Risk and Reliability Analysis

Other Titles of Interest

Acceptable Risk Processes: Lifeline and Natural Hazards, edited by Craig Taylor and Erik VanMarcke (TCLEE Monograph). Report explaining the use of acceptable risk processes to develop risk reduction strategies and earthquake mitigation plans. (ISBN 978-0-7844-0623-6)

Appraisal, Risk and Uncertainty, by Nigel Smith (Thomas Telford, Ltd.). An integrated approach to appraisal, risk, and uncertainty, especially when operating in ambiguous situations. (ISBN 978-0-7277-3185-2)

Artificial Neural Networks in Water Supply Engineering, edited by Srinivasa Lingreddy and Gail Brion (ASCE Committee Report). Compilation of information concerning successful applications of Artificial Neural Networks (ANN) for water supply engineering problems. (ISBN 978-0-7844-0765-3)

Degrees of Belief: Subjective Probability and Engineering Judgment, by Steven Vick (ASCE Press). Examination of the intersection of probability and risk analysis with professional judgment and expertise from a geotechnical perspective. (ISBN 978-0-7844-0598-7)

Infrastructure Risk Management Processes: Natural, Accidental, and Deliberate Hazards, edited by Craig Taylor and Erik VanMarcke (ASCE Monograph). Quantifies exposure and vulnerability of complex systems to derive estimates of local and system-wide potential losses under various hazard situations. (ISBN 978-0-7844-0815-5)

Risk-Based Decision Making in Water Resources X, edited by Yacov Haimes, et al. (ASCE Proceedings). Continuation of series focusing on risk and vulnerability of homeland water resource systems to terrorism. (ISBN 978-0-7844-0694-6)

Strategic Risk: A Guide for Directors (Thomas Telford, Ltd.). Explanation of the recommended approach to successfully managing strategic risk. (ISBN 978-0-7277-3467-9)

Risk and Reliability Analysis

A Handbook for Civil and Environmental Engineers

Vijay P. Singh

Sharad K. Jain

Aditya Tyagi

ASCE PRESS

Library of Congress Cataloging-in-Publication Data

Singh, V. P. (Vijay P.)

Risk and reliability analysis : a handbook for civil and environmental engineers / Vijay P. Singh, Sharad K. Jain, Aditya K. Tyagi.

p. cm.

Includes bibliographical references and index.

ISBN-13: 978-0-7844-0891-9

ISBN-10: 0-7844-0891-2

1. Engineering—Management—Handbooks, manuals, etc. 2. Reliability (Engineering)—Handbooks, manuals, etc. 3. Risk assessment—Handbooks, manuals, etc. I. Jain, S. K. (Sharad Kumar), 1960- II. Tyagi, Aditya K. III. Title.

TA190.S594 2007

620'.00452—dc22

2006038853

Published by American Society of Civil Engineers
1801 Alexander Bell Drive
Reston, Virginia 20191
www.pubs.asce.org

Dedicated to our families:

Anita, Vinay, and Arti Singh
Shikha, Sanchit, and Surabhi Jain
Deepa, Surabhi, and Uday Tyagi

Table of Contents

Part III   Uncertainty Analysis

Part IV   Risk and Reliability Analysis

Preface

Risk is an inherent part of any decision-making process. Risk and uncertainty are unavoidable in the planning, design, construction, and management of engineering systems. Until recently, a common way to account for risk and uncertainty in engineering design was through a factor of safety. However, this does not provide any quantitative idea of the risk in a particular situation and most decision makers these days want to explicitly know the risk involved while making a decision as well as the risk of operating an existing structure or project and the consequences of a management action. Answers to these questions require an understanding of the behavior of inputs to the system under study as well as the consequences of a management action. Despite the importance and relevance of the subject, there are very few universities and institutions where risk and reliability analysis is taught or where it is part of either civil and environmental engineering curricula or environmental and watershed sciences curricula. Commonly, these topics form the subject matter of a course in statistics or systems analysis. A review of the literature in civil and environmental engineering shows that there are few books providing a comprehensive discussion of relevant issues related to uncertainty, risk, and reliability. This constituted the motivation for the proposed book.

The subject matter of the book is divided into four parts. The first part, termed Preliminaries, contains three chapters. Introducing the basic theme of the book, Chapter 1 provides a broad overview of the art of decision making under uncertainty. Chapters 2 and 3 provide a preliminary background of probability and random variables and moments and expectations of data that are needed to grasp the topics that follow.

The second part of the book deals with Probability Distributions and Parameter Estimation, which is a bouquet of techniques organized into six chapters. Discrete and continuous probability distributions form the subject matter of Chapter 4. The distributions discussed include Bernoulli, binomial, geometric, and negative binomial distributions. Another set of distributions is used when interest lies in the number of times a specified event occurs over a certain duration. Poisson, exponential, and gamma distributions are used in such cases and these are also described in Chapter 4. Chapter 5 deals with limit distributions and other continuous distributions. Two frequently used distributions that form the backbone of statistical analysis—the normal and log-normal distributions—are discussed. These distributions are based on an important theorem in statistical analysis, the central limit theorem, which is discussed next. Distributions of extremes and other distributions found useful in environmental and water engineering, such as uniform, triangular, beta, Pareto, logistic, Pearson type III, and log-Pearson type III distributions, are also presented. Many environmental processes can be described by using the concepts of probability and physical laws. Chapter 6 focuses on impulse response functions as probability distributions. These distributions have a quasi-physical basis. The concepts discussed include impulse response of a linear reservoir, cascade of linear reservoirs, the Muskingum model, the diffusion model, and the linear channel downstream model. These concepts are widely used in hydrologic systems analysis. Many real-world decisions frequently involve more than one variable and there may not be a one-to-one relationship among them. In such a situation, an analysis of the joint probabilistic behavior of the variables involved would be desirable. Chapter 7 presents multivariate distributions, with particular attention to bivariate distributions. A newly emerging copula methodology is presented and several bivariate distributions are discussed using this methodology. The concept of return period is extended to more than one variable.

In statistical analysis, a considerable effort is devoted to deriving parameters of a distribution, which constitutes the subject matter of Chapter 8. Many techniques are available for this purpose; these include the method of moments, the method of maximum likelihood, the method of probability weighted moments, *L*-moments, and the method of least squares. Discussion of these methods is followed by a treatment of the problems of parameter estimation. Besides point estimates, interval estimation of parameters is carried out to determine the confidence that can be placed in the point estimates and this chapter also includes a description of interval estimates. The last chapter in the second part, Chapter 9, deals with entropy. Originating in thermodynamics, the principle of information theoretic entropy has found applications in many branches of engineering, including civil and environmental engineering. The fundamental concepts of the Shannon entropy theory are discussed. The methodology to derive parameters of normal and gamma distributions by following the Lagrange multiplier method and the parameter-space expansion method is described. This chapter also provides a discussion of the fields where the entropy concept has proved to be useful.

Part 3 of the book, comprising four chapters, deals with Uncertainty Analysis. Chapter 10 discusses the concepts of error and uncertainty analysis. The focus of this chapter is on a treatment of the types of uncertainties and analysis of errors. The Monte Carlo method, a powerful tool to solve a range of problems, is discussed in Chapter 11. Generation of random numbers comprises an important part of the Monte Carlo method. Therefore, this chapter provides a discussion of many techniques that can be used to generate random numbers that follow a given distribution. Several examples help illustrate the application of Monte Carlo methods. Because many environmental processes are stochastic and can be treated as stochastic processes, Chapter 12 gives a preliminary treatment of this topic. It includes a discussion of mean, variance, covariance, correlation, stationarity, correlogram, and spectral density of stochastic processes. The chapter is concluded with a discussion of time series analysis. Description of many environmental processes requires the use of stochastic differential equations, which are presented in Chapter 13. Several examples are presented to demonstrate the application of these equations and the techniques to solve them.

The fourth and last part of the book, encompassing three chapters, focuses on Risk and Reliability Analysis. The first chapter of this section, Chapter 14, describes the various reliability measures, such as time to failure and the hazard function. This is followed by a discussion of two concepts that are widely used in reliability analysis: margin of safety and factor of safety. The highlight of this chapter is the discussion on methods of reliability analysis, such as the first-order approximation, the first-order second moment (FOSM) method, the mean-value FOSM method, and point estimation methods. A number of real-world examples are discussed to illustrate the concepts of reliability analysis and estimation.

Chapter 15 is related to risk analysis and management. In this chapter, a broad view of risk is explained wherein risk is considered as a triplet involving answering three questions: What can go wrong? What is the probability of things going wrong? What are the consequences if something goes wrong? Concepts of reliability and failure analysis and event and fault-tree analysis are explained using suitable examples.

An important area that has drawn much attention and application of the principles of reliability analysis is the design of water distribution networks (WDNs). The final chapter of the book, Chapter 16, begins with a basic description of analysis of WDNs, followed by hydraulic reliability analysis of a WDN for a range of conditions. The entropy method can be useful in hydraulic reliability analysis of a WDN and the entropy-based methodology is described in this chapter.

The chapters of this book are arranged by keeping in view the requirements of a typical engineering student, who hopefully has a fundamental knowledge of mathematics and statistics. This book is intended for senior undergraduate and beginning graduate students as well as water resources practitioners. Those who have an adequate background in probability and statistical analysis can skip the first section of the book and do a quick reading of the first two chapters of the second section. Numerous examples have been solved step by step and this should help with understanding of computational procedures. Besides the book being of value to students, it should also be useful to faculty members and practitioners working in the fields of civil and environmental engineering, watershed sciences, and biological and agricultural engineering. Much of the material has been used for teaching a course on risk and reliability analysis in civil and environmental engineering.

—*Vijay P. Singh, College Station, Texas*

—*Sharad K. Jain, Roorkee, India*

—*Aditya Tyagi, Austin, Texas*

Acknowledgments

Dr. Lan Zhang and Mr. Hemant Chowdhary of the Civil and Environmental Engineering Department at Louisiana State University reviewed most of the chapters of the book and provided critical and constructive comments. Inclusion of these comments significantly improved the manuscript. Dr. Zhang also helped with final preparation of several chapters. We are deeply grateful to Dr. Zhang and Mr. Chowdhary. Mr. Jose Villalobos-Enciso of the Civil and Environmental Engineering Department at Louisiana State University and Dr. P. K. Bhunya from the National Institute of Hydrology, India, also read several chapters and provided critical comments. We are grateful to both Mr. Villalobos-Enciso and Dr. Bhunya. There are tens of scientists whose works have been inspiring. This book draws upon the fruits of their labor. We have tried to make our acknowledgments as specific as possible. Any omission on our part has been entirely inadvertent and we offer our sincere apologies in advance. We would be most grateful if readers discovering any discrepancies, errors, or misprints would bring them to our attention.

Our families provided unwavering support and help, without which this book would not have been completed. We would like to take this opportunity to express our deep appreciation and dedicate the book to them.

Part I

Preliminaries

Chapter 1

Rational Decision Making Under Uncertainty

The process of decision making can be traced to the beginning of human civilization. However, the nature of problems requiring decisions, the type of decisions, and the decision making tools have undergone dramatic changes over time. People's intuitive judgment and cognitive ability; the availability of data; access to computational tools; environmental and ecological considerations; and social, political, and economic constraints all influence the process of decision making and the ensuing decisions. Most day-to-day decisions involve a certain amount of risk, which is factored, either knowingly or unknowingly, into the decision-making process.

Planning, design, operation, and management of civil and environmental engineering systems are greatly affected by the vagaries of nature or the uncertainty of natural events. Nature has immense variability, and the information available to quantify this variability is usually limited. Nevertheless, decisions have to be made and implemented. Decision theory attempts to provide a systematic approach to making rational decisions. Haimes and Stakhiv (1985) have aptly summarized the overall philosophy of decision making as shown in Fig. 1-1. This philosophy presents decision making through a triangle whose three vertices are occupied by benefit–cost theory, decision theory, and sustainability theory. As shown in the figure, risk and reliability analysis occupies a central place in the interaction of certainty and uncertainty; efficiency and equity; and single decision making and collective decision making. The relative importance of the vertices of

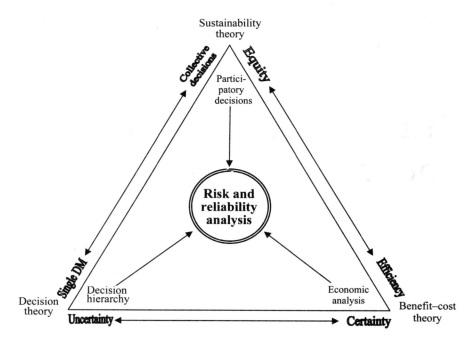

Figure 1-1 *The decision triangle for environmental systems*
[adapted from Haimes and Stakhiv, 1985].

the triangle in the figure changes with social evolution and the development stage of the society. These days, most societies attach utmost importance to sustainability and equity, and decision making therefore is becoming participatory. In line with the modern-day philosophy and development paradigms, principles of sustainability, equity, and participatory decision making are placed at the apex of the decision triangle.

Central to rational decision making and risk assessment is uncertainty. Klir (1991) and Shackle (1961) have argued that the necessity of decision making results totally from uncertainty. In other words, if the uncertainty did not exist, there would be no need for decision making or decision making would be relatively simple and straightforward. One of the main causes of uncertainty in natural systems is the unpredictability of system behavior. For example, flow in a river varies in time and experiences highs and lows each year. If one were to consider the lowest flow in each year for a number of years, a series of low flows would be the result. Prediction of these flows cannot be made with certainty. The same would apply to the highest yearly flows. Another example is the unpredictability of rainfall or for that matter forecasting of climate. Prediction of earthquakes also entails a very high degree of uncertainty, as does prediction of tornadoes.

A risky decision exposes the decision maker to the possibility of some type of loss but there are many situations when such a decision has to be made. The foremost cause involves the vagaries of nature. For example, a decision to build a project may be risky, because a large flood or hurricane or earthquake might occur and endanger the structure with resulting loss of life and property. In addition, a decision may be risky because the natural phenomena are not clearly understood. Sometimes a risky choice has to be made if the cost of an alternative that can control the risk is more than the ability or willingness to pay for it. The ultimate goal is to reduce uncertainty and thereby risk.

1.1 Problems Requiring Decision Making

Most problems in environmental and water resources, as shown in Fig. 1-2, can be classified in six categories: (1) prediction (i.e., the system output is unknown but the system geometry and governing equations as well as system input are known), (2) forecasting (i.e., for a given system input and system geometry, the system output is forecasted in real time; this is different from prediction, where a specific time is not of concern), (3) detection (i.e., system output and system equations are known but system input is unknown; this is also referred to as an instrumentation problem), (4) identification (i.e., the system input and output are known but system parameters are unknown; this is an identification problem), (5) design (i.e., the system input is known and output is either known or assumed but the system is constructed based on certain hypotheses and then the desired system output is predicted; if the predicted output is acceptable, the constructed system is acceptable; otherwise the system needs to be reconstructed and the cycle needs to be repeated), and (6) simulation or modeling (i.e., a combination of categories (1) and (4) in which first the system is identified and then prediction is performed). Illustrative examples of these problems are as follows: A highway engineer may be assigned the task of designing a highway bridge (design problem), a water resources manager may be interested in predicting the peak flow of the Mississippi River at Baton Rouge (prediction problem), a hydrometeorologist may be asked to forecast rainfall on a particular day next week (forecasting problem), an environmental engineer may want to analyze the water quality of the Amite River at Denham Springs or calibrate a water treatment system for performance evaluation (analysis and identification problem), a groundwater engineer may want to determine parameters of an aquifer using pumping data (identification problem), a soil physicist may want to calibrate a neutron probe for measuring soil moisture (detection problem), a geotechnical engineer may want to simulate the behavior of a pile foundation when a goods train passes over the bridge (simulation or modeling problem), and a watershed manager may want to develop a watershed model for simulating the impact of land use changes on the watershed (modeling or simulation problem).

An environmental or water resources system can be represented by (1) the system geometry, (2) the equations governing the system, (3) the sources and sinks to which the system is subjected, (4) initial and boundary conditions that

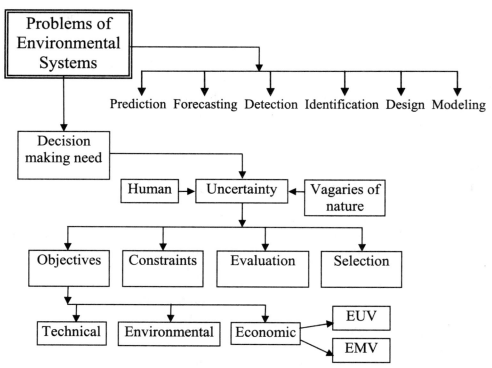

Figure 1-2 *Conceptual depiction of decision making under uncertainty. EMV = effective monetary value, and EUV = effective utility value.*

the system must satisfy, and (5) the system output or response. As an example, consider a watershed with the goal of predicting runoff from the watershed as a function of time for a given rainfall event. Thus, the watershed is the system here. Rainfall is the input or source for the watershed. Infiltration and evaporation are the sinks of the watershed. The watershed has a certain topography and channel network, which, in turn, define the watershed geometry. Hydraulic equations of flow over land areas and in channels are the equations governing the flow in the watershed. Runoff is generated based on the initial state of the watershed (antecedent moisture condition) and the upstream and downstream boundaries. The governing equations must satisfy these conditions. Solution of these equations yields runoff as a function of time. This is a typical prediction problem.

Environmental and water resources systems are subject to uncertainties that are due to natural randomness (or caused by the vagaries of nature) as well as human-induced errors or factors. For example, consider the problem of predicting runoff from a watershed for a given rainfall event. In this problem, there is uncertainty in the areal mapping of rainfall, because rainfall varies spatially. In practice, rainfall is measured only at a point and the point measurement is used to represent rainfall over an area. Because of inherent randomness in the rainfall field, there are uncertainties in rainfall measurements caused by wind, angle of incidence, raindrop size, and so on. There may also be errors in rainfall measurements resulting from instrumental defects, improper rain gauge location, etc. Similarly, infiltration and evaporation have uncertainties. The watershed geometry also has less than certain elements. The governing equations, expressed as partial differential equations (PDEs), may themselves be in error. The resistance parameter, such as Manning's friction factor, in the momentum equation is spatially variable but only an average value is used. Thus it is also subject to uncertainty. Furthermore, round-off and truncation errors may arise in computations. These uncertainties in virtually every component of the prediction problem will introduce uncertainty in the predicted runoff.

This discussion shows that solutions of the aforementioned types of problems are subject to uncertainty. This leads to the necessity of making decisions under uncertainty. To make a decision, the problem is to be solved nevertheless. In the event of uncertainty, a typical problem-solving approach entails preparing a model of the system, as shown in Fig. 1-3. The model variables are considered random and are described by laws of probability or probability distribution functions. Then, parameters of these distributions need to be estimated. One can then compute the error and thereby the associated risk and reliability of the model output or the solution of the problem.

It may be noted that the uncertainty resulting from natural causes can be reduced to some extent by collecting more comprehensive data and using improved models. Even then, it is not possible to remove the uncertainty beyond a limit, because nature has immense variability and any model used will be a simplified depiction of reality. Thus, the objective should be to understand the causes and sources of uncertainty, deal rationally with uncertainty, and integrate it with decision making. Klemes (1971) appropriately noted: "Nowadays in hydrology, and the more so in engineering, uncertainty is still regarded as a regrettable imperfection in the body of knowledge, as it was in the 19th century physics. As in physics, it also seems that in hydrology and engineering, progress lies not in trying to remove the uncertainty at any cost but in learning how to make it one of the legitimate elements of our concepts." Of course, this was the view about 30 years ago, but the perceptions about uncertainty have now begun to change.

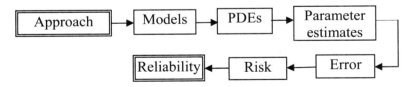

Figure 1-3 *An approach for decision making under uncertainty.*

1.2 Concept of Rationality

The presence of uncertainty notwithstanding, decisions about environmental and water engineering systems, such as reservoirs, irrigation systems, water purification systems, flood control, water diversion, water supply, land reclamation, hydropower generation, drainage, transportation, environmental pollution, and so on, should be rational. To judge whether a decision is rational or not, the decision-making process itself needs to be considered along with criteria of rationality. The basic question is "What is rationality?" In social sciences, Boudon (2003) reviews rational choice theory and presents a broader concept of rationality. He quotes Rescher (1975): "[R]ationality is in its very nature teleological and ends-oriented." He goes on: "Cognitive rationality is concerned with achieving true beliefs. Evaluative rationality is concerned with making correct evaluation. Practical rationality is concerned with the effective pursuit of appropriate objectives." Boudon (2003) emphasizes that "teleological" is not synonymous with "instrumental" or "consequential." These forms of rationality are goal oriented but the nature of the goals can be diverse. From an engineering standpoint, four criteria seem to emanate from these forms of rationality that may essentially constitute the concept of rationality: (1) the objective to be achieved, (2) the identification of alternatives to achieve the objective, (3) the evaluation of alternatives, and (4) the selection of an alternative using objective criteria.

First, the decision must be aimed at an objective that is to be achieved when solving a problem. Consider, for example, that the life of an overhead water tank is over and it needs to be replaced. The problem is to design a tank. The objective is to design a tank that meets some projected demands, is economical, is durable, and has a pleasing appearance. Thus, the objective includes the technical solution plus some additional aspects. Failure to meet any of these objectives in a new design may make it unacceptable and the underlying decision will be irrational. Consider another example in which an old bridge over a stream is to be replaced. The objective is not just to design any bridge but a bridge that will meet the increased traffic demands, is designed for the heaviest vehicles anticipated, is economical and durable, and has a pleasing appearance. Failure to meet any of these objectives in the new bridge design will make the decision to build the bridge irrational.

Second, there might be many designs that would, to a greater or lesser degree, meet the stated objective. This leads to the second criterion of rationality, which states that the decision maker must identify and study enough alternatives to ensure that the best alternative is among them. This does not imply an exhaustive search of all possible alternatives, but one must have an open mind for alternative solutions. For example, the tank designer may be a specialist in steel construction but that should not prevent the designer from considering the advantages that reinforced concrete might offer. In the case of a bridge design, it is appropriate to consider alternative designs based on steel, concrete, or timber. Each alternative design has its pros and cons and has a certain amount of risk.

Third, a choice has to be made among alternatives by following an objective evaluation process. Each alternative has its consequences and associated benefits and costs. In the tank or bridge design example, some designs have a lower cost, some have a longer useful life, some require more maintenance and some less, some look better in appearance than others, some can be built using local labor and locally available material, some can be built faster than others and with little interruption in services, and so on. All the consequences must be taken into account when evaluating and ranking different alternatives. For the evaluation process to be objective, one must employ objective criteria. For example, one can express each consequence in terms of a monetary value measured in terms of a currency, say, the U.S. dollar. A difficulty, however, is encountered with the evaluation of those aspects for which there is no market value, for example, service interruption or aesthetic value or loss of life. The answers in such cases are not very precise and have an element of subjectivity. Nevertheless, one can make a guess and obtain some upper and lower limits that will suffice to rank alternatives and find the best alternative among them. One then adds up the value of each of the consequences and arrives at the relative value of the alternatives.

Fourth, the relative values of different alternatives are then compared and ranked. The best alternative is thus selected. However, this selection procedure should be employed qualitatively and with a sense of judgment. There may be other considerations that should also be taken into account. For example, the relative value of one alternative may be lower than that of another alternative but may still be preferable. The higher-value alternative may cost more than a lower-value alternative and may therefore not be affordable.

Another way to select an alternative from a number of alternative decisions may be to employ two steps. Step one is to do the initial screening of all alternatives and eliminate the inferior alternatives. This will permit only a few worthy alternatives for further consideration. Step two is to consider all the available information and choose the alternative that is expected to have the highest value or the least risk. Indeed this two-step procedure is normally the one employed in hiring people. Of course, there is no guarantee that this alternative will be the best; there might be situations wherein the selected alternative will not be up to the expectation.

Another consideration in selecting an alternative is the constituency for which the decision is to be made. For example, a decision about water supply for agricultural irrigation must take into account farmers' concerns and attitudes relative to crops, productivity, soil and water management, environmental quality, and so on. Another example is making a decision on evacuating people before the arrival of a likely hurricane. Different people have different attitudes toward evacuation, tempered by the level of risk and damage. Any evacuation plan must consider people's aspirations, attitudes, and concerns. In such cases people's participation is of considerable value.

In engineering design, especially structural, geotechnical, and hydraulic design, it is normally argued that the factor of safety adequately accounts for the adverse impacts of uncertainty (e.g., environmental, engineering, social, or economic). However, this factor can, at best, be considered sufficient only with respect to the longevity of the structure against uncertainty. Engineering design and operation of projects require an explicit risk analysis and management. This is particularly important when dealing with low-probability high-consequence events, such as large floods, dam failure, levee rupture, subsidence, hurricanes, earthquakes, and large-scale drought.

In real life, personal considerations play a significant role in choosing an alternative and these personal considerations vary with the size of the project or the nature of the problem. Many times people make decisions based on just one factor. For example, for small projects, such as building a house, personal likes or dislikes of a particular architectural design may be the determinant factor. In other words, the scale of a project and personal likes and dislikes must be taken into consideration.

1.3 Evaluation of Alternatives

There are many criteria by which alternatives, taking uncertainty into account, can be evaluated and selected: economic, risk, safety, environmental, and so on. Here we discuss only a simple economic criterion and defer discussion of other criteria to Chapter 15. One economic measure in which the uncertainty can be accounted for is the expected monetary value (EMV) or the expected utility value (EUV). Consider, for example, alternatives and their consequences or outcomes. It is assumed that possible outcomes expressed as profit or loss are associated with appropriate probability values. For each alternative, its outcomes are then weighted with the corresponding probability values, and the sum of the weighted outcomes of each alternative decision is then computed. The weighted sum determines the EMV of each decision. The decision with the greatest EMV may be the preferred decision. In this manner, EMV attempts to maximize the expected benefit or minimize the expected cost.

Let us now consider that a house is to be constructed for specified requirements on a purchased parcel of land. Several alternative house designs can be considered. Let us consider how an alternative house design can be evaluated. The initial cost of the house, operational (heating, cooling, and so on) costs, maintenance requirements and costs, and useful life can be expressed in dollars. Likewise, it is relatively easy to calculate the cost of painting the house, redoing the patio, or replacing certain components, such as a heater, an air conditioner, a ventilating system, a fireplace, windows, and doors. However, there are other considerations that are also important and need to be considered: vulnerability to unexpected natural events, such as extreme winds, flooding, and fire; the

possibility for further extension; and safety. It is not easy to express these considerations in economic terms but there are indirect ways to accomplish this. For example, vulnerability can be expressed in terms of the cost of insurance that one may purchase. It should, however, be noted that expenditures, such as the initial cost, are not always immediate but occur at specified intervals. For example, the house loan may be for a period of 10 years; the cost of the house will then be paid during this period. This points to the effective value of money, say, in terms of dollars (i.e., because of inflation the value of a dollar at a future date is not the same as it is today). To account for the change in the value of a dollar with time, the money markets have established interest rates that express this change in value. Consider, for example, an interest rate of 5% for the next 10 years. If an amount of $100,000 is invested today, then 10 years from now, it will amount to $100,000(1 + 0.05)^{10} = $162,889$. In other words, $162,889 invested 10 years from now will have the same value as $100,000 has now. There are, however, consequences for which there is no market value, as, for example, interruption in residency during major repair work, inconvenience, emotional value, etc. It is difficult to evaluate such consequences. The question then arises as to the worth of these nonquantifiable consequences. How much are the people willing to pay for less interruption, reduced inconvenience, more emotional value, etc.? It is difficult to get precise numbers for such consequences, for they are subjective. Nevertheless, one can at least specify some upper and lower limits for the monetary value that might suffice to rank alternative designs and enable selection of the best alternative.

As an example, consider the case of a house design where three alternative designs are to be evaluated. These alternatives are designated as I, II, and III. Assume that each design is to be evaluated by considering three aspects: foundation A, material B, and labor C. For design I, IA denotes the foundation for design I, IB denotes the material needed for design I, and IC denotes the labor for design I. Each aspect has an effective cost based on probabilistic considerations. For design I, IA has effective cost ECIA, IB has effective cost ECIB, and IC has effective cost ECIC. In a similar manner, designs II and III are represented. These designs can be represented as a decision tree, as schematically shown in Fig. 1-4. Associated with each branch representing a design is a set of consequences, which are entered into the calculation of the relative value of the alternative. The relative value is then used for ranking alternative designs.

This exercise essentially comprises a rational planning process consisting of defining an objective, identifying alternative means of achieving the objective, and applying a ranking procedure to determine the best alternative. Although it is conceptually quite simple, in the real world often little planning and little rationality are employed even for important decisions. A common occurrence is not to consider alternatives. Quite frequently, the decision is made based on precedent, tradition, lack of preparation, personal bias, prejudice, or shortsightedness. Many a time, there is a deliberate effort to postpone making decisions, however consequential they are. The result is that no time is left for anything but

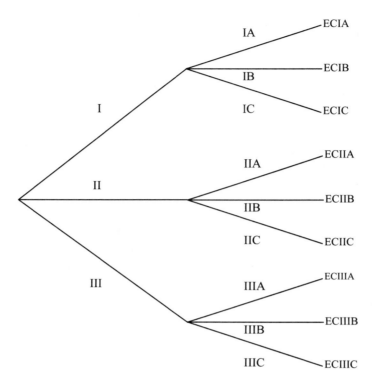

Figure 1-4 *Decision tree. I, II, and III are alternatives; letters A, B, and C associated with the alternatives denote the consequences of the respective alternatives in terms of foundation, material, and labor; and ECIs, ECIIs and ECIIIs are costs of the consequences.*

to continue the present practice. Another common error is the presumption in value judgment. The one making the decision may have his or her own ideas as to what is important and what is not important and may not bother to support others in participating in the decision-making process. For rational decision making, decision makers must transcend their own value judgment and keep the judgment of the client or the community being served as paramount. Examples of different types of errors abound and they will be discussed in Chapter 10. Each error introduces an element of uncertainty.

Three types of situations can arise in a decision-making process: decision making under certainty, decision making under uncertainty, and decision making under risk. When a decision is to be made under certainty, the input data and their relations are known. The preference of the decision maker is expressed through an objective function, which is defined or known in the given circumstances. A common objective function is the benefit–cost ratio. The problem is solved using an optimization technique, such as linear programming, dynamic programming, or goal programming. Then the best decision is determined based on the value of the objective function. Additionally, one may employ intuitive techniques to solve the problem. For example, Taha (2003) describes an analytical hierarchical approach for such problems. In this case, decision making is analytical and relatively simple.

In the problems related to decision making under uncertainty or risk, the benefits and costs associated with each decision are commonly expressed using probability distributions. In the absence of one definite outcome, an expected value criterion may be adopted for comparing decisions. Based on the optimization of the expected profit or expected loss, decisions are then evaluated. When the number of alternatives is small, decision-tree analysis can be used to find the best alternative or decision.

1.4 Dealing with Uncertainty

There are many types of uncertainties, some quantifiable and some not. For example, some people may have difficulty making up their minds, some people may have their own personal biases or preferences, there may be delays in the transmission of pertinent information, there may be mismanagement, and so on. All these may create considerable uncertainty in decision making and are difficult to quantify. Our objective here is not to deal with every kind of uncertainty but only with the kind of uncertainty that can be measured quantitatively, at least in principle. For example, if a coin is tossed, there is no way to know in advance whether heads will turn up. But the event that heads will turn up has a probability of 50% each time the coin is tossed. Similarly, a probability value can be assigned to the event that the peak flow in the Amite River at Denham Springs in Louisiana in any given year will exceed 5,000 m^3/s. One can also determine the probability that the number of rain-free days in the month of August in Baton Rouge will exceed 25 or that the number of westbound cars that will cross the toll bridge at the Mississippi River in New Orleans will exceed 10,000 on any given working day. This kind of uncertainty is the uncertainty associated with the randomness of the event.

One can go a step further and investigate the uncertainty in the conclusions about uncertain events. For example, one may calculate the probability p that in any given year the peak flow in the Amite River exceeds 5,000 m^3/s and conclude that p is 2%. But there is an element of uncertainty in the computed value of p that depends on the amount and quality of the data as well as the method used for its determination. Thus, the following question arises: What is the probability that p lies within the given range $p - \Delta p$ to $p + \Delta p$? Statistical procedures can be employed to analyze random events and make probability assessments. In this manner, one can deal rationally with uncertainty but cannot eliminate it.

In principle, it is not too difficult to deal with uncertain events in the decision-making process. Events that are certain to occur, or conclusions that are certainly true, must, of course, be fully taken into account. Corresponding to the probability of 1 assigned to certain events, these certain events are given the weight of 1. Impossible events, in contrast, are disregarded in decision making and given the weight of 0, corresponding to the probability of 0 assigned to all impossible events. These two types of events define the length of the scale. Any event in between is given a weight equal to the probability of its occurrence. Thus, the more likely an event, the more weight it gets and the greater its relative effect on the outcome or the decision. A simple illustration of this decision policy is considered in Example 1.1.

Example 1.1 A farmer in India owns an agricultural farm adjacent to a canal. He can use the canal water for irrigation, which is very cheap, but the supply is available with only a 70% reliability. The cost of canal water per year is estimated at 2,000 rupees (Rs). The farmer has an option to install a small well costing Rs 54,000, which he can use when water from the canal is not available. The farmer estimates that he will be spending about Rs 13,000 each year to meet the running expenditure of this well to supply water and the life of this well is about 3 years. Alternately, he may construct a big well to meet his entire farm requirement and be free from dependence on the canal water. This big well will cost Rs 195,000 to construct and will last for about 15 years. The well operation will cost nearly Rs 17,000 per year. If adequate water is available, the value of production from the farm is Rs 120,000 per year. What should the farmer do?

Solution Let us consider the three options that the farmer has. Based on these options he can make a rational decision.

(i) If the farmer is completely dependent on the canal water, the expected value of production will be

$$Rs\ 120,000 \times 0.7 - Rs\ 2,000 = Rs\ 82,000\ \text{per year}$$

(ii) If the farmer decides to construct a small well, the expenditure toward the well will be Rs 12,000 + Rs 54,000/3 = Rs 30,000. Thus, his net benefit will be

$$Rs\ 120,000 - Rs\ 2,000 - Rs\ 30,000 = Rs\ 88,000\ \text{per year}$$

(iii) If he constructs a large well, the annual expenditure toward the well will be Rs 17,000 + Rs 195,000/15 = Rs 30,000. Therefore, his net benefit will be

$$Rs\ 120,000 - Rs\ 30,000 = Rs\ 90,000\ \text{per year}$$

Each option now has a monetary value. Option iii has the largest expected benefit and option i has the lowest expected benefit. However, the construction of a large well involves a heavy expenditure in the beginning, whereas a small well is considerably cheaper to construct. Of course, the small well will require installation expenditure every three years. Many farmers will have to borrow money to construct a large well and the interest rate will be an important parameter in decision making. Clearly, it is better for the farmer to construct a well. Whether he goes for a large well or a small well largely depends upon his paying capacity and the willingness to spend the needed sum of money. If his paying capacity is limited, the large well may not be a viable option.

Many farmers in countries with a large population and limited agricultural land have small holdings. If the value of the farm produce is Rs 80,000 per annum then one gets a different trade-off. For the first option, his net benefit will be Rs 80,000 × 0.7 − Rs 2,000 = Rs 54,000 per annum and for the second option, it will be Rs 80,000 − Rs 2,000 − Rs 30,000 = Rs 48,000 (assuming that the running expenditure remains the same). For the third option, the net benefit will be Rs 80,000 − Rs 30,000 = Rs 50,000. Thus, this farmer will be worse off if he decides to construct a well and opt for the second option. He may elect to stay with the first option. Even though the third option ranks second in terms of the net expected benefit it may not be worth considering for a small farmer because of his limited paying capacity.

Example 1.2 A paper factory is being planned in an area. As paper making requires a considerable amount of water (about 40 m^3 of water is needed per ton of paper produced), the owner prefers a site near a river. One such site that is also close to an interstate highway is available but the river runs dry for three months each year. This problem can be overcome by constructing a small reservoir but the river water necessarily requires treatment before use. The cost of raw water will be 50 cents per m^3 but the treatment cost and expenditure for the impoundment will make the cost nearly 110 cents per m^3. The factory owner can also get a permit to pump water from an aquifer at a depth of 325 m and this water will not require any treatment but the cost will be about 120 cents per m^3 of water. What is the best course of action to meet the water supply demand of the factory? Discuss the likely answer qualitatively without doing any calculation.

Solution A first inspection of the data shows that it will be better for the factory owner to opt for surface water for the plant. However, there might be some other factors that may influence the owner's final decision. Additional land will be required for impoundment and treatment facilities. Since the river will be dry for three months, the size of impoundment should be sufficiently large to meet the demand during this period. There might be significant losses of water owing to evaporation and seepage. Another important point is that the groundwater availability usually has a high reliability, whereas there may be instances when the river is dry for more than three months. After all, rainfall is highly uncertain and prolonged droughts are not uncommon as evidenced in recent years.

The quality of paper produced depends upon the quality of water used. Since the marginal difference in the price of water is only 10 cents per ton of paper produced, many decision makers may base their decisions on other factors.

The purpose of this example is to qualitatively illustrate that many real-life problems do not have a straightforward answer and a number of related factors require a careful examination before making a decision. Different people have different risk perceptions. Consequently, the whole process may be quite subjective, particularly when two or more options are equally attractive.

Example 1.3 Suppose a contractor in Louisiana has bid for a job of repairing a highway in the month of June. For doing the repair work, the contractor needs a certain number of rain-free days. To ensure completion of the work in time, there is a clause in the contract that the contractor forfeits her payment if the work is not completed in time. The contractor enters a bid for $5,000,000. What is the reasonable course of action?

Solution There are several alternatives that the contractor initially considers. After initial screening she eliminates what she considers inferior alternatives and finally decides on evaluating only three alternatives for completing the work. The first alternative is based on the calculation that she can complete the work in 20 days at a cost of $3,000,000 with her own equipment. Analysis of rainfall data reveals that there is a 30% chance of having fewer than 20 rain-free days in June.

The second alternative is that the contractor can buy additional equipment and can then finish the work in 15 days at a cost of $3,500,000. There is, however, a 10% probability that there will be fewer than 15 rain-free days in June.

The third alternative is that the contractor can partner with another contractor and finish the work in 10 days at a cost of $4,000,000. Analysis of rainfall data shows that there is virtually no chance of having fewer than 10 rain-free days in June.

To make a rational decision, one can consider the decision tree with the three alternatives and their associated consequences or outcomes. Rainfall in the month of June is a random variable and clearly influences the consequences of the three alternative courses of action and the resulting outcomes. The possible outcomes, profit or loss, are associated with appropriate probability values. These outcomes must be weighted with the corresponding probability values, and the sum of the weighted outcomes of each decision is then computed. The weighted sum determines the EMV of each decision. The decision with the greatest EMV may be the preferred decision but may not necessarily be the best decision.

Let us now compute the EMV of each alternative decision. In the first alternative, the profit will be $2,000,000 with a probability of 0.7 and the loss will be $3,000,000 with a probability of 0.3. Therefore,

$$EMV = \$2,000,000 \times 0.7 - \$3,000,000 \times 0.3 = \$500,000$$

For the second alternative, the profit will be $1,500,000 with a probability of 0.9 and the loss will be $3,500,000 with a probability of 0.1. Therefore,

$$EMV = \$1,500,000 \times 0.9 - \$3,500,000 \times 0.1 = \$1,000,000$$

For the third alternative, the profit will be $1,000,000 with a probability of 1 and the loss will be $4,000,000 with a probability of 0. Thus,

$$EMV = \$1,000,000 \times 1 - \$4,000,000 \times 0 = \$1,000,000$$

These three courses of action are depicted in Fig. 1-5. In terms of the EMVs, alternatives can be ranked as alternative C, alternative B, and alternative A.

options designed to protect the area. Use both methods to select the flood protection scheme and compare the EMVs.

Assume the following construction costs: $40 million for a dam, $30 million for levees, $24 million for a drainage system, and $20 million for land use. Each scheme will reduce the damage differently and also change the stage–discharge curve differently. The reduced damage will be $20 million for the dam, $25 million for the levees, $36 million for the drainage system, and $40 million for the land use.

Table 1-4 contains data on annual peak discharge, stage, and flood damage for the area under consideration in Louisiana. For the data the relationship between discharge Q ft^3/s (cfs) and stage H (ft) may be expressed by the following equation:

$$H = 8.0451 \ln Q - 52.619 \text{ or } Q = 692.63 \times \exp\left(\frac{H}{8.0451}\right)$$

The relationship between stage and damage D (in millions of dollars) may be expressed by the following equation conditioning on the stage greater than or equal to 31 ft:

$$H = 1.191 \ln D + 19.987 \text{ or } D = 5.55 \times 10^{-8} \times \exp\left(\frac{H}{1.191}\right)$$

One does not necessarily have to use these relationships. Other suitable relationships can be derived if so desired.

Solution The first option for evaluating flood protection schemes is without considering uncertainty; that is, determine the EMV of each scheme. To that end, if the dam is constructed it will have a benefit of $80 million. The net benefit will be $80 million − $40 million = $40 million. This is also the EMV of the flood benefit of the dam scheme. The benefit–cost ratio is 1. For the levees, the damage is reduced to $25 million. The benefit is $100 million − $25 million = $75 million. The net benefit is $75 million − $30 million = $45 million. This is also the EMV of the flood benefit of the levee scheme. This gives a benefit–cost ratio of 1.5. For the drainage system, the damage is reduced to $36 million, so the benefit is $100 million − $36 million = $64 million. The net benefit is $64 million − $24 million = $40 million, giving a benefit–cost ratio of 1.67. The EMV of the drainage system scheme is also $40 million. For the land use option, the benefit is $100 million − $40 million = $60 million. The net benefit is $60 million − $20 million = $40 million, yielding a benefit–cost ratio of 2. The EMV of the land use scheme is $40 million. Based on the benefit–cost ratio, one may want to select the land use option because it has the highest benefit–cost ratio.

Now the second option is examined. This option considers evaluating the EMV of flood damage for each flood protection scheme under uncertainty. Note that flood peak Q is a random variable, and so are floods on the Amite River occurring each year. Likewise, flood damage is a random variable and its value

Table E1-4 *Discharge, stage, and damage data.*

Discharge (cfs)	Stage (ft)	Damage ($)
91,282	39.27	1.16×10^7
39,348	32.5	3.94×10^4
44,335	33.46	8.83×10^4
63,340	36.33	9.82×10^5
60,418	35.95	7.14×10^5
50,265	34.47	2.06×10^5
50,768	34.55	2.20×10^5
47,531	34.02	1.41×10^5
55,452	35.26	4.00×10^5
53,755	35.01	3.24×10^5
64,693	36.5	1.13×10^6
114,313	41.08	5.30×10^7
33,394	31.18	1.30×10^4
66,321	36.7	1.34×10^6
60,493	35.96	7.20×10^5
37,861	32.19	3.04×10^4
98,473	39.88	1.93×10^7
43,300	33.27	7.52×10^4
79,419	38.15	4.53×10^6
61,555	36.1	8.10×10^5
60,268	35.93	7.02×10^5
43,679	33.34	7.98×10^4
54,023	35.05	3.35×10^5
44,778	33.54	9.44×10^4
35,184	31.6	1.85×10^4
81,317	38.34	5.31×10^6
48,426	34.17	1.60×10^5

changes from year to year. The other variable is water level or stage, because flood damage is related to it rather than to flow. Thus, three random variables are to be dealt with: the flood peak discharge Q, the corresponding flood stage H, and the corresponding flood damage D. To compute flood damage, three relationships are needed: (1) a relationship between peak flow and the corresponding water level (i.e., a rating curve) at the nearby gauge, (2) a relationship between stage and the corresponding flood damage, and (3) a relationship between the dollar value and damage. From the data available at the nearby gauge, the rating curve (the relationship between H and Q) is given as shown in Fig. 1-7a. The second relationship between the flood stage H and the flood damage D is obtained from the data on actual damage figures and water level observations and is also given as shown in Fig. 1-7b. Analytical expressions for these relationships are given.

In evaluating flood protection schemes under uncertainty, the first step is to perform a frequency analysis of historical flood peak data (Table 1-4) and construct a cumulative distribution function (CDF) of Q, $F(Q)$, as shown in Fig. 1-7b. Likewise, a CDF of flood damage D, $F(D)$, is constructed as shown in Fig. 1-7d.

Not all flood peaks cause damage; only some do and these usually are in the upper 20% range (i.e., flood peaks in this range contain all damaging floods). This part of the CDF is plotted separately as shown in Fig. 1-7c. The second step is to construct a CDF of annual flood damage as shown in Fig. 1-7d (for flood discharge values, selected ones are in the upper 20th percentile). Each year the flood peak remains below 70,000 cfs with a probability of $F(Q) = 0.7$ or 70%. This flood discharge corresponds to a stage of 37.1 ft. Correspondingly, there is a 70% probability that the flood damage D will remain below $4.53 million each year. In other words, $F(D)$ for $D = \$4.53$ million is equal to 0.70 or 70%. In this way the function $F(D)$ can be constructed.

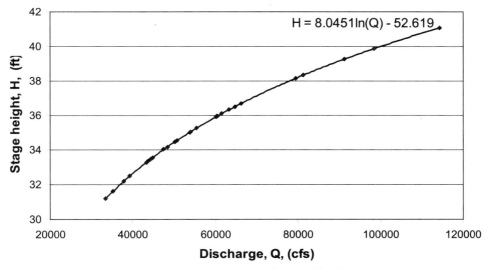

Figure 1-7a *Stage (H)-discharge (Q) relation.*

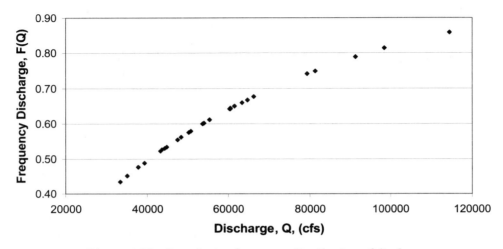

Figure 1-7b Cumulative frequency distribution of discharge.

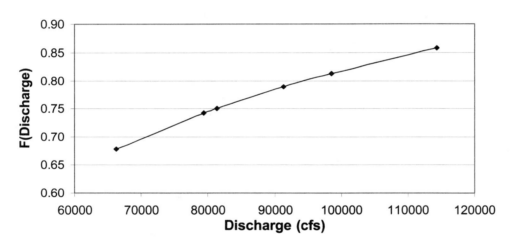

Figure 1-7c Frequency distribution of upper 20% damage-causing floods.

To that end, the CDF of D, $F(D)$, as shown in Fig. 1-7d, is employed for computing the probabilities of flood-damage values. For calculating the EMV of the annual flood damage, consider an interval between $D' - (\frac{1}{2})\, dD$ and $D' + (\frac{1}{2})\, dD$ as shown in Fig. 1-8. The probability that D lies in this interval is equal to $dF(D)$, the interval on the $F(D)$ axis that corresponds to the interval dD on the D axis. The incremental value of EMV of the damage in this interval can be denoted by dEMV and can be written as the magnitude of the damage times the probability of its occurrence:

$$d\text{EMV} = D \times dF(D)$$

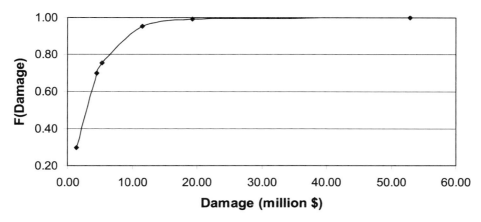

Figure 1-7d Frequency distribution of damage due to upper 20% floods.

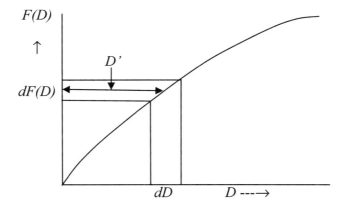

Figure 1-8 Calculation of EMV of the annual flood damage.

Therefore, EMV is obtained by integrating

$$\text{EMV} = \int_0^1 D\, dF(D) \tag{1.1}$$

The integral of Eq. 1.1 defines the area between the $F(D)$ axis and the CDF curve of flood damage in Fig. 1-8. The integration can be performed numerically by taking probability intervals that are sufficiently small for determining the average value of the damage in the interval. Multiplying the average damage with the probability interval and adding the values for all intervals gives the EMV of the annual flood damage. Since costs of alternatives will be spread over time, the present worth of the EMVs can then be determined by using a given interest rate. The final value is the EMV of the flood damage. This can be done for each flood protection scheme and then a flood protection scheme can be selected on this basis. For the flood-damage curve, the EMV is computed as

$$EMV = (1.34 \times 0.30) + (4.53 \times 0.70) + (5.31 \times 0.75)$$
$$+ (11.6 \times 0.95) + (19.3 \times 0.99) + (53 \times 1.00)$$

$$= \$90.8 \text{ million}$$

A comment regarding computation of EMV for each option is in order. Each flood protection scheme affects the EMV of flood damage in its own way. If, for example, levees are to be used for a flood protection scheme, then levees eliminate the flood damage up to the level where they control the channel flooding and therefore would change the rating curve at the gauging site. This in turn would change the damage curve. In a similar manner, each flood protection scheme would lead to a modified EMV of flood damage. Without a rating curve and a damage curve for each option, it is not possible to compute EMVs of these options and make a statement as to which should be the preferred option.

Example 1.5 Tresimeno Lake in central Italy is used for irrigation and recreational purposes. The water level in the lake reservoir needs to be regulated through control works at the outlet and outlet channel. The channel allows large discharges to pass without raising the lake levels too high, whereas the control works, such as a gate, permit holding water back when the water level would be low. The release of water from the lake for irrigation purposes is also controlled by the gate. The lake water level is to be regulated to make it attractive for recreation on the one hand and satisfy the irrigation water supply need on the other hand. There may be a conflict in satisfying these two objectives. The question to be addressed is how best to regulate the lake water level. Analyze conceptually the approach for determining the best solution in these circumstances. The lake level is a random variable and can be denoted as X. The annual benefit to be derived from recreation and irrigated agriculture is also a random variable and can be denoted as Y.

Solution There can be several alternative proposals to achieve the twin objectives. One alternative may be to emphasize the recreational benefits only. The other alternative is to emphasize irrigation benefits only. There may be several alternatives combining the two objectives in different ways. Regardless of the strategy to regulate the lake, one can hypothesize that the best solution is the one for which the net benefits are maximized. The determination of the best solution then requires, for each proposal, the calculation of the cost and the benefits. Were there no uncertainties, one can first calculate the benefits as the EMV of the recreation and irrigation benefits under natural conditions. Second, one can calculate the EMV of the recreation benefits with the various proposals for lake regulation. Then, the difference between the two EMVs can be computed for each proposal and a decision can be made. However, the lake level is a random variable and thus calculations of EMVs should be done under uncertainty.

To illustrate the procedure under uncertainty, we consider the average lake level during the months of June and July as the basic random variable X. There is a relationship between lake level and possible benefits, that is, between X and Y. For example, if the lake level is very high, the available beach area may be reduced and more water may have to be released than what is needed for irrigation. High lake levels also cause beach erosion and may cause damage to cottages and boats and other tourist facilities, especially when coupled with strong winds. However, low lake levels expose mudflats and cause difficulties with boating on the lake, such as shoals and docks that are too high. Desired releases of water for irrigation may not be permissible. This means that a monetary value must be assigned to each possible lake level. To that end, a survey can be conducted and on that basis a kind of utility function of the lake levels can be prepared. This function quantitatively expresses the monetary worth of the average lake level during the months of June and July. Figure 1-9 shows this hypothetical utility function. It is admitted that this utility function is imprecise, subjective, and difficult to determine. It involves value judgment and attempts to express intangibles, such as recreation benefits, in terms of monetary value. To make a rational decision, the worth of benefits has to be expressed in quantitative terms. It is assumed that the probability density function (PDF) of X is known or can be obtained from data, as shown in the bottom of Fig. 1-9. Then the CDF of X is determined from its PDF. Similarly, the PDF and CDF of Y can be derived from the knowledge of its relationship with X or independently from data, as shown in Fig. 1-9. It should be noted that the same value of Y can occur for two different values of X. If Y is set at a given arbitrary value y_1, then it can be stated that Y is smaller than y_1 if and only if either X is smaller than x_1 or X is larger than x_2. Thus,

$$P(Y < y_1) = P(X < x_1) + P(X > x_2) \qquad (1.2)$$

The two probabilities, $P(X < x_1)$ and $P(X > x_2)$, in Eq. 1.2 can be read from the CDF of X, as shown by a and b in Fig. 1-9. Also,

$$F(y) = P(Y < y) = P(Y < y_1) = a + b \qquad (1.3)$$

In this manner, every point of the CDF of Y can be determined, which is the CDF of the annual benefits. Then EMVs can be computed and in this way different options for lake regulation can be evaluated.

Example 1.6 A chemical plant produces wastewater in a city at a rate of about 2,000 m^3/day and a price has to be paid for this pollution. The owner wants to know whether he should install a pollution abatement device at the plant that will eliminate pollution or simply pay the penalty each year. Another option could be that he pays a small amount of penalty and abates pollution to some extent. However, all this depends on the weather also. In bad weather, the production will have to be reduced, which leads to less pollution. It is assumed that the probability of concentration of the wastewater exceeding the related standard is 0.8. The owner has the following three options: (1) He can install a pollution abatement device at the plant, which will eliminate pollution. The cost for the construction is about $20,000, and the cost for abatement is about $3/$m^3$.

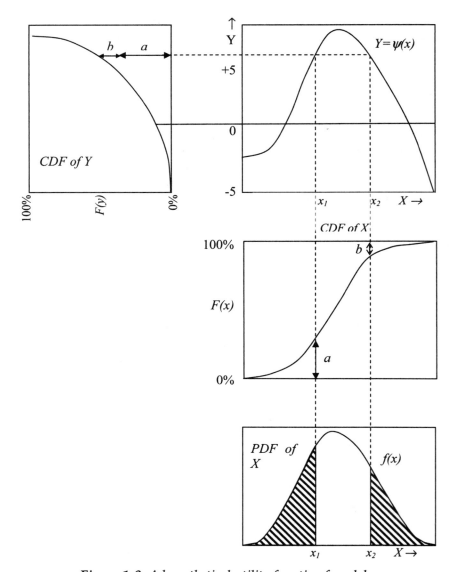

Figure 1-9 *A hypothetical utility function for a lake.*

(2) The owner can just pay the penalty each year. The penalty is about $25/m^3$.
(3) The owner can pay a small amount of penalty and abate pollution to some extent. It is assumed that the owner decides to treat about 70% of the wastewater and pays a penalty for the remaining 30%.

From the perspective of the plant owner, the less the cost is, the better. Other factors, such as environmental and social consequences, are not as important as economic ones. Find the least-cost solution.

Solution The least-cost solution or option can be determined using EMV. To that end, let us consider each option one by one.

Option A: For this option there are two consequences. First, if the pollutant concentration is below the discharge standard, the cost is $20,000, which is the construction cost of the abatement device. If the concentration is above the discharge standard, the cost will be the cost of construction and treatment = $20,000 + $3 × 2,000 × 0.8 = $24,800. Therefore, the cost of this consequence will be $24,800.

Option B: If the concentration is below the discharge standard, the owner has to pay nothing with a probability of 1 − 0.8 = 0.2. Thus, the cost of this consequence is $0. If the concentration is above the discharge standard, the cost will be the penalty, which is $25 × 2,000 = $50,000, with a probability of 0.8. Thus, the effective cost will be $50,000 × 0.8 = $40,000.

Option C: If the concentration is below the discharge standard, the cost will be the cost of the abatement device construction, which is $20,000. Thus, the effective cost of this consequence will be $20,000 + 0.8 × 2,000 × [($3 × 0.7) + ($25 × 0.3)] = $35,360, as the random part is only the level of pollution exceeding the standard with a probability of 0.8 and 70% of the wastewater to be treated (at $3/m^3) and the penalty is to be paid for the rest (at $25/m^3). All the costs are listed in Table 1-6a.

Now we calculate the EMV of each option as follows:

$$EMV_A = \$20{,}000 + (\$3 \times 2{,}000 \times 0.8) = \$24{,}800$$

$$EMV_B = \$25 \times 2{,}000 \times 0.8 = \$40{,}000$$

$$EMV_C = \$20{,}000 + 0.8 \times 2{,}000 \times [(\$3 \times 0.7) + (\$25 \times 0.3)] = \$35{,}360$$

Thus $EMV_A < EMV_C < EMV_B$.

If we consider the nonlinearity of gains and losses to improve decision making and assume that a curve similar to Fig. 1-6 is applicable in this case, we can construct a table of real and utility dollars (Table 1-6b).

Then we can calculate the corresponding EUVs:

$$EUV_A = (\$30{,}000 \times 0.8) + (\$22{,}000 \times 0.2) = \$28{,}400$$

$$EUV_B = (\$66{,}000 \times 0.074) + (\$48{,}000 \times 0.926) = \$49{,}332$$

$$EUV_C = (\$48{,}000 \times 0.71) + (\$30{,}000 \times 0.29) = \$42{,}780$$

One can also compute the EUVs directly by converting the total value into utility dollars. Thus $EUV_A < EUV_C < EUV_B$.

From the EMVs and EUVs, it is seen that option A costs the least and option B costs the most. Thus, it is concluded that option A is the best choice for the plant owner.

Table E1-6a Costs associated with different alternatives.

Alternatives	Cost of maintaining the desired level of concentration or paying penalty ($)		Fixed cost for abatement device ($)	Total value ($)
	C < standard ($p = 0.2$)	C > standard ($p = 0.8$)		
Option A	$2{,}000 \times 0.2 \times 0 = 0$	$2{,}000 \times 3 \times 0.8 = 4{,}800$	20,000	24,800
Option B	$20{,}000 \times 0.2 \times 0 = 0$	$2{,}000 \times 25 \times 0.8 = 40{,}000$	0	40,000
Option C	$2{,}000 \times 0.2 \times 0 = 0$	$2{,}000 \times (3 \times 0.7 + 25 \times 0.3) \times 0.8 = 15{,}360$	20,000	35,360

Table E1-6b Real and utility dollars.

Real dollars	20,000	26,000	39,200	50,000
Utility dollars	22,000	30,000	48,000	66,000

1.6 Sources of Uncertainties

There are many causes of uncertainty in a system that interacts with nature. The major cause of uncertainty in the behavior of environment-related systems involve rapid fluctuations in space and time, which are inherent features of natural processes. Meteorological, hydrological, environmental, social, and economic processes are highly influenced by random factors whose interaction affects the performance of a project. Consequently, inputs and outputs of natural and engineering systems, such as a catchment, a river reach, an aquifer, the foundation of a skyscraper, or a bridge, are highly random. Although this topic will be covered in greater detail in subsequent chapters, it is worthwhile to provide a synopsis here.

Another major source of uncertainty is data. A detailed risk analysis requires collection and analysis of huge volumes of quality data for a number of variables. In most studies, such data are not available because of bad equipment or human malfeasance, or the quality of data is inadequate owing to measurement errors, inadequate sampling, or variability and complexity of the underlying systems. In the absence of the required quality data, simplifying assumptions are often made and thus uncertainty creeps into the analysis.

Since the systems are highly complex, the models that are used to represent these systems are not perfect. For example, the response of a catchment is a nonlinear function of many input variables but in a particular study these nonlinearities may be ignored. Such simplifications are typical sources of uncertainty. It is important to recall that most natural and socio-economic systems have yet to be fully understood.

It is instructive here to note the following pertinent observations of Cornell (1972): "It is important in engineering applications that we avoid the tendency to model only those probabilistic aspects that we think we know how to analyze. It is far better to have an approximate model of the whole problem than an exact model of only a portion of it."

In an ideal situation, one should derive joint probability distributions for all the sources of uncertainty that significantly influence the behavior of the system. This however is not possible, except in simple cases, owing to the involvement of a large number of variables and their interactions in a nonlinear manner. The techniques that are most commonly used for uncertainty analysis are Monte Carlo simulation (MCS), the mean-value first-order second-moment (MFOSM) method, and the advanced first-order second-moment (AFOSM) method. In Monte Carlo simulation, long and multiple series of input random variables are generated according to the distributions that they follow. Next, these are input to the model of the system and the output is monitored. Statistical analysis of this output yields measures of its behavior and the probability distribution.

In the mean-value first-order second-moment method, the Taylor series expansion of the system performance function is truncated after the first terms. The use of the term "mean-value" signifies that the expansion is about the mean value of the variable. Further, only the first two moments of the variable are needed. This makes the method easy to apply and simplifies the calculations. However, if the linearization about the central value is not an adequate representation of the true behavior of the variable, the method will not give acceptable results. Another criticism of this method arises from the fact that most engineering systems do not fail near the average of the performance function. Rather, failure occurs at some extreme value.

In the AFOSM method, the Taylor series expansion of the performance function is taken at a likely failure point. Thus, the key to a successful application of AFOSM is the determination of the likely failure point. In AFOSM, the reliability index is the shortest distance between the mean state of the system and the failure surface. This index can be found either by applying some nonlinear optimization procedure or by following an iterative scheme.

While comparing the reliability analysis methods with specific reference to watershed models, Melching (1995) noted that the AFOSM method displayed good agreement with MCS and better agreement for the tails of the probability distributions. Expectedly, he noted that when the nonlinearities were not large, MFOSM performed as accurately as (and sometimes better than) AFOSM. Note that the results of MCS for a large number of runs formed the standard against which the performance of the other method was compared.

1.7 Rational Decision Making

Bouchart and Goulter (1998) have reviewed rational decision making in the context of the management of irrigation reservoirs. They examined a number of models that permit rational decision making under risk. Rational decision making under risk or uncertainty involves identifying (1) the actual decisions to be taken in a particular situation (optimal policy) and (2) the models that are employed to select the optimal policy. The models they discussed include expectation objective, expectation-variance objective, safety-first rule, utility function, stochastic dominance, and risk curves.

The mathematical expectation of the risk curve is the simplest approach to making a rational decision. This is not suitable for management of irrigation reservoirs, because it is risk neutral. The expectation-variance approach bases preferences on mean and variance of the net returns or the outcome is normally distributed. These assumptions are quite restrictive. The safety-first rule avoids risk and is therefore not a valid approach. The utility function measures the utility level of each decision or action. It is quite difficult to quantify the utility level. Stochastic dominance or stochastic efficiency is an efficiency measure of different decisions. It lacks the ability to identify a strategy for comparing risk curves. It is difficult to analytically define risk curves of different decisions. One of the main problems of these decision models is that they are only capable of partially accommodating the concerns of the decision maker. For rational decision making Bouchart and Goulter (1998) proposed a methodology using neural networks.

1.8 Questions

1.1   An urban area gets flooded and flooding needs to be mitigated. To that end, one can consider construction of a detention pond and attenuate flood peaks using the pond. Another option could be upgrading or enhancement of the existing drainage system, which will suffice to carry greater runoff. Still another option could be construction of additional drainage channels. Proper land use management can be another option. There can be other options also. Since flooding is a random variable, a decision has to be made under uncertainty. Analyze the urban flooding and discuss conceptually which way is the most rational way to mitigate flooding.

1.2   Consider the problem of water supply to a city. There can be several ways by which water can be supplied to the city. Water can be supplied from a nearby river. Of course, the river flow is subject to uncertainty. Water can be supplied from groundwater, which is also subject to uncertainty. Water can be supplied using a combination of surface and groundwater sources. Still another source can be water harvesting. In any case, a decision has to be made under uncertainty. What is the rational decision for supplying water to the city? Discuss it conceptually.

1.3   Consider a problem of solid waste disposal. An alternative is to burn it in the open or incinerate it mechanically. One can also landfill it. Or one can use both options. Still another way is to haul it away to another place or dump it in the nearby sea. There are many options, but each option has an element of uncertainty. Discuss conceptually the rational decision for disposing of the solid waste.

1.4   A farmer has to make a rational decision about selecting a crop for production in a season. There are cash crops and noncash crops. There is, however, a question of the sensitivity of crops to weather, which is uncertain. Thus, the risk of having a crop come to fruition varies with the type of crop. The farmer would like to make a rational decision about growing crops on his farm. Discuss conceptually what the rational decision should be.

1.5   A farmer has a steady water supply from a reservoir. She has to make a decision about growing those crops that can bring her the most returns. Each crop has advantages and disadvantages. Crops can be rice, sugarcane, vegetables, etc. How should she go about deciding which crops to grow?

1.6   A chemical plant in a city produces different types of chemicals. These chemicals lead to different profit margins, depending on the market demand. Since the chemicals are produced following different production procedures, the level of resulting pollution varies with the type of the chemical produced. The chemical plant owner would like to minimize the pollution and of course the owner's main objective is to make money or maximize profit. He is considering different options. First, he can produce only one chemical that leads to the least pollution. Second, he can install pollution devices and produce a variety of chemicals. Third, he can produce the chemical that brings him the most money, even if it produces the most pollution. He can combine some of these options. Qualitatively analyze the rational decision the plant owner can make.

1.7   Coastal areas often suffer from erosion. Erosion can be controlled by using a sea wall, by coastal management, or by installing wave breakers. Each option is subject to uncertainty. The question is one of selecting the most rational option. Discuss qualitatively the rational decision for controlling coastal erosion.

1.8   A new urban area is to be developed. This area is to be provided with adequate drainage facilities. The developer has to ask whether to provide adequate drainage or to provide a little less than adequate drainage and bear the cost of flooding of houses if it occurred. Drainage systems can be in different forms. Each option is subject to uncertainty. Discuss qualitatively the rational decision that the builder can make.

1.9   A reservoir is employed for both recreation and water supply. The reservoir water level has to be monitored accordingly. The higher reservoir water level is better for recreation but may cause other problems, such as beach erosion. When the reservoir water level is low, one has to decide about irrigation and recreation, which may be in conflict. Each option is subject to uncertainty. What is the rational decision for operating the reservoir? Discuss it qualitatively.

1.10   An industrial plant generates considerable waste and the owner wants to determine the best way to dispose of it. There can be many options. The plant is located near a river. The plant can dump the waste in the river but the river water quality has to be maintained. This means that the entire waste may not be dumped. The owner can construct a storage facility and treat the waste or use a combination of both. Discuss conceptually the rational decision that can be made for disposing of the waste.

1.11   A family is building a new home in the New Orleans area. The base cost of the home is $100,000, but the family can choose the degree of wind resistance of the structure. For the base price, the house is designed to withstand wind gusts of 90 miles per hour (mph). For an additional cost of $2,000, the house will withstand winds of 100 mph, and for an additional $4,000 the house can withstand winds of 110 mph. The expected cost of repairing wind damage is $10,000, and the family hopes to not have to pay for damage during the first 10 years of living in the house. From the FEMA Multihazard Loss Estimation Methodology Hurricane Model HAZUS-MH Technical Manual, the return periods of various wind gusts can be estimated as shown below.

Wind gust (mph)	Return period (years)	Annual probability (per year)
90	15	0.067
100	23	0.043
110	35	0.029

From these data and the building costs given, one can determine the expected net benefit for different options. What should be the rational decision for the family for building its home?

1.12   Following catastrophic Hurricane Katrina, nearly 1.5 million people had to be evacuated from the Louisiana–Mississippi Gulf coast. Clean up operations will take about a year. Assume that 500,000 people will be forced to live in temporary housing for a year. Supplying adequate drinking water is an important aspect of maintaining safe housing development. Relief officials have three options to choose from: (1) shipping bottled water, (2) pumping well water, and (3) on-site chlorination of surface water. Assume that the cost of bottled water is $1.89 per liter and the demand can be met with certainty. Per capita consumption can be assumed to be 2 liters per day. It can be assumed that pumping groundwater will cost $0.02912 per liter but there is a 75% chance that the groundwater supply will be exhausted before the end of the year. The cost of chlorination is $0.035071 per liter and there is a 90% chance that there will be sufficient surface water and chlorination to meet the water demand. What is the rational decision that the relief officials should make for meeting the water demand of the evacuees in temporary housing?

1.13   A system is to be designed for transporting semirefined uranium ore for a distance of 100 miles in the Athabasca Basin region of Saskatchewan, Canada. It is found that a 30-mile distance is over continuous muskeg or peat bog. Muskeg itself consists of dead plants in various stages of decomposition. The water level in muskeg is usually at or near the surface. Proposed alternative configurations could be monorail, truck or train on all-season roads, ropeway, or any other transportation technology. What should be the rational decision for designing the transportation system? Discuss conceptually.

Chapter 2

Elements of Probability

Most natural processes, including environmental, hydrologic, and hydraulic processes, are not deterministic; that is, they are unpredictable or stochastic. For example, the next state of the environment is not fully determined by the previous state of the environment. Uncertainty is introduced into engineering systems through the variation inherent in nature, lack of understanding of causes and effects in physical systems, and lack of sufficient data or inaccuracy thereof. As an example, consider the prediction of the maximum flood in a river in the next 10 years. One cannot predict the magnitude of this flood with certainty even with a long record of data. The uncertainty in flood prediction results from natural variation. Thus, we must consider the "possibility" of occurrence of such events and determine the likelihood of their occurrence. Information on their occurrence will form an input while mitigating the ensuing undesirable consequences that might arise from risk and uncertainty.

In engineering design and analysis, many problems involve a study of mathematical models of some random phenomena. Dealing with uncertainty of an engineering random phenomenon has been and always will be a challenge. Some things can never be predicted with absolute certainty. A random phenomenon is defined as a phenomenon that obeys probabilistic, rather than deterministic, laws. Statistics plays a major role in engineering by offering probabilistic concepts to deal with random phenomena.

To apply probability theory to an engineering process, we study the observed data of that process. The collection of all possible observations of a process is called a *statistical population*. The population itself often cannot be totally observed, because the process is time dependent and infinitely long and thus cannot be observed in a limited time frame. Generally the number of observations is limited by the availability of money, time, and/or period of interest under study. Most often, only a portion of the population is observed, which is called a *sample*.

Statistics deals with the computation of statistical characteristics using sampled data, and probability deals with the prediction of chance or likelihood of occurrence of an event from sampled data. Frequency analysis deals with estimating future probabilities of occurrence of some specific events, such as floods, droughts, rainfall, water pollution, air quality, sediment load, depletion of ozone layer, occurrence of earthquakes, and snow avalanches, by analyzing past sampled data or records of interest.

2.1 Random Variables

An observation (e.g., daily rainfall amount or number of cars passing through an intersection every hour) or the outcome of an experiment (such as analyzing stream samples for water quality variables, e.g., dissolved oxygen, bio-oxygen demand, pH, total phosphorus, or total suspended solids) or a mathematical model of a random phenomenon (e.g., a dissolved oxygen–bio-oxygen demand model or a rainfall–runoff model) can be characterized by one or more variables that are, to a certain degree, unpredictable. Yet there is frequently a degree of consistency in the factors governing the outcome that exhibits a statistical regularity. This statistical regularity is expressed through a probability distribution defined on the probability space. Such variables are called *random variables* or *stochastic variables*.

The term "experiment" is used here in a general sense. An experiment may be counting the times the water level at a certain river cross section exceeds a defined threshold, or measuring the discharge at that section. In Fig. 2-1 daily observed flow of the Hillsborough River downstream of the Tampa Dam is given, and flow of more than 2,000 cubic feet per second (cfs) is of interest because of flooding issues. In this case, the measurements of daily discharges during, say, January 1990 to December 2000 will constitute the experiment's sample space. This space consists of a set S containing sample points, each of which is associated with one and only one distinguishable outcome. An event is a collection of sample points (for instance, the collection of observed daily discharges in Fig. 2-1) in the sample space of an experiment.

It is important to precisely specify the conditions under which a variable can be regarded as random. The qualification "random" implies the absence of

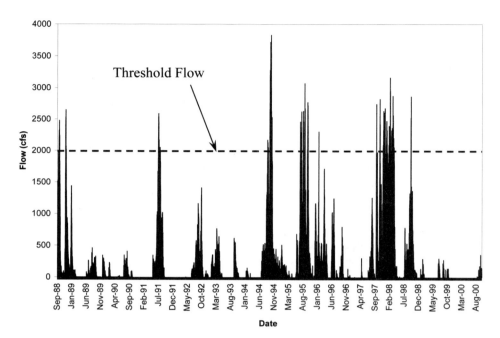

Figure 2-1 *Daily discharge of the Hillsborough River downstream of the Tampa Dam.*

discernable systematic variations, such as those caused by cycles, trends, or changes in the factors governing the outcome of an event. The variables, such as the daily temperature in a greenhouse, outflows from a controlled reservoir, roughness of a steel pipe, and many others, cannot be considered random. The systematic variations in these variables would render them incompatible with the qualification "random."

Now consider the number of eclipses occurring in a year as a variable. Astronomers can accurately calculate the number of eclipses in any year. Therefore this number is not a random variable for them. However, someone not familiar with the laws governing the occurrence of eclipses could well regard it as a random variable and use the observations of many years to estimate the probability associated with the occurrences of eclipses. There may be similar variables involving cycles and trends that are commonly regarded as random because not enough is known about them, but they may not be truly random. Thus, the question whether or not a real-life variable is random must then be addressed, based on observations and whether the conditions of randomness are met. Random variables are well-defined objects in a probability space.

To define a random variable, its probability space must be fully described. A probability space is composed of the following three elements:

1. A sample space S can be any set and can usually be defined as the set of the most elementary events that allows us to describe a given process.

2. A set of events E, defined over the sample space S, is a collection of subsets of the sample space S.

3. A measure of probability P is a function that assigns probability to each member of event set E.

The outcome of an experiment need not be a number. For example, the outcome of an event can be a "head" or a "tail" when a coin is tossed; "pass" or "fail" in a classroom test; "spade," "heart," "diamond," or "club" in a game of cards; or "dry" or "wet" for weather on a day; and so on. However, we often want to represent outcomes as numbers. A random variable is obtained by assigning, without prejudice, a numerical value to each outcome of a particular experiment. Random variables may be positive, negative, or zero. When outcomes are not already numerical values, then scores or numerical values are assigned to make them so. The value of a random variable will vary from trial to trial as the experiment is repeated. For example, one may associate number 1 with pass and number 0 with fail, one may associate the numbers 1 to 4 with the four suits in a deck of cards, and so on. Such numbers are then, of course, purely conventional. In customary nomenclature, random variables are usually represented by uppercase letters, such as $X, Y, Z, ...$, and their values by lowercase letters, such as $x, y, z, ...$ The notation $X(s) = x$ means that x is the value associated with the outcome of the random variable X at s. A random variable X may have values such as $x = 1.5$, $x = 2.7$, and $x = -3.9$, etc. Frequently, the values of 0 and 1 are assigned to a random variable for representing success or failure of an event. Any random variable whose possible values are either 0 or 1 is called a Bernoulli random variable

2.1.1 Types of Random Variables

In the preceding discussion we defined a random variable to be a function from the sample space to the real numbers. Based on the type and number of values it takes on, a random variable can thus be assumed to be either a discrete or continuous random variable. If assigned values for the random variable constitute a finite set or a countably infinite set, it is called a *discrete variable*: associated with discrete random events or outcomes. The roll of dice is one classical example of a discrete random variable. Other examples of discrete random variables include the number of rainy days or number of cloudy days in a month, the number of violations of a water quality standard, and the number of exceedances of a threshold flow at a given site. *Continuous random variables* are those associated with a continuous interval where any event can occur within the interval. For example, a random variable that can take on any value between 0 and 1 is a continuous random variable. Other examples of continuous random variables include river stage or flow, reservoir level, temperature, rainfall amount, and dissolved oxygen (DO) or bio-oxygen demand (BOD) of river water. Consider, for example, the data on DO concentration taken six times a day in a month as presented in Fig. 2-2. In this example, the DO concentration is a continuous

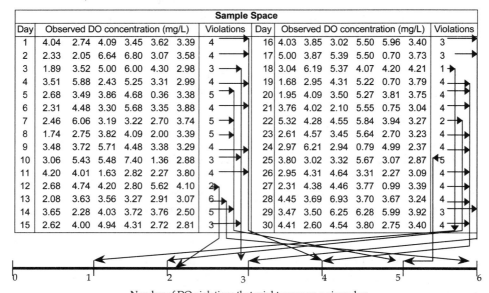

Figure 2-2 *Random variable X (number of DO violations) as a function mapping the elements of the sample space onto the real line.*

variable, whereas the number of violations of the DO standard is a discrete variable. A DO violation is said to occur when the observed DO concentration is found to be less than 4.0 mg/L.

Figure 2-3 depicts the random variable X, the daily violation of the DO standard, with a bar chart. Mathematically, a random variable X can be considered as a real-valued function that maps all elements of the sample space into points on the real number line (R). In other words, in this example, the random variable is a numerical characteristic that depends on the daily sample with daily sample size $n = 6$. Each of 30 elementary events is represented by a string of the actual DO concentration.

For discrete random variables the probability is concentrated in single points of the probability space. The number of possible outcomes with a discrete random variable is not necessarily finite. The number of occurrences of a discrete random variable is called *frequency*. For example, if, in an experiment, a coin is tossed repeatedly, one may ask what the probability is that heads will turn up for the first time on the nth toss. The variable that heads will turn up, X, is then a discrete random variable that has no upper bound if the number of tosses is infinite. With continuous random variables no single point carries a measurable probability (i.e., $P(X = x) = 0$) and, consequently, measurable probability is only associated with nondegenerate intervals (i.e., a nonzero-width interval). The number of possible outcomes of a continuous variable is always infinite where its set of possible values consists of an entire interval in R.

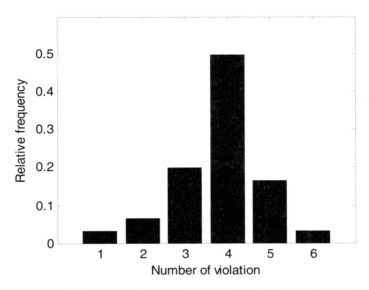

Figure 2-3 *Discrete random variable* X *(number of DO violations).*

For a discrete variable, there is at most a finite number of possible outcomes in any finite interval. For example, the number of floods per year at a point on a river can take on only integer values, which is a discrete event, so it is a discrete random variable. For a discrete random variable having an infinite number of possible outcomes, these outcomes must be enumerable. This means that the outcomes can be arranged in a sequence so that there is a one-to-one correspondence between the elements in the sequence and the positive integer numbers 1, 2, 3, etc. Stated another way, the possible outcomes can be counted even if counting would take an infinitely long time. However, this is evidently not possible with a continuous variable. For example, considering the discharge of the Amite River at Denham Springs, we see that any value satisfied by the continuous variables can occur in R^+ (where R^+ is a set of positive real numbers). The annual flood peaks would be infinite in number; even between any two peak discharge values, there will be an infinite number of flood peaks.

A random variable is continuous if its possible values span an entire interval in R. A continuous variable takes on values within a continuum of values. The discharge at a point on a river is a continuous variable. A random variable may be discrete or continuous, depending on the interval under consideration. Annual river flows are evidently represented by continuous variables. Temperature, vapor pressure, relative humidity, etc. are other examples of continuous variables. However, there may be a measurable probability that the flow in any particular year is zero. At point zero, the variable is then discrete; everywhere else it is continuous.

Here are some other examples:

1. A coin is tossed 20 times. In this case, the random variable X is the number of tails that are noted. X can take on only values such as 0, 1, ..., 20, so X is a discrete random variable.
2. A light bulb is lit until it burns out. In this example, the random variable X is its lifetime in hours. X can take on any positive real value, so X is a continuous random variable.

2.1.2 Graphical Description of Random Variables

When studying a random variable, the first step is to characterize it by using the collected observed data. The data screening or exploratory data analysis can be done by representing the data set using visual techniques, such as bar graphs, pie graphs, histograms, or other kinds of pictorial portrayals. Graphical descriptors are useful to grasp the data characteristics and for determining the family to which the random variable under study belongs. For discrete random variables, it is often possible to determine the appropriate family from a physical description of an engineering system. In a continuous case, it is generally more difficult to determine the family of a random variable. The graphical description of a random variable is very helpful to get an idea of the shape of its probability distribution. There are several graphical methods used to describe the data of a random variable. Here we will discuss two commonly used graphical descriptors to display a random variable: histograms and stem-and-leaf diagrams.

2.1.2.1 Histogram

Histogramming is a method of discretization (encoding a data set using bins or classes) wherein a continuous data set is converted into discrete data. A histogram is constructed by dividing the observed data into bins (or classes) such that for the first bin, $x_1 \le X < x_2$, for the second bin, $x_2 \le X < x_3$, etc. The DO concentration values presented in Fig. 2-2 are divided into bins and then a plot is made of the number of observations in each bin versus the value of X as shown in Fig. 2-4.

The appropriate width of a bin or class interval and the number of total bins depend on the number of data points, the minimum and maximum values of data, and the overall behavior of the data. Often, 5 to 15 bins are sufficient for most practical applications. After categorizing the data into bins or classes, the number of observations in each bin is determined. As a rule of thumb, the number of classes is approximately equal to $\sqrt{n}$, where n is the number of observations. The best histogram is obtained by using the Sturges rule:

$$N = 1 + 3.3\log_{10}n \qquad (2.1)$$

where N is the number of classes and n is the total number of observations in the data set.

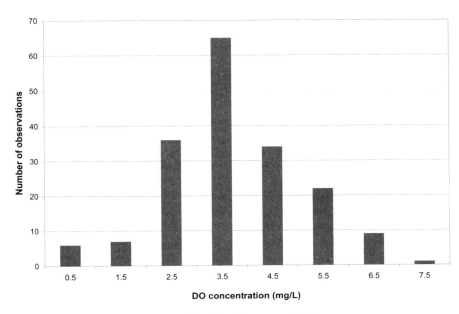

Figure 2-4 *Absolute frequency histogram.*

The purpose of a histogram is to graphically summarize the distribution of a univariate data set that includes the following information: center (i.e., the location of gravity) of the data, spread (i.e., the scale) of the data, skewness of the data, presence of outliers, and presence of multiple modes in the data. These features provide strong indications of a proper statistical model for the data. The most common form of the histogram is obtained by splitting the range of the data into equal-sized bins (called classes). Then for each bin, the number of points from the data set that fall into each bin is counted. The histogram can be normalized by plotting the relative frequency of each class versus the value of X in two ways:

1. The *normalized count* (relative frequency) is the count in a class divided by the total number of observations. In this case the relative counts are normalized to sum to one (or 100 if a percentage scale is used). This is the intuitive case where the height of the histogram bar represents the proportion of the data in each class, as given in Fig. 2-5a.
2. The *relative frequency* is the count in the class divided by the number of observations times the class width. For this normalization, the area (or integral) under the histogram is equal to one. From a probabilistic point of view, this normalization results in a relative histogram that is most akin to the probability density function as given in Fig. 2-5b.

The frequency polygon is obtained by joining the midpoint of each class matching the class frequency. The frequency histogram provides a meaningful arrangement of observed data and is considered as the foundation of any

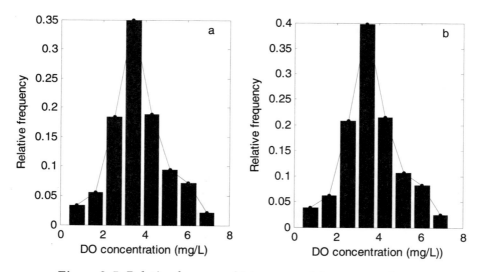

***Figure* 2-5** *Relative frequency histogram and frequency polygon.*

statistical analysis. But it must define the number of observations in each bin or class. In a comparison of two data sets, this will only be useful if they have the same number of observations. To compare groups of different sizes, the histogram must be modified by plotting the relative frequency of each class versus the value of X, as given in Fig. 2-5. The relative frequency of a class is determined by dividing the number of observations in that class by the total number of observations in the data set. An alternative display is the frequency polygon, shown for the same data in Fig. 2-5. The frequency polygon is obtained by joining the midpoint of each class matching the class frequency.

In many applications regulatory requirements are in terms of the number of exceedances or nonexceedances, as, for example, the number of violations of the critical dissolved oxygen level in a given reach of stream or the number of exceedances above a given flow or stage at a given cross section of a river. In these cases, it is advantageous to make another transformation of class frequencies to obtain a *cumulative frequency* plot. Cumulative relative frequency is determined by adding the frequency of each class to the sum of the frequencies for the lower classes by considering the relative frequency constructed by approaches (1) and (2). Table 2-1 provides cumulative frequencies for the DO data presented in Fig. 2-2. Figure 2-6 presents the cumulative frequency plot for the DO data. Based on this figure, one can make statements related to the chances of exceedance of a certain DO level, such as chances of exceedance of 3.0 mg/L = 1 − 0.27 = 0.73 (i.e., 73%).

Example 2.1 Table E2-1a gives the annual peak flow at the USGS 08075000 site on Brays Bayou River in downtown Houston, Texas. Develop a frequency histogram and a cumulative frequency plot for the peak flow.

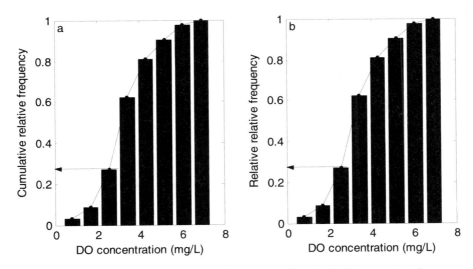

Figure 2-6 *Cumulative relative frequency plots for DO concentration.*

Table 2-1 *Development of histogram for the data of DO concentration.*

Lower	Upper	Midpoint	Frequency	Relative frequency		Cumulative frequency	
				(1)	(2)	(1)	(2)
0.36	1.24	0.8	6	0.033	0.038	0.033	0.033
1.24	2.12	1.68	10	0.056	0.063	0.089	0.089
2.12	3.00	2.56	33	0.183	0.208	0.272	0.272
3.00	3.88	3.44	63	0.350	0.398	0.622	0.622
3.88	4.76	4.32	34	0.189	0.215	0.811	0.811
4.76	5.64	5.2	17	0.094	0.107	0.905	0.906
5.64	6.52	6.08	13	0.072	0.082	0.977	0.978
6.52	7.4	6.96	4	0.022	0.025	1.000	1.000

Solution In the data from the table, $n = 66$. Using the Sturges rule, $N = 1 + 3.3 \log_{10}(66) = 7$. Then the bin width for each class is 4,030.6 cfs. Thus seven bins were selected and frequencies were determined for each bin. By dividing the frequency of each class by n, the relative frequency was determined and then the cumulative frequency was evaluated as given in Table E2-1b (where (1) and (2) denote the two approaches outlined earlier). Figures 2-7 and 2-8 show the relative frequency and cumulative frequency histograms obtained by both normalization approaches.

Table E2-1a

Year	1929	1936	1937	1938	1939	1940	1941	1942	1943	1944	1945
Peak flow	11100	6600	1270	4530	6800	1340	6460	4590	6280	8120	5590
Year	1946	1947	1948	1949	1950	1951	1952	1953	1954	1955	1956
Peak flow	3880	4360	1440	2340	5340	786	1850	3580	3680	3300	1180
Year	1957	1958	1959	1960	1961	1962	1963	1964	1965	1966	1967
Peak flow	4660	5100	7760	12600	6320	7720	8300	4060	3160	9400	4730
Year	1968	1969	1970	1971	1972	1973	1974	1975	1976	1977	1978
Peak flow	12000	9240	11500	15500	11700	24800	8660	18000	29000	8710	6260
Year	1979	1980	1981	1982	1983	1984	1985	1986	1987	1988	1989
Peak flow	25500	11300	25400	17700	29000	8640	12300	17300	22400	8290	21500
Year	1990	1991	1992	1993	1994	1995	1996	1997	1998	1999	2000
Peak flow	10400	19800	23000	16000	16600	27000	17700	23400	25500	16700	7640

Table E2-1b

Lower	Upper	Mid value	Frequency	Relative frequency		Cumulative frequency	
				(1)	(2)	(1)	(2)
786	4816.6	2801.3	18	0.27	6.77×10^{-5}	0.27	0.27
4816.6	8847.1	6831.9	18	0.27	6.77×10^{-5}	0.55	0.55
8847.1	12878	10862	9	0.14	3.38×10^{-5}	0.68	0.68
12878	16908	14893	5	0.08	1.88×10^{-5}	0.76	0.76
16908	20939	18924	5	0.08	1.88×10^{-5}	0.83	0.83
20939	24969	22954	5	0.08	1.88×10^{-5}	0.91	0.91
24969	29000	26985	6	0.09	2.26×10^{-5}	1	1

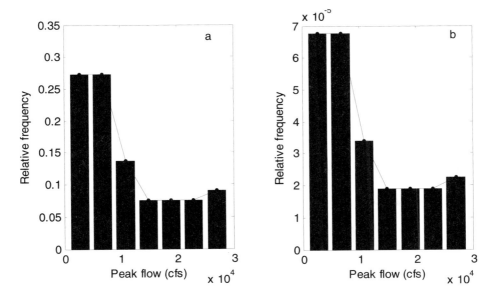

Figure 2-7 *Frequency histogram plot for the Brays Bayou River peak flow data.*

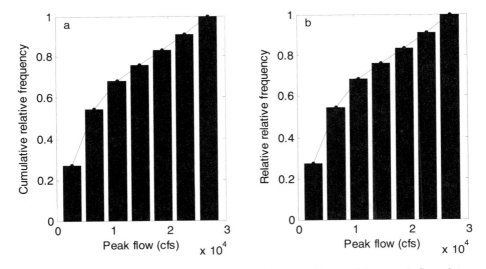

Figure 2-8 *Cumulative frequency plot for the Brays Bayou River peak flow data.*

2.1.2.2 Stem-and-Leaf Plot

To get additional information about the distribution of a random variable, the histogram can be modified into a ratio or interval variable to obtain a stem-and-leaf plot. The display of this plot provides particulars of the typical or representative value, the extent of spread about the typical value, the presence of gaps, the extent of symmetry, and the number and location of peaks.

The stem-and-leaf plot consists of a series of columns of numbers. Each column is labeled via a number called its *stem*, which is generally the first one or two digits of the numbers in the data set. The other numbers in the columns are called *leaves*, which are the last digits of the data. The following steps are used to construct a stem-and-leaf plot:

Step 1: Partition each data value into two parts.

Step 2: Based on the least and greatest data values, choose some convenient numbers to serve as stems. Stems may consist of one or more initial digits of the data values.

Step 3: Reproduce the data set graphically by recording the digits following the stem as a leaf. A leaf may consist of one or more of the remaining digits of the data values.

Figure 2-9 depicts a stem-and-leaf plot prepared using the DO data presented in Fig. 2-2. It is obvious from Fig. 2-9 that a stem-and-leaf plot does resemble a histogram. The stem values represent the intervals of a histogram while the leaf values represent the frequency for each interval. An advantage of the stem-and-leaf plot over the histogram is that the stem-and-leaf plot displays not only the frequency for each interval, but also displays all of the individual values within that interval.

2.2 Elements of Probability

People have an intuitive appreciation of the concept of probability, although they may have difficulty defining it precisely. Some events are known to be downright impossible, others quite unlikely or improbable, still others likely, and some almost certain. The occurrence of events may be almost unlikely, likely, very likely, almost certain, or even certain. Thus, people have an appreciation of a probability scale, but such an intuitive, nonscientific probability assessment is too vague for rational decision making.

Classical probability theory has two cornerstones. The first was the equal likelihood of all possible outcomes in a game of chance. For example, drawing the ace of hearts from a deck of 52 cards is neither more nor less likely than drawing any other card. In the toss of a true coin, any particular outcome is not favored over any other outcome: A head is just as likely an outcome as a tail.

The second cornerstone of classical theory is related to the relative frequency with which an event tends to occur. In the long run the relative frequency tends to approach the ratio of the number of successes (s) to the total number of trials (n), s/n. This is the number adopted to define the probability of an event. Consider, for example, the throw of two dice. If two dice are cast a very large number of times, the sum of the spots will be larger than 10 in approximately $1/12$ of the total number of throws. (In a throw of two dice, 36 outcomes are possible. Out of

	STEM							
	0	1	2	3	4	5	6	7
	0.36	0.36	0.00	0.02	0.00	0.00	0.00	0.40
	0.70	0.63	0.05	0.02	0.01	0.00	0.06	
	0.70	0.68	0.08	0.04	0.02	0.22	0.19	
	0.75	0.74	0.10	0.04	0.03	0.25	0.21	
	0.79	0.89	0.27	0.06	0.03	0.27	0.25	
	0.99	0.95	0.27	0.07	0.04	0.32	0.28	
			0.28	0.07	0.07	0.37	0.64	
			0.31	0.07	0.09	0.39	0.80	
			0.31	0.09	0.09	0.43	0.93	
			0.33	0.19	0.09	0.48		
			0.37	0.22	0.10	0.50		
			0.43	0.23	0.20	0.50		
			0.46	0.24	0.20	0.55		
			0.50	0.27	0.20	0.62		
			0.60	0.27	0.21	0.64		
			0.61	0.29	0.28	0.67		
			0.62	0.30	0.30	0.68		
			0.68	0.31	0.31	0.71		
			0.68	0.31	0.31	0.84		
			0.70	0.32	0.31	0.88		
			0.70	0.35	0.38	0.96		
			0.72	0.38	0.41			
			0.74	0.38	0.45			
			0.75	0.39	0.46			
			0.75	0.39	0.48			
L			0.80	0.39	0.48			
E			0.81	0.40	0.54			
A			0.82	0.40	0.55			
V			0.87	0.45	0.57			
E			0.88	0.45	0.64			
S			0.91	0.47	0.68			
			0.94	0.48	0.74			
			0.95	0.49	0.94			
			0.95	0.50	0.99			
			0.97	0.50				
			0.98	0.51				
			0.99	0.52				
				0.56				
				0.58				
				0.62				
				0.63				
				0.65				
				0.67				
				0.69				
				0.70				
				0.72				
				0.72				
				0.73				
				0.74				
				0.75				
				0.76				
				0.76				
				0.77				
				0.79				
				0.80				
				0.80				
				0.80				
				0.81				
				0.82				
				0.85				
				0.86				
				0.87				
				0.88				
				0.92				
				0.94				

Figure 2-9 Stem-and-leaf plot of DO concentration data.

these, three sums are larger than 10. A sum of 11 can be obtained in two ways: 6 on the first die and 5 on the second or vice versa, whereas 12 is possible only when both dice have 6. Hence, the probability of getting a sum larger than 10 is $3/36 = 1/12$.) The approximation tends to get better as the number of throws increases.

When a coin is tossed, there are two equally likely possible outcomes; that is, there is mutual symmetry of all possible outcomes. In any toss, therefore, we have a number (say, n) of cases that are equally likely. Consider, for example, an event E defined as getting more than 3 heads in a single throw of 5 coins. The cases that favor the event E or successes (see Fig. 2-10) are (H, H, H, H, H) and (H, H, H, H, T). Note that the concern here lies only with the total number of heads or tails in a single throw. The total number of possible outcomes are (H, H, H, H, H), (T, H, H, H, H), (T, T, H, H, H), (T, T, T, H, H), (T, T, T, T, H), and (T, T, T, T, T). If the number of cases favorable to the event (or number of successes) is denoted by s then in this example, $s = 2$ and $n = 6$. Similarly, let event E be defined as throwing more than 10 with two dice in one single throw. The cases that favor this event or successes are $(6, 6)$, $(6, 5)$, and $(5, 6)$. In this example, $s = 3$ and $n = 36$.

2.2.1 Definition of Probability

The ratio s/n is a measure of the likelihood that event E will occur. If s/n is zero, there are no cases favorable to the event or no successes; the event is clearly impossible. If s/n is one, all possible cases are favorable to the event—the event is clearly a certain event. In the theory of probability, the ratio s/n has been adopted as the definition of probability. Thus, the probability of the occurrence of a given event is equal to the ratio s/n, where s is the number of cases favorable to the event or successes and n is the total number of possible cases, provided that all n cases are equally likely and mutually symmetric.

If the number of observations approaches infinity, the frequency ratio would approach a constant number P in a mathematical sense. If an event occurred s times in n observations, then P is defined as

$$P = \lim_{n \to \infty} \frac{s}{n} \qquad (2.2)$$

The stability of the relative frequency number is an empirical observation. Observed values must be sufficient to get an acceptable estimate of the probability P. Since Eq. 2.2 employs a frequency-type definition of probability, it is assumed that in a random phenomenon there is an underlying statistical regularity, which demonstrates itself in the long run. Figure 2-11 depicts this trend of the underlying statistical regularity in a random experiment of tossing a coin. A bent coin might show a different frequency ratio of "heads" but the ratio would still converge to an approximately constant number.

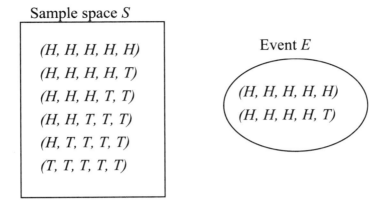

Figure 2-10 Set of all possible outcomes and subset E.

2.2.2 Distinction Between Probability and Frequency

When differentiating between probability and frequency, two schools of thought are distinguished. The objectivists view probability as something that is the result of repetitive experiments—it is something external. In contrast, the subjectivists view that probability is an expression of an internal state of the system and is basically a state of knowledge. Kaplan and Garrick (1981) termed the former interpretation as "frequency." Of course, the estimation of probability has to be objective, because it should not depend upon the person who is determining it: Two different persons should arrive at the same result given the same observed data. To elaborate, probability is a numerical measure of a state of knowledge or a degree of belief whereas frequency refers to the outcome of some experiment that involves repeated trials. Thus frequency is a hard measurable number even though the experiment may not be really performed—it may be conducted in future or only in "thoughts." Appropriately, the statistical analysis of floods is termed as "flood frequency analysis" because observations of floods can be perceived as repeated experiments and actual observations are employed in frequency analysis.

One should note that probability and frequency are closely connected. Frequency is used to calibrate the probability scale and, after the calibration is over, we use probability to express our state of confidence or knowledge in those areas where we may not have any information about frequency. Furthermore, relative frequency is expressed in terms of probability or percent of chance. Therefore, frequency and probability are often used interchangeably.

2.2.3 Random Events

The collection of all possible random events that might arise from a random experiment constitutes a set of elementary events. The sets of combinations of

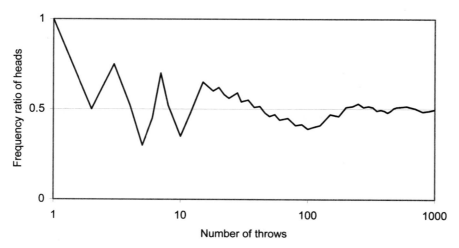

Figure 2-11 *Frequency ratio for "heads" in a sequence of coin throws.*
Note the abscissa has logarithmic scale.

elementary events are called composite events. Together, all elementary and composite events constitute the sample space. A sample event can also be an event consisting of a single sample point, and a compound event is made up of two or more sample points or elementary outcomes of an experiment. The complement A^c of an event A consists of all sample points in the sample space of the experiment not included in the event A. Therefore, the complement of an event is also an event. If A is a certain event (i.e., if A is the collection of all sample points in the sample space S), then its complement A^c will be the null event (i.e., it will contain no sample events).

   If two events contain no sample points in common, the events are said to be mutually exclusive or *disjointed* (see Fig. 2-12). If two events A and B are not mutually exclusive, the set of points that they have in common is called their intersection, denoted as AB. Figure 2-13 shows this concept using the Venn diagram. If the intersection of two events A and B is equivalent to one of the events, say, A, then event A is said to be contained in event B and is written as $A \subset B$. The union of two events A and B is the event that is the collection of all sample points that occur at least once in either A or B and is written as $A \cup B$.

   Another sample space of interest is conditional sample space. For example, a hydrologist might be interested in floods exceeding a certain threshold event denoted as A. The set of events exceeding event A can be considered as a new, reduced sample space. Only the sample events associated with the sample points in that reduced space, which is conditional on A, are possible outcomes of the experiment. The reduced sample space is the conditional sample space and can be represented as {events | events $\geq A$}.

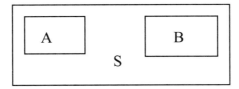

Figure 2-12 *Two mutually exclusive events* A *and* B *in sample space* S.

Figure 2-13 *Venn diagram showing intersection and union of two events.*

2.2.4 Mathematical Representation of Events

The city of New Orleans in Louisiana is surrounded by lakes on three sides and the Mississippi River passes through it. Located on the northern side is Lake Pontchartrain, the largest lake. The city is protected by dykes from likely water level rise in this lake. Consider a case where the dykes need to be raised along the lake so that dyke elevations would protect the city with a probability of more than 99% in any given year. In this case, the interest is really in the maximum lake water level (H). Other related aspects of interest might be the duration of the maximum water level, the time intervals between two high-water levels, or the cause of the water level rise. If the maximum water level rise is considered to be of main interest in the present case, then it can be used to differentiate among events. This is accomplished by representing the maximum water level in a one-dimensional space, the probability space, which is simply a straight line with an origin, a scale, and a positive direction. Any single lake water level that has occurred in the past or that may occur in the future can be represented by a single point in this probability space. Conversely, any point on the line can be thought of as representing a lake level, provided, of course, that impossible events, that is, events with a probability of zero, are included in the description.

Not all events can be represented by single points on the probability scale. In fact, when water levels that will overtop a dyke are considered, then all water levels equal to or greater than the particular maximum level that will overtop the dyke are considered. That event, say, $H > 10$ m, is represented in the probability space by an interval, namely, the infinite. Here H is above a certain threshold. One interval in the probability space represents an event.

It should be emphasized that not every event may be represented in the probability space by a single interval. Consider, for example, an event that the maximum lake level is not between 5 and 7 m. It must be then either smaller than 5 m or greater than 7 m. Any point in the interval to the left of 5 m or to the right of 7 m satisfies this condition. The event thus specified is represented by the sum of two intervals, namely the interval from $-\infty$ to 5 m and the interval from 7 m to $+\infty$. The sum of two intervals may be an interval, but it is not necessarily an interval as this example illustrates.

Any event can be represented in the probability space by an interval or by the sum of a number of intervals. All intervals can be arranged in a sequence so that there is a one-to-one correspondence between the elements of the sequence and the set of positive integer numbers 1, 2, 3, etc. Figure 2-14 shows examples representing flood peak events on a one-dimensional probability scale. Three kinds of events can be distinguished:

Event A: Peak between 6,000 and 7,000 cubic meters per second (cumecs).

Event B: Peak larger than 10,000 cumecs.

Event C: Peak either between 2,000 and 3,000 cumecs or between 4,000 and 5,000 cumecs.

The boundaries of the intervals need attention. When the flow between 6,000 and 7,000 cumecs specifies event A, the boundaries of 6,000 and 7,000 cumecs may or may not be included, depending on how the event is defined. In practice one can never determine whether or not the flow is exactly 7,000 cumecs. When variables take on only a limited number of specific values, the inclusion of the boundaries becomes important. Consider, for instance, the throwing of a dice having six faces wherein only six outcomes, including 1, 2, 3, 4, 5, and 6, are possible. If an event is defined such that the outcome must be larger than 5 then the value 5 is specifically excluded.

In real life there are many cases where more than one aspect of an event is of interest, as, for example, the annual peak flow and the total annual flow in a river, pollution concentration and pollutant loading, or the number of air quality violations and the time interval between violations. If river flows have been recorded for a number of years, then each year of record gives a recorded event characterized by a pair of observations: the annual flood peak (say, in cumecs) and the annual volume of flow (say, in cubic meters). The behavior of the river is now described by two variables and their representation consequently needs a two-dimensional probability space.

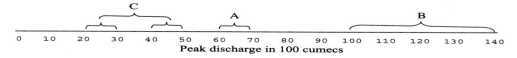

Peak discharge in 100 cumecs

Figure 2-14 *Representation of events on a one-dimensional probability scale.*

Figure 2-15 shows a two-dimensional probability space in which event A represents all years in which the annual flow volume is between 60 and 70 million m^3 while the peak discharge is between 120,000 and 130,000 cumecs. Similarly, the concept of probability space can easily be extended to events characterized by three or more parameters. In each case any event can be represented by the sum of an enumerable number of intervals.

2.2.5 Axioms of Probabilities

The preceding discussion shows that an event can be an empty (or null) set, a subset of a sample space, or the sample space itself. In other words, events are sets and therefore the usual set operators, including union, intersection, and complement, are applicable. These operators help define axioms of probability. To that end, consider two events A and B. If A and B both belong to set S then the union of A and B, $(A + B) = C$, will be denoted by $A \cup B = C$. Clearly, C is the set of events that belong to either A or B or both A and B if they are not mutually exclusive (or disjointed) events. The intersection of A and B is denoted by $A \cap B$. If D is the intersection of A and B, then it is the set of all events that belong to both A and B (i.e., $D = A \cap B$). The notation $(A - B)$ will designate the set of all events that belong to A and do not belong to B. The notation A^C will designate the complementary event of A, that is, the set of all events that belong to S but do not belong to A. Thus $A \cup A^C = S$. If the occurrence of event A depends on the occurrence of event B, then these events are called conditional events, and they are denoted as $A \mid B$. Figure 2-16 illustrates some of these familiar concepts by means of a Venn diagram. Each event defined on the probability space must be assigned a definite probability number.

Before we discuss axioms of probability, it may be instructive to write the four frequently used rules for set operations, considering three events A, B, and C:

1. Commutative rule: $A \cup B = B \cup A$, $A \cap B = B \cap A$.
2. Associative rule: $(A \cup B) \cup C = A \cup (B \cup C)$, $(A \cap B) \cap C = A \cap (B \cap C)$.
3. Distributive rule: $A \cap (B \cup C) = (A \cap B) \cup (A \cap C)$, $A \cup (B \cap C)$ $= (A \cup B) \cap (A \cup C)$.
4. De Morgan's rule: $(A \cup B)^C = A^C \cap B^C$, $(A \cap B)^C = A^C \cup B^C$.

The notation $P[A]$ is used to denote the probability of a random event A. For the complementary event, one can write

$$P(A) = 1 - P(A^C) \tag{2.3}$$

Axiom 1: The probability of an event A is a number greater than or equal to zero but less than or equal to unity:

$$0 \leq P[A] \leq 1 \tag{2.4a}$$

Axiom 2: The probability of an event A, whose occurrence is certain, is unity:

$$P[A] = 1 \tag{2.4b}$$

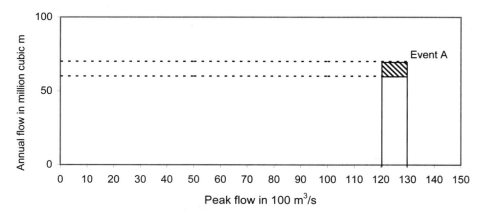

Figure 2-15 *Depiction of an event in a two-dimensional probability space.*

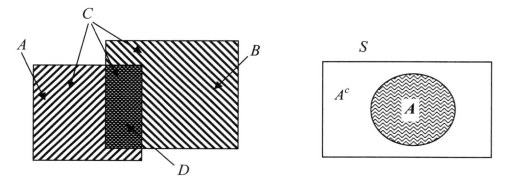

Figure 2-16 *Venn diagram.*

where A is the event associated with all sample points in the sample space.

Axiom 3: The probability of an event that is the union of two events is

$$P[A \text{ or } B] = P[A \cup B] = P[A] + P[B] - P[A \cap B] \tag{2.5}$$

where $A \cup B$ denotes the union of events A and B, which means that either event A occurs or event B occurs, and $A \cap B$ denotes the intersection of event A and event B. Equation 2.5 can be extended to the union of n events. If A and B are two mutually exclusive (disjointed) events, the probability of A and B, $P[A \cap B]$, will be zero and Eq. 2.5 becomes

$$P[A \text{ or } B] = P[A] + P[B] \tag{2.6a}$$

Equation 2.6a can be generalized: If E_1, E_2, ..., E_n, are n mutually exclusive events then

$$P(E_1 \text{ or } E_2 \text{ ... or } E_n) = P(E_1 \cup E_2 \cup E_3 \cup ... E_n) = \sum_{i=1}^{n} P(E_i) \tag{2.6b}$$

Axiom 4: The probability of two (statistically) independent events occurring simultaneously or in succession is the product of individual probabilities:

$$P[E_1 \cap E_2] = P[E_1] \times P[E_2] \tag{2.7a}$$

Statistical independence implies that the occurrence of E_1 has no influence on the occurrence of event E_2. Equation 2.7a can be extended to n independent events, E_i, $i = 1, 2, ..., n$, as

$$P[\bigcap_{i=1}^{n} E_i] = P(E_1) \times P(E_2) \times ... \times P(E_n) = \prod_{i=1}^{n} P(E_i) \tag{2.7b}$$

2.2.6 Probabilities of Simple Events

Since a simple event is associated with one or more sample points and simple events are mutually exclusive by the construction of the sample space, the probability of any event is the sum of the probabilities assigned to the sample points with which it is associated. If an event consists of all sample points with nonzero probabilities, its probability is one and it is certain to occur. If an event is impossible (i.e., it cannot occur), then the probabilities of occurrence of all the sample points associated with the event are zero.

2.2.7 Conditional Probability

Many times the occurrence of one event depends on the occurrence of another event. The conditional probability of an event A, given that event B has occurred, is denoted by $P[A \mid B]$. Knowing that event B has occurred reduces the sample space for determining $P[A]$ to B. By applying the definition of probability, the number of ways in favor of A and B, $s = P[A \cap B]$, and the total number of ways B can occur, $n = P[B]$, $P[A \mid B]$ is defined as (see Fig. 2-17)

$$P[A \mid B] = s/n = P[A \cap B]/P[B] \tag{2.8a}$$

or

$$P[A \cap B] = P[A \mid B] \times P[B] \tag{2.8b}$$

In a similar manner, one can also write

$$P[B \mid A] = P[A \cap B]/P[A] \text{ or } P[A \cap B] = P[B \mid A] \times P[A] \tag{2.9}$$

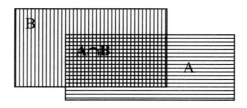

Figure 2-17 *Conditional probability.*

This leads to

$$P[A \cap B] = P[A \,|\, B] \times P[B] = P[B \,|\, A] \times P[A] \qquad (2.10)$$

If the two events A and B are statistically independent, then

$$P[A \,|\, B] = P[A] \qquad (2.11a)$$

$$P[B \,|\, A] = P[B] \qquad (2.11b)$$

$$P[A \cap B] = P[A] \times P[B] \qquad (2.12)$$

which is the same as Eq. 2.7a.

Taking advantage of Eq. 2.9, one can express the joint occurrence of n dependent events as

$$P[\bigcap_{i=1}^{n} E_i] = P(E_1) \times P(E_2 | E_1) \times P(E_3 | E_2, E_1) \ldots \times P(E_n | E_{n-1}, E_{n-2}, \ldots, E_1) \qquad (2.13)$$

Example 2.2 Consider a river passing through an urban area that reaches a flood stage each summer with a relative frequency of 0.1. Power failures in an industrial complex along the river occur with a probability of 0.2. Experience shows that when there is a flood, the chances of a power failure are raised to 0.4. Determine the probability of flooding or power failure.

Solution We are given the following:

$$P[\text{Flood}] = P[F] = 0.1$$

$$P[\text{Power failure}] = P[P_P] = 0.2$$

$$P[\text{Power failure} \,|\, \text{flood occurs}] = P[P_P | F] = 0.4$$

Therefore,

$$P[\text{no flood}] = P[\overline{F}] = 1.0 - 0.1 = 0.90$$
$$P[\text{no power failure}] = P[\overline{P_F}] = 1.0 - 0.2 = 0.80$$

If flooding and power failure were independent, the joint probability would be

$$P\left[F \cap P_P\right] = 0.1 \times 0.2 = 0.02$$
$$P\left[F \cap \bar{P}_P\right] = 0.1 \times 0.8 = 0.08$$
$$P\left[\bar{F} \cap P_P\right] = 0.9 \times 0.2 = 0.18$$
$$P\left[\bar{F} \cap \bar{P}_P\right] = 0.9 \times 0.8 = 0.72$$

The probability of a flood or a power failure during summer would be the sum of these first three joint probabilities:

$$P\left[F \cup P_P\right] = P\left[F \cap P_P\right] + P\left[F \cap \bar{P}_P\right] + P\left[\bar{F} \cap P_P\right]$$
$$= 0.02 + 0.08 + 0.18 = 0.28$$

The events are, however, dependent. When a flood occurs with $P[F] = 0.1$, a power failure occurs with probability $P[P_P|F] = 0.4$. Therefore, the true joint probability is

$$P[F \cap P_P] = P[F] \times P[P_P|F] = 0.1 \times 0.4 = 0.04$$

The union of probabilities is

$$P[F \cup P_P] = P[F] + P[P_P] - P[F \cap P_P] = 0.1 + 0.2 - 0.04 = 0.26$$

Note the contrast:

$$P[F \cup P_P] = 0.3 \text{ for mutually exclusive events}$$

$$P[F \cup P_P] = 0.28 \text{ for joint but independent events}$$

$$P[F \cup P_P] = 0.26 \text{ otherwise}$$

Example 2.3 Consider the design of an underground utility system for an industrial park containing six similar building sites. The sites have not yet been located and hence their nature is not yet known. If the power and water are provided in excess of demand, there will be wastage of client's capital. However, if the facilities prove inadequate, expensive changes will be required. For simplicity of numbers, let us consider a particular site where the electric power required by the occupant will be either 5 or 10 units while the water capacity demand would be either 1 or 2 units. It is assumed that the probability of electric power demand being 5 units and water demand being 1 unit is 0.1, the probability of electric power demand being 5 units and water demand being 2 units is 0.2, the probability of electric power demand being 10 units and water demand being 1 unit is 0.1, and the probability of electric power demand being 10 units and water demand being 2 units is 0.6. Calculate the probabilities of water or power demands.

Solution First, we define four associated events. Let us denote the following: event $W1$ = the water demand is 1 unit, $W2$ = the water demand is 2 units, $E5$ = the electricity demand is 5 units, and $E10$ = the electricity demand is 10 units.

Note that water and power demands occur simultaneously. In other words, they both need to be satisfied simultaneously. For example, the water demand of 1 unit can occur either with electric power demand of 5 units or with 10 units. Then the sample experimental space associated with a single occupant consists of four points as shown in Table E2-3.

Table E2-3

Water demand	Electric power demand	
	E5 = 5 units	E10 = 10 units
W1 = 1 unit	P[E5W1] = 0.1	P[E10W1] = 0.1
W2 = 2 units	P[E5W2] = 0.2	P[E10W2] = 0.6

Thus, the probability of event *W1* is

$$P[W1] = P[E5W1] + P[E10W1] = 0.1 + 0.1 = 0.2$$

Likewise, the probability of event *W2* is the sum of the probabilities of the corresponding mutually exclusive simple events:

$$P[W2] = P[E5W2] + P[E10W2] = 0.2 + 0.6 = 0.8$$

Similarly, the probability of event *E5* is

$$P[E5] = P[E5W1] + P[E5W2] = 0.1 + 0.2 = 0.3$$

and the probability of electric power demand being 10 units, *E10*, is

$$P[E10] = P[E10W1] + P[E10W2] = 0.1 + 0.6 = 0.7$$

The probability that either the water demand is 2 units or the power demand is 10 units may be calculated as

$$P[W2 \cup E10] = P[W2] + P[E10] - P[W2 \cap E10] = 0.8 + 0.7 - 0.6 = 0.9$$

Alternatively, one can obtain this probability as

$$P[W2 \cup E10] = P[E10W2] + P[E5W2] + P[E10W1] = 0.6 + 0.2 + 0.1 = 0.9$$

The probability that a site with a power demand of *E10* will also require a water demand *W2* is

$$P[W2 \mid E10] = P[W2 \cap E10]/P[E10] = P[E10W2]/P[E10] = 0.6/0.7 = 0.86$$

Example 2.4 In a survey, a number of firms in similar industrial parks are sampled. It is found that there is no apparent relationship between their electricity demand and their water demand. A high electricity demand does not always seem to be correlated with a high water demand. Based on this information, probabilities are assigned, as listed in Table E2-4a. Find the probabilities for the joint or simultaneous occurrence of events denoted by water demand and electricity demand.

Solution Adopting the assumption of independence of events of power and water demands, one can calculate the probabilities for the joint occurrence or simple events. The results are listed in Table E2-4b.

Example 2.5 For simplicity, let us calculate the probabilities of water demand only and investigate the design capacity of a pair of similar sites in the industrial park. The occupancies of the two sites represent two repeated trials of the previous experiment. Denote by $W1W1$ the events that the demand of each firm is one unit and by $W1W2$ the event that the demand of the first firm is one unit and that of the second firm is two units and so on. Find the probabilities for the joint occurrence of water demands at two sites.

Solution Assuming independence of demands for water from the two sites, one calculates the values in Table E2-5.

If the demands of all six sites are mutually independent, the probability that all sites will demand two units of water is

$$P[W2W2...W2] = P[W2]P[W2]...P[W2] = 0.7^6 = 0.117$$

Table E2-4a

	Event	Estimate of probability
Electricity demand	$E5$	0.2
	$E10$	0.8
	Sum	1.0
Water demand	$W1$	0.3
	$W2$	0.7
	Sum	1.0

Table E2-4b

$P[E5W1] = P[E5]P[W1]$	0.2×0.3	0.06
$P[E5W2] = P[E5]P[W2]$	0.2×0.7	0.14
$P[E10W1] = P[E10]P[W1]$	0.8×0.3	0.24
$P[E10W2] = P[E10]P[W2]$	0.8×0.7	0.56
	Total	1.00

Table E2-5

$P[W1W1] = P[W1]P[W1]$	0.3×0.3	0.09
$P[W1W2] = P[W1]P[W2]$	0.3×0.7	0.21
$P[W2W1] = P[W2]P[W1]$	0.7×0.3	0.21
$P[W2W2] = P[W2]P[W2]$	0.7×0.7	0.49
	Total	1.00

2.3 Conditional Probability and Independence

Probability values may change when new information is gathered. For example, consider drawing the ace of hearts in a single draw from a complete deck of 52 playing cards. Clearly the probability of drawing the ace of hearts is 1/52. Now, suppose that it is somehow known that the card that was drawn is a heart. What is the probability that it is the ace of hearts? Evidently, the probability would be 1/13.

The probability, thus revised with the availability of extra information, is called *conditional probability*, for it is conditional upon the knowledge that another event has occurred, for instance that the card drawn was a heart. If the event of drawing a heart is denoted by H and the event of drawing the ace of hearts by A_h then the conditional event will be designated by $A_h | H$ and the conditional probability by $P(A_h | H)$. In this example, event A_h is no longer one particular outcome out of 52 equally possible outcomes, since event H has a probability of 1/4.

Consider another experiment. We again draw a card from a deck of 52 and we ask the following: What is the probability that it is an ace? The answer is 1/13, because there are four aces in the 52-card deck: $P(A) = 4/52$. Now if it is revealed that the card that was drawn is a heart then we ask the following: What is the probability that the card is an ace? Using the definition of conditional probability one can write $P(A | H) = P(A \cap H)/P(H)$. But since $P(A \cap H) = 1/52$ and $P(H) = 1/4$, the answer is still 1/13. In this case, the probability value before and after the information is the same. In other words, knowing that event H has occurred does not tell us anything about the probability that event A will occur. This is logical since the suits have the same configuration. We call such two events A and H independent.

To extend the discussion, consider throwing two dice and the probability of the occurrence of 4 or 5 is to be determined. The domain of an event A, which is the occurrence of either 4 or 5, can be sketched as shown in Fig. 2-18.

Example 2.6 Consider a standard deck of playing cards from which we remove all hearts except the ace. We now have 40 cards left. What is the probability that a card drawn at random is both a heart and an ace?

Solution The event of drawing both a heart and an ace can be regarded as the intersection of two events: event A, meaning the card is an ace, and event H, meaning that the card is a heart. Since there are four aces, the probability of drawing one out of a deck of 40 cards is $4/40 = 0.10$:

$$P(A) = 0.10$$

Since there is only one heart, the probability of drawing it out of a deck of 40 is $1/40 = 0.025$:

$$P(H) = 0.025$$

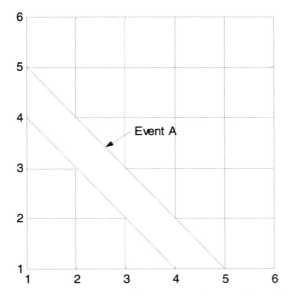

Figure 2-18 *Domain of event* A *in throw of two dice.*

Because events A and H are both independent, the multiplication rule would lead to

$$P(A \cap H) = P(A) \times P(H) = 0.10 \times 1/40 = 1/400$$

This is obviously wrong.

To do the calculation correctly, we need to use

$$P(A \cap H) = P(A \mid H) \times P(H)$$

$P(A \mid H)$ is the conditional probability that the card is an ace if it is known to be a heart. This event is certain; it has a probability of one. Therefore

$$P(A \cap H) = 1 \times 1/40 = 1/40$$

Alternatively, we can use

$$P(A \cap H) = P(H \mid A) \times P(A)$$

$P(H \mid A)$ is the conditional probability that the card is a heart if it is known to be an ace. This probability is evidently 1/4. Therefore

$$P(A \cap H) = (1/4) \times 1/10 = 1/40$$

Example 2.7 A contractor wants to construct a tunnel below the bed of a river in connection with the development of a transportation project. The period of construction will take two years. The building site must be surrounded by a cofferdam to keep the water out against a 10-year flood. What is the probability that the site will be flooded during construction? The water level in the river is considered as a random variable.

Solution The 10-year flood is the flood that has a 10% probability of being exceeded every year. This means that the contractor accepts a 10% probability that the site will be flooded in the year the tunnel is constructed. It is assumed that in each of the two years the probability of flooding remains at 10% and that the events in the two years are independent. One might be tempted to argue that doubling the time will double the risk, since extending the time will increase the risk of flooding. This clearly is not the case, since the answer would then be that the probability of flooding is 20%. By the same reasoning one would conclude that flooding would occur with absolute certainty if the construction period would be extended to 10 years. If the period of construction were extended to 12 years, this reasoning would lead to an answer of 120%. This answer is patently wrong.

To analyze this problem, we need to only consider the maximum water level during the first year and the maximum water level during the second year, because our interest is in the question of whether the level is higher or lower than the critical level that would cause the cofferdam to be overtopped. The water levels in successive years are assumed to be independent. The probability of no flooding in the first year is 0.9; likewise, the probability of no flooding in the second year is also 0.9. The probability that there is no flooding in either year is equal to the product of the probabilities, 0.9 × 0.9 = 0.81. Similarly, probabilities can be assigned to the other three possible points. In this case, the sum of all the probabilities adds up to 1.00.

In a similar manner, no flooding in the first year and no flooding in the second year is the intersection of two events defined as a single point that will have a probability of 0.81. Likewise, flooding in the first year and flooding in the second year is the intersection of two events, say, event A and event B, and results in a single point that will have a probability of 0.1 × 0.1 = 0.01. In the probability space we can arbitrarily give the water levels lower than the critical level the value 0 and those higher than the critical level the value 1. This gives four possible points in probability space, as shown in Fig. 2-19, where we now have a discrete distribution.

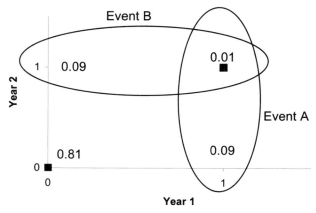

Figure 2-19 *Probability space of Example 2.7.*

The event of interest is flooding during the construction period. This means flooding during the first year or flooding during the second year. This event is the union of the two events A and B:

$$P(A \cup B) = P(A) + P(B) - P(A \cap B)$$

$$P(A \cup B) = 0.10 + 0.10 - (0.10 \times 0.10) = 0.19$$

The probability associated with the event $(A \cup B)$ is 0.19. It then follows that the probability is 19% that the site will be flooded during construction.

Alternatively, it is even simpler to observe that the described event "flooding during years 1 and 2" has a complementary event: "no flooding during year 1 and no flooding during year 2." The latter is the intersection of two independent events A^c and B^c, each having a probability of 0.9 [i.e., $P(A^c) = P(B^c) = 1 - 0.1 = 0.9$]. The probability of the intersection is $P(A^c \cap B^c) = P(A^c) \times P(B^c) = 0.9 \times 0.9 = 0.81$. The probability of the complementary event that the site will be flooded at least once in two years is $1.0 - 0.81 = 0.19$.

Example 2.8 A highway culvert is designed for a 10-year flood. What is the probability that the design flood will be exceeded in the next 20 years?

Solution The event "exceedance during the 20 years" is the union of 20 events that have many points in common, namely, all the possible multiple occurrences of exceedance during the 20-year period. Applying the addition rule thus becomes quite awkward. However, the complementary event has a probability that can be evaluated easily. The complementary event of exceedance at any time during the period of 20 years is evidently "no exceedance at all." That means "no exceedance in the first year," "no exceedance in the second year," "no exceedance in the third year," and so on. This is the intersection of 20 events, each having a probability of 0.9. Assuming statistical independence and applying the multiplication rule gives a probability equal to $(0.9)^{20} = 0.12$. It then follows that the probability of exceedance during the 20 year period is $1.0 - 0.12 = 0.88$.

2.4 Total Probability and Bayes's Formula

Sometimes, it is difficult to directly determine the probability of occurrence of an event. Such an event may occur along with other events, called attribute events. These events are exclusive and mutually exhaustive. In such cases the sample space can be divided into a number of mutually exclusive and collectively exhaustive subsets, $S_1, S_2, \ldots, S_n$, each subset corresponding to an attribute event. For an event A taking place in this sample space as shown in Fig. 2-20, one can write

$$A = A \cap S = A \cap (S_1 \cup S_2 \cup \ldots S_n) = (A \cap S_1) \cup (A \cap S_2) \cup \ldots \cup (A \cap S_n) \quad \textbf{(2.14)}$$

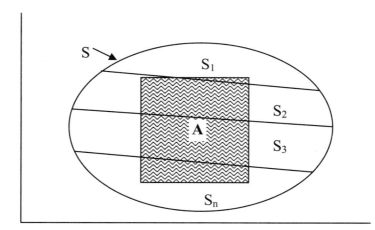

Figure 2-20 *Venn diagram of a set* S *consisting of subsets and an event* A.

By invoking the properties of mutually exclusive events, Eq. 2.14 can be simplified to

$$P(A) = P(A \cap S_1) + P(A \cap S_2) + \ldots + (A \cap S_n) \tag{2.15}$$

Using Eq. 2.9, $P(A \cap B) = P[A \mid B] \times P[B]$, allows one to write Eq. 2.15 as

$$P[A] = P[A \mid S_1] \times P[S_1] + P[A \mid S_2] \times P[S_2] + \ldots + P[A \mid S_n] \times P[S_n]$$

$$= \sum_{i=1}^{n} P[A \mid S_i] \times P[S_i] \tag{2.16}$$

Equation 2.16 is also known as the theorem of total probability. This gives the probability of event A, regardless of the attributes.

Rewriting Eq. 2.10, one gets

$$P[A \cap S_i] = P[A \mid S_i] \times P[S_i] = P[S_i \mid A]P \times [A] \tag{2.17}$$

or

$$P[S_i \mid A] = P[A \mid S_i] \times P[S_i]/P[A]$$

Substituting the value of $P[A]$ from Eq. 2.16, one obtains

$$P[S_i \mid A] = \frac{P[A \mid S_i] \times P[S_i]}{\displaystyle\sum_{i=1}^{n} P[A \mid S_i] \times P[S_i]} \tag{2.18}$$

Equation 2.18 is known as Bayes's theorem and follows from the definition of conditional probability; it is regarded as a fundamental theorem to revise the probability value through evidence. Bayes's theorem involves a prior (or *a priori*) distribution that contains all the relevant information about the variable before additional data become available. Given the *a priori* distribution, the *posteriori* distribution can be evaluated when a new set of data becomes available.

Example 2.9 A contractor must complete a highway section in a period of 2 weeks. If it does not rain at all during that time he determines that he has a 90% chance of completing the work on time. Two rainy days will reduce this probability to 70%, three rainy days will reduce this probability to 50%, and four rainy days will mean only a 10% chance of finishing on time. With more than four rainy days he cannot make it. From the U.S. National Weather Service he obtains the following information about the probability of rain during the construction period: There is a 40% chance of zero days of rain, 30% chance of two days of rain, 20% chance of three days of rain, and 10% chance of four days of rain. The probability of more than four days of rain is negligible. What is the probability that the contractor will finish the work on time? This example is based on Booy (1990).

Solution Consider the number of rainy days as a variable and denote it by X. On the probability space one can now distinguish events A_1, A_2, A_3, A_4, and A_5 corresponding to 0, 2, 3, 4, and more than 4 rainy days. Also, one can define event R corresponding to "being on time" and event Q to "not being on time." Events A_1, A_2, A_3, A_4, and A_5 are independent, mutually exclusive, and collectively exhaustive; that is, they have no common points and their union contains all points on the probability space. Conditional probabilities $P(R|A_1)$, $P(R|A_2)$, $P(R|A_3)$, $P(R|A_4)$, and $P(R|A_5)$ are also given. It is then possible to construct a simple two-dimensional probability space in which the probabilities associated with events A_1, A_2, A_3, A_4, and A_5 can be given in the first row and conditional probabilities $P(R|A_1)$, $P(R|A_2)$, $P(R|A_3)$, $P(R|A_4)$, and $P(R|A_5)$ in the second row. These are listed in Table E2-9.

Table E2-9

	Probability space X→ Number of days of rain				
	A_1	A_2	A_3	A_4	A_5
	0	2	3	4	> 4
Chance of rain [P(X=Ai)]	40%	30%	20%	10%	0%
Chance of completing work (given) [P(R|X=Ai)]	90%	70%	50%	10%	0%

From the total probability theorem, the probability of completing the work on time can be expressed as

$$P(R) = \sum_{i=1}^{5} P(R \cap A_i) = \sum_{i=1}^{5} P(R|A_i)P(X = A_i)$$

One can now determine the probabilities associated with the intersections of events A_1, A_2, A_3, A_4, and A_5 with events R and Q using the multiplication rule:

$$P(R \cap A_i) = P(R \mid A_i)P(A_i)$$

As an example, let us consider the case $A_1 = 0$. Then,

$$P(R \cap A_1 = 0) = P(R \mid A_1 = 0)P(A_1 = 0) = P(R \mid A_1 = 0)P(A_1 = 0) = 0.9 \times 0.4 = 0.36$$

Thus,

$$P(R) = \sum_{i=1}^{5} P(R \cap A_i)$$

$$= \sum_{i=1}^{5} P(R \mid A_i)P(X = A_i)$$

$$= 0.9 \times 0.4 + 0.7 \times 0.3 + 0.5 \times 0.2 + 0.1 \times 0.1 = 0.68$$

It is seen that the probability of finishing the work on time, now called the total probability, is equal to 36% + 21% + 10% + 1% = 68%. One can visualize that each of the events A_1, A_2, A_3, A_4, and A_5 carries part of this total probability and a percentage of each corresponding to the conditional probability is associated with R and the rest is associated with Q.

Example 2.10 Assume that the engineer who calculated the probabilities shown in Example 2.9 was away when the work was being done. She did not know any-thing about the weather at the site but she learned that the contractor was able to finish in time. Did it rain? And if it did, did it rain for two, three, or four days? So she decided to calculate the probabilities of rain during zero, two, three, and four days, knowing that the contractor was able to finish on time. It should be noted that the engineer did have the information on the chances of rainy days and those of completing the work on time. This example is based on Booy (1990).

Solution The probabilities to be calculated are the probabilities of rain for a given number of days given that the work has been completed on time. Thus these are conditional probabilities. These probabilities are not the probabilities of number of rainy days, which are already given. Therefore, to distinguish between the probabilities of number of rainy days, which are given, and the probabilities to be computed, the latter are denoted as R_1, R_2, R_3, and R_4. For computing the probability that it did not rain given that the work was com-pleted one can write

$$R_i = P(A_i \mid R) = P(A_1 \cap R)/P(R)$$

From Example 2.9, $P(A_i \cap R)$ is known. Therefore,

$$R_1 = P(A_1 \mid R) = 0.36/0.68 = 0.53$$

$$R_2 = P(A_2 \mid R) = 0.21/0.68 = 0.31$$

$$R_3 = P(A_3 \mid R) = 0.10/0.68 = 0.15$$

$$R_4 = P(A_4 \mid R) = 0.01/0.68 = 0.01$$

It is worth noting that before the engineer knew that the contractor had finished on time she would have rated the probabilities of no rain, two rainy days, three rainy days, and four rainy days as 0.40, 0.30, 0.20, and 0.10.

2.5 Probability Distribution

We have discussed that any event can be specified on the probability space either as an interval or as the (enumerable) sum of intervals, or the probability of any event can be determined as part of the total probability of one, which is located in the intervals that represent the event. How is the probability of one distributed over the probability space? Once the distribution is known, the probability that each interval carries can be determined.

The probability distribution function gives a complete characterization of its random variable. It allows us to determine various moments of the random variable through the theory of statistical expectation, which will be covered in the next chapter. Further, knowing the probability distributions of two random variables will allow us to examine whether the two random variables are identical and if they are not, how they differ from one another.

2.5.1 Probability Mass Function

The probability distribution associated with a discrete random variable may be defined by means of probability mass function (PMF). It can be a bar graph displaying the distribution or simply a list of probabilities associated with points of the probability space. It can also be a formula that relates probability to values of the variable. For example, the probability that X takes on a value x may be specified by the equation

$$f(x) = P(X = x) = 0.9 \times (0.1)^{x-1}, x = 1, 2, 3, \ldots, \infty \qquad \textbf{(2.19)}$$

where $f(x)$ is the notation for PMF, which is equal to the probability that the random variable X will take on the value x.

The probability mass function of a discrete variable X is shown in Fig. 2-21, in which the values of the variable are plotted on the X axis and the probability that the discrete random variable X takes on values $x_1, x_2, \ldots, x_n$ is given on the Y axis.

The PMF function $f(x)$ is not continuous, which means that we may not draw the function as a curve by joining the spikes in Fig. 2-21. The function is defined only for specified values of X, and it is not defined if the discrete random variable X cannot take on the value x.

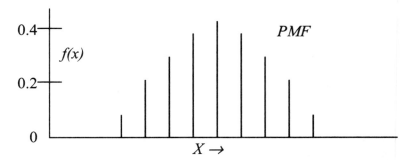

Figure 2-21 *Diagram showing the probability of obtaining various values of a variable* X.

2.5.1.1 Properties of the Probability Mass Function

The probability mass function has the following two properties:

1. $0 \le f(x) \le 1$ for all x
2. $\sum f(x) = 1$

Example 2.11 A regulatory agency has issued a water use permit to a water supply utility for a typical year according to the rules in Table 2-11. Develop a probability mass function to show the availability of surface water in a typical year.

Solution The water supply is arranged in increasing order and Table 2-11 is prepared. Then, the bar chart of $f(x)$ versus X is plotted as given in Fig. 2-22. This bar chart is the PMF.

Example 2.12 Develop a PMF for the data presented in Table 2-1.

Solution Assume that the sample data are sufficient to characterize the PMF of the DO violation. In such a case, the relative frequency corresponding to a given number of DO violations can be regarded as its probability. Therefore, Fig. 2-6 can be assumed to represent the PMF of DO violations.

Example 2.13 A subdivision has a provision for water supply from four water supply utilities. Based on the past history of these individual utilities, it has been noted that 95% of the time these utilities are able to meet the demand but, because of maintenance, drought conditions, or other reasons, 5% of the time they fail to meet the required demand. Develop a probability mass function to show the distribution of probability with respect to the number of ways the subdivision will meet its demand.

Solution Let X be the random variable representing the number of ways the subdivision will meet its demand. The possible values that X can take on are

Table E2-11

Month	Available surface water, x (million gallons)	Probability, $f(x)$
Jan	484	0.120
Feb	277	0.123
Mar	395	0.118
Apr	793	0.100
May	839	0.098
Jun	1,390	0.077
Jul	2,353	0.040
Aug	3,202	0.017
Sep	2,631	0.039
Oct	1,936	0.064
Nov	1,235	0.094
Dec	770	0.110

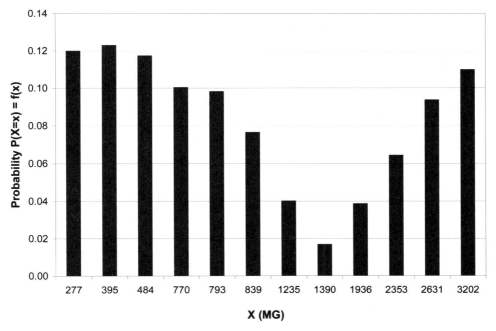

Figure 2-22 PMF for availability of surface water.

0, 1, 2, 3, and 4. To determine the probability corresponding to these values, the following calculation is performed:

$f(0)$	$P(X = 0)$	$= 0.05 \times 0.05 \times 0.05 \times 0.05$	0.00003
$f(1)$	$P(X = 1)$	$= 4 \times 0.05 \times 0.05 \times 0.05 \times 0.95$	0.0005
$f(2)$	$P(X = 2)$	$= 4 \times 0.05 \times 0.05 \times 0.95 \times 0.95$	0.0090
$f(3)$	$P(X = 3)$	$= 4 \times 0.05 \times 0.95 \times 0.95 \times 0.95$	0.1715
$f(4)$	$P(X = 4)$	$= 0.95 \times 0.95 \times 0.95 \times 0.95$	0.8145

Total = 1.00

The resulting PMF is plotted in Fig. 2-23.

Example 2.14 If two dice are thrown, compute the probability of obtaining different values as the sum of the points on their top faces.

Solution Each die has six faces and each face has a unique number of dots, varying from 1 to 6. Hence, if two dice are thrown, the minimum that one can get is $1 + 1 = 2$ and the maximum is $6 + 6 = 12$. In all, there are $6 \times 6 = 36$ possible outcomes. Some numbers can occur in more than one way. For example, one can obtain a sum of 6 in five ways: $1 + 5$, $2 + 4$, $3 + 3$, $4 + 2$, and $5 + 1$. Clearly, the probability of obtaining a sum of 6 is $5/36 = 0.139$. A complete calculation is shown in Table 2-14.

The data generated for X are summarized in Table 2-14.

Now, the PMF of the random variable X, which is the total number of points obtained when throwing two ordinary dice, is plotted in Fig. 2-24. It is seen from the figure that number 7 has the highest probability of occurrence.

Table E2-14a

Die #1	Die #2					
	1	2	3	4	5	6
1	2	3	4	5	6	7
2	3	4	5	6	7	8
3	4	5	6	7	8	9
4	5	6	7	8	9	10
5	6	7	8	9	10	11
6	7	8	9	10	11	12

2.5.2 Probability Density Function

For a continuous random variable, no single point carries a measurable probability. Therefore, each point is associated with a probability per unit basis on the

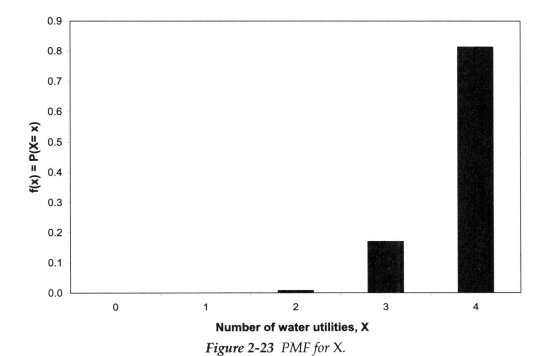

Figure 2-23 PMF *for* X.

Table 2-14b *Frequency summary and calculation of PMF for X.*

X	Frequency	PMF = *f*(*x*)
2	1	0.03
3	2	0.06
4	3	0.08
5	4	0.11
6	5	0.14
7	6	0.17
8	5	0.14
9	4	0.11
10	3	0.08
11	2	0.06
12	1	0.03
Sum =	36	1

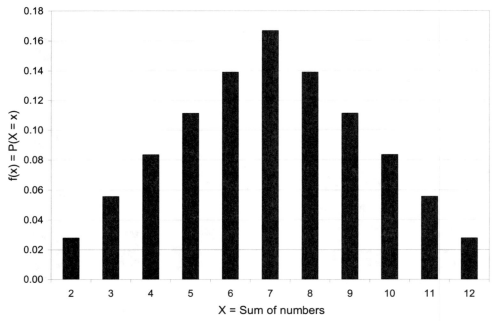

Figure 2-24 *PMF for the sum of numbers obtained by throwing two dice.*

probability space and this ratio is called the *probability density*. For a single random variable X, the probability density is a function of the value of X, x, and this function is called the probability density function (PDF); for two variables, X and Y, it is a function of x and y, and so on. The PDF is denoted by the symbol $f(x)$. Figure 2-25 shows a graph of a typical probability density function. Since a continuous random variable X is defined on a particular interval, it may take on any value in that particular interval. For example, if we say that X is defined on any arbitrary interval between points a and b, then the probability that X lies within this interval (a, b) is equal to the area of the PDF, $f(x)$, intercepted by $X = a$ and $X = b$:

$$P(a \leq x \leq b) = P(a < x < b) = \int_{a}^{b} f(x)dx \tag{2.20}$$

2.5.2.1 Properties of Probability Density Function

To be a probability density function, a function $f(x)$ must satisfy the following two properties:

1. $f(x) \geq 0$ for all x

2. $\int_{-\infty}^{\infty} f(x)dx = 1$

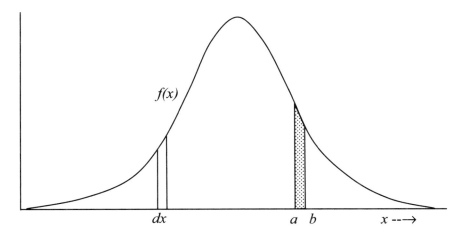

Figure 2-25 *Probability density function.*

That means the PDF, $f(x)$, is a non-negative function and the total area under the PDF is equal to unity.

Example 2.15 Evaluate constant c for the following expression to be considered as a PDF:

$$f(x) = c(x-1) \text{ , for } 0 \le X \le 10$$

Solution Using the first property of the PDF, $f(x) \ge 0$ for all x, the range of X that X can take on is determined. Thus, for $f(x)$ to be a non-negative function, $X \ge 1$. Therefore, the range on which X is defined is for $1 \le X \le 10$. Now, using the second property of the PDF,

$$\int_{-\infty}^{\infty} f(x)dx = 1$$

we obtain

$$\int_{1}^{10} c(x-1)dx = 1$$

$$c\left(\frac{x^2}{2} - x\right)\Big|_{1}^{10} = 1$$

$$c = 2/81$$

Example 2.16 Plot the PDF of X given in Example 2.15 and determine the following: (a) probability $X \le 3$, (b) probability $X \ge 9$, and (c) probability $4 \le X \le 8$.

Solution The function $f(x) = 2(x-1)/81$ is evaluated for various values of X ranging from 1 to 10 and is plotted in Fig. 2-26.

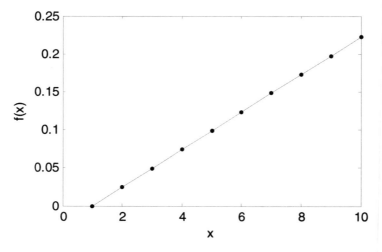

Figure 2-26 PDF of X.

(a) Probability $X \leq 3 = P(X \leq 3) = \int_{1}^{3} \frac{2(x-1)}{81} dx = 0.0494$

(b) Probability $X \geq 9 = 1 - P(X \leq 9) = 1 - \int_{1}^{9} \frac{2(x-1)}{81} dx = 0.21$

(c) Probability $4 \leq X \leq 8 = P(4 \leq X \leq 8) = \int_{4}^{8} \frac{2(x-1)}{81} dx = 0.493$

2.5.3 Joint Probability Distribution

When we want to express the joint behavior of more than one variable, joint probability distributions are needed. For example, consider coastal land loss along the Gulf of Mexico in the United States. The objective is to determine the relationship between the annual land loss and the severity of hurricane activity in the Gulf region. Both annual land loss (Y) and hurricane severity (X) are considered random variables. It is logical to state that there would be a significant degree of correlation between hurricane severity and annual land loss. A joint probability distribution would be needed to express the joint behavior of these variables. This can be expressed in the shape of the contours of the joint PDF, as shown in Fig. 2-27 for a hypothetical case. This figure shows that large annual land losses are more likely to occur when hurricane activity is high and low land losses are more likely when hurricane activity is low.

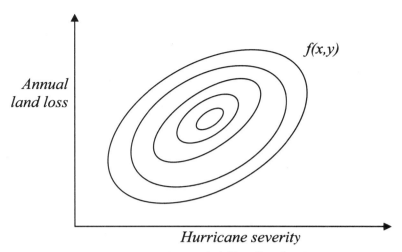

Figure 2-27 Joint probability distribution.

2.5.3.1 Joint Probability Density Function

For two random variables X and Y, the PDF is a surface lying above the probability space such that the elevation of the surface at any point indicates the probability density at that point. The shape of the surface can be constructed by using contours of equal probability density as shown in Fig. 2-28. The probability associated with any interval ($a < X < b$, $c < Y < d$) can be expressed as a double integral:

$$P(a < x < b, c < y < d) = \int\limits_{c}^{d} \int\limits_{a}^{b} f(x,y) \, dx \, dy \qquad (2.21)$$

Just as in the case of a univariate, continuous random variable where the probability is interpreted as an area, the probabilities in the bivariate case are represented by volumes under the PDF and the total volume under the PDF is equal to 1.

If the random variables are discrete, the corresponding bivariate PMF is given as

$$f(x,y) = P(X = x \text{ and } Y = y) \qquad (2.22)$$

In real life, most engineering problems contain more than one random variable to define the process of an engineering system. For example, the groundwater level in a phreatic aquifer depends on withdrawal by pumping, rainfall, evaporation, and inflow or outflow from other surface water bodies. Another example is one of a reservoir in which the volume of water and reservoir level depend upon input from its contributing streams, outflow to downstream reach, water supply withdrawal, losses from evaporation and seepage, etc. Further, if these random variables are statistically dependent, analysis related to these

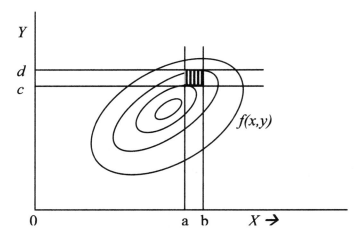

Figure 2-28 Contours of equal probability.

systems becomes mathematically challenging, depending on the mathematical functions involved in the formulation of a particular problem, because in many cases it is not possible to analytically integrate the bivariate or multivariate probability density function. Simplifying assumptions are therefore made to solve a given problem and the implications of these assumptions are analyzed numerically.

Example 2.17 The following bivariate probability density function is given: $f(x, y) = ax^2y^2$, $0 \leq X \leq 2$, $0 \leq Y \leq 2$. Evaluate constant a so that $f(x, y)$ may be considered as a bivariate PDF. Determine the probability $P(0.5 \leq X \leq 1.5, 0 \leq Y \leq 1)$.

Solution Using the property of a bivariate PDF,

$$\int_{-\infty}^{\infty} \int_{-\infty}^{\infty} f(x,y)dxdy = 1$$

we write

$$\int_{0}^{2}\int_{0}^{2} ax^2y^2 dydx = 1$$

$$\int_{0}^{2} \frac{8a}{3}x^2 dx = 1$$

$$a = 9/64$$

$$P(0.5 \leq X \leq 1.5, 0 \leq X \leq 1) = \frac{9}{64}\int_{0.5}^{1.5}\int_{0}^{1} x^2y^2 dydx = \frac{3}{64}\int_{0.5}^{1.5} x^2 dx = 0.0508$$

2.5.3.2 Marginal Probability Density Function

The marginal probability density function can be derived from the bivariate probability density function. This can be illustrated as follows. Consider two random variables X and Y. To determine the probability associated with $X = x$, we consider the interval between $x - (1/2)dx$ and $x + (1/2)dx$; this interval is taken to define an event. To determine the probability of this event, the probability mass in a strip of thickness dx parallel to the Y axis in the X–Y plane must be computed; that is, the value of Y that is associated with a particular value of X is not counted. Mathematically speaking, this process of adding up the probabilities of all events in the strip dx implies integration of $f(x, y)$ over all values of Y. The probability of the event defined by the interval equals $f(x)dx$ and can be expressed as

$$f(x)dx = \int_{-\infty}^{+\infty} dx\, f(x,y)\, dy \qquad (2.23)$$

In Eq. 2.23 dx is infinitesimally small but remains invariant with y. Therefore, dividing by dx gives

$$f(x) = \int_{-\infty}^{+\infty} f(x,y)\, dy \qquad (2.24)$$

Equation 2.24 shows that the marginal distribution of X, $f(x)$, is obtained by integrating $f(x, y)$ over all possible values of Y. Similarly,

$$f(y) = \int_{-\infty}^{+\infty} f(x,y)\, dx \qquad (2.25)$$

Thus, the marginal distribution of Y, $f(y)$, is obtained by integrating $f(x, y)$ over all possible values of X. The separate distributions of X and Y are called the *marginal distributions* of X and Y and the process of obtaining them is illustrated in Fig. 2-29.

Example 2.18 The joint probability distribution of x and y is given as

$$f(x,y) = \frac{3(x-y)^2}{8}, \quad -1 \le x \le 1; -1 \le y \le 1$$

$$= 0 \text{ otherwise}$$

Find the marginal probability density function of X.

Solution Applying Eq. 2.24,

$$f(x) = \int_{-\infty}^{+\infty} f(x,y)\, dy,$$

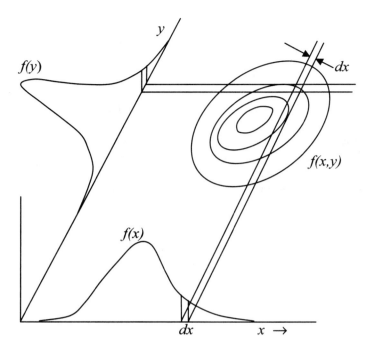

Figure 2-29 *Joint and marginal distributions.*

gives

$$f(x) = \frac{3}{8} \int_{-1}^{1} (x-y)^2 dy = \frac{3}{8} \int_{-1}^{1} \left(x^2 - 2xy + y^2 \right) dy$$

$$f(x) = \frac{3}{8} \left[x^2 y - xy^2 + \frac{y^3}{3} \right]_{-1}^{1} = \frac{1}{4} \left(3x^2 + 1 \right)$$

So the marginal distribution of X is

$$f(x) = \frac{3x^2 + 1}{4} \, , -1 \leq x \leq 1$$

$$= 0, \text{otherwise}$$

In the same way one can determine the marginal distribution of Y:

$$f(y) = \frac{3y^2 + 1}{4} \, , -1 \leq y \leq 1$$

$$= 0, \text{otherwise}$$

2.5.3.3 Conditional Probability Density Functions

Frequently, we are interested in determining the probability distribution of a random variable given a specific value of another related random variable. For example, we may want to determine the probability distribution of rainfall depth for rainfall duration equal to 6 hours, or the probability distribution of annual river flow for winter snowfall of 100 cm, or the probability distribution of drought length for annual rainfall equal to 500 mm in a drainage basin, and so on. Such determinations are made using conditional probability distributions.

For two random variables X and Y, we denote the PDF of the conditional probability distribution of Y for a given value of $X = x$ by the symbol $f(y \mid x)$. The probability that Y is in the interval $y - (1/2)dy$ to $y + (1/2)dy$ for a given value of X is given by the expression $f(y \mid x)dy$; this will be true in every case where the probability distribution is specified by a PDF. This is a conditional probability, conditional, namely, upon X being in the interval, say, between $x_* - (1/2)dx$ and $x_* + (1/2)dx$.

Recalling the definition of conditional probability,

$$P(A \mid B) = \frac{P(A, B)}{P(B)}$$

we thus have

$$f(y \mid x)dy = \frac{f(x_*, y)dx \, dy}{f(x_*)dx} \tag{2.26}$$

or

$$f(y \mid x) = \frac{f(x_*, y)}{f(x_*)} \tag{2.27}$$

and

$$f(x \mid y) = \frac{f(x, y_*)}{f(y_*)} \tag{2.28}$$

Note that x_* is a constant in Eq. 2.27 and y_* is a constant in Eq. 2.28. Since x_* is a constant, the expression $f(x_*, y)$ signifies the function of $f(x, y)$ for a constant value of x. On a two-dimensional probability space, this is the cross section of the plane $X = x_*$ with the probability density surface $f(x, y)$, which is a curve, not a surface. The curve $f(x_*, y)$, shown in Fig. 2-30, measures the way the probability density changes with Y for a constant X but is not a proper PDF since its area will not be equal to 1. However, the area can be computed by integrating $f(x, y)dy$ over the entire range of Y for constant X. This is the same way that the joint probability density function of X is obtained. This will yield the area under the curve $f(x_*, y)$ as equal to $f(x_*)$, as shown in Fig. 2-30. Thus, dividing the function $f(x_*, y)$ by $f(x_*)$ would make the area under the curve equal to 1.

Equations 2.27 and 2.28 show the relationships among conditional, joint, and marginal distributions, which can be clarified by considering their geometric

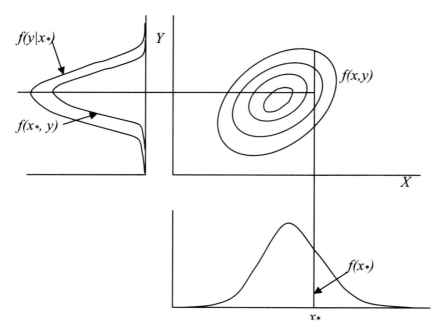

Figure 2-30 *Joint and marginal distributions.*

representations. Figure 2-30 shows the joint distribution and the marginal distribution of X.

Example 2.19 Consider the joint PDF $f(x,y)$ given in Example 2.18. Determine the conditional distribution of X given $Y = y$.

Solution Applying Eq. 2.28,

$$f(x \mid y) = \frac{f(x,y)}{f(y)} \; ,$$

we obtain the conditional distribution of X given y:

$$f(x \mid y) = \frac{3(x-y)^2 / 8}{[3y^2 + 1]/4} = \frac{3(x-y)^2}{6y^2 + 2}$$

In the same way one can determine the conditional distribution of Y given x:

$$f(y \mid x) = \frac{3(x-y)^2}{6y^2 + 2}, \; -1 \le y \le 1$$

$$= 0, \text{ otherwise}$$

2.5.4 Cumulative Distribution Function

For continuous random variables, probabilities are calculated by integration of the PDF over certain intervals. To avoid complications with integration, we introduce the concept of the cumulative distribution function (CDF). The CDF of a random variable X is denoted by the symbol $F(x)$ and is defined as the probability that X will be equal to or less than a given value x:

$$F(x) = P(X \le x) \tag{2.29}$$

$F(x)$ is a function of x. The CDF is calculated by the integral

$$F(x) = \int_{-\infty}^{x} f(x)\,dx \tag{2.30}$$

Once the CDF is determined, the probability associated with any interval can be determined as

$$P(a < X \le b) = F(b) - F(a) \tag{2.31}$$

For a discrete variable, $F(x)$ is calculated by summation of probabilities:

$$F(x) = \sum p(x_i), \quad \text{for all } x_i \le x \tag{2.32}$$

Figure 2-31a and Fig. 2-31b show the probability density functions and cumulative distribution functions for continuous and discrete variables. The CDF is a function that starts with zero somewhere on the left-hand side and increases till it reaches one on the right-hand side. $F(x)$ is the total probability to the left of x and in point x itself. For a continuous distribution, the inclusion of point x makes no difference since the probability in each single point is zero.

Example 2.20 Develop a cumulative frequency distribution for Example 2.13.

Solution Using the PMF data developed in Example 2.14, the CDF values corresponding to each value of X were determined by using Eq. 2.32, as shown in Table E2-20. The resulting CDF is presented in Fig. 2-32.

Example 2.21 Plot the CDF of X given in Example 2.15 and determine (a) probability $X \le 3$, (b) probability $X \ge 9$, and (c) probability $4 \le X \le 8$.

Solution By using the PDF and Eq. 2.30, the CDF values are calculated at several values of X. Then these points are joined by a smooth curve. The obtained curve is the required CDF as shown in Fig. 2-33.
   Using the plot in Fig. 2-33 one can read the probability corresponding to any interval. The answers to the posed questions are (a) probability $X \le 3 = 0.05$, (b) probability $X \ge 9 = 1 - 0.79 = 0.21$, and (c) probability $4 \le X \le 8 = 0.60 - 0.11 = 0.49$.

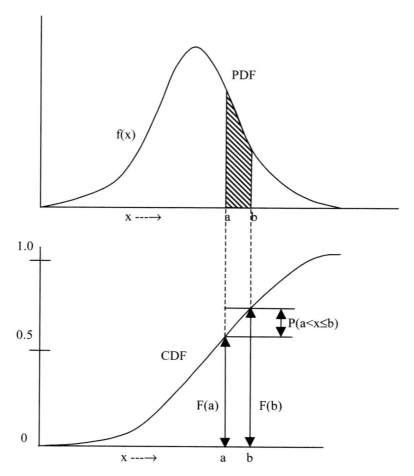

Figure 2-31a PDF and CDF of a continuous distribution.

2.6 Function of a Random Variable

There are many cases where we are interested in determining the probability distribution of a random variable, say, Y, that is a function of another random variable, say, X. For example, we may be interested in the probability of flood damage, which may be a function of water level in the river, or in the probability of traffic interruption on rainy days, which may be expressed as a function of water depth on the highway, or in the probability of beach erosion during a hurricane, which may be a function of tidal currents, and so on. To express the probability algebraically, the basic random variable may be X and its probability distribution may be determined from observations. Our interest is in Y, which is expressed as a function of X as $Y = w(x)$. If this function is known analytically

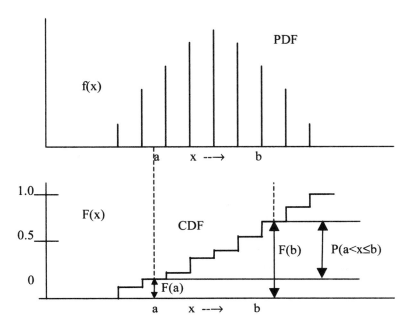

Figure 2-31b PDF and CDF of a discrete distribution.

Table E2-20

X	Frequency	PMF = *f(x)*	CDF = *F(x)*
2	1	0.03	0.03
3	2	0.06	0.08
4	3	0.08	0.17
5	4	0.11	0.28
6	5	0.14	0.42
7	6	0.17	0.58
8	5	0.14	0.72
9	4	0.11	0.83
10	3	0.08	0.92
11	2	0.06	0.97
12	1	0.03	1.00
Sum =	36	1	

then the probability distribution of Y can be determined. There is a one-to-one relationship between X and Y.

If there is a functional relationship between X and Y, say, $Y = w(X)$, the PDF of Y can be determined through the PDF of X, as shown in Fig. 2-34. From this, one notes that Y lies in the interval between $y + (\frac{1}{2})dy$ and $y - (\frac{1}{2})dy$ if and only

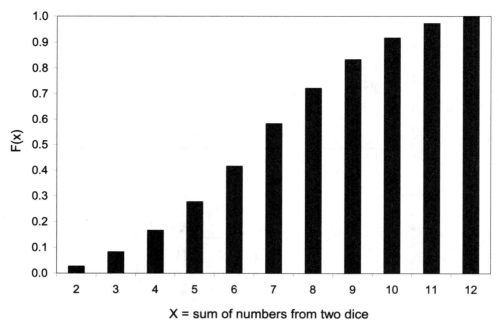

Figure 2-32 *CDF for the sum of numbers obtained by throwing two dice.*

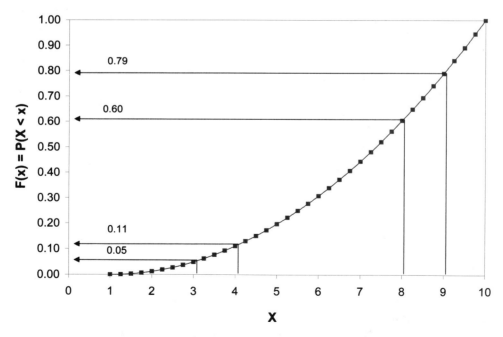

Figure 2-33 *CDF of* X.

if x lies in the interval between $x + (\frac{1}{2})dx$ and $x - (\frac{1}{2})dx$, such that $y = w(x)$. This means that the probability that Y lies in the interval between $y + (\frac{1}{2})dy$ and $y - (\frac{1}{2})dy$ is equal to the probability that X lies between $x + (\frac{1}{2})dx$ and $x - (\frac{1}{2})dx$. In other words,

$$f(y)dy = f(x)dx \tag{2.33}$$

The differential quotients dy/dx and dx/dy can be positive as well as negative and can be determined by differentiation. The ratio between two positive intervals corresponding to events on the probability space can be obtained as

$$f(y) = f(x)\left|\frac{dx}{dy}\right| \tag{2.34}$$

If Y is an analytical function of X then one must first determine the inverse and obtain the ratio by differentiation with respect to Y. Referring to Fig. 2-34, we can obtain the ratio $|dx/dy|$ by drawing the tangent of the function at any particular point and by measuring the line segments cut by the tangent from the positive axes. This procedure involves (a) the determination of the interval dy corresponding to dx at any point x and (b) making the areas under the PDFs corresponding to the intervals the same.

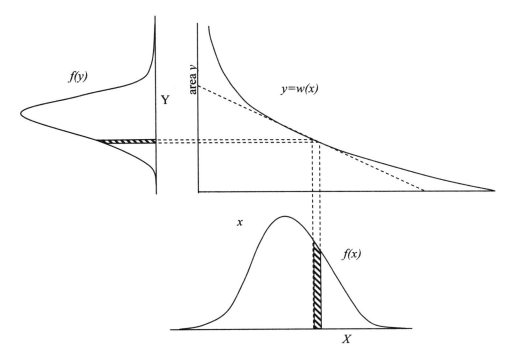

Figure 2-34 *Schematic for deriving PDF of Y as function of X.*

Example 2.22 Assume X is a continuous variable defined in the interval $0 < X < 2$ and characterized by the following PDF:

$$f(x) = \frac{3}{8}x^2$$

If $y = x^3$, find the probability $0 < Y < 5$.

Solution Applying Eq. 2.34, $f(y) = f(x)\, dx/dy$, and differentiating the relationship $y = x^3$, $dy = 3x^2 dx$, we get $dx/dy = 1/3x^2$ and so

$$f(y) = f(x)\frac{dx}{dy} = \frac{3}{8}x^2\frac{1}{3x^2} = \frac{1}{8}$$

As X is defined from 0 to 2, the range over which Y is defined is 0 to 2^3 (i.e., 0 to 8). Thus, Y is characterized by a uniform distribution defined in the interval $0 < y < 8$. The probability $(0 < Y < 5)$ = area of the rectangle having length 5 and height $1/8 = 5 \times 1/8 = 5/8$.

Example 2.23 A number of network modeling packages assume that chlorine decay follows first-order kinetics. The chlorine concentration, C (mg/L), at any time t, is given by the following equation:

$$C = C_0 \exp(-xt)$$

The decay parameter X is characterized by the following PDF:

$$f(x) = \frac{\lambda e^{-\lambda x} x^{\alpha-1}}{\Gamma(\alpha)}$$

Determine the PDF of C.

Solution The distribution of C can be obtained by applying Eq. 2.34:

$$f(c) = f(x)\frac{dx}{dc}$$

If x is rewritten as

$$x = -\frac{1}{t}\ln\left(\frac{C}{C_0}\right)$$

then

$$\frac{dx}{dC} = \frac{d\left[-\frac{1}{t}\ln\left(\frac{C}{C_0}\right)\right]}{dC} = \frac{1}{-tC_0 C}$$

Now, replacing x in $f(x)$ by

$$x = -\frac{1}{t}\ln\left(\frac{C}{C_0}\right)$$

and substituting dx/dc and simplifying the expression gives the distribution of chlorine residual as

$$f(c) = \frac{\lambda\left(\dfrac{C}{C_0}\right)^{\lambda/t}\left[\ln\left(\dfrac{C}{C_0}\right)\right]^{\alpha-1}}{\Gamma(\alpha)C \cdot C_0(-t)^{\alpha}}$$

2.7 Questions

2.1 Obtain daily temperature data for the city you live in for a number of years, say, 30 years or more. Plot the temperature data against time. Is daily temperature random? Obtain the maximum temperature and the minimum temperature for the month of August for each year. Plot this temperature as a function of year. Compute the mean temperature for August and plot it. Discuss whether the maximum temperature and minimum temperatures are random variables.

2.2 Obtain rainfall data for a city of your choice for several years, say, 30 years or more. Compute yearly rainfall and the long-term yearly mean. Plot yearly rainfall as well as the mean. Is yearly rainfall a random variable? Now obtain the rainfall data for the month of August for each year. Also obtain the long-term mean rainfall for the month of August. Plot the August rainfall data as a function of year. Also plot the mean. Is rainfall for the month of August a random variable?

2.3 Obtain the yearly maximum wind velocity data for a city of your choice for a number of years. Plot the wind velocity as a function of time and discuss whether this velocity is a random variable. Also plot the wind velocity mean.

2.4 Obtain the instantaneous maximum discharge data for the Amite River at Darlington, Louisiana. Plot the maximum discharge as a function of year and show if the discharge can be considered as a random variable. Also plot the mean discharge. Also compute the time between two consecutive maximum discharge values, called the interarrival time. Compute the average value of this time. Can the interarrival time be considered random?

2.5 Each year air quality standards established by the EPA are violated for a certain number of times in Baton Rouge, Louisiana. Obtain the number of violations occurring in Baton Rouge for several years. Can the number of violations be considered a random variable? Compute the period of each violation as well as the time between violations. Can these be considered random? Compute their mean values.

2.6 The following experiments involve a random variable or nonrandom variable:

(a) Roll two fair dice at a time.

(b) Measure the time between consecutive plane arrivals.

(c) Toss a coin five times.

(d) Take a penalty shot on goal.

(e) Measure the number of days an air quality standard is violated in a year.

(f) Roll a die and determine whether it is a 6 or not.

(g) Test a randomly selected circuit to see whether it is defective.

(h) Determine the number of vehicles on a bridge at a particular period of time.

(i) Determine whether there was flooding this year in New Orleans.

(j) Find out the flow rate in a stream.

(k) Determine when a comet appears in the sky.

(l) Request a person's highest educational level.

Indicate which of the variables is random, and if it is, then determine which type of random variable it is (e.g., discrete, continuous, or Bernoulli).

2.7 A concrete culvert is to be designed such that it can carry a predicted flow. Discharge measurements are irregular, and the engineer assigns estimates of annual maximum flow rates and their likelihoods of occurrences (assuming that a maximum 20 cfs is possible) as follows: event A = [10 to 17] with $P[A] = 0.6$; event B = [13 to 20] with $P[B] = 0.6$, and event $C = [A \cap B]$ with $P[A] = 0.7$.

(a) Construct the sample space. Indicate events A, B, C, $A \cap C$, $A \cap B$, and $A^c \cap B^c$ on the sample space.

(b) Find $P[A \cap B]$, $P[A^c]$, and $P[B \cap A^c]$.

(c) Find $P[A|B]$, $P[B|A]$, and $P[B|A^c]$.

2.8 It is assumed that earthquakes and high wind speeds are unrelated. At a particular location the probability of "high" wind speed occurring throughout any single minute is 10^{-6} and the probability of a "moderate" earthquake during any single minute is 10^{-9}.

(a) Find the probability of the joint occurrence of the two events during any minute.

(b) Find the probability of the occurrence of one or the other or both during any minute.

(c) If the events in succeeding minutes are mutually independent, what is the probability that there will be no moderate earthquakes in a year near this location? What is the probability in 5 years?

2.9 Consider the possible failure of a water supply system to meet demand during any given dry-season day.

(a) Use the total probability (Eq. 2.16) to determinate the probability that the supply will be insufficient if the probabilities are as listed in Table Q2-9.

Table Q2-9

	Demand level (gal/day)	P [level]	P [inadequate supply level]
D_1	50,000	0.55	0
D_2	100,000	0.35	0.1
D_3	150,000	0.10	0.5
		1.00	

(b) Determine the probability that a demand level of 100,000 gal/day was the "cause" of the system's failure to meet demand if an inadequate supply was observed.

2.10 Consider the continuous PDF $f(x) = 0.25$ for $0 < X < a$.

(a) What is a?

(b) What is $P [X < a/2]$?

(c) What is $P [X > a/2 \mid X > a/4]$?

(d) What is $P [X > a/2 \mid X < a/4]$?

2.11 Assume that in earthquake-resistant design, the following relations take place: $Y = ce^x$, where Y is the ground-motion intensity at the building site, X is the magnitude of an earthquake, and c is related to the distance between the site and center of the earthquake. If X is exponentially distributed $f(x) = le^{-lx}$ with $x \geq 0$. Show that the CDF is $F(y) = 1 - (y/c)^{-l}$ with $y \geq c$. Sketch the distribution.

2.12 Table Q2-12 gives the annual peak flow at the USGS gauge station 05578500 on Salt Creek near Rowell, Illinois. Develop a stem-and-leaf diagram for the peak flow.

Table Q2-12

Year	1943	1944	1945	1946	1947	1948	1949	1950	1951	1952	1953
Peak flow	12400	8850	1380	3040	1600	2480	1810	6890	9390	3170	2210
Year	1954	1955	1956	1957	1958	1959	1960	1961	1962	1963	1964
Peak flow	829	1320	10300	7950	5730	7500	2290	10300	4110	6050	10600
Year	1965	1966	1967	1968	1969	1970	1971	1972	1973	1974	1975
Peak flow	1830	1090	1040	24500	1700	5060	1310	2020	7270	8060	3920

2.13   Develop a frequency histogram and a cumulative frequency plot for the peak flow of Question 2.12 using different numbers of bins. Evaluate the impact of the number of bins on both the frequency histogram and cumulative frequency plots. What frequency bin gives you the most appropriate results?

2.14   If a highway bridge is constructed on Salt Creek near Rowell (Question 2.12) for a 15-year flood, what is the probability that the design flood will be exceeded in the next 30 years?

2.15   Assume X is a continuous variable defined in the interval $0 < X < 1$ and characterized by the following PDF:

$$f(x) = 4x^2 + 2x + 1$$

If $y = x^{3/2}$, find the probability $0 < Y < 1/2$.

2.16   The joint probability distribution of random variables X and Y is given as

$$f(x,y) = \frac{3(x+y)^2}{8}, \quad -1 \le x \le 1; -1 \le y \le 1$$

$$= 0 \text{ otherwise}$$

Find the marginal probability density function of X.

2.17   If the dynamic head and discharge relationship in a given pipe system is described by the following relationship:

$$h = aq^b$$

where h is the dynamic head and q is the flow described by the following distribution function:

$$f(q) = \frac{1}{\sigma\sqrt{2\pi}} \exp\left[-\frac{1}{2}\left(\frac{q-\mu}{\sigma}\right)^2\right]$$

determine the probability distribution function of the dynamic head h.

2.18 Based on soil data it was found that the concentration c (µg/L) of tetrachloroethylene is described by the following log-normal distribution:

$$f(c) = \frac{1}{1.2c\sqrt{2\pi}} \exp\left[-\frac{1}{2}\left(\frac{\ln c - 6.8}{1.2}\right)^2\right]$$

Further experiments were conducted to determine the amount of solute sorbed onto the soil. The sorption of tetrachloroethylene onto this soil is described by the Freundlich isotherm given as $q = 0.054c^{0.86}$, where q is mass of solute sorbed per unit of soil in µg/g. Find the distribution of the solute sorbed on the soil. Further, determine the following:

(a) probability [$q < 0.50$ µg/g],
(b) probability [$2 < q < 20$ µg/g], and
(c) probability [$q > 3000$ µg/g].

2.19 Consider a random variable X described by a uniform distribution $f(x) = 1/(b-a), a \leq x \leq b, a > 0$. If X is related to another random variable Y with the following relationship:

$$y = c\exp(-kx)$$

derive a formula for the probability density function of Y.

2.20 Let X be a random variable described as

$$f(-2) = 1/10, \quad f(-1) = 2/10, \quad f(0) = 3/10, \quad f(1) = 2/10, \quad f(3) = 1.5/8, \text{ and}$$
$$f(4) = 1/8.$$

(a) Let R be a random variable defined by the equation $R = (X-1)^2$. Find the distribution function of R.
(b) Let S be a random variable defined by the equation $S = X^3$. Find the distribution function of S.
(c) Let T be a random variable defined by the equation $T = \exp(2X + 5)$. Find the distribution function of T.

2.21 Let X and Y be two random variables described by a joint probability distribution function given as

$$f(x,y) = k\left(x^2y + xy^2 + xy + x + y + c\right), 0 \leq x \leq 1, 0 \leq y \leq 1.$$

(a) If $c = 0$, determine the value of k so that $f(x,y)$ is a valid bivariate distribution.
(b) Determine the marginal distributions of X and Y.
(c) Determine the values of k and c for which X and Y are statistically independent.

2.22   Let X and Y be two random variables described by their probability density functions $f(x)$ and $f(y)$ as

$$f(x) = e^{-(2x+1)} \ (x \geq 0) \text{ and } f(y) = e^{-(2y-1)} \ (y \geq 0)$$

and define a random variable $Z = (X + Y)^2$.

(a) Determine $f(z)$ at $z = 0$.

(b) Sketch the probability distribution of Z and find the value of Z corresponding to the maximum of $f(z)$.

(c) Find the probability that Z is greater than 2.0.

2.23   A contaminant is discovered in a sample taken from a stream used as the main drinking water source by a city. You are told that the test used to detect the contaminant is extremely reliable. It is 100% sensitive (i.e., it is always correct if contamination exists). But this test gives false results about 0.01% of the time. Determine the probability of being contamination free given a positive test result.

2.24   Consider a bivariate distribution $f(x,y)$ given as

$$f(x,y) = \frac{1}{24\pi} e^{-\frac{1}{2}\left[\left(\frac{x^2}{9} + \frac{y^2}{16}\right)\right]}$$

Determine the following:

(a) probability $[X \leq 2 \text{ and } Y \leq 3]$,

(b) probability $[-1 \leq X \leq 3 \text{ and } 2 \leq Y \leq 5]$, and

(c) probability $[X \geq 4 \text{ and } Y \geq 6]$.

2.25   Consider the bivariate distribution of Question 2.24. Determine the marginal distributions of X and Y. Use marginal distributions to determine whether X and Y are independent random variables.

2.26   A preliminary groundwater drilling was conducted with an assumed prior probability of 0.81. The electrical resistivity method might be used to locate the drilling locations. This method gives favorable results for about 78% of applications where water was known to be present and 97% unfavorable results where water was not found. Determine (a) the probability of finding water given a favorably result and (b) the probability of finding water given an unfavorable result.

Chapter 3

Moments and Expectation

As mentioned in Chapter 2, the probabilistic approach plays an important role and offers meaningful measures for decision making related to planning, design, analysis, management, and regulatory compliance of engineering projects. Further, parameters having significance in the design or analysis are frequently subject to significant variability and are taken as random variables. Instead of having precise single values, random variables assume a range of values in accordance with their probability mass or probability density functions. The probability distribution of a random variable quantifies the likelihood that its value lies in any given interval. But the mathematical form of the probability distribution for the complete characterization of a random variable relevant to a real-life engineering system is often not known, as most generally only a sample datum is known. In many cases, descriptive parameters—called moments—are specified to approximately define the distribution. These parameters are estimated from the available data to extract valuable information about distributions of relevant random variables. In this chapter, we first introduce the theory of statistical expectations and moments. Then, we discuss the specific moments and expectations that are most commonly utilized as descriptive parameters in most civil engineering-related projects.

3.1 Expectation

Let X be a random variable characterized with a probability density function (PDF) or probability mass function (PMF), $f(x)$. Further, let $g(x)$ be another function of x defining a given system. The expectation of a function $g(x)$, denoted $E[g(x)]$, is defined as

$$E[g(x)] = \int_{-\infty}^{\infty} g(x)f(x)dx, \text{ if } X \text{ is a continuous random variable} \qquad (3.1a)$$

$$E[g(x)] = \sum_{\text{all } x} g(x)f(x), \text{ if } X \text{ is a discrete random variable} \qquad (3.1b)$$

In other words, the expectation of a function $g(x)$ is a weighted average of its possible values determined at various X values that it can take. Each $g(x)$ value is weighted by its corresponding probability.

If $g(x) = x$, Eq. 3.1 can be rewritten as

$$E[X] = \int_{-\infty}^{\infty} xf(x)dx, \text{ if } X \text{ is a continuous random variable} \qquad (3.2a)$$

$$E[X] = \sum_{\text{all } x} xf(x), \text{ if } X \text{ is a discrete random variable} \qquad (3.2b)$$

3.1.1 Properties of Expectation

From the definition of expectation it is easy to prove the following properties:

1. The expectation of a constant c is the constant itself; that is,

    $$E[X] = c, \text{ if } X = c$$

2. The expectation of a product of a constant c and X is equal to the constant multiplied by the expectation of X; that is,

    $$E[c\,X] = c\,E[X]$$

    Further, if we extend this rule to a linear sum of functions and let a, b, ... be constants, then

    $$E[a\,g_1(X) + b\,g_2(X) + ...] = a\,E[g_1(X)] + b\,E[g_2(X)] + ...$$

3. The expectation of the sum of n random variables is equal to the sum of the expectaion of the n individual random variables:

    $$E[X_1 + X_2 + ... + X_n] = E[X_1] + E[X_2] + ... + E[X_n]$$

4. The expectation of the multiplication of n independent random variables is equal to the product of the expectation of the n individual random variables:

$$E[X_1 \times X_2 \times X_3 \times ... \times X_n] = E[X_1] \times E[X_2] \times [E_3] \times ... \times E[X_n]$$

5. $|E[X]| \leq E[|X|]$
6. $|E[X]| \leq c$ if $P(|X| \leq c) = 1$

The expectation of common random variables can also be calculated by using Eqs. 3.1 and 3.2.

Example 3.1 Suppose that a discrete random variable X has the following PMF:

X	0	1	2	3	4	5	6
$f(x)$	0.05	0.1	0.2	0.3	0.2	0.1	0.05

Calculate the following:

   (i) $E[X]$
   (ii) $E[10X]$
   (iii) $E[g(x)]$, where $g(x) = (10X + 2)/3$.

Solution

   (i) From Eq. 3.2b, we have

$$E[X] = \Sigma x \, f(x) = 0(0.05) + 1(0.1) + 2(0.2) + 3(0.3) + 4(0.2) + 5(0.1) + 6(0.05) = 3$$

   (ii) Because of the basic property of expectation, we know that $E[cX] = c \, E[X]$. Thus,

$$E[10X] = 10E[X]$$

Substituting $E[X] = 3$ as calculated in (i), we get

$$E[10X] = 10(3) = 30$$

   (iii) $E[g(x)] = E[(10X+2)/3] = E[(10X+2)]/3 = \{10 \times E[X] + E[2]\}/3 = \{10 \times E[X] + 2\}/3$. Substituting $E[X] = 3$ as calculated in (i), we get

$$E[g(x)] = (10 \times 3 + 2)/3 = 32/3$$

Example 3.2 Calculate (i) $E[X]$, (ii) $E[2X]$, (iii) $E[2X+8]$, if X is uniformly distributed over (0, 1).

Solution In this example, we have to determine the mathematical form of the distribution $f(x)$. Because X is distributed uniformly, $f(x)$ is parallel to the X axis. Let $f(x) = c$. Moreover, $f(x)$ is defined only over (0, 1). For $f(x)$ to be a distribution, its area (a rectangle with base length = 1 and width = c) should be 1. So, $1 \times c = 1$, giving $c = 1$. So, $f(x) = 1$ is defined over $(0, 1)$. Knowing $f(x)$, one can calculate the required expectations as given in the following.

(i) Using Eq. 3.2a, we have

$$E[X] = \int_{-\infty}^{\infty} xf(x)dx = \int_{0}^{1} xf(x)dx = \int_{0}^{1} xdx = \frac{1}{2}\left[x^2\right]_0^1 = \frac{1}{2}$$

(ii) From the basic properties of expectation, we know $E[cX] = cE[X]$, so
$E[2X] = 2\,E[X] = 2(1/2) = 1$

(iii) $E[2X + 8] = 2\,E[X] + 8 = 2(1/2) + 8 = 9$

Example 3.3 Let X be a random variable defined by the following PDF:

$$f(x) = \frac{4}{3}\left(1 - x^3\right) \text{ for } 0 < x \le 1$$

$$= 0 \text{ elsewhere}$$

Find (a) $E[X]$ and (b) $E[3X + 2]$.

Solution

(a) $\quad E[X] = \int_{-\infty}^{\infty} xf(x)dx = \int_{0}^{1} xf(x)dx = \frac{4}{3}\int_{0}^{1} x\left(1 - x^3\right)dx = \frac{4}{3}\left[\frac{x^2}{2} - \frac{x^5}{5}\right]_0^1 = \frac{4}{10}$

(b) $E[3X + 2] = 3\,E[X] + E[2]$, using the second property of expectation
$\qquad = 3\,E[X] + 2$, using the first property of expectation
$\qquad = 3 \times 4/10 + 2$ (substituting the result $E[X] = 4/10$)
$\qquad = 16/5$

Example 3.4 A water resources engineer is interested in determining the average annual concentration of total phosphorus (*TP*) in a stream draining the runoff of a local watershed into a lake. This lake is used as a water supply and the state agency is concerned with eutrophication resulting from increased levels of *TP* loading. The engineer has collected water quality samples during several storm events and has found that the *TP* concentration has the following relationship with stream flow:

$$TP = 0.40 \ln(Q) + 2.07$$

where *TP* is in mg/L and *Q* is in cfs. A USGS site upstream of the lake records daily flow, and data of last five years are available. The engineer is interested in using the flow information to determine the average annual *TP* concentration. The empirical frequency distribution based on five years of daily flow is tabulated in Table 3-4 and is presented graphically in Fig. 3-1.

Table E3-4 Observed frequency distribution of flow.

Lower class interval	Upper class interval	Midpoint	Absolute frequency	Relative frequency
0	2	1	20	0.011
2	4	3	64	0.035
4	8	6	226	0.124
8	16	12	280	0.153
16	32	24	320	0.175
32	64	48	335	0.183
64	128	96	292	0.160
128	256	192	187	0.102
256	512	384	75	0.041
512	1024	768	18	0.010
1024	2048	1536	9	0.005

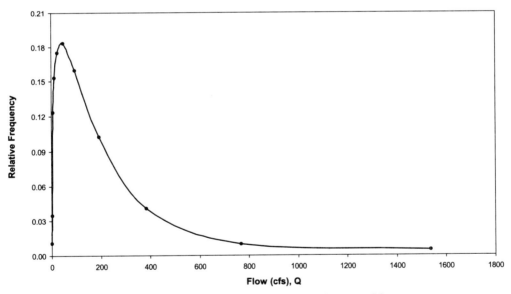

Figure 3-1 Observed frequency distribution of flow.

Solution By using the relationship between *TP* and flow, the midpoints of the flow classes are converted into the corresponding *TP* concentrations. Then, the expected value of the *TP* concentration can be calculated by applying Eq. 3.1. For example, for the second class the midpoint is 3 cfs. The corresponding *TP* concentration is 2.51 mg/L. The relative frequency of *TP* will be equal to the corresponding relative frequency of flow in the same class interval. Thus *TP* = 2.51 has a relative frequency of 0.04. Therefore, the *TP* and its relative frequency are given as follows:

TP (mg/L)	2.07	2.51	2.79	3.06	3.34	3.62	3.90	4.17	4.45	4.73	5.00
Relative frequency	0.01	0.04	0.12	0.15	0.18	0.18	0.16	0.10	0.04	0.01	0.00

Applying Eq. 3.1 gives the expected value of the TP concentration:

$$E[TP] = 2.07 \times 0.01 + 2.51 \times 0.04 + 2.79 \times 0.12 + 3.06 \times 0.15 + 3.34 \times 0.18 + 3.62 \times 0.18 +$$
$$3.9 \times 0.16 + 4.17 \times 0.1 + 4.45 \times 0.04 + 4.73 \times 0.01 + 5 \times 0.0 = 3.43 \text{ mg/L}$$

3.2 Moments

The moments of a distribution comprise a special class of expectations that can be used to compare distributions and derive properties of the distributions. In many cases, the moments of a distribution are used as a way of summarizing the important characteristics of a distribution as single numbers, without entailing too much detail. There are several types of moments in the statistical literature, but we most commonly deal with two general types of moments: moments about the origin (or regular moments) and moments about the centroid (or central moments).

3.2.1 General Moments or Noncentral Moments

The most commonly used moments are the moments about the origin, called noncentral moments or general moments. For any random variable X defined by PDF (or PMF) $f(x)$, and any positive integer k, the expectation $E[X^k]$ is called the kth noncentral moment of X and is given by

$$\mu'_k = E\left[X^k\right] = \int_{-\infty}^{\infty} x^k f(x)dx \text{ for the continuous case} \qquad (3.3a)$$

$$\mu'_k = E\left[X^k\right] = \sum_{\text{all } x} x^k f(x) \text{ for the discrete case} \qquad (3.3b)$$

It is clear from Eq. 3.3 that the zeroth noncentral moment is the integration of the PDF or PMF itself, giving $\mu'_0 = E[1] = 1$. Further, for $k = 1$ Eq. 3.3 gives the first noncentral moment, which is equal to the mean of the distribution $f(x)$, that is, $\mu'_1 = E[X] = \mu$. In general, the first noncentral moment provides a measure of the central location of a distribution. We will discuss measures of central location in detail later in this chapter.

3.2.2 Central Moments

The central moments are used to measure various aspects of a distribution with respect to its mean μ. For any random variable X defined by PDF or PMF $f(x)$ and any positive integer k, the expectation $E[(X- \mu)^k]$ is called the kth central moment of X and is given by

$$\mu_k = E\left[(X-\mu)^k\right] = \int_{-\infty}^{\infty} (x-\mu)^k f(x)dx \text{ for the continuous case} \qquad (3.4a)$$

$$\mu_k = E\left[(X-\mu)^k\right] = \sum_{\text{all } x} (x-\mu)^k f(x) \text{ for the discrete case} \qquad (3.4b)$$

Again for $k = 0$, $\mu_0 = E[1] = 1$.

For $k = 1$, $\mu_1 = E\left[(X-\mu)\right] = E[X] - E[\mu] = \mu - \mu = 0$; that is, the first central moment is always zero. The second central moment obtained by substituting $k = 2$ in Eq. 3.4 is $E[(X-\mu)^2]$, which measures the spread of a distribution about its mean, also known as the variance of the distribution. After the mean, the variance is the most important moment of a distribution. Its unit is the square of the unit of the random variable and hence it is always positive. A zero variance thus implies a deterministic variable. We will discuss variance in more detail later in the chapter.

3.2.3 Relationship Between Central and Noncentral Moments

Rewriting Eq. 3.4, we can express the kth central moment of X as

$$\mu_k = E\left[(X-\mu)^k\right] = E\left[\sum_{i=0}^{k} \binom{k}{i} X^i (-\mu)^{k-i}\right] = \sum_{i=0}^{k} \binom{k}{i}(-\mu)^{k-i} E\left[X^i\right] = \sum_{i=0}^{k} \binom{k}{i}(-\mu)^{k-i} \mu_i' \quad (3.5)$$

By substituting $k = 1, 2, 3$, and 4, the relationship between the first four central and noncentral moments are derived as given in the following. For $k = 1$, the relationship is

$$\mu_1 = \sum_{i=0}^{1} \binom{1}{i}(-\mu)^{1-i} \mu_i' = \binom{1}{0}(-\mu)^1 \mu_0' + \binom{1}{1}(-\mu)^0 \mu_1'$$

We know that $\mu_0' = E[1] = 1$ and $\mu_1' = E[X] = \mu$. Substituting these relations into the previous expression gives

$$\mu_1 = \binom{1}{0}(-\mu) + \binom{1}{1}\mu = \mu - \mu = 0 \qquad (3.5a)$$

Thus, the first central moment of any random variable is always zero. Now, for $k = 2$,

$$\mu_2 = \sum_{i=0}^{2}\binom{2}{i}(-\mu)^{2-i}\mu_i' = \binom{2}{0}(-\mu)^2\,\mu_0' + \binom{2}{1}(-\mu)^1\,\mu_1' + \binom{2}{2}(-\mu)^0\,\mu_2' = \mu^2 - 2\mu^2 + \mu_2' = \mu_2' - \mu^2$$

Thus,

$$\mu_2 = \mu_2' - \mu^2 \tag{3.5b}$$

Now, substituting $k = 3$ in Eq. 3.5, one gets

$$\mu_3 = \sum_{i=0}^{1}\binom{3}{i}(-\mu)^{3-i}\mu_i' = \binom{3}{0}(-\mu)^3\,\mu_0' + \binom{3}{1}(-\mu)^2\,\mu_1' + \binom{3}{2}(-\mu)^1\,\mu_2' + \binom{3}{3}(-\mu)^0\,\mu_3'$$

$$= -\mu^3 + 3\,\mu^3 - 3\,\mu\,\mu_2' + \mu_3'$$

$$= \mu_3' - 3\,\mu\,\mu_2' + 2\,\mu^3 \tag{3.5c}$$

Similarly, for $k = 4$ one can show that

$$\mu_4 = \mu_4' - 4\,\mu\,\mu_3' + 6\,\mu^2\mu_2' - 3\,\mu^4 \tag{3.5d}$$

Example 3.5 A random variable X is defined by the following PDF:

$$f(x) = x \,, 0 \le X \le \sqrt{2}$$

(i) Determine the first four moments about the origin, and (ii) use the non-central moments to determine the central moments.

Solution

(i) Using Eq. 3.3a gives the first moment of X about the origin:

$$\mu_1' = \int_{-\infty}^{\infty} xf(x)dx = \int_{0}^{\sqrt{2}} x^2 dx = \frac{1}{3}\Big[x^3\Big]_0^{\sqrt{2}} = \frac{2\sqrt{2}}{3}$$

The second moment of X about the origin is

$$\mu_2' = \int_{-\infty}^{\infty} x^2 f(x)dx = \int_{0}^{\sqrt{2}} x^3 dx = \frac{1}{4}\Big[x^4\Big]_0^{\sqrt{2}} = 1$$

The third moment of X about the origin is

$$\mu_3' = \int_{-\infty}^{\infty} x^3 f(x)dx = \int_{0}^{\sqrt{2}} x^4 dx = \frac{1}{5}\Big[x^5\Big]_0^{\sqrt{2}} = \frac{4\sqrt{2}}{5}$$

The fourth moment of X about the origin is

$$\mu_4' = \int_{-\infty}^{\infty} x^4 f(x) dx = \int_0^{\sqrt{2}} x^5 dx = \frac{1}{6}\left[x^6\right]_0^{\sqrt{2}} = \frac{4}{3}$$

(ii) Using Eq. 3.5a, one gets $\mu_1 = 0$. The second central moment can be calculated by using Eq. 3.5b:

$$\mu_2 = \mu_2' - \mu^2 = 1 - \frac{8}{9} = \frac{1}{9}$$

Using Eq. 3.5c gives the third central moment:

$$\mu_3 = \mu_3' - 3\mu\mu_2' + 2\mu^3 = \frac{4\sqrt{2}}{5} - 3\left(\frac{2\sqrt{2}}{3}\right)(1) + 2\left(\frac{2\sqrt{2}}{3}\right)^3 = -\frac{2\sqrt{2}}{135}$$

Using Eq. 3.5d gives the fourth central moment:

$$\mu_4 = \mu_4' - 4\mu\mu_3' + 6\mu^2\mu_2' - 3\mu^4 = \frac{4}{3} - 4\left(\frac{2\sqrt{2}}{3}\right)\left(\frac{4\sqrt{2}}{5}\right) + 6\left(\frac{2\sqrt{2}}{3}\right)^2(1) - 3\left(\frac{2\sqrt{2}}{3}\right)^4 = \frac{4}{135}$$

3.3 Moment-Generating Functions

A moment-generating function (mgf) is the expectation of a very special function of X used in many areas of probability and statistics from which all kinds of moments of a random variable are obtained. An mgf offers shortcuts for finding the expected value, variance, and higher order moments. Further, an mgf uniquely identifies its corresponding distribution if the distribution has the mgf.

Let X be a random variable with PDF $f(x)$. Then the mgf of X is defined as the expectation of $g(X) = e^{tX}$, where t is any real number:

$$M_x(t) = E\left[e^{tX}\right] = \begin{cases} \sum_{\text{all } x} e^{tx} f(x) & \text{for X discrete} \\ \int_{-\infty}^{\infty} e^{tx} f(x) dx & \text{for X continuous} \end{cases} \tag{3.6}$$

Using the mgf $M_X(t)$, we can define the kth moment about the origin as the value of the kth derivative with respect to t, evaluated at $t = 0$:

$$\mu_k' = E\left[X^k\right] = \frac{d^k M_X(t)}{dt^k}\bigg|_{t=0} \tag{3.7}$$

The expectation of X is said to exist if the integral or infinite series given in Eq. 3.6 converges absolutely. Thus, a random variable may not possess a finite mean, variance, or moment-generating function.

Example 3.6 Determine the moment-generating function for a discrete random variable X whose PMF is as follows:

x	-10	-5	0	5	10	100
$F(x)$	$1/20$	$1/15$	$1/10$	$1/5$	$1/4$	$1/3$

Using the mgf determine the first noncentral moment (i.e., the mean of X).

Solution Using Eq. 3.6 gives the mgf as

$$M_X(t) = E\left[e^{tx}\right] = \sum e^{tx} f(x) = \frac{1}{20}e^{-10t} + \frac{1}{15}e^{-5t} + \frac{1}{10} + \frac{1}{5}e^{5t} + \frac{1}{4}e^{10t} + \frac{1}{3}e^{100t}$$

To determine the first moment of X, we need to first determine the first derivative of the $M_X(t)$ with respect to t. Thus,

$$\frac{dM_X(t)}{dt} = -\frac{1}{2}e^{-10t} - \frac{1}{3}e^{-5t} + e^{5t} + \frac{5}{2}e^{10t} + \frac{100}{3}e^{100t}$$

Now, the first noncentral moment of X can be determined by evaluating $\frac{dM_X(t)}{dt}$ at $t = 0$:

$$E[X] = -\frac{1}{2}e^{-10t} - \frac{1}{3}e^{-5t} + e^{5t} + \frac{5}{2}e^{10t} + \frac{100}{3}e^{100t}\bigg|_{t=0} = -\frac{1}{2} - \frac{1}{3} + 1 + \frac{5}{2} + \frac{100}{3} = \frac{58}{3}$$

Example 3.7 A random variable X is defined by the normal distribution with parameters mean μ and standard deviation σ with the following PDF:

$$f(x) = \frac{1}{\sigma\sqrt{2\pi}}e^{-\frac{(x-\mu)^2}{2\sigma^2}}, \quad -\infty < x < \infty$$

Determine the mgf.

Solution Using Eq. 3.6 we have

$$M_X(t) = \int_{-\infty}^{\infty} e^{tx} f(x)dx = \int_{-\infty}^{\infty} e^{tx} \frac{1}{\sigma\sqrt{2\pi}}e^{-\frac{(x-\mu)^2}{2\sigma^2}} dx$$

Letting $z = (x - \mu)/\sigma$ and then rearranging terms gives

$$M_X(t) = \frac{e^{t\mu + \frac{1}{2}t^2\sigma^2}}{\sqrt{2\pi}} \int_{-\infty}^{\infty} e^{-\frac{(z-t\sigma)^2}{2}} dz = e^{t\mu + \frac{1}{2}t^2\sigma^2} \left[\frac{1}{\sqrt{2\pi}} \int_{-\infty}^{\infty} e^{-\frac{(z-t\sigma)^2}{2}} dz \right] = e^{t\mu + \frac{1}{2}t^2\sigma^2}$$

3.4 Characteristic Functions

As mentioned earlier, the moment-generating functions do not always exist. The most commonly used alternative to the mgf is the characteristic function (cf). It is worth mentioning that a random variable always possesses a characteristic function. Furthermore, there is a one-to-one correspondence between distribution functions and characteristic functions. The characteristic function of X is defined as

$$\varphi_X(t) = E\left[e^{itX}\right] = \int_{-\infty}^{\infty} e^{itx} f(x) dx \tag{3.8}$$

where i is an imaginary number and t is a real number. Using the characteristic function $\varphi_X(t)$, we define the kth moment of X about the origin as the value of the kth derivative with respect to t, evaluated at $t = 0$ and divided by i^k:

$$\mu'_k = E\left[X^k\right] = \frac{1}{i^k} \frac{d^k \varphi_X(t)}{dt^k} \bigg|_{t=0} \tag{3.9}$$

Example 3.8 Find the characteristic function for the random variable X defined in Example 3.2. Using the obtained characteristic function, find the expected value of X.

Solution The PDF of X is $f(x) = 1$ for $0 \le x \le 1, = 0$ elsewhere
The characteristic function of X is

$$\varphi_X(t) = E\left[e^{itX}\right] = \int_{-\infty}^{\infty} e^{itx} f(x) dx = \int_0^1 e^{itx}(1) dx = \frac{e^{itx}}{it} \bigg|_0^1 = \frac{1}{it}\left(e^{it} - 1\right)$$

So, this uniform distribution can be uniquely represented by the characteristic function $\left(e^{it} - 1\right)/it$. Now the expected value of X is

$$\mu'_1 = E[X] = \frac{1}{i} \frac{d\varphi_X(t)}{dt} \bigg|_{t=0} = \frac{1}{i} \frac{\left(i^2 t e^{it} - i e^{it} + i\right)}{i^2 t^2} \bigg|_{t=0} = -\frac{1}{t^2}\left[e^{it}(it - 1) + 1\right]\bigg|_{t=0}$$

Expanding the exponential term up to second order and then taking the limit at $t = 0$, one obtains

$$E[X] = -\frac{1}{t^2}[(1 + it + \frac{(it)^2}{2!})(it - 1) + 1]\Big|_{t=0} = -\frac{1}{t^2}[-\frac{t^2}{2} - \frac{it^3}{2}]\Big|_{t=0} = \frac{1}{2}$$

Example 3.9 Find the characteristic function of X characterized by the following PDF:

$$f(x) = e^{-x} \quad x > 0$$
$$= 0 \quad x < 0$$

Solution The characteristic function of X is

$$\varphi_X(t) = E[e^{itX}] = \int_{-\infty}^{\infty} e^{itx} f(x)dx = \int_0^{\infty} e^{(it-1)x} dx = \frac{e^{(it-1)x}}{(it - 1)}\Big|_0^{\infty} = \frac{1}{(1 - it)}$$

3.5 Characterization of a Single Random 1Variable

Let the random variable be designated as X. It is assumed that the probability distribution of X is given explicitly in the form of a PMF, a PDF, or a cumulative distribution function (CDF). Based on this knowledge, the distribution parameters are defined.

3.5.1 Mean

The most important parameter of a distribution is the mean of a random variable. The mean is defined as the first moment of the probability distribution about the origin of the probability space and is usually designated by the Greek letter μ. For a continuous distribution, μ is defined as

$$\mu = \int_{-\infty}^{+\infty} x f(x)dx = \int_0^1 x dF(x) = \int_{-\infty}^{\infty} [1 - F(x)]dx \tag{3.10}$$

For a discrete distribution, μ is defined as

$$\mu = \sum_{i=1}^{n} x_i p(x_i) \tag{3.11}$$

where x is the value of the random variable X, n is the number of observations, x_i is the ith observation of X, and $i = 1, 2, 3, \ldots, n$. The concept of mean is explained in Fig. 3-2a and Fig. 3-2b for continuous and discrete cases.

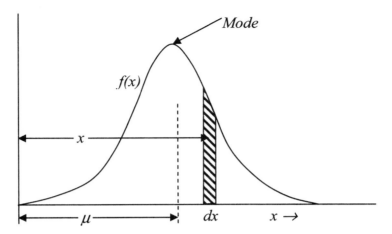

Figure 3-2a *The mean for a continuous case.*

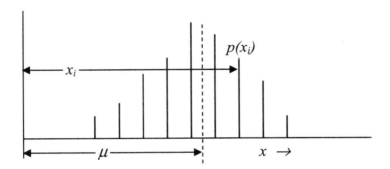

Figure 3-2b *The mean for a discrete case.*

The probability distribution is the distribution of a probability mass of unity (here one-dimensional) over the probability space. The mean gives the distance of the center of gravity of the probability mass from the origin and describes the location of the probability mass on the probability space. Other parameters can also be used to determine the location. One of them is the mode, which is the value of X for which $f(x)$ is a maximum. The mode can be obtained by solving

$$\left[\frac{\partial f(x)}{\partial x}\right]_{x=x_{mo}} = 0$$

where x_{mo} is the mode of X.

Another important parameter is the median, which is the value of X with a 50% probability of being exceeded and therefore also a 50% probability of not being exceeded. In other words, the median divides the distribution into two equal halves and represents the 50th quantile of X:

$$F(x_{med}) = \int_{-\infty}^{x_{med}} f(x)dx = 0.5$$

In engineering, however, the mean is by far the most useful parameter of location. The expected value, designated by $E(x)$, is also given by the mean. The expected monetary value (EMV) can also be interpreted in this sense.

3.5.1.1 Sample Mean

To understand how the mean μ of a random variable X and the sample mean m are related, the concept of a sample distribution is introduced. For each observation of X, the relative frequency can be calculated and plotted. There will probably be n clearly distinguishable points so that each point gets a relative frequency of $1/n$. If points happen to overlap, k observations might fall on the same point x. The relative frequency assigned to that point then becomes k/n. The relative frequency distribution is in all respects identical to the probability distribution for a discrete random variable. The sum of all relative frequencies is equal to 1.

Suppose one wants to determine the mean saturated hydraulic conductivity of soil of a watershed. Let the saturated hydraulic conductivity be the random variable X. It is random because of the variability of soil characteristics, such as texture, particle size distribution, organic matter, land use, and vegetation. Since the probability distribution of X is not known, one must resort to measuring it experimentally. It is necessary to take a large number of soil samples and measure the conductivity for each sample. Let the sample be of size n, and let the saturated hydraulic conductivity value for each sample be x_i. The mean or average saturated hydraulic conductivity is then determined by adding all the test results and dividing them by n:

$$m = \frac{1}{n}\sum_{i=1}^{n} x_i \tag{3.12}$$

Similarly, one may want to determine the mean elevation of the water level during February in a river at a specified location. For this purpose, measurements of the river water level can be carried out each day and their mean can then be computed.

But why do we need to take a sample? Taking a sample allows us to estimate the probability associated with each event in the probability space. Consider an example of cement strength, where one may wish to estimate the probability that the strength of a batch of cement to be used in a structure falls below a specified strength. To compute this probability, a sample of observations of cement strength is collected. If, in the sample, say, 10% of the observations of X were equal to or less than x then it is expected that the probability that X will be equal to or less than x for the batch of cement to be manufactured will also be 10%. In other words, $F(x) = 10\%$. Similarly, for computing the probability of discharge exceeding a certain value we need observations of river flow; for computing the probability of rainfall amount exceeding a certain value we need observations of rainfall; for computing the probability of water quality violations we need observations of water quality constituents.

One can place reasonable confidence in such probability estimates if the sample is representative of the expected hydraulic conductivity or of the concrete strength. Therefore, special precautions are taken to ensure that the sample will be free of bias (i.e., one should not choose all "good" or "bad" samples). If possible, numerous test observations are taken so that exceptionally high or low values have less influence on the final result. Ideally, one should take samples such that the entire range of variable X is covered, as shown in Fig. 3-3. The sampling should be such that the distribution of relative frequency over the sample space is representative of the distribution of probability over probability space.

Since the mean of X lies at the center of gravity of the probability mass, one may expect the center of gravity of the relative frequency mass to be close to the mean. The latter is calculated by Eq. 3.3. The sample mean is denoted by m and is an estimator of the mean of the random variable X, which has been designated by μ.

It is important to keep the following points in mind: (1) The sample distribution is always a discrete distribution, even though the random variable may be continuous. (2) The sample distribution is representative of a set of observations. (3) The sample distribution tends to vary from sample to sample. (4) The sample mean m itself is a random variable. It varies from sample to sample and each additional observation tends to change it. (5) The sample mean is usually referred to as a statistic rather than a parameter. To distinguish it from the mean of the random variable, μ, it is designated by the letter m or by $\bar{x}$. (6) The larger the sample, the better the agreement between m and μ.

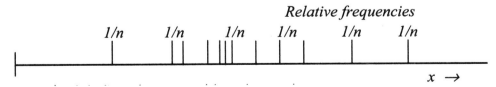

Figure 3-3 Sample space with n *observations.*

3.5.1.2 Arithmetic Mean

Example 3.10 The number of passengers who arrived at a railway terminal was counted from Monday to Friday and the following values were obtained, respectively: 10,371, 8,448, 9,165, 8,974, and 10,739. Find the average number of passengers arriving at the railway terminal each day.

Solution The average number of passengers is

$$m = (10{,}371 + 8{,}448 + 9{,}165 + 8{,}974 + 10{,}739)/5 = 9{,}539.$$

This gives the arithmetic mean of the variable. Besides arithmetic mean, there are also other types of means.

3.5.1.3 Geometric Mean

Another type of mean of data is the geometric mean. The geometric mean is used when the data consist of rates of change or are distributed exponentially. The geometric mean is determined as

$$\bar{x}_g = \left(x_1 x_2 \cdots x_n\right)^{1/n} = \exp\left(\frac{1}{n}\sum_{i=1}^{n} \log x_i\right) = \prod_{i=1}^{n} x_i^{1/n} \tag{3.13}$$

Note that if any one of the observations is zero, the geometric mean will be zero. Further, if any of the observations is less than zero, the geometric mean cannot be computed. Logarithms are helpful in computations when more than three observations are involved. Most often water quality, air pollution, and soil contaminant data are handled by transforming the raw data by taking logarithms (i.e., these data are log-normally distributed). Further, in real-life problems, decisions are taken using sample data as the data collection effort is extremely costly and time consuming. For example, a primary concern in risk-based corrective action (RBCA) is the decision criterion for evaluating attainment of cleanup objectives. The statistic for comparison with a risk-based cleanup objective should be accurate and stable. It has been observed that in such a situation the geometric mean is a more accurate and stable estimator of the true average concentration and its confidence intervals.

Example 3.11 Analysis of demographic data for a country showed that its population growth rate from 1970 to 1980 was 1.25%; from 1980 to 1990, it was 1.22%, and from 1990 to 2000; the rate was 1.15%. Find the average growth rate for the period 1970 to 2000.

Solution The average growth rate can be obtained by taking the geometric mean of given rates:

$$\text{average growth rate} = (1.25 \times 1.22 \times 1.15)^{1/3} = 1.206\%$$

Example 3.12 Consider the case of river pollution. Careful observations showed that the river's pollutant concentration near an industrial town increased by 20% in the year 2002. The next year the pollutant concentration increased by 60%. Compute the average rate of increase in pollutant concentration.

Solution The average rate of increase in pollutant concentration can be determined by using the geometric mean. Thus,

$$\bar{x}_g = (x_1 x_2 \cdots x_n)^{1/n} = (20 \times 60)^{1/2} = 34.64$$

The average rate of increase in pollutant concentration is 34.64%. If we use the arithmetic mean, the average rate will be $(20 + 60)/2 = 40$.

3.5.1.4 Harmonic Mean

When variables are expressed as ratios of two quantities (e.g., kilometers per hour) another type of mean, called the harmonic mean, is useful. The definition of the harmonic mean is

$$\bar{x}_h = \frac{1}{\left(\dfrac{1}{n}\right)\left[\left(\dfrac{1}{x_1}\right)+\left(\dfrac{1}{x_2}\right)+\cdots+\left(\dfrac{1}{x_n}\right)\right]} = \frac{n}{\displaystyle\sum_{i=1}^{n} 1/x_i} \tag{3.14}$$

Equation 3.14 may yield erratic and misleading results if an observation is negative or zero and therefore the harmonic mean should be taken only when the quantities are positive. The kind of bias associated with the harmonic mean is opposite to that of the arithmetic mean. The use of inverses allows a smaller observation to get a larger weight and a larger observation to get a smaller weight. This reduces the value of the average and hence the harmonic average is always smaller than the arithmetic average. The harmonic mean involves a sort of weighting system different from the arithmetic mean and may be more informative when weighting is useful. For this reason the harmonic mean has been used as an aggregation tool to aggregate several water quality subindices to give an overall water quality indicator. This removes one of the subjective aspects of indicator development (i.e., assignment of weights to subindices). Further, it has been observed that the harmonic mean is more sensitive to the subindex with the lowest score.

Example 3.13 Consider a rainfall event with an intensity of 2.5 cm/h that produced 5 cm of rainfall. Another event had an intensity of 4 cm/h and produced 7 cm. Compute the mean rainfall intensity.

Solution The time for the first event is $5/2.5 = 2$ h. The time for the second event is $7/4 = 1.75$ h. Thus, the total time is $2 + 1.75 = 3.75$ h. The total rainfall amount is simply $5 + 7 = 12$ cm. Therefore the average (arithmetic) rainfall intensity is $12/3.75 = 3.2$ cm/h.

The harmonic mean of rainfall intensities is

$$\bar{x}_h = \frac{1}{\left(\frac{1}{2}\right)\left[\left(\frac{1}{x_1}\right)+\left(\frac{1}{x_2}\right)\right]} = \frac{1}{\left(\frac{1}{2}\right)\left[\left(\frac{1}{2.5}\right)+\left(\frac{1}{4}\right)\right]} \approx 3.08 \text{ cm}$$

Example 3.14 The discharge along a reach of the Song River was 256 m^3/s on November 14. Measurements showed that the cross-sectional areas of flow at five locations were 103, 96, 114, 107, and 91 m^2. Find the mean velocity in the reach.

Solution Velocity is the ratio of discharge to cross-sectional area. The harmonic mean of the flow areas is

$$v_m = \frac{1}{\left(\frac{1}{5}\right)\left[\frac{1}{v_1}+\frac{1}{v_2}+\frac{1}{v_3}+\frac{1}{v_4}+\frac{1}{v_5}\right]}$$
$$= \frac{1}{\left(\frac{1}{5}\right)\left[\frac{103}{256}+\frac{96}{256}+\frac{114}{256}+\frac{107}{256}+\frac{91}{256}\right]}$$
$$= 2.504 \text{ m/s}$$

One can also compute the harmonic mean velocity by obtaining the harmonic mean area first as

$$\bar{A}_h = \frac{1}{\left(\frac{1}{5}\right)\left[\frac{1}{103}+\frac{1}{96}+\frac{1}{114}+\frac{1}{107}+\frac{1}{91}\right]} = 101.56 \text{ m}^2$$

so that the mean velocity is 256 × 101.56 = 2.52 m/s.

In contrast, if the arithmetic mean is employed, then for each cross section, the average velocity (v) can be computed as v_1 = 256/103 = 2.486 m/s, v_2 = 256/96 = 2.667 m/s, v_3 = 256/114 = 2.245 m/s, v_4 = 256/107 = 2.392 m/s, and v_5 = 256/91 = 2.813 m/s. The mean velocity is therefore (2.486 + 2.667 + 2.245 + 2.392 + 2.813)/5 = 2.521 m/s.

Example 3.15 A man makes a round-trip drive to a location 60 km away from his home. He drives at a speed of 60 km/h while going and at 30 km/h while returning. Find his average speed for the entire trip.

Solution The average speed computed by the arithmetic mean is (60 + 30)/2 = 45 km/h. But the total journey time is 1 h (while going) + 2 h (while returning) = 3 h. Therefore, he should have covered a distance of 45 × 3 = 135 km, but actually he did only 120 km. Obviously, there is a discrepancy somewhere.

Taking the harmonic mean, one obtains

$$\bar{x}_h = \frac{1}{\left(\dfrac{1}{2}\right)\left[\left(\dfrac{1}{30}\right) + \left(\dfrac{1}{60}\right)\right]} = \frac{1}{\left(\dfrac{1}{2}\right)\left(\dfrac{3}{60}\right)} = 40 \text{ km/h}$$

This is the correct answer. It is observed that the involvement of inverses results in the assignment of weights to data that are inversely proportional to the magnitudes of the data. In this example, the driver traveled at a speed of 60 km/h for one hour and at 30 km/h for two hours. This leads to the suggestion that if the denominator (hours in this case) varies, compute the harmonic mean, and if the numerator (kilometers in this case) varies, use the arithmetic mean.

3.5.1.5 Comparison of the Three Means

If none of the observations is zero or negative (and of course they are not equal), then

$$\bar{x} > \bar{x}_g > \bar{x}_h \tag{3.15}$$

Referring to Example 3.12, we see that the difference in the numbers was purposely kept large to illustrate the relative importance of the numbers and why the behavior given by Eq. 3.15 is noted. The second number (60) is three times the first and its influence on the arithmetic average is also three times larger than the former. When the geometric mean is taken, the influence is not three times larger because the differential in logarithms is not that large. With the harmonic mean, the relative weights are reversed—the contribution of the smaller number to the mean is more. In summary, the size matters in the arithmetic mean, the size of logarithms matters in the geometric mean, and the reciprocals determine the relative importance in the harmonic mean.

3.5.2 The Median and the Mode

Another measure of the central tendency is the median. The median of a distribution is the value that divides the members of the distribution such that half of them are larger and half of them are smaller. Thus, the median divides a distribution into its two halves. Sometimes the median is also called the second quartile. When observations are tied or have the same value, the median may not exist. When observations are ungrouped and are arranged in either descending or ascending order, the median is the middle observation. In the case of an even number of observations, the median is the average of the two central observations.

The mode is the value of the variable that occurs most often or is the value at which the frequency is the maximum. For ungrouped data it is the value that occurs most frequently. Sometimes two or more modes may appear.

The relative positions of the arithmetic mean, the median, and the mode depend on the symmetry of the distribution. If the distribution is symmetric, these measures of central tendency are equal. If the distribution is asymmetric, they take different positions. If the distribution is not unimodal, the simple relationships between them may not be valid.

3.5.3 Variance

Variance measures the variability of a random variable and is the second most important descriptor of its probability distribution. A small variance of a variable indicates that its values are likely to stay near the mean value whereas a large variance implies that the values have large dispersion around the mean. If the stage of a river at a gauging station is independently measured in a quick succession a number of times in a survey, then there will likely be variability in the stage measurements. The magnitude of the variability is a measure of the natural variation and the measurement error.

Variance, designated by the Greek letter σ^2, measures the deviation from the mean and is universally accepted as given by the second moment of the probability mass about the mean. Sometimes, the notation $VAR(x)$ or $var(x)$ is also used. For a continuous variable X variance is expressed as

$$\sigma^2 = \int_{-\infty}^{+\infty} (x-\mu)^2 \, f(x)dx \tag{3.16}$$

and for a discrete variable

$$\sigma^2 = \sum_{i=1}^{n} (x_i-\mu)^2 \, f(x_i) \tag{3.17}$$

In structural engineering, the variance is the moment of inertia of the probability mass about the center of gravity. Figure 3-4 shows PDFs of three probability distributions; the random variables X_1 and X_2 have the same variance but different means; variables X_2 and X_3 have the same mean but different variances.

The variance has a dimension that is equal to the square of the dimension of the random variable. If X is in m^3/s, then σ^2 is in $(m^3/s)^2$. This makes it quite difficult to visualize the degree of variability associated with a given value of the variance. For this reason, the positive square root of the variance, called the standard deviation, denoted by σ, is often used. Its mechanical analogy is the radius of gyration. Figure 3-4 also shows the standard deviation for the three probability distributions.

Variance has four important properties:

1. The variance of a constant is zero: $var[a] = 0$, where a = constant.
2. The variance of X multiplied by a constant a is equal to the variance of X multiplied by the square of a: $var[aX] = a^2 \times var[X]$.

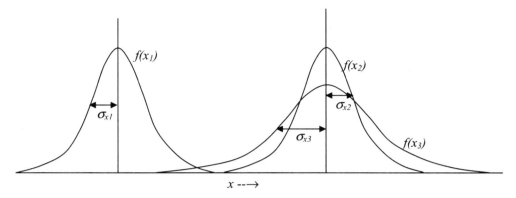

Figure 3-4 *PDFs and standard deviations of three probability distributions.*

3. The variance of X is the difference between the second moment of X about the origin and the second moment of X about the centroid:

$$\text{var}[X] = \mu_2' - \mu^2.$$

4. The variance of the sum of n independent variables is equal to the sum of variances of the n individual random variables:

$$\text{var}[X_1 + X_2 + X_3 + ... + X_n] = \text{var}[X_1] + \text{var}[X_2] + \text{var}[X_3] + ... + \text{var}[X_n]$$

3.5.3.1 Sample Variance

The sample variance is the moment of inertia of the relative frequency mass about its center of gravity on the sample space. Using Eq. 3.17 for discrete variables and observing that $p(x_i)$ is to be replaced by the relative frequency $1/n$, one gets

$$s^2 = \frac{1}{n} \sum_{i=1}^{n} (x_i - m)^2 \tag{3.18}$$

With the sample mean calculated first, the sample variance can be determined from Eq. 3.18. The sample variance serves as an estimator of the variance of the random variable X. The sample variance is also referred to as a statistic, rather than a parameter. It is itself a random variable that is likely to change when additional observations are made. To distinguish sample variance from the variance of the random variable X, it is denoted as s^2. When the number of samples $n \leq 30$, an unbiased estimate of variance is obtained by

$$s^2 = \frac{1}{n-1} \sum_{i=1}^{n} (x_i - m)^2 \tag{3.19}$$

The sample variance can also be computed by

$$s^2 = \frac{1}{n}\sum_{i=1}^{n} x_i^2 - m^2 \qquad (3.20)$$

It is easier to use Eq. 3.20 because one need not subtract m from all values of x before squaring. Instead, x^2 can be computed at the time when computing the mean.

3.5.3.2 Coefficient of Variation

A dimensionless measure of dispersion is the coefficient of variation, c_v, which is computed as the ratio of standard deviation and mean:

$$c_v = s/m \qquad (3.21)$$

When the mean of the data is zero, c_v is undefined. This coefficient is useful in comparing different populations or their distributions. For example, if two samples of aggregates of water quality are analyzed, the one with larger c_v will have more variation.

If each value of a variable is multiplied by a constant α, the mean, variance, and standard deviation are obtained by multiplying the original mean, variance, and standard deviation by α, α^2, and α, respectively; the coefficient of variation remains unchanged. If a constant α is added to each value of the variable, the new mean is equal to the old mean + α; the variance and the standard deviation remain unchanged; the coefficient of variation changes because the unchanged standard deviation is divided by the new mean.

3.5.3.3 Standard Error of Estimate

For a statistical parameter, the standard deviation of its sampling distribution is known as its standard error. The standard error of the mean is $\sigma/\sqrt{n}$ and the standard error of the standard deviation is $\sigma/\sqrt{2n}$.

Example 3.16 For a lake, water levels have been observed for 10 years. The maximum level (x) for each year is listed in Table E3-16. From these data, estimate the mean and the standard deviation of the maximum lake levels.

Solution The computations are demonstrated in Table E3-16.
From the table, we have

$$m = (1/10)\Sigma\, x_i = (1/10) \times 2{,}180 = 218.0 \text{ m}$$

$$s^2 = (1/9)\Sigma\,(x_i - m)^2 = 13.24/9 = 1.47 \text{ m}^2$$

$$s = \sqrt{1.47} = 1.21 \text{ m}$$

Table E3-16

Year	x (meters)	x − m	(x − m)²
1971	217.5	− 0.5	0.25
1972	218.8	0.8	0.64
1973	216.0	− 2.0	4.00
1974	217.8	− 0.2	0.04
1975	220.0	2.0	4.00
1976	218.2	0.2	0.04
1977	217.2	− 0.8	0.64
1978	218.5	0.5	0.25
1979	219.3	1.3	1.69
1980	216.7	− 1.3	1.69
Sum	2,180.0	0.0	13.24

The standard error of $m = 1.21/\sqrt{10} = 0.383$. The standard error of $s = 1.21/\sqrt{2 \times 10} = 0.271$.

Example 3.17 The discharge and stage values of the Amite River near Darlington, Louisiana, are given in Table 3-17a. Compute the mean, median, mode, mean deviation, standard deviation, coefficient of variation, and ratio of standard deviation to the mean deviation of the discharge and stage values.

Solution The sum of discharge and stage values is given in the last row of Table 3-17a. We have 48 sets of data. Thus the mean discharge is $1,376,440/48 = 28,675.83$ cfs. The statistical properties of discharge (cusecs) and stage (ft) data are computed and given in Table 3-21b.

3.5.3.4 Interpretation of Variance as Expectation

For a continuous distribution, the variance is defined by the integral

$$\text{var}(X) = \int_{-\infty}^{+\infty} \{x - E(x)\}^2 f(x) dx \tag{3.22}$$

Variance can also be written as an expectation:

$$\text{var}(X) = E[X - E(X)]^2 \tag{3.23}$$

Expanding the square on the right side yields

$$\text{var}(X) = E\{X^2 - 2XE(X) + [E(X)]^2\}$$
$$= E(X^2) - 2[E(X)]^2 + [E(X)]^2 = E(X^2) - [E(X)]^2 \tag{3.24}$$

Table E3-17a

Year	Discharge (cubic ft/sec)	Stage (ft)
1949	20,000	17.79
1950	43,400	20.2
1951	31,600	19.04
1952	3,180	11.22
1953	18,900	17.63
1954	3,280	11.57
1955	55,700	21.17
1956	20,400	17.84
1957	20,200	17.81
1958	6,900	18.05
1959	9,800	14.83
1960	37,900	15.9
1961	15,400	19.69
1962	4,530	17.06
1963	44,500	12.92
1964	44,500	19.4
1965	20,000	19.37
1966	39,300	18.97
1967	8,000	14.13
1968	8,600	13.82
1969	36,300	9.26
1970	10,100	14.44
1971	45,500	19.43
1972	62,100	20.19
1973	22,400	17.13
1974	40,700	18.98
1975	7,660	12.35
1976	76,400	21.76
1977	30,500	18.09
1978	43,400	19.25
1979	47,500	19.59
1980	8,320	12.85
1981	18,100	16.03

Table E3-17a *(Continued)*

Year	Discharge (cubic ft/sec)	Stage (ft)
1982	63,300	20.29
1983	13,000	14.56
1984	8,970	12.73
1985	17,500	15.92
1986	21,200	16.68
1987	22,000	16.69
1988	16,000	15.66
1989	104,000	22.05
1990	19,500	16.1
1991	26,900	17.46
1992	19,400	16.08
1993	60,800	20.17
1994	23,300	16.95
1995	16,200	15.21
1996	39,300	18.8
Sum	1,376,440	813.11

Table E3-17b

Parameter	Discharge	Stage
Mean	28,675.83 cusec	16.94 ft
Standard deviation	21,117.14 cusec	2.937 ft
Median	20,800 cusec	17.295 ft
Mode	20,000 cusec	17.13 ft
Mean deviation (mean of absolute deviations from the mean)	16,744.13 cusec	2.366 ft
Coefficient of variation	0.736	0.173
Ratio of standard deviation to mean deviation	1.261	1.241

Equation 3.24 is often used to calculate the sample variance. Equations 3.23 and 3.24 can also be employed to calculate the variance of simple functions of X. For example, the variance of the linear functions of $Y = a + bX$, where a and b are constants, can be determined as

$$\begin{aligned}
\text{var}(a + bX) &= E[(a + bX)^2] - [E(a + bX)]^2 \\
&= E[a^2 + 2abX + b^2X^2] - [a + bE(X)]^2 \\
&= a^2 + 2abE(X) + b^2E(X^2) - a^2 - 2abE(X) - b^2[E(X)]^2 \\
&= b^2\{E(X^2) - [E(X)]^2\} = b^2 \,\text{var}(X) \quad\quad\text{(3.25)}
\end{aligned}$$

Notice that adding a constant to a variable does not change its variance. The variance of the variable multiplied by constant amounts to the variance of the variable multiplied by the constant squared. Evidently, taking the variance is not a linear operation.

3.5.4 Skewness

Probability distributions are not usually symmetrical about their mean. This property of being asymmetrical is commonly referred to as the skewness of the distribution. The degree of skewness is measured by the third moment of the probability mass about the mean. For symmetrical distributions, the third moment is zero because the contributions of the probabilities on either side of the mean have opposite signs and cancel each other in the integral. The more asymmetrical the distribution is, the greater will be the absolute value of the third moment. The third moment about the mean can be positive or negative, corresponding to a positive or negative skew. Figure 3-5 shows the possible cases.

The third moment about the mean for a continuous distribution can be calculated as

$$\mu_3 = \int_{-\infty}^{+\infty} (x - \mu)^3 \, f(x)\,dx \quad\quad\text{(3.26)}$$

and the formula for a discrete distribution is

$$\mu_3 = \sum_{i=1}^{n} (x_i - \mu)^3 \, f(x_i) \quad\quad\text{(3.27)}$$

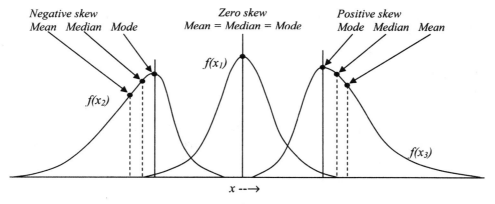

Figure 3-5 Symmetrical and skewed distributions.

3.5.4.1 Coefficient of Skewness

The third moment has a dimension equal to the dimension of the random variable cubed. This makes it awkward to compare the degrees of skewness of different populations or their distributions. For this reason it is customary to divide the third moment by the standard deviation cubed. The resulting parameter is nondimensional and is called the coefficient of skewness:

$$\gamma = C_s = \frac{\mu_3}{\sigma^3} \tag{3.28a}$$

The common notation for the coefficient of skewness is γ or C_s. The sign of the skewness coefficient can be used to denote the degree of symmetry of the probability distribution function. If γ is zero, the distribution is symmetric about its mean. If γ is greater than zero, the distribution is positively skewed or the distribution has a long tail to the right. If γ is negative then the distribution is negatively skewed or the distribution has a long tail to the left.

Another measure of asymmetry is the Pearson skewness coefficient γ_1 expressed as

$$\gamma_1 = \frac{\mu - x_{\text{mod}}}{\sigma} \tag{3.28b}$$

Clearly this does not involve computation of moments higher than two and is thus less susceptible to error.

3.5.4.2 Sample Skewness

An estimate of the skewness of a random variable is obtained from a sample of observations. The third moment of the relative frequency is taken about the sample mean and is divided by the cube of the sample standard deviation. The common notation for sample skewness coefficient is g or G. An unbiased estimate of g, for small samples, is obtained by

$$g = \frac{n \sum_{i=1}^{n} (x_i - \bar{x})^3}{(n-1)(n-2)s^3} \tag{3.29}$$

Example 3.18 Compute the coefficient of skewness of the discharge of the Amite River, given in Example 3.17.

Solution For the data (48 values), the mean and standard deviation were computed as 28,675.83 cusec and 2,117.14 cusec, respectively. The summation term in Eq. 3.29 comes out to be 5.6729×10^{14}.
Hence,

$$g = 48 \times 5.6729 \times 10^{14} / [47 \times 46 \times 2,117.14^3] = 1.34$$

Example 3.19 At a number of recording stations the annual number of severe storms was observed during a series of years. There were 360 observations (the number of stations times the number of years, assuming independence among observations of different stations). This is equivalent to 360 years of observations at one recording station. Each observation recorded the annual number of storms observed, whose number varied from zero to five. The frequencies of the number of storms from zero to five are listed in Table 3-19. Call the number of storms per year the random variable X and estimate the mean, the standard deviation, and the coefficient of the skewness of this random variable.

Solution From the table, the mean is

$$m = \Sigma \, x_i \, f(x_i) = 1.183 \text{ storms per year}$$

The standard deviation is

$$s = [(x_i - m)^2 f(x_i)]^{0.5} = \sqrt{1.0994} = 1.05 \text{ storms per year}$$

and the skewness is

$$g = \Sigma \, (x_i - m)^3 f(x_i)/s^3 = 1.1022/1.05^3 = 0.958$$

A close examination of these calculations reveals that the coefficient of skewness is quite sensitive to incidental variations in the frequency of relatively rare events.

3.5.5 Shape Factors

The idea of plotting dimensionless shape factors to compare distributions was advanced by Nash (1959). A dimensionless moment of order R is defined as the Rth moment about the center of area divided by the first moment about the origin raised to the power R:

$$m_R = \frac{\mu_R}{(\mu_0')^R} \tag{3.30}$$

Table E3-19

X_i	Frequency	Relative frequency $f(x_i)$	$x_i \, f(x_i)$	$(x_i - m)^2 f(x_i)$	$(x_i - m)^3 f(x_i)$
0	102	0.283	0.0	0.3963	− 0.4689
1	144	0.400	0.400	0.0134	− 0.0025
2	74	0.206	0.412	0.1374	+0.1122
3	28	0.078	0.234	0.2574	+0.4677
4	10	0.028	0.112	0.2221	+0.6257
5	2	0.005	0.025	0.728	+0.2780
Sum	360	1.000	1.183	1.0994	+1.1022

For a distribution, higher order moments can be expressed as a function of lower order moments. For instance, the coefficient of skewness $c_s = m_3/m_2^{1.5}$ can be expressed as a unique function of the coefficient of variation $c_v = m_2^{0.5}/m_1'$. Here m_1' is the sample moment about the origin, m_2 is the second sample moment about the mean, and m_3 is the third moment about the mean.

3.5.6 Higher Order Moments

The sampling variance of a moment depends on the population. It becomes very large for higher moments even when the sample size is large. For this reason, higher order moments have limited applications in practical cases and are not commonly used in analysis of civil and environmental engineering systems. Higher order moments are related to the properties of the probability distribution or mass function. Higher order moments describe more subtle properties, such as the symmetry of a PDF or whether its mass is centered around the mean or distributed toward the margins. To derive higher order moments, moment-generating or characteristic functions can be used.

Kurtosis is an indicator of the degree of peakedness of a probability distribution function and is related to the fourth-order central moment as

$$K = \frac{\mu_4}{\mu_2^2} = \frac{\int\limits_{-\infty}^{\infty} (x - \mu)^4 f(x)\,dx}{\sigma^4} \tag{3.31a}$$

with $K > 0$. For the normal distribution the value of K is 3. This value is used as a reference to indicate the degree of peakedness. The coefficient of excess, C_e, is then defined as $K-3$. If the value of K is greater than 3 or $C_e > 0$, then the distribution is called leptokutic. If K is less than 3 or $C_e < 0$, then the distribution is platykutic. Stuart and Ord (1987) presented an inequality that must be satisfied by all plausible distributions:

$$\gamma^2 + 1 \leq K \tag{3.31b}$$

3.5.7 Moment Ratio Diagrams

The moment ratios can be defined in different ways. Johnson and Kotz (1985) defined the Rth moment ratio, m_R, as

$$m_R = \frac{\mu_R}{(\mu_2)^{R/2}} \tag{3.32}$$

If $R = 3$ then one gets the coefficient of skewness and if $R = 4$ then one gets the coefficient of kurtosis.

Now two ratios are defined as

$$\beta_1 = \frac{\mu_3}{\mu_2^{3/2}} \tag{3.33a}$$

$$\beta_2 = \frac{\mu_4}{\mu_2^2} \tag{3.33b}$$

The classical form of the moment ratio diagram (MRD) is a graphical plot of β_1 and β_2 for a specific distribution or a group of distributions, as shown in Fig. 3-6, for a number of distributions. Usually, β_1 is on the abscissa and β_2 is on the ordinate but with its values increasing downward. Pearson (Johnson and Kotz 1985) has shown that for all distributions the following must be satisfied:

$$\beta_2 - \beta_1 - 1 \geq 0 \tag{3.34}$$

When these ratios are plotted, one can discern the impossible region on the graph.

Bobee et al. (1993) have described two kinds of moment ratio diagrams and their applications in hydrology. Ashkar et al. (1988) have plotted the MRD for a number of distributions. These can be employed to distinguish families of distributions, such as the Pearson system of distributions, the Johnson family of distributions, and so on. They can also be used to classify distributions. They allow us to distinguish distributions into three categories: those represented by a point, those represented by a curve, and those represented by a region. For example, the normal, exponential, and uniform distributions are represented by a point, because these distributions do not have a shape parameter but have only a scale or location parameter. The gamma, log-normal, and Student distributions are represented by a curve, because they have one shape parameter. In contrast, the beta distribution has two shape parameters and therefore is represented by a region. In this manner, an MRD permits a comparison of distributions in terms of their flexibility, because the more flexible shape of the distribution occupies a greater portion of the diagram. An MRD also aids in selecting a probability distribution to represent a given sample. This is done by computing the values of β_1 and β_2 from the sample and then plotting the resulting point on the MRD. One then selects the distribution that seems to best reflect the position of this point on the MRD. It must however be emphasized that the sampling variance associated with skewness and kurtosis may be large, especially with small samples, and one may end up selecting a wrong distribution as a result.

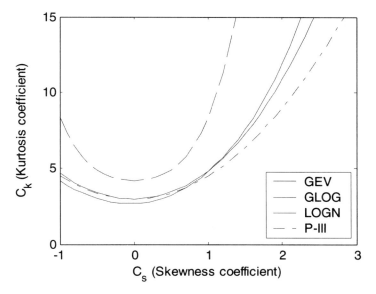

Figure 3-6 *C_s-C_k moment ratio diagram (GEV: generalized extreme value; GLOG: Generalized Logistic) (LOGN: three parameter Lognormal; P-III: Pearson type III)*

3.6 Two Random Variables

The measures of location, dispersion, and asymmetry, namely the mean, variance or standard deviation, and coefficient of skewness, introduced for a single variable, can also be defined for the joint distribution of two or more random variables. Because these descriptive parameters define properties of marginal distributions, they do not shed any light on the joint behavior of the variables: the way the outcome of one variable is influenced by the values the other variable may assume. To that end, parameters describing the degree of dependence that may exist between the random variables are introduced.

3.6.1 Mean and Variance

Consider two random variables X and Y whose joint PDF and marginal PDFs are shown in Fig. 3-7. The centroid is defined by two coordinates designated as μ_x and μ_y. Calculation of μ_x, for example, involves first the determination of the probability mass in a strip of thickness dx, parallel to the Y axis. This elementary mass is determined by the integral

$$dx \int_{-\infty}^{+\infty} f(x,y)dy \tag{3.35}$$

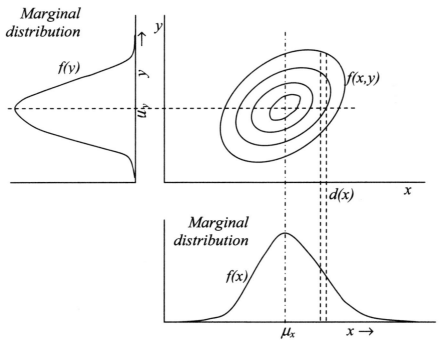

Figure 3-7 *Joint and marginal distributions of two random variables.*

The product of the probability and the distance to the Y axis, x, is then integrated over the entire reach of X, resulting in the double integral

$$\mu_x = \int\limits_{-\infty}^{+\infty} \int\limits_{-\infty}^{+\infty} x f(x,y)\,dx\,dy \tag{3.36}$$

Similarly,

$$\mu_y = \int\limits_{-\infty}^{+\infty} \int\limits_{-\infty}^{+\infty} y f(x,y)\,dx\,dy \tag{3.37}$$

Recall that the marginal distributions are defined by the same process of summing the probability located in small strips parallel to the coordinate axes. Thus, one can write

$$f(x) = \int\limits_{-\infty}^{+\infty} f(x,y)\,dy \tag{3.38}$$

It follows that the double integral in Eq. 3.36 can be rewritten as

$$\mu_x = \int\limits_{-\infty}^{+\infty} f(x)\,dx$$

This is the mean of the random variable X.

For discrete distributions, double integrals are replaced by summations and discrete probabilities replace the term $f(x,y)dxdy$. The coordinates of the centroid corresponding to the means of the marginal distributions remain the same.

Example 3.20 From the measured rainfall data in a catchment, information about rainfall duration, amount, and frequency was derived and is presented in Table 3-20. Compute the marginal probabilities of rainfall duration and amounts. Also determine the conditional probability of 15 mm of rainfall if the storm duration is 3 h.

Table E3-20

Rainfall duration x (h) and amount y (mm)		Relative frequency
x	y	
1	10	0.188
1	15	0.109
1	20	0.086
2	10	0.168
2	15	0.106
2	20	0.057
3	10	0.069
3	15	0.083
3	20	0.044
4	10	0.052
4	15	0.027
4	20	0.011
	Sum	1.000

Solution The marginal probability for various rainfall durations (X) can be computed by

$$f_X(x) = \sum_{y_i} f_{X,Y}(x, y_i)$$

Thus, for the various values of X, the marginal probabilities are calculated as follows:

$$P(X = 1) = 0.188 + 0.109 + 0.086 = 0.382$$

$$P(X = 2) = 0.168 + 0.106 + 0.057 = 0.331$$

$$P(X = 3) = 0.069 + 0.083 + 0.044 = 0.196$$

$$P(X = 4) = 0.052 + 0.027 + 0.011 = 0.090$$

Similarly, for rainfall amount Y, the marginal probabilities are

$$P(Y = 10) = 0.188 + 0.168 + 0.069 + 0.052 = 0.477$$

$$P(Y = 15) = 0.109 + 0.106 + 0.083 + 0.027 = 0.325$$

$$P(Y = 10) = 0.086 + 0.057 + 0.044 + 0.011 = 0.198$$

The conditional probability of rainfall duration of 3 h and with an amount of 15 mm can be computed by

$$p_{\text{intensity} \,|\, \text{duration}} (15 \,|\, 3) = f_{Y|X}(15 \,|\, 3) = \frac{f_{X,Y}(3,15)}{f_X(3)} = \frac{0.083}{0.196} = 0.423$$

3.6.2 Covariance

The marginal distributions can be readily determined from the joint distribution of two variables but the converse is not true. The joint distribution also depends on the degree of dependence and the nature of dependence existing between the random variables. This dependence is the reason for the study of the joint distribution. The covariance is a second moment about the centroidal axes and is defined as follows:

$$\text{cov}(x,y) = \sigma_{xy} = E[(x - \mu_X)(Y - \mu_Y)]$$

$$= \int_{-\infty}^{+\infty} \int_{-\infty}^{+\infty} (x - \mu_x)(y - \mu_y) f(x,y) \, dx \, dy \qquad (3.39)$$

$$= E[XY] - \mu_x \mu_y$$

or

$$\text{cov}(x,y) = \sigma_{xy} = \sum_{i=1}^{n} \sum_{j=1}^{m} (x_i - \mu_x)(y_j - \mu_y) p(x_i, y_j) \qquad (3.40)$$

Figure 3-8 shows the elements of the integration for the continuous case. If random variables X and Y are standardized as

$$X_* = \frac{X - \mu_x}{\sigma_x}, \qquad Y_* = \frac{Y - \mu_y}{\sigma_y}$$

then the standardized variables have zero mean and unity variance. It can be shown that the covariance of standardized X and Y is equal to the correlation coefficient between nonstandardized X and Y. Standardization of a random variable does not influence its skewness coefficient or kutosis.

The centroidal axes divide the probability space into four quadrants, as marked by the Roman numerals I to IV in Fig. 3-8. The probability masses in the first and third quadrants make positive contributions to the value of the covariance, since the product $(x - \mu_x)(y - \mu_y)$ is positive. The probability masses in the

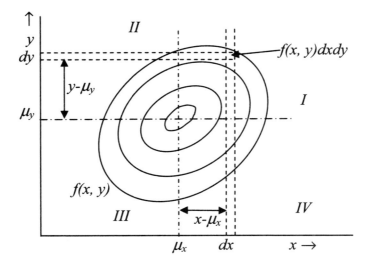

Figure 3-8 *Computation of covariance for a continuous case.*

second and fourth quadrants make negative contributions to the value of the covariance. Therefore, if values larger than the average value of X are associated with larger than the average values of Y, and conversely, values smaller than the average value of X occur simultaneously with values smaller than the average value of Y, then a relatively large part of the probability mass is located in the first and third quadrant and the covariance will be positive. However, if values larger than the average value of Y occur with values smaller than the average value of X and vice versa, then the covariance will be negative. If X and Y are unrelated then the positive and negative contributions cancel each other and the covariance will be zero.

For independent random variables X and Y,

$$f(x, y)dxdy = f(x)dx\, f(y)dy \tag{3.41}$$

Substitution of Eq. 3.32 in Eq. 3.30 permits writing the double integral as the product of two integrals. For independent variables,

$$\text{cov}(x,y) = \int_{-\infty}^{+\infty} (x - \mu_x) f(x)dx \int_{-\infty}^{+\infty} (y - \mu_y) f(y)dy$$

since

$$\mu_x = \int_{-\infty}^{+\infty} x f(x)dx \quad \text{and} \quad \mu_y = \int_{-\infty}^{+\infty} y f(y)dy$$

It follows that both integrals in the product are equal to zero. Thus, the covariance of independent variables is zero.

3.6.3 Correlation Coefficient

The covariance has a dimension equal to the product of the dimensions of the random variables. For that reason the magnitude of the parameter does not reveal much about the degree of dependence between the random variables. It is, therefore, customary to divide the covariance by the product of the standard deviations of the respective random variables. This results in a dimensionless parameter, called the correlation coefficient, denoted by ρ:

$$\rho = \frac{\sigma_{xy}}{\sigma_x \sigma_y} \tag{3.42a}$$

For computation, it is convenient to write

$$\rho = \frac{n \sum x_i y_i - (\sum x_i)(\sum y_i)}{\sqrt{n \sum x_i^2 - (\sum x_i)^2} \sqrt{n \sum y_i^2 - (\sum y_i)^2}} \tag{3.42b}$$

The correlation coefficient lies between -1 and $+1$. The variables having either of the two extreme values of the correlation coefficient are said to be highly correlated. However, a high correlation does not mean that the variables have a cause-and-effect relationship. A value of zero is obtained when the variables are independent but uncorrelated variables are not necessarily independent.

3.6.4 Evaluation of Regression Models

When a regression line is fitted to data on X and Y, the regression line always passes through the point defined by the mean of the x values ($\bar{x}$) and the mean of the y_p values predicted by the regression line ($\bar{y}$). One can relate the dispersion of the actual y values about their mean, denoted as SS or $n\sigma_0^2$ or $n\sigma_y^2$, to the sum of squares of regression (SSR or $n\sigma_r^2$ or $n\sigma_{y_p}^2$) and the sum of squares of error (SSE or $n\sigma_e^2$). These are expressed algebraically as

$$SS = n\sigma_y^2 = \sum_{i=1}^{n}(y_i - \bar{y}) \tag{3.43a}$$

$$SSR = n\sigma_{y_p}^2 = \sum_{i=1}^{n}(y_{pi} - \bar{y})^2 \tag{3.43b}$$

$$SSE = n\sigma_e^2 = \sum_{i=1}^{n}(y_i - y_{pi})^2 \tag{3.43c}$$

SSR is a measure of dispersion of the y_p values predicted by the regression line about the mean of the observed y values. SSE is a measure of dispersion of the actual or observed y values about their corresponding y_p values predicted by the regression line. It can be shown that

$$SS = SSR + SSE \qquad \text{(3.43d)}$$

This relationship shows that the dispersion of the observed y values about their mean is equal to the sum of the dispersion of the predicted y values about that mean and the dispersion of the actual y values about their corresponding predicted y values.

The measure of goodness of regression is the standard error of regression, s_e, computed as

$$s_e = \sqrt{\frac{1}{n}\sum_{i=1}^{n}(y_i - y_{pi})^2} = \sqrt{\frac{SSE}{n}} = \sqrt{\frac{S_e}{n}} = \sqrt{n}\sigma_e \qquad \text{(3.44a)}$$

where y_i and y_{pi} represent the ith observed and predicted values of y, respectively, S_e is the sum of squares of errors, and n is the number of observations. The standard error of regression, s_e, quantifies the spread of data around the regression line of fit and can be referred to as the unexplained sum of squares. Thus it is the standard deviation of the errors of estimation. It is worth mentioning that if any of the regression assumptions (independence, zero mean error, or common variance) concerning the residual error ($e_i = y_i - y_{pi}$) are incorrect, then s_e may not be a useful estimate of scale or dispersion for the residual errors. Clearly, a smaller value of s_e indicates that points lie closer to the regression line. If all points lie on the regression line then $s_e = 0$. For a large sample (i.e., large n) two-thirds of the errors $|y - y_p|$ will be less than s_e and about one-third of the errors will exceed it. Thus, by drawing two lines parallel to the regression line and at a vertical distance equal to s_e from it, one can draw a region within which about two-thirds of the sample points will fall. Likewise, one can show that 95% of the sample points will fall within the region bounded by two lines parallel to the regression line at a vertical distance of twice s_e from it.

3.6.5 Coefficient of Determination

The square of the coefficient of correlation is known as the coefficient of determination, C_d, and is a measure of the degree to which the variance in the dependent variable is explained by the linear regression relation between the two variables. The coefficient of determination can be related to SSE or σ_e^2 and SS or σ_o^2 as

$$C_d = 1 - \frac{\sigma_e^2}{\sigma_o^2} = 1 - \Delta = \frac{\sigma_o^2 - \sigma_e^2}{\sigma_o^2} = \frac{\text{fraction of } \sigma_o^2 \text{ explained by regression}}{\text{variance of the observed data}} \qquad \text{(3.44b)}$$

where σ_0^2 is the variance of the observed data, which is a measure of the variability associated with the dependent variable before regression. This coefficient is computed as $(1 - \Delta)$, where Δ is the difference of the variance of the observed values of the dependent variable and the variance of the values of the dependent variable that have been computed using the regression relation divided by the variance of the observed values. Clearly, as Δ becomes smaller, the coefficient of determination becomes larger or the regression "improves." Thus, it is a useful measure of the goodness of fit for evaluating simple regression models.

As an important note about application of Eq. 3.44b, its useful characteristics are dependent on the partitioning of the variance of the observed data into error and regression components by using the minimization criteria of ordinary least square. If either of these criteria is not used, then the interpretations associated with Eq. 3.44b are no longer valid.

3.6.6 Evaluation of Nonregression Models

While evaluating the efficacy of nonregression models, such as mechanistic hydrologic/water quality models, characteristics of the goodness of fit are determined by analyzing the differences between observed and predicted values. Although the linear-regression concepts are still useful, much of the interpretative power is lost because minimization criteria are no longer valid. The most commonly used error statistic is the root mean square error (RMSE), which may be an adequate statistic for summarizing the predictive accuracy of models for the same data set. RMSE describes the magnitude of the direct error and hence is used in decision making when one needs to know the implication of a model's uncertainty. The root mean square error is given as

$$RMSE = \sqrt{\frac{\sum_{i=1}^{n}(O_i - P_i)^2}{n}} \tag{3.45}$$

where n is the number of observations, O_i is the ith observed value, and P_i is the ith predicted value. RMSE should be as small as possible. Note that the RMSE is affected by the units used for expressing the parameter of concern; consequently, it cannot be used to compare a model's efficacy across parameters having different units of measurement. To overcome this shortcoming, normalized error statistics are used. One of the normalized error statistics is the normalized mean square error (NMSE), defined as (Gershenfeld and Weigend 1993)

$$NMSE = \frac{\sum_{i=1}^{n}(O_i - P_i)^2}{\sum_{i=1}^{n}(O_i - \bar{O})^2} \tag{3.46}$$

where $\bar{O}$ is the mean of the observed data. NMSE ranges from 0 to $+\infty$. When NMSE is zero, the model is perfect. When NMSE is 1, the model is as good as the observed mean value. When NMSE is greater than 1, the model is poor.

The other measure of goodness of fit that has been widely used to evaluate the performance of hydrologic/water quality models is the coefficient of efficiency developed by Nash and Sutcliffe (1970). Mathematically, the coefficient of efficiency is defined as

$$\eta_1 = 1 - \frac{\sum_{i=1}^{n}(O_i - P_i)^2}{\sum_{i=1}^{n}(O_i - \bar{O})^2} = 1 - NMSE \tag{3.47}$$

The Nash–Sutcliff coefficient ranges from $-\infty$ to 1.0. The model performance is measured as follows:

- If $\eta_1 < 0$, the model is poor and the observed mean is better than the model predictions.
- If $\eta_1 = 0$, the model is as good as the observed mean value.
- If $\eta_1 = 1$, the model is perfect.

The decision is subjective when η_1 ranges between 0 and 1, depending upon what is considered acceptable. Both NMSE and η_1 contain square terms that make them sensitive to outliers. Willmott et al. (1985) suggested a modified coefficient of efficiency, η_2, also known as the index of agreement. Mathematically, η_2 is defined as

$$\eta_2 = 1 - \frac{\sum_{i=1}^{n}|O_i - P_i|}{\sum_{i=1}^{n}|O_i - \bar{O}|} \tag{3.48}$$

Interpretations similar to those for η_1 can be made for η_2.

Example 3.21 An engineer developed a mechanistic lake model and used it to predict the water quality of a lake. The engineer is interested in knowing whether the developed model is efficient enough to be used for the TMDL (total maximum daily load) process. The observed and model predicted water quality parameters are listed in Table 3-21a. *TP* stands for total phosphorus, *TN* for total nitrogen, and (*Chl-a*) for chlorophyll-a.

Solution By applying Eq. 3.45 to Eq. 3.48, the various error statistics are determined (see Table 3-21b).

Based on the RMSE given in Table 3-21b, one cannot draw conclusions about the model's efficacy with respect to *Chl-a*, *TP*, and *TN* because the RMSE value is affected by scale. This is why a relative measure is needed to know how this model predicts various parameters. The relative measures NMSE , η_1, and

Table E3-21a

Date	Observed (O)			Predicted (P)		
	Chl-a (µg/L)	TP (mg/L)	TN (mg/L)	*Chl-a* (µg/L)	TP (mg/L)	TN (mg/L)
05/21/97	29.16	0.04	0.38	51.03	0.11	1.15
06/09/97	23.87	0.11	0.61	32.12	0.09	0.78
07/01/97	29.64	0.06	0.48	33.27	0.07	0.62
08/04/97	26.94	0.08	0.51	48.18	0.10	0.88
09/18/97	39.98	0.07	0.49	40.35	0.09	0.84
06/17/98	23.73	0.07	0.31	26.03	0.06	0.51
07/07/98	21.73	0.07	0.30	38.02	0.08	0.68
08/05/98	18.57	0.06	0.44	46.20	0.10	0.87
04/29/99	9.00	0.09	0.22	14.40	0.05	0.47
05/18/99	29.42	0.07	0.24	18.02	0.05	0.50
06/09/99	26.09	0.05	0.21	27.42	0.05	0.49
07/21/99	44.83	0.06	0.24	36.41	0.07	0.64
08/04/99	43.28	0.09	0.54	52.40	0.10	0.86
01/11/01	37.33	0.08	1.29	1.89	0.04	1.00
04/30/01	44.24	0.13	0.75	26.98	0.07	0.82
05/10/01	32.06	0.11	0.48	27.92	0.06	0.76
05/30/01	45.48	0.10		28.50	0.06	0.70
06/12/01	22.97	0.09	0.91	33.46	0.07	0.70
06/26/01	24.14	0.09	1.07	34.84	0.07	0.70
07/09/01	44.90	0.09	0.41	39.51	0.08	0.76
07/23/01	61.73	0.11	1.22	39.91	0.08	0.79
08/06/01	45.44	0.09	1.19	40.32	0.08	0.80
08/20/01	38.51	0.10	1.08	49.66	0.10	0.91
09/05/01	35.36	0.10	1.16	47.78	0.10	0.96
09/18/01	44.50	0.10	0.88	38.72	0.09	0.89
10/03/01	59.54	0.06	1.06	28.38	0.08	0.73
11/07/01	59.90	0.11	1.25	21.30	0.06	0.60

Table E3-21b

Statistic	Segment 1-4			Segment 14-15		
	Chl-a	TP	TN	Chl-a	TP	TN
RMSE	16.11	0.03	0.35	19.26	0.05	0.26
NMSE	1.47	1.97	0.89	2.26	1.12	1.05
Nash–Sutcliff coefficient, η_1	−0.47	−0.97	0.11	−1.25	−0.12	−0.04
Index of agreement, η_2	−0.16	−0.30	0.08	−0.56	0.00	0.03

η_2 indicate that the model performs poorly for predicting *Chl-a* and *TP* in both lake segments, indicating that the corresponding observed average levels of *Chl-a* and *TP* are far better than the model predictions. For *TN*, the model predictions are improved marginally in segment 1-4, whereas its predictions are poor in segment 14-15.

3.6.7 Sample Statistics

The parameters of a two-dimensional joint distribution can be estimated from a sample by considering the relative frequency distribution of the observations on the sample space. This leads to a discrete distribution of the relative frequencies. Treating each relative frequency as if it were a probability, one can apply the formulas for the parameters to obtain the corresponding statistics.

Example 3.22 The precipitation (in millimeters) and runoff (in millimeters) for a catchment for the month of July are given in Table 3-2. Compute the coefficient of correlation of the data.

Solution The various variables required to calculate the coefficient of correlation are computed in Table 3-22. Here, $\bar{x} = 687.05/16 = 42.94$ and $\bar{y} = 234.04/16 = 14.63$.
Now

$$\sigma_{xy} = \frac{1}{n}\sum(x - \bar{x})(y - \bar{y}) = 369.423/16 = 23.09$$

$$s_x = (570.056/16)^{0.5} = 5.97$$

$$s_y = (363.07/16)^{0.5} = 4.76$$

Hence,

$$\rho = 23.09/(5.97 \times 4.76) = 0.81$$

$$\text{coefficient of determination } (R^2) = \rho^2 = 0.81^2 = 0.656.$$

Table E3-22 *Calculations for correlation coefficient*

SN	Year	Precipitation (x)	Runoff (y)	$x - \bar{x}$	$y - \bar{y}$	$(x - \bar{x}) \times (y - \bar{y})$	$(x - \bar{x})^2$	$(y - \bar{y})^2$
1	1953	42.39	13.26	− 0.55	− 1.37	0.7535	0.3025	1.8769
2	1954	33.48	3.31	− 9.46	− 11.32	107.0872	89.4916	128.1424
3	1955	47.67	15.17	4.73	0.54	2.5542	22.3729	0.2916
4	1956	50.24	15.50	7.3	0.87	6.3510	53.2900	0.7569
5	1957	43.28	14.22	0.34	− 0.41	− 0.1394	0.1156	0.1681
6	1958	52.60	21.20	9.66	6.57	63.4662	93.3156	43.1649
7	1959	31.06	7.70	− 11.88	− 6.93	82.3284	141.1344	48.0249
8	1960	50.02	17.64	7.08	3.01	21.3108	50.1264	9.0601
9	1961	47.08	22.91	4.14	8.28	34.2792	17.1396	68.5584
10	1962	47.08	18.89	4.14	4.26	17.6364	17.1396	18.1476
11	1963	40.89	12.82	− 2.05	− 1.81	3.7105	4.2025	3.2761
12	1964	37.31	11.58	− 5.63	− 3.05	17.1715	31.6969	9.3025
13	1965	37.15	15.17	− 5.79	0.54	− 3.1266	33.5241	0.2916
14	1966	40.38	10.40	− 2.56	− 4.23	10.8288	6.5536	17.8929
15	1967	45.39	18.02	2.45	3.39	8.3055	6.0025	11.4921
16	1968	41.03	16.25	− 1.91	1.62	− 3.0942	3.6481	2.6244
	Total	687.05	234.04	0.01	− 0.04	369.4230	570.0559	363.0714

3.6.8 Interpretation of Covariance as Expectation

For two continuous random variables X and Y, the covariance is defined as

$$\text{cov}(X,Y) = \int\limits_{-\infty}^{+\infty} \int\limits_{-\infty}^{+\infty} \{x - E(X)\}\{y - E(Y)\} f(x,y) \, dx \, dy$$

This expression can be written as an expectation:

$$\text{cov}(X,Y) = E\{[X - E(X)][Y - E(Y)]\} \tag{3.49}$$

An alternative expression for the covariance can be obtained as

$$\begin{aligned}
\text{cov}(X, Y) &= E[XY - Y \times E(X) - X \times E(Y) + E(X) \times E(Y)] \\
&= E(XY) - E(X)E(Y) - E(Y)E(X) + E(X)E(Y) \\
&= E(XY) - E(X)E(Y) \tag{3.50}
\end{aligned}$$

Equation 3.51 is used for calculating the sample covariance:

$$s_{x,y} = \frac{1}{n} \sum_{i=1}^{n} x_i \, y_i - m_x \, m_y \tag{3.51}$$

Equation 3.51 can be used to derive an important theorem about the variance of the sum of random variables:

$$\text{var}(X + Y) = E[(X + Y)]^2 - [E(X + Y)]^2$$

$$= E(X^2) + E(Y^2) + 2E(XY) - [E(X)]^2 - [E(Y)]^2 - 2E(X)E(Y)$$

$$= E(X^2) - [E(X)]^2 + E(Y)^2 - [E(Y)]^2 + 2E(XY) - 2E(X)E(Y)$$

$$= \text{var}(X) + \text{var}(Y) + 2\text{cov}(X,Y) \tag{3.52}$$

The covariance can be written as the product of the standard deviations and the correlation coefficient. From Eq. 3.42a,

$$\sigma_{x,y} = \rho \sigma_x \sigma_y$$

Hence, Eq. 3.53 can be expressed as

$$\text{var}(X+Y) = \sigma_x^2 + \sigma_y^2 + 2\rho\sigma_x\sigma_y \tag{3.53}$$

Equations 3.52 and 3.53 are equivalent and show that the variance of the sum of two random variables is equal to the sum of their variances if the variables are independent; otherwise one must add twice the covariance.

3.7 Functions of Random Variables

First, consider a few useful rules that govern the expectation of functions of random variables. Let X be a random variable and $Y = \varphi(X)$. It can be shown that the mean or expectation of Y is

$$E(Y) = \int_{-\infty}^{+\infty} y\, f(y)\, dy$$

If the distribution of X is continuous and there is a one-to-one relationship between X and Y then one can write

$$f(y)dy = f(x)dx$$

Thus, one can also write

$$E(Y) = \int_{-\infty}^{+\infty} y\, f(x)\, dx \tag{3.54}$$

$$E\{\varphi(x)\} = \int_{-\infty}^{+\infty} \varphi(x) f(x) dx \tag{3.55}$$

$$E\{\varphi(x)\} = \sum_{i=1}^{n} \varphi(x) p(x_i) \tag{3.56}$$

which immediately gives Eq. 3.54 from the definition of $E(Y)$.

For two-dimensional cases,

$$E\{\varphi(X,Y)\} = \int_{-\infty}^{+\infty} \int_{-\infty}^{+\infty} \varphi(x,y) f(x,y) dx\, dy \tag{3.57}$$

$$E\{\varphi(X,Y)\} = \sum_{i=1}^{n} \sum_{j=1}^{m} \varphi(x,y) p(x_i, y_j) \tag{3.58}$$

It can be seen from Eq. 3.55 to Eq. 3.58 that calculation of the expectation of a random variable is a linear operation. This means that the expectation of the sum of two random variables, or the expectation of the sum of two functions of random variables, is equal to the sum of the expectations. This follows immediately from the fact that when $\varphi(X)$ or $\varphi(X, Y)$ can be written as a sum, the integral can be written as the sum of two separate integrals. For example,

$$E\{\varphi_1(X) + \varphi_2(X)\} = E\{\varphi_1(X)\} + E\{\varphi_2(X)\} \tag{3.59}$$

For two random variables that may or may not be statistically independent,

$$E\{\varphi_1(X) + \varphi_2(Y)\} = E\{\varphi_1(X)\} + E\{\varphi_2(Y)\} \tag{3.60}$$

The expectation of a constant C is, of course, equal to the constant itself:

$$E(C) = \int_{-\infty}^{+\infty} C f(x) dx = C \int_{-\infty}^{+\infty} f(x) = C \tag{3.61}$$

Appendix 3A

To prove that

$$s^2 = \frac{\sum (x_i - \bar{X})^2}{n-1} \tag{3A.1}$$

is an unbiased estimator of

$$\sigma^2 = \frac{\sum (x_i - \mu)^2}{n} \tag{3A.2}$$

we write

$$\sum (x_i - \mu)^2 = \sum [(x_i - \bar{X}) + (\bar{X} - \mu)]^2$$
$$= \sum [(x_i - \bar{X})^2 + (\bar{X} - \mu)^2 + 2(x_i - \bar{X})(\bar{X} - \mu)]$$

Since $\sum (x_i - \bar{X}) = 0$, we have

$$\sum (x_i - \mu)^2 = \sum (x_i - \bar{X})^2 + n(\bar{X} - \mu)^2$$
$$= (n-1)s^2 + n(\bar{X} - \mu)^2$$

Moving s^2 to the left-hand side, we get

$$s^2 = \frac{1}{n-1} \sum (x_i - \mu)^2 - \frac{n}{n-1}(\bar{X} - \mu)^2$$

Taking the expectation gives

$$E(s^2) = \frac{1}{n-1} \sum E(x_i - \mu)^2 - \frac{n}{n-1} E\{(\bar{X} - \mu)^2\}$$

or

$$E(s^2) = \frac{1}{n-1} \sum \mathrm{var}(X) - \frac{n}{n-1} \mathrm{var}(\bar{X})$$
$$= \frac{1}{n-1} n\sigma^2 - \frac{n}{n-1} \frac{\sigma^2}{n}$$
$$= \frac{n}{n-1} \sigma^2 - \frac{1}{n-1} \sigma^2 = \sigma^2$$

which proves the assumption.

3.8 Questions

3.1   Compute the mean, standard deviation, coefficient of variation, coefficient of skewness, and coefficient of kurtosis for the temperature, rainfall, wind velocity, and discharge data that you have used in the previous chapter [Questions 2.1 to 2.5]. Also compute shape factors and moment ratios.

3.2   Using the data from Question 3.1, plot histograms of temperature, rainfall, wind velocity, and discharge.

3.3 Represent each histogram by a probability distribution plotted in Question 3.2, based on the histogram shape. Do not perform any numerical fitting. Compute the area under this probability distribution and show that this is or is not a probability distribution.

3.4 What would be the consequence of adding a constant to each observation in Question 3.1 on the mean, standard deviation, coefficient of variation, and coefficient of skewness?

3.5 What would be the consequence of multiplying each observation in Question 3.1 by a constant on the mean, standard deviation, coefficient of variation, and coefficient of skewness?

3.6 Divide the data of Question 3.1 in two groups, one bigger than the other. How do the grouped data mean and standard deviation compare to the ungrouped mean and standard deviation? Which estimate would you prefer?

3.7 Find the moment-generating function of the Bernoulli distribution with $f(x) = p^x(1-p)^{1-x}$, where $x = 0,1$.

3.8 Find the moment-generating function of $f(x) = xe^{-x}$, where $x \geq 0$. Hint: Use the change of variable technique to integrate with respect to $w = x(1-x)$ instead of x.

3.9 If a random variable X is defined by the normal distribution with parameters μ (mean) and σ (variance), then find the distribution of $Y = aX + b$. Remember that $M_X(t) = \exp[t\mu + (t^2\sigma^2/2)]$.

3.10 A discrete random variable X has the following PMF:

X	0	1	2	3	4	5	6	7	8	9	10
f(x)	0.00	0.04	0.08	0.12	0.16	0.20	0.16	0.12	0.08	0.04	0.00

Calculate the following: (i) $E[X]$, (ii) $E[2X]$, (iii) $E[2X+2]$, and (iv) $E[g(X)]$, where $g(X) = (X^2 - 2X + 4)$.

3.11 Calculate (i) $E[X]$, (ii) $E[2X]$, (iii) $E[2X + 2]$, and (iv) $E[g(X)]$, where $g(X) = (X^2 - 2X + 4)$ if X is described by (a) a uniform distribution in the range $(0, 10)$ [i.e., $f(X) = 1/10$], (b) a triangular distribution given in Fig. Q3-11, and (c) an exponential distribution defined as $f(X) = 0.25 \exp(-0.25X)$.

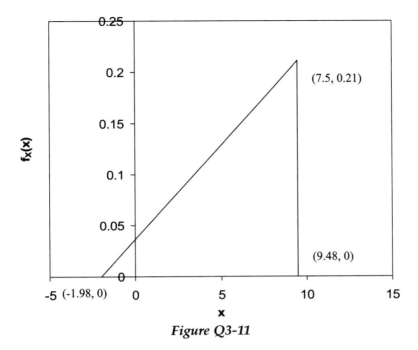

Figure Q3-11

3.12 Let X be a random variable defined by the following PDF:

$$f(x) = c\left(1 - x^3\right) \quad \text{for } 0 < x \le 1$$

$$= 0 \text{ elsewhere}$$

(a) Find the value of c so that $f(x)$ is a valid distribution. (b) Determine the first four moments of X about the origin. (c) Use the noncentral moments to determine the first four central moments.

3.13 Find the characteristic function for the random variable X defined in Example 3.2. Using the characteristic function thus obtained, find the expected value of X.

3.14 Find the moment-generating function of X characterized by the following PDF:

$$f(x) = \frac{a^x \exp(-b)}{x!}, \quad \text{where } x \text{ is a positive integer}$$

Determine the mean and variance of X using this moment-generating function.

3.15 Let X be defined by the following gamma distribution function:

$$f(x) = \frac{\beta^\alpha}{\Gamma(\alpha)} x^{\alpha-1} e^{-\beta x}, \quad x, \alpha > 0$$

Find the moment-generating function of X and determine the mean, variance, skewness, and kurtosis of X using this function.

3.16 A water resources engineer is interested in determining the average annual concentration of total nitrogen (*TN*) in the inflow to a lake. This lake is used as a water supply and the state agency is concerned with eutrophication in the lake from increased levels of *TN* loading. It was determined that the inflow follows a log-normal distribution given as

$$f(Q) = \frac{1}{9Q\sqrt{2\pi}} e^{-\frac{(\ln Q - 4)^2}{18}}, \quad 0 < Q < \infty$$

The relationship between inflow (Q) and *TN* is given as

$$TN = 1.5 \times 10{-}4Q^2 - 1.5 \times 10^{-2}Q + 0.15$$

where *TN* is in mg/L and Q is in cfs. Determine the distribution of *TN* and find its moment-generating function and evaluate the mean, variance, skewness, and kurtosis of *TN*.

3.17 Determine the moment-generating function for a discrete random variable X given in Question 3.10. Using this moment-generating function, determine the first four moments of X about the mean.

3.18 A random variable X is defined by the following PDF:

$$f(x) = \frac{1}{2} e^{-\frac{|x|}{2}}, \quad -\infty < x < \infty$$

Determine its moment-generating function and the first three central and noncentral moments. Determine its characteristic functions.

3.19 A random variable X is defined by the normal distribution with parameters μ and σ with the following PDF:

$$f(x) = \frac{1}{\sigma\sqrt{2\pi}} e^{-\frac{(x-\mu)^2}{2\sigma^2}}, \quad -\infty < X < \infty$$

Determine its characteristic function.

3.20 Let R, S, T, U, and V be random variables with the first four noncentral moments given in Table Q3-20. Using the relationship between central and noncentral moments, determine the mean, variance, skew, and kurtosis of these random variables.

Table Q3-20

Moment order, k	1	2	3	4
$E[R^k]$	2.00E-01	4.04E-02	8.24E-03	1.70E-03
$E[S^k]$	1.49E-01	5.24E-03	2.13E-04	9.41E-06
$E[T^k]$	5.00E-02	2.53E-03	1.29E-04	6.63E-06
$E[U^k]$	2.02E-01	4.15E-02	8.63E-03	1.82E-03
$E[V^k]$	1.01E-02	1.05E-04	1.11E-06	1.20E-08

3.21 Let X be a random variable with CDF defined as

$$F(x) = \begin{cases} 0, & \text{if } x < -5 \\ 0.25, & \text{if } -5 \le x < 5 \\ 0.45, & \text{if } 5 \le x < 10 \\ 0.65, & \text{if } 10 \le x < 15 \\ 0.85, & \text{if } 15 \le x < 20 \\ 1.0, & \text{if } x > 20 \end{cases}$$

(a) Determine the PDF $f(X)$.

(b) Compute $E(X)$, $E(X^2)$, and $E(X^3)$.

(c) Compute the variance, skew, and kurtosis of X.

3.22 The discharge and stage values of Salt Creek at the USGS gauge near Rowell, Illinois, are given in Table Q3-22. Compute the mean, median, mode, mean deviation, standard deviation, coefficient of variation, and ratio of standard deviation to the mean deviation of the discharge and stage values.

3.23 An engineer developed a mechanistic watershed model and used it to predict the water quality of a stream. The engineer is interested in knowing whether the developed model is efficient enough to be used for the total maximum daily load computation. The observed and model predicted water quality parameters are listed in Table Q3-23.

3.24 An engineer performed watershed modeling and conducted a statistical error analysis to determine the model's efficacy. The various statistics calculated are shown in Table Q3-24. Comment on the efficacy of this modeling exercise with respect to both flow and water quality.

Table Q3-22

Year	Stage (ft)	Flow (cfs)	Year	Stage (ft)	Flow (cfs)
1968	29.21	24,500	1952	19.29	3,170
1943	24.77	12,400	1946	18.83	3,040
1964	24.71	10,600	1976	19.51	2,910
1956	23.67	10,300	1948	18.7	2,480
1961	23.72	10,300	1960	18.74	2,290
1951	23.23	9,390	1953	18.35	2,210
1944	22.94	8,850	1972	18.67	2,020
1974	22.92	8,060	1965	17.83	1,830
1957	22.45	7,950	1949	17.76	1,810
1959	24.84	7,500	1969	17.59	1,700
1973	22.48	7,270	1947	17.36	1,600
1950	21.77	6,890	1945	16.58	1,380
1963	21.87	6,050	1955	16.75	1,320
1958	21.1	5,730	1971	17.4	1,310
1970	21.1	5,060	1966	16.07	1,090
1962	19.98	4,110	1967	15.85	1,040
1975	20.33	3,920	1954	14.8	829

Table Q3-23

Observed water quality						Model-calculated water quality					
DO[a]	FCOLI[b]	NO3[c]	NH3[d]	TSS[e]	TEMP[f]	DO	FCOLI	NO3	NH3	TSS	TEMP
3.0	5.	0.0	0.0	11.0	46.0	6.0	158.	0.1	0.0	6.5	41.7
6.6	50.	0.1	0.0	12.0	48.2	9.2	180.	0.1	0.0	6.7	52.4
6.8	62.	0.1	0.0	13.0	52.7	8.0	124.	0.1	0.0	44.5	50.8
6.8	92.	0.1	0.0	14.0	55.0	7.0	335.	0.1	0.0	23.2	54.0
7.5	100.	0.1	0.0	15.0	56.7	8.3	196.	0.3	0.0	47.7	55.3
7.5	108.	0.1	0.0	15.0	59.2	9.1	170.	0.3	0.1	29.8	53.1
7.7	120.	0.1	0.0	16.0	59.9	7.7	89.	0.3	0.0	32.0	56.5
7.7	122.	0.1	0.0	18.0	59.9	7.7	155.	0.3	0.0	51.1	46.9
7.7	125.	0.1	0.0	19.0	60.6	7.8	124.	0.1	0.0	5.5	55.7
7.8	148.	0.1	0.0	20.0	61.0	7.6	180.	0.1	0.0	25.1	61.4
7.8	160.	0.1	0.0	22.0	61.0	8.1	135.	0.2	0.0	74.5	56.4
7.9	165.	0.1	0.0	23.0	61.9	8.8	40.	0.1	0.0	101.0	59.3
8.0	170.	0.1	0.0	25.0	63.3	7.8	819.	0.3	0.0	114.0	60.3

Table Q3-23 *(Continued)*

Observed water quality						Model-calculated water quality					
DO[a]	FCOLI[b]	NO3[c]	NH3[d]	TSS[e]	TEMP[f]	DO	FCOLI	NO3	NH3	TSS	TEMP
8.0	215.	0.1	0.0	25.0	66.2	10.2	150.	0.1	0.0	112.0	60.4
8.1	220.	0.1	0.0	31.0	67.6	8.6	54.	0.1	0.0	15.0	63.6
8.2	225.	0.1	0.0	33.0	69.3	10.1	95.	0.1	0.0	55.8	69.5
8.4	240.	0.1	0.0	44.0	69.6	8.9	208.	0.3	0.0	34.0	69.1
8.5	250.	0.1	0.0	44.0	70.5	9.4	148.	0.3	0.0	53.2	60.7
8.5	280.	0.2	0.0	47.0	72.1	9.6	48.	0.5	0.0	109.0	69.8
8.9	300.	0.2	0.0	48.0	72.9	9.5	77.	0.4	0.0	11.2	71.4
9.1	310.	0.2	0.0	56.0	73.4	8.9	50.	0.3	0.1	29.4	69.6
9.2	340.	0.2	0.0	57.0	74.5	10.6	317.	0.3	0.0	52.9	71.1
9.2	350.	0.2	0.0	59.0	75.0	9.7	40	0.6	0.0	36.8	81.2
9.3	370.	0.2	0.0	64.0	75.7	7.9	124.	0.1	0.0	26.6	74.4
9.4	468.	0.2	0.0	67.0	76.1	9.2	77.	0.2	0.0	1.4	77.8
9.6	980.	0.2	0.0	69.0	77.0	10.9	820.	0.2	0.0	56.6	78.7
9.6	2320.	0.2	0.0	73.0	77.0	9.2	146.	0.1	0.0	67.1	75.2

[a]Dissolved oxygen (mg/L). [b]Numbers of fecal coliform (#/100 mL). [c]Nitrate (mg/L). [d]Ammonia (mg/L). [e]Total suspended solids (mg/L). [f]Temperature (°F).

Table Q3-24

Data	Root mean square error (RSME)	Normalized mean square error (MSE*)	Coefficient of determination (CD)
Flow assessment			
Daily flow (cfs)	47.01	1.06	− 0.06
Monthly flow (cfs)	598.94	1.95	− 0.95
10% Highest flow (cfs)	133.06	1.55	− 0.55
10% Lowest flow (cfs)	3	21.03	− 20.03
Water quality assessment			
TSS (mg/L)	217.43	1.45	− 0.45
TP (mg/L)	0.23	1.66	− 0.66
Zn (mg/L)	56.29	3.6	− 2.6

Part II

Probability Distributions and Parameter Estimation

Chapter 4

Discrete and Continuous Probability Distributions

In earlier chapters we studied characterization of random variables through their probability distributions along with their expectation and moments without referring to a specific distribution. In this chapter we describe several probability distributions that are commonly used for performing reliability, risk, and uncertainty analysis of various engineering systems. As mentioned in earlier chapters, the probability distributions can be classified into discrete and continuous distributions, based on the nature of the random variable of interest. First we will discuss discrete distributions and some continuous distributions derived from discrete distributions.

Many engineering problems relate to the number of times a particular event may be observed in a series of repeated observations in a certain space or in a certain interval of time. For example, in the quality control of manufactured products a number of samples may be subjected to testing to determine the number of times a sample fails to meet a predetermined quality criterion. In the lining of irrigation canals, interest may be in the number of flaws that may occur in, say, 1000 m of lining. A dam designer may want to know the number of times the design flood is likely to be exceeded during the economic life of the dam. Studies on railway transportation safety may be concerned with the number of railway accidents that occur annually in a region. A flood management officer may want to know how many times the water level in a given river would be higher than the danger level in a given year. In flood protection work, interest may lie in knowing the number of levee sections liable to break. In such examples, an event may be characterized as simply "failure" or "no failure," "success" or "no success," "win" or "loss," "flaw" or "no flaw," "accident" or "no accident," and so on. In other words, the specified event may be described by "occurrence" or "nonoccurrence." It is further assumed that the probability of "occurrence" (and of "nonoccurrence") is the same from trial to trial, or per unit of space, or per unit of time, as the case may be. Each time a lined canal is analyzed, the probability of leakage is assumed to be the same. Each time a water distribution network is analyzed, the probability of failure of pipes is assumed to be the same. The probability of encountering a flaw is assumed to be the same for each meter of canal lining. The probability of accident per unit of time in the case of transportation is assumed to be the same for each unit of time. Furthermore, it is assumed that the occurrence or nonoccurrence of the specified event is independent of the previous occurrences or nonoccurrences.

The number denoting the occurrence of the event or the success is regarded as the random variable, denoted as X, whose probability distribution is of interest. The distribution of X is obviously discrete. Depending on whether the occurrences are considered in a fixed number of observations or in a fixed continuum of space or time, the random variable X will follow a binomial or a Poisson distribution if three conditions are met: (i) Only two discrete states of the random variable are possible, (ii) the probability of occurrence is constant for each trial, and (iii) the occurrence of the specified event is independent of the previous occurrences. The binomial and the Poisson distributions are the most commonly used discrete probability distributions.

4.1 Simple Discrete Random Trials

The state of many engineering systems can be classified into two exclusive categories: success or failure, satisfactory functioning or unsatisfactory functioning, working in compliance with the regulatory standards or violating the regulatory standards, winning or losing, exceeding or not exceeding standards, etc. Under such conditions the following distributions arise: the Bernoulli distribution, the binomial distribution, the geometric distribution, and the negative binomial distribution.

4.1.1 Bernoulli Distribution: Single Trial

Let a random variable be denoted by X, which takes on any one of two possible values: one associated with success and the other with failure. When we consider whether it is a wet or dry day, hot or cold, flooded or not, windy or tranquil, day or night, sunny or cloudy, clear or hazy, foggy or not foggy, urbanized or rural, rich or poor, etc., it is seen that only two mutually exclusive or collectively exhaustive events are possible outcomes. Such a variable is defined as the Bernoulli random variable. Now for the random variable X, we assign a value of zero for "nonoccurrence" of the specified event and a value of one for its "occurrence." Thus, all possible events associated with the experiment can be defined on the probability space as shown in Fig. 4-1.

Let the probability of "occurrence" be p, and let the probability of "nonoccurrence" be $q = 1 - p$. If a success is observed, $x = 1$, and if a failure is observed, $x = 0$. The PMF of X, denoted as $f_X(x)$, is simply

$$f_X(x) = \begin{cases} p & x = 1 \\ q = 1 - p & x = 0 \end{cases} \tag{4.1a}$$

Alternatively, Eq. 4.1a can be written as

$$f_X(x) = p^x (1-p)^{1-x} \text{ for } x = 0, 1 \tag{4.1b}$$

where p is the probability of success. Evidently, $p + q = 1$. The mean of the random variable X can be determined by taking the first moment about the origin and is written as

$$E(X) = \Sigma x f_X(x) = (1) p + (0) (1 - p) = p \tag{4.2}$$

Similarly, the variance of X can be determined by taking the second moment about the mean:

$$\sigma_X^2 = \sum (x - \mu)^2 f_X(x) = p^2 q + q^2 p = pq(p+q) = pq = p(1-p) \tag{4.3}$$

For a Bernoulli variable, the probability of occurrence of the event in each trial is the same from trial to trial and the trials are statistically independent. We use the notation X ~ Bernoulli (p), which reads as X is characterized by a Bernoulli distribution with parameter p. The Bernoulli distribution is useful for modeling an experiment or an engineering process that results in exactly one of two mutually exclusive outcomes. The experiments involving repeated sampling of a Bernoulli random variable are frequently called Bernoulli trials [e.g., tossing a coin repeatedly and observing the outcomes (heads or tails)].

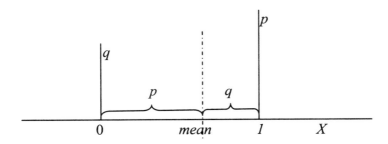

Figure 4-1 *Probability space in a Bernoulli trial.*

Several commonly used discrete distributions arise from examining the results of Bernoulli trials repeated several times. Three basic questions come to mind when we observe a set of Bernoulli trials: (1) How many successes will be obtained in a fixed number of trials? (2) How many trials must be performed until we observe the first success? (3) How many trials must be performed until we observe the kth success? Answering these three questions motivates the development of the binomial, geometric, and negative binomial distributions, respectively.

Example 4.1 If the occurrence of rainfall in Baton Rouge, Louisiana, follows a Bernoulli distribution with probability of occurrence X on any given day as 0.2, find the mean of the random variable and its variance.

Solution The mean of $X = 0.2$. The variance of $X = 0.2 \times 0.8 = 0.16$.

4.1.2 Binomial Distribution: Repeated Trials

A binomial random variable represents the number of successes obtained in a series of n independent and identical Bernoulli trials, the number of trials is fixed, and the number of successes varies from experiment to experiment. Consider a sequence of Bernoulli trials, where the outcomes of the experiment are mutually independent and the probability of success remains unchanged. For example, for a sequence of n years of flood data, the maximum annual flood magnitudes are independent and the probability of occurrence, p, of a flood in any year remains unchanged throughout the period of n years. If the random variable is whether the flood occurs or not, then the sequence of n outcomes is Bernoulli trials. Let the random variable be designated by Y and its specific value by y. We wish to determine the probability of exactly y occurrences (the number of successes) in n Bernoulli trials. To that end, let the probability of success (occurrence of flood) be p. First, consider a simple case of $n = 3$. If there are no successes or all trials lead to failure (nonoccurrence of flood), then $y = 0$. This event has a probability of

$$(1 - p)(1 - p)(1 - p) = (1 - p)^3$$

If there is one success (denoted by 1) and two failures (denoted by 0) in each of the three trials (i.e., $y = 1$), then the following sequence is possible:

Trial 1	Trial 2	Trial 3
1	0	0
0	1	0
0	0	1

Each sequence is an event with the probability of occurrence as $p(1 - p)^2$. Therefore, the probability of event $y = 1$ or $f[y = 1] = 3p(1 - p)^2$, since the sequences are mutually exclusive events. Similarly, for the event $y = 2$, the mutually exclusive occurring sequences are

Trial 1	Trial 2	Trial 3
1	1	0
1	0	1
0	1	1

Each occurs with probability $p^2(1 - p)$, leading to $y = 2$. Hence,

$$f[y = 2] = 3p^2 (1 - p)$$

Likewise, for $y = 3$, $P[y = 3] = p^3$, since only one sequence corresponds to $y = 3$. In summary

$$f_Y(0) = (1 - p)^3$$

$$f_Y(1) = 3p (1 - p)^2$$

$$f_Y(2) = 3p^2 (1 - p)$$

$$f_Y(3) = p^3$$

One can write

$$f_Y(y) = \binom{3}{y} p^y (1 - p)^{3-y}, \quad y = 0, 1, 2, 3$$

where $\binom{3}{y}$ the Binomial coefficient, equals $3!/[(y! \, (3 - y)!)]$, the number of ways that exactly y successes can be found in a sequence of three trials.

To generalize, if there are n Bernoulli trials, the probability mass function of the total number of successes y is given as

$$f_Y(y) = \binom{n}{y} p^y (1-p)^{n-y} = \frac{n!}{y!(n-y)!} p^y (1-p)^{n-y}, \quad y = 0, 1, 2, 3, \ldots, n \qquad (4.4)$$

$$= B(n, p)$$

Here n must be an integer and $0 \le p \le 1$. Equation 4.4 defines the *binomial distribution* of Y for given values of p and n. The distribution is called binomial because the coefficients are the well-known binomial coefficients that arise when the series $(a+b)^n$ is expanded using the binomial theorem. The binomial distribution has two parameters: the number of trials and the probability of occurrence of the specified event in a single trial. In an abbreviated form, this is referred to as $B(n, p)$. The shape of $B(n, p)$ depends on parameters n and p. The probability of each sequence is equal to $p^y q^{n-y}$. With use of Eq. 4.4, the probabilities that Y will take on the values of 0, 1, 2, ..., n, which exhaust all possibilities, can be calculated. The binomial coefficients can be calculated or obtained from mathematical tables when Y and n are large.

Many everyday situations entail events that have just two possibilities. A highway bridge may or may not be flooded in the next year, an area may or may not get flooded this year, it may or may not rain today, it may be windy or may not be windy next week, it may snow or may not snow next week, it may be cloudy or sunny tomorrow, a car accident may or may not occur next week, a column may or may not buckle, an excavator may or may not cease to operate in the next week, and so on.

The mean of Y can be determined by taking the first moment of Eq. 4.4 about the origin:

$$E[Y] = \sum_{y=-\infty}^{n} y\, f_Y(y) = \sum_{y=0}^{n} y \binom{n}{y} p^y (1-p)^{n-y}$$

$$= \sum_{y=1}^{n} y \binom{n}{y} p^y (1-p)^{n-y} = np \sum_{y=1}^{n} \frac{(n-1)!}{(y-1)!(n-y)!} p^{y-1} (1-p)^{n-y} \quad (4.5)$$

Let $u = y - 1$. Then, Eq. 4.5 can be written as

$$E[Y] = np \sum_{u=0}^{n-1} \frac{(n-1)!}{u!(n-1-u)!} p^u (1-p)^{n-1-u} = np \quad (4.6)$$

because the term after the summation will add up to unity. Similarly, the variance of Y can be obtained as

$$\mathrm{var}[Y] = E[Y^2] - [E(Y)]^2$$

To express the variance, $E[Y^2]$ needs to be specified. This term can be derived as

$$E[Y^2] = \sum_{y=-\infty}^{\infty} y^2 f_Y(y) = \sum_{y=-\infty}^{n} y^2 \binom{n}{y} p^y (1-p)^{n-y} = \sum_{y=1}^{n} y^2 \binom{n}{y} p^y (1-p)^{n-y}$$

$$= \sum_{y=1}^{n} y^2 \frac{n!}{y!(n-y)!} p^y (1-p)^{n-y} = np \sum_{y=1}^{n} y \frac{(n-1)!}{(y-1)!(n-y)!} p^{y-1} (1-p)^{n-y}$$

$$= np \left[\sum_{y=1}^{n} \frac{(y-1+1)(n-1)!}{(y-1)!(n-y)!} p^{y-1} (1-p)^{n-y} \right]$$

$$= np[\sum_{y=1}^{n} \frac{(y-1)(n-1)!}{(y-1)!(n-y)!} p^{y-1}(1-p)^{n-y} + \sum_{y=1}^{n} \frac{(n-1)!}{(y-1)!(n-y)!} p^{y-1}(1-p)^{n-y}]$$

$$= np[\sum_{y=2}^{n} p\frac{(n-1)(n-2)!}{(y-2)!(n-y)!} p^{y-1}(1-p)^{n-y} + \sum_{y=1}^{n} \frac{(n-1)!}{(y-1)!(n-y)!} p^{y-1}(1-p)^{n-y}]$$

Let $u = y - 2$ for the first summation and $v = y - 1$ for the second summation in this equation. Then one obtains

$$E[Y^2] = np\{p(n-1)[\sum_{u=0}^{n} \frac{(n-2)!}{u!(n-2-u)!} p^{u+1}(1-p)^{n-2-u} + \sum_{v=0}^{n} \frac{(n-1)!}{v!(n-1-v)!} p^{v}(1-p)^{n-1-v}]$$

$$= np\{p(n-1) \times 1 + 1\} = np\{np - p + 1\} = n^2 p^2 - np^2 + np$$

Therefore,

$$\text{var}[Y] = E[Y^2] - [E(Y)]^2 = [n^2 p^2 - np^2 + np] - n^2 p^2 = np(1-p) \quad (4.7)$$

It should be noted that the total number of successes in n trials Y can be interpreted in terms of the success or failure in each Bernoulli trial. Let the ith trial be denoted by X_i. Then, $X_i = 1$ if success occurs and $X_i = 0$ if failure occurs. Then, Y can be written as the sum

$$Y = X_1 + X_2 + X_3 + \dots + X_n \quad (4.8)$$

of n independent identically distributed Bernoulli random variables. The mean and variance of Y can be written as

$$E[Y] = \sum_{i=1}^{n} E[X_i] = n E[X_i] = np$$

$$\text{var}[Y] = \sum_{i=1}^{n} Var[X_i] = n Var[X_i] = np(1-p)$$

This also shows that the sum of two binomial random variables, $B(n_1, p)$ and $B(n_2, p)$, also has a binomial distribution, $B(n_1 + n_2, p)$, as long as p remains constant. As p tends to 0.5 and n tends to a large number, $B(n, p)$ tends to a normal distribution (see the plots in Example 4.2), which will be discussed in the next chapter.

Example 4.2 Consider the binomial distribution with parameters n and p. Graph the binomial distribution for the following parameter sets: (1) $n = 5$, $p = 0.1$; (2) $n = 5$, $p = 0.25$; (3) $n = 5$; $p = 0.5$; (4) $n = 15$, $p = 0.25$; (5) $n = 15$, $p = 0.5$; (6) $n = 30$, $p = 0.25$, (7) $n = 30$, $p = 0.5$; and $n = 50$, $p = 0.5$.

Solution The plots of the binomial distribution for different combinations of parameters are given in Fig. 4-2.

Example 4.3 Daily rainfall data for Baton Rouge, Louisiana, is available for the years 1948 to 1990. Consider the rainfall data for the month of September. Assuming that the occurrence of rainfall on any day is an independent event, compute the probability of 2 rainy days, 4 rainy days, and 10 rainy days in September. From the data, the total number of rainy days is 380.

Solution The total number of days in the month of September from 1948 to 1990 will be $43 \times 30 = 1,290$. Thus, the probability of rain on any given day in September is $380/1,290 = 0.2978$. The probability of the occurrence of a given number of rainy days in a month follows a binomial distribution. If there are n Bernoulli trials, the probability mass function of the total number of successes y is given by Eq. 4.4. Here, the number of trials n is the number of days in September (30), and p is 0.2978. Hence, the probability of 2 successes, $y = 2$, is

$$P_2 = \binom{30}{2} 0.2978^2 (1 - 0.2978)^{30-2} = 0.001938$$

Similarly, for $y = 4$,

$$P_4 = \binom{30}{4} 0.2978^4 (1 - 0.2978)^{30-4} = 0.022$$

and for $y = 10$,

$$P_{10} = \binom{30}{10} 0.2978^{10} (1 - 0.2978)^{30-10} = 0.14$$

The probability of a different number of rainy days in September is plotted in Fig. 4-3.

Example 4.4 Using the data of Example 4.3, find the probability of 2, 3, and 5 consecutive rainy days in the month of September.

Solution Since the probability of rain falling on a day is independent of the rain on the previous day, the probability of 2 consecutive rainy days in the month of September is

$$P_2 = (p)^2 = (0.2978)^2 = 0.0887$$

The probability of 3 consecutive rainy days in the month of September is

$$P_3 = (p)^3 = (0.2978)^3 = 0.0264$$

The probability of 5 consecutive rain days in the month of September is

$$P_5 = (p)^5 = (0.2978)^5 = 0.0023$$

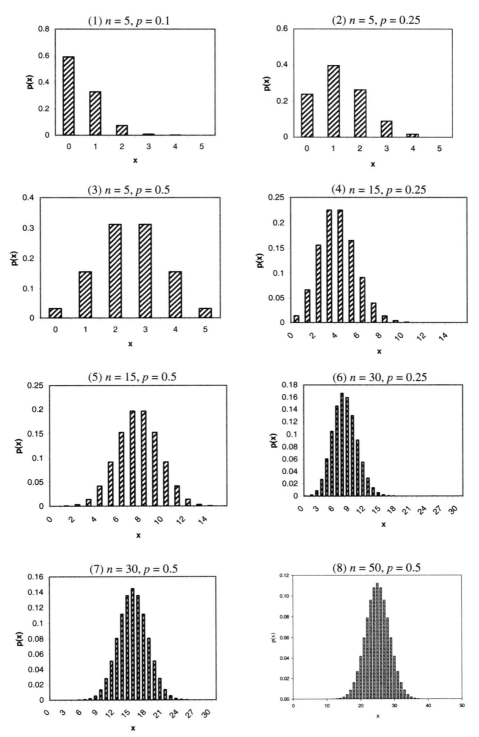

Figure 4-2 *Plots of binomial distribution for various combinations of* n *and* p.

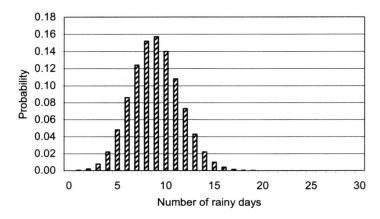

Figure 4-3 Probability of given number of rainy days in September for Baton Rouge.

Example 4.5 Consider the annual peak discharge series for the Amite River near Darlington, Louisiana. Assume that the annual peak values are independent. If the probability of a given T-year flood is constant from year to year, then the successive years represent independent Bernoulli trials. What is the probability that a 50-year flood will occur at least once during 20 years? Compute the probability of a 100-year flood occurring at least once in 20 years.

Solution When a return period (T) for an event is given, the probability p is defined as (the definition of return period is provided later) p = the probability of occurrence of a T-year flood in any given year = $1/T$ = $1/50$ = 0.02. Let x = the number of occurrences of the 50-year flood. $X \sim B(n, p = 0.02)$. The probability that the 50-year flood will occur in 20 years will be

$$P[x \geq 1] = 1 - f_x(0) = 1 - \binom{20}{0}(0.02)^0 (1 - 0.02)^{20} = 0.332$$

The probability that the 100-year flood ($p = 0.01$) will occur in 20 years will be

$$P[x \geq 1] = 1 - f_x(0) = 1 - \binom{20}{0}(0.01)^0 (1 - 0.01)^{20} = 0.182$$

Example 4.6 A factory produces plastic pipes and an inspection showed that 10% of the pipes produced are defective. Prepare the PMF of the number of defective pipes encountered in a sample of 10. Assume that the number of defective pipes follows a binomial distribution.

Solution Let X be the number of defective items. Here $p = 0.1$ and $n = 10$. A sample calculation for $x = 2$ is

$$P(x = 2) = \binom{10}{2}(0.1)^2(0.9)^8 = 45 \times 0.0043047 = 0.1937$$

The binomial distribution is $X \sim B(10,0.10)$. The PMF is listed in Table E4-6. (Note that $0! = 1$.)

Table E4-6

X	0	1	2	3	4	5	6	7	8	9	10
$\binom{n}{x}$	1	10	45	120	210	252	210	120	45	10	1
$p_X(x)$	0.3487	0.3874	0.1937	0.0574	0.0112	0.0015	0.0001	0.000	0.0	0.0	0.0

Example 4.7 Using the data of Example 4.6, compute the probability that in one particular sample of 10 plastic pipes, one would find three or more defective pipes.

Solution This probability can be calculated in two ways. First, one can add up the probabilities of $x = 3$, $x = 4$, $x = 5$, etc. up to $x = 10$ defective pipes. Second, a more efficient way is to consider the complementary event: less than three defective pipes. This means adding up

$$P(x = 0) = 0.3487 + [P(x = 1) = 0.3874] + [P(x = 2) = 0.1937] = 0.9298$$

The probability of finding more than three defective pipes is, therefore, $1.0 - 0.9298 = 0.0702$.

4.1.3 Geometric Distribution: Repeated Trials

In the preceding cases, we focused on the number of successes occurring in a fixed number of Bernoulli trials. Here we focus on the question of determining the number of trials when the first success would occur. For example, how many days would pass before the next rain if the probability of occurrence of rain on any day is p? What would be the year when a flood would occur if the probability of occurrence of flood in any year is p? When would the next accident occur? When would the next hurricane strike the Louisiana coast? When would the next earthquake hit the Los Angeles area? When would the next snowfall occur in Denver? Thus a geometric random variable represents the number of trials needed to obtain the first success.

If we assume the independence of trials and a constant value of p, the distribution of N, the number of trials to the first success, can be found as follows. The first success would occur on the nth trial if and only if (1) the first $(n-1)$ trials are failures, which occur with probability $(1-p)^{n-1}$, and (2) the nth trial is a success, which occurs with probability p. That is,

$$P[N=n]=P_N(n)=(1-p)^{n-1}p \ , \quad n=1,2,3,... \tag{4.9}$$

This is the geometric distribution with parameter p and is denoted as $G(p)$. The cumulative distribution function is

$$F_N(n) = \sum_{j=1}^{n} P_N(j) = \sum_{j=1}^{n} (1-p)^{j-1} p = 1 - (1-p)^n \qquad \textbf{(4.10)}$$

Alternatively, we could observe directly that the probability that $N \leq n$ is the probability that there is at least one occurrence in n trials, or P (no occurrence in n trials) $= (1-p)^n$. Thus, the probability of at least one occurrence is

$$P \text{ (at least one occurrence in } n \text{ trials)} = 1 - (1-p)^n$$

which is the same as Eq. 4.10.

The mean and variance of N can be expressed as

$$E[N] = \sum_{n=1}^{\infty} np(1-p)^{n-1} = \frac{1}{p} \qquad \textbf{(4.11)}$$

$$Var[N] = \frac{1-p}{p^2} \qquad \textbf{(4.12)}$$

Example 4.8 Plot the geometric distribution for various values of p and n, taking n from 0 to 20, and p from 0 to 1.

Solution The distribution has been plotted in Fig. 4-4.

4.1.3.1 Concept of Return Period

Most often hydrologic events (flood, low flow, drought, etc.) are described in terms of *return period*. The return period of a given event under study is defined as the expected time to obtain the first success (i.e., its first occurrence). Thus, the return period can be characterized by a geometric distribution. Mathematically, the return period T can be defined as the first moment about the origin:

$$T = E[N] = \sum_{n=1}^{\infty} np_N(n) = \sum_{n=1}^{\infty} npq^{n-1} = p(1 = 2q + 3q^2 + \ldots)$$

Using the algebra of infinite series we get

$$\left(1 + 2q + 3q^2 + \ldots\right) = 1/\left(1-q\right)^2 = 1/p^2$$

Thus,

$$T = \frac{p}{p^2} = \frac{1}{p}$$

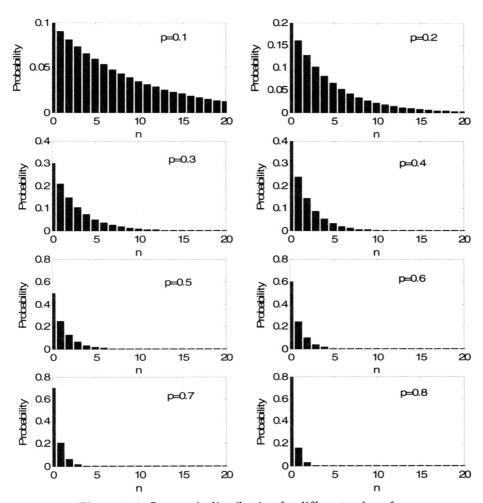

Figure **4-4** *Geometric distribution for different value of* p.

Example 4.9 Using the data in Example 4.3, compute the probability of the first rainy day in September using the geometric distribution. Also, compute the probability of the first two consecutive rainy days.

Solution The probability of the first occurrence of rain in September (first success) can be obtained using Eq. 4.9:

$$P(n = 1) = (1 - p)^{n-1} p = (1 - 0.2978)^{1-1} (0.2978) = 0.2978$$

The probability of the occurrence of the first two consecutive rainy days in September can be obtained as follows: The probability of two consecutive rainy days is

$$P = (0.2978)(0.2978) = 0.089$$

Then, the probability of occurrence of the first two consecutive rainy days in September can be obtained again by substituting this value in Eq. 4.9:

$$P = (1 - 0.089)^{1-1}(0.089) = 0.089$$

Example 4.10 Using the data of Example 4.5, compute the probability that (a) a 50-year flood will occur at least once during 20 years, (b) a 100-year flood will occur at least once during 20 years, (c) the number of years for the first occurrence of the 50-year flood is greater than 10 years, (d) the number of years for the first occurrence of the 50-year flood is greater than 30 years, (e) the number of years for the first occurrence of the 100-year flood is greater than 10 years, (f) the number of years for the first occurrence of the 100-year flood is greater than 20 years, (g) there will be no floods greater than the 50-year flood in 50 years, and (h) there will be no floods greater than the 100-year flood in 100 years.

Solution

(a) For a 50-year flood, the probability in any year is $p = 1/50 = 0.02$. The probability that a 50-year flood will occur at least once during 20 years is one minus the probability that it will not occur in 20 years:

$$P[x \geq 1] = 1 - P(0) = 1 - \binom{20}{0}(0.02)^0 (1 - 0.02)^{20} = 0.332$$

(b) The probability of a 100-year flood in any year is $p = 1/100 = 0.01$. The probability that a 100-year flood will occur at least once during 20 years is

$$P[x \geq 1] = 1 - P(0) = 1 - \binom{20}{0}(0.01)^0 (1 - 0.01)^{20} = 0.182$$

(c) The probability that the number of years for the first occurrence of the 50-year flood is greater than n years is

$$P(N > n) = 1 - [1 - (1 - p)^n]$$

For $p = 0.02$, $n = 10$,

$$P(N > 10) = 1 - [1 - (1 - 0.02)^{10}] = 0.817$$

(d) For $p = 0.02$, $n = 30$,

$$P(N > 30) = 1 - [1 - (1 - 0.02)^{30}] = 0.545$$

(e) If this is a 100-year flood, $p = 0.01$. For $p = 0.01$, $n = 10$,

$$P(N > 10) = 1 - [1 - (1 - 0.01)^{10}] = 0.904$$

(f) For $p = 0.01$, $n = 20$,

$$P(N > 10) = 1 - [1 - (1 - 0.01)^{20}] = 0.818$$

(g) The probability that there will be no floods greater than the 50-year flood in 50 years is

$$P[x = 0] = f_x(0) = \binom{50}{0}(0.02)^0 (1 - 0.02)^{50} = 0.364$$

(h) The probability that there will be no floods greater than the 100-year flood in 100 years is

$$P[x = 0] = f_x(0) = \binom{100}{0}(0.01)^0 (1 - 0.01)^{100} = 0.366$$

4.1.4 Negative Binomial Distribution

The negative binomial random variable represents the number of trials needed to obtain exactly k successes. Here the number of successes (k) is fixed and the number of trials varies from experiment to experiment. For this reason it is thought of as a reversal of the binomial distribution, because the number of successes and number of trials are reversed. Each trial has two possible outcomes—success or failure—and the probability of success is constant from one trial to another. As mentioned earlier, in the binomial case the number of trials is fixed and the number of successes varies. Consider a random variable W_k that is the sum of random variables $N_1, N_2, ..., N_k$, where N_i is the number of trials between $(i-1)$th and ith successes. Thus,

$$W_k = N_1 + N_2 + N_3 + ... + N_k \tag{4.13}$$

where N_i ($i = 1, 2, ..., k$) are mutually independent random variables, each with a common geometric distribution with parameter p, the probability of success in a trial. The distribution of k successes in w trials can be derived as

$$P_{W_k}(w) = \binom{w-1}{k-1}(1-p)^{w-k}p^k, \quad w = k, k+1, ... \tag{4.14}$$

where W_k is the trial number at which the kth success occurs. Equation 4.14 implies that $k-1$ successes in the preceding $w-1$ trials have already occurred. The probability of $k-1$ successes in $w-1$ trials is obtained from the binomial distribution.

This is the *negative binomial distribution*, also called the Pascal distribution, with parameters k and p, and is denoted as $NB(k,p)$. Note that $P_{W_k}(w) = 0$ for $w < k$. Interestingly, the sum of two negative binomial random variables is also a negative binomial random variable; that is, $NB(k_1, p) + NB(k_2, p)$ is also $NB(k_1 + k_2, p)$. Parameters of the negative binomial distribution are given as

$$E(w) = \frac{k}{p}$$

$$Var(w) = \frac{k(1-p)}{p^2}$$

Example 4.11 Consider the negative binomial distribution with parameters k and p. Graph the distribution for the following parameter sets, where p is the number of trials: (1) $p = 0.25$, $k = 2$; (2) $p = 0.5$, $k = 2$; (3) $p = 0.25$, $k = 3$; and (4) $p = 0.5$, $k = 3$.

Solution The graph in Fig. 4-5 shows the negative binomial distribution for the desired combination of parameters.

Example 4.12 Using the data of Example 4.3, compute the probability that the second rainy day will occur on the 10th day of September. Also compute the probability that the third rainy day will occur on the 15th day of the month using the negative binomial distribution.

Solution The probability of the kth success occurring at the wth trial can be calculated by the negative binomial distribution. Here, $k = 2$, $w = 10$. Hence,

$$P_{wk}(10) = \binom{10-1}{2-1}(1-0.2978)^{10-2}(0.2978)^2 = 0.047$$

In the second case, $k = 3$, $w = 15$. Hence,

$$P_{wk}(15) = \binom{15-1}{3-1}(1-0.2978)^{15-3}(0.2978)^3 = 0.04$$

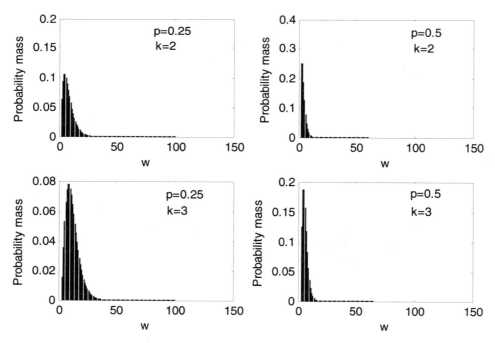

Figure 4-5 *Negative binomial distribution for different combinations of* p *and* k.

Example 4.13 Consider a 50-year flood. There is 1 chance in 50 that a flood greater than the critical value will occur in any particular year. What is the probability that at least one 50-year flood will occur during the 30-year economic lifetime of a proposed flood-control system?

Solution Let X equal the number of 50-year floods in 30 years. Then, X is $B(30, 0.02)$ and

$$p[X \geq 1] = 1 - f_X(0) = 1 - \binom{30}{0}(0.02)^0 (0.98)^{30} = 1 - (0.98)^{30} = 0.45$$

One can also use the binomial theorem and get the same answer:

$$P[X \geq 1] = 1 - (0.98)^{30} = 1 - (1 - 0.02)^{30}$$

$$= 1 - \sum_{i=0}^{30} \binom{30}{i}(0.02)^i$$

$$= 1 - [1 - 30(0.02) + \frac{(30)(29)}{2}(0.02)^2 - \frac{(30)(29)(28)}{(2)(3)}(0.02)^3 + ...]$$

$$= 0.6 - 0.174 + 0.01624 \approx 0.45$$

If this risk is too large, the design capacity is increased such that the magnitude of the critical flood would be exceeded with an acceptable probability of, say, 0.01 in any one year. Then, X is $B(30, 0.01)$, and $P(X \geq 1) = 1 - f_X(0) = 0.26$. The risk is lowered, but one must weigh the initial cost of the system versus the decreased risk of incurring the damage associated with the failure of the system to contain a large flood. The number of years N to the first occurrence of the critical flood is a random variable with a geometric distribution, $G(0.01)$, in the latter case. The probability that it is greater than 10 years is

$$P(N > 10) = 1 - F_N(10) = 1 - [1 - (1 - p)^{10}] = (1 - p)^{10} = 0.90$$

Suppose now the probability that $N > 30$ is to be computed. This is the probability that there are no floods in 30 years; that is, $X = 0$, where X is $B(30, 0.01)$. Thus,

$$P(N > 30) = P[X = 0] = 0.74$$

In a similar manner, one can compute the average return period or the expected value of N, which simply is

$$m_N = \frac{1}{p} = \frac{1}{0.01} = 100 \text{ years}$$

This is the average number of trials (years) to the first flood of magnitude greater than the critical flood after its last occurrence. This is referred to as the return period or recurrence interval in hydrology. Since X is $B(m, 1/m)$, the probability that there will be no floods greater than the m-year flood in m years is

$$P(X=0)=\left(1-\frac{1}{m}\right)^m = \sum_{i=0}^{m} \binom{m}{i}\left(\frac{1}{m}\right)^i$$

$$=1-m\left(\frac{1}{m}\right)+\frac{m(m-1)}{2}\left(\frac{1}{m}\right)^2 \cdots$$

$$=1-\frac{u}{1!}+\frac{u^2}{2!}-\cdots-=e^{-u}, \quad u=m(1/m)=1 \qquad (4.15)$$

For large m,

$$P[X=0]\approx e^{-1}\approx 0.368$$

This states that the probability that one or more m-year events will occur in m years is approximately $1 - e^{-1} = 0.632$. Thus, a system designed for the "m-year flood" will be inadequate, because the m-year flood will occur with a probability of about 2/3 at least once during the period of m years.

4.2 Models for Random Occurrences

In some cases, the number of trials is infinitely large, for example, events occurring at any instant over an interval of time or at any location along the length of a line or on the area of a surface that may be large. In such cases, it is difficult to identify discrete trials at which different events (or successes) might have occurred. Then the occurrences of events may be more appropriately modeled by a Poisson process rather than a binomial process.

4.2.1 Poisson Distribution: Counting Events

The binomial distribution is used when the random variable X is the number of times a specified event occurs in a fixed number of trials. When our interest is in the number of times a specified event occurs in a certain length of time, such as a given monitoring period, or how often the event is observed in a continuum of space, such as the length of a highway, an area of land, etc., and the number of trials is not specified, then the binomial distribution cannot be used. In such cases, it is more appropriate to use the Poisson distribution. Of course, the number of times a specified event occurs in a given continuum of space or time can be counted, but it makes little sense to specify the number of times the event did not occur. To illustrate this point, consider an example of thunder and lightning.

On a given day, it is easy to count the number of times the thundering and lightning occurred, but it makes little sense to state the number of times it did not occur. Similarly, the number of flaws in a 1000 m of a water supply pipe can be counted but the number of nonflaws cannot be stated. Thus, instead of defining the probability of "occurrence" for the specified event in a single trial, as for the binomial distribution, what is defined here is the probability of occurrence per unit of time or of space. For example, the probability that lightning in New Orleans in the month of May will occur may be 0.025 per day. The probability that a flaw occurs in a water supply pipe may be 0.000045 per meter of pipe length or the probability of flooding in an urban area may be 0.01 per year. It is assumed that these probabilities are the same for every day, every meter, or every year. It is further assumed that the occurrences and the nonoccurrences are independent along the continuum. The difference between these binomial and Poisson distributions can be summarized by noting that both the occurrences and nonoccurrences can be specified for the binomial distribution, whereas they cannot be for the Poisson distribution.

The binomial and Poisson distributions share some similarities. The probability distribution of the number of occurrences X in a given continuum of time or space can be treated as a special case of the binomial distribution under two conditions: (1) The number of trials becomes infinitely large, and (2) the average number of occurrences defined by np remains constant. By dividing the continuum into small intervals, the problem can be reduced to one of "occurrence" and "nonoccurrence" of the specified event in any of these intervals, provided these intervals are made so small that the probability of getting two or more "occurrences" in any interval is negligible. To that end, consider a fixed interval of time, say, t. Assume that the probability of an event occurring at any instant is p (and it is assumed here that the probability of two or more events occurring at any one instant is negligible). Then, the total number of events X in the $n = t$ (assumed) independent trials is binomial, $B(n, p)$:

$$B(n,p) = f_X(x) = \binom{n}{x} p^x (1-p)^{n-x}, \quad x = 0, 1, 2, ..., n \tag{4.16}$$

If an individual trial is represented by a smaller and smaller time duration, the number of trials n increases and the probability of success, p, on any one trial decreases, but the expected number of events in the total interval must remain constant at np. Let $v = np$ and let the trial duration tend to zero:

$$n \to \infty, \; p \to 0, \; np = v$$

Substituting for $p = v/n$ in the PMF of X and rearranging gives

$$
\begin{aligned}
f_X(x) &= \frac{n!}{x!(n-x)!}\left(\frac{v}{n}\right)^x\left(1-\frac{v}{n}\right)^{n-x} \\
&= \frac{v^x}{x!}\left(1-\frac{v}{n}\right)^n \frac{n!}{(n-x)!}\frac{1}{n^x\left(1-\frac{v}{n}\right)^x} \\
&= \frac{v^x}{x!}\left(1-\frac{v}{n}\right)^n\left\{\frac{n(n-1)(n-2)...(n-x+1)}{\left[n\left(1-\frac{v}{n}\right)\right]^x}\right\}
\end{aligned}
\tag{4.17}
$$

When n becomes very large compared to x, the product in the numerator of the large fraction approaches n^x and the term $[1 - (v/n)]^x$ will approach unity. Thus, as n tends to infinity,

$$
\left\{\frac{n(n-1)(n-2)...(n-x+1)}{\left[n\left(1-\frac{v}{n}\right)\right]^x}\right\} = 1; \quad \left(1-\frac{v}{n}\right)^n \Rightarrow e^{-v}
\tag{4.18}
$$

Thus, from Eq. 4.17,

$$
f_x(x) = \frac{v^x e^{-v}}{x!}, \quad x = 0, 1, 2, ..., \infty
\tag{4.19}
$$

This is the Poisson distribution. This distribution has one parameter and is entirely specified by the average number of occurrences of the specified event over the interval of time or space in question. It is denoted by $X \sim P(v)$.

The average of X can be expressed as

$$
\begin{aligned}
E[X] &= \sum_{x=0}^{\infty} x\frac{v^x e^{-v}}{x!} = v\sum_{x=1}^{\infty}\frac{v^{x-1}e^{-v}}{(x-1)!} = ve^{-v}\sum_{x=1}^{\infty}\frac{v^{x-1}}{(x-1)!} \\
&= ve^{-v}e^v = v
\end{aligned}
\tag{4.20}
$$

In a similar manner, one obtains the variance of X, $\text{var}(X) = v$.

The sum of two Poisson random variables with parameters v_1 and v_2 must again be a Poisson random variable with parameters $v = v_1 + v_2$. The distributions with the property that the sum of independent random variables has the same distribution are said to be *regenerative*. If p remains the same, then binomial and negative binomial distributions are also regenerative.

It is worth mentioning here the difference between the binomial and Poisson processes. The binomial process is concerned with the number of successes and failures (two events) in a fixed number of trials, whereas in the Poisson process only one event (rather than two) is of concern. The Poisson distribution is most commonly used in waiting time evaluations and reliability analysis (e.g., the number of arrivals of vehicles at a highway toll booth in a given hour, the number of times a water or air or noise quality standard is violated at a given site during a given monitoring period, the number of windy days in a given period, or the number of snowfalls in a month).

Example 4.14 Consider the Poisson distribution with parameter v. Graph the Poisson distribution for v = 0.5, 1.0, 2.0, 5.0, and 10.0.

Solution The shape of the distribution for a given value of parameter v is given in Fig. 4-6.

Example 4.15 Determine the probability of occurrence of a storm with a return period of 25 years in 5 years using the Poisson distribution.

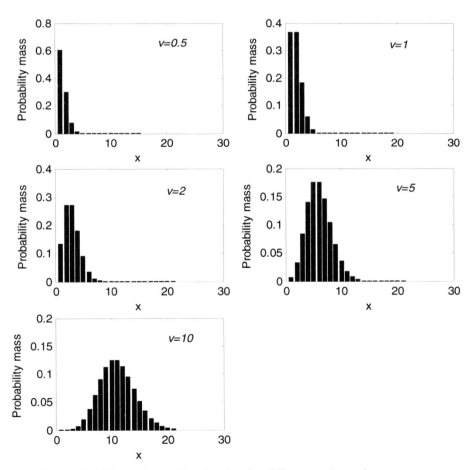

Figure 4-6 *The Poisson distribution for different values of parameter* v.

Solution The probability of occurrence of a 25-year storm in a year is $1/25 = 0.04$. Thus,

$$v = 5 \times 0.04 = 0.20$$

$$P(x) = 0.20^1 \times e^{-0.2}/1! = 0.2 \times 0.819 = 0.164$$

Example 4.16 A left turn lane is to be designed at an intersection to accommodate left-turning cars that arrive during 30 seconds when the traffic light does not permit a left-hand turn. Since an insufficient length of traffic lane would cause cars to interfere with through traffic, it is decided that the probability of having too many left-turning cars arrive during the 30 seconds should be kept to 1%. Left-turning cars arrive at an average rate of 6/min. (a). Draw the PMF of the number of left-turning cars during the 30-second period that the light is red for them. (b). Determine the number of cars the left turn lane must be designed for. This example was discussed by Booy (1990).

Solution Here, the random variable is the number of left-turning cars arriving in the 30-second period. The average number of such cars is equal to 3. Therefore, $X \sim P(3)$ and

$$f_X(x) = \frac{3^x}{x!} e^{-3} = \frac{3^x}{x!} (0.04979)$$

Successive values of X can be calculated from this formula.

Alternatively, one can use the probabilities listed in Table E4-16. From the table it may be seen that more than 8 cars arrive in the left turn lane within 30 seconds with a probability of 1% or less ($< 0.4\%$ to be exact). Thus, a capacity to accommodate 8 cars would be sufficient for the left turn lane, as per the allowed design exceedance probability of 1%.

Table E4-16

$X =$	0	1	2	3	4	5	6
$P(x)$	0.0498	0.1494	0.2240	0.2240	0.1680	0.1008	0.0504
$P(X<x)$	0.0498	0.1991	0.4232	0.6472	0.8153	0.9161	0.9665
$P(X>x)$	0.9502	0.8009	0.5768	0.3528	0.1847	0.0839	0.0335

$X =$	7	8	9	10	11	12	13
$P(x)$	0.0216	0.0081	0.0027	0.0008	0.0002	0.0001	0.0000
$P(X<x)$	0.9881	0.9962	0.9989	0.9997	0.9999	1.0000	1.0000
$P(X>x)$	0.0119	0.0038	0.0011	0.0003	0.0001	0.0000	0.0000

Example 4.17 Irrigation canals were lined using two types of lining: concrete and brick. It was observed after a number of years that 5 out of 70 brick-lined canal reaches leaked and only 1 out of 50 concrete-lined canal reaches showed cracks and leakage. Assuming that there should be no difference in the durability of the

linings, what is the probability that a difference equal to or greater than the observed difference in the number of leaks (= 4) would occur?

Solution On the assumption of no difference in durability, there are 120 lined canal reaches, including leaking lined canals. The probability of a leak per single canal can be estimated at $6/120 = 5\%$. There are then two samples, one of 70 and one of 50 canal reaches, and the distribution of the number of defects in each can be determined. Let these numbers be designated as X_{70} and X_{50}. If they follow a binomial distribution, then

$$X_{70} \sim B(70, 0.05)$$

$$X_{50} \sim B(50, 0.05)$$

The variances of the two variables are $70(0.05)(0.95) = 3.33$ and $50(0.05)(0.95) = 2.38$, respectively.

The Poisson distribution can be expressed with $v = 70 \times 0.05 = 3.5$ and $50 \times 0.05 = 2.5$ as $X_{70} \sim P(3.5)$ and $X_{50} \sim P(2.5)$.

To determine the probability that $D = X_{70} - X_{50}$ is equal to or larger than 4, the type of distribution of the difference between two Poisson or binominal distributions is required, which is not known. A simple way out of this difficulty is as follows. Event D equal to or larger than 4 can result from a rather limited number of combinations of X_{70} and X_{50}, namely,

$$X_{50} = 0 \text{ and } X_{70} \text{ equal to or larger than } 4$$

$$X_{50} = 1 \text{ and } X_{70} \text{ equal to or larger than } 5$$

$$X_{50} = 2 \text{ and } X_{70} \text{ equal to or larger than } 6$$

etc. These probabilities can be computed and the total probability can be determined for the Poisson distributed variables as listed in Table E4-17a.

Similarly, the probabilities computed and the total probability for binomial-distributed variables are given as listed in Table E4-17b. The calculated values given in the table indicate that the probability obtained converges to zero quickly. So it is safe to calculate the probability until $X_{50} = 6$, which results in $P(D4) = 0.1492$ if it is assumed that the variables are Poisson distributed, and 0.1438 if it is assumed that the variables are binomial distributed. This means that the evidence of greater durability is by no means conclusive.

Example 4.18 At a meteorological station, Sombor, in the region of Backa, Yugoslavia, during a period of 39 years, the number of years without drought was 4, the total number of droughts was 70, and the longest drought lasted for 62 days. The period considered in a year was the growing season, April 1 through September 30, which was 183 days long. Suppose an irrigation project is to be built to serve agriculture for 50 years and we want to examine the drought situation during the lifetime of the project. Compute the probability of 1, 2, 3, 4, 5, and 6 droughts occurring in the growing season.

Table E4-17a

X_{50}	$P(X_{50})$	$P(X_{70} \geq X_{50} + 4)$	$P(X_{50}) \times P(X_{70} \geq X_{50} + 4)$
0	0.0821	0.4634	0.0380
1	0.2052	0.2746	0.0563
2	0.2565	0.1424	0.0365
3	0.2138	0.0653	0.0140
4	0.1336	0.0267	0.0036
5	0.0668	0.0099	0.0007
6	0.0278	0.0033	0.0001
7	0.0099	0.001	1.01×10^{-5}
8	0.0031	0.0003	8.98×10^{-7}
9	0.0009	0.00008	6.55×10^{-8}
10	0.0002	0.00002	4×10^{-9}
		Sum	14.92%

Table E4-17b

X_{50}	$P(X_{50})$	$P(X_{70} \geq X_{50} + 4)$	$P(X_{50}) \times P(X_{70} \geq X_{50} + 4)$
0	0.077	0.47	0.036
1	0.202	0.27	0.055
2	0.261	0.14	0.036
3	0.219	0.06	0.013
4	0.136	0.023	0.0032
5	0.066	0.008	0.00053
6	0.026	0.0025	6.42×10^{-5}
7	0.0086	0.00068	5.88×10^{-6}
8	0.0024	0.00017	4.18×10^{-7}
9	0.00059	3.94×10^{-5}	2.35×10^{-8}
10	0.00013	8.27×10^{-6}	1.07×10^{-9}
		Sum	14.38%

Solution For the Poisson distribution, parameter λ is the mean number of droughts in the growing season: $\lambda = np = 70/39 = 1.79$. The probability of x (the number of droughts) will be

$$p[x] = \frac{\lambda^x e^{-\lambda}}{x!}$$

Hence,

$$p[1] = \frac{1.79^1 e^{-1.79}}{1!} = 0.299$$

$$p[2] = \frac{1.79^2 e^{-1.79}}{2!} = 0.267$$

In a similar manner, we find $p[3] = 0.160$, $p[4] = 0.071$, $p[5] = 0.026$, and $p[6] = 0.0076$. These probabilities are plotted in Fig. 4-7.

4.2.2 Exponential Distribution: Time Between Events

The exponential distribution has wide application in engineering evaluations. The main applications of the exponential distribution include characterization of the time interval between successive events (e.g., distribution of time between rainfall events) occurring in a Poisson process, the distribution of interarrival times of floods, the distribution of interarrival times of droughts, the determination of time to failure for a certain engineering system, and the length of time between customers arriving at a shop. Whereas the Poisson distribution provides the number of events in a fixed span of time, the exponential distribution treats the span itself as the random variable, thus making it continuous.

Now consider the length of time interval between events at a point. Let us assume that the events follow a Poisson arrival process. Let the average number of events per unit time be λ. If the interval is too short, it will cause the events to merge with the stream of events or to interrupt the stream. For example, if the time interval between two flood events is very short, the events may merge. Let the random variable T denote the time to the first arrival. Then, the probability that T exceeds some time t is equal to the probability that no events occur in that time interval of length t. The number of events in time interval t is λt. Then $1 - F_T(t)$ is the former probability and $P_X(0)$ is the latter probability—the probability that a Poisson random variable X with parameter λt is zero. Thus, the CDF of the exponential distribution is the result:

$$1 - F_T(t) = \frac{(\lambda t)^0 e^{-\lambda t}}{0!} = e^{-\lambda t}, \quad t \geq 0$$
$$F_T(t) = 1 - e^{-\lambda t}, \quad t \geq 0 \tag{4.21}$$

Differentiation of Eq. 4.21 yields the PDF of the exponential distribution:

$$f_T(t) = \frac{dF_T(t)}{dt} = \lambda e^{-\lambda t}, \quad t \geq 0 \tag{4.22}$$

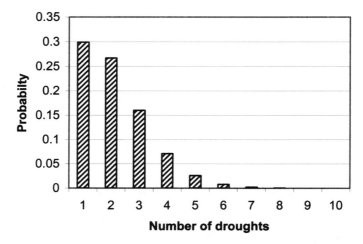

Figure 4-7 *The probability of number of droughts.*

which is the exponential distribution, denoted as $EX(\lambda)$. This distribution describes the time to the first occurrence of a Poisson event and is a continuous analog of the geometric distribution.

The Poisson process follows the property of independence and stationarity. The term $e^{-\lambda t}$ is the probability of no events in any interval of time t, whether or not it begins at time 0. If we use the arrival time of the nth event as the beginning of the time interval, the term $e^{-\lambda t}$ is the probability that the time to the $(n+1)$th event is greater than t. In short, the interarrival times of a Poisson process are independent and exponentially distributed.

The mean of the exponential distribution is

$$E[T] = \int_0^\infty t\lambda e^{-\lambda t} dt = \frac{1}{\lambda} \tag{4.23}$$

where $1/\lambda$ denotes the average time between arrivals, and λ is the average number of events per unit time.

The variance of T is given as

$$\sigma_T^2 = \frac{1}{\lambda^2} \tag{4.24}$$

The coefficient of variation of T is

$$c_v = \frac{E[T]}{\sigma_T} = \frac{1/\lambda}{1/\lambda} = 1 \tag{4.25}$$

for any value of parameter λ.

Example 4.19 Consider the exponential distribution with parameter λ. Graph the exponential distribution for the ratio of time t to the mean interarrival time $1/\lambda$ (i.e., for λt).

Solution The exponential distribution for various values of λ is plotted in Fig. 4-8.

Example 4.20 In Example 4.18, the mean and standard deviation of the drought duration (beyond a threshold value of 15 days) were 8.06 and 8.575 days, respectively. The drought duration was found to follow an exponential distribution. The maximum observed value of this duration (counted beyond the threshold value) during the course of 39 years was 47 days. Compute the probability of drought duration to be less than 5, 10, 15, 20, 25, 30, 35, 40, and 45 days.

Solution The mean of exponential distribution is $E[T] = 1/\lambda = 8.06$; thus $\lambda = 0.124$. Therefore,

$$F_T(5) = 1 - e^{-(0.124)5} = 0.462$$

$$F_T(10) = 1 - e^{-(0.124)10} = 0.711$$

In a similar manner, we obtain $F_T(15) = 0.844$, $F_T(20) = 0.9167$, $F_T(25) = 0.955$, $F_T(30) = 0.976$, $F_T(35) = 0.987$, $F_T(40) = 0.993$, and $F_T(45) = 0.996$. These probabilities have been plotted in Fig. 4-9.

4.2.3 Gamma Distribution: Time to the kth Event

Consider the distribution of X_k, the time to the kth arrival of a Poisson process. The times between arrivals, T_i, $i = 1, 2, \ldots, k$, are independent and have exponential distributions with common parameter λ. X_k is the sum $T_1 + T_2 + \ldots + T_K$. Its distribution is given by the repeated application of the convolution integral. For any $k = 1, 2, 3, \ldots$.

$$f_{Xk}(x) = \frac{\lambda(\lambda x)^{k-1} e^{-\lambda x}}{(k-1)!}, \quad x \geq 0, \ \lambda > 0, \ k = 1, 2, \ldots \tag{4.26}$$

Equation 4.26 is the gamma distribution of X_k or X, denoted as $G(k, \lambda)$ with k and λ as parameters.

Here, X is gamma distributed and is the sum of k independent exponentially distributed random variables. More generally, k need not be integer valued; then

$$f_X(x) = \frac{\lambda(\lambda x)^{k-1} e^{-\lambda x}}{\Gamma(k)}, \quad x \geq 0, \ \lambda > 0, \ k > 0 \tag{4.27}$$

$$\Gamma(k) = \int_0^\infty e^{-u} u^{k-1} du, \ k > 0 \tag{4.28}$$

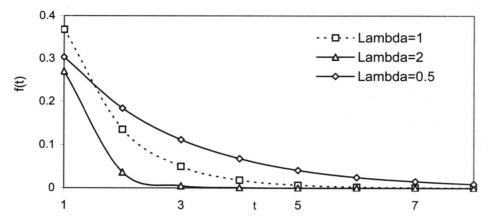

Figure 4-8 *Exponential distribution for various values of lambda (λ).*

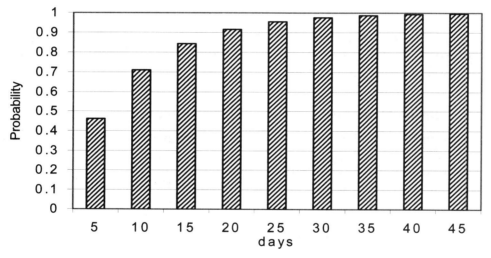

Figure 4-9 *Probabilities of drought durations.*

is the gamma function. Here, k is a shape parameter and λ is a scale parameter. The cumulative distribution function can be computed by

$$F_X(x) = \int_0^x \frac{\lambda^k x^{k-1} e^{-\lambda x}}{\Gamma(k)} dx \qquad (4.29)$$

This function can be evaluated by using the table of incomplete gamma functions. If k is an integer, the cumulative distribution function can be computed as

$$F_X(x) = 1 - e^{-\lambda x} \sum_{j=0}^{k-1} \frac{(\lambda x)^j}{j!} \qquad (4.30)$$

Also, $\Gamma(k+1) = k\Gamma(k)$, $k > 0$; $\Gamma(k) = \Gamma(k+1)/k$, $k < 1$; and $\Gamma(2) = \Gamma(1) = 1$, $\Gamma(1/2) = \sqrt{\pi}$. The incomplete gamma function is defined as

$$\Gamma(k,x) = \int_0^x e^{-u} u^{k-1} du \qquad (4.31)$$

Abramowitz and Stegun (1965) have given the following numerical approximation to evaluate the gamma function:

$$\Gamma(k+1) = 1 + b_1 y + b_2 y^2 + b_3 y^3 + \ldots + b_8 y^8 + \varepsilon(y) \qquad (4.32)$$

where $0 \le y \le 1$. The coefficients are $b_1 = -0.577191652$, $b_2 = 0.988205891$, $b_3 = -0.897056937$, $b_4 = 0.918206857$, $b_5 = -0.756704078$, $b_6 = 0.482199394$, $b_7 = -0.193527818$, and $b_8 = 0.035868343$. The absolute error in the approximation is $|\varepsilon(y)| \le 3 \times 10^{-7}$.

It is easy to see that the exponential distribution is a special case of the gamma distribution where $k = 1$. In the field of water resources, the gamma distribution is used to model the instantaneous unit hydrograph and to perform flood frequency analysis; it is also used in many other problems.

The mean and variance of the variable X can be expressed as

$$E[X] = \frac{k}{\lambda} \qquad (4.33)$$

$$\sigma_X^2 = \frac{k}{\lambda^2} \qquad (4.34)$$

The coefficient of skewness is defined as

$$\gamma = \frac{E\left[(x - m_x)^3\right]}{\sigma_x^3} = \frac{2}{\sqrt{k}} \qquad (4.35)$$

Example 4.21 Graph the gamma distribution with parameters k and λ for $\lambda = 2.2$, $k = 1, 2, 3, 4, 5$, and 10. Take the X axis as λx.

Solution The shape of the gamma distribution for different values of parameter k is given in Fig. 4-10.

Example 4.22 Markovic (1965) used the gamma distribution to model the maximum annual river flows in the Weldon River at Mill Grove, Missouri, based on the data for 1930 to 1960. He found $k = 1.727$ and $\lambda = 0.00672$ (cfs)$^{-1}$. Determine the probability that the maximum flow is less than 400 cfs in any year.

Solution The mean is

$$m_X = \frac{k}{\lambda} = \frac{1.727}{0.00672} = 256.7 \text{ cfs}$$

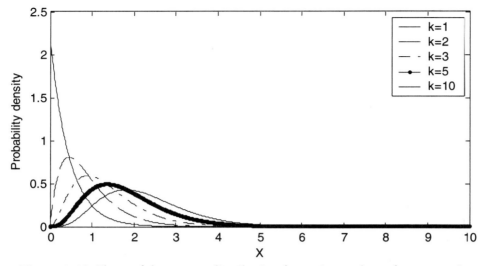

Figure 4-10 *Shape of the gamma distribution for various values of parameter* k.

with a standard deviation of

$$\sigma_X = \frac{\sqrt{k}}{\lambda} = 195.6 \;\; \text{cfs}$$

To compute the probability that the maximum flow (x) is less than 400 cfs in any year (with $\lambda \times x = 0.00672 \times 400 = 2.69$), Eq. 4.29 is used as

$$F_x(400) = \frac{\displaystyle\int_0^{2.69} y^{(1.727-1)} e^{-y} dy}{\Gamma(1.727)} = \Gamma(2.69, 1.727) = 0.809$$

Note that $\Gamma(2.69, 1.727)$ denotes the incomplete gamma function as

$$\Gamma(x, a) = \frac{1}{\Gamma(a)} \int_0^x t^{a-1} e^{-t} dt$$

Example 4.23 The mean and the standard deviation of the annual peak flow data for the Amite River near Darlington, Louisiana, are 28,675.833 cfs and 21,117.138 cfs. Assuming that the peak flow data follow a two-parameter gamma distribution, determine the parameters of the distribution. What is the probability that the maximum flow is less than 100,000 cfs, less than 80,000 cfs, and less than 50,000 cfs in any year? What is the return period of each of these flows? What is the probability that peak flow will occur in any year between one standard deviation on either side of the mean, and between two standard deviations on either side of the mean?

Solution For a two-parameter gamma distribution,

$$m_x = \frac{k}{\lambda} = 28675.833$$

$$\sigma_x = \frac{\sqrt{k}}{\lambda} = 21117.138$$

Thus,

$$k \approx 1.844$$

$$\lambda \approx 6.43 \times 10^{-5}$$

The probability that the maximum flow is less than 100,000 cfs in any year is (with $\lambda \times 100{,}000 = 6.43$)

$$F_x(100000) = \Gamma(6.43, 1.844) = 0.9907$$

The probability that the maximum flow is less than 80,000 cfs in any year is (with $\lambda \times 80{,}000 = 5.144$)

$$F_x(80000) = \Gamma(5.144, 1.844) = 0.971$$

The probability that the maximum flow is less than 50,000 cfs in any year is (with $\lambda \times 50{,}000 = 3.215$):

$$F_x(50000) = \Gamma(3.215, 1.844) = 0.863$$

The return period can be calculated from

$$m_N = \frac{1}{1 - F_x(x)}$$

For a maximum flow of 100,000 cfs, the return period is $1/(1 - 0.9907) = 107.78$ years, for a maximum flow of 80,000 cfs, the return period is $1/(1 - 0.971) = 34.48$ years, and for a maximum flow of 50,000 cfs, the return period is $1/(1 - 0.863) = 7.30$ years.

The probability that peak flow will occur within one standard deviation of the mean is

$$F_x(m_x - \sigma < x < m_x + \sigma) = F_x(m_x + \sigma) - F_x(m_x - \sigma)$$

$$= F_x(49792.97) - F_x(7558.695) = 0.856 - 0.112 = 0.744$$

The probability that peak flow will occur within two standard deviations of the mean is

$$F_x(m_x - 2\sigma < x < m_x + 2\sigma) = F_x(m_x + 2\sigma) - F_x(m_x - 2\sigma)$$

$$= F_x(70910.11) - F_x(-13558.4) = 0.953 - 0 = 0.953$$

The distribution is plotted in Fig. 4-11.

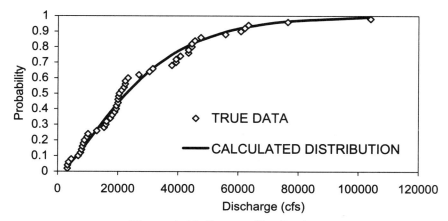

Figure 4-11 *Gamma distribution.*

4.3 Summary of Distributions

The distributions presented in this chapter are summarized in Table 4-1. This table gives the distribution name, mathematical form, random variable and its range, parameters, and the applications for which the distribution is used.

4.4 Questions

4.1 Using daily rainfall data for the month of January in Houston, Texas, compute the probability of 2, 5, and 10 rainy days in January in Houston. Use the binomial distribution. What will be the probability of having two consecutive rainy days, three consecutive rainy days, four consecutive rainy days, and five consecutive rainy days? What will be the probability that a 50-year rainfall (in terms of amount) will occur at least once in 20 years? A 100-year rainfall in 20 years?

4.2 Compute the probability of the first occurrence of rain in January in Houston. Use the geometric distribution.

4.3 Compute the probability that the third rainy day will occur on the 10th day or on the 15th day in August in Houston. You can use the negative binomial distribution.

Table 4-1 *Summary of probability distributions.*

Distribution name and symbol	Mathematical form	Range of random variable	Parameters	Applications
Bernoulli distribution	$f_X(x) = p^x(1-p)^{1-x}$	$X = [0,1]$ $0 \le \rho \le 1$	$E(X) = p$ $\sigma_X^2 = pq$	To model the behavior of a random variable that can take on any one of the two values: success or failure, rain or dry, etc.
Binomial distribution, $B(n, p)$	$f_Y(y) = \dfrac{n!}{y!(n-y)!} p^y (1-p)^{1-y}$	$y = 0, 1, 2, 3, \ldots, n$ $0 \le \rho \le 1$	$E(Y) = np$ $\mathrm{var}[Y] =$ $np(1-p)$	To model events whose outcomes are mutually independent and for which the probability of success or failure is fixed.
Geometric distribution, $G(p)$	$P[N=n] = (1-p)^{n-1} p$	$n = 1, 2, 3, \ldots$	$E[N] = 1/p$ $\mathrm{var}[N] = \dfrac{1-p}{p^2}$	Used when the question is to find the number of trials before the first success occurs.
Negative binomial distribution, $NB(k, p)$	$f_{W_k}(w) = \dbinom{w-1}{k-1}(1-p)^{w-k} p^k$	$w = k, k+1 \ldots$ $k = 1, 2, 3, \ldots$ $0 \le \rho \le 1$	$E(w) = \dfrac{k}{p}$ $\mathrm{var}(w) = \dfrac{k(1-p)}{p^2}$	Used to model the number of trials, w, to obtain k successes or k successes in w trials.
Poisson distribution, $P(v)$	$p_x(x) = \dfrac{v^x e^{-v}}{x!}$	$X = 0, 1, 2, \ldots$ ∞	$E(X) = v$ $\mathrm{var}(X) = v$	Used to model the number of times a specified event occurs in a fixed span of time; used for waiting time evaluation and reliability analysis.
Exponential distribution, $EX(\lambda)$	$f_T(t) = \lambda e^{-\lambda t}$	$t > 0$	$E[T] = \dfrac{1}{\lambda}$ $\mathrm{var}[T] = \dfrac{1}{\lambda^2}$	To characterize the time interval between successive events (e.g., distribution of time between rainfall events) occurring in a Poisson process, to determine time to failure for certain engineering systems and length of time between customers arriving at a shop.
Gamma distribution, $G(k, \lambda)$	$f_{X_k}(x) = \dfrac{\lambda(\lambda x)^{k-1} e^{-\lambda x}}{(k-1)!}$	$x \ge 0, \lambda > 0$ $k > 0$	$E[X] = \dfrac{k}{\lambda}$ $\sigma_X^2 = \dfrac{k}{\lambda^2}$	Used to model the time to the kth event wherein times between arrivals of events are independent and exponentially distributed. It is also used to model the instantaneous unit hydrograph. Many distributions from the gamma family are employed in flood frequency analysis.

4.4   Count the number of wet periods in Houston for the data you have. Define a wet period by the sum of consecutive months each having rainfall equal to or greater than 5 inches. The number of wet periods can be described by the Poisson distribution. Compute the probability that Houston will have 2, 4, 5, 6, and 10 wet periods.

4.5   Count the number of dry periods in Houston each year for the data you have. Define a dry period as the sum of months each having rainfall less than or equal to 2 inches. The number of dry periods can be described by the Poisson distribution. Compute the probability that Houston will have 2, 4, 5, 6, and 10 dry periods in a year.

4.6   Compute the time interval between wet periods as defined in Question 4.4. This should follow an exponential distribution. Compute the probability that the time interval would be less than 2, 3, 4, and 5 months.

4.7   Compute the time interval between dry periods as defined in Question 4.5. This should follow an exponential distribution. Compute the probability that the time interval would be less than 2, 3, 4, and 5 months.

4.8   Compute the maximum monthly rainfall for each year. Determine the probability that the maximum monthly rainfall is less than 10 inches. You can use the gamma distribution here.

4.9   The daily concentration of a pollutant in a stream follows an exponential distribution and is independent from day to day.

   (a) If the mean daily concentration of the pollutant is 2 mg/L, estimate the parameter λ of the exponential distribution given by $f(x) = \lambda e^{-\lambda x}$.
   (b) Pollution is a problem if the concentration exceeds 6 mg/L. What is the probability of a pollution problem on any particular day?
   (c) What is the return period in days of the pollution problem?
   (d) What is the probability of a pollution problem in at most 1 day in any 3 consecutive days?
   (e) If instead of being exponentially distributed the pollution level is described by a gamma distribution with the same mean and variance as the exponential distribution, then what is the probability of a pollution problem on any particular day?

4.10   The time between rainstorms is thought to be exponentially distributed with a mean of 5 days. What would you expect the distribution of the time for the occurrence of 10 rainstorms to be? What would you expect for the values of the parameters of this distribution?

4.11   The probability of flooding in a given low-lying roadway crossing of a small river in any given year is 0.51. Consider X as a random variable representing the number of yearly floods at this location in a period of 15 consecutive years.

(a) Determine the first four moments of X.

(b) Determine the mean and variance of X.

(c) Determine the probability that there will be no floods in the 15-year period.

(d) Determine the probability that there will be exactly 3 floods in the 15-year period.

(e) Determine the probability that there will be at least 2 floods in the 15-year period.

(f) Determine the probability that there will be at most 7 floods in the 15-year period.

(g) Determine the probability that there will be between 3 and 7 floods in the 15-year period.

4.12   To monitor the current water quality status in a river basin, automatic water quality monitoring stations will be installed. In the initial phase, it has been found that about 85% of all the monitoring sites operate correctly at a given time after installation. The rest require some adjustments. About 10 monitoring sites are installed in a given month. Determine the probability that at least 9 of the automatic monitoring sites operate correctly upon installation. Consider 6 consecutive months in which 10 monitoring sites are established. What is the probability that at least 9 monitoring sites operate correctly in each of the 6 months?

4.13   It is possible for a piece of monitoring equipment to not detect a particular pollutant. This error is called the eclipsing error. Equipment is defective when it introduces an eclipsing error with probability 0.1. The equipment is used 20 times during a given week.

(a) Find the probability that no eclipsing error occurs.

(b) Find the probability that at least one such error occurs.

(c) Would it be unusual for more than five such errors to occur? Explain, based on the calculation of probability involved.

4.14   Based on water quality monitoring data, it was found that a particular industry releases a detectable amount of benzene once in a month, on average, in a river used as a raw water supply for drinking purposes. Find the probability that there will be at most three such releases during a month. What is the expected number of releases during a three-month period? If, in fact, 10 or more releases are detected during a three-month period, do you think that there is a reason to suspect the reported average figure of once a month? Explain, on the basis of calculations.

4.15   Let X be binomial with $n = 25$ and $p = 0.04$. Find $P[X = 0, 1, 2, 3, 4,$ and $5]$ using the binomial probability density function and compare your answer to that obtained using the Poisson approximation to this probability density function. Comment on the error involved in the approximation for each value of X. Further, determine the mean and variance of X using both distributions.

4.16   At a given location extreme temperature has been observed in some September months. On average, there are about 7 years between each of these warm events. If these warm events were to be randomly distributed in time, determine the following:

(a) the probability of having no warm September in 10 years

(b) the probability of having 2 events in 10 years

(c) the probability of more than 3 events in 10 years

4.17   During a 4-month hurricane season at a given coastal town, on average severe hurricane events occur with 1 event/month. Last year, 11 hurricanes occurred and the news media blamed the climate change on the greenhouse effect. From a statistical point of view, how unusual are the seasons of this or higher severity? Would you agree or disagree with the news media that this was an exceptionally severe season? Support your arguments with the calculation of probability value involved.

4.18   A new filtration cartridge is being studied. It is thought that the cartridge will treat at least 70,000 gallons of water on 90% of the filtration units in which it is used. Laboratory trials are conducted to simulate 100 filtration units using this type of cartridge. Assume X is a random variable representing the number of filtration units whose cartridge must be replaced before treating 70,000 gallons of water.

(a) What is the distribution of X? Determine the expected value of X.

(b) What distribution can be used to approximate probabilities for X?

(c) If out of 100 filtration units 21 or more need replacement, what is the probability that you will conclude that the 90% figure is correct?

4.19   A survey of 70 systems found about 2,100 pressure-reduction incidents caused by water main breaks in a one-year period. Water main breaks not only cause substantial loss of water but also cause pressure reductions in the water distribution system that enable backflow of contaminants into the system. Assume that city Z has very similar characteristics to those cities included in the survey. City Z wants to allocate some emergency budget for maintenance of its water distribution system, particularly pressure-reduction incidents.

(a) What is the probability that city Z will be having exactly one main break in a given year?

(b) What is the probability that city Z will be having at least one main break in a given year?

(c) What is the probability that city Z will be having no breaks in 10 years?

(d) What is the probability that city Z will be having between 3 to 7 main breaks in 10 years?

4.20 A drinking water company produces 10,000 bottles per day. Each bottle has a 0.001 chance of being affected by some contaminant. Assume that the chance of a bottle being affected by contamination is independent of the daily supply orders.

(a) What is the most appropriate distribution for the number of bottles affected by contamination?

(b) If your answer is the Poisson distribution, provide your justification for selecting the Poisson distribution as an acceptable approximation.

(c) What is the probability that 21 bottles turn out to be affected by contamination?

4.21 What design return period should be used to ensure a 90% chance that the design will not be exceeded in a 50-year period? What design return period should be used to ensure an 85% chance of no more than 1 exceedance in 25 years?

4.22 Two widely separated watersheds are selected for a study on peak discharges. If the occurrence of flood flows on the two basins can be considered as independent events, what is the probability of experiencing a total of ten 25-year events on the two watersheds in a 50-year period?

4.23 A scientist has predicted that during a certain 5-year period a severe drought will occur in the high plateau of Mexico. She made this prediction based on her observation of sunspot activity. If the probability of a drought is 0.18 in any year, what is the probability that the scientist's prediction will come true if the occurrence of a drought was a strictly random phenomenon unrelated to sunspot activity?

4.24 On average how many times will a 5-year flood occur in a 50-year period? What is the probability that exactly this number of 5-year floods will occur in a 50-year period?

4.25 What is the probability that exactly 4 years will elapse between occurrences of a 5-year event?

4.26 A binomial random variable has a mean of 20 and a variance of 16. Find the values of n and p that characterize the distribution of this random variable.

4.27   Assume that California is hit by approximately 500 earthquakes that are large enough to be felt every year. However, those of destructive magnitudes occur on average once every year. Find the probability that California will experience at least two destructive earthquakes during a 3-year period.

4.28   Assume that the probability of a rainy day is $p = 0.25$ in a specific watershed. What is the probability that the next year would have at least 125 rainy days?

4.29   Suppose that the arrivals of small aircraft at a certain airport follow a Poisson process, with a rate of 5 per hour (then $\lambda = 5t$).

   (a) What is the probability that exactly 3 small aircraft arrive during a 1-hour period?

   (b) What is the probability that at least 3 small aircraft arrive during a 1-hour period?

   (c) If a working day has 12 hours, what is the probability that at least 50 small aircraft arrive during a day?

Chapter 5

Limit and Other Distributions

In real-life civil and environmental engineering projects, situations arise that warrant a proper characterization of the sample mean (e.g., several samples are tested to determine the average concentration of a chemical discharged from a point source, the average export coefficient for a nutrient from a given land use, the event mean concentration, the average flow of a stream at a given site, the average compressive strength of a cement concrete specimen, or the average shear strength of a soil sample). There are so many procedures, both deterministic and statistical, that are based on the sample mean. In the deterministic area, engineering design and analysis utilize mean values of the input parameters. Statistical methods include parameter estimation using the method of moments, determination of bias, making statistical inferences, and calculation of confidence intervals. Because of economic difficulties, lack of access to the site, lack of time, or other reasons, it is not always possible to collect enough samples to sufficiently characterize the population frequency distribution. In these situations, we are not sure about the type of the distribution possessed by the sample mean and we have many questions in our mind (e.g., Does the sample mean follow some distribution, such as the uniform, triangular, normal, log-normal, or gamma?). To help us answer these questions, we first introduce the highly celebrated central limit theorem, which is one of the most significant theorems in the field of probability theory. Then we will discuss the most widely used continuous probability distribution, known as the Gaussian or normal distribution. The normal distribution is based on sound theoretical principles, is capable of representing many population frequency distributions (both PDFs and PMFs), and helps provide a convenient approximation to reality. Then, we will discuss normal approximations to several PMFs. In the end we will discuss other commonly used PDFs along with their applications in the area of civil and environmental engineering.

5.1 Normal Distribution: Model of Sums

The normal distribution is also known as Gaussian distribution after the famous German mathematician Karl Gauss who widely used it. It is also called the law of errors. The name "normal" became popular because it was believed that most random phenomena could be described by this distribution. When the random variation in a phenomenon arises from a number of additive variations, then it can be described by the normal distribution.

For a random variable X, the normal distribution can be expressed as

$$f_X(x) = \frac{1}{\sigma_X \sqrt{2\pi}} \exp\left[-\frac{1}{2}\left(\frac{x - \mu_X}{\sigma_X}\right)^2\right], \quad -\infty \leq x \leq \infty \tag{5.1}$$

where $f_X(x)$ is the PDF of X, μ_X is the mean value of X, and σ_X is the standard deviation of X. The normal distribution is denoted as $N(\mu, \sigma^2)$, which means that X is normally distributed with a mean of μ and a variance of σ^2; these are also known as scale and shape parameters of this two-parameter continuous distribution. It can be shown that $f(x)$ is symmetrical about the mean and that it decreases on either side of the mean without ever reaching zero. The distribution has a characteristic bell shape. The range of a normally distributed variable is from $-\infty$ to $+\infty$ but most engineering variables vary from 0 to some high positive value. Strictly speaking, such variables cannot be said to follow a normal distribution. However, if μ_X of a random variable is more than 3 times σ_X, the probability of the variable acquiring negative values is very small and hence the normal distribution can be applied without incurring unacceptable error.

The effect of different values of the parameters on the shape of the distribution is shown in Fig. 5-1. If only μ_X of a random variable changes but σ_X remains the same, the distribution just gets shifted, but if μ_X remains the same and σ_X changes, the spread of the distribution changes.

Example 5.1 Consider the normal distribution with parameters μ and σ. Graph the normal distribution for the following sets of parameters: (1) $\mu = 0$, $\sigma = 1$; (2) $\mu = 1$, $\sigma = 1$; (3) $\mu = 1$, $\sigma = 0.5$; (4) $\mu = 1$, $\sigma = 1.5$; and (5) $\mu = 1$ and $\sigma = 2.5$.

Solution The shape of the distribution for the desired combinations of parameters can be seen in Fig. 5-2.

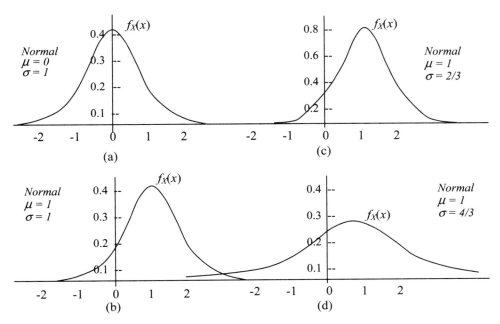

Figure 5-1 *Shape of normal distribution for various values of parameters.*

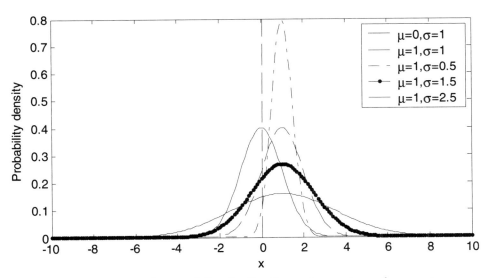

Figure 5-2 *Shape of normal distribution for various values of parameters.*

The cumulative distribution function (CDF), $F(x)$, can be expressed as

$$F_X(x) = \int_{-\infty}^{x} \frac{1}{\sigma_X \sqrt{2\pi}} \exp\left[-\frac{1}{2}(\frac{x - \mu_X}{\sigma_X})^2 \right] dx, \quad -\infty \leq X \leq \infty \tag{5.2}$$

Equation 5.2 cannot be evaluated in closed form and must be evaluated by expressing it in an infinite series and integrating term by term, or by expressing it in the standardized form; the values of the cumulative distribution are obtained from the standard normal distribution tables.

The standardized variable $U = (X - \mu_x)/\sigma$ has mean 0 and standard deviation 1 and is normally distributed, $N(0,1)$. Thus, the PDF of U is

$$f_U(u) = \frac{1}{\sqrt{2\pi}} e^{-u^2/2}, \quad -\infty \leq u \leq \infty \tag{5.3a}$$

$$f_X(x) = \frac{1}{\sigma_X} f_U\left(\frac{x - \mu_X}{\sigma_X} \right), \quad -\infty \leq x \leq \infty \tag{5.3b}$$

The mean and variance of U can be easily shown to be

$$E[U] = \frac{1}{\sigma_X}[E(X) - \mu_X] = \frac{1}{\sigma_X}[\mu_X - \mu_X] = 0 \tag{5.4}$$

and

$$\text{var}[U] = \frac{1}{\sigma_X^2}[Var(x)] = \frac{1}{\sigma_X^2}\sigma_X^2 = 1 \tag{5.5}$$

Thus, the parameters of U are now fixed and the PDF and CDF are functions of U only. For $X = 1$, $\mu_X = 1$, and $\sigma_X = 4/3$,

$$f_X(x) = \frac{1}{\sigma_X} f_U\left(\frac{1-1}{4/3} \right) = \frac{3}{4} f_U(0) = \frac{3}{4} \cdot \left(\frac{1}{\sqrt{2 \times 3.1415}} \right) = 0.2660$$

The cumulative distribution function can be expressed as

$$F_X(x) = P[X \leq x] = P\left[U \leq \frac{x - \mu_X}{\sigma_X} \right]$$

$$= F_U\left(\frac{x - \mu_X}{\sigma_X} \right) = F_U(u) \tag{5.6}$$

$$= \frac{1}{\sqrt{2\pi}} \int_{-\infty}^{u} \exp\left[-\frac{1}{2}u^2 \right] du, \quad -\infty \leq u \leq \infty$$

The quantity $u = (x - \mu_x)/\sigma_x$ can be interpreted as the number of deviations by which x differs from the mean. The values of f_U are listed in tables in statistical books for $u \geq 0$. By virtue of symmetry, $f_U(u)$ for $u \leq 0$ can be obtained directly.

A numerical approximation of $f(u)$ was given by Abramowitz and Stegun (1965) as

$$f(u) = (2.5052367 + 1.2831204u^2 + 0.2264718u^4 +$$
$$0.1306469u^6 - 0.0202490u^8 + 0.003932u^{10})^{-1} \qquad (5.7)$$

The error of this approximation is $< 2.3 \times 10^{-4}$. An approximate formula to calculate $F(u)$ is

$$F(u) = 0.5 + \{0.5 + 0.064 \times [\exp(-0.4 \times u^2)]\} \times \{1. - \exp[-(u^2/2)]\}^{0.5} \qquad (5.8)$$

However, an improved approximation for positive values of u can be obtained by computing t as

$$t = 1.0/(1.0 + u \times 0.2316419)$$

and then

$$b = 0.3989423 \times \exp(-u \times u/2.0)$$

Next, compute a variable m as

$$m = (\{(1.330274429 \times t - 1.821255978) \times t + 1.781477937$$
$$\times t - 0.356563782\} \times t + 0.31938153) \times t$$

Finally,

$$F(u) = 1.0 - m \times b$$

Example 5.2 Determine the value of $F(u)$ for $u = 2.5$ from Eq. 5.6 and Eq. 5.8 and compare the results with tabulated values.

Solution Using Eq. 5.6, we get $F(2.5) = 0.994029$ whereas Eq. 5.8 gives $F(2.5) = 0.99379$. The value obtained from standard tables is also 0.99379. While writing programs to compute $F(u)$ using either Eq. 5.6 or Eq. 5.8, it is advisable to use double precision.

The range of a normally distributed random variable is $(-\infty$ to $+\infty)$ and hence many hydrologic variables, such as rainfall, discharge, or storage in a reservoir, cannot be strictly normal. But for the random variable whose mean is quite high, the probability of acquiring a negative value is negligible and the normal distribution can still be applied to such variables.

Example 5.3 A random variable X has a mean of 3,000 and a standard deviation of 400. Compute the probability that this variable will have a value less then 4,000.

Solution We evaluate

$$F_X(x) = \frac{1}{400} \frac{1}{\sqrt{2\pi}} \int_{-\infty}^{4000} \exp\left[-\frac{1}{2}\left(\frac{x-3000}{400}\right)^2\right] dx$$

$$= F_U\left(\frac{4000-3000}{400}\right) = F_U(2.5)$$

$$= \frac{1}{\sqrt{2\pi}} \int_{-\infty}^{2.5} \exp\left[-\frac{1}{2}u^2\right] du = 0.99379$$

Note that $F_U(-u) = 1 - F_U(u)$.

Example 5.4 Compute the probability that the random variable X in the previous example will be less than 3,400.

Solution The probability is given by

$$P[X \le 3400] = F_X(3400) = F_U\left(\frac{3400-3000}{400}\right)$$

$$= F_U(1) = 0.8413$$

5.1.1 Properties of the Normal Distribution

Sometimes, $1 - F_U(u)$ is given in tables for $u \ge 0$ and sometimes $1 - 2F_U(-u)$, $u \ge 0$ is tabulated. The latter tables are useful when we want to determine the probability that a normal variate will fall within, say, r standard deviations of its mean:

$$P[\mu - r\sigma \le X \le \mu + r\sigma] = P\left[-r \le \frac{X-\mu}{\sigma} \le r\right]$$

$$= F_U(r) - F_U(-r)$$

$$= 1 - F_U(-r) - F_U(-r) = 1 - 2F_U(-r)$$

One can easily compute higher moments of $N(\mu, \sigma)$. Since $N(\mu, \sigma)$ is symmetric, this implies that all odd-ordered central moments (and the skewness coefficient) are zero. The even-ordered moments are functions of the mean and standard deviations. Thus, one obtains

$$m_n = E\left[(X - \mu_X)^n\right] = \frac{n!\sigma^n}{n2^{n/2}\left(\frac{n}{2}\right)!}, \qquad n = 2, 4,\ldots \tag{5.9a}$$

Note that

$$m_4 = \frac{4!\sigma^4}{2^2 2!} = 3\sigma^4 \tag{5.9b}$$

Hence, the coefficient of kurtosis is

$$\gamma = \frac{m_4}{(m_2)^2} = \frac{3\sigma^4}{(\sigma^2)^2} = 3 \tag{5.10}$$

If a distribution has $\gamma < 3$, it is considered flatter than the standard normal distribution. Thus, one defines the coefficient of excess as $\gamma - 3$. Negative values of this coefficient imply flatter and positive values more peaked distributions than normal.

The sum of two independent normal random variables would also be distributed as normal. Thus, let

$$Z = X + Y, \quad \begin{matrix} X \approx N(\mu_X, \sigma_X^2) \\ Y \approx N(\mu_Y, \sigma_Y^2) \end{matrix} \tag{5.11}$$

Then,

$$\mu_Z = \mu_X + \mu_Y \tag{5.12a}$$

$$\sigma_Z = \sqrt{\sigma_X^2 + \sigma_Y^2} \tag{5.12b}$$

Thus,

$$Z \approx N(\mu_X + \mu_Y, \sqrt{\sigma_X^2 + \sigma_Y^2}) \tag{5.13}$$

Example 5.5 Assuming that the data of Example 4.18 follow a normal distribution, compute the probability that the peak flow will be less than 100,000 cfs, less than 80,000 cfs, and less than 50,000 cfs in any year. What is the return period of each of these flows? Compare these probabilities and return periods with those computed in Example 4.18. Which probabilities and return periods are more realistic? What is the probability that the peak flow will occur in any year between one standard deviation on either side of the mean and between two standard deviations on either side of the mean?

Solution If the sample data follow a normal distribution with mean m_X and standard deviation s_X, then

$$F_x(x) = P[x \le x] = \frac{1}{2\pi} \int_{-\infty}^{u} \exp\left[-\frac{1}{2}u^2\right] du$$

where $u = (x - m_x)/s_x = (x - 28{,}675.833)/21{,}117.138$.

The return period can be calculated according to the following equation:

$$R_N = \frac{1}{1 - F_X(x)}$$

The results are given in Table E5-5.

The probability that the peak flow will be between ±1 and ±2 standard deviations will be

$$F_X(\mu - \sigma < x < \mu + \sigma) = F_X(-1 < U < 1) = 2 \times 0.3414 = 0.6828$$

$$F_X(\mu - 2\sigma < x < \mu + 2\sigma) = F_X(-2 < U < 2) = 2 \times 0.4772 = 0.9544$$

The CDF of the actual data and the normal distribution are plotted in Fig. 5-3.

To compare the distributions, we can define an index C by taking the sum of squares of the differences between calculated and true values of P and then dividing by $(n-1)$ as

$$C = \frac{\sum\limits_{i=1}^{n}(P_{\text{calculated}} - P_{\text{true}})^2}{n-1} \tag{5.14}$$

The values for gamma and normal distributions are $C_{\text{gamma}} = 0.001369$ and $C_{\text{normal}} = 0.005775$. Hence, the estimates of the gamma distribution appear to be more realistic.

Table E5-5

Peak flow (cfs)	u	Probability	Return period (years)
100,000	3.377	0.9996	2500
80,000	2.43	0.9920	125
50,000	1.009	0.8621	7.25

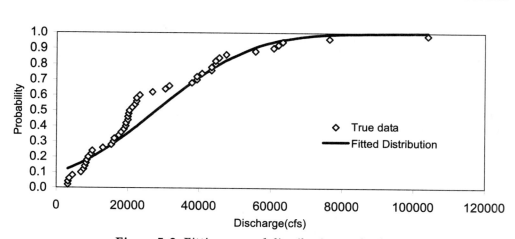

Figure 5-3 Fitting normal distribution to the data.

5.1.2 Frequency Factors for the Normal Distribution

Some problems require estimation of that particular value of a random variable that can occur with a known probability. A random variate X can be expressed as the sum of its mean value plus the departure of the variate from the mean:

$$X = m_X + \Delta X \tag{5.15}$$

The departure ΔX depends on the statistical characteristics of the distribution and can be assumed to be some multiple of its standard deviation s and a factor K:

$$X = sK \tag{5.16}$$

Equation 5.16 can now be written as

$$X/m_X = 1 + (s/m_X)K \tag{5.17}$$

or

$$X/m_X = 1 + C_v K \tag{5.18}$$

Chow (1951) proposed Eq. 5.18 as the general equation for frequency analysis and coined the term *frequency factor* for K. In addition to the statistical characteristics, the frequency factor also depends upon the recurrence interval. For a particular distribution, the relationship between K and recurrence interval (T) can be presented through tables or graphs.

For a normal distribution

$$K = (X - m_X)/s \tag{5.19}$$

the exceedance probability can be computed by

$$P(X \geq x) = 1.0 - \frac{1}{\sigma_X \sqrt{2\pi}} \int_{-\infty}^{Ks+m_X} \exp(-K^2/2)dK \tag{5.20}$$

and the recurrence interval is

$$T = 1/P(X \geq x) \tag{5.21}$$

5.1.3 Approximation of the Binomial Distribution by the Normal Distribution

The binomial distribution has two parameters: n and p. For smaller values of n and p this distribution is quite asymmetric, as shown in Fig. 5-4. For constant p, if the sample size n is increased then the distribution becomes progressively more symmetrical. It can be shown that in the limit as $n \to \infty$ the shape of the histogram depicting the PMF of X will approach the shape of the PDF of a normally distributed variable. The rapidity with which the binomial distribution approaches the normal distribution depends on the magnitude of p; the

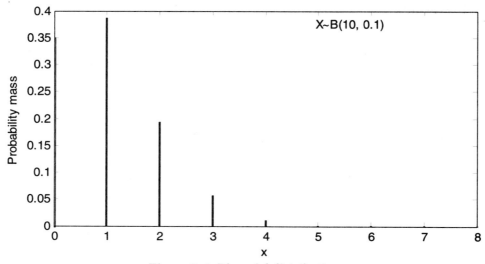

Figure 5-4 *Binomial distribution.*

approximation is quicker when p is closer to 0.5. The approximation is satisfactory when the product, np or nq, is larger than 5. This criterion can be employed to determine the usefulness of the normal distribution as a useful approximation or when the actual evaluation of the binomial probabilities would entail an excessive amount of work.

Example 5.6 Consider a binomial random variable X with $n = 20$ and $p = 0.4$. Evaluate the probability $5 < X \leq 7$ using binomial as well as normal approximation of the binomial distribution. What will be the answer when $n = 30$?

Solution If X follows a binomial distribution, $P(5 < X \leq 7)$ can be computed by

$$P(5 < X \leq 7) = \sum_{x=6}^{7} \binom{20}{x} p^x q^{n-x}$$

$$= \binom{20}{6} 0.4^6 \times 0.6^{20-6} + \binom{20}{7} 0.4^7 \times 0.6^{20-7}$$

$$= 0.124 + 0.166 = 0.39$$

As n becomes large, the standard normal variate is computed as

$$Z = \frac{(X - \mu)}{\sigma} = \frac{(X - np)}{\sqrt{np(1-p)}}$$

By approximating with the normal distribution, the requisite probability is

$$P(5.5 < X < 7.5) = P\left(\frac{5.5 - 20 \times 0.4}{\sqrt{20 \times 0.4 \times 0.6}} < Z < \frac{7.5 - 20 \times 0.4}{\sqrt{20 \times 0.4 \times 0.6}}\right)$$

$$= P(-1.14 < Z < -0.23) = 0.375 - 0.087 = 0.288$$

When $n = 30$,

$$P(5 < X \leq 7) = \binom{30}{6} 0.4^6 \times 0.6^{30-6} + \binom{30}{7} 0.4^7 \times 0.6^{30-7}$$
$$= 0.0115 + 0.0263 = 0.0388.$$

Approximating with the normal distribution, one gets

$$P(5.5 < X < 7.5) = P\left(\frac{5.5 - 30 \times 0.4}{\sqrt{30 \times 0.4 \times 0.6}} < Z < \frac{7.5 - 30 \times 0.4}{\sqrt{30 \times 0.4 \times 0.6}}\right)$$

$$= P(-2.42 < Z < -1.67) = 0.4922 - 0.4525 = 0.0397$$

Obviously, the approximation gets better as n increases.

5.1.4 Approximation of the Poisson Distribution by the Normal Distribution

The Poisson distribution can also be approximated by the normal distribution if parameter m (which happens to be not only the mean but also the variance of the distribution) is equal to or larger than 9. The normal approximation is particularly useful if one considers functions of the random variable. Any linear function of one or more normally distributed variables is itself normally distributed. Neither the binomial nor the Poisson distribution shares this general "regenerative property." For the Poisson distribution, one can state

$$\text{if } X_1 \sim P(m_1) \text{ and } X_2 \sim P(m_2) \text{ then } (X_1 + X_2) \sim P(m_1 + m_2)$$

provided that X_1 and X_2 are independent. For the binomial distribution one can say

$$\text{if } X_1 \sim B(n_1, p) \text{ and } X_2 \sim B(n_2, p) \text{ then } (X_1 + X_2) \sim B(n_1 + n_2, p)$$

provided that X_1 and X_2 are independent and that p is the same throughout. If both variables are distributed as binomial or Poisson, then the difference between them does not necessarily follow the same distribution. However, in case of a difference between random variables that follow either a binomial or a Poisson distribution, a normal approximation is often acceptable even if the variables themselves deviate quite a bit from normality.

Example 5.7 Referring to Example 4.18, one sees that the difference D may be approximately normally distributed even if X_{70} and X_{50} are not. Thus, alternative calculations can be made to do these calculations.

Solution The means and the variances of X_{70} and X_{50} have already been calculated:

$$E(X_{70}) = 3.5, E(X_{50}) = 2.5$$

therefore,

$$E(D) = 3.5 - 2.5 = 1.00$$

$$\text{var}(X_{70}) = 3.33, \text{var}(X_{50}) = 2.38$$

so

$$\text{var}(D) = 3.33 + 2.38 = 5.7$$

It is now assumed that $D \sim N(1.0, 2.39)$. To compute the probability of $D \geq 4$, the correction for continuity that makes $D \geq 4$ equivalent to $D > 3.5$ is noted. The latter value corresponds to $U > 1.05$. This probability is 14.69%, quite close to the accurate value obtained in Example 4.18.

5.2 Central Limit Theorem

Let us consider $X_1, X_2,..., X_n$ to be a sequence of n independent identically distributed random variables each having mean μ and standard deviation σ. Then the distribution of the sum, $S_n = X_1 + X_2 + ... + X_n$, tends to a normal distribution with mean $n\mu$ and standard deviation $\sigma \sqrt{n}$, if n tends to infinity.

The mean and variance of random variable S_n can be derived as

$$E[S_n] = E[X_1 + X_2 + X_3 + ... + X_n] = E[X_1] + E[X_2] + ... + E[X_n] = n\mu$$

$$\text{var}[S_n] = \text{var}[X_1 + X_2 + ... + X_n] = n\text{var}[X] = n\sigma^2$$

Thus the standard deviation of S_n is $\sigma \sqrt{n}$.

Mathematically, the central limit theorem indicates that

$$\lim_{n \to \infty} P[S_n \leq s] = \lim_{n \to \infty} P[\frac{S_n - n\mu}{\sigma \sqrt{n}} \leq \frac{y - n\mu}{\sigma \sqrt{n}}] = \frac{1}{2\pi} \int_{-\infty}^{z} e^{-z^2/2} dz \qquad \text{(5.22a)}$$

where y is a particular value of S_n and

$$z = \frac{y - n\mu}{\sigma \sqrt{n}}$$

Further, if there exists a constant A, such that $|X_n| \leq A$ for all n, then for $a < b$,

$$\lim_{n \to \infty} P\left[a \leq \frac{S_n - n\mu}{\sigma\sqrt{n}} \leq b\right] = \frac{1}{2\pi}\int_a^b e^{-z^2/2}dz \qquad (5.22b)$$

Similarly, one can obtain that the average $\overline{X} = (X_1 + X_2 + ... + X_n)/n$ tends toward a normal distribution function with mean μ and standard deviation $\sigma/\sqrt{n}$.

The central limit theorem helps approximate the sampling distribution of the sum by an appropriate normal curve regardless of the form of the parent PDF from which individual observations were derived. To explain it further, let us consider measurements of discharge of a river being made at a gauging station. The technician takes note of the computed discharge after rounding it off to the nearest integer. These data are subsequently used in hydrologic analysis and design. For example, monthly flow at the station is obtained by adding daily values. What is the rounding error in the sum of n measurements? The rounding error in a single measurement display is a random variable, called X. It is assumed that this variable is uniformly distributed. Suppose now that n rounding errors are summed. The sum is a random variable, denoted as S_n. Quite likely, the values of X will be clustered around zero since positive and negative values in the individual measurements tend to cancel each other out to some extent. Values near the extremes will have a very low probability density because outcomes near extremes would occur if all or most individual measurements have large rounding errors of the same sign. To demonstrate this trend, PDFs of S_1, S_2, S_3, and S_4 are drawn as shown in Fig. 5-5. It is evident that the PDF of S_n rapidly approaches a characteristic bell shape.

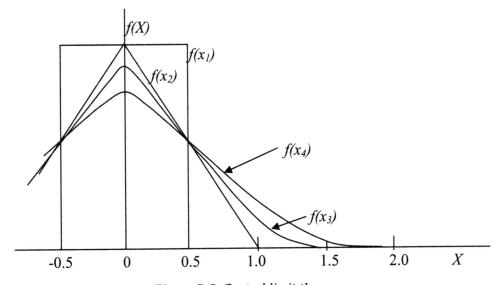

Figure 5-5 *Central limit theorem.*

It can be shown that when the number n becomes large, the PDF of S_n will approach

$$f_{S_n}(s) = k\,e^{-cs^2} \tag{5.23}$$

where k and c are constants and must be determined for each value of n such that the area under the curve is equal to one. (The correct mean of zero value is already ensured by the symmetry of the function about the $f(x)$ axis.)

The observation that the sum of a number of uniformly distributed random variables approaches the normal distribution is a special case of a more general law. The sum of triangularly distributed variables would also approach a normally distributed variable. The general law is called the *central limit theorem*. Accordingly, under very general conditions the distribution of the sum of n random variables approaches a normal distribution when n is large, regardless of the shape of the distribution of the contributing variables. The theorem applies even if the number of variables is only moderately large, as long as they are not highly dependent and as long as each contribution to the sum is relatively small; that is, there must not be one or two dominating contributing variables. The approximation improves with an increasing number of variables and is better near the center than near the tails of the distribution. If contributing variables were symmetrical then one needs fewer variables to obtain a good approximation than if they were asymmetrical. If the random variables already have close to normal distributions, then the approximation will be very rapid. In a similar vein, the sum of two or more normally distributed variables is also normally distributed.

In natural phenomena it is not uncommon to frequently observe near-normal distributions. This may be partly because variations in observed data can often be regarded as the sum of variations in additive contributing factors. For example, the total annual flows of a river result from the runoff caused by many rainstorms over its drainage basin. One would expect that the annual flow may be approximately normally distributed. This is indeed the case. Even when there is little reason to regard the random variables as the sum of many contributing random variables, the distribution may still be approximately normal. However, the agreement between the normal distribution and the probabilities encountered in empirical observations cannot be expected to be perfect but the discrepancies in probability are small.

The central limit theorem holds for most physically meaningful random variables: (1) independent and identically distributed variables, (2) independent but not identically distributed variables, and (3) not independent but weakly dependent variables. It is applicable without the knowledge of (1) the marginal distributions of the contributing random variables, (2) their number, or (3) their joint distribution. Examples 5.6 and 5.7 explain how the central limit theorem works.

Example 5.8 Choose n random numbers from a uniform PDF in the interval $[0, 1]$ and obtain the distribution of the sum S_n. Plot and compare the resulting PDFs.

Solution The PDF of the uniform distribution is

$$f_X(x) = 1 \text{ for } 0 \le X \le 1$$

with

$$\mu = E[X] = \int_0^1 x f_X(x) dx = 1/2$$

$$\sigma^2 = \text{var}[X] = \int_0^1 x^2 f_X(x) dx = 1/12$$

Based on the central limit theorem, the sampling distribution of the sum $S_n = X_1 + X_2 + \ldots + X_n$ tends to a normal distribution with mean $n\mu = n/2$ and standard deviation $\sigma\sqrt{n} = \sqrt{n}/12$. Table E5-8 compares the statistics (mean and standard deviation) of the actual S_n obtained by adding the n uniformly distributed numbers with the statistics obtained by applying the central limit theorem. It can be noticed that the statistics of S_n matches very closely.

It is noted from Fig. 5-6a that the probability density function of S_n tends to have a bell shape but is centered at $n/2$. To compare the shapes of these probability density functions for various values of n, we standardize S_n. The standardized parameter is defined as $S_n^* = (S_n - n\mu)/\sigma\sqrt{n}$. Figure 5-6b depicts the distribution of S_n^* for various values of n.

Example 5.9 Choose n random numbers from an exponential PDF in the interval $[0, +\infty]$ with the parameter $\lambda = 1$ and obtain the distribution of the sum S_n. Plot and compare the resulting PDFs.

Table E5-8 (Top) Statistics of S_n based on the chosen numbers from the uniform distribution. (Bottom) Statistics of S_n based on the central limit theorem.

N	2	4	8	16	32
Mean	1.0	2.0	4.0	8.0	16.0
St. dev.	0.41	0.58	0.81	1.16	1.63
CV	0.41	0.29	0.20	0.14	0.10
Min	0.00	0.26	1.14	2.94	10.02
Max	1.98	3.85	6.92	12.66	21.92

Mean = $n/2$	1	2	4	8	16
St. dev. = $\sqrt{n}/12$	0.41	0.58	0.82	1.15	1.63

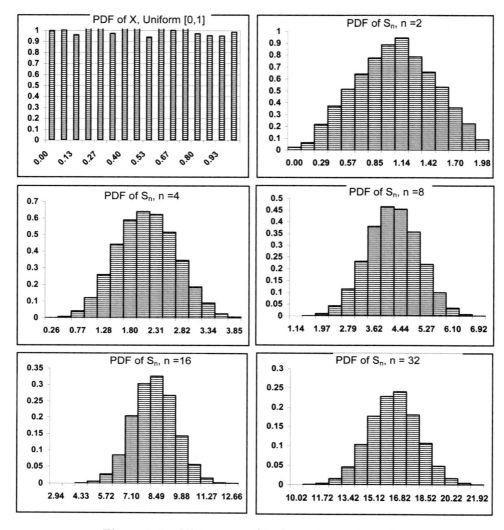

Figure 5-6a *Histograms of* S_n *for various values of* n.

Solution The PDF of the exponential distribution is

$$f_X(x) = \lambda e^{-\lambda x}, \lambda > 0$$

The mean of the exponential distribution is $\mu = 1/\lambda = 1$, and the variance is $\sigma^2 = 1/\lambda^2 = 1$. Therefore, the mean and standard deviation of the sum S_n of n random numbers from this distribution are n and $\sqrt{n}$, respectively. In Table E5-9, statistics of the actual S_n are compared with the statistics obtained by applying the central limit theorem. Note that the actual mean and standard deviations match very closely with the corresponding mean and standard deviation values obtained by applying the central limit theorem.

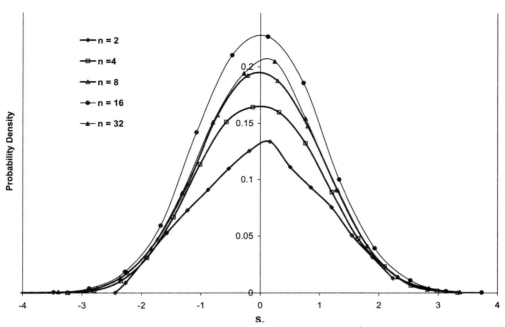

Figure 5-6b Comparison of probability density functions of S_n^* for various values of n.

Table E5-8

Statistics of S_n based on the chosen numbers from the exponential distribution					
n	2	4	8	16	32
Mean	2.02	4.01	8.01	16.02	32.04
St. dev.	1.41	1.99	2.79	4.01	5.65
CV	70.0	49.7	34.9	25.0	17.6
Min	0.0	0.3	1.4	5.5	14.4
Max	10.5	14.1	22.1	33.3	55.4

Statistics of S_n based on the central limit theorem					
Mean = n	2	4	8	16	32
St. dev. = $\sqrt{n}$	1.41	2.00	2.83	4.00	5.66

Figure 5-7a compares the histograms of S_n for various values of n. To compare the shapes of these probablility density functions for various values of n, we standardize S_n. The standardized parameter is defined as $S_n^* = (S_n - n\mu)/\sigma\sqrt{n}$. Figure 5-7b depicts the distribution of S_n^* for various values of n.

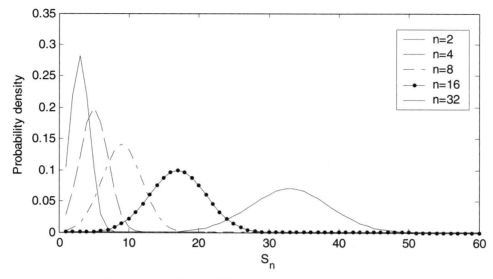

Figure 5-7a *Probability density function plots of* S_n.

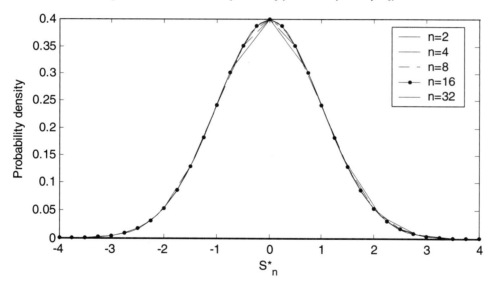

Figure 5-7b *Comparison of density functions of* S_n^* *for various values of* n.

5.3 Log-Normal Distribution: Product of Variables

Many probability distributions encountered in engineering practice are skewed. The distribution of annual flood peaks is an example. In his study of flood peaks Hazen (1914) found that the frequency curves of such peaks generally showed a marked upward curvature. This corresponds to a positive skewness, as is shown in the PDF in Fig. 5-8.

Hazen also found that the frequency curve could be straightened out in most cases if a logarithmic scale, instead of a linear scale, was used along the horizontal axis. Using a logarithmic scale means, of course, plotting the logarithms of the data instead of the data themselves. If this results in a straight line on a probability graph then the logarithms of X are normally distributed. A wide range of variables, such as daily stream flows, annual flood peaks, earthquake magnitudes, particle size in a soil sample, hydraulic conductivity of geologic formations, and strength of some materials, follow the log-normal distribution. Another reason for the popularity of this distribution is that it avoids negative values of variables.

Now consider the distribution of a phenomenon that occurs as a result of a multiplicative action (mechanism) of a number of factors. An example is the evaporation of water into the atmosphere. Evaporation depends on temperature, radiation, relative humidity, wind velocity, sunshine hours, etc. For evaporation, a product type relationship holds. In such cases, the variable of interest can be expressed as a product of a large number of variables, each of which, in itself, is difficult to study and describe. By taking the natural logarithm of the product, we obtain the sum of logarithms. By the central limit theorem, the sum will be normally distributed. Another example is the sediment particle size that results from a number of collisions of particles of many sizes traveling at different velocities. Each collision reduces the particle size by a random proportion of its size at the time. Thus, the size of a randomly chosen particle after n collisions, X_n, is the product of its size prior to the collision, X_{n-1}, and the random reduction factor, w_i. One can then write

$$X_n = X_{n-1}w_n = X_{n-2}w_{n-1}w_n = ... = X_0 \prod_{i=1}^{n} w_i$$

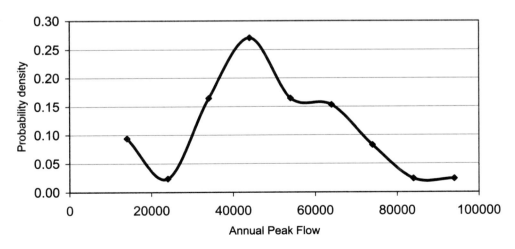

Figure 5-8 *Probability density plot of annual peak flow of a river.*

Similarly, discharge on any day k, Q_k, on the recession hydrograph can be expressed as

$$Q_k = Q_0 \prod_{i=1}^{k} K_i$$

where K_i is the recession factor on the ith day and Q_0 is the peak discharge or the discharge at the beginning of recession. In a similar manner autoregressive processes used in hydrology can be expressed in product form. A random variable X is said to be log-normally distributed if its logarithm, $Y = \ln(X)$, can be characterized by a normal distribution with parameters μ_Y and σ_Y. Thus, by using Eq. 5.1 the distribution of Y can be written as

$$f_Y(y) = \frac{1}{\sigma_Y \sqrt{2\pi}} \exp\left[-\frac{1}{2}\left(\frac{y - \mu_Y}{\sigma_Y} \right)^2 \right], \quad -\infty \le Y \le \infty \qquad (5.24)$$

where $\mu_Y = \mu_{\ln(y)}$ and $\sigma_Y = \sigma_{\ln(y)}$. Parameters μ_Y and σ_Y can be estimated by the transformed sample data by using a logarithmic transformation such that $y_i = \ln(x_i)$. The sample mean and standard deviation are used as the estimates of μ_Y and σ_Y. Thus,

$$\mu_Y \approx \bar{y} = \frac{\sum y_i}{n}$$

and

$$\sigma_Y \approx s_Y = \sqrt{\frac{(y_i - \bar{y})^2}{n - 1}}$$

One can determine the distribution of X by the technique of variable transformation explained in Chapter 2: One takes

$$f_X(x) = f_Y(y) \left| \frac{dy}{dx} \right|$$

Differentiating $Y = \ln(X)$ with respect to X, one gets

$$\frac{dY}{dX} = \frac{1}{X}$$

Substituting this relationship and Eq. 5.24 into the preceding relation gives the PDF of X as

$$f_X(x) = \frac{1}{x\sigma_Y \sqrt{2\pi}} \exp\left[-\frac{1}{2}\left(\frac{\ln x - \mu_Y}{\sigma_Y} \right)^2 \right], \quad x \ge 0 \qquad (5.25)$$

Equation 5.25 represents the log-normal distribution. Note that the range of Y is $-\infty$ to $+\infty$ whereas that of X is 0 to ∞.

It is worth mentioning that the mean of X, μ_X, should not be interpreted as a 50% probable value. Instead the median value is the 50% probable value of a log-normally distributed variable X on either side of which half of the distribution lies. Let M_X be the median value and the geometric mean of X. Thus, we can write $P(X \le M_X) = 0.5$. Further, based on the definition of the normal distribution, this relationship can be rewritten as

$$F_u\left(\frac{\ln M_X - \mu_Y}{\sigma_Y}\right) = 0.5$$

Thus,

$$\frac{\ln M_X - \mu_Y}{\sigma_Y} = F_u^{-1}(0.5) = 0$$

Therefore, one can write

$$\mu_Y = \ln(M_X) \tag{5.26}$$

Equation 5.25 can be rewritten in terms of the median value of X:

$$f_X(x) = \frac{1}{x\sigma_Y\sqrt{2\pi}} \exp\left[-\frac{1}{2}\left\{\frac{1}{\sigma_Y}\ln\left(\frac{x}{M_X}\right)\right\}^2\right], \quad x \ge 0 \tag{5.27}$$

denoted as $LN(M_X, \sigma_{\ln X})$, M_X with $\sigma_{\ln X}$ as parameters.

One can use normal tables as follows. If U is a standardized $N(0,1)$ variable, then

$$f_X(x) = \frac{1}{x\sigma_{\ln X}} f_U\left(\frac{\ln(x/M_X)}{\sigma_{\ln X}}\right)$$

In other words,

$$f_X(x) = \frac{1}{x\sigma_{\ln X}} f_U(u) \tag{5.28a}$$

where

$$u = \frac{1}{\sigma_{\ln X}}\ln\frac{x}{M_X} \tag{5.28b}$$

Tables of the function $f_U(u)$ are widely available in statistical textbooks or in mathematical handbooks.

The CDF of X is easily calculated from the tables of the normal distribution:

$$F_X(x) = P[X \le x] = P[\ln X \le \ln x]$$
$$= P[Y \le \ln x] = F_Y[\ln x]$$

where Y is $N(\mu_Y, \sigma_Y^2)$ or $N(\ln M_X, \sigma_Y^2)$. Hence,

$$F_X(x) = F_U\left(\frac{\ln x - \ln M_X}{\sigma_{\ln X}}\right) = F_U\left(\frac{\ln\left(y/M_X\right)}{\sigma_{\ln X}}\right) = F_U(u) \qquad (5.29)$$

where u is defined as before.

Example 5.10 Consider the log-normal distribution with parameters mean and standard deviation of Y. Graph the log-normal distribution for different values of the parameters. Take the values of standard deviation as 0.1, 0.25, 0.5, 0.75, 1.0, 1.5, 2.5, and 5.0.

Solution The log-normal distribution for two cases is plotted in Fig. 5-9. The mean and variance of Y can be estimated without transforming the data using the following relations:

$$\mu_Y = \frac{1}{2}\ln\left(\frac{\mu_X^2}{1+CV_X^2}\right) \qquad (5.30)$$

$$\sigma_Y^2 = \ln\left(1+CV_X^2\right) \qquad (5.31)$$

where CV_X is the coefficient of variation of X. The mean, variance, and the coefficient of variation of the log-normal distribution are

$$\mu_X = \exp\left(\mu_Y + \frac{\sigma_Y^2}{2}\right) \qquad (5.32)$$

$$\sigma_X^2 = \mu_X^2\left[\exp\left(\sigma_Y^2\right) - 1\right] \qquad (5.33)$$

$$CV_X = \left[\exp\left(\sigma_{\ln Y}^2\right) - 1\right]^{1/2} \qquad (5.34)$$

The coefficient of skewness is

$$\gamma_X = 3CV_X + CV_X^3 \qquad (5.35)$$

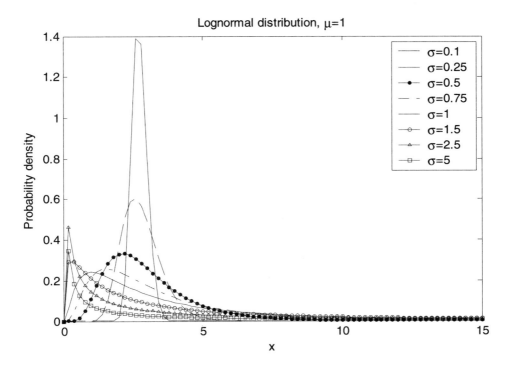

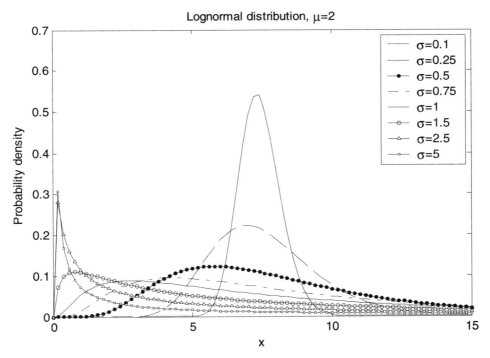

Figure 5-9 *Log-normal distribution for various combinations of parameters.*

Example 5.11 Assume that the peak flow data of Example 4.18 follow a two-parameter log-normal distribution. Compute the parameters of the log-normal distribution. Compute the probability that the peak flow will be less than 100,000 cfs, less than 80,000 cfs, and less than 50,000 cfs in any year. Compare these probabilities and return periods with those computed in Example 4.18. Which probabilities and return periods are more realistic? What is the probability that peak flow will occur in any year between one standard deviation on either side of the mean and between two standard deviations on either side of the mean?

Solution The mean and standard deviations of the data are 28,675.833 cfs and 21,117.138 cfs, respectively. Hence,

$$CV_x = s_x/m_x = 21{,}117.138/28{,}675.833 = 0.736$$

and

$$s_{ln(x)} = \sqrt{\ln(CV_x^2 + 1)} = \sqrt{\ln(0.736^2 + 1)} = 0.658$$

From Eq. 5.32, one obtains

$$m_{\ln x} = \ln m_X - \frac{s_{\ln x}^2}{2} = \ln 28675.833 - \frac{0.658^2}{2} = 10.264 - 0.216 = 10.048$$

Now,

$$P(x < 100{,}000) = P\left(u < \frac{\ln(100000) - 10.048}{0.658}\right) = P(u < 2.226) = 0.9868$$

$$P(x < 80{,}000) = P\left(u < \frac{\ln(80000) - 10.048}{0.658}\right) = P(u < 1.887) = 0.9706$$

$$P(x < 50{,}000) = P\left(u < \frac{\ln(50000) - 10.048}{0.658}\right) = P(u < 1.173) = 0.879$$

The return periods are

$$R(x < 100{,}000) = 1/(1 - 0.9868) = 75.75 \text{ years}$$

$$R(x < 80{,}000) = 1/(1 - 0.9706) = 34.01 \text{ years}$$

$$R(x < 50{,}000) = 1/(1 - 0.879) = 8.26 \text{ years}$$

The log-normal distribution is plotted in Fig. 5-10.

The performance index C in Eq. 5.14 was computed for gamma and log-normal distributions and the values were 0.001369 and 0.003222, respectively. Hence, the gamma distribution appears to better represent the data.

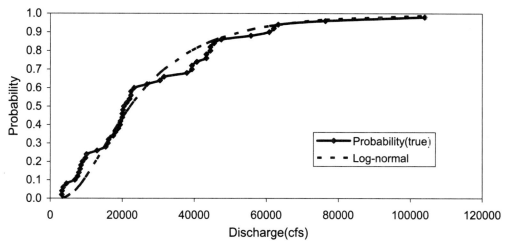

Figure 5-10 *Theoretical log-normal distribution and the curve of example data.*

5.3.1 Frequency Factor for the Log-Normal Distribution

Here X is normally distributed while Y (= exp$X$) is log-normally distributed. Chow (1951) presented the frequency factor as

$$K = \frac{\exp(s_x K_x - s_x^2/2) - 1}{\sqrt{\exp(s_x^2 - 1)}}$$ (5.36)

where $K_X = (X - m_X)/s_X$. For a given value of X, K_X and thereby K can be calculated. The coefficient of variation and the coefficient of skewness are computed as

$$C_v = [\exp(\sigma_x^2) - 1]^{0.5}$$ (5.37)

$$C_s = 3C_v + C_v^3$$ (5.38)

Using the computed value of C_s and exceedance probability allows K to be read from tables of frequency factors for the log-normal distribution.

5.4 Distribution of Extremes

Frequently in the design and management of environmental systems, our concern resides with the largest or the smallest value of a number of random variables. For example, the largest flood is critical to the design of dams, spillways, highway bridge openings, drainage systems, and levees; the maximum load is critical to the safety of a bridge, etc. Sometimes the adequacy of a system is judged by the smallest features: The capacity of a canal reach depends on the section with the smallest conveyance, the capacity of a highway depends on the narrowest section, etc. If Y is the maximum of n random variables X_1, X_2, ..., X_n, then the probability

$$F_Y(y) = P[Y \le y] = P[\text{All } n \text{ random variables } X_i \le y]$$

If random variables X_is are independent,

$$\begin{aligned} F_Y(y) &= P[X_1 \le y]P[X_2 \le y]...P[X_n \le y] \\ &= F_{X_1}(y) \cdot F_{X_2}(y) \cdot ... \cdot F_{X_n}(y) \end{aligned}$$

(5.39)

If all the X_is are identically distributed with CDF $F_X(x)$, then

$$F_Y(y) = \{F_X(y)\}^n$$

(5.40)

Assuming X_i to be continuous random variables with PDF $f_X(x)$, one obtains

$$f_Y(y) = \frac{d}{dy}[F_Y(y)]^n = n[F_X(y)]^{n-1} f_X(y)$$

(5.41)

Equations 5.40 and 5.41 can be used to determine the distribution of Y if the X_is are mutually independent and identically distributed. Three limiting forms of $f_Y(y)$ for large values of n are found depending on (1) the interest in the largest or smallest value and (2) the behavior of the appropriate table of X_i.

5.4.1 Extreme Value Type 1 Distribution: Distribution of the Largest Value

In many applications, our interest lies in knowing the limiting distribution of the largest of n values of X_i as n becomes large. Suppose that all the variables have the lower positive bound ($X_i \ge 0$) and in the upper tail the common CDF is of the exponential type:

$$F_X(x) = 1 - e^{-g(x)}$$

(5.42)

where $g(x)$ is an increasing function of x. If a linearly increasing function is chosen, $g(x) = \lambda x$, then

$$F_X(x) = 1 - e^{-\lambda x}$$

(5.43)

Equation 5.43 represents the negative exponential distribution. The normal and gamma distributions are also of this type.

Now consider Y as the largest of the independent random variables whose upper tail follows a general exponential-type distribution. Then, the distribution of Y can be derived as follows:

$$F_Y(y) = \exp\left[-e^{-\alpha(y-u)}\right], \qquad -\infty \le y \le \infty$$

(5.44)

and

$$f_Y(y) = \alpha \exp\left[-\alpha(y-u) - e^{-\alpha(y-u)}\right] \tag{5.45}$$

where α and u are parameters. This is the extreme value type I (EVI) distribution, also known as the Gumbel distribution. It is represented as $EV_{I,L}(u,\alpha)$. Parameter u is the mode of the distribution and parameter α is a measure of dispersion. The moments of the distribution are

$$m_Y = u + \frac{\gamma}{\alpha} \cong u + \frac{0.5772}{\alpha}, \quad \gamma = \text{Euler's constant} = 0.5772 \tag{5.46}$$

$$\sigma_Y{}^2 = \frac{\pi^2}{6\alpha^2} \cong \frac{1.645}{\alpha^2} \tag{5.47}$$

$$\sigma_Y = \frac{\pi}{\sqrt{6}\alpha} = \frac{1.282}{\alpha} \tag{5.48}$$

$$\gamma_1 = 1.1396 \text{ (coefficient of skewness)} \tag{5.49}$$

In analogy with the standardized variate for the normal distribution, a reduced variate is defined as

$$R = \alpha\,(Y-u) \tag{5.50}$$

In terms of the reduced variate,

$$F_R(r) = e^{-e^{-r}} \tag{5.51}$$

$$F_Y(y) = F_R\left[(y-u)\alpha\right] \tag{5.52}$$

$$f_Y(y) = \alpha f_r\left[(y-u)\alpha\right] \tag{5.53}$$

The frequency factor of the EVI distribution is given by

$$K = \frac{\sqrt{6}}{\pi}\left[0.5772 + \ln\left\{\ln\left(\frac{T}{T+1}\right)\right\}\right] \tag{5.54a}$$

Referring to Eq. 5.18, when $x = m_X$ (the average of x), one sees that $K = 0$. This condition from Eq. 5.55 gives $T = 2.33$ years. This value (2.33 years) is considered to be the recurrence interval of the mean annual flood.

Gumbel (1958) showed that asymptotically for large x the following holds:

$$T(x) \propto -\frac{f'(x)}{[f(x)]^2} \tag{5.54b}$$

This yields the tail behavior of EVI.

Example 5.12 Plot the extreme value type I distribution, with parameters u and α, for both the largest and the smallest values.

Solution The EVI distribution for the largest values:

$$f(y) = \alpha \exp\left[-\alpha(y-u) - \exp(-\alpha(y-u))\right]$$

When $u = 0$,

$$f(y) = \alpha \exp\left[-\alpha y - \exp(-\alpha y)\right]$$

The distribution for various values of α is plotted in Fig. 5-11.
    For the smallest values, the EVI distribution is

$$f(y) = \alpha \exp\left[\alpha(y-u) - \exp(-\alpha(y-u))\right]$$

When $u = 0$,

$$f(y) = \alpha \exp\left[\alpha y - \exp(-\alpha y)\right]$$

The shape is plotted in Fig. 5-12.

Example 5.13 Assume that the peak flow data of Example 4.18 follows a two-parameter extreme value type I distribution. Compute the parameters of the distribution and the probability that the peak flow in any year will be less than 100,000 cfs, less than 80,000 cfs, and less than 50,000 cfs. Compare these probabilities and return periods with those computed in Example 4.18. Which probabilities and return periods are more realistic? What is the probability that peak flow will occur in any year between one standard deviation on either side of the mean and between two standard deviations on either side of the mean?

Solution For the EVI distribution,

$$\alpha = \frac{1.282}{\sigma_y} = \frac{1.282}{21117.138} = 6.07 \times 10^{-5}$$

and

$$u = 28675.833 - \frac{0.5772}{\alpha} = 19171.474$$

Hence, the nonexceedance probabilities are

$$P(X < x) = \exp(-e^{-\alpha(x-u)}) = \exp[-e^{-0.0000607(x - 19171.5)}]$$

The probabilities and return periods are shown in Table E5-13.
    The probability that the peak flow will be within one or within two standard deviations is

$$P_x(\mu - \sigma < u < \mu + \sigma) = P_x(x < 49{,}792.97) - P_x(x < 7{,}558.695) = 0.853 - 0.132 = 0.721$$

$$P_x(\mu - 2\sigma < u < \mu + 2\sigma) = P_x(x < 70{,}910.11) - P_x(x < -13{,}558.4) = 0.958 - 0.041 = 0.917$$

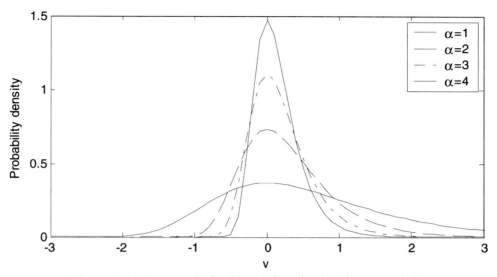

Figure 5-11 *Extreme Value Type I distribution (largest value).*

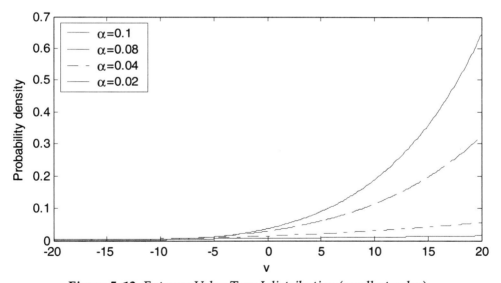

Figure 5-12 *Extreme Value Type I distribution (smallest value).*

Table E5-13

Peak flow (cfs)	Probability	Return period (years)
100,000	0.993	142.8
80,000	0.975	40
50,000	0.857	7

Figure 5-13 shows the distribution. The performance index for EVI turns out to be 0.0022. Here too, the gamma distribution (C_{gamma} = 0.001369) appears to better represent the data.

Example 5.14 For the peak annual flow in a small stream it is found that $m_Y = 200$ m³/s and $\sigma_Y = 100$ m³/s. Compute the PDF of the extreme value type I distribution for the given data.

Solution We first determine parameters α and u:

$$\alpha = \frac{1.282}{\sigma_Y} = \frac{1.282}{100} = 0.01282.$$

$$u = m_Y - \frac{0.5772}{\alpha} = 200 - \frac{0.5772}{0.01282} = 154.98 \text{ m}^3/\text{s}$$

Thus,

$$F_Y(y) = \exp\left[-e^{-0.01282(y-154.98)}\right]$$

One can compute the probability that the peak flow in a particular year will exceed a given value, say, 400 m³/s as

$$P[Y \geq 400] = 1 - F_Y(400) = 1 - e^{-e^{-0.01282(400-154.98)}} = 0.0423$$

Tables of $F_R(r)$ and $f_R(r)$ and R are available. Then, the values of Y and $f_Y(y)$ are easily computed. Note that

$$R = \alpha(Y - u)$$

Hence,

$$Y = u + \frac{R}{\alpha}$$

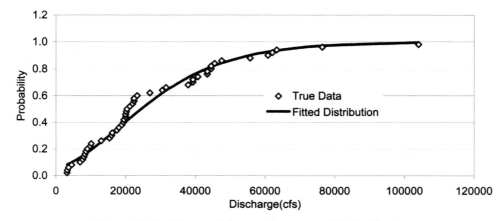

Figure 5-13 *Observed data and exponential distribution.*

5.4.2 Extreme Value Type I: Distribution of the Smallest Value

Sometimes our interest lies in knowing the distribution of the smallest of many independent variables with a common unlimited distribution having an exponential-like lower tail. In a manner similar to that for the largest of independent variables discussed in the preceding section, the distribution of the smallest of independent variables can be found. To that end, let Z be the lowest of n random variables. Then, the asymptotic distribution of Z when n is large is

$$F_Z(z) = 1 - \exp\left(-e^{-\alpha(z-u)}\right), \qquad -\infty \leq z \leq \infty \tag{5.55}$$

$$f_Z(z) = \alpha \exp\left[\alpha(z-u) - e^{-\alpha(z-u)}\right] \tag{5.56}$$

where

$$m_Z = u - \frac{\gamma}{\alpha} = u - \frac{0.5772}{\alpha} \tag{5.57}$$

$$\sigma_Z = \frac{\pi}{\sqrt{6}\alpha} = \frac{1.282}{\alpha} \tag{5.58}$$

$$\gamma_1 = -1.1396 \tag{5.59}$$

In view of the asymmetry of the distribution for the largest and the smallest values, the tables of the EVI distribution for the largest values can also be used for the smallest values. In terms of the reduced variate R,

$$F_Z(z) = P[Z \leq z] = P[R \geq -(z-u)\alpha] = 1 - F_R[-(z-u)\alpha] \tag{5.60}$$

$$f_Z(z) = \alpha f_R(-(z-u)\alpha), \qquad -\infty \leq z \leq \infty \tag{5.61}$$

Example 5.15 Consider the same mean and variance values as used in Example 5.13. For minimum annual flow in a large stream, EVI might be an appropriate model.

Solution The mean and the standard deviation are

$$m_z = 200 \text{ m}^3/\text{s}$$

$$\sigma_z = 100 \text{ m}^3/\text{s}$$

The parameters are

$$\alpha = \frac{1.282}{\sigma_Z} = \frac{1.282}{100} = 0.01282$$

$$u = m_Z + \frac{0.5772}{\alpha} = 200 + \frac{0.5772}{0.01282} = 1245.02 \text{ m}^3/\text{s}$$

The PDF of Z can be constructed from a table of PDF of R by noting that

$$R = -\alpha(Z - u)$$

Example 5.16 The distribution of the largest drought in Example 4.18 can be given by the extreme value type I distribution. Compute the probability of the largest drought being 5, 10, 15, 20, 25, 30, 35, 40, 45, and 50 days. From these probabilities, compute the return period of 10-, 20-, 30-, 40-, 50-, 60-, 70-, and 80-day droughts.

Solution For the EVI distribution,

$$\alpha = \frac{1.282}{\sigma_y} = \frac{1.282}{8.575} = 0.150$$

and

$$u = m_y - \frac{0.5772}{\alpha} = 8.06 - \frac{0.5772}{0.150} = 4.213$$

The probabilities of the largest drought being 5, 10, 15, 20, 25, 30, 35, 40, 45, and 50 days are tabulated as follows and shown in Fig. 5-14:

Y	5	10	15	20	25	30	35	40	45	50
$F_y(y)$	0.411	0.657	0.820	0.911	0.957	0.979	0.990	0.995	0.998	0.999

To compute the return period of droughts, the probability of the largest drought being 10, 20, 30, 40, 50, 60, 70, and 80 days is to be computed first according the EVI distribution. The parameters of EVI are

$$\alpha = \frac{1.282}{\sigma_y} = \frac{1.282}{8.575} = 0.150$$

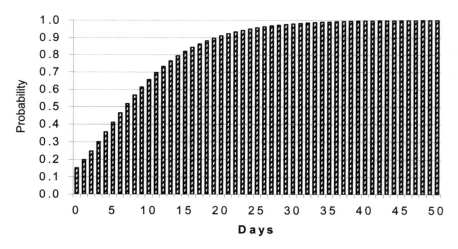

Figure 5-14 Probabilities of duration of the largest drought.

and

$$u = m_y - \frac{0.577}{\alpha} = 8.06 - \frac{0.577}{0.150} = 4.213$$

We first compute

$$F_y[y] = \exp\left[-e^{-\alpha(y-u)}\right]$$

The average return period is computed by

$$m_N = \frac{1}{1 - F_y(y)}$$

The results are listed in Table E5-16. The return periods are plotted in Fig. 5-15.

Table E5-16

y	$F_y(y)$	Return period (days)
10	0.657198	2.92
20	0.910589	11.18
30	0.979318	48.35
40	0.995348	214.94
50	0.99896	961.57
60	0.999768	4,307.73
70	0.999948	19,304.16
80	0.999988	86,513.50

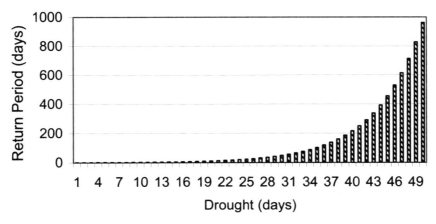

Figure 5-15 Return periods of droughts of various durations.

5.4.3 Extreme Value Type II Distribution

The extreme value type II (EVII) distribution arises as the limiting distribution of the largest value of many independent, identically distributed random variables. Each of the variables has a distribution limited on the left at zero but unlimited to the right in the tail of interest. The tail is such that the CDF of X_i has the form

$$F_X(x) = 1 - \beta \left(\frac{1}{x} \right)^k, \qquad x \geq 0 \tag{5.62}$$

Let the largest of many X_i be denoted by Y. The asymptotic distribution of Y is of the form

$$F_Y(y) = \exp[-(u/y)^k], \qquad y \geq 0 \tag{5.63}$$

$$f_Y(y) = \frac{k}{u} \left(\frac{u}{y} \right)^{k+1} \exp\left(-(u/y)^k \right), \qquad y \geq 0 \tag{5.64}$$

The distribution is designated as $EX_{II,L}(u,k)$.

For $j \geq k$ the moments of order j of Y do not exist; they do for $j < k$ and are

$$E\left[Y^j \right] = u^j \Gamma \left(1 - \frac{j}{k} \right) \tag{5.65}$$

$$m_Y = u \Gamma \left(1 - \frac{1}{k} \right), \qquad k > 1 \tag{5.66}$$

$$\sigma_Y^2 = u^2 \left[\Gamma \left(1 - \frac{2}{k} \right) - \Gamma^2 \left(1 - \frac{1}{k} \right) \right], \qquad k > 2 \tag{5.67}$$

If the coefficient of variation is represented by V_Y, then

$$V_Y^2 = \sigma_Y^2 / m_Y^2$$

or

$$V_Y^2 = \frac{\Gamma \left(1 - \frac{2}{k} \right)}{\Gamma^2 \left(1 - \frac{1}{k} \right)} - 1, \qquad k > 2 \tag{5.68}$$

The relation between this distribution and type I is the same as that between log-normal and normal distributions. If Y has a type II distribution with parameters u and k, then $Z = \ln Y$ has the type I distribution with parameters $u^0 = \ln u$ and $\alpha = k$. It follows that

$$F_Y(y) = F_Z(\ln y) \tag{5.69}$$

$$f_Y(y) = \frac{1}{y} f_Z(\ln y) \tag{5.70}$$

where Y is $EX_{II,L}(u,k)$ and Z is $EX_{I,L}(\ln u,k)$. In terms of tabulated values of the reduced variable R of EVI, one has

$$F_Y(y) = F_W\left[(\ln y - \ln u)k\right] \tag{5.71}$$

$$f_Y(y) = \frac{k}{y} f_W\left[(\ln y - \ln u)k\right] \tag{5.72}$$

The EVII distribution has also been used to model the annual maximum wind velocity.

Example 5.17 For the extreme value type II distribution, plot the coefficient of variation versus k. Graph the distribution for various values of the parameters.

Solution The graph of CV versus k for the extreme value type II distribution, based on Eq. 5.68, is shown in Fig. 5-16. The shape of the distribution for various combinations of parameters is given in Fig. 5-17.

Example 5.18 From the measured wind data at an airport location, the mean and standard deviation of the maximum annual wind velocity were $m_Y = 60$ km/hour and $\sigma_Y = 12.6$ km/hour, respectively. Find the wind velocity that will be exceeded with a probability of 0.05 in any year.

Solution To determine parameters u and k, first calculate

$$CV = \sigma_Y / m_Y = 12.6/60 = 0.21$$

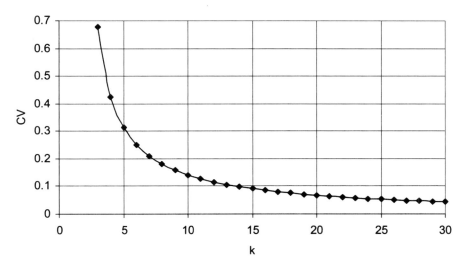

Figure 5-16 Coefficient of variation versus k for EVII distribution.

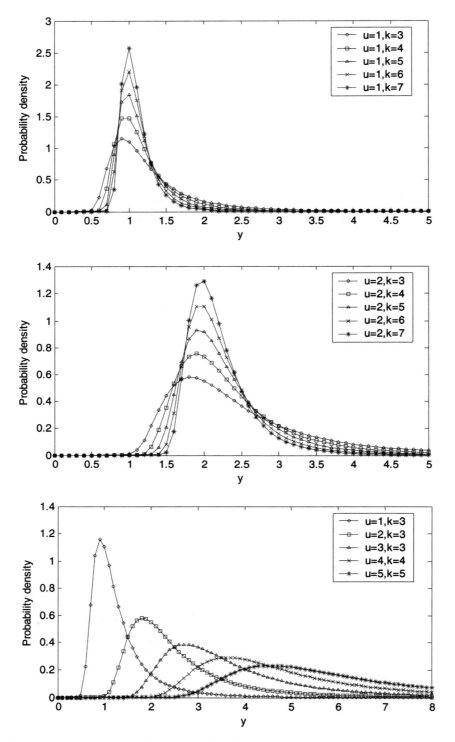

Figure 5-17 *Extreme value type II distribution for various values of parameters.*

From the graph of the CV function, for $CV = 0.21$, $k = 6.4$. Then

$$u = \frac{60}{\Gamma\left(1 - \dfrac{1}{6.4}\right)} = \frac{60}{\Gamma(0.844)} = \frac{60}{1.118} = 53.67 \text{ km/hour}$$

$$F_Y(y) = \exp\left[-\left(\frac{53.67}{y}\right)^k\right]$$

The velocity y, which will be exceeded with probability $0.05 = 1/20$ in any year, is found as

$$P[Y \geq y] = 1 - P[Y \leq y] = 1 - F_Y(y)$$

$$\exp\left[-\left(\frac{53.67}{y}\right)^{6.4}\right] = 0.95$$

Solving for y, one gets $y = 84.76$ km/hour.

Example 5.19 The data for the maximum annual wind velocity in the Baton Rouge area are given in Table E5-19. Compute the mean and standard deviation of the maximum annual wind velocity. Assume that the peak wind velocity follows an extreme value type II distribution. Compute the 20-, 30-, 50-, 80-, and 100-year wind velocity.

Solution Table E5-19 gives the maximum annual wind velocity and direction from 1980 to 2000. The coefficient of variation is

$$CV = \frac{8.83}{47.38} = 0.186$$

From the CV function, $k = 8.5$. Then,

$$u = \frac{47.38}{\Gamma\left(1 - \dfrac{1}{8.5}\right)} = 43.74$$

Thus

$$F_Y(y) = \exp\left[-\left(\frac{43.74}{y}\right)^{8.5}\right]$$

Table E5-19

Year	Month	Day	Speed (mph)	Wind direction (degrees)
1980	7	18	41	30
1981	3	22	29	210
1982	5	22	35	110
1983	1	31	35	270
1984	2	12	44	270
1985	4	5	51	330
1986	8	10	41	180
1987	12	14	48	270
1988	7	4	54	180
1989	5	16	51	220
1990	5	31	48	360
1991	4	25	46	270
1992	8	26	70	130
1993	7	3	60	110
1994	3	27	43	180
1995	12	18	52	300
1996	3	18	49	290
1997	7	10	47	280
1998	2	10	51	170
1999	5	10	52	250
2000	8	30	48	130
		Mean	47.38	216.19
		Standard deviation	8.83	83.99

For a 20-year wind, $F_Y(Yy) = 1 - 0.05 = \exp[-(43.74/y)^{8.5}]$, or $y = 62.03$ mph. Similarly, for a 30-year wind, $y = 65.96$ mph, for a 50-year wind, $y = 69.22$ mph, for an 80-year wind, $y = 73.19$ mph, and for a 100-year wind, $y = 75.15$ mph.

5.4.4 Extreme Value Type III Distribution

This distribution arises when the underlying distributions are limited in the tail of interest. For the largest values of interest X_i falls off to some maximum value, w, such that the CDF holds near w:

$$F_X(x) = 1 - c(w - x)^k, \qquad x \le w, \quad k > 0 \tag{5.73}$$

The distribution of Y, the largest of X_i, is

$$F_Y(y) = \exp\left[-\left(\frac{w-y}{w-u}\right)^k\right], \qquad y \leq w \tag{5.74}$$

$$f_Y(y) = \frac{k}{w-u}\left(\frac{w-y}{w-u}\right)^{k-1} \exp\left[-\left(\frac{w-y}{w-u}\right)^k\right], \qquad y \leq w \tag{5.75}$$

where u and k are parameters of the distribution, k is the scale parameter and w is the location parameter, and u is the lower limit of x. When $k = 2$, this results in a triangular distribution. Most useful applications of this distribution deal with the smallest values. Hence, the left-hand tail of the PDF of X_i satisfies $X \geq \varepsilon$ such that near $X = \varepsilon$, the CDF has the form

$$F_X(x) = c(x - \varepsilon)^k, \qquad x \geq \varepsilon$$

where ε is the lower limit of x. For $\varepsilon = 0$, the gamma distribution acquires this form. For independent and identically distributed X_i, the distribution of Z is

$$F_Z(z) = 1 - \exp\left[-\left(\frac{z-\varepsilon}{u-\varepsilon}\right)^k\right], \qquad z \geq \varepsilon \tag{5.76}$$

$$f_Z(z) = \frac{k}{u-\varepsilon}\left(\frac{z-\varepsilon}{u-\varepsilon}\right)^{k-1} \exp\left[-\left(\frac{z-\varepsilon}{u-\varepsilon}\right)^k\right], \qquad z \geq \varepsilon \tag{5.77}$$

The moments of the distribution are

$$m_Z = \varepsilon + (u - \varepsilon)\Gamma\left(1 + \frac{1}{k}\right) \tag{5.78}$$

$$\sigma_Z^2 = (u - \varepsilon)^2\left[\Gamma\left(1 + \frac{2}{k}\right) - \Gamma^2\left(1 + \frac{1}{k}\right)\right] \tag{5.79}$$

$$E\left[(Z - \varepsilon)^j\right] = (u - \varepsilon)^j\Gamma\left(1 + \frac{j}{k}\right) \tag{5.80}$$

This distribution gives rise to the Weibull distribution, commonly employed in studies on reliability of the lifetimes of components, rainfall analysis, and so on.
If $\varepsilon = 0$,

$$F_Z(z) = 1 - \exp\left(-(z/u)^k\right), \qquad z \geq 0 \tag{5.81}$$

with

$$m_Z = u\Gamma\left(1 + \frac{1}{k}\right) \tag{5.82}$$

$$\sigma_Z^2 = u^2\left\{\Gamma\left[1 + \frac{2}{k}\right] - \Gamma^2\left[1 + \frac{1}{k}\right]\right\} \tag{5.83}$$

If the coefficient of variation is represented by CV_Z, one has

$$CV_Z^2 = \sigma_Z^2 / m_Z^2$$

$$CV_Z^2 = \frac{\Gamma\left(1 + \frac{2}{k}\right)}{\Gamma^2\left(1 + \frac{1}{k}\right)} - 1 \tag{5.84}$$

A logarithmic transformation permits the use of EVI tables. If Z is EVIII distributed with parameters ε, u, k [denoted as $EX_{III,s}(\varepsilon, u, k)$], then $X = \ln(Z - \varepsilon)$ is EVI distributed with parameters $u_0 = \ln(u - \varepsilon)$ and $\alpha = k$; that is, x is $EX_{1,s}[\ln(u - \varepsilon), k]$. Thus,

$$\begin{aligned} F_Z(z) &= F_X\left[\ln(z - \varepsilon)\right] \\ &= 1 - F_W\left\{-k\left[\ln(z - \varepsilon) - \ln(u - \varepsilon)\right]\right\} \quad Z \geq \varepsilon \end{aligned} \tag{5.85}$$

where W is the $EV_{LL}(0,1)$ variable whose values are tabulated. Consequently,

$$\begin{aligned} f_Z(z) &= \frac{1}{z - \varepsilon} f_x\left[\ln(z - \varepsilon)\right] \\ &= \frac{k}{z - \varepsilon} f_W\left(-k\ln\frac{z - \varepsilon}{u - \varepsilon}\right), \quad z > \varepsilon \end{aligned} \tag{5.86}$$

When $\varepsilon = 0$, the extreme value type III distribution is the gamma distribution. The probability density function of the gamma distribution is

$$f(x; \alpha, \beta) = \frac{1}{\beta^\alpha \Gamma(\alpha)} x^{\alpha-1} e^{-\frac{x}{\beta}}$$

The standard gamma distribution is

$$f(x; \alpha) = \frac{x^{\alpha-1} e^{-x}}{\Gamma(\alpha)}$$

When $\alpha = 1$, we get the exponential distribution with $\lambda = 1/\beta$. The distribution is plotted in Fig. 5-18.

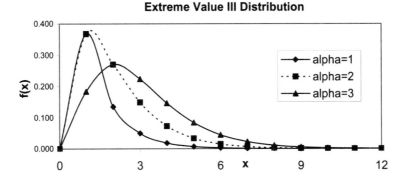

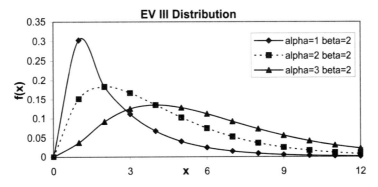

Figure 5-18 EV III distribution.

Example 5.20 Assume that the minimum annual low flows follow an extreme value type III distribution. For the Amite River at Darlington, Louisiana, compute the mean and standard deviation of the low flow data. Also, compute the distribution parameters.

Solution From the data of annual minimum low flow one obtains

$$\text{mean} = 106.3235 \text{ cfs}$$

$$\text{standard deviation} = 28.03755 \text{ cfs}$$

$$CV = \frac{28.03755}{106.3235} = 0.264$$

From the *CV* function or Eq. 5.84, $k = 4.278$. Now, Eq. 5.82 gives

$$u = \frac{106.3235}{\Gamma\left(1 + \dfrac{1}{4.278}\right)} = 116.852$$

The distribution of this annual low flow data is plotted in Fig. 5-19.

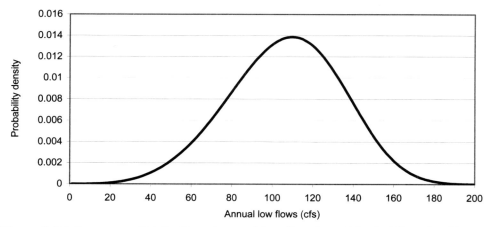

Figure 5-19 *Frequency distribution of annual low flow data of Amite River, Darlington.*

5.4.5 Generalized Extreme Value Distribution

The generalized extreme value (GEV) distribution was introduced by Jenkinson (1955, 1969) and recommended by the Natural Environment Research Council (1975) of Great Britain. The GEV distribution is a three-parameter distribution and is the most widely used distribution in the United Kingdom for analyzing frequencies of flood peaks and has also become popular elsewhere. For a random variable X, the PDF of the GEV is expressed as

$$f_X(x) = \frac{1}{a}\left(1 - \frac{b}{a}(x - c)\right)^{\frac{1-b}{a}} \exp\left[-\left(1 - \frac{b}{a}(x - c)\right)^{1/b}\right] \qquad (5.87)$$

where $a > 0$ and c are, respectively, the scale and location parameters, and b is a shape parameter. The range of X depends on the value of b; it is bounded by $c + (a/b)$ from above for $b > 0$ [i.e., $-\infty < x < c + (a/b)$] and is bounded from below for $b < 0$ [i.e., $c+(a/b) < x < \infty$]. Depending on the value of b, different extreme value distributions are represented by Eq. 5.87. For example, the GEV distribution corresponds to the Gumbel distribution (or extreme value type I) for $b = 0$, the extreme value type II distribution for $b < 0$, and the extreme value type III distribution for $b > 0$. Equation 5.87 gives rise to a reverse Raleigh distribution for $b = 2$ and to a reverse exponential distribution for $b = 1$. It can also be shown that the Weibull distribution is a reverse GEV distribution.

The CDF of the GEV distribution can be expressed as

$$F_X(x) = \exp\left[-\left(1 - \frac{b}{a}(x - c)\right)^{\frac{1}{b}}\right] \qquad (5.88)$$

Sometimes Eq. 5.87 is also expressed as

$$f_X(x) = \frac{1}{a(1-z)} \exp[-y - \exp(-y)] \tag{5.89}$$

where

$$y = -\frac{1}{b} \ln(1-z) \tag{5.90}$$

$$z = \frac{b}{a}(x-c) \tag{5.91}$$

The three parameters, a, b, and c, of the GEV distribution can be estimated by using the method of moments. For $b < 0$ (extreme value type II distribution) the first three moments, by using the transformation

$$y = \left[1 - \frac{b}{a}(x-c) \right]^{\frac{1}{b}} \tag{5.92}$$

are found to be

$$M_1^0 = c + \frac{1}{b}[1 - \Gamma(1+b)] \tag{5.93}$$

$$M_2 = \frac{a^2}{b^2}[\Gamma(1+2b) - \Gamma^2(1+b)] \tag{5.94}$$

$$M_3 = \frac{a^3}{b^3}[-\Gamma(1+3b) + 3\Gamma(1+b)\Gamma(1+2b) - 2\Gamma^3(1+b)] \tag{5.95}$$

where M_1^0, M_2, and M_3 are, respectively, the first moment about the origin and the second and third moments about the centroid. The value of b is computed numerically from the relationship to the coefficient of skewness C_s, defined as

$$C_s = \frac{M_3}{M_2^{3/2}} \tag{5.96}$$

Example 5.21 Assume that annual peak flows follow the GEV distribution for the Amite River at Darlington, Louisiana. Compute the mean and standard deviation of the peak flow data. Also, compute the distribution parameters.

Solution For the peak flow data at Darlington, Louisiana, one has mean = 20,371 cfs, standard deviation = 20,643 cfs, $C_v = 0.7542$, $C_s = 1.37$.

Thus parameter b can be obtained by solving the following equation (Rao and Hamed 2000):

$$b = 0.2858221 - 0.357983C_s + 0.116659C_s^2 - 0.022725C_s^3 + 0.002604C_s^4$$
$$- 0.000161C_s^5 + 0.000004C_s^6$$
$$= -0.04$$

Now we have that

$$\Gamma(1 + \hat{b}) = \Gamma(0.96) = 1.022$$
$$\Gamma(1 + 2\hat{b}) = \Gamma(0.92) = 1.047$$

Then from Eq. 5.94 and Eq. 5.95 we have $a = 16{,}462$, $c = 20{,}370$. The distribution of this annual peak flow data is plotted in Fig. 5-20.

5.5 Other Distributions

In this section, we discuss some other distributions that are useful in water and environmental engineering and risk analysis.

5.5.1 Uniform Distribution

If a random variable X is equally likely to assume any value in the interval 0 and 1, its probability density function is constant over the interval:

$$f_X(x) = 1, 0 \le x \le 1 \tag{5.97}$$

Equation 5.98 defines the *uniform distribution*, which is also known as the rectangular distribution. Its cumulative distribution function is triangular in shape:

$$F_X(x) = \begin{cases} 0 & x < 0 \\ x & 0 \le x \le 1 \\ 1 & x > 1 \end{cases} \tag{5.98}$$

The mean and variance of the rectangular distribution are

$$\mu_X = \frac{1}{2} \tag{5.99}$$

$$\sigma_X^2 = \frac{1}{12} \tag{5.100}$$

This rectangular distribution can be generalized to any arbitrary range a to b. Then, the PDF becomes

$$f_X(x) = \begin{cases} \dfrac{1}{b - a} & a \le x \le b \\ \\ 0 & \text{elsewhere} \end{cases} \tag{5.101}$$

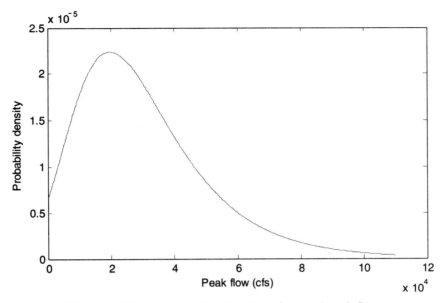

Figure 5-20 *The GEV distribution of annual peak flow.*

The mean and variance become

$$\mu_X = \frac{a+b}{2} \tag{5.102}$$

$$\sigma_X^2 = \frac{(b-a)^2}{12} \tag{5.103}$$

The uniform distribution is shown in Fig. 5-21.

5.5.2 Beta Distribution

For a random variable X, $0 \le X \le 1$, the beta distribution takes the form

$$f_X(x) = \frac{1}{B} x^{r-1}(1-x)^{t-r-1}, \qquad \begin{matrix} r > 0 \\ t-r > 0 \end{matrix} \tag{5.104}$$

where B is the normalizing constant defined as

$$B = \frac{(r-1)!(t-r-1)!}{(t-1)!} \tag{5.105}$$

if r and $t-r$ are integer valued, or

$$B = \frac{\Gamma(r)\Gamma(t-r)}{\Gamma(t)}, \quad B(m,n) = \frac{\Gamma(m)\Gamma(n)}{\Gamma(m+n)} = \int_0^1 x^{m-1}(1-x)^{n-1}dx \tag{5.106}$$

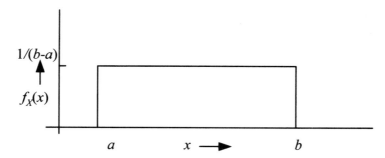

Figure 5-21 *Generalized uniform or rectangular distribution.*

if r and $t - r$ can take on noninteger values. The mean and variance of the beta distribution are

$$m_X = \frac{r}{t} \tag{5.107}$$

$$\sigma_X^2 = \frac{m_X(1 - m_X)}{t + 1} = \frac{r(t - r)}{t^2(t + 1)} \tag{5.108}$$

The skewness coefficient γ is

$$\gamma = \frac{1}{\sigma_x^3} \frac{r}{t} \left[\frac{(r + 2)(r + 1)}{(t + 2)(t + 1)} - \frac{3r(r + 1)}{t(t + 1)} + \frac{2r^2}{t^2} \right] \tag{5.109}$$

The beta distribution can assume a wide variety of shapes for different values of its parameters r and t. It reduces to the uniform distribution for $r = 1$, $t = 2$ and to the triangular distribution for $t = 3$, $r = 1$ or 2. It is symmetrical about $x = 0.5$ if $r = t/2$. It is skewed to the right if $r < t/2$ and to the left if $r > t/2$. It is U–shaped if $r < 1$ and $t \le 2r$. It is J-shaped if $r < 1$ and $t > r + 1$ or if $r > 1$ and $t < r + 1$. It is unimodal and bell shaped (generally skewed) if $r > 1$ and $t > r + 1$ with the mode at $x = (r - 1)/(t - 2)$. The distribution is plotted in Fig. 5-22 for several combinations of parameters.

For integer values of t and r, the tables of the binomial distribution can be used to evaluate $f_X(x)$ and $F_X(x)$. The binomial distribution for Y as a function of n and p can be written as

$$P_Y(y) = \frac{n!}{y!(n - y)!} p^y (1 - p)^{n - y} \tag{5.110}$$

Recall that the beta distribution is

$$f_X(x) = \frac{(t - 1)!}{(r - 1)!(t - r - 1)!} x^{r-1} (1 - x)^{t - x - 1} \tag{5.111}$$

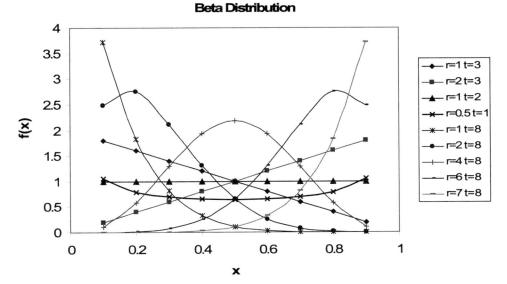

Figure 5-22 Shape of beta distribution for selected combination of parameters.

Comparing the two distributions, we get

$$f_X(x) = (t-1)p_Y(y) \tag{5.112}$$

if $P_Y(y)$ is evaluated at $n = t - 2$ and $y = r - 1$ for various values of $p = x$.
For $a \le x \le b$, the beta distribution can be generalized to

$$f_Y(y) = \frac{1}{B(b-a)^{t-1}}(y-a)^{r-1}(b-y)^{t-r-1}, \qquad a \le y \le b \tag{5.113}$$

when

$$m_Y = a + \frac{r}{t}(b-a) \tag{5.114}$$

$$\sigma_Y^2 = (b-a)^2 \frac{r(t-r)}{t^2(t+1)} \tag{5.115}$$

There is a simple relation among Y, $BT(a, b, r, t)$, and X, which is a $BT(0, 1, r, t)$ variable. In particular,

$$Y = a + (b-a)X \tag{5.116}$$

Thus,

$$F_Y(y) = F_X\left(\frac{y-a}{b-a}\right) \tag{5.117}$$

$$f_Y(y) = \frac{1}{b-a}f_X\left(\frac{y-a}{b-a}\right) \tag{5.118}$$

Example 5.22 To investigate dispersion of pollutants, the prevailing wind direction and the random variations in the direction are critical. The wind direction Y at a major oil refinery near Baton Rouge, Louisiana, measured from north has mean and standard deviation of about $m_Y = 200°$ and $\sigma_Y = 100°$, respectively. Understandably, the range of wind direction is limited between $0°$ and $360°$. Thus, assuming that a beta distribution is appropriate,

$$a = 0, \quad b = 360°$$

$$\frac{r}{t} 360 = m_Y = 200°$$

$$(360)^2 \frac{r(t-r)}{t^2(t+1)} = \sigma_y{}^2 = 10000$$

Solution Solving for r and t, one gets

$$r = 1.22, t = 2.2$$

Thus, Y is $BT(0, 360, 1.25, 2.2)$, and its PDF is

$$f_Y(y) = \frac{1}{360} f_X\left(\frac{y}{360}\right)$$

where X is a beta-distributed variable, $BT(1.22, 2.2)$

Example 5.23 The data on wind direction at the Baton Rouge airport measured in degrees from north are given in Example 5.19. Compute the mean and standard deviation of the wind direction in degrees. Assume the beta distribution is appropriate for modeling the wind direction and compute the distribution parameters. Compute the probability of the wind direction (measured from north) exceeding $45°$, $60°$, $90°$, $120°$, $145°$, $180°$, $200°$, $245°$, $290°$, $320°$, and $345°$. Compute the recurrence intervals of these wind directions.

Solution From the data, $m_y = 216.19$ and $\sigma_y = 83.99$. The direction is limited between $0°$ and $360°$. Thus, for a beta distribution,

$$a = 0, b = 360°$$

$$\frac{r}{t} \times 360 = m_y = 216.19$$

$$(360)^2 \times \frac{r(t-r)}{t^2(t+1)} = \sigma_y^2 = 83.99^2$$

Solving for r and t, one gets

$$r = 2.05, t = 3.41$$

To compute the probability of the wind direction exceeding the given degrees, one calculates

$$B = \frac{\Gamma(r)\Gamma(t-r)}{\Gamma(t)} = \frac{\Gamma(2.05)\Gamma(3.41-2.05)}{\Gamma(3.41)} = 0.3$$

$$f_Y(y) = \frac{1}{0.3 \times (360-0)^{3.41-1}} (y-0)^{2.05-1} (360-y)^{3.41-2.05-1}$$
$$= 2.302 \times 10^{-6} y^{1.05} (360-y)^{0.36}$$

$$F_Y(y) = \int_0^y f_Y(y) dy$$

For the probability of exceedance of the given degree,

$$F_Y'(y) = 1 - F_Y(y) = \int_y^{360} f_Y(y) dy$$

The results are shown in Table E5-23.

Table E5-23

y (degrees)	$F_Y'(y)$	Recurrence (years)
45	0.984	1.02
60	0.967	1.03
90	0.918	1.09
120	0.850	1.18
145	0.782	1.28
180	0.669	1.49
200	0.597	1.68
245	0.419	2.39
290	0.233	4.29
320	0.115	8.70
345	0.032	31.25

5.5.3 Three-Parameter Pearson Distribution

The probability density function of a three-parameter Pearson distribution for variable X is given by

$$f_X(x) = \frac{1}{\alpha\Gamma(\beta)}\left(\frac{x-\gamma}{\alpha}\right)^{\beta-1}\exp\left(-\frac{x-\gamma}{\alpha}\right) \qquad \gamma < x < \infty \qquad (5.119)$$

where α (scale), β (shape), and γ (location) are parameters. Parameter α can be positive or negative; for negative values of α, the distribution has an upper bound and thus it is not useful for analysis of maximum extreme events, such as flood flows. The rth moment of Eq. 5.120 about the point γ can be written as

$$M_r^c = \int_c^\infty \frac{(x-\gamma)^r}{\alpha\Gamma(\beta)}\left(\frac{x-\gamma}{\alpha}\right)^{\beta-1}\exp\left(-\frac{x-\gamma}{\alpha}\right)dx \qquad (5.120)$$

The parameter of the distribution can be obtained by using the method of moments as follows:

$$\hat{\beta} = (2/C_S)^2 \qquad (5.121)$$

$$\hat{\alpha} = \sqrt{(m_2/\hat{\beta})} \qquad (5.122)$$

$$\hat{\gamma} = m_1' - \sqrt{m_2\hat{\beta}} \qquad (5.123)$$

If $\log x$ follows a Pearson type III distribution then x follows a log-Pearson type 3 distribution whose density function is given by

$$f(x) = \frac{1}{\alpha x\Gamma(\beta)}\left(\frac{\log(x)-\gamma}{\alpha}\right)^{\beta-1}\exp\left(-\frac{\log(x)-\gamma}{\alpha}\right) \qquad \gamma < x < \infty \qquad (5.124)$$

If $\gamma = 0$, the resulting equation represents a two-parameter gamma distribution. Furthermore, if a new variable Y, defined as

$$Y = (x - \gamma)/\alpha$$

is substituted in Eq. 5.109, a one-parameter gamma distribution is obtained:

$$f_Y(y) = \frac{1}{\Gamma(b)}y^{b-1}e^{-y} \qquad (5.125)$$

$$F_Y(y) = \frac{1}{\Gamma(b)}\int_0^y y^{b-1}e^{-y}dy \qquad (5.126)$$

This distribution was described in Chapter 4.

Example 5.24 The data of annual maximum stage and discharge for the Amite River at Darlington have been given in Example 3.17. For the discharge data, $n = 48$, mean $m_X = 28{,}676$ cusecs, coefficient of variation $c_v = 0.736$, and coefficient of skewness $c_s = 1.34$. For these data, compute the parameters of the gamma distribution.

Solution We first compute β using Eq. 5.121:

$$\beta = (2/1.34)^2 = 2.228$$

Now, from Eq. 5.122, we get

$$\alpha = \sqrt{(0.736 \times 28676)^2 / 2.228} = 14139.65$$

Finally, from Eq. 5.123, we have

$$\gamma = 28676 - \sqrt{(0.736 \times 28676)^2 \times 2.228} = -2827.15$$

5.5.4 Log Pearson Type 3 Distribution

If the variable $\ln X$ follows a Pearson type 3 distribution then the variable X will have a log-Pearson type 3 (LP3) distribution. The PDF of LP3 is given by

$$f(x) = \frac{1}{\alpha x \Gamma(\beta)} \left(\frac{\ln x - \gamma}{\alpha} \right)^{\beta-1} e^{-\left(\frac{\ln x - \gamma}{\alpha} \right)} \tag{5.127}$$

where $\alpha > 0$, $\beta > 0$, and $0 < \gamma < \ln X$ are parameters. The PDF of the LP3 distribution may be J-shaped, reverse J-shaped, U-shaped, inverted U-shaped, inverted U with inflexions, bell shaped with an upper bound, bell shaped with a lower bound, etc. The behavior of this distribution depends upon the skewness coefficient to a large extent. Let y be defined as $y = (\ln x - \gamma)/\alpha$. Then, the distribution function of LP3 is given by

$$F(y) = \frac{1}{\Gamma(\beta)} \int_0^y y^{\beta-1} e^{-y} dy \tag{5.128}$$

which is the same as Eq. 5.126.

The LP3 distribution has been extensively investigated after it was recommended by the U.S. Water Resources Council for flood frequency analysis in the United States.

5.5.5 Pareto Distribution

A new distribution for flood frequency analysis, namely the Wakeby distribution was proposed by Houghton (1978). If a random variable X follows the Wakeby distribution then

$$x = \varepsilon + (\alpha/k)[1 - (1 - F)^k] - (\gamma/\delta)[1 - (1 - F)^{-\delta}] \tag{5.129}$$

where $F = F(x) = P(X \le x)$. For this distribution, explicit expressions of the probability density function or cumulative distribution function are not available. This distribution is not commonly used because its application requires estimation of five parameters.

The Pareto distribution is a special case of the Wakeby distribution. If in Eq. 5.130, $\gamma = 0$, we get

$$x = \varepsilon + (\alpha / k)[1 - (1 - F)^k] \tag{5.130}$$

From Eq. 5.130, explicit expression for $F(x)$ can be written as

$$F(x) = \begin{cases} 1 - \left(1 - a\dfrac{x-c}{b}\right)^{\frac{1}{a}} , & a \ne 0 \\[2ex] 1 - \exp\left(-\dfrac{x-c}{b}\right), & a = 0 \end{cases} \tag{5.131}$$

and the probability density function is

$$f(x) = \begin{cases} \dfrac{1}{b}\left(1 - a\dfrac{x-c}{b}\right)^{\frac{1}{a}-1} , & a \ne 0 \\[2ex] \dfrac{1}{b}\exp\left(\dfrac{x-c}{b}\right), & a = 0 \end{cases} \tag{5.132}$$

The range of X depends upon a. When $a \le 0$, $c \le X \le \infty$, and for $a > 0$, $c \le X \le c + b/a$. When $a = 0$ in Eq. 5.123, this yields the exponential distribution. When $a = 1$ in Eq. 5.132, the uniform distribution with the range $[c, c + \alpha]$ is obtained. Figure 5-23 graphs the PDF for different values of parameter a.

The parameters of the distribution are related to the moments as follows:

$$\bar{x} = c + \frac{b}{1+a} \tag{5.133}$$

$$\mu_2 = \frac{b^2}{(1+a)^2(1+2a)} \tag{5.134}$$

$$C_s = \frac{2(1-a)\sqrt{(1+2a)}}{(1+3a)} \tag{5.135}$$

Example 5.25 For the Amite River data used in Example 5.24, important parameters are $n = 48$, mean $m_x = 28{,}676$ cusecs, coefficient of variation $c_v = 0.736$, and coefficient of skewness $c_s = 1.34$. For these data, find the parameters of the Pareto distribution.

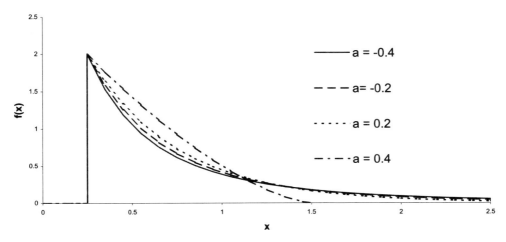

Figure 5-23 *Probability density function of the three-parameter generalized Pareto distribution for different values of the shape parameter, a.*

Solution From Eq. 5.135, we have

$$1.34 = \frac{2(1-a)\sqrt{(1+2a)}}{(1+3a)}$$

Solving this gives $a = 0.205$. From Eq. 5.134, we get

$$b = \sqrt{\mu_2 \times (1+a)^2(1+2a)}$$
$$= \sqrt{(0.736 \times 28676)^2 \times (1+0.205)^2 \times (1+2 \times 0.205)}$$
$$= 30199$$

Finally, Eq. 5.133 yields

$$c = 28{,}676 - 30{,}199/1.205 = 3{,}614.59$$

5.5.6 Logistic Distribution

The probability density function of the logistic distribution is given by

$$f(x) = \frac{1}{\alpha} e^{-\left(\frac{x-m}{\alpha}\right)} \left[1 + e^{-\left(\frac{x-m}{\alpha}\right)} \right]^{-2} \qquad (5.136)$$

The cumulative distribution function is given by

$$F(x) = \left[1 + e^{-\left(\frac{x-m}{\alpha}\right)} \right]^{-1} \qquad (5.137)$$

By using the method of moments, parameters of the logistic distribution can be estimated as

$$\alpha = \frac{\sqrt{3}}{\pi}\sigma \qquad (5.138)$$

and

$$m = \bar{x} \qquad (5.139)$$

The frequency factor for the distribution is given by

$$K_T = \frac{\sqrt{3}}{\pi}\log(T-1)$$

and the T-year return period flood is

$$x_T = m + \alpha \log(T-1)$$

Example 5.26 For the Amite River data given in Example 5.24, find the parameters of the logistic distribution and a 100-year return period quantile.

Solution The parameters of the distribution are estimated as

$$\alpha = \frac{\sqrt{3}}{\pi}(0.736 \times 28676) = 11636$$

$$m = 28,676$$

The 100-year flood will be

$$X100 = 28,676 + 11,636 \log(100-1) = 51,897 \text{ cusec}$$

5.5.7 Triangular Distribution

In many hydrologic and environmental engineering applications, a triangular distribution is used for the sake of convenience. This distribution is fully defined by three points: minimum, most likely (mode), and maximum values. The probability density function $f_X(x)$ for a triangular distribution is given by

$$f_X(x) = \begin{cases} \dfrac{2(x-a)}{(b-a)(c-a)} & \text{for } a \le x \le c \\ \dfrac{2(b-x)}{(b-a)(b-c)} & \text{for } c \le x \le b \end{cases} \qquad (5.140)$$

where a, b, and c are the minimum, maximum, and mode values of X as defined in Fig. 5-24.

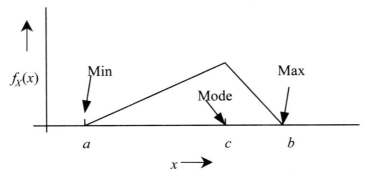

Figure 5-24 *Probability density function for a triangular distribution.*

The cumulative density function $F_X(x)$ for this triangular distribution is given by

$$F_X(x) = \begin{cases} \dfrac{(x-a)^2}{(b-a)(c-a)} & \text{for } a \le x \le c \\[2mm] 1 - \dfrac{(b-x)^2}{(b-a)(b-c)} & \text{for } c \le x \le b \end{cases} \qquad (5.141)$$

Equations 5.140 and 5.141 are convenient to use when parameters a, b, and c for the population of X are known. Generally, it is very unlikely to have the population data of a random variable. We most often only have the sample data and their characteristics, such as the mean (μ_x), coefficient of variance (CV_x), and skew (γ_x). Thus, it is necessary to determine parameters a, b, and c, given the values of μ_x, CV_x, and γ_x. The sample characteristics μ_x, CV_x, and γ_x can be represented in terms of a, b, and c as

$$\mu_X = \frac{1}{3}(a+b+c) \qquad (5.142)$$

$$\text{var}(X) = \sigma^2 = \frac{1}{18}\left[a^2 + b^2 + c^2 - (ab + bc + ca)\right] \qquad (5.143)$$

$$CV_X = \frac{1}{\sqrt{2}}\frac{\sqrt{a^2 + b^2 + c^2 - (ab + bc + ca)}}{a+b+c} \qquad (5.144)$$

$$\gamma_X = \frac{\sqrt{2}}{5}\frac{2\left(a^3 + b^3 + c^3\right) - 3\left[ab(a+b) + bc(b+c) + ca(c+a)\right] + 12abc}{\left[a^2 + b^2 + c^2 - (ab + bc + ca)\right]^{3/2}} \qquad (5.145)$$

If the values of μ_x, CV_x, and γ_x are known then a unique triangle can be delineated by determining its parameters a, b, and c. These parameters can be obtained by using the following equation (Tyagi, 2000):

$$\bar{a} = \mu_X \left\{ 1 + 2\sqrt{2} CV_X \cos\left[\frac{2\pi n}{3} + \frac{1}{3} \cos^{-1}\left(\frac{5}{2\sqrt{2}} \gamma_x \right) \right] \right\} \tag{5.146}$$

where $\bar{a}$ = a vector containing b, a, and c that can be obtained by substituting $n = 0$, 1, and 2, respectively, in Eq. 5.137. Equation 5.146 shows that the maximum value of coefficient of skew for a triangular distribution is $2\sqrt{2}/5$. To determine higher order moments of X defined by a generalized triangular distribution, first determine the moments about the origin, $E[X^r]$, and then use the relationship between central and noncentral moments as given in Eq. 3.5 of Chapter 3 to determine the central moments. The rth moment of X about the origin is given as (Tyagi, 2000)

$$\mu'_r = E\left[X^r\right] = \frac{2\left[(b-c)a^{r+2} + (c-a)b^{r+2} + (a-b)c^{r+2}\right]}{(r+1)(r+2)(b-c)(c-a)(b-a)} \tag{5.147}$$

For a symmetrical triangle, $\gamma_X = 0$ and the parameters a, b, and c can be obtained corresponding to $n = 1$, 0, and 2. The obtained c is the μ_X and the parameters a and b are the same as those obtained using the method of moments. The estimates of a and b are given as

$$\hat{a} = \mu_X \left(1 - \sqrt{6} CV_X\right) \tag{5.148}$$

$$\hat{b} = \mu_X \left(1 + \sqrt{6} CV_X\right) \tag{5.149}$$

Using Eqs. 5.147, 5.148, and 5.149 one can rewrite the expression for $E[X^r]$ as

$$E\left[X^r\right] = \frac{\mu_X^r}{6(r+1)(r+2)CV_X^2} \left[\left(1 + CV_X \sqrt{6}\right)^{r+2} + \left(1 - CV_X \sqrt{6}\right)^{r+2} - 2 \right] \tag{5.150}$$

Example 5.27 Determine the parameter of a random variable X if it is defined by a triangular distribution with mean, CV, and skew of 10, 0.5, and 0.4, respectively. Sketch the obtained distribution.

Solution Substituting $\mu_x = 10$, $CV_x = 0.5$, and $\gamma_x = 0.4$ in Eq. 5.147, for various values of n, gives the following parameters: When $n = 0$, $b = 23.66$. When $n = 1$, $a = 0$. When $n = 2$, $c = 6.34$, and further using Eq. 5.140 one gets $f_X(c) = 0.031$. The obtained distribution is depicted in Fig. 5-25.

Example 5.28 For the previous example determine the various noncentral and central moments.

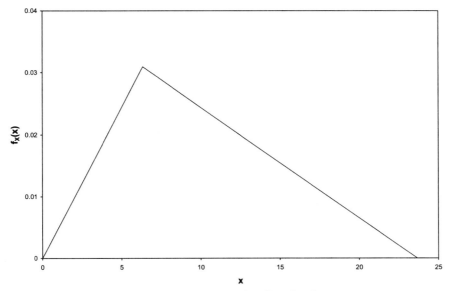

Figure 5-25 Triangular distribution.

Solution Substituting values of *a*, *b*, and *c* in Eq. 5.147 gives the first four moments about the origin. Then, the relationship between the central and non-central moments is used to determine the moments about the origin. The obtained results are as follows:

Item		$r = 1$	$r = 2$	$r = 3$	$r = 4$
Moments about the origin $\mu'_r = E\left[X^r\right]$		10	125	1,800	28,500
Moments about the mean $\mu_r = E\left[(X - \mu)^r\right]$		0	25	50	1,500

5.5.8 Halphen Distributions

Perreault et al. (1999a,b) have provided a comprehensive discussion of the Halphen distribution system, including mathematical and statistical properties, parameter and quantile estimation, and application to flood data. The discussion here is excerpted from their work. The Halphen system comprises three distributions: Halphen type A, Halphen type B, and Halphen type IB. The PDF of the Halphen type A distribution for a random variable $X > 0$ can be expressed as

$$f_A(x) = \frac{1}{2m^v K_v(2\alpha)} x^v \exp[-\alpha(\frac{x}{m} + \frac{m}{x})] \tag{5.151}$$

where $m > 0$ is a scale parameter, $\alpha > 0$ and υ are shape parameters, and K_u is the modified Bessel function of the second kind of order υ. Equation 5.151 contains three parameters and is a reparameterized form of the generalized inverse Gaussian distribution (Good 1953). Taking $\upsilon = 0$ specializes Eq. 5.151 into the harmonic distribution.

The PDF of the Halphen type B distribution can be expressed as

$$f_B(x) = \frac{2}{m^{2\upsilon} ef_\upsilon(\alpha)} x^{2\upsilon-1} \exp[-(\frac{x}{m})^2 + \alpha(\frac{x}{m})] \qquad (5.152)$$

where $m > 0$ is a scale parameter, α and $\upsilon > 0$ are shape parameters, and $ef_\upsilon(\alpha)$ is the exponential factorial function (a normalizing function). The Halphen type B distribution is used for modeling smaller values in data sets.

The PDF of the Halphen inverse B distribution can be written as

$$f_{IB}(x) = \frac{2}{m^{-2\upsilon} ef_\upsilon(\alpha)} x^{-2\upsilon-1} \exp[-(\frac{x}{m})^2 + \alpha(\frac{m}{x})] \qquad (5.153)$$

where parameters have the same connotation as in the case of the type B distribution. The relation between the type B distribution and type IB distribution is seen by noting that if X follows a type B distribution then $1/X$ follows a type IB distribution.

For the tail behavior of these distributions on the relationship between return period and quantile, Morlat (1956) and Ouarda et al. (1994) reported the following:

1. For the Halphen type A, the gamma, and the Gumbel distributions, $x \propto \ln T$.

2. For the Halphen B and normal distributions, $x \propto \sqrt{\ln T}$.

3. For the Halphen type IB, $x \propto T^{1/2\upsilon}$. This relation depends on parameter υ.

The log-normal distribution has $x \propto \exp(\sqrt{\ln T})$ and the inverse gamma has $x \propto T^{1/\lambda}$, in which case the relation depends on parameter λ.

Table 5-1 Summary of probability distributions.

Distribution	Mathematical form	Range of random variable	Parameters	Applications
Normal distribution	$f_X(x) = \dfrac{1}{\sigma_X \sqrt{2\pi}} \exp\left[-\dfrac{1}{2}\left(\dfrac{x-\mu_X}{\sigma_X}\right)^2\right]$	$-\infty \le x \le \infty$	$E(X) = \bar{x}$ $\sigma_X^2 = \text{variance}$	Used to model phenomena in which random variations arise from a number of additive variations; the most widely used distribution.
Log-normal distribution	$f_X(x) = \dfrac{1}{x\sigma_Y \sqrt{2\pi}} \exp\left[-\dfrac{1}{2}\left(\dfrac{\ln x-\mu_Y}{\sigma_Y}\right)^2\right]$	$Y = \ln(x)$ $x \ge 0$	$\mu_Y = \dfrac{1}{2}\ln\left(\dfrac{\mu_X^2}{1+CV_X^2}\right)$ $\sigma_Y^2 = \ln\left(1+CV_X^2\right)$	Used to model phenomena that occur as a result of a multiplicative mechanism among many factors (e.g., evaporation of water, stream-flow, particle size, or damage).
Extreme value type I distribution: largest values	$f_Y(y) = \alpha \exp\left[-\alpha(y-u) - e^{-\alpha(y-u)}\right]$	$-\infty \le y \le \infty$	$\sigma_Y^2 = \dfrac{1.645}{\alpha^2}$ $u = \dfrac{0.5772}{\alpha} - m_Y$	Used where interest lies in modeling the behavior of the largest values (e.g., flood peaks).
Extreme value type I distribution: smallest values	$f_Z(z) = \alpha \exp\left[\alpha(z-u) - e^{-\alpha(z-u)}\right]$	$-\infty \le z \le \infty$	$u = m_Z + \dfrac{0.5772}{\alpha}$ $\alpha = \dfrac{1.282}{\sigma_Z}$	Used when interest lies in knowing the distribution of the smallest of many independent variables with a common unlimited distribution (e.g., low-flow analysis).
Generalized extreme value distribution	$f(x) = \dfrac{1}{\alpha}\left[1 - k\left(\dfrac{x-u}{\alpha}\right)\right]^{1/k-1} \exp\left\{-\left[1 - k\left(\dfrac{x-u}{\alpha}\right)\right]^{1/k}\right\}$ $(u+\alpha/k) < x < \infty$		$u = m_1 - \dfrac{\alpha}{k}[1-\Gamma(1+k)]$ $C_S = \dfrac{k}{\|k\|}\dfrac{\left[-\Gamma(1+3k)+3\Gamma(1+k)\Gamma(1+2k)-2\Gamma^3(1+k)\right]}{\left[\Gamma(1+2k)-\Gamma^2(1+k)\right]^{3/2}}$	Used for frequency analysis of largest values (e.g., floods).

Table 5-1 Summary of probability distributions. (Continued)

Distribution	Mathematical form	Range of random variable	Parameters	Applications
Beta distribution	$f_X(x) = \dfrac{1}{B} x^{r-1}(1-x)^{t-r-1}, \quad \begin{array}{l} r > 0 \\ t-r > 0 \end{array}$	$0 \le X \le 1$	$m_X = \dfrac{r}{t}$ $\sigma_X^2 = \dfrac{r(t-r)}{t^2(t+1)}$	Used to model the behavior of a variable that is bounded (e.g., stream flows, duration of an activity, or strength of a metal).
Log-Pearson type 3 distribution	$f_X(x) = \dfrac{1}{\alpha \Gamma(\beta)} \left(\dfrac{x-\gamma}{\alpha}\right)^{\beta-1} \exp\left(-\dfrac{x-\gamma}{\alpha}\right)$	$0 < x < \infty$	$\hat{\beta} = (2/C_S)^2$ $\hat{\alpha} = \sqrt{(m_2/\hat{\beta})}$ $\hat{\gamma} = m_1' - \sqrt{m_2 \hat{\beta}}$	Used to model extreme values of a variable (e.g., flood frequency); use has grown tremendously after the U.S. Water Resources Council recommended its adoption as the standard distribution for flood frequency analysis.
Pareto distribution	$x = \varepsilon + (\alpha/k)[1 - (1-F)^k]$ $-(\gamma/\delta)[1 - (1-F)^{-\delta}]$	If $a \le 0, c \le X \le \infty$ If $a > 0, c \le X \le c + b/a$	$\bar{x} = c + \dfrac{b}{1+a}$ $C_s = \dfrac{2(1-a)\sqrt{(1+2a)}}{(1+3a)}$	Used for frequency analysis of large floods.
Logistic distribution	$f(x) = \dfrac{1}{\alpha} e^{-\left(\frac{x-m}{\alpha}\right)} \left[1 + e^{-\left(\frac{x-m}{\alpha}\right)}\right]^{-2}$	$-\infty < x < \infty$	$m = \bar{x}$ $\alpha = \dfrac{\sqrt{3}}{\pi} \sigma$	Used for flood frequency analysis.

5.6 Questions

5.1   Obtain one-day maximum rainfall for each year for Houston, Texas, for a period of 30 to 50 years. Compute the mean, standard deviation, and coefficient of variation of the daily maximum values. Fit a suitable distribution. Give the parameter values. Do the same for two-day and three-day maximum rainfall.

5.2   Use the maximum yearly discharge values for the Amite River at Darlington for as long a period as you can get. Compute the mean, standard deviation, and coefficient of variation of the instantaneous maximum discharge values. Fit a suitable distribution.

5.3   A random variable X has a mean of 5,000 and a standard deviation of 1,000. Compute the probability that this variable will have a value less then 7,000. Compute the probability that the random variable X will be less than 3,000.

5.4   Assuming that the data of Example 4.18 follow a normal distribution, compute the probability that the peak flow will be less than 50,000 cfs, less than 30,000 cfs, and less than 20,000 cfs in any year. What is the return period of each of these flows? What is the probability that the peak flow will occur in any year between one standard deviation, between two standard deviations, and between three standard deviations on either side of the mean?

5.5   Consider a binomial random variable X with $n = 30$ and $p = 0.3$. Evaluate the probability $3 < X \le 8$ using the binomial as well as the normal approximation of the binomial distribution. What will be the answer when $n = 50$?

5.6   Assume that the peak flow data of Example 4.18 follow a two-parameter log-normal distribution. Compute the parameters of the log-normal distribution. Compute the probability that the peak flow will be less than 150,000 cfs, less than 100,000 cfs, and less than 70,000 cfs in any year. What is the probability that peak flow will occur in any year between one standard deviation, between two standard deviations, and between three standard deviations on either side of the mean?

5.7   Assume that the peak flow data of Example 4.18 follow a two-parameter extreme value type I distribution. Compute the parameters of the distribution and the probability that the peak flow in any year will be less than 150,000 cfs, less than 100,000 cfs, and less than 70,000 cfs. What is the probability that peak flow will occur in any year between one standard deviation, between two standard deviations, and between three standard deviations on either side of the mean?

5.8 For the peak annual flow in a small stream it is found that $m_Y = 500$ m³/s and $\sigma_Y = 200$ m³/s. Compute the PDF of the extreme value type I distribution for peak annual flow.

5.9 From the measured data of wind at an airport location, the mean and standard deviation of the maximum annual wind velocity were $m_Y = 50$ km/hour and $\sigma_Y = 15$ km/hour, respectively. Find the wind velocity that will be exceeded with a probability of 0.01 in any year.

5.10 For the data of annual maximum discharge for a river, $n = 50$, mean $m_X = 30,000$ cusecs, coefficient of variation $c_v = 0.50$, and coefficient of skewness $c_s = 2.00$. For these data, compute the parameters of the gamma distribution.

5.11 Consider the normal distribution with parameters μ and CV. Graph the normal distribution for $\mu = 1$, $CV = 0.01$. Keep the value of μ constant but increase CV by 25% in each step. Perform this calculation for about 20 steps and plot both the PDF and cumulative PDF. For each step also calculate the $P(x < 0)$ and plot it with respect to CV. What conclusion can you draw from this plot about the nature of the probability distributions of hydrologic and water quality variables?

5.12 A random variable X has a mean of 300 and a standard deviation of 100. Compute the following probabilities:

(a) X will have a value less than 50.

(b) X will have a value more than 550.

(c) X will be between 50 and 550.

5.13 Consider a binomial random variable X with $n = 22$ and $p = 0.7$. Evaluate the probability $4 < X \leq 8$ using the binomial distribution, the Poisson distribution, and the normal approximation of the binomial distribution. Repeat the same calculation for $n = 30$, $n = 50$, and $n = 100$. Comment on your results about the approximation of a binomial distribution by the Poisson and normal distributions.

5.14 Choose n random numbers $u_1, u_2, \ldots, u_n$ from a uniform PDF in the interval [0, 10] and obtain the distribution of the mean value $\mu_n = (u_1 + u_2 + \ldots + u_n /n)$. Plot and compare the resulting PDFs of μ_n.

5.15 Choose n random numbers from an exponential PDF in the interval [0, $+\infty$] with the parameter $\lambda = 10$ and obtain the distribution of the mean μ_n. Plot and compare the resulting PDFs.

5.16 Assume that the peak flow data of Example 3.17 follow a two-parameter log-normal distribution. Compute the parameters of the log-normal distribution. Compute the probability that the peak flow will be less than 10,000 cfs, less than 8,000 cfs, and less than 15,000 cfs in any year. What is the probability that peak flow will occur in any year between one standard deviation on either side of the mean and between two standard deviations on either side of the mean?

5.17 Assume that the peak flow data of Example 3.17 follow a two-parameter extreme value type I distribution. Compute the parameters of the distribution and the probability that the peak flow in any year will be less than 10,000 cfs, less than 8,000 cfs, and less than 15,000 cfs. Compare these probabilities and return periods with those computed in Question 5.16. Which distribution is a better choice to characterize the peak flow? What is the probability that peak flow will occur in any year between one standard deviation on either side of the mean and between two standard deviations on either side of the mean?

5.18 Assume that the peak flow data of Example 3.17 follow a two-parameter gamma distribution. Compute the parameters of the distribution and the probability that the peak flow in any year will be less than 10,000 cfs, less than 8,000 cfs, and less than 15,000 cfs. Compare these probabilities and return periods with those computed in Question 5.17. Do you think a gamma distribution is a better choice to characterize the peak flow? What is the probability that peak flow will occur in any year between one standard deviation on either side of the mean and between two standard deviations on either side of the mean?

5.19 Repeat Question 5.18 using the log Pearson type III distribution.

5.20 For the peak annual flow in a stream, the mean and standard deviation of the peak flow are 2,000 and 1,200 m^3/s, respectively. Compute the PDF of the peak flow distribution under the following assumptions:

(a) Peak flow is described by an extreme value type I distribution.

(b) Peak flow is described by a log-normal distribution.

(c) Peak flow is described by a gamma distribution.

(d) Peak flow is described by log Pearson type III distribution.

5.21 For National Pollutant Discharge Elimination System permits it is common practice to evaluate the water quality status during low flow conditions. For the Chattahoochee River downstream of the Buford Dam at Lake Lanier, the 7-day minimum flow is given in Table Q5-21. Using these data, select the most appropriate distribution from various candidate distributions, such as the extreme value type I, log-normal, gamma, log Pearson type III, etc. Explain, using the fitting characteristics, why you feel that your selected distribution is the most appropriate choice. Further, compute the probability of the low flows for the return periods of 10, 20, 30, 40, 50, 60, 70, and 90 years.

5.22 From the measured data of air temperature at a given location, the mean and standard deviations of the maximum temperature were 50°F and 15.5°F, respectively. Find the temperature that will be exceeded with probability of (a) 0.05°F in any year, (b) 5°F in any year, and (c) 0°F in any year.

Table Q5-21

Year	7-day min flow (cfs)	Year	7-day min flow (cfs)	Year	7-day min flow (cfs)	Year	7-day min flow (cfs)
1958	12.43	1970	15.29	1982	17.57	1994	31.86
1959	9.43	1971	26.14	1983	14.00	1995	14.00
1960	13.00	1972	14.29	1984	23.29	1996	24.29
1961	16.14	1973	28.00	1985	25.00	1997	12.43
1962	13.86	1974	19.86	1986	12.29	1998	13.86
1963	18.14	1975	41.43	1987	11.00	1999	10.87
1964	25.00	1976	18.71	1988	10.01	2000	7.36
1965	16.71	1977	17.71	1989	21.86	2001	12.57
1966	28.29	1978	9.21	1990	18.86	2002	5.47
1967	26.43	1979	15.86	1991	25.29	2003	24.71
1968	25.43	1980	22.43	1992	20.86	2004	41.14
1969	24.43	1981	6.20	1993	14.29		

5.23 Assume that annual peak flow of Salt Creek at the USGS gauge near Rowell, Illinois, is described by the GEV distribution as given in Question 3.22. Compute the mean and standard deviation of the peak flow data. Also, compute the parameter of the distribution parameters.

5.24 The data of annual maximum stage for Salt Creek near Rowell, Illinois, has been given in Question 3.22. For the stage data, $n = 34$, $m_X = 20.34$ ft, $CV = 0.16$, and $c_s = 0.53$. For these data, compute the parameters of the gamma distribution.

5.25 Repeat Question 5.24 using the Pareto distribution.

5.26 Determine the parameter of a random variable X if it is defined by a triangular distribution with mean, CV, and skew of 100, 0.33, and 0.1, respectively. Sketch both the frequency and cumulative distribution functions. Further, sketch both the frequency and cumulative distribution functions by increasing the CV by 10% and varying the skew between -0.55 and 0.55. For each case determine the probability of $X < 0$. What is the maximum skew that a triangular distribution can describe?

5.27 Assume that peak discharge Q is exponentially distributed with mean μ_Q and variance σ_Q^2. What is the probability distribution of stage S? Suppose stage and discharge are related by $Q = aS^b$.

5.28 A set of data having a mean of 6.5 and a standard deviation of 2.5 is thought to follow the extreme value type I distribution for minima. What proportion of observations from this distribution would exceed 7.0? Plot the probability density function.

5.29   From a data series of minimum annual discharges on a stream, one obtained an average of 175 cfs, a standard deviation of 65 cfs, and a coefficient of skew of 1.5. Using both extreme value type I (for) minimum and type III (for minimum) distributions, evaluate the probability of an annual minimum flow being less than 125 cfs.

5.30   If flood flows from a large watershed have an average value of 1,500 cms with a variance of 53,500 $(cms)^2$, what is the probability that a flood will be equal to or exceed 2,000 cms, if the Gumbel distribution is used?

Chapter 6

Impulse Response Functions as Probability Distributions

A multitude of environmental and hydrologic processes embody both the elements of chance and the descriptive laws of physics. A finer process description at one scale is lost through the processes of integration in time and space and through averaging. This justifies simplification in representation of the processes. It is hypothesized that if an environmental process is described by a linear or linearized governing equation, then the solution of this equation for a unit impulse (or Dirac delta) function can be interpreted as a probability density function for describing the probabilistic properties of the process. This hypothesis is tantamount to mapping from the unit impulse response (UIR) function, $h(t)$, to the probability density function (PDF), $f(x)$, where h is the UIR as a function of time or space variable denoted by t and f is the PDF as a function of the random variable of the process. For example, the impulse response of a diffusion equation for pollutant transport described by the space–time variation of concentration can be used as a probability distribution for pollutant concentration in a medium, such as a river, a lake, conduit storm water, soil, or a saturated geologic formation. Likewise, the impulse response of a linearized diffusion model of channel flow can be interpreted as a probability distribution for frequency analysis of extreme values (such as floods, droughts, hurricanes, earthquakes, and so on). Similarly, the impulse response of a linear reservoir can be used as an exponential probability distribution model. The impulse response of a cascade of equal linear reservoirs is the gamma distribution, which has a number of applications in environmental and water resources data analysis. In this vein, a number of impulse responses of physically based equations that apply to environmental and hydrologic processes and data are discussed and illustrated by using field or laboratory data.

Environment can be defined as a continuum of three components: air, water, and land. The processes dealing with these components and their dynamic interactions constitute environmental processes. Examples of such processes are solute transport in a river, a lake, storm water, soil, or an aquifer; flood; drought; rainfall; erosion; sediment transport by storm water or in a river; air pollution; depletion in the ozone layer; glacial movement and melting; climate change; occurrence of an epidemic resulting from pollution; seawater rise; and salt-water intrusion. A quantitative description of these processes involves a determination of the space, time, or space–time history of flux, concentration, the peak, the average, or the volume of the process variable. For example, to describe the transport of a pollutant in a river, the pollutant flux or concentration as a function of space and time may be selected. One may also select the pollutant load passing through a given point on the river over a selected period of time, say, a month or a year. Similarly, to describe the quality of air in an urban area, one may select the ozone level and determine its variability in time.

Because environmental processes frequently embody both the elements of chance and the descriptive laws of physics, environmental variables cannot be completely described either by deterministic means or by stochastic means alone. Rather, a better approach has to be a combination of both the stochastic and the deterministic means. Considerable simplification is usually needed to view the variables deterministically. This can be justified in light of the observation that excessive process description at one scale is lost through the operations of integration in time and averaging. Furthermore, the governing equations themselves have inherent limitations with regard to accuracy. Even more stark is the state of data acquisition and processing.

When the stochastic aspect of the environmental variables is considered, the element of chance is attributed to a complex mix of factors, such as inherent stochasticity in environmental processes owing to interactions of the environmental continuum components, human–environment interaction, and our limitations to observe and quantify spatial and temporal variability of environmental variables. A stochastic description usually includes analysis of variance, time series analysis, or frequency analysis. The type of analysis that is needed depends on the demand of the problem. For example, a frequency analysis is needed for planning and design. A time series analysis is needed for operation and management. A regression analysis is needed for prediction, extrapolation, or interpolation. Analogous to a deterministic description, a stochastic analysis also involves simplifications that can be justified on the basis of lack of data or lack of adequate knowledge of processes to be modeled as well as of the methodologies for modeling of nonlinear and non-Gaussian processes, requirement of simplicity, parsimony of parameters, and so on.

An environmental process in nature exhibits itself in many ways and a variable selected to describe some aspect of this process must obey the commands of the process. The variable has an intrinsic nature and its characterization and analyses, whether deterministic or stochastic or a combination thereof, is important. This means that there must be an inherent connectivity among these analyses. In environmental science and engineering, this connection is frequently observed. For example, a regression analysis without error analysis in statistics is no different from curve-fitting techniques in mathematics. Indeed, regression analysis techniques are often employed to find a best-fit curve for a given set of data, and the connection between these types of techniques is well known. Another example is the autoregressive (AR) technique in time series analysis. When an AR model is applied to, say, daily, monthly, or annual river flow, the coefficients associated with the autoregressive variables are nothing but the ordinates of linear kernel of the flow variable. Since the AR technique is linear, it is equivalent to the impulse response function of a linear flow process. In hydrology, this is known as the unit hydrograph method. One also finds a connection between the unit hydrograph method and spectral analysis. This means that certain linear time series analysis techniques are equivalent to linear response functions of environmental processes. However, the connection between frequency analysis methods and deterministic methods is not clear yet. This may be because frequency, by definition, is stochastic in nature and finding a deterministic equivalent seems somewhat contradictory in terms. However, our objective here is to find a connection through techniques of analysis, not through concept. This constitutes the subject matter of this chapter.

6.1 Hypothesis

It is hypothesized that if an environmental variable is described by a linear or linearized governing equation, then the solution of this equation for a unit impulse (or Dirac delta) function (UIF) can be interpreted as a PDF for describing the probabilistic properties of the random variable, say, X. The solution for the UIF can be characterized as the UIR or $h(t)$. If the UIR is a function of time t, then the PDF is a mapping from the (h, t) plane to the (f, x) plane, where x is the value (or quantile) of the random variable X for which $h(x, t)$ is desired, and f is the PDF.

There are many environmental variables that can be reasonably well described linearly. If some of the variables cannot be described linearly in the real domain, then they can be described linearly in the logarithmic domain or in an appropriate transformed domain. Examples of linear approximation are surface runoff from rainfall excess, river flow, monthly sediment discharge, and solute concentration in a tube or soil. Thus, their UIRs can be considered as their PDFs. It is not surprising that several probability distributions have found their niche in linear environmental analyses. This hypothesis will be explored in what follows.

6.2 Impulse Responses of Linear Systems

6.2.1 Linear Reservoir

The simplest linear system in hydrology is probably a linear reservoir (or storage element), shown in Fig. 6-1, and is described by the spatially lumped form of the continuity equation

$$\frac{dS(t)}{dt} = I(t) - Q(t) \tag{6.1}$$

and a storage–discharge type relation

$$S = k\,Q \tag{6.2}$$

where $I(t)$ is the rate of inflow to the reservoir at time t, $Q(t)$ is the rate of outflow from the reservoir at time t, $S(t)$ is the storage in the reservoir at time t, and k is the storage coefficient or average travel (or residence or lag) time. Substitution of Eq. 6.2 in Eq. 6.1 yields

$$I(t) - Q(t) = \frac{dS(t)}{dt} = k\frac{dQ}{dt} \tag{6.3}$$

Solution of Eq. 6.3 gives

$$Q_t = I[1 - \exp(-t/k)] \quad t \le D \tag{6.4}$$

and

$$Q_t = Q_P \exp[-(t-D)/k] \quad t \ge D \tag{6.5}$$

where D is the inflow duration, and Q_P is the peak of outflow hydrograph given by

$$Q_P = I[1 - \exp(-D/k)] \tag{6.6}$$

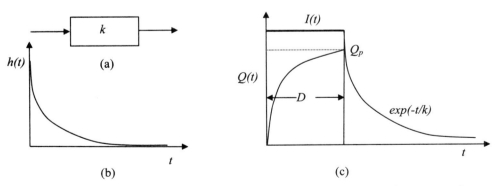

Figure 6-1 *Depiction of linear reservoir concept: (a) lag time, (b) IUH, and (c) hydrograph due to a pulse of* D *hour duration.*

The linear reservoir has been used for rainfall–runoff modeling either by itself or as an element of a network model. For instantaneous inflow that fills the reservoir in time $t = 0$,

$$Q = \frac{S}{k}\exp(-t/k) \tag{6.7}$$

If $I(t)$ is denoted by a unit delta function $\delta(t)$, then the UIR of the linear reservoir, $h(t)$, is

$$h(t) = \frac{\exp(-t/k)}{k} \tag{6.8}$$

In hydrology, $h(t)$ is known as the instantaneous unit hydrograph (IUH). The determination of $h(t)$, the impulse response function (or the kernel function or Green's function) of a system from input and output data, is known as system identification. Convolution of the impulse response function with the system inputs gives the system output. Then, the PDF of a variable described by a linear reservoir becomes

$$f(x) = \frac{\exp(-x/k)}{k} \tag{6.9}$$

where x is the quantile of the variable X described by the linear reservoir and k is a parameter. Thus, it is seen that $h(t)$ is mapped onto $f(x)$. Equation 6.9 is an exponential density function and is widely used in environmental and water resources. For example, if an environmental process is described by the Poisson process then the interarrival times follow an exponential distribution. Interarrival times of floods can be modeled by using Eq. 6.9. Rainfall depth, intensity, and duration have been modeled with Eq. 6.9. It should be noted that k in $f(x)$ represents the average of X and hence its interpretation from Eq. 6.8 remains unchanged under mapping of $h(t)$ onto $f(x)$.

Another modification of the linear reservoir involves restating the unit delta function $\delta(t)$ as $\delta(t - t_0)$, where t_0 is the time at which the function occurs. In that case, $h(t)$ of Eq. 6.8 becomes

$$h(t) = \frac{\exp[-(t - t_0)/k]}{k} \tag{6.10}$$

Equation 6.10 is the UIR of a lag and route linear reservoir system in which t_0 is the amount of lag time before water is released from the reservoir. This is equivalent to a linear reservoir and a linear channel, connected in series. By mapping Eq. 6.7 onto the probability plane, the PDF becomes

$$f(x) = \frac{\exp[-(x - x_0)/k]}{k} \tag{6.11}$$

where x_0 is the threshold of X, $x \geq x_0$. The threshold is the minimum value of X. This is useful in frequency analysis of environmental data.

Example 6.1 Consider a linear reservoir with a lag time of 10 hours. This reservoir receives a pulse of 10 m³/s for a duration of 5 hours. Determine the peak outflow and graph the outflow hydrograph. Also graph the outflow hydrograph when lag time is 5 hours and compare the two hydrographs.

Solution The peak of the hydrograph can be computed from Eq. 6.6:

$$Q_P = I[1 - \exp(-D/k)] = 10[1 - \exp(-5/10] = 3.93 \text{ cumec}$$

The hydrograph for both cases is plotted in Fig. 6-2. Notice that the peak is higher when the lag time is smaller and the recession is slower and lasts longer when k is larger. Clearly, a larger catchment will have a longer lag time. This hydrograph can also be considered as a probability density function of peak discharge exceeding a given threshold.

6.2.2 Muskingum Model

The Muskingum model is described by Eq. 6.1 and the Muskingum hypothesis:

$$S(t) = K\left[\alpha I(t) - (1 - \alpha)Q(t)\right] \tag{6.12}$$

where K is the average reach travel time and α is a parameter or a weighting coefficient. The unit impulse response of the Muskingum model (see Fig. 6-3), is given by

$$h(t) = -\frac{\alpha}{1 - \alpha}\delta(t) + \frac{1}{K(1-\alpha)^2}\exp\left[-\frac{t}{K(1-\alpha)}\right] \tag{6.13}$$

It has been shown that modeling flood routing along a short reach of a lowland river may result in the negative value of α and

$$0 \leq \left(-\frac{\alpha}{1 - \alpha}\right) \leq 1 \tag{6.14}$$

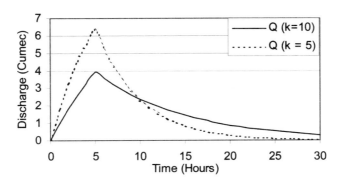

Figure 6-2 *Outflow hydrograph from the linear reservoir for two values of lag time.*

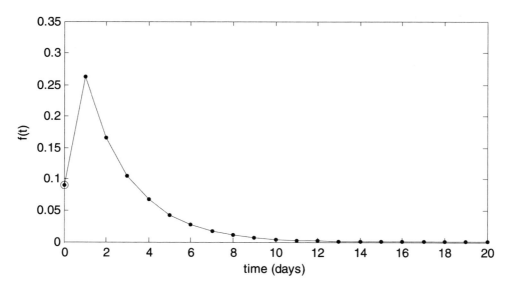

Figure 6-3 *Muskingum reach impulse response function.*

Denoting $\dfrac{1}{1-\alpha} = \beta$ and $\dfrac{\beta}{K} = \gamma$ and renaming t as x, one gets a two-parameter probability distribution function:

$$f(x) = (1 - \beta)\delta(x) + \beta\gamma \exp(-\gamma x) \tag{6.15}$$

The PDF given by Eq. 6.15 is a weighted sum of two functions: a delta function and an exponential function. It is interesting to note that in this function parameter β is a weighting factor and parameter $K = \beta/\gamma$ becomes the average of X. Thus, the original expressions of the weighting factor and the average travel time are modified under mapping, but the conceptual meaning of the modified expressions remains more or less intact. Equation 6.15 is useful for frequency analysis of floods with zero values as well as flood damage.

Example 6.2 Let the average travel time of a Muskingum lowland reach, K, be 2 days, and the weighting coefficient parameter be −0.1. Determine the impulse response function of reach outflow.

Solution According to Eq. 6.15,

$$\beta = \frac{1}{1+0.1} = \frac{1}{1+0.1} = 0.91, \gamma = \beta/K = 0.91/2 = 0.455$$

Then

$$f(t) = 0.09\delta(t) + 0.414\exp(-0.455t)$$

6.2.3 Cascade of Linear Reservoirs

If an environmental system is represented by a cascade of n equal linear reservoirs, then its UIR becomes

$$h(t) = \frac{1}{k\,\Gamma(n)}\left(\frac{t}{k}\right)^{n-1} \exp(-t/k) \tag{6.16}$$

where k is the storage parameter of each reservoir and $\Gamma(n)$ is the gamma of n. Since there are n reservoirs, nk represents the total lag time (or the average residence time) of the system. By mapping onto the probability plane, the PDF becomes

$$f(x) = \frac{1}{k\,\Gamma(n)}\left(\frac{x}{k}\right)^{n-1} \exp(-x/k) \tag{6.17}$$

where k and n are parameters. Equation 6.17 is the gamma probability density function. The gamma distribution results from the sum of exponentials, where n is the number of exponentials. In deterministic parlance, each exponential represents a linear reservoir. Thus, the deterministic interpretation of parameters is carried over through mapping. The gamma distribution is one of the most commonly used probability distributions for environmental frequency analysis.

If an environmental system satisfies the requirement that $h(t) > 0$ if $t \geq t_0$ then the UIR becomes

$$h(t) = \frac{1}{k\,\Gamma(n)}\left(\frac{t - t_0}{k}\right)^{n-1} \exp[-(t - t_0)/k)] \tag{6.18}$$

The interpretation of t_0 is that the cascade of equal linear reservoirs retains water for time t_0 before it starts to release it. Mapping onto the probability plane transforms Eq. 6.18 into

$$f(x) = \frac{1}{k\,\Gamma(n)}\left(\frac{x - x_0}{k}\right)^{n-1} \exp[-(x - x_0)/k)] \tag{6.19}$$

which represents the three-parameter Pearson type III probability density function. This is equivalent to a cascade of linear reservoirs and channels connected in series. This is one of the most widely used frequency distributions in hydrology and environmental sciences. Note that Eq. 6.17 is a special case of Eq. 6.19. Here parameter x_0 is the lowest value or threshold of the variable X. Although these parameters, k, n, and x_0, can be interpreted by using the deterministic analogy, their optimal values are better found by curve fitting. This means that under mapping onto the probability plane, the interpretation of the parameters may be somewhat transformed.

Example 6.3 Take n as 3 and k as 6 hours. Compute the probability density function of peak discharge. Assume the lowest value or threshold of discharge as 100 cumecs.

Solution Substituting $n = 3$, $k = 6$, and $x_0 = 100$ into Eq. 6.19, we have the probability density function of discharge as

$$f(x) = \frac{1}{6\Gamma(3)} \left(\frac{x-100}{6} \right)^2 \exp[-(x-100)/6]$$

The probability density function is plotted in Fig. 6-4.

6.2.4 Linear Downstream Channel Routing Model

One of the most important problems in one-dimensional flood routing analysis is the downstream problem (i.e., the prediction of flood characteristics at a downstream section on the basis of the knowledge of flow characteristics at an upstream section). By using the linearization of the Saint-Venant equation, the solution of the upstream boundary problem was derived by Deymie (1939), Masse (1937), Dooge and Harley (1967), and Dooge et al. (1987a,b), among others; a discussion of this problem is presented in Singh (1996a). The solution is a linear, physically based model with four parameters dependent on the hydraulic characteristics of the channel reach at the reference level of linearization. However, the complete linear solution is complex in form and is relatively difficult to compute (Singh 1996a). Two simpler forms of the linear channel downstream response are recognized in the hydrologic literature and are designated as the linear diffusion (LD) analogy model and the linear rapid flow (LRF) model. These correspond to the limiting flow conditions of the linear channel response, that is, where the Froude number is equal to zero (Hayami 1951; Dooge 1973) and where it is equal to one (Strupczewski and Napiorkowski 1980c).

The complete linearized Saint-Venant equation is of hyperbolic type and may be written as

$$a\frac{\partial^2 Q}{\partial y^2} + b\frac{\partial^2 Q}{\partial y \partial t} + c\frac{\partial^2 Q}{\partial t^2} = d\frac{\partial Q}{\partial y} + e\frac{\partial Q}{\partial t} \tag{6.20}$$

where Q is the perturbation of flow about an initial condition of steady-state uniform flow, y is the distance from the upstream boundary, t is the elapsed time, and a, b, c, d, and e are parameters as functions of channel and flow characteristics at the reference steady-state condition. A number of models of simplified forms of the complete Saint-Venant equation have been proposed in the hydrological literature.

If all three second-order terms on the left-hand side of Eq. 6.20 are neglected, the linear kinematic wave model is obtained. Expressing the second and the third second-order terms in terms of the first on the basis of the linear kinematic wave approximation leads to a parabolic equation (Dooge 1973), in contrast with

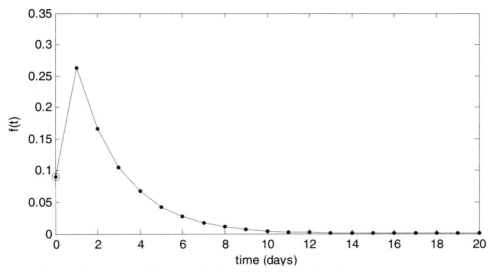

Figure 6-4 *Probability density function of cascade of reservoirs in series.*

the original Eq. 6.20, which is hyperbolic. Its solution for a semi-infinite channel, known under the name of the linear diffusion analogy (LDA) or the convective–diffusion solution, has the form

$$h(y,t) = \frac{x}{\sqrt{4\pi Dt^3}} \exp\left[-\frac{(y - ut)^2}{4Dt}\right] \tag{6.21}$$

where y is the length of the channel reach, t is the time, u is the convective velocity, and D is the hydraulic diffusivity. Both u and D are functions of channel and flow characteristics at the reference steady-state condition. Besides flood routing, LDA has been applied by Moore and Clarke (1983) and Moore (1984) as a transfer function of a sediment routing model.

The function given by Eq. 6.21 is rarely quoted in statistical literature. It was derived by Cox and Miller (1965, p. 221) as the probability density function of the first passage time T for a Wiener process starting at 0 to reach an absorbing barrier at a point x, where u is the positive draft and D is the variance of the Wiener process. Tweedie (1957) termed the density function of Eq. 6.21 as an inverse Gaussian PDF, Johnston and Kotz (1970) summarized its properties, and Folks and Chhikara (1978) provided a review of its development. The function in Eq. 6.21 has been applied by Strupczewski et al. (2001) as a flood frequency model expressed as

$$f(x) = \frac{\alpha}{\sqrt{\pi x^3}} \exp\left[-\frac{(\alpha - \lambda x)^2}{x}\right] \tag{6.22}$$

where $\alpha = x/\sqrt{4D} > 0$, $\beta = xu/(4D) > 0$, and $\gamma = \beta/\alpha$. Equation 6.22 can be extended to a three-parameter PDF by introducing a location parameter or a lower bound for x as

$$f(x) = \frac{\alpha}{\sqrt{\pi x^3}} \exp\left[-\frac{(\alpha - \lambda[x - \varepsilon])^2}{[x - \varepsilon]}\right] \tag{6.23}$$

where ε is the lower bound of x.

If the diffusion term is expressed in terms of two other terms by using the kinematic wave solution, one gets the rapid flow (RF) equation, which is of parabolic-like form (Strupczewski and Napiorkowski 1990a). Therefore, it filters out the downstream boundary condition. It provides the exact solution for a Froude number equal to one and consequently can be used for large values of the Froude number. If the alternative approach is taken by expressing all the second-order terms as cross-derivatives, one gets the equation representing the diffusion of kinematic waves (Lighthill and Witham 1955; Strupczewski and Napiorkowski 1989). The kinematic diffusion (KD) model, being of parabolic-like form, satisfactorily fits the solution of the complete linearized Saint-Venant equation only for small values of the Froude number and slow rising waves.

Although RF and KD models correspond to quite different flow conditions, the structure of their impulse response is similar (Strupczewski et al. 1989; Strupczewski and Napiorkowski 1990c). In both cases, the impulse response is

$$h(x,t) = P_0(\lambda)\delta(t - \Delta) + \sum_{i=1}^{\infty} P_i(\lambda) \cdot h_i\left(\frac{t - \Delta}{\alpha}\right) \cdot 1(t - \Delta) \tag{6.24}$$

where

$$P_i(\lambda) = \frac{\lambda^i}{i!} \exp(-\lambda) \tag{6.24a}$$

is the Poisson distribution,

$$h_i\left(\frac{t}{\alpha}\right) = \frac{1}{\alpha(i-1)!}\left(\frac{t}{\alpha}\right)^{i-1} \exp\left(-\frac{t}{\alpha}\right) \tag{6.24b}$$

is the gamma distribution, and $1(t)$ is the unit step function. Parameters α, λ, and Δ are functions of both channel geometry and flow conditions, which are different for the two models. Furthermore, there is no time lag (Δ) in the impulse response function of the KD model.

Both models can be considered as hydrodynamic and conceptual. Note that the solution of both models can be represented in terms of basic conceptual elements used in hydrology, namely, a cascade of linear reservoirs and a linear channel in case of the RF model. The upstream boundary condition is delayed by a linear channel with time lag Δ divided according to the Poisson distribution with mean, and then transformed by parallel cascades of equal linear reservoirs (with time constant α) of varying lengths. Note that λ is the average number of reservoirs in a cascade. Strupczewski et al. (1989) and Strupczewski and Napiorkowski (1989) have derived the distributed Muskingum model from the multiple Muskingum model and have shown its identity to the KD model. Similarly, the RF model happens to be identical to the distributed delayed Muskingum model (Strupczewski and Napiorkowski 1990c).

Einstein (1942) introduced the function given by Eq. 6.21 to hydrology as the mixed deterministic–stochastic model for the transportation of bed load. It has also been used as the PDF of the total rainfall depth derived from the assumption of a Poisson process for storm arrivals and an exponential distribution for storm depths (Eagleson 1978). The function in Eq. 6.24 is considered to be a flood frequency model. An example of such a model is shown in Fig. 6-5.

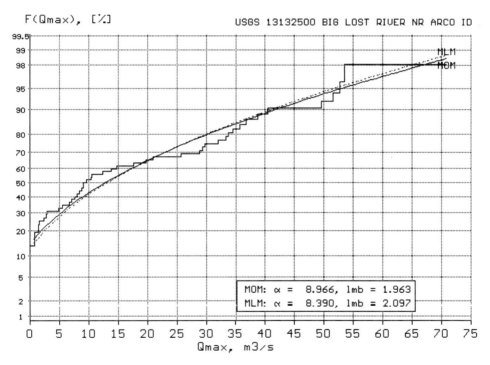

Figure 6-5 *Empirical and two theoretical KD cumulative distribution functions for the Big Lost River, Arco, Idaho. MOM and MLM estimated parameters are shown. Solid line: MOM estimated CDF, dotted line: MLM estimated CDF.*

The RF model can be employed for modeling samples censored by the value. If the delay is equated to zero and t is renamed as x then Eq. 6.24 yields a two-parameter probability distribution of the form

$$f(x) = P(z = 0)\delta(x) + \sum_{i}^{\infty} P(z = i) + h_i(\frac{x}{\alpha}) \cdot I(x) \tag{6.25}$$

where z is the Poisson-distributed random variable

$$P(z = i) = P_i(\lambda) = \frac{\lambda^i}{i!} \exp(-\lambda) \tag{6.25a}$$

and x is the gamma-distributed variable

$$h_i(\frac{x}{\alpha}) = \frac{1}{\alpha(i-1)!} (\frac{x}{\alpha})^{i-1} \exp(-\frac{x}{\alpha}) \tag{6.25b}$$

$I(x)$ is the unit step function. Equation 6.25 differs from Eq. 6.24 since its second term cannot be expressed as the product of the probability of nonzero value (i.e., $(1 - P_0(\lambda))$) and the conditional PDF (i.e., $f_1(x, g)$ with β not included in g, where $\beta = \exp(-\lambda)$ and $g = [\alpha, \beta]$ in the KD distribution function). The second term of the PDF,

$$f_c = \sum_{i}^{\infty} P_i(\lambda) h_i(\frac{x}{\alpha}) \tag{6.26}$$

can be expressed by the first-order modified Bessel function of the first type,

$$I_1(z) = \sum_{i}^{\infty} \frac{1}{(i-1)! i!} (\frac{z}{2})^{2(i-1)+1} \tag{6.27}$$

as

$$f_c(x) = \exp(-\lambda - \frac{x}{\alpha}) \sqrt{\frac{\lambda}{\alpha x}} I_1(2\sqrt{\frac{\lambda x}{\alpha}}) I(x) \tag{6.28}$$

Thus, Eq. 6.25 can be written as

$$f(x) = P_0(\lambda)\delta(x) + \exp(-\lambda - \frac{x}{\alpha}) \sqrt{\frac{\lambda}{\alpha x}} I_1 \times (2\sqrt{\frac{\lambda x}{\alpha}}) I(x) \tag{6.29}$$

which is the KD–PDF.

6.2.5 Diffusion Equation

Many natural processes are diffusive in nature and can, therefore, be modeled using diffusion-based equations. Examples of such processes are dye transport in a container of water, contaminant mixing in rivers and estuaries, transport of sediment in rivers, and migration of microbes. Thus, the probability distributions based on random diffusion processes may be better suited to represent the data of diffusion-driven processes. To illustrate this concept, a dye diffusion equation is considered to describe the concentration distribution produced from an injection of a mass of dye that is introduced as a plane source located at coordinate x_0 at time zero into a liquid-filled, semi-infinite tube of cross section A. The tube is closed on the left end and extends to the right to infinity. The one-dimensional diffusion equation for a dye of mass M introduced at time $t = 0$ into a liquid-filled tube of cross section A that extends from $x = 0$ to infinity is

$$\frac{\partial C}{\partial t} = D \frac{\partial^2 C}{\partial x^2} + \left(\frac{M}{A} \right) \delta(y - y_0) \delta(t) \tag{6.30}$$

where D is the diffusion coefficient, $\delta(y - y_0)$ is a Dirac delta function of $(y - y_0)$, $\delta(t)$ is a Dirac delta function of t, and y_0 is the location where the mass is inserted at time $t = 0$. A Dirac delta function has the property that it is equal to zero if the argument is nonzero; when the argument is zero, the Dirac delta function becomes infinite. The definition of the Dirac delta requires that the product $\delta(x) \partial(x)$ is dimensionless. Thus, the units of the Dirac delta are the inverse of those of the argument $\overline{x}$. That is, $\delta(x)$ has units meters^{-1}, and $\delta(t)$ has units sec^{-1} (Scott 1955).

The first boundary condition states that there is no diffusion of dye through the closed left end of the tube at $y = 0$:

$$\frac{\partial C}{\partial y} = 0, \quad \text{at } y = 0 \tag{6.31}$$

$$C(y, 0) = 0 \tag{6.32}$$

The second boundary condition states that the concentration and the concentration flux are zero at infinity. (More generally, all of the terms in the Taylor series expansion of the concentration are zero at infinity.) The second boundary condition is stated for the terms of the Taylor series of concentration as

$$C = 0, \quad \frac{\partial C}{\partial y} = 0, ..., \quad \frac{\partial^n C}{\partial y^n} = 0, ..., \quad \text{as } x \to \infty \tag{6.33}$$

The initial condition states that there is no dye in the tube at time zero.

Using the integral transform method gives the solution of Eq. 6.30 subject to Eq. 6.31 to Eq. 6.33 (Özisik 1968; Cleary and Adrian 1973):

$$C(x,t) = \frac{M}{A} \frac{1}{\sqrt{4\pi D t}} \left(\exp\left[-\frac{(x - x_0)^2}{4 D t} \right] + \exp\left[-\frac{(x + x_0)^2}{4 D t} \right] \right) \tag{6.34}$$

We now reduce the number of terms in Eq. 6.34, normalize the equation so that it represents a unit mass injected over a unit area, and map onto the probability plane by introducing a frequency term instead of concentration. These changes make Eq. 6.34 resemble a probability distribution. The term "$4Dt$" appears together in Eq. 6.34. We define a new term

$$\sigma^2 = 2Dt \tag{6.35}$$

In addition, the mass and cross-sectional area are combined with concentration, σ is held constant so it is treated as a parameter, and a new term $f(x; x_0, \sigma)$ is introduced, so that $f(x; x_0, \sigma) = AC(x, t)/M$, which has units length^{-1}. The result is the equation

$$f(x; \sigma, x_0) = \frac{1}{\sqrt{2\pi}\sigma}\left\{\exp\left[-\left(\frac{x - x_0}{\sqrt{2}\sigma}\right)^2\right] + \exp\left[-\left(\frac{x + x_0}{\sqrt{2}\sigma}\right)^2\right]\right\} \tag{6.36}$$

which is now interpreted as a probability distribution that is bounded by $x = 0$ on the left side and extends to infinity on the right. The term σ represents the spread of the probability distribution, and x_0 usually represents the location of the peak frequency, although it is possible that if σ is large, the peak frequency may not be located at x_0 but may be located at $x = 0$. The distribution is called a two-parameter semi-infinite Fourier distribution as it was developed from the diffusion (Fourier) equation using semi-infinite Fourier transforms.

Equation 6.36 is limited to application to data that are distributed along the positive x axis. However, if the restriction on x only being able to represent distance is relaxed, so that x can represent any dimension that is appropriate for a frequency distribution, then the number of applications of Eq. 6.36 can increase. For example, if one is interested in the frequency distribution of a chemical, such as manganese concentration in a river, then x could have units of milligrams per liter. The units of σ are the same as the units of x.

6.3 Application

In the frequency analyses of environmental (say, hydrological) data in arid and semiarid regions, one often encounters data series that contain several zero values with zero being the lower limit of the variability range. From the viewpoint of probability theory, the occurrence of zero events can be expressed by placing a nonzero probability mass on a zero value (i.e., $P(X = 0) \neq 0$, where X is the random variable and P is the probability mass). Therefore, the distribution functions from which such hydrological series were drawn would be discontinuous with discontinuity at the zero value having a form

$$f(x) = (1 - \beta)\delta(x) + f_c(x; \mathbf{h}) \cdot 1(x) \tag{6.37}$$

where $(1-\beta)$ denotes the probability of the zero event, that is, $1-\beta = P(X = 0)$, $f_c(x;\mathbf{h})$ is the continuous function such that $\int_0^\infty f_c(x;\mathbf{h})dx = \beta$, $\mathbf{h}$ is the vector of parameters, $\delta(x)$ is the Dirac delta function, and $1(x)$ is a unit step function.

The estimation procedures for hydrologic samples with zero events have been a subject of several publications. The theorem of total probability has been employed (Jennings and Benson 1969, Woo and Wu 1989, Wang and Singh 1995) to model such series. Then, Eq. 6.37 takes the form

$$f(x) = (1-\beta)\delta(x) + \beta f_1(x;\mathbf{g}) \cdot 1(x) \quad \beta \notin g \tag{6.38}$$

where $f_1(x; \mathbf{g})$ is the conditional probability density function (CPDF), that is, $f_1(x;\mathbf{g}) \equiv f_1(x;\mathbf{g}|X > 0)$, which is continuous in the range $(0, +\infty)$ with a lower bound of zero value. Wang and Singh (1995) estimated β and the parameters of the CPDF separately by considering the positive values as a full sample. Having estimated g and β allows one to transform the conditional distribution to the marginal distribution [i.e., to $f(x)$] by Eq. 6.33. Among several PDFs with zero lower bound recognized in flood frequency analysis (FFA) (e.g., Rao and Hamed 2000), the gamma distribution given by Eq. 6.17 was chosen by Wang and Singh (1995) as an example of a CPDF and four estimation methods were applied: the maximum likelihood method (MLM), the method of moments (MOM), probability weighted moments (PWM), and the ordinary least-squares method. By using monthly precipitation and annual low-flow data from China, and annual maximum peak discharge data from the United States, the suitability of the distribution and the estimation methods was assessed. The histogram and the estimated PDF of all three series indicated a reverse J-shape without mode, whereas the value of the coefficient of variation of $f_1(x; \mathbf{g})$ was close to one, pointing out a good fit of data to the Muskingum-originated PDF given by Eq. 6.10.

Among positively skewed distributions, it is the log-normal (LN) distribution, which together with the gamma, is most frequently used in environmental frequency analysis. The LN distribution has been found to describe hydraulic conductivity in a porous medium (Freeze 1975), annual peak flows, raindrop sizes in a storm, and other hydrologic variables. Chow (1954) reasoned that this distribution is applicable to hydrologic variables formed as the product of other variables since if $X = X_1 \cdot X_2 \cdot \ldots X_n$ then $Y = \log X$ tends to the normal distribution for large n provided that the X_i are independent and identically distributed.

Kuczera (1982d) considered six alternative PDFs and found the two-parameter LN to be most resistant to an incorrect distributional assumption in at-site analysis and also while combining site and regional flood information. Strupczewski et al. (2001) fitted seven two-parameter distribution functions—namely, normal, gamma, Gumbel (extreme value type I), Weibull, log-Gumbel, and log-logistic—to thirty-nine 70-year long annual peak flow series of Polish rivers. The criterion of the maximum log-likelihood value was used for the best model choice. From these competing models, the log-normal was selected in 32 cases out of 39, the gamma in 6 cases, the Gumbel in one case, and the remaining four were not identified as the best model even in a single case.

The LDA model shows a similarity to the LN model (Strupczewski et al. 2001). Only for large values of the coefficient of variation do the LDA lines deviate apparently from straight lines on a log-normal probability plot. A comparison of the maximum likelihood (ML) values of both distributions of the thirty-nine 70-year long annual peak flow series of Polish rivers has shown that in 27 out of 39 series the LDA model provided a better fit to the data than did the LN model (Strupczewski et al. 2001). For Polish rivers, the average value of the ratio of the skewness coefficient (c_s) to the coefficient of variation (c_v) equals 2.52. This value is only just closer to the ratio of the LDA model, where $c_c/c_v = 3$, than to that of the LN model, where $c_s = 3c_v + c_v^3$. Moreover it is interesting to learn that the LDA model represents flood frequency characteristics quite well when the LDA model is likely to be better than other linear models (i.e., for lowland rivers; Fig. 6-6).

Comparing the potential for applicability of the two distributions, one should also take into account real conditions, where the hypothetical PDF differs from the true one. Applying the ML method, one gets unbiased moment estimates if instead of the LN model the LD model is used for the LN-distributed data, whereas in the opposite case the ML estimate of the variance is biased. Therefore, if both models show an equally good fit to the data it seems reasonable to select the LD model if the ML method is to be applied. One should also be aware that the LN distribution is not uniquely determined by its moments (Kendall and Stuart 1969, p. 179).

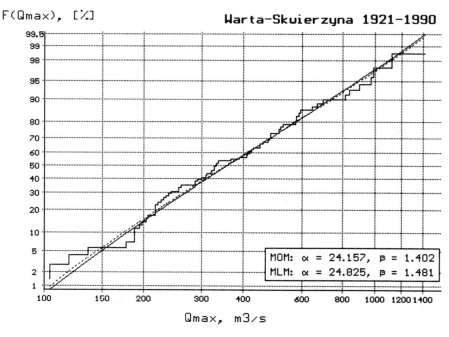

Figure 6-6 *Empirical and two theoretical LD cumulative distribution functions for the Warta River, Skwierzyna cross-section data. MOM and MLM estimated parameters are shown. Solid line: MOM estimated CDF, dotted line: MLM estimated CDF.*

Application of the function given by Eq. 6.24 to flood data reveals that for $\lambda < 2$, which corresponds to the probability of the zero event being equal to $P_0(\lambda = 2) = \exp(-2) = 0.135$, the PDF has a reverse J-shape without mode. By modeling longer time series, it may be reasonable to introduce a third parameter to the model of Eq. 6.22, making the shape of the continuous part of the distribution independent of the probability of zero event:

$$f(x) = \beta \cdot \delta(x) + \frac{1-\beta}{1-e^{-\lambda}} \sum_{i=1}^{\infty} P_i(\lambda) \cdot h_i\left(\frac{x}{\alpha}\right) \cdot 1(x) \tag{6.39}$$

where $P_i(\lambda)$ and $h_i\left(\dfrac{x}{\alpha}\right)$ are defined by Eq. 6.25a and Eq. 6.25b.

6.4 Summary

To summarize, one can conclude the following:

1. The unit impulse response functions of linear or linearized physically based models form suitable models for environmental frequency analysis.

2. Many of the unit response functions are found to be the same as those that have been used in statistics for a long time.

3. The use of the UIR functions can provide a physical basis to many of the statistical distributions.

4. The UIR approach provides a hope for linking deterministic and stochastic frequency models.

6.5 Questions

6.1   Consider a linear reservoir whose impulse response function can be defined as

$$h(t) = \frac{1}{k}\exp(-t/k)$$

where $h(t)$ is the impulse response at time t and k is a parameter, called the lag time. Integrate $h(t)$ over time and show that it can be considered as a probability density function regardless of the value of k. Plot $h(t)$ (1/hour). In the probability domain, $h(t)$ will assume the role of a probability density function and t will be the random variable and its value will be a quantile.

6.2   Integration of the impulse response function in Question 6.1, $U(t)$, can be
       expressed as

$$U(t) = 1 - \exp(-t/k)$$

Here $U(t)$ takes on the role of the cumulative distribution function in the
probability domain. Plot $U(t)$ for $k =1, 2, 5, 10$, and 15 hours and interpret
these plots physically.

6.3   The impulse response function for an aquifer interacting with a stream
       can be expressed as

$$h(t) = \frac{1}{S_c} \exp[-a(t - t_0)/S_c]$$

where S_c is the storage coefficient and is equal to specific yield for unconfined aquifers, a (1/time) is the subsurface flow constant, and t_0 is the
initial time. In the probability domain this can also be considered as a
probability density function with parameters two parameters and a
threshold value of t_0. Plot the UIR for different values of S_c: 0.01, 0.05, 0.1,
0.2, 0.3, 0.4, and 0.5. Take the value of a as 1/day, 0.5/day, and 0.2/day.

6.4   The impulse response of a time-variant linear reservoir can be expressed as

$$h(t) = \frac{1}{k(t)} \exp[-\int_\tau^t \frac{dw}{k(w)}]$$

where $k(t)$ is the time-varying lag time and is the time at which input to
the reservoir is applied. Plot $h(t, \tau)$ assuming $k(t) =10 + t$. Take τ as 0, 1, 2,
3, 4, and 5 hours. Interpret these graphs physically and discuss the kind
of probability density function these graphs look like.

6.5   Plot the impulse response function using Eq. 6.17 for different values of
       n as 1, 2, 3, 4, and 5 and k as 2, 4, 6, 8, and 10 hours. Interpret these plots
       physically. Now use the impulse response function as a probability den-
       sity function of peak discharge and then compute the probability of dis-
       charge equal to or exceeding 1,000 cumecs.

6.6   Consider the impulse response function of Eq. 6.18. Then compute the
       probability of discharge equal to or exceeding 1,000 cumecs if the lowest
       value or threshold of discharge is 100 cumecs.

6.7   Consider a linear reservoir with a lag time of 15 hours. This reservoir
       receives a pulse of 600 m^3/s for a duration of 6 hours. Determine the
       peak outflow and graph the outflow hydrograph. Also graph the out-
       flow hydrograph for lag times as 10 and 20 hours and compare all three
       hydrographs and comment on your results.

6.8   A reservoir releases a peak flow of 531 m^3/s. It was determined that the lag time for this reservoir was 12 hours. It receives an inflow pulse for a duration of 6 hours. Determine the magnitude of the inflow pulse.

6.9   Determine the reservoir storage coefficient K and inflow pulse duration D using the following data:

Inflow (m^3/s)	27	82	109	136	190	218	245	272	299	326	354	381	408	435	463	490	517	544
Peak outflow (m^3/s)	14	43	57	71	99	113	128	142	156	170	184	198	213	227	241	255	269	283

6.10   Let the average travel time of a Muskingum lowland reach K be 5 days, and let the weighting coefficient parameter be -0.15. Determine the impulse response function of reach outflow.

6.11   Take n as 5 and k as 9 hours. Compute the probability density function of peak discharge. Assume the lowest value or threshold of discharge as $300\ m^3/s$.

6.12   Select a watershed in your area that has two USGS gauging stations on a stream. Use the discharge data at the two stations to determine the Muskingum method routing coefficients.

6.13   Rework Question 6.12 using a linear reservoir and determine the lag K and pulse duration D parameters.

6.14   What is a response function? Discuss its applications and advantages.

Chapter 7

Multivariate Probability Distributions

Although univariate frequency analysis is often employed for hydrologic and hydraulic design, in reality it might not suffice to meet the design needs and to calculate the corresponding risk appropriately. More than one hydrologic and hydraulic variable may be needed. For example, for designing a drainage system, design peak discharge of a given return period is commonly used in practice, but the total volume and duration of the peak discharge event are also needed. Similarly, for design of a detention reservoir not only is the flood peak discharge needed but also the total volume and duration of the flood.

Similar to hydrologic and hydraulic design, univariate analysis may not always be appropriate in other fields. For example, the transportation engineer designing a highway requires knowledge of peak traffic time as well as peak traffic duration. In environmental engineering, multivariate analysis is needed for water quality analysis. For example, when studying sediment and pollutant loading, frequency analysis of discharge or velocity may also be needed. Therefore this chapter discusses multivariate distributions based on the conventional approach as well as the copula method.

7.1 Existing Multivariate Distributions

7.1.1 Multivariate Normal Distribution

Let random variables $X_1, X_2, ..., X_n$ each be normally distributed. Before presenting the multivariate normal distribution, the bivariate normal distribution is given first.

7.1.1.1 Bivariate Normal Distribution

For the bivariate normal distribution, let X_1 and X_2 be two normally distributed variables as $X_1 \sim N(\mu_1, \sigma_1)$ and $X_1 \sim N(\mu_2, \sigma_2)$, then the joint PDF of X_1 and X_2 is expressed as

$$f_{X_1, X_2}(x_1, x_2) = \frac{1}{2\pi\sigma_1\sigma_2\sqrt{1-\rho^2}} \exp\left[-\frac{z}{2(1-\rho^2)}\right] \qquad (7.1a)$$

where

$$z \equiv \frac{(x_1 - u_1)^2}{\sigma_1^2} - \frac{2\rho(x_1 - \mu_1)(x_2 - \mu_2)}{\sigma_1\sigma_2} + \frac{(x_2 - \mu_2)^2}{\sigma_2^2} \qquad (7.1b)$$

where μ_1, μ_2, σ_1, and σ_2 are the means and standard deviations of variables X_1 and X_2, respectively, and ρ is the coefficient of correlation between variables X_1 and X_2. Symbol $X_1 \sim N(\mu_1, \sigma_1)$ means that X_1 is normally distributed with mean μ_1 and variance σ_1 and the same applies to other variables.

Example 7.1 Let the bivariate variables listed in Table E7-1 be normally distributed after the Box–Cox transformation. What is the joint probability density function of these two variables?

Solution The values of the random variable X_1 representing peak discharge (in cfs) and X_2 representing the corresponding flow volume (in cfs·day), after the Box–Cox transformation, are given as follows:

1. To calculate the first two moments of variables X_1 and X_2, one evaluates

$$m_1 = \bar{X}_1 = \frac{\sum_{i=1,n} x_1(i)}{n} = \frac{2477.3}{62} = 39.36; \quad \sigma_1 = S_1 = \sqrt{\frac{\sum_{i=1,n}(x_1(i) - \bar{X}_1)^2}{n-1}} = \sqrt{\frac{2887.9}{61}} = 6.88$$

$$m_2 = \bar{X}_2 = \frac{\sum_{i=1,n} x_2(i)}{n} = \frac{98034}{62} = 1581.2; \quad \sigma_2 = S_2 = \sqrt{\frac{\sum_{i=1,n}(x_2(i) - \bar{X}_2)^2}{n-1}} = \sqrt{\frac{20314000}{61}}$$

$$= 577.08$$

Table E7-1

No.	X_1	X_2	No.	X_1	X_2	No.	X_1	X_2	No.	X_1	X_2
1	31.583	733.08	17	33.17	1465.8	33	54.272	2513.8	49	44.654	1698
2	33.882	1698.3	18	35.414	1631.1	34	40.018	1442.3	50	46.771	1858.4
3	36.174	1902.4	19	35.275	1095.5	35	48.448	2304.2	51	44.917	2192.9
4	31.649	863.53	20	44.676	1568.3	36	47.678	2773.2	52	42.209	1514.2
5	42.55	2098.4	21	44.808	1548.4	37	31.042	674.27	53	28.181	316.14
6	30.831	1506.1	22	25.374	533.26	38	37.442	1267.6	54	47.874	1860
7	31.248	1111.2	23	42.627	1648.5	39	54.506	2848.8	55	33.935	1096.8
8	33.614	1156.9	24	44.852	1822.4	40	37.325	998	56	31.714	764.94
9	38.867	1828.8	25	42.42	2345.7	41	38.374	1427.4	57	48.086	1915.1
10	43.761	2142.9	26	44.385	2013.3	42	43.474	1855.2	58	39.57	1361.1
11	39.207	2164.7	27	26.725	582.37	43	40.018	1830.3	59	37.246	1452.1
12	42.704	1634.3	28	37.088	1215.3	44	42.831	2064.8	60	45.782	1969.6
13	41.668	1581.5	29	36.556	840.95	45	39.406	1429	61	36.388	1193
14	28.637	596.3	30	31.908	902.43	46	52.622	2565.5	62	50.563	2106.4
15	48.173	2267.2	31	45.261	1984.6	47	41.887	1806.2			
16	33.449	843.3	32	46.942	1971.8	48	44.565	1635.7			

Thus, $X_1 \sim N(39.36, 6.88^2)$ and $X_2 \sim N(1581.2, 577.8^2)$ as shown in Fig. 7-1.

2. To calculate the correlation coefficient r, one evaluates

$$\rho = \frac{\text{cov}(x_1, x_2)}{\sigma_1 \sigma_2} = \frac{E\left[(X_1 - \mu_{X_1})(X_2 - \mu_{X_2})\right]}{\sigma_1 \sigma_2} = \frac{3460.6}{6.88 \times 577.08} = 0.87$$

where cov (x_1, x_2) is the covariance of two variables.

3. Then the joint probability density function, plotted in Fig. 7-2, is

$$f_{X_1, X_2}(x_1, x_2) = \frac{1}{12300} \exp\left[-\frac{z}{0.4862}\right]$$

7.1.1.2 Multivariate Normal Distribution

The bivariate normal distribution can be extended to the multivariate normal distribution (dimension $N \geq 3$), which can be expressed as

$$f(X) = \frac{\exp\left(-\frac{1}{2}(\vec{X} - \vec{\mu})' \Sigma^{-1}(\vec{X} - \vec{\mu})\right)}{\sqrt{(2\pi)|\Sigma|}} \tag{7.2}$$

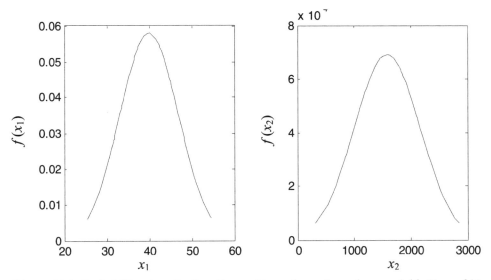

Figure 7-1 *Probability density functions of transformed random variable* X_1 *and* X_2.

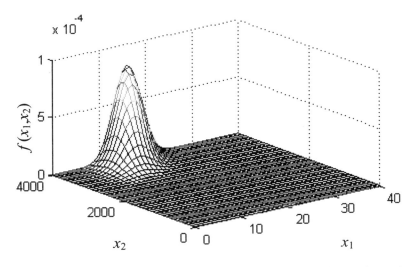

Figure 7-2 *Joint normal probability density function of transformed random variables* X_1 *and* X_2.

where $|\Sigma|$ denotes the determinant of the covariance matrix of the random variable vector $\vec{X}$, with Σ positive definite, and $\vec{\mu}$ denotes the mean vector of the random variable vector $\vec{X}$. Consider the case of $N = 2$. Then the covariance matrix Σ is expressed as

$$\Sigma = \begin{vmatrix} \sigma_1^2 & \rho\sigma_1\sigma_2 \\ \rho\sigma_1\sigma_2 & \sigma_2^2 \end{vmatrix}$$

Similarly, for $N = 3$, the covariance matrix is expressed as

$$\Sigma = \begin{vmatrix} \sigma_1^2 & \rho_{12}\sigma_1\sigma_2 & \rho_{13}\sigma_1\sigma_3 \\ \rho_{12}\sigma_1\sigma_2 & \sigma_2^2 & \rho_{23}\sigma_2\sigma_3 \\ \rho_{13}\sigma_1\sigma_3 & \rho_{23}\sigma_2\sigma_3 & \sigma_3^2 \end{vmatrix}$$

Example 7.2 Similar to Example 7.1, let the trivariate variables—peak discharge (in cfs), volume (in cfs·day), and duration (in days)—be normally distributed after the Box–Cox transformation (see Table E7-2). What is the joint probability density function?

Table E7-2

No.	X_1	X_2	X_3	No.	X_1	X_2	X_3	No.	X_1	X_2	X_3
1	31.58	733.1	4.696	22	35.28	1096	3.887	43	31.04	674.27	4.696
2	33.88	1698	7.229	23	44.68	1568	3.887	44	37.44	1267.6	5.379
3	36.17	1902	7.341	24	44.81	1548	5.217	45	54.51	2848.8	5.379
4	31.65	863.5	4.105	25	25.37	533.3	5.379	46	37.33	998	4.311
5	42.55	2098	7.45	26	42.63	1649	5.379	47	38.37	1427.4	6.251
6	30.83	1506	7.229	27	44.85	1822	5.217	48	43.47	1855.2	5.379
7	31.25	1111	6.115	28	42.42	2346	4.696	49	40.02	1830.3	5.535
8	33.61	1157	6.115	29	44.39	2013	5.379	50	42.83	2064.8	6.251
9	38.87	1829	6.251	30	26.73	582.4	4.696	51	39.41	1429	5.379
10	43.76	2143	5.05	31	37.09	1215	5.379	52	52.62	2565.5	5.217
11	39.21	2165	7.341	32	36.56	841	3.15	53	41.89	1806.2	3.887
12	42.7	1634	5.686	33	31.91	902.4	4.105	54	44.57	1635.7	4.508
13	41.67	1582	6.251	34	45.26	1985	5.05	55	50.56	2106.4	5.976
14	28.64	596.3	4.105	35	46.94	1972	4.696	56	48.09	1915.1	3.887
15	48.17	2267	5.833	36	36.39	1193	4.876	57	47.87	1860	4.105
16	33.45	843.3	4.105	37	39.57	1361	5.379	58	44.65	1698	4.876
17	45.78	1970	5.686	38	33.94	1097	5.686	59	46.77	1858.4	6.638
18	37.25	1452	5.686	39	54.27	2514	5.686	60	44.92	2192.9	5.217
19	31.71	764.9	4.105	40	40.02	1442	4.876	61	42.21	1514.2	5.833
20	33.17	1466	6.115	41	48.45	2304	4.508	62	28.18	316.14	2.555
21	35.41	1631	6.383	42	47.68	2773	6.512				

Solution The first two moments of variables X_1 and X_2 are calculated from Example 7.1 as

$$m_1 = 39.36; \; \sigma_1 = 6.88; \; m_2 = 1581.2; \; \sigma_2 = 577.08$$

The first two moments of X_3 are calculated as

$$m_3 = \frac{\sum\limits_{i=1:n} x_3(i)}{n} = \frac{327.77}{62} = \bar{X}_3 = 5.29; \; \sigma_3 = S_3 = \sqrt{\frac{\sum\limits_{i=1,n}\left[x_3(i) - \bar{X}_3\right]^2}{n-1}} = \sqrt{\frac{65.77}{61}} = 1.04$$

Thus, $X_1 \sim N(39.36, 6.88^2)$, $X_2 \sim N(1{,}581.2, 577.8^2)$, and $X_3 \sim N(5.29, 1.04^2)$ as shown in Fig. 7-3.

The covariance matrix of random variables X_1, X_2, X_3 is determined from

$$\Sigma = \begin{bmatrix} 37.34 & 3460.6 & 0.81 \\ 3460.6 & 33302 & 257.81 \\ 0.81 & 257.81 & 1.08 \end{bmatrix}, \mu = [39.36 \quad 1{,}581.2 \quad 5.29], \; |S| = 2.17 \times 10^6$$

Then the joint probability density function is expressed as

$$f(X) = \frac{\exp\left(-\frac{1}{2}(\bar{X} - [39.36 \; 1581.2 \; 5.29])' \begin{bmatrix} 37.34 & 3460.6 & 0.81 \\ 3460.6 & 33302 & 257.81 \\ 0.81 & 257.81 & 1.08 \end{bmatrix}^{-1} (\bar{X} - [39.36 \; 1581.2 \; 5.29])\right)}{\sqrt{(2\pi) \times 2.07 \times 10^6}}$$

7.1.2 Bivariate Exponential Distribution

To apply the bivariate exponential distribution, each marginal needs to be exponentially distributed. It is possible that the bivariate random variables may be either positively or negatively distributed. Then there are two types of bivariate exponential distributions.

Type 1 model: The bivariate exponential distribution, proposed by Marshall and Ingram (1967), Singh and Singh (1991), and Bacchi et al. (1994), is expressed as

$$f(x_1, x_2) = \alpha\beta[(1 + \alpha\delta x_1)(1 + \beta\delta x_2) - \delta]\exp(-\alpha x_1 - \beta x_2 - \alpha\beta\delta x_1 x_2) \quad (7.3)$$

where variable X_1 is exponentially distributed with parameter α as

$$f_{X_1}(x_1) = \alpha\exp(-\alpha x_1)$$

variable X_2 is also exponentially distributed with parameter β as

$$f_{X_2}(x_2) = \beta\exp(-\beta x_2)$$

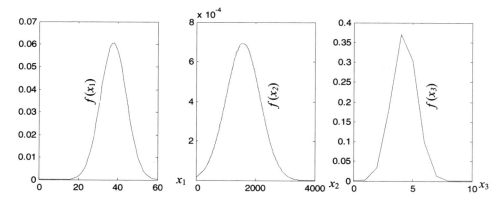

Figure 7-3 *Probability density functions of transformed random variables X_1, X_2, and X_3.*

and δ represents the correlation between the two variables, which is in the range of [0, 1]. This parameter δ is defined through the correlation ρ of two variables as

$$\rho = -1 + \int_0^\infty \frac{1}{1+\delta x} \exp(-x)dx \qquad (7.3a)$$

Note that this bivariate exponential distribution is only valid when $\rho \in [-0.404, 0]$; the relationship of ρ and δ plotted in Fig. 7-4.

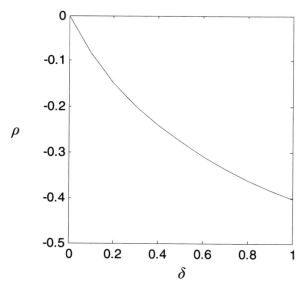

Figure 7-4 *Relationship between correlation coefficient and parameter of bivariate exponential distribution (Model I).*

Type 2 model: The Nagao–Kadoya (NK) model is another bivariate exponential distribution. The NK model relaxes the restriction of the type 1 bivariate exponential distribution, which may be applied to both negatively and positively correlated bivariate random variables. The NK model is expressed as

$$f(x_1, x_2) = \frac{\alpha\beta}{(1-\rho)} \exp\left(-\frac{\alpha x_1}{(1-\rho)} - \frac{\beta x_2}{(1-\rho)}\right) I_0\left(2\frac{\sqrt{\rho}}{1-\rho}\sqrt{x_1 x_2 \alpha\beta}\right) \tag{7.4}$$

where α denotes the parameter of variable X_1, which is exponentially distributed, β denotes the parameter of variable X_2, which is also exponentially distributed, ρ denotes the correlation coefficient of variables X_1 and X_2, and I_0 denotes the modified first-kind Bessel function of zero order expressed as

$$I_0(z) = \sum_{k=0}^{\infty} \frac{(z^2/4)^k}{(k!)^2} \tag{7.4a}$$

Example 7.3 Suppose two low-flow random variables X_1 (duration in days) and X_2 (discharge in cfs) are exponentially distributed, with values as given in Table E7-3. What is the joint bivariate exponential distribution?

Table E7-3

No.	X_1	X_2	No.	X_1	X_2	No.	X_1	X_2	No.	X_1	X_2
1	50	2011	6	39	5166	11	65	2360	16	50	1866
2	40	5199	7	39	1421	12	53	797	17	52	5280
3	49	4930	8	60	1303	13	29	4813	18	40	2455
4	38	2547	9	64	1266	14	36	469	19	51	3041
5	47	1526	10	50	1839	15	48	2312	20	45	698

Solution A calculation of the correlation coefficient gives $\rho = -0.28$. Thus both bivariate exponential distributions may be applied. Parameters of the marginal exponential distributions are estimated for two random variables as

$$\alpha = \frac{1}{\bar{X}_1} = \frac{1}{47.25} = 0.021; \quad \beta = \frac{1}{\bar{X}_2} = \frac{1}{2564.9} = 0.0004$$

The marginal probability density functions are given in Fig. 7-5.

- Type 1 model: The parameter δ is determined by solving

$$\rho = -1 + \int_0^{\infty} \frac{1}{1+\delta x} \exp(-x)\,dx = -0.28,$$

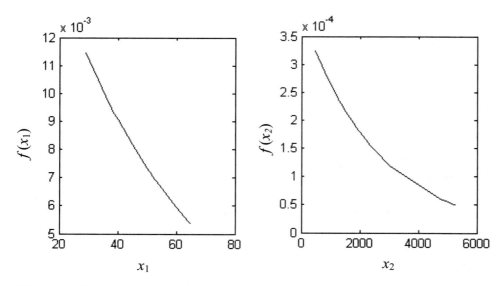

Figure 7-5 *Exponential probability density functions of random variable X_1 and X_2.*

which yields $\delta = 0.508$. Then the joint probability density function, plotted in Fig. 7-6(a), is

$$f(x_1,x_2) = 8.4 \times 10^{-6} \left[(1+0.11x_1)(1+2.03 \times 10^{-4}x_2) - 0.508 \right]$$
$$\times \exp(-0.021x_1 - 0.0004x_2 - 4.27 \times 10^{-6} x_1 x_2)$$

- Type 2 model: For parameters α, β, ρ determined earlier, the bivariate distribution by the NK model, plotted in Fig. 7-6(b), is expressed as

$$f(x_1,x_2) = \frac{8.4 \times 10^{-6}}{0.72} \exp\left(-\frac{0.021x_1}{0.72} - \frac{0.0004x_2}{0.72} \right) I_0 \left(0.0058 \frac{\sqrt{-0.28}}{0.72} \sqrt{x_1 x_2} \right)$$

Notice that, for the modified Bessel function, the variable can be complex.

7.1.3 Bivariate Gumbel Mixed Distribution

Gumbel (1960) proposed the Gumbel mixed distribution with standard Gumbel marginal probability distributions. The general formulation of the Gumbel mixed distribution (cumulative density function) is expressed as

$$F_{X_1,X_2}(x_1,x_2) = F_{X_1}(x_1)F_{X_2}(x_2)\exp\left\{ -\theta \left[\frac{1}{\ln F_{X_1}(x_1)} + \frac{1}{\ln F_{X_2}(x_2)} \right]^{-1} \right\}, 0 \le \theta \le 1 \quad \textbf{(7.5)}$$

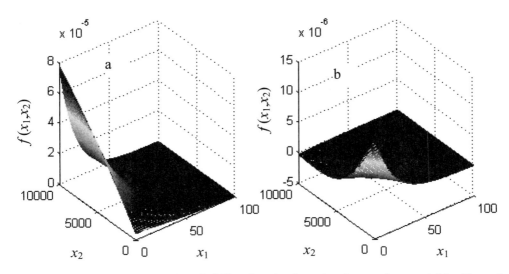

Figure 7-6 *Joint exponential probability density function for random variables X_1 and X_2. (a: Type I model; b: Type II model).*

where $F_{x_1}(x_1)$ and $F_{x_2}(x_2)$ are the Gumbel marginal distributions of random variables X_1 and X_2, $F_{X_1,X_2}(x_1,x_2)$ denotes the joint distribution of two random variables X_1 and X_2, and θ denotes the correlation between the two random variables, which can be estimated as

$$\theta = 2\left[1 - \cos(\pi\sqrt{\rho/6})\right], \quad \text{for} \ \ 0 \le \rho \le 2/3 \tag{7.5a}$$

where ρ is the correlation coefficient of the two random variables.

Example 7.4 Suppose two low-flow random variables X_1 (discharge in cfs) and X_2 (volume in cfs·day) follow the Gumbel distribution, with the data given in Table E7-4. What is the joint cumulative probability distribution?

Table E7-4

No.	X_1	X_2	No.	X_1	X_2	No.	X_1	X_2	No.	X_1	X_2
1	610	35600	6	1100	37213	11	1360	48790	16	1470	38634
2	934	39744	7	1130	49226	12	1370	38682	17	1490	57769
3	949	33010	8	1170	42497	13	1380	45263	18	1490	55766
4	968	58538	9	1210	74840	14	1420	60824	19	1500	41943
5	993	36882	10	1330	47627	15	1460	50895	20	1530	60767

Solution First one determines the parameters of the Gumbel marginal distributions. The probability distribution function of the Gumbel variable is

$$F(x) = \exp\left[-\exp\left(-\frac{x-\beta}{\alpha}\right)\right] \tag{7.5b}$$

Using the method of moments, one obtains

$$\alpha = \sqrt{6}S / \pi \text{ and } \beta = M - 0.577\alpha \tag{7.5c}$$

where M and S are the sample mean and sample standard deviation, respectively, given by

$$M_1 = 1423.2; \ S_1 = 253.08; \ M_2 = 47726; \ S_2 = 10905$$

Then, the distribution parameters are

$$\alpha_1 = 197.32, \ \beta_1 = 1129.3 \ ; \ \alpha_2 = 8502.4 \ , \ \beta_2 = 42820$$

The Gumbel marginal distributions are plotted in Fig. 7-7.

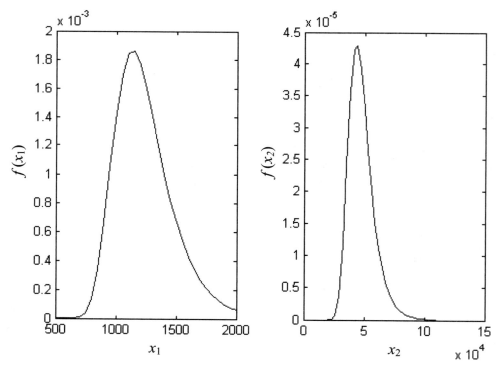

Figure 7-7 *Gumbel probability density functions of random variables* X_1 *and* X_2.

Now one must determine the parameter θ. The value of $\rho = 0.407$ is in the range $[0, 2/3]$. Thus the Gumbel mixed distribution is valid and we have

$$\theta = 2\left[1 - \cos(\pi\sqrt{0.407/6})\right] = 0.68$$

Then the bivariate Gumbel mixed distribution for the two variables, plotted in Fig. 7-8, is

$$F_{X_1,X_2}(x_1,x_2) = F_{X_1}(x_1)F_{X_2}(x_2)\exp\left\{-0.68\left[\frac{1}{\ln F_{X_1}(x_1)} + \frac{1}{\ln F_{X_2}(x_2)}\right]^{-1}\right\}$$

with

$$F_{X_1}(x) = \exp\left[-\exp\left(-\frac{x_1 - 1129.3}{197.32}\right)\right]$$

and

$$F_{X_2}(x) = \exp\left[-\exp\left(-\frac{x_2 - 42820}{8502.4}\right)\right]$$

7.1.4 Bivariate Gumbel Logistic Distribution

Similar to the bivariate Gumbel mixed distribution, the Gumbel logistic distribution is also considered as the bivariate extreme value distribution with Gumbel marginals and may be applied for the representation of joint distribution of extreme hydrologic events.

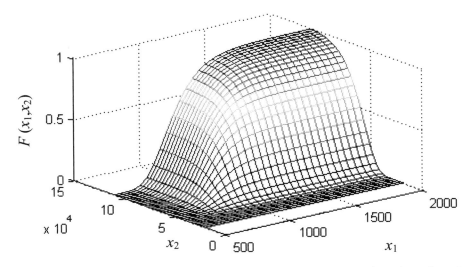

Figure 7-8 *Bivariate Gumbel mixed joint probability distribution function of random variables X_1 and X_2.*

The bivariate Gumbel logistic distribution is expressed as

$$F_{X_1,X_2}(x_1,x_2) = \exp\{-[(-\ln F_{X_1}(x_1))^m + (-\ln F_{X_2}(x_2))^m]^{1/m}\}, m \geq 1 \qquad (7.6)$$

where m is the parameter and Eq. 7.5b gives the marginal distribution representation of variables X_1 and X_2. It will be seen later that Eq. 7.6 has the same form as the Gumbel–Hougaard copula formula.

Example 7.5 Using data in Example 7.4, if the Gumbel logistic distribution is applied for the bivariate frequency analysis of two variables X_1 and X_2, determine its joint distribution.

Solution The Gumbel marginal distributions of two variables were already obtained through Example 7.4. Now one must determine parameter m in the bivariate Gumbel logistic model. This parameter can be estimated through the correlation coefficient of two variables as

$$m = \frac{1}{\sqrt{1-\rho}} \qquad (7.6a)$$

The restriction of Eq. 7.6a is that correlated variables need to be positively correlated.

The correlation coefficient $\rho = 0.41$; then from Eq. 7.6a one gets $m = 1.3$. To this end, the bivariate Gumbel logistic distribution, plotted in Fig. 7-9, is expressed as

$$F_{X_1,X_2}(x_1,x_2) = \exp\{-[(-\ln F_{X_1}(x_1))^{130} + (-\ln F_{X_2}(x_2))^{130}]^{1/130}\}, m \geq 1$$

with the marginal distribution of $F_{X_1}(x_1)$ and $F_{X_2}(x_2)$ given in Example 7.4.

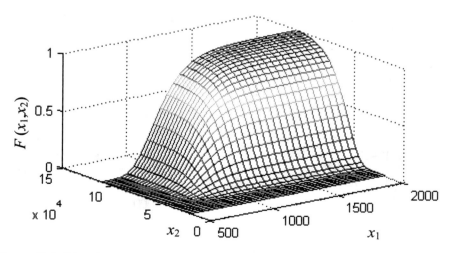

Figure 7-9 *Bivariate Gumbel logistic distribution of random variables X_1 and X_2.*

7.1.5 Bivariate Gamma Distribution

Yue et al. (2001) discussed that the bivariate gamma distribution with the gamma marginals might be useful for bivariate hydrologic frequency analysis.

7.1.5.1 Izawa Model

This bivariate gamma model, developed by Izawa (1953), has the following form:

$$f_{X_1,X_2} = \frac{(x_1 x_2)^{(n-1)/2} x_1^m \exp\left(-\dfrac{\alpha_{X_1} x_1 + \alpha_{X_2} x_2}{1-\eta}\right)}{\Gamma(n)\Gamma(m)(\alpha_{X_1}\alpha_{X_2})^{-(n+1)/2}\alpha_{X_1}^{-m}(1-\eta)\eta^{(n-1)/2}}$$

$$\times \int_0^1 (1-t)^{(n-1)/2} t^{m-1} \exp\left[\alpha_{X_1}\eta x_1 t/(1-\eta)\right] I_{n-1}\left(\frac{2\sqrt{\alpha_{X_1}\alpha_{X_2}\eta x_1 x_2 (1-t)}}{1-\eta}\right) dt \tag{7.7}$$

$$\eta = \rho\sqrt{\beta_{X_1}/\beta_{X_2}} \quad (0 \le \rho \le 1, 0 \le \eta \le 1) \tag{7.7a}$$

$$n = \beta_{X_2}, m = \beta_{X_1} - \beta_{X_2} = \beta_{X_1} - n, n \ge 0, m \ge 0 \tag{7.7b}$$

where $I_{n-1}(.)$ is the modified Bessel function of the first kind, η is the association parameter between random variables X_1 and X_2, and ρ is Pearson's correlation coefficient.

7.1.5.2 Smith–Adelfang–Tubbs (SAT) Model

This bivariate gamma model with gamma marginals was developed by Smith et al. (1982). The joint probability density function has the following form:

$$f_{X_1,X_2}(x_1,x_2) = \begin{cases} \dfrac{K_1}{K_2}\displaystyle\sum_{j=0}^{\infty}\sum_{k=0}^{\infty} c_{jk}(\alpha_{X_1}x_1)^j(\eta\alpha_{X_2}x_2)^{j+k}, & 0 < \eta < 1 \\ f_{X_1}(x_1,\alpha_{X_1},\beta_{X_1})f_{X_2}(x_2,\alpha_{X_2},\beta_{X_2}) \end{cases} \tag{7.8}$$

where $\beta_{X_2} \ge \beta_{X_1}$,

$$K_1 = (\alpha_{X_1}x_1)^{\beta_{x_1}-1}(\alpha_{X_2}x_2)^{\beta_{x_2}-1}\exp\left(-\frac{\alpha_{X_1}x_1+\alpha_{X_2}x_2}{1-\eta}\right) \tag{7.8a}$$

$$K_2 = (1-\eta)^{\beta_{x_1}}\Gamma(\beta_{X_1})\Gamma(\beta_{X_2}-\beta_{X_1}) \tag{7.8b}$$

$$c_{jk} = \frac{\eta^{j+k}\Gamma(\lambda_{X_2}-\lambda_{X_1}+k)}{(1-\eta)^{2j+k}\Gamma(\lambda_{X_2}+j+k)j!k!} \tag{7.8c}$$

and

$$\eta = \rho\sqrt{\beta_{X_1}/\beta_{X_2}} \tag{7.8d}$$

Example 7.6 Suppose the correlated rainfall variables intensity and depth have the gamma marginals with the data given in Table E7-6. Determine the joint probability distribution of rainfall intensity and depth.

Table E7-6

No.	X_1	X_2	No.	X_1	X_2	No.	X_1	X_2	No.	X_1	X_2	No.	X_1	X_2	No.	X_1	X_2	No.	X_1	X_2
1	2.65	7.95	9	1.09	2.17	17	2.47	7.42	25	3.2	6.5	33	1.2	6.2	41	1.3	6.5	49	3.04	6.08
2	1.8	3.59	10	1.12	4.49	18	2.25	4.5	26	3.5	6.9	34	1.9	3.8	42	4.5	4.5	50	1.09	5.43
3	3.03	9.1	11	3.03	3.03	19	2.78	5.55	27	4.2	8.4	35	1.6	8.2	43	2.1	4.2	51	2.65	10.6
4	1.26	2.51	12	3.25	3.25	20	2.24	2.24	28	5.3	5.3	36	3.1	12	44	1.1	5.5	52	1.89	3.77
5	2.91	2.91	13	1.6	3.19	21	0.72	2.87	29	1.3	4	37	1	7.2	45	3.2	9.7	53	1.21	4.84
6	2.65	5.3	14	2.54	5.08	22	3.26	6.52	30	2.7	11	38	1.3	6.4	46	3.2	9.5	54	2.2	15.4
7	4.41	4.41	15	8.05	8.05	23	2.43	4.85	31	1.7	5.2	39	1	3.8	47	1.7	5.2	55	2.74	5.48
8	3.69	7.38	16	2.24	2.24	24	1.11	4.43	32	3.2	13	40	1.3	6.5	48	7.3	7.3			

Solution

1. First one needs to determine the marginal distribution of rainfall intensity X_1 and depth X_2. Consider the gamma probability density function is expressed as

$$f(x,\alpha,\beta)=\frac{1}{\Gamma(\beta)}\alpha^\beta x^{\beta-1}\exp(-\alpha x)$$

   Following the maximum likelihood method, one can obtain the parameters of rainfall intensity X_1 and depth X_2:

$$X_1 \sim \Gamma(1.5, 3.81) \text{ and } X_2 \sim \Gamma(0.85, 5.09)$$

   with a correlation coefficient equal to 0.28. The probability density functions are plotted and shown in Fig. 7-10.

2. Using the Izawa bigamma model for determination of parameters, one obtains

$$n = \beta_{X_1} = 3.81; m = \beta_{X_2} - \beta_{X_1} = 5.09 - 3.81 = 1.28; \eta = 0.28\sqrt{\beta_{X_1}/\beta_{X_2}} = 0.32$$

   By substituting these parameters into Eq. 7.7, the joint probability density function of the Izawa bigamma model is thus obtained. The results are plotted in Fig. 7-11.

3. Using the SAT model, one obtains $K_2 = 0.98$ by Eq. 7.8b and η is obtained from the previous model. Substituting the parameters and Eqs. 7.8a to 7.8d into Eq. 7.8 gives the joint probability density function of the SAT model (see Fig. 7-12).

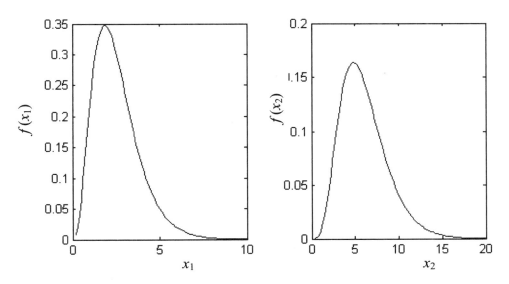

Figure 7-10 *Gamma probability density functions of random variables* X_1 *and* X_2.

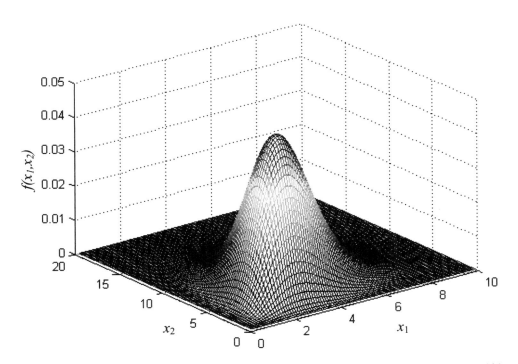

Figure 7-11 *Bivariate gamma distribution (Izawa model) of random variables* X_1 *and* X_2.

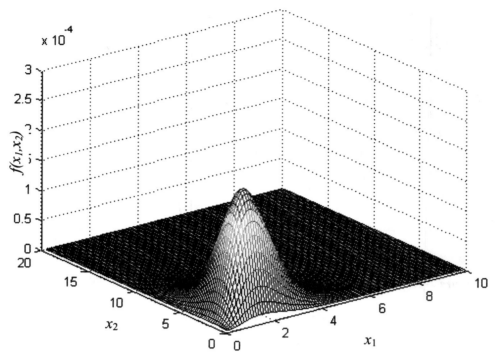

Figure 7-12 *Bivariate gamma distribution (SAT model) of random variables X_1 and X_2.*

7.1.6 Bivariate Log-Normal Distribution

Yue (2000) discussed the application of the bivariate log-normal distribution in hydrological frequency analysis. Consider the univariate log-normal distribution expressed as

$$f_X(x) = \frac{1}{x\sqrt{2\pi\sigma_Y^2}} \exp\left[-\frac{1}{2}\left(\frac{\ln x - \mu_Y}{\sigma_Y}\right)^2\right] \tag{7.9}$$

where $Y = \ln X$ and μ_Y and s_Y are the mean and the standard deviation, respectively, of Y.

The bivariate log-normal density function with log-normal marginals is expressed as

$$f_{X_1,X_2}(x_1,x_2) = \frac{1}{2\pi\sigma_{Y_1}\sigma_{Y_2}\sqrt{1-\rho^2}} \exp\left(-\frac{z}{2}\right)$$

$$z = \frac{1}{1-\rho^2}\left[\left(\frac{\ln x_1 - \mu_{Y_1}}{\sigma_{Y_1}}\right)^2 - 2\rho\left(\frac{\ln x_1 - \mu_{Y_1}}{\sigma_{Y_1}}\right)\left(\frac{\ln x_2 - \mu_{Y_2}}{\sigma_{Y_2}}\right) + \left(\frac{\ln x_2 - \mu_{Y_2}}{\sigma_{Y_2}}\right)^2\right] \tag{7.10}$$

$$\rho = \frac{E\left[\left(Y_1 - \mu_{Y_1}\right)\left(Y_2 - \mu_{Y_2}\right)\right]}{\sigma_{Y_1}\sigma_{Y_2}} \qquad (7.10a)$$

Example 7.7 Suppose the correlated rainfall variables $X_1 \sim$ rainfall intensity (in inches/day) and $X_2 \sim$ depth (in inches) follow the log-normal distribution, with the data given in Table E7-7. Determine the joint distribution of rainfall intensity and depth.

Table E7-7

No.	X_1	X_2	No.	X_1	X_2	No.	X_1	X_2
1	4.3	4.30	11	1.5	7.72	21	2.03	6.10
2	4.95	4.95	12	2.2	11.17	22	7.15	7.15
3	1.24	3.73	13	1.3	5.21	23	2.24	6.72
4	2.5	7.51	14	3.1	3.13	24	2.25	4.50
5	1.7	5.11	15	0.9	5.31	25	1.5	6.00
6	3.04	9.11	16	2.2	8.68	26	2.3	6.91
7	6.11	6.11	17	2.2	4.42	27	1.08	4.31
8	0.74	2.96	18	0.9	5.37	28	2.06	14.40
9	1.67	8.36	19	1.9	5.54	29	1.92	7.67
10	2.95	11.79	20	2.8	8.52			

Solution

1. To determine the parameters of the log-normal (LN2) marginal distribution of rainfall intensity X_1 and depth X_2, one takes the logarithm of variables X_1 and X_2. For rainfall intensity, one gets

$$\mu_{Y_1} = \frac{\sum_{i=1}^{29} \ln(x_1(i))}{29} = \frac{21.56}{29} = 0.74; \quad \sigma_{Y_1} = \left(\frac{\sum_{i=1}^{29}\left(\ln(x_1(i)) - 0.74\right)^2}{28}\right)^{0.5} = 0.54$$

So, $X_1 \sim$ LN2($0.74, 0.54^2$). Similarly, parameters for rainfall depth are $\mu_{Y_2} = 1.82$; $\sigma_{Y_2} = 0.38$. Thus, $X_2 \sim$ LN2($1.82, 0.38^2$). The log-normal marginals are shown in Fig. 7-13.

2. To determine the parameters needed for the bivariate log-normal distribution, one first calculates the correlation coefficient of the log transformed

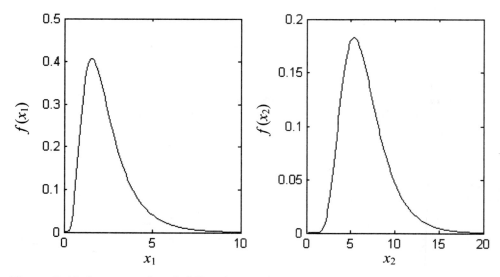

Figure 7-13 *Log-normal probability density functions of random variables* X_1 *and* X_2.

rainfall variables: $\rho = 0.25$. The corresponding bivariate log-normal distribution is then obtained as

$$f_{X_1, X_2}(x_1, x_2) = \frac{1}{1.25} \exp\left(-\frac{z}{2}\right)$$

$$z = \frac{1}{0.94}\left[\left(\frac{\ln x_1 - 0.74}{0.54}\right)^2 - 0.5\left(\frac{\ln x_1 - 0.74}{0.54}\right)\left(\frac{\ln x_2 - 1.82}{0.38}\right) + \left(\frac{\ln x_2 - 1.82}{0.38}\right)^2\right]$$

These results are plotted in Fig. 7-14.

7.1.7 Box–Cox Transformation

The Box–Cox transformation is used to transform the non-normal-distributed random variables to normally (N) distributed random variables using the following equation:

$$Y = \begin{cases} \dfrac{X^\lambda - 1}{\lambda}, & \lambda \neq 0 \\ \ln(X), & \lambda = 0 \end{cases} \tag{7.11}$$

Example 7.8 Consider the rainfall intensity variable in Example 7.7. Determine parameter λ for the Box–Cox transformation.

Solution The rainfall intensity variable is expressed as random variable X_1. Parameter λ for the Box–Cox transformation is determined through the

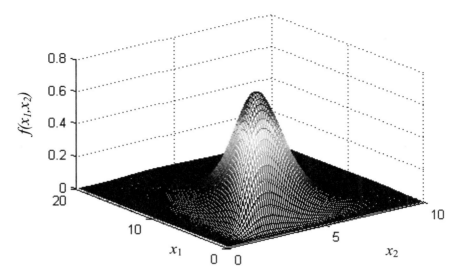

Figure 7-14 *Bivariate log-normal density function of random variables* X_1 *and* X_2.

maximum likelihood method as $\lambda = -0.07$. The first four moments of the transformed variable are

$$m = 0.76; \quad S = 0.495; \quad \text{skewness} = 0.005; \quad \text{kurtosis} = 2.56$$

Thus, it is safe to say that the transformed variable by the Box–Cox transformation follows the normal distribution $\sim N(0.76, 0.495)$.

7.1.8 Conditional Distributions

7.1.8.1 Conditional Bivariate Distributions

Let $f_{X_1,X_2}(x_1, x_2)$ be the joint PDF of random variables X_1 and X_2, and let $f_{X_1}(x_1)$ and $f_{X_2}(x_2)$ be the marginal PDFs of X_1 and X_2, respectively.

1. The conditional PDF of X_1 given $X_2 = x_2$, $f_{X_1|X_2=x_2}(x_1 \mid x_2)$, can be written as

$$f_{X_1|X_2=x_2} = \frac{f_{X_1,X_2}(x_1, x_2)}{f_{X_2}(X_2 = x_2)} \qquad (7.12)$$

and the conditional cumulative distribution function (CCDF) of X_1 given $X_2 = x_2$ can be expressed as

$$F_{X_1|X_2=x_2}(X_1 \le x_1 \mid X_2 = x_2) = \int_{-\infty}^{x_1} f_{X_1|X_2=x_2}(u \mid x_2)du = \frac{\displaystyle\int_{-\infty}^{x_1} f_{X_1,X_2}(u_1, x_2)du}{f_{X_2}(x_2)} \qquad (7.12a)$$

2. The conditional cumulative distribution function of $X_1 \le x_1$ given $X_2 \le x_2$, $F'_{X_1|X_2 \le x_2}(x_1|x_2)$, is given as

$$F'_{X_1|X_2 \le x_2}(X_1 \le x_1 | X_2 \le x_2) = \frac{F_{X_1,X_2}(X_1 \le x_1, X_2 \le x_2)}{F_{X_2}(X_2 \le x_2)} = \frac{F_{X_1,X_2}(x_1,x_2)}{F_{X_2}(x_2)} \quad \textbf{(7.12b)}$$

3. Special case: If random variables X_1 and X_2 are independent, then the conditional bivariate distribution based on $X_2 = x_2$ and $X_2 \le x_2$ is the same and is

$$F_{X_1|X_2=x_2} = F'_{X_1|X_2 \le x_2} = F_{X_1}(x_1) \quad \textbf{(7.12c)}$$

7.1.8.2 Conditional Trivariate Distributions

The conditional trivariate distribution can be derived in a manner similar to that used for the conditional bivariate distribution. Let $f_{X_1,X_2,X_3}(x_1,x_2,x_3)$ be the joint PDF of random variables X_1, X_2, and X_3, and let $f_{X_1}(x_1)$, $f_{X_2}(x_2)$, and $f_{X_3}(x_3)$ be the marginal PDFs of X_1, X_2, and X_3, respectively.

1. The conditional PDF of X_1 and X_2, given $X_3 = x_3$, is expressed as

$$f_{X_1,X_2|X_3=x_3} = \frac{f_{X_1,X_2,X_3}(x_1,x_2,x_3)}{f_{X_3}(X_3 = x_3)} \quad \textbf{(7.12d)}$$

2. The conditional PDF of X_1, given $X_2 = x_2$ and $X_3 = x_3$, is expressed as

$$f_{X_1|X_2=x_2,X_3=x_3} = \frac{f_{X_1,X_2,X_3}(x_1,x_2,x_3)}{f_{X_2,X_3}(X_2 = x_2, X_3 = x_3)} \quad \textbf{(7.12e)}$$

3. The conditional probability distribution of X_1 and X_2, given $X_3 \le x_3$, is expressed as

$$F_{X_1,X_2|X_3=x_3} = \frac{F_{X_1,X_2,X_3}(x_1,x_2,x_3)}{F_{X_3}(X_3 \le x_3)} = \frac{F_{X_1,X_2,X_3}(x_1,x_2,x_3)}{F_{X_3}(x_3)} \quad \textbf{(7.12f)}$$

4. The conditional probability distribution of X_1, given $X_2 \le x_2$ and $X_3 \le x_3$, is expressed as

$$F'_{X_1|X_2 \le x_2,X_3 \le x_3} = \frac{F_{X_1,X_2,X_3}(x_1,x_2,x_3)}{F_{X_2,X_3}(X_2 \le x_2, X_3 \le x_3)} = \frac{F_{X_1,X_2,X_3}(x_1,x_2,x_3)}{F_{X_2,X_3}(x_2,x_3)} \quad \textbf{(7.12g)}$$

5. Special case: Let X_1, X_2, and X_3 be independent variables. Then the conditional distributions of X_1 and X_2, given $X_3 = x_3$, and of X_1 and X_2, given $X_3 \le x_3$, are the same and are expressed as

$$F_{X_1,X_2|X_3=x_3} = F_{X_1,X_2|X_3 \leq x_3} = F_{X_1,X_2}(x_1,x_2) = F_{X_1}(x_1)F_{X_2}(x_2) \qquad (7.12h)$$

The conditional distributions of X_1, given $X_2 = x_2$ and $X_3 = x_3$, and of X_1, given $X_2 \leq x_2$ and $X_3 \leq x_3$, are the same and are expressed as

$$F_{X_1|X_2=x_2,X_3=x_3} = F_{X_1|X_2 \leq x_2, X_3 \leq x_3} = F_{X_1}(x_1) \qquad (7.12i)$$

In Example 7.9, only bivariate conditional distributions are considered. Trivariate conditional distribution can be obtained in a similar manner.

Example 7.9 Consider Example 7.1. What is the probability density function of $f_{X_1|X_2}(x_1 \mid X_2 = 2000)$ and $f_{X_1|X_2}(x_1 \mid X_2 \leq 2000)$?

Solution $f_{X_1|X_2}(x_1 \mid X_2 = 2000)$ can be solved by using Eq. 7.12. The joint distribution and each marginal have already been calculated in Example 7.1. Then

$$f_{X_1|X_2}(x_1 \mid X_2 = 2000) = \frac{f_{X_1,X_2}(x_1, x_2 = 2000)}{f_{X_2}(x_2 = 2000)}$$

$$f_{X_1|X_2}(x_1 \mid X_2 \leq 2000) = \frac{\partial}{\partial x_1}\left(\frac{F_{X_1,X_2}(x_1, x_2 \leq 2000)}{F_{X_2}(x_2 \leq 2000)}\right)$$

The conditional probability density functions are plotted in Fig. 7-15.

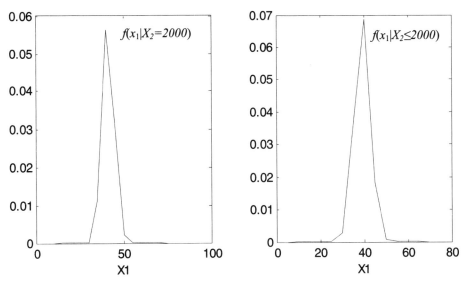

Figure 7-15 Conditional probability density functions.

7.2 Return Period

7.2.1 Bivariate Return Period

Let random variables X_1 and X_2 have the bivariate distribution $F_{X_1,X_2}(x_1,x_2)$. Then the joint return period can be expressed as

$$T_{X_1,X_2}(x_1,x_2) = \frac{1}{1 - F_{X_1,X_2}(x_1,x_2)} \qquad (7.13)$$

The conditional bivariate return period of $X_1 \geq x_1$, given $X_2 = x_2$, is expressed as

$$T_{X_1|X_2=x_2}(x_1 \mid X_2 = x_2) = \frac{1}{1 - F_{X_1|X_2=x_2}(x_1 \mid x_2 = x_2)} \qquad (7.13a)$$

The conditional bivariate return period of $X_1 \geq x_1$, given $X_2 \leq x_2$, is expressed as

$$T_{X_1|X_2=x_2}(x_1 \mid X_2 \leq x_2) = \frac{1}{1 - F_{X_1|X_2 \leq x_2}(x_1 \mid x_2 \leq x_2)} \qquad (7.13b)$$

7.2.2 Trivariate Return Period

1. Let random variables X_1 X_2, and X_3 have the trivariate distribution $F_{X_1,X_2,X_3}(x_1,x_2,x_3)$. Then the joint return period can be expressed as

$$T_{X_1,X_2,X_3}(x_1,x_2,x_3) = \frac{1}{1 - F_{X_1,X_2,X_3}(x_1,x_2,x_3)} \qquad (7.13c)$$

2. The conditional trivariate return period of $X_1 \geq x_2$, or $X_2 \geq x_2$, given $X_3 = x_3$, is expressed as

$$T_{X_1,X_2|X_3=x_3} = \frac{1}{1 - F_{X_1,X_2|X_3=x_3}(x_1,x_2 \mid X_3 = x_3)} \qquad (7.13d)$$

3. The conditional trivariate return period of $X_1 \geq x_2$, or $X_2 \geq x_2$, given $X_3 \leq x_3$, is expressed as

$$T_{X_1,X_2|X_3 \leq x_3} = \frac{1}{1 - F_{X_1,X_2|X_3=x_3}(x_1,x_2 \mid X_3 \leq x_3)} \qquad (7.13e)$$

4. The conditional trivariate return period of $X_1 \geq x_1$, given $X_2 = x_2$ and $X_3 = x_3$, is expressed as

$$T_{X_1 | X_2 = x_2, X_3 = x_3} = \frac{1}{1 - F_{X_1 | X_2 = x_2, X_3 = x_3}(x_1 \mid X_2 = x_2, X_3 = x_3)} \qquad \text{(7.13f)}$$

5. The conditional trivariate return period of $X_1 \geq x_1$, given $X_2 \leq x_2$ and $X_3 \leq x_3$, is expressed as

$$T_{X_1 | X_2 \leq x_2, X_3 \leq x_3} = \frac{1}{1 - F_{X_1 | X_2 \leq x_2, X_3 \leq x_3}(x_1 \mid X_2 \leq x_2, X_3 \leq x_3)} \qquad \text{(7.13g)}$$

In Example 7.10, only bivariate conditional return periods are considered. The trivariate conditional return period can be obtained in a similar manner.

Example 7.10 Consider Example 7.4. Determine the conditional return period given $X_2 \leq$ 50,000, 60,000, and 90,000 cfs·day.

Solution In Example 7.4, marginal distributions and joint distribution were calculated. Thus the conditional return period can be estimated by using Eq. 7.12 as

$$F_{X_1 | X_2}(x_1 \mid X_2 \leq x_2) = \frac{F_{X_1, X_2}(x_1, x_2)}{F_{X_2}(x_2)}$$

and the corresponding conditional return period is directly expressed from Eq. 7.13b. Substituting the results in Example 7.4 into these equations one can obtain the conditional return period as plotted in Fig. 7-16.

7.3 Derivation of Multivariate Distributions

The problem with the conventional approach to deriving multivariate probability distributions is that the marginal distributions need to be of the same type and if the marginal distributions are of the same type, then the Box–Cox transformation is needed to apply the multivariate normal distribution. This problem can be circumvented by applying the copula method.

7.3.1 Copula Method

7.3.1.1 Definition of Copula

Consider multivariate random variables $X_1, X_2, \ldots, X_N$ with marginal distribution functions as $F_{X_i}(x) = P_{X_i}(X_i < x_i)$, where N is the number of random variables and x_i is the value of random variable X_i $(X_1, \ldots, X_N)$. The joint distribution of multivariate random variables $X_1, X_2, \ldots, X_N$ is then expressed as

$$F_{X_1, \ldots, X_N}(x_1, x_2, \ldots, x_N) = H = P[X_1 \leq x_1, X_2 \leq x_2, \ldots, X_N \leq x_N] \qquad \text{(7.14)}$$

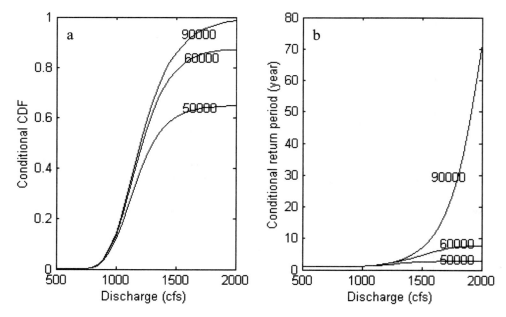

Figure 7-16 *Conditional probability distribution of discharge (a) and conditional return period (b) given different flood volumes (cfs·day).*

Copulas are functions that connect multivariate probability distributions to their one-dimensional marginal probability distributions. Thus, the multivariate probability distribution F is expressed in terms of its marginals and the associated dependence function C as

$$C\left[F_{X_1}(x_1), F_{X_2}(x_2), ..., F_{X_N}(x_N)\right] = F_{X_1, X_2, ...X_N}(x_1, x_2, ..., x_N)$$

where the copula C is a mapping uniquely determined on the unit square whenever $F_{X_i}(x_i)$, $i = 1, 2, ..., N$, are continuous. A copula captures the essential features of the dependence between the random variables and is essentially a function that connects the multivariate probability distribution to its marginals. Thus the problem of determining H reduces to determining C. There are a variety of copulas that can be employed for deriving multivariate distributions. De Michele and Salvadori (2003) applied the copula method for rainfall analysis. Favre et al. (2004) applied elliptical and Archimedean copulas and copulas with quadratic section for frequency analysis of flood peaks as well as that of peak flows and volumes. Salvadori and De Michele (2004) employed copulas for bivariate frequency analysis of hydrological events. De Michele et al. (2005) used a two-dimensional copula to derive a bivariate probability distribution for evaluating the adequacy of dam spillways. Zhang and Singh (2006) used Archimedean copulas for flood frequency analysis. The Archimedean copula family is perhaps the most popular and commonly used family in hydrology and environmental engineering.

7.3.1.2 Archimedean Copula

The Archimedean copulas are found to be perhaps the most important copulas for hydrologic analysis for the following reasons:

1. They can be easily constructed.
2. A large variety of copula families belong to this class (see Nelson 1999).
3. The Archimedean copulas have desirable mathematical properties, such as convexity of generating function, convexity of level curves, continuity, and generation of copula by nonparametric methods.

7.3.1.3 Two-Dimensional Archimedean Copula

Let φ be a continuous, strictly decreasing function from I to $[0, \infty]$ such that $\varphi(1) = 0$, defined as

$$\varphi\left[C_\theta(u_1, u_2)\right] = \varphi(u_1) + \varphi(u_2) \tag{7.15}$$

Let $\varphi^{[-1]}$ be the pseudo-inverse of φ. Then, C is the function from I^2 to I, where $I \in [0, 1]$ and is given as

$$C_\theta(u_1, u_2) = \varphi^{-1}\left[\varphi(u_1) + \varphi(u_2)\right] \tag{7.16}$$

where $\varphi(\bullet)$ is the generating function of the Archimedean copula, θ is the copula parameter, which is hidden in the generating function, and C_θ denotes the representation of the copula. Thus, the Archimedean copula is determined from Eq. 7.15. Furthermore, the N-dimensional Archimedean copulas can be defined as described in the following section.

7.3.1.4 *N*-Dimensional Archimedean Copula

Following the two-dimensional Archimedean copula, one can express the N-dimensional copula as

$$C_\theta^N(\mathbf{u}) = \varphi^{-1}\left[\varphi(u_1) + \varphi(u_2) + \ldots + \varphi(u_N)\right] \tag{7.17}$$

Following Nelson (1999), one has that the functions C^N in Eq. 7.17 are the serial iterates of the two-dimensional Archimedean copulas generated by φ. Then for $N \geq 3$,

$$C_\theta^N(u_1, u_2, \ldots, u_N) = C_\theta\left[C_\theta^{N-1}(u_1, u_2, \ldots, u_{N-1}), u_N\right] \tag{7.18}$$

7.3.1.5 Archimedean Copula Families

There exists a large variety of Archimedean copula families that are used for constructing copulas to represent multivariate distributions. Here the most widely used one-parameter Archimedean copulas are introduced. The two-dimensional widely used Archimedean copulas are given first.

Gumbel–Hougaard Archimedean Copula

The Gumbel–Hougaard Archimedean copula was first introduced by Gumbel (1960). Nelson (1999) discussed that the Gumbel–Hougaard copula can be considered as the representation of the bivariate extreme value distribution. By virtue of this characteristic, the Gumbel–Hougaard Archimedean copula might be a suitable candidate for multivariate hydrologic frequency analysis of extreme hydrological events (i.e., peak discharge and the corresponding volume and duration). This copula can be expressed as

$$C_\theta(u_1, u_2) = C_\theta\left[F_{X_1}(x_1), F_{X_2}(x_2) \right] = F_{X_1, X_2}(x_1, x_2)$$
$$= \exp\left\{ -\left[(-\ln u_1)^\theta + (-\ln u_2)^\theta \right]^{1/\theta} \right\}, \ \theta \in [1, \infty) \qquad (7.19)$$

where θ is a parameter of the generating function $\varphi(t) = (-\ln t)^\theta$, with $t = u_1$ or u_2 as a uniformly distributed random variable varying from 0 to 1, and $\tau = 1 - \theta^{-1}$, which is Kendall's coefficient of correlation between random variables X_1 and X_2. Note that $\varphi(u_1) = (-\ln u_1)^\theta$ and $\varphi(u_2) = (-\ln u_2)^\theta$ in this equation. Parameters θ and τ will have the same connotation in the following three copulas.

Example 7.11 Determine the joint distribution of random variables X_1 and X_2 using the Gumbel–Hougaard copula, where the marginal distributions of X_1 and X_2 are given as $X_1 \sim \exp(\lambda)$, and $X_2 \sim N(\mu, \sigma^2)$.

Solution The marginal distributions of two random variables X_1 and X_2 are

$$u_1 = F_{X_1}(x_1) = 1 - \exp(-\lambda x_1); \ \ u_2 = F_{X_2}(x_2) = \Phi(\frac{x_2 - \mu}{\sigma})$$

Substituting the marginal distribution obtained into Eq. 7.17, one can express the joint distribution through the Gumbel–Hougaard copula as

$$F(x_1, x_2) = C_\theta(u_1, u_2) = \exp\left(-\left\langle \left\{ -\ln\left[1 - \exp(-\lambda x_1) \right] \right\}^\theta + \left\{ -\ln\left[\Phi(\frac{x_2 - \mu}{\sigma}) \right] \right\}^\theta \right\rangle^{1/\theta} \right)$$

where $F_{X_2}(x_2)$ is the cumulative probability of X_2.

Ali–Mikhail–Haq Archimedean Copula

The Ali–Mikhail–Haq Archimedean copula was developed by Ali et al. (1978). This Archimedean copula family was developed based on the concept of the univariate logistic distribution, which may be specified by considering a suitable form for the odds in favor of a failure against survival. The parameter of this copula is a measure of departure from independence or measure of the association between two variables. This copula can be expressed as

$$C_\theta(u_1, u_2) = C_\theta\left[F_{X_1}(x_1), F_{X_2}(x_2)\right] = F_{X_1, X_2}(x_1, x_2)$$

$$= \frac{u_1 u_2}{1 - \theta(1 - u_1)(1 - u_2)}, \theta \in [-1, 1) \tag{7.20}$$

with

$$\varphi(t) = \ln\frac{1 - \theta(1 - t)}{t}, \quad \tau = \left(\frac{3\theta - 2}{\theta}\right) - \frac{2}{3}\left(1 - \frac{1}{\theta}\right)^2 \ln(1 - \theta) \tag{7.20a}$$

Example 7.12 Determine the joint distribution of random variables X_1 and X_2 using the Ali–Mikhail–Haq copula, where the marginal distributions of X_1 and X_2 are given as $X_1 \sim \exp(\lambda)$, and $X_2 \sim$ Gumbel(α, β). The marginal distribution of X_1 and X_2 is given as

$$u_1 = F_{X_1}(x_1) = 1 - \exp(-\lambda x_1); \ u_2 = \exp\left[-\exp(-\frac{x_2 - \beta}{\alpha})\right]$$

Substituting u_1 and u_2 into Eq. 7.20, one can obtain the joint distribution through the Ali–Mikhail–Haq copula as

$$F_{X_1, X_2}(x_1, x_2) = C_\theta(u_1, u_2) = \frac{\left[1 - \exp(-\lambda x_1)\right]\left\{\exp\left[-\exp(-\frac{x_2 - \beta}{\alpha})\right]\right\}}{1 - \theta \exp(-\lambda x_1)\left\{1 - \exp\left[-\exp(-\frac{x_2 - \beta}{\alpha})\right]\right\}}$$

Frank Archimedean Copula

The Frank Archimedean copula was developed by Frank (1979). The Frank copula satisfies all the conditions for the construction of bivariate distributions with fixed marginals. It is absolutely continuous and has full support on the unit square. This copula can be expressed as

$$C_\theta(u_1, u_2) = C_\theta\left[F_{X_1}(x_1), F_{X_2}(x_2)\right] = F_{X_1, X_2}(x_1, x_2)$$

$$= \frac{1}{\theta}\ln\left\{1 + \frac{\left[\exp(\theta u) - 1\right]\left[\exp(\theta v) - 1\right]}{\exp(\theta) - 1}\right\}, \theta \neq 0 \tag{7.21}$$

with

$$\varphi(t) = \ln\left[\frac{\exp(\theta t) - 1}{\exp(\theta) - 1}\right], \quad \tau = 1 - \frac{4}{\theta}[D_1(-\theta) - 1] \tag{7.21a}$$

where D_1 is the first-order Debye function D_k, which is defined as

$$D_k(\theta) = \frac{k}{x^k} \int_0^\theta \frac{t^k}{\exp(t)-1} dt, \ \theta > 0 \qquad (7.21b)$$

The Debye function D_k with negative argument can be expressed as

$$D_k(-\theta) = D_k(\theta) + \frac{k\theta}{k+1} \qquad (7.21c)$$

Example 7.13 Determine the joint distribution of random variables X_1 and X_2 by the Frank copula, where the marginal distributions of X_1 and X_2 are given as $X_1 \sim \exp(\lambda)$, and $X_2 \sim \text{Gumbel}(\alpha, \beta)$.

Solution Again for two random variables X_1 and X_2, we have

$$u_1 = F_{X_1}(x_1) = 1 - \exp(-\lambda x_1); \ u_2 = \exp\left[-\exp\left(-\frac{x_2 - \beta}{\alpha}\right)\right]$$

Substituting u_1 and u_2 into Eq. 7.21, one can obtain the joint distribution through the Frank copula as

$$F_{X_1,X_2}(x_1,x_2) = C_\theta(u_1,u_2)$$

$$= \frac{1}{\theta} \ln\left(1 + \frac{\left\langle \exp\{\theta[1 - \exp(\lambda x_1)]\} - 1\right\rangle \exp\left\{\theta \exp\left[\exp\left(-\frac{x_2 - \beta}{\alpha}\right)\right] - 1\right\}}{\exp(\theta) - 1}\right)$$

Cook–Johnson (Clayton) Archimedean Copula

This copula can be used for modeling nonelliptically symmetric (non-normal) multivariate data (Nelson 1999). When $\theta = 0$, this copula represents the bivariate logistic distribution. The formulation of this copula is expressed as follows:

$$C_\theta(u_1,u_2) = C_\theta\left[F_{X_1}(x_1), F_{X_2}(x_2)\right] = F_{X_1,X_2}(x_1,x_2) = \left(u_1^{-\theta} + u_2^{-\theta} - 1\right)^{-1/\theta}, \ \theta \geq 0 \ (7.22)$$

with

$$\varphi(t) = t^{-\theta} - 1, \ \tau = \frac{\theta}{\theta + 2} \qquad (7.22a)$$

Example 7.14 Determine the joint distribution of random variables X_1 and X_2 by the Cook–Johnson Archimedean copula, where the marginal distributions of X_1 and X_2 are given as $X_1 \sim \exp(\lambda)$ and $X_2 \sim \text{Gumbel}(\alpha, \beta)$.

Solution The marginal distributions of random variables X_1 and X_2 are given in Example 7.13. With this in hand and substituting the marginal distributions into

Eq. 7.22, one can obtain the joint distribution through the Cook–Johnson (Clayton) copula as

$$F_{X_1,X_2}(x_1,x_2) = C_\theta(u_1,u_2)$$

$$= \left\{ \left[1-\exp(-\lambda x_1)\right]^{-\theta} + \exp\left[-\theta\exp(-\frac{x_2-\beta}{\alpha})\right] - 1 \right\}^{-1/\theta}$$

Joe Archimedean Copula

The Joe Archimedean copula was first introduced by Joe (1993). When $\theta = 1$, this copula represents the bivariate distribution of two independent variables. Again this copula is applicable to extreme value analysis and is independent of univariate marginals. This copula can be expressed as

$$C_\theta(u_1,u_2) = C_\theta\left[F_{X_1}(x_1), F_{X_2}(x_2)\right] = H(x_1,x_2)$$
$$= 1 - \left[(1-u_1)^\theta + (1-u_2)^\theta - (1-u_1)^\theta(1-u_2)^\theta\right]^{1/\theta} \qquad (7.23)$$

with

$$\varphi(t) = -\ln\left[1-(1-t)^\theta\right], \ \theta \geq 1 \qquad (7.23a)$$

Example 7.15 Determine the joint distribution of random variables X_1 and X_2 by the Joe Archimedean copula, where the marginal distributions of X_1 and X_2 are given as $X_1 \sim \exp(\lambda)$, and $X_2 \sim \text{Gumbel}(\alpha, \beta)$.

Solution The marginal distribution of exponential- and Gumbel-distributed random variables are given in the previous examples, and thus by substituting these marginal distributions into Eq. 7.23, one can obtain the joint distribution using the Joe copula as

$$F_{X_1,X_2}(x_1,x_2) = C_\theta(u_1,u_2) = 1 - \left\langle \begin{array}{c} \exp(-\lambda x_1)^\theta + \left\{1-\exp\left[\exp(-\frac{x_2-\beta}{\alpha})\right]\right\}^\theta \\ -\exp(-\lambda x_1)^\theta \left\{1-\exp\left[\exp(-\frac{x_2-\beta}{\alpha})\right]\right\}^\theta \end{array} \right\rangle^{1/\theta}$$

Survival Copulas Associated with Gumbel's Bivariate Exponential Distribution

As its name implies, this family is the survival copula, which is actually the survival probability distribution of the Gumbel bivariate exponential distribution. It can be expressed as

$$C_\theta(u_1,u_2) = C_\theta\left[F_{X_1}(x_1), F_{X_2}(x_2)\right] = F_{X_1,X_2}(x_1,x_2)$$
$$= u_1 u_2 \exp(-\theta \ln u_1 \ln u_2) \qquad (7.24)$$

with

$$\varphi(t) = \ln(1 - \theta \ln t) \qquad \text{(7.24a)}$$

Example 7.16 Determine the joint distribution of random variables X_1 and X_2 by the survival copula, where the marginal distributions of X_1 and X_2 are given as $X_1 \sim \exp(\lambda)$ and $X_2 \sim \text{Gumbel}(\alpha, \beta)$.

Solution The marginal distributions of exponential- and Gumbel-distributed random variables are given in the previous examples, and thus by substituting these marginal distributions into Eq. 7.24, one can obtain the joint distribution using survival copulas associated with Gumbel's bivariate exponential distribution as

$$F_{X_1,X_2}(x_1, x_2) = C_\theta(u_1, u_2)$$
$$= \left[1 - \exp(\lambda x_1)\right] \exp\left[-\exp(-\frac{x-\beta}{\alpha})\right]$$
$$\exp\left\{\theta \ln\left[1 - \exp(\lambda x_1)\right] \exp\left(-\frac{x-\beta}{\alpha}\right)\right\}$$

There are still other Archimedean copulas (e.g., the copula proposed by Genest and Ghoudi (1994)) in the Archimedean copula family. By using the same procedure as for the generation of two-dimensional Archimedean copulas, N-dimensional Archimedean copulas can be generated and expressed as

$$C_\theta^N(\mathbf{u}) = \varphi^{-1}\left[\varphi(u_1) + \varphi(u_2) + \dots + \varphi(u_N)\right]$$

where superscript N denotes the dimension of the copula and u denotes the variable vector.

Following Nelson (1999), we obtain the copula function C_θ^N, the serial iterate of the Archimedean two-dimensional Archimedean copula generated by j, which can be expressed as

$$C_\theta^N(u_1, u_2, \dots, u_N) = C_\theta\left[C_\theta^{N-1}(u_1, u_2, \dots, u_{N-1}), u_N\right]$$

but this procedure may not always succeed. Thus only the Gumbel–Hougaard, Frank, Cook–Johnson (Clayton), and Ali–Mikhail–Haq multivariate copulas are considered here. These N-dimensional Archimedean copulas can be represented as follows:

- Gumbel–Hougaard multivariate Archimedean copula:

$$C_\theta^N(\mathbf{u}) = C_\theta^N\left[F_{X_1}(x_1), F_{X_2}(x_2), \dots, F_{X_N}(x_N)\right] = F_{X_1, X_2 \dots X_N}(x_1, x_2, \dots, x_N)$$
$$= \exp\left\{-\left[(-\ln u_1)^\theta + (-\ln u_2)^\theta + \dots + (-\ln u_n)^\theta\right]^{1/\theta}\right\} \qquad \text{(7.25)}$$

with generating function $\varphi(t) = (-\ln t)^\theta$.

- Frank multivariate Archimedean copula:

$$C_\theta^N(\mathbf{u}) = C_\theta^N\left[F_{X_1}(x_1), F_{X_2}(x_2), ..., F_{X_N}(x_N)\right] = F_{X_1,X_2...X_N}(x_1, x_2, ..., x_N)$$

$$= -\frac{1}{\theta}\ln\left[1 + \frac{(e^{-\theta u_1} - 1)(e^{-\theta u_2} - 1)...(e^{-\theta u_N} - 1)}{(e^{-\theta} - 1)^{N-1}}\right] \qquad (7.26)$$

with $\varphi(t) = -\ln\left((e^{-\theta t} - 1)/(e^{-\theta} - 1)\right)$.

- Cook–Johnson (Clayton) multivariate Archimedean copula:

$$C_\theta^N(\mathbf{u}) = C_\theta^N\left[F_{X_1}(x_1), F_{X_2}(x_2), ..., F_{X_N}(x_N)\right] = F_{X_1,X_2...X_N}(x_1, x_2, ..., x_N)$$

$$= (u_1^{-\theta} + u_2^{-\theta} + ... + u_N^{-\theta} - N + 1)^{1/\theta} \qquad (7.27)$$

with generating function $\varphi(t) = t^{-\theta} - 1, \theta > 0$.

- Ali–Mikhail–Haq multivariate Archimedean copula:

$$C_\theta^N(\mathbf{u}) = C_\theta^N\left[F_{X_1}(x_1), F_{X_2}(x_2), ..., F_{X_N}(x_N)\right] = F_{X_1,X_2...X_N}(x_1, x_2, ..., x_N)$$

$$= \frac{u_1 u_2 ... u_N}{1 - \theta(1 - u_1)(1 - u_2)...(1 - u_N)} \qquad (7.28)$$

7.3.1.6 Estimation of Copula Parameter

The copula parameter may be estimated either semiparametrically by the maximum likelihood method for both two-dimensional or N-dimensional Archimedean copulas or by the nonparametric method through Kendall's τ (tau) for two-dimensional copulas.

Semiparametric Method

To estimate the copula parameter θ, two conditions may be considered. First, if appropriate marginals are already available, then one simply expresses the likelihood function for the copula. The resulting estimate of θ would then be marginal dependent; the same maximum likelihood methodology, which is usually applied for estimation of parameters of univariate probability distributions, is indirectly effected for the copula method. This is a semiparametric method. Second, if nonparametric estimates are contemplated for the marginals, the estimation of the copula parameter θ will be marginal free.

The semiparametric estimation can be expressed step by step as follows:

1. Let a random sample $\{(X_{1k}, ..., X_{Nk}) : k = 1, ..., n\}$ be given from the distribution

$$H(x_1, ..., x_N) = C_\theta^N\left[F_1(x_1), ..., F_N(x_N)\right] \qquad (7.29)$$

2. Write the log-likelihood function for the copula in Eq. 7.29 as

$$L(\theta) = \sum_{k=1}^{n} \log\left\{c_{\theta}^{N}\left[F_{1n}(X_{1k}),...,F_{Nn}(X_{Nk})\right]\right\} \tag{7.30}$$

where F_{in} denotes $n/(n + 1)$ times the marginal empirical distribution function of the ith variable (Genest and Rivest 1993). This rescaling avoids the difficulty of the potential unboundedness of $\log[c_{\theta}(u_1,...,u_N)]$, since some of the u_is tend to 1. The term c_{θ} denotes the probability density function of the copula, which has the same meaning as the probability density function of a univariate random variable.

3. According to the property of the semiparametric estimator, θ is consistent and asymptotically normal under the same conditions as the maximum likelihood estimation (Genest, et al. 1995), which is an asymptotic property. To maximize the preceding log-likelihood function the following step is needed:

$$\frac{1}{n}\frac{\partial}{\partial\theta}L(\theta) = \frac{1}{n}\sum_{k=1}^{n} l_{\theta}\left[\theta, F_{1n}(X_{1k}), F_{2n}(X_{2k}),...F_{Nn}(X_{Nk})\right] = 0 \tag{7.31}$$

where L denotes the log-likelihood function and l_{θ} denotes the derivative of L with respect to parameter θ.

Nonparametric Method

Genest and Rivest (1993) described a procedure to identify a copula function based on a nonparametric estimation for bivariate Archimedean copulas. It is assumed that a random sample of bivariate observations (x_{11},x_{21}), (x_{12},x_{22}),..., (x_{1n},x_{2n}) is available and that its underlying distribution function $F_{X_1,X_2}(x_1,x_2)$ has an associated Archimedean copula C_{θ}, which also can be regarded as an alternative expression of F. Then the following steps are followed to identify the appropriate copula:

- Determine Kendall's τ (the dependence structure of the bivariate random variables) from observations as

$$\tau_n = \binom{n}{2}^{-1} \sum_{i<j} \text{sign}[(x_{1i} - x_{1j})(x_{2i} - x_{2j})] \tag{7.32}$$

where n is the number of observations; sign = 1 if $x_{1i} \leq x_{1j}$ and $x_{2i} \leq x_{2j}$ and, otherwise, sign = –1; i, j = 1, 2, ..., n; and τ_n is the estimate of τ from the observations.

- Determine the copula parameter θ from this value of τ according to the relationship between Kendall's τ and copula parameter θ (i.e., for the Gumbel–Hougaard copula, the relationship between Kendall's τ and the copula parameter θ is given by Eq. 7.19 and is similar for other Archimedean copula families).

- Obtain the generating function of each copula, ϕ, by inserting the parameter θ obtained.

- Obtain the copula from its generating function ϕ.

Example 7.17 Consider the bivariate flow variables X_1 (volume in cfs·day) and X_2 (duration in days) with the sample data for each random variable given in Table E7-17. Suppose that the Gumbel–Hougaard copula family can be applied to this bivariate data set. Determine the parameter by the semiparametric method for this bivariate case and the parameter by the nonparametric method for correlated variables X_1 and X_2.

Table E7-17

No.	X_1	X_2	No.	X_1	X_2	No.	X_1	X_2	No.	X_1	X_2
1	16664	12	16	21173	12	31	42881	17	46	85723	17
2	36987	19	17	37707	11	32	23467	15	47	38543	18
3	42365	14	18	41328	15	33	58831	14	48	44991	14
4	54302	21	19	8259	15	34	36481	16	49	71320	15
5	58371	22	20	40544	14	35	68179	14	50	46025	12
6	34125	20	21	52301	13	36	78119	16	51	67491	26
7	38884	18	22	79864	15	37	13775	12	52	38752	12
8	19104	32	23	53831	12	38	23484	16	53	52107	19
9	64769	18	24	5263	15	39	95828	10	54	65776	18
10	18644	12	25	28363	18	40	16513	11	55	26842	17
11	23249	22	26	17538	8	41	33681	9			
12	31075	17	27	18926	11	42	46742	22			
13	13246	12	28	49351	15	43	22192	11			
14	12854	8	29	44354	12	44	48304	15			
15	29882	13	30	41115	15	45	63794	18			

Solution

1. Estimation by the semiparametric method for the bivariate variables:

   (a) Determine the marginal probabilities: For simplicity, empirical probabilities are obtained from the Gringorten plotting-position formula as $P_i = (I - 0.44)/(n + 0.12)$ in which n is the sample size and i is the rank. To avoid the possibility of some of the u_is tending to 1, we need to take $n/(n + 1)$ times the marginal probability denoted as P_{ic} [i.e., $P_{ic} = nP_i/(n + 1)$].

(b) Obtain the probability density function of the Gumbel–Hougaard copula: Similar to the conventional approach for the derivation of a joint probability density function from its probability distribution functions, the density function represented by the copula is

$$c(u_1,u_2) = \frac{\partial^2 C(u_1,u_2)}{\partial u_1 \partial u_2} \tag{7.33}$$

in which u_1 and u_2 denote the marginal probabilities for random variable X_1 and X_2, respectively.

(c) Obtain the log-likelihood function as in Eq. 7.31.

(d) Take the derivative with respect to parameter θ of the log-likelihood function obtained in step (c) and set the equation equal to 0.

(e) Then, parameter θ is calculated numerically.

(f) Following the preceding steps, we obtain $\theta = 1.32$.

2. Estimation by nonparametric estimation of variables X_1 and X_2:

(a) Calculate Kendall's τ from the data set by Eq. 7.32: $\tau_n = 0.21$.

(b) Obtain the copula parameter θ from the relationship between Kendall's τ and the copula parameter. For the Gumbel–Hougaard copula $(\tau = 1 - \theta^{-1})$, we have $\theta = 1.25$.

7.3.1.7 Identification of the Copula by the Nonparametric Approach

The nonparametric approach employs the Q–Q plot, which can be applied for the identification of an appropriate copula. This involves the following steps:

1. Define an intermediate random variable $Z = H(X_1, X_2)$ that has a distribution function $K(z) = P(Z \le z)$.

2. Construct a nonparametric estimate of $K(z)$ as follows:

(a) Obtain $z_i = \{$number of (x_{1j}, x_{2j}) such that $x_{1j} < x_{1i}$ and $x_{2j} < x_{2i}\}/(n-1)$ for $i = 1, \ldots, n$.

(b) Construct an estimate of $K(z)$ as $K_n(z) = $ the proportion of z_is $\le z$.

3. Construct a parametric estimate of $K(z)$ using

$$K_C(t) = t - \frac{\varphi(t)}{\varphi'(t)} \tag{7.34}$$

4. Construct a plot of nonparametrically estimated $K_n(z)$ versus parametrically estimated $K(z)$ by using Eq. 7.31. This plot is referred to as the Q–Q plot. If the plot is in agreement with a straight line passing through the origin at a 45° angle then the generating function is satisfactory. The 45°

$$H_{Y|X}(y|x) = Q\left\{ \frac{Q^{-1}[G(y)] - \gamma Q^{-1}[F(x)]}{\sqrt{1-\gamma^2}} \right\} \quad (7.48)$$

The corresponding density function can be obtained by using Eq. 7.41b and

$$h_{Y|X}(y|x) = \frac{h(x,y)}{f(x)} \quad (7.49)$$

Likewise, the conditional distributions $H_{X|Y}(x|y)$ and $h_{X|Y}(x|y)$ can be derived.

The bivariate distribution Φ, the marginal distributions Φ_X and Φ_Y, and the conditional distributions $\Phi_{X|Y}$ and $\Phi_{Y|X}$ can be determined from the eight elements in Eq. 7.44. The advantages of the meta-Gaussian distribution are (1) F and G can be of any form; (2) H and h can be expressed analytically; (3) it is easy to determine Q and Q^{-1}; and it is easy to estimate γ.

7.4 Questions

7.1 Get instantaneous yearly peak discharge and associated volume data for a period of at least 30 years for a gauging station near your town. Check whether the frequency distributions of peak discharge and volume are normal. If not, then use the Box–Cox transformation to transform them to normal. Then sketch the joint probability density function of these two variables.

7.2 Obtain the duration data for the flood events in Question 7.1. If the flood duration is not normally distributed, then use the Box–Cox transformation to transform it to normal. Then sketch the trivariate probability density function.

7.3 Obtain data for two low-flow duration and discharge data for a gauging station near your town. Assume that they are exponentially distributed. Then sketch their joint distribution.

7.4 Suppose the low-flow variables in Question 7.3 follow the Gumbel distribution. Sketch the joint Gumbel distribution.

7.5 Suppose the low-flow variables in Question 7.3 follow the Gumbel logistic distribution. Then sketch the bivariate distribution.

7.6 Obtain data on rainfall intensity and depth for a rain gauge station nearby. Assume that intensity and depth have gamma distributions. Sketch the joint probability distribution of rainfall intensity and depth.

7.7 Suppose rainfall intensity and X_2 follow a log-normal distribution. Sketch the joint distribution of rainfall intensity and depth.

7.8   Consider rainfall intensity in Question 7.7. Determine parameter λ for the Box–Cox transformation.

7.9   Consider Question 7.1 and take a value of the conditioning variable as, say, 5,000. Then determine the probability density function if the conditioning variable (X_2) takes on a value equal to the given value and also when it is less than or equal to the given value.

7.10   Consider Question 7.4. Determine the conditional return period given $X_2 \leq$ 5,000, 10,000, and 500,000 cfs·day.

7.11   Determine the joint distribution of random variables X_1 and X_2 using the Gumbel–Hougaard copula, where the marginal distribution of X_1 is given by a gamma distribution and the marginal distribution of X_2 is given by a normal distribution.

7.12   Determine the joint distribution of random variables X_1 and X_2 using the Ali–Mikhail–Haq copula, where the marginal distribution of X_1 is gamma and the marginal distribution of X_2 is normal.

7.13   Determine the joint distribution of random variables X_1 and X_2 by the Frank copula, where the marginal distribution of X_1 is gamma and the marginal distribution of X_2 is Gumbel.

7.14   Determine the joint distribution of random variables X_1 and X_2 by the Cook–Johnson Archimedean copula, where the marginal distribution of X_1 is gamma and the marginal distribution of X_2 is Gumbel.

7.15   Determine the joint distribution of random variables X_1 and X_2 by the Joe Archimedean copula, where the marginal distribution of X_1 is gamma and the marginal distribution of X_2 is Gumbel.

7.16   Determine the joint distribution of random variables X_1 and X_2 by the survival copula, where the marginal distribution of X_1 is gamma and the marginal distribution of X_2 is Gumbel.

7.17   Obtain data for flow variables X_1 (volume in cfs·day) and X_2 (duration in days) from a nearby gauging station. Suppose that the Gumbel–Hougaard copula family would be applied to this bivariate data set. Determine the parameter by the semiparametric method for this bivariate case and the parameter by the nonparametric method for correlated variables X_1 and X_2.

7.18   Considering Question 7.17, obtain the Q–Q plot for the identification of the Gumbel–Hougaard copula.

Chapter 8

Parameter Estimation

To conduct a probabilistic analysis (e.g., deriving probabilistic information and performing relevant statistical inferences) for a given real-world engineering project or system, most often probability density functions (as described in Chapters 4 and 5) are used. After selecting a suitable probability model (PDF in continuous and PMF in discrete cases) for the random variable of interest, the parameters (i.e., the variables that govern the characteristics of the PDF or PMF) need to be determined. The most common procedure for determining parameters of a PDF or PMF is performed by fitting the known mathematical function to the observed data. Once the parameters of the fitted distribution are determined, this can serve as a reasonable model for the phenomenon under consideration. Fitting a mathematical model to match the experimental data falls in the domain of parameter estimation. In other words, the problem of parameter estimation can be stated as the problem of determining the probability density function for the random variable X given a set of observations of X. It is of particular practical importance, because most parameters cannot be measured directly.

When discussing various methods of parameter estimation, it is worth discussing the accuracy of estimated parameters. An estimator $\hat{a}$ of a known or unknown quantity a is a function of the observed random sample $x_1,..., x_n$ from the population X that is available to estimate the value of a. It is necessary to emphasize that the estimator $\hat{a}$ is itself a random variable since its value depends on the sample values of the random variable that are observed.

This chapter presents an overview of various methods for parameter estimation and the most commonly used techniques for accuracy assessment of estimated parameters.

8.1 Methods of Parameter Estimation

A number of methods are available for estimating parameters of probability distributions. Some of the popular methods used in hydrology include (1) the method of moments, (2) the method of probability-weighted moments, (3) the method of mixed moments, (4) L-moments, (5) maximum likelihood estimation, and (6) the least-squares method. A brief review of these methods is given here.

8.1.1 Method of Moments

This method was developed by Karl Pearson in 1902 based on the premise that, when the parameters of a probability distribution are estimated correctly, the moments of the probability density function are equal to the corresponding moments of the sample data. The method of moments is frequently utilized to estimate parameters of linear hydrologic models (Nash 1959, Dooge 1973, Singh 1988). Nash (1959) developed the theorem of moments relating the moments of input, output, and impulse response functions of linear hydrologic models. Moments of functions are amenable to the use of standard methods of transform, such as the Laplace and Fourier transforms. The method of moments has been used to estimate parameters of frequency distributions. Wang and Adams (1984) reported on parameter estimation in flood frequency analysis. Ashkar et al. (1988) developed a generalized method of moments and applied it to the generalized gamma distribution. Kroll and Stedinger (1996) estimated moments of a log-normal distribution using censored data.

In the method of moments, distribution moments are equated to the sample moments to estimate distribution parameters. The advantage of the method of moments is that moments are simple to understand and interpret, and they are easy to calculate. But the disadvantage is that they are often not available for all the probability distribution functions and they do not have the desirable optimality properties of other methods, such as the maximum likelihood and least-squares estimators. Another important application of the method of moments is to provide starting values for the more precise estimates. The main steps for parameter estimation for a distribution having k parameters using the method of moments are as follows:

1. Compute the algebraic expressions for the first k moments from the assumed distribution.
2. Compute the numerical values for the first k sample moments from the sample data.

3. Equate each algebraic expression of the assumed distribution moment to the corresponding numerical value of the sample moment. This will form k algebraic equations.

4. Solve these equations to determine the k unknown parameters.

8.1.1.1 Computation of Algebraic Expressions for Moments

We have discussed central moments (moments about the mean) and moments about the origin in Chapter 3. Here we will further generalize the concept of statistical moments about any generic point. Let X be a continuous variable (which may or may not be a random variable) and $f(x)$ be its function satisfying some necessary conditions. The rth moment of $f(x)$ about an arbitrary point is denoted as μ_r^a (f). Here μ denotes the moment, the subscript ($r \geq 0$) denotes the order of the moment, the superscript denotes the point about which the moment is taken, and the quantity within the parentheses denotes the function, in normalized form, whose moment is to be taken. Then, the rth moment of a function $f(x)$ can be defined as

$$\mu_r^a(f) = \int_{-\infty}^{\infty} (x-a)^r f(x) dx$$

(8.1)

This is the definition normally used for functions representing PDFs. If the area enclosed by the function $f(x)$ does not add to unity (i.e., $f(x)$ is not a PDF), then the definition of Eq. 8.1 is modified as

$$\mu_r^a(f) = \frac{\int_{-\infty}^{\infty} (x-a)^r f(x) dx}{\int_{-\infty}^{\infty} f(x) dx}$$

(8.2)

As the denominator in Eq. 8.2 defines the area under the curve, which is usually unity or made to unity by normalization, the two definitions are numerically the same. In this text the definition of Eq. 8.1 is used with $f(x)$ normalized beforehand. It is assumed here that the integral in Eq. 8.1 converges. There are some functions that possess moments of lower order; some do not possess any moment except of zero order. However, if a moment of higher order exists, then moments of all lower orders must exist. Figure 8-1 shows the concept of moment of a function about an arbitrary point located at a distance a from the origin.

Moments are statistical descriptors of a distribution and reflect on its quantitative properties. For example, if $r = 0$ then Eq. 8.1 yields

$$\mu_0^a(f) = \int_{-\infty}^{\infty} (x-a)^0 f(x) dx = \int_{-\infty}^{\infty} f(x) dx = 1$$

(8.3)

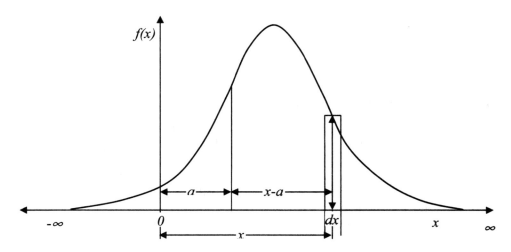

Figure 8-1 *Concept of moment of a function f(x) about an arbitrary point.*

Thus, the zero-order moment is the area under the curve defined by $f(x)$ subject to $-\infty < x < \infty$.

If $r = 1$, then Eq. 8.1 yields

$$\mu_1^a(f) = \int_{-\infty}^{\infty} (x - a)f(x)dx = \mu - a \tag{8.4}$$

where μ is the centroid of the area or mean. Thus, the first moment is the weighted mean about the point a. If $a = 0$, it is called the first moment about the origin. It gives the mean and is represented by

$$\mu_1' = \int_{-\infty}^{\infty} xf(x)dx \tag{8.5a}$$

When $a = \mu$, the rth moment about the mean is represented by μ_r and is called central moment:

$$\mu_r = \int_{-\infty}^{\infty} (x - \mu)^r f(x)dx = \int_{-\infty}^{\infty} (x - a)^r f(x)dx \tag{8.5b}$$

The descriptive properties of the moments with respect to a specific function can be summarized as follows:

μ_0 = area

$\mu_1' = \mu$ = mean

μ_2 = variance, a measure of dispersion of the function about the mean

μ_3 = measurement of skewness of the function

μ_4 = kurtosis, a measure of the peakedness of the function

The second moment about the mean is known as variance and is a measure of how the data are scattered about the mean value. The third moment is a measure of symmetry; it indicates whether the data are evenly distributed about the mean or the mode is to the left or right of the mean. If the mode is to the left of the mean, the data are said to have positive skew and if it is to the right, the data are said to have negative skew. The coefficient of skewness is a quantitative measure of the skewness in the data. The fourth moment is a measure of peakedness, which is explained through kurtosis. Kurtosis is the peakedness or flatness of data with respect to the normal distribution.

8.1.1.2 Determination of Sample Moments

Let the sample data be represented by a discrete function f_j, $j = -\infty, \ldots, -1, 0, 1, \ldots, \infty$. The rth moment about any arbitrary point can be defined in a manner analogous to that for continuous functions. When the arbitrary point is the origin, the rth moment is defined as

$$M_r = \sum_{x=-\infty}^{\infty} x^r f_x \tag{8.6}$$

when f_x is normalized:

$$\sum_{x=-\infty}^{\infty} f_x = 1 \tag{8.7}$$

Otherwise,

$$M_r = \sum_{x=-\infty}^{\infty} x^r f_x / \sum_{x=-\infty}^{\infty} f_x \tag{8.8}$$

It is thus seen that Eq. 8.6 and Eq. 8.8 are analogous to Eq. 8.1 and Eq. 8.2. Figure 8-2 explains the concept of moment of a discrete function.

Sample moments are often biased owing to the small size of the sample. Commonly, the first four moments are used in parameter estimation. Selected central moments or moments about sample mean $(\bar{x})$ are

$$M_2^\mu = \frac{1}{n} \sum_{i=1}^{n} (x_i - \bar{x})^2 \tag{8.9}$$

$$M_3^\mu = \frac{n}{(n-1)(n-2)} \sum_{i=1}^{n} (x_i - \bar{x})^3 \tag{8.10}$$

$$M_4^\mu = \frac{n^2}{(n-1)(n-2)(n-3)} \sum_{i=1}^{n} (x_i - \bar{x})^4 \tag{8.11}$$

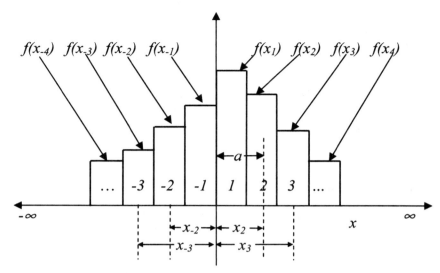

Figure 8-2 *Concept of moment of a discrete function about an arbitrary point.*

The ratios of moments summarize useful information about a probability distribution. The commonly used ratios are identified by their popular names:

$$\text{coefficient of variation} = c_v = \sqrt{M_2^\mu / M_1}$$

$$\text{coefficient of skewness} = c_s = M_3^\mu / (M_2^\mu)^{1.5}$$

$$\text{coefficient of kurtosis} = c_k = M_4^\mu / (M_2^\mu)^2$$

Example 8.1 The histogram of annual flows of the Sabarmati River in India is given in Table 8-1. Find the mean and variance of the sample data and suggest the candidate distribution(s) using the method of moments.

Solution Summing the frequencies gives

$$6 + 11 + 9 + 19 + \ldots + 2 + 0 + 0 = 98$$

The first moment of the data is

$$[(150 \times 6) + (250 \times 9) + (350 \times 11) + \ldots + (1{,}750 \times 1)]/98 = 664.29 \text{ cumecs}$$

This is the mean of the data.
    The second moment about the mean gives the variance:

$$\text{second moment} = [(150 - 664)^2 \times 6 + (250 - 664)^2 \times 9 + \ldots + (1{,}750 - 664)^2$$
$$\times 1]/98 = 120{,}000 \text{ cumecs}^2$$

Table 8-1 *Data description of annual flows of the Sabarmati River in India*

Discharge range (m³/s)	Average (m³/s)	Frequency	Discharge range (m³/s)	Average (m³/s)	Frequency
100–200	150	6	200–300	250	9
300–400	350	11	400–500	450	9
500–600	550	9	600–700	650	9
700–800	750	19	800–900	850	6
900–1,000	950	6	1,000–1,100	1,050	1
1,100–1,200	1,150	5	1,200–1,300	1,250	2
1,300–1,400	1,350	3	1,400–1,500	1,450	0
1,500–1,600	1,550	2	1,600–1,700	1,650	0
1,700–1,800	1,750	1			

Hence, the standard deviation is $(120,000)^{0.5} = 346.41$ cumecs. One then obtains the following results:

coefficient of variation $= 346.41/664.29 = 0.52$

third moment about the mean $= [(150 - 664)^3 \times 6 + (250 - 664)^3 \times 9 + \ldots + (1,750 - 664)^3 \times 1] \times 98/(97 \times 96) = 31,148,196$

coefficient of skewness $= 31,148,196/(120,000)^{1.5} = 0.75$

fourth moment about the mean $= [(150 - 664)^4 \times 6 + (250 - 664)^4 \times 9 + \ldots + (1750 - 664)^4 \times 1] \times 98 \times 98/(97 \times 96 \times 95) = 50,810,846,446$

coefficient of kurtosis $= 50,810,846,446/(120,000)^2 = 3.53$

Note that the normal distribution has zero skewness and its kurtosis is 3.

Based on sample moments, one can select several candidate distributions for determining their parameters. As an example, we selected the normal, the log-normal, and the gamma distributions as candidate distributions. Now, we will use the method of moments to determine their parameters.

As described in Chapter 5, the normal distribution is given as

$$f_X(x) = \frac{1}{x\sigma_y\sqrt{2\pi}} \exp\left[-\frac{1}{2}\left(\frac{\ln x - \mu_Y}{\sigma_Y}\right)^2\right], \quad -\infty \leq x \leq \infty$$

It has two parameters, μ_X and σ_X. By definition these are the first and second moments of the normal distribution, that is, $E[X] = \mu_X$ and $E\left[(X - \mu_X)^2\right] = \sigma_X^2$.

Thus, by using the method of moments the parameter estimates for μ_X and σ_X are given as

$$\hat{\mu}_X = \text{the first sample moment} = 664.29$$

$$\hat{\sigma}_X^2 = \text{the second sample moment} = 346.41^2$$

The log-normal distribution (see Chapter 5) is given as

$$f_X(x) = \frac{1}{x\sigma_y\sqrt{2\pi}}\exp\left[-\frac{1}{2}\left(\frac{\ln x - \mu_Y}{\sigma_Y}\right)^2\right], \qquad x \geq 0$$

The log-normal distribution can be treated in exactly the same way as the normal distribution by transforming the random variable X into another random variable Y using the transformation $Y = \ln(X)$. As described in Chapter 5, the estimates of the mean and variance of Y can be estimated using the following relations:

$$\hat{\mu}_Y = \frac{1}{2}\ln\left(\frac{\hat{\mu}_X^2}{1+\hat{C}V_X^2}\right) = 6.5$$

$$\hat{\sigma}_Y^2 = \ln\left(1+\hat{C}V_X^2\right) = 0.49^2$$

The estimate for the coefficient of skew is

$$\hat{\gamma}_X = 3\hat{C}V_X + \hat{C}V_X^3 = 1.71$$

The PDF of the gamma distribution with k and λ as parameters has been described in Chapter 4 as

$$f_{Xk}(x) = \frac{\lambda^k x^{k-1} e^{-\lambda x}}{\Gamma(k)}, \qquad x \geq 0, \ \lambda > 0, \ k > 0$$

The algebraic expressions for the mean and variance of X are expressed as

$$E[X] = \frac{k}{\lambda}$$

and

$$\sigma_X^2 = \frac{k}{\lambda^2}$$

Equating the distribution moments of X with its sample moments, one gets

$$\frac{\hat{k}}{\hat{\lambda}} = 664.29$$

and

$$\frac{\hat{k}}{\hat{\lambda}^2} = 346.41^2$$

On solving these two equations, one gets $\hat{k} = 3.677$ and $\hat{\lambda} = 0.006$. The fitted gamma distribution has a coefficient of skewness of 1.04.

The plot in Fig. 8-3 compares the PDFs of various fitted distributions with the observed data relative frequency distributions. It appears that out of these three candidate distributions, the gamma distribution is a better choice.

Example 8.2 If one wants to fit an exponential distribution to the data given in Example 8.1, what will be its parameter based on the method of moments?

Solution As given in Chapter 4, the PDF of an exponential distribution is

$$f_x(x) = \lambda e^{-\lambda x}, x \geq 0$$

It has only one parameter, λ. The first distribution moment of X is

$$E[X] = \int_0^\infty x \lambda e^{-\lambda x} dx = \frac{1}{\lambda}$$

Based on the method of moments, the first distribution moment of X = the first sample moment of X, that is, $1/\hat{\lambda} = 664.29$ or $\hat{\lambda} = 0.0015$.

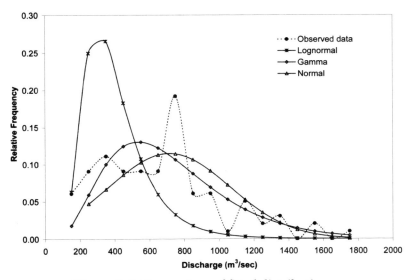

Figure 8-3 Comparison of fitted distributions.

8.1.2 Method of Maximum Likelihood

The maximum likelihood estimation (MLE) method is widely accepted as one of the most powerful parameter estimation methods. Asymptotically, the maximum likelihood (ML) parameter estimates are unbiased, have minimum variance, and are normally distributed, whereas in some cases these properties hold for small samples. The MLE method has been extensively used for estimating parameters of frequency distributions as well as fitting conceptual models.

Let $f(x; a_1, a_2, ..., a_m)$ be a PDF of the random variable X with parameters a_i, $i = 1, 2, ..., m$, to be estimated. For a random sample of data $x_1, x_2, ..., x_n$, drawn from this probability density function, the joint PDF is defined as

$$f(x_1, x_2, ..., x_n; a_1, a_2, ..., a_m) = \prod_{i=1}^{n} f(x_i; a_1, a_2, ..., a_m) \tag{8.12}$$

Interpreted conceptually, the probability of obtaining a given value of X, say x_1, is proportional to $f(x_1; a_1, a_2, ..., a_m)$. Likewise, the probability of obtaining the random sample $x_1, x_2, ..., x_n$ from the population of X is proportional to the product of the individual probability densities or the joint PDF, owing to the independence among $x_1, x_2, ..., x_n$. This joint PDF is called the likelihood function, denoted

$$L = \prod_{i=1}^{n} f(x_i; a_1, a_2, ..., a_m) \tag{8.13}$$

where parameters a_i, $i = 1, 2, ..., m$, are unknown.

By maximizing the likelihood that the sample under consideration is the one that would be obtained if n random observations were selected from $f(x; a_1, a_2, ..., a_m)$, the unknown parameters are determined, hence giving rise to the name of the method. The values of parameters so obtained are known as ML estimators. Since the logarithm of L attains its maximum for the same values of a_i, $i = 1, 2, ..., m$, as does L, the MLE function can also be expressed as

$$\ln L = L^* = \ln \prod_{i=1}^{n} f(x_i; a_1, a_2, ..., a_m) = \sum_{i=1}^{n} \ln f(x_i; a_1, a_2, ..., a_m) \tag{8.14}$$

Frequently $\ln[L]$ is maximized, for it is much easier to find the maximum of the logarithm of the maximum likelihood function than that of L itself.

The procedure for estimating parameters or determining the point where the MLE function achieves its maximum involves differentiating L or $\ln L$ partially with respect to each parameter and equating each differential to zero. This results in as many equations as the number of unknown parameters. For m unknown parameters, we get

$$\frac{\partial \ln L(a_1, a_2, ..., a_m)}{\partial a_1} = 0, \frac{\partial \ln L(a_1, a_2, ..., a_m)}{\partial a_2} = 0, ..., \frac{\partial \ln L(a_1, a_2, ..., a_m)}{\partial a_m} = 0 \tag{8.15}$$

These m equations in m unknowns are then solved for the m unknown parameters. The parameters determined using the method of maximum likelihood are efficient; that is, in a large sample, they attain minimum variance and are asymptotically unbiased. A drawback of this method is that it requires that the underlying distribution be known. Many times, this distribution is not known. Furthermore, there may not be an analytical solution of the ML equations to estimate the parameters in terms of sample statistics. Consequently, one may have to resort to a numerical solution.

Example 8.3 Using the maximum likelihood estimation procedure, find the parameter α of the exponential distribution for the data of the Sabarmati River in India given in Example 8.1.

Solution The probability density function of the one-parameter exponential distribution is given by

$$f_X(x) = \alpha \exp(-\alpha x) \tag{8.16}$$

The likelihood function is given by

$$L(\alpha) = \prod_{i=1}^{n} \alpha \exp(-\alpha x_i) = \alpha^n \exp\left(-\alpha \sum_{i=1}^{n} x_i\right) \tag{8.17}$$

This can be used to form the log-likelihood function:

$$\ln L(\alpha) = n \ln(\alpha) - \alpha \left(\sum_{i=1}^{n} x_i\right) \tag{8.18}$$

where n is the sample size. Differentiating Eq. 8.18 with respect to α gives

$$\frac{d \ln L(\alpha)}{d\alpha} = \frac{n}{\alpha} - \sum_{i=1}^{n} x_i = 0$$

This yields

$$\alpha = n / \left(\sum_{i=1}^{n} x_i\right) = \frac{1}{\bar{x}} \tag{8.19}$$

In Example 8.1, the mean of the data was found to be 664.9 cumecs. This will give the estimate of α as

$$\alpha = 1/664.29 = 1.51 \times 10^{-3} \text{ cumec}^{-1}$$

In Fig. 8-4, the likelihood function for a typical case is plotted. The maximum likelihood estimation tries to find that value of parameter that gives the maximum value of the likelihood function (or its logarithm). Thus, in the present case, a value of α that is equal to the reciprocal of the mean ($1/\bar{x}$ or $1/m_x$) is most likely to be the true value of the parameter.

Example 8.4 The probability density function of a two-parameter exponential distribution is given by

$$f(x) = \frac{1}{\alpha} \exp\left(-\frac{x-\varepsilon}{\alpha}\right), \quad \varepsilon < x < \infty \tag{8.20}$$

Estimate the parameters of this distribution using the data of the Amite River at Darlington, Louisiana (Example 3.6), using the method of maximum likelihood estimation.

Solution The likelihood function is formed as

$$L = \frac{1}{\alpha^n} \exp\left[-\frac{1}{\alpha} \sum_{i=1}^{n} (x_i - \varepsilon)\right] \tag{8.21}$$

The log-likelihood function becomes

$$\log L = -n \log \alpha - \frac{1}{\alpha} \sum_{i=1}^{n} (x_i - \varepsilon) \tag{8.22}$$

Differentiating this function with respect to α and equating the derivative to zero, one obtains

$$\frac{\partial \log L}{\partial \alpha} = -\frac{n}{\alpha} + \frac{1}{\alpha^2} \sum_{i=1}^{n} (x_i - \varepsilon) = 0$$

$$\alpha = \frac{1}{n} \sum_{i=1}^{n} (x_i - \varepsilon) = m_x - \varepsilon$$

As x is bounded as $\varepsilon < x < \infty$, parameter ε cannot be greater than x_{min}. However, equating ε with x_{min} gives a biased estimator of ε. Rao and Hamed (2000) proposed the following equations to compute the unbiased and minimum variance estimates of parameters α and ε.

$$\hat{\alpha} = \frac{n(m_x - x_1)}{(n-1)} \tag{8.23}$$

$$\hat{\varepsilon} = \frac{n x_1 - m_x}{(n-1)} \tag{8.24}$$

where x_1 is a minimum value.

For the Amite River data given in Example 3.6, $n = 48$, mean = 28,676, and $x_1 = 3,180$. Therefore,

$$\alpha = 48(28{,}676 - 3{,}180)/(48 - 1) = 26{,}038.0$$

$$\varepsilon = [(48 \times 3{,}180) - 28{,}676]/(48 - 1) = 2{,}637.5$$

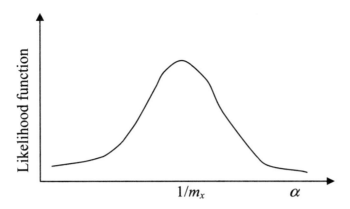

Figure 8-4 *A typical likelihood function.*

8.1.3 Method of Probability-Weighted Moments

Greenwood et al. (1979) introduced the method of probability-weighted moments (PWM) and showed its usefulness in deriving explicit expressions for parameters of distributions whose inverse forms $X = X(F)$ can be explicitly defined. They derived relations between parameters and PWM for generalized lambda, Wakeby, Weibull, Gumbel, logistic, and kappa distributions. Hosking (1986) developed the theory of probability-weighted moments and applied it to estimate parameters of several distributions. For flood frequency analysis, Haktanir (1996) modified the conventional method of probability-weighted moments for estimation of parameters of any distribution without the need to use a plotting position formula. Wang (1997) defined partial PWM and derived them for extreme value type I and III distributions. He applied these moments to lower bound censored samples. Singh (1998) employed PWM to estimate parameters of a number of distributions used in hydrology.

Let a probability distribution function be denoted as $F = F(X) = P[X \le x]$. The PWM of this function can be defined as

$$M_{i,j,k} = E[x^i F^j (1-F)^k] = \int_0^1 [x(F)]^i F^j (1-F)^k \, dF \tag{8.25}$$

where $M_{i,j,k}$ is the probability-weighted moment of order (i, j, k), E is the expectation operator, and i, j, and k are real numbers. If $j = k = 0$ and i is a non-negative integer then $M_{i,0,0}$ represents the conventional moment of order i about the origin. If $M_{i,0,0}$ exists and X is a continuous function of F, then $M_{i,j,k}$ exists for all non-negative real numbers j and k.

For non-negative integers j and k, we can express

$$M_{i,0,k} = \sum_{j=0}^{k} \binom{k}{j}(-1)^j M_{i,j,0} \tag{8.26}$$

$$M_{i,j,0} = \sum_{k=0}^{j} \binom{j}{k}(-1)^k M_{i,0,k} \tag{8.27}$$

If $M_{i,0,k}$ exists and X is a continuous function of F then $M_{i,j,0}$ exists. When the inverse $X = X(F)$ of the distribution $F = F(X)$ cannot be analytically defined, it may, in general, be difficult to derive $M_{i,j,k}$ analytically. We normally work with the moments $M_{i,j,k}$ into which x enters linearly. In particular, if we consider an ordered sample in which $x_1 \leq x_2 \leq \ldots \leq x_n$, the PWM for hydrologic applications are defined as

$$M_{1,0,s} = a_s = \frac{1}{n}\sum_{i=1}^{n}\binom{n-i}{s}x_i \bigg/ \binom{n-1}{s} \tag{8.28}$$

$$M_{1,0,r} = b_r = \frac{1}{n}\sum_{i=1}^{n}\binom{i-1}{r}x_i \bigg/ \binom{n-1}{r} \tag{8.29}$$

where $n > r$ and $n > s$.

In general, a_s and b_r are functions of each other as

$$a_s = \sum_{k=0}^{s}(-1)^k \binom{s}{k}b_k \tag{8.30}$$

$$b_r = \sum_{k=0}^{r}(-1)^k \binom{r}{k}a_k \tag{8.31}$$

Therefore,

$$
\begin{array}{ll}
a_0 = b_0, & b_0 = a_0 \\
a_1 = b_0 - b_1, & b_1 = a_0 - a_1 \\
a_2 = b_0 - 2b_1 + b_2, & b_2 = a_0 - 2a_1 + a_2 \\
a_3 = b_0 - 3b_1 + 3b_2 - b_3, & b_3 = a_0 - 3a_1 + 3a_2 - a_3
\end{array} \tag{8.32}
$$

A complete set of these a or b probability-weighted moments characterizes a distribution. The theory of PWM parallels the theory of conventional moments. The main advantage of PWM over conventional moments is that, because PWM are linear functions of the data, they suffer less from the effects of sampling variability: PWM are more robust than conventional moments to outliers in the data, enable more secure inferences to be made from small samples about an underlying probability distribution, and they frequently yield more efficient parameter estimates than the conventional moment estimates.

Example 8.5 Using the data of Example 8.4, find the parameters of a two-parameter exponential distribution using the PWM method.

Solution For the given data, the first PWM are determined as

$$a_0 = 28,675.8$$

$$a_1 = 8,724.1$$

The inverse form of a two-parameter exponential distribution as given by Eq. 8.20 is written as

$$x = \varepsilon - \alpha \ln(1 - F)$$

By using Eq. 8.25, expressions for the first two PWM are written as

$$a_0 = \int_0^1 x\,df = \int_0^1 [\varepsilon - \alpha \ln(1 - F)]\,dF$$

$$a_1 = \int_0^1 x(1 - F)\,df = \int_0^1 [\varepsilon - \alpha \ln(1 - F)](1 - F)\,df$$

Solving these integrals one gets

$$a_0 = \hat{\varepsilon} + \hat{\alpha}, \quad a_1 = \hat{\varepsilon}/2 + \hat{\alpha}/4$$

Substituting the sample estimates of the PWM, a_0 and a_1 in the preceding equations, one gets

$$\hat{\varepsilon} + \hat{\alpha} = 28675.9, \quad 2\hat{\varepsilon} + \hat{\alpha} = 34896$$

Now solving for ε and α_1 one gets $\hat{\varepsilon} = 6220.6$ and $\hat{\alpha} = 22455.2$.

8.1.4 Method of *L*-Moments

The probability-weighted moments characterize a distribution but are not meaningful by themselves. *L*-moments were developed by Hosking (1986) as functions of PWM that provide a descriptive summary of the location, scale, and shape of the probability distribution. These moments are analogous to ordinary moments and are expressed as *linear* combinations of order statistics, hence giving rise to their name. They can also be expressed by linear combinations of probability-weighted moments. Thus, the ordinary moments, the probability-weighted moments, and L-moments are related to each other. L-moments are known to have several important advantages over ordinary moments. L-moments have less bias than ordinary moments because they are linear combinations of ranked observations. As an example, the variance (second moment) and skewness (third moment) involve squaring and cubing of observations, respectively, which compel them to give greater weight to the observations far from the mean. As a result, they result in substantial bias and variance.

If X is a real-value ordered random variate of a sample of size n, such that $x_{1:n} \le x_{2:n} \le \ldots \le x_{n:n}$ with the cumulative distribution $F(x)$ and quantile function $x(F)$, then the rth L-moment of X (Hosking 1990) can be defined as a linear function of the expected order statistics as

$$L_r = \frac{1}{r}\sum_{k=0}^{r-1}(-1)^k \binom{r-1}{k} E\{X_{r-k:r}\}, \quad r = 1,2,\ldots \tag{8.33}$$

where $E\{\bullet\}$ is the expectation of an order statistic and is equal to

$$E\{X_{j:r}\} = \frac{r!}{(r-j)!j!}\int x\{F(x)\}^{j-1}\{1-F(x)\}^{r-j}\,dF(x) \tag{8.34}$$

As noted by Hosking (1990), the natural estimator of L_r, based on an observed sample of data, is a linear combination of the ordered data values (i.e., an L-statistic). Substituting Eq. 8.34 in Eq. 8.33, expanding the binomials of $F(x)$, and summing the coefficients of each power of $F(x)$, one can write

$$L_r = E[xP^*_{r-1}\{F(x)\}] = \int_0^1 x(F)P^*_{r-1}(F)dF, \quad r = 1,2,\ldots \tag{8.35}$$

where $P^*_r(F)$ is the rth shifted Legendre polynomial expressed as

$$P^*_r(F) = \sum_r^k (-1)^{r-k}\binom{r}{k}\binom{r+k}{k}F^k \tag{8.36}$$

Equation 8.36 can simply be written as

$$P^*_r(F) = \sum_{k=0}^r P_{r,k}F^k \tag{8.37}$$

and

$$P_{r,k} = (-1)^{r-k}\binom{r}{k}\binom{r+k}{k} \tag{8.38}$$

The shifted Legendre polynomials are related to the ordinary Legendre polynomials $P_r(u)$ as $P^*_r(u) = P_r(2u - 1)$ and are orthogonal on the interval $(0,1)$ with a constant weight function.

The first four *L*-moments are

$$L_1 = E(x) = \int x\, dF \tag{8.39}$$

$$L_2 = \frac{1}{2} E(x_{2:2} - x_{1:2}) = \int x(2F - 1)\, dF \tag{8.40}$$

$$L_3 = \frac{1}{3} E(x_{3:3} - 2x_{2:3} + x_{1:3}) = \int x\left(6F^2 - 6F + 1\right) dF \tag{8.41}$$

$$L_4 = \frac{1}{4} E(x_{4:4} - 3x_{3:4} + 3x_{2:4} - x_{1:4}) = \int x\left(20F^3 - 30F^2 + 12F - 1\right) dF \tag{8.42}$$

The *L*-moments can be defined in terms of PWM α and β as

$$L_{P,r+1} = (-1)^r \sum_{k=0}^{r} P_{r,k}\alpha_k = \sum_{k=0}^{r} P_{r,k}\beta \tag{8.43}$$

These can be written as

$$L_{P,1} = a_0 = b_0$$

$$L_{P,2} = a_0 - 2a_1 = 2b_1 - b_0 \tag{8.44}$$

$$L_{P,3} = a_0 - 6a_1 + 6a_2 = 6b_2 - 6b_1 + b_0$$

$$L_{P,4} = a_0 - 12a_1 + 30a_2 - 20a_3 = 20b_3 - 30b_2 + 12b_1 - b_0$$

Parallel to conventional moment ratios, *L*-moment ratios are defined by

$$T_1 = L_{P,2}/L_{P,1} \tag{8.45}$$

$$T_k = L_{P,k}/L_{P,2}, \ k \geq 3 \tag{8.46}$$

Example 8.6 Using the data of Example 8.4, find the parameters of exponential distribution using the *L*-moments.

Solution The parameter estimation using *L*-moments is very similar to the method of moments. As described earlier, in the method of moments we equate the first k conventional moments of the distribution to the first k conventional sample moments, whereas in the *L*-moment method, the first k *L*-moments of the distribution are equated to the first k *L*-moments of the sample data.

We determined the first two PWM in Example 8.5 as $a_0 = 28{,}675.8$ and $a_1 = 8{,}724.1$. Using Eq. 8.4 we obtain the first two *L*-moments based on the sample data:

$$\text{sample } L_{P,1} = a_0 = 28{,}675.8$$

$$\text{sample } L_{P,2} = a_0 - 2a_1 = 28{,}675.8 - (2 \times 8{,}724.1) = 11{,}227.6$$

Now we will determine the algebraic expression for the first two L-moments of the exponential distribution for which the parameters are to be determined. The inverse form of exponential distribution given by Eq. 8.20 is written as

$$x = \varepsilon - \alpha \ln(1 - F) \tag{8.47}$$

Using Eq. 8.25, we can write expressions for PWM as

$$a_0 = \int_0^1 x df = \int_0^1 [\varepsilon - \alpha \ln(1 - F)] dF$$

Solving this yields

$$a_0 = \varepsilon + \alpha \tag{8.48}$$

$$a_1 = \int_0^1 x(1 - F) dF = \int_0^1 [\varepsilon - \alpha \ln(1 - F)](1 - F) dF$$

$$= \varepsilon/2 + \alpha/4 \tag{8.49}$$

Hence, the first two L-moments of the distribution are

$$\text{distribution } L_{P1} = a_0 = \varepsilon + \alpha$$

$$\text{distribution } L_{P2} = a_0 - 2a_1 = (\varepsilon + \alpha) - 2(\varepsilon/2 + \alpha/4) = \alpha/2$$

Now equating the distribution moments with the sample moments, one can find the estimators of ε and α. Thus,

$$\text{distribution } L_{P,1} = \text{sample } L_{P,1}: \varepsilon + \alpha = 28{,}675.8$$

$$\text{distribution } L_{P,2} = \text{sample } L_{P,2}: \hat{\alpha}/2 = 11{,}227.6$$

yielding

$$\alpha = 2 \times 11{,}227.6 = 22{,}445.2$$

$$\varepsilon = 28{,}675.8 - 22{,}445.2 = 6{,}230.6$$

8.1.5 Method of Ordinary Least Squares

The ordinary least-squares parameter-estimation method (MOLS) is a variation of the probability plotting methodology in which one mathematically fits the best straight line or curve to a set of data points in an attempt to estimate the parameters. The method of least squares requires that a straight line be fitted to a set of data points such that the sum of the squares of the deviations from the observed data points to the assumed line is minimized.

MOLS is one of the most frequently used parameter-estimation methods in hydrology. Natale and Todini (1974) presented a constrained MOLS for linear models in hydrology. Williams and Yeh (1983) described MOLS and its variants for use in rainfall–runoff models. Jones (1971) linearized weight factors for least squares (LS) fitting. Shrader et al. (1981) developed a mixed-mode version of MOLS and applied it to estimate parameters of the log-normal distribution. Snyder (1972) reported on fitting of distribution functions by nonlinear least squares. Stedinger and Tasker (1985) performed regional hydrologic analysis using ordinary, weighted, and generalized least squares.

MOLS is quite good for mathematical functions that can be linearized. Most of the distributions used in engineering analysis can be linearized rather easily. For these distributions, the calculations are relatively easy and straightforward. Further, this technique provides a good measure of the goodness of fit of the chosen distribution in the form of R-square value (coefficient of determination). MOLS is generally best used with complete data sets containing no censored or interval data.

Let $Y = f(X; a_1, a_2, \ldots, a_m)$ be a linearized form of a distribution function, where a_i, $i = 1, 2, \ldots, m$, are parameters to be estimated. The method of least squares involves estimating parameters by minimizing the sum of squares of all deviations between observed and computed values of Y. Mathematically, this sum S can be expressed as

$$S = \sum_{i=1}^{n} d_1^2 = \sum_{i=1}^{n} \left[y_0(i) - y_c(i) \right]^2 = \sum_{i=1}^{n} \left[y_0(i) - f(x; a_1, a_2, \ldots, a_m) \right]^2 \qquad (8.50)$$

where $y_0(i)$ is the ith observed value of Y, $y_c(i)$ is the ith computed value of Y, and $n > m$ is the number of observations. The minimum of S in Eq. 8.50 can be obtained by differentiating S partially with respect to each parameter and equating each differential to zero:

$$\frac{\partial \sum_{i=1}^{n} \left[y_0(i) - f(x_i; a_1, a_2, \ldots, a_m) \right]^2}{\partial a_i} = 0 \qquad (8.51)$$

This leads to m equations, usually called the normal equations, which are then solved to estimate the m parameters. This method is used to estimate parameters of a linear regression model. For instance, suppose a linear equation of the type

$$y_i = a + bx_i \qquad (8.52)$$

is to be fitted. The regression coefficients (a and b) are estimated by minimizing the sum of squares of deviations of y_i from the regression line. For a point x_i, the corresponding $\hat{y}_i$ computed by the regression equation will be

$$\hat{y}_i = a + bx_i \qquad (8.53)$$

The residual error at this point is $e_i = y_i - \hat{y}_i$, which is a measure of how well the least-squares line conforms to the raw data. If the line passes through each sample point, the error e_i would be zero. The sum of the square of the errors is

$$S_{se} = \sum_{i=1}^{n} e_i^2 = \sum_{i=1}^{n} (y_i - y_i)^2 \qquad (8.54)$$

Minimizing S_{se} leads to the following values of parameters:

$$b = S_{xy}/S_{xx}, \quad a = \bar{y} - b\bar{x} \qquad (8.55)$$

where

$$S_{xx} = \sum_{i=1}^{n} (x_i - \bar{x})^2, \quad S_{yy} = \sum_{i=1}^{n} (y_i - \bar{y})^2, \quad S_{xy} = \sum_{i=1}^{n} (x_i - \bar{x})(y_i - \bar{y}) \qquad (8.56)$$

Example 8.7 The precipitation and runoff for a catchment for the month of July are given in Table E8-7. The relationship between rainfall and runoff follows a linear relation of the form $y = a + bx$, where y represents runoff and x precipitation. Estimate parameters a and b using MOLS.

Table E8-7 Precipitation runoff data and calculations

SN	Year	Precipitation (x)	Runoff (y)	$x - \bar{x}$	$y - \bar{y}$	$(x - \bar{x}) \times (y - \bar{y})$	$(x - \bar{x})^2$	$(y - \bar{y})^2$
1	1953	42.39	13.26	−0.55	−1.37	0.75	0.3025	1.8769
2	1954	33.48	3.31	−9.46	−11.32	107.08	89.49	128.14
3	1955	47.67	15.17	4.73	0.54	2.55	22.37	0.29
4	1956	50.24	15.50	7.3	0.87	6.35	53.29	0.76
5	1957	43.28	14.22	0.34	−0.41	−0.14	0.1156	0.1681
6	1958	52.60	21.20	9.66	6.57	63.47	93.32	43.16
7	1959	31.06	7.70	−11.88	−6.93	82.33	141.13	48.02
8	1960	50.02	17.64	7.08	3.01	21.31	50.13	9.06
9	1961	47.08	22.91	4.14	8.28	34.28	17.14	68.56
10	1962	47.08	18.89	4.14	4.26	17.64	17.14	18.15
11	1963	40.89	12.82	−2.05	−1.81	3.71	4.20	3.28
12	1964	37.31	11.58	−5.63	−3.05	17.17	31.69	9.30
13	1965	37.15	15.17	−5.79	0.54	−3.13	33.52	0.29
14	1966	40.38	10.40	−2.56	−4.23	10.83	6.55	17.89
15	1967	45.39	18.02	2.45	3.39	8.31	6.00	11.49
16	1968	41.03	16.25	−1.91	1.62	−3.09	3.65	2.62
	Total	687.05	234.04	0.01	−0.04	369.42	570.06	363.07

Solution Parameters a and b are computed using Eq. 8.55. To that end, $\bar{x} = 687.05/16 = 42.94$, $\bar{y} = 234.04/16 = 14.63$. The various other quantities, such as S_{xy} and S_{xx}, required to calculate a and b are computed in Table E8-7. Thus,

$$b = S_{xy}/S_{xx} = 369.423/570.0559 = 0.648$$

$$a = \bar{y} - b\bar{x} = 14.63 - (0.648 \times 42.94) = -13.195$$

Hence, the relationship between runoff and rainfall is $y = -13.195 + 0.648x$.

Example 8.8 Solve Example 8.4 using the method of ordinary least squares.

Solution The probability density function of a two-parameter exponential distribution is given by

$$f_X(x) = \frac{1}{\alpha}\exp\left(-\frac{x-\varepsilon}{\alpha}\right) \qquad \varepsilon < x < \infty$$

The cumulative distribution function for this two-parameter exponential distribution is given by

$$F_X(x) = \int_0^x f_X(x)dx = \frac{1}{\alpha}\int_0^x \exp\left(-\frac{x-\varepsilon}{\alpha}\right)dx = 1 - \exp\left(-\frac{x-\varepsilon}{\alpha}\right)$$

On further simplification, one can write

$$1 - F_X(x) = \exp\left(-\frac{x-\varepsilon}{\alpha}\right)$$

Taking logarithm of both sides gives

$$\ln\left[1 - F_X(x)\right] = \frac{\varepsilon}{\alpha} - \frac{1}{\alpha}x$$

This equation is the linearized form of the two-parameter exponential distribution as it matches with Eq. 8.52 in which $y = \ln[1 - F_X(x)]$, $a = \varepsilon/\alpha$, and $b = -1/\alpha$. Now, we can use MOLS as described earlier. It is easier to perform the calculations of the required quantities in a tabular form as given in Table E8-8.

Now using Eq. 8.55, one has

$$\hat{b} = \frac{-870741.27}{2.10 \times 10^{10}} = -0.00004$$

and

$$\hat{a} = \bar{y} - \hat{b}\bar{x} = -0.96 - (-0.00004)28675.83 = 0.23$$

Table E8-8 *Calculations of the required quantiles for Example 8.8*

Rank i	X	Plotting position $F_X(x) = i/(N+1)$	$y_i = \ln[1-F_X(x)]$	$(x_i-M_x)^2$	$(x_i-M_x)(y_i-M_y)$
1	3180	0.02	−0.02	6.50×10^8	−23978.69
2	3280	0.04	−0.04	6.45×10^8	−23349.97
3	4530	0.06	−0.06	5.83×10^8	−21681.38
4	6900	0.08	−0.09	4.74×10^8	−19074.67
5	7660	0.10	−0.11	4.42×10^8	−17936.65
6	8000	0.12	−0.13	4.27×10^8	−17171.14
7	8320	0.14	−0.15	4.14×10^8	−16426.40
8	8600	0.16	−0.18	4.03×10^8	−15716.67
9	8970	0.18	−0.20	3.88×10^8	−14940.43
10	9800	0.20	−0.23	3.56×10^8	−13833.25
11	10100	0.22	−0.25	3.45×10^8	−13130.87
12	13000	0.24	−0.28	2.46×10^8	−10662.88
13	15400	0.27	−0.31	1.76×10^8	−8666.63
14	16000	0.29	−0.34	1.61×10^8	−7917.85
15	16200	0.31	−0.37	1.56×10^8	−7431.28
16	17500	0.33	−0.40	1.25×10^8	−6323.30
17	18100	0.35	−0.43	1.12×10^8	−5658.38
18	18900	0.37	−0.46	9.56×10^7	−4919.99
19	19400	0.39	−0.49	8.60×10^7	−4364.19
20	19500	0.41	−0.52	8.42×10^7	−4006.07
21	20000	0.43	−0.56	7.53×10^7	−3483.33
22	20000	0.45	−0.60	7.53×10^7	−3167.81
23	20200	0.47	−0.63	7.18×10^7	−2774.90
24	20400	0.49	−0.67	6.85×10^7	−2384.84
25	21200	0.51	−0.71	5.59×10^7	−1849.13

Table E8-8 *Calculations of the required quantiles for Example 8.8 (Continued)*

Rank i	X	Plotting position $F_X(x) = i/(N+1)$	$y_i = \ln[1 - F_X(x)]$	$(x_i - M_x)^2$	$(x_i - M_x)(y_i - M_y)$
26	22000	0.53	−0.76	4.46×10^7	−1367.13
27	22400	0.55	−0.80	3.94×10^7	−1006.24
28	23300	0.57	−0.85	2.89×10^7	−611.85
29	26900	0.59	−0.90	3.15×10^6	−115.47
30	30500	0.61	−0.95	3.33×10^6	25.05
31	31600	0.63	−1.00	8.55×10^6	−117.95
32	36300	0.65	−1.06	5.81×10^7	−743.31
33	37900	0.67	−1.12	8.51×10^7	−1458.51
34	39300	0.69	−1.18	1.13×10^8	−2365.54
35	39300	0.71	−1.25	1.13×10^8	−3098.53
36	40700	0.73	−1.33	1.45×10^8	−4397.93
37	43400	0.76	−1.41	2.17×10^8	−6564.03
38	43400	0.78	−1.49	2.17×10^8	−7845.20
39	44500	0.80	−1.59	2.50×10^8	−9939.50
40	44500	0.82	−1.69	2.50×10^8	−11606.74
41	45500	0.84	−1.81	2.83×10^8	−14321.83
42	47500	0.86	−1.95	3.54×10^8	−18537.98
43	55700	0.88	−2.10	7.30×10^8	−30779.10
44	60800	0.90	−2.28	1.03×10^9	−42444.66
45	62100	0.92	−2.51	1.12×10^9	−51620.70
46	63300	0.94	−2.79	1.20×10^9	−63434.74
47	76400	0.96	−3.20	2.28×10^9	−106785.67
48	104000	0.98	−3.89	5.67×10^9	−220753.04
N=48		$M_X = 28675.83$	$M_y = -0.96$	$S_{XX} = 2.10 \times 10^{10}$	$S_{XY} = -870741.27$

It is known that $a = \varepsilon/\alpha$ and $b = -1/\alpha$. The estimates of α and ε are given as

$$\hat{\alpha} = -\frac{1}{\hat{b}} = \frac{-1}{-0.0004} = 24070.15$$

$$\hat{\varepsilon} = \hat{a}\hat{\alpha} = 0.23 \times 24070.15 = 5541.68$$

Figure 8-5 compares observed and fitted distributions using MOLS.

8.2 Assessment of Parameter Accuracy

The parameters of probability distribution functions can be estimated by various methods. The choice of the method to be used depends on the statistic to be estimated. It is desired that the method provides an unbiased estimate of the statistic under consideration with as small a variance as possible. However, it is not always possible to obtain an estimator that is both unbiased and efficient (with minimum variance). Because estimators can be derived in a variety of ways, their error characteristics must always be analyzed and compared. In practice, many problems and the estimators derived for them are sufficiently complicated to render analytic studies of the errors difficult, if not impossible. Instead, numerical simulation and comparison with lower bounds on the estimation error are frequently used to assess the estimator performance.

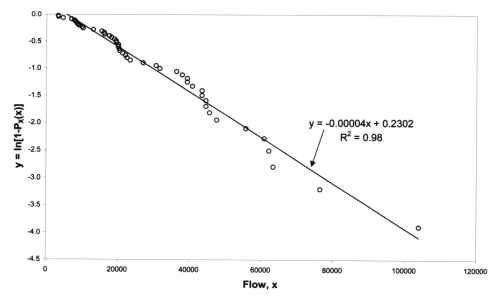

Figure 8-5 *Observed and fitted distribution using MOLS.*

As previously mentioned, the parameters obtained by maximizing the likelihood function are estimators of the true value. It is clear that the sample size determines the accuracy of an estimator. If the sample size equals the whole population, then the estimator is the true value. Estimators have properties, such as unbiasedness, sufficiency, consistency, and efficiency. Standard statistical books deal with these properties and this coverage lies beyond the scope of this reference. However, we would like to briefly address unbiasedness and consistency here. The parameters of a distribution function are estimated from sample values and these can, of course, be obtained in myriad ways. The sample data may contain errors, the hypotheses underlying the method of parameter estimation may not yield accurate estimates, and there may be truncation and round-off errors. These sources of errors may result in errors in parameter estimates. Each estimate of a parameter is a function of sample values that are observations of a random variable. Thus, the parameter estimate itself is a random variable having its own sampling distribution. An estimate obtained from a given set of values can be regarded as an observed value of the random variable. Thus, the goodness of an estimate can be judged from its distribution.

Some important questions arise here. How should we best use the data to form estimates? What do we mean by the best estimates? Are these estimates unique? How do we select the best parameter estimator if there is one? A number of statistical properties are available by which one can address these questions about the appropriateness of a parameter estimation method. These are briefly discussed next.

8.2.1 Bias

Bias measures how close an estimator is on average to the true parameter value. Let the parameter be a and its estimate $\hat{a}$. The estimate $\hat{a}$ is called an unbiased estimate of a if the expected value of the estimate equals the true value of the parameter (i.e., $E[\hat{a}] = a$). Otherwise, the estimate is said to be biased (i.e., $E[a_c] \neq a$). Since the parameters and estimators were known, their bias can be calculated by

$$Bias(\hat{a} \mid a) = \frac{1}{m} \sum_{j=1}^{m} \hat{a}_j - a = E[\hat{a}] - a \tag{8.57}$$

where $j = 1,..., m$ are the numbers of samples, $\hat{a} \in \{\hat{a}_1, \hat{a}_2, \hat{a}_3, ...\}$, and $a \in \{a_1, a_2, a_3, ...\}$. An unbiased estimate has a probability distribution where the mean equals the actual value of the parameter. For the sake of convenience, we will write bias($\hat{a} \mid a$) as bias($\hat{a}$). Obviously, bias($\hat{a}$) = 0 for an unbiased estimate. It should, however, be noted that an individual $\hat{a}$ may not be equal to or even close to a even if bias(a) = 0. It simply implies that the average of many independent estimates of parameter a will be equal to its true value. The bias($\hat{a}$) is usually considered to be additive, so that bias($\hat{a}$) = $E[\hat{a}] - a$. When we have a biased estimate, the bias usually depends on the number of observations, n. An estimate is said to be asymptotically unbiased if the bias tends to zero for large n; that is $\lim_{n\to\infty} bias(a) \to 0$. An estimate's variance equals the mean-squared estimation error only if the estimate is unbiased.

The bias in a given quantity is usually measured in dimensionless terms and is often referred to as standardized bias. It is particularly important when bias is compared with respect to several parameters. Thus, the dimensionless measure of bias is defined as

$$\text{bias} = \frac{E(\hat{a}) - a}{SD(\hat{a})} \tag{8.58}$$

where $SD(\hat{a})$ is the standard deviation of $\hat{a}$. The following are the properties of unbiased estimators:

1. They are not unique. For example, let $x_1, x_2, \ldots, x_n$ constitute a random sample from a uniform distribution with the range defined by parameters a_1 and a_2. Then, $[(n+1)/n]y_n$ is an unbiased estimator of a_2, where $y_n = \max(x_1, x_2, \ldots, x_n)$ is the largest sample value. Further, $2\bar{x}$ is also an unbiased estimator of a_2. This shows that unbiased estimates are not unique.
2. If $\tilde{a}$ is an unbiased estimator of a, it does not necessarily follow that $f(\tilde{a})$ is an unbiased estimator of $f(a)$, where $f(.)$ is any mathematical function operating on parameter a. For example, the square root of the sample variance is not an unbiased estimator of the standard deviation.

Should the lack of bias be considered a desirable property? If many unbiased estimates are computed from statistically independent sets of observations having the same parameter value, the average of these estimates will be close to the true parameter value. This property does not mean that the estimate has less error than a biased one; there exist biased estimates whose mean-squared errors are smaller than unbiased ones. In such cases, the biased estimate is usually asymptotically unbiased. Lack of bias is good, but that is just one aspect of how we evaluate estimators.

8.2.2 Efficiency

Efficiency refers to the variance of an estimator. An efficient estimate $\hat{a}$ of a has to satisfy two conditions: (1) It must be unbiased, and (2) its variance must be at least as small as that of any other unbiased estimate of a. If there are two estimates of a, say, a_1 and a_2, then the relative efficiency of a_2 with respect to a_1 is defined as the ratio of their variances (i.e., $\text{var}(\hat{a}_1)/\text{var}(\hat{a}_2)$). Mathematically, it is given as

$$e = \frac{E[\hat{a}_1 - a]^2}{E[\hat{a}_2 - a]^2} \tag{8.59}$$

If $e < 1$, then $\hat{a}_1$ is more efficient than $\hat{a}_2$. Only an efficient estimate has $e = 1$. If an efficient estimate exists, it may be approximately obtained by the use of MLE or the entropy method.

An *efficient* estimate has a mean-squared error that equals a particular lower bound known as the Cramer–Rao bound. If an efficient estimate exists (the Cramer–Rao bound being the greatest lower bound), it is optimum in the mean-squared sense, meaning that no other estimate has a smaller mean-squared error. If is an unbiased estimator $\tilde{a}$ of parameter a exists, then under some very general conditions var($\tilde{a}$) is given by the Cramer–Rao inequality as

$$\mathrm{var}(\tilde{a}) \geq \frac{1}{nE\left[\left(\dfrac{\partial \ln f(X)}{\partial a}\right)^2\right]}$$

(8.60)

where $f(X)$ is the probability density function of random variable X. If var($\tilde{a}$) is equal to the right-hand side of the inequality in Eq. 8.60, then $\tilde{a}$ is a minimum variance unbiased estimator (MVUE) of parameter a.

Example 8.9 Consider a normally distributed random variable X with mean μ and standard deviation σ. A sample of size $(2n + 1)$ is selected randomly from its population and the sample mean and median are estimated. What is the efficiency of the median relative to the mean? Further, determine which parameter is more efficient when sample size is large.

Solution We know $E[X] = \mu$ and var$[X] = \sigma^2$. Let the mean and median be denoted by $\bar{X}$ and $\tilde{X}$, respectively

Thus, $\mathrm{var}(\bar{X}) = \sigma^2/(2n+1)$ and the sample median $\mathrm{var}(\tilde{X}) = \pi\sigma^2/4n$ are the unbiased estimates of variance of $\bar{X}$ and $\tilde{X}$. The efficiency of median relative to mean is var($\bar{X}$)/var($\tilde{X}$). Thus, by using Eq. 8.60 the efficiency of the median with respect to the mean is $e = 4n/\pi(2n+1)$, which is a function of the sample size as shown in Fig. 8-6.

For large samples, the asymptotic efficiency is calculated as

$$e = \lim_{n\to\infty} \frac{4n}{\pi(2n+1)} = \lim_{n\to\infty} \frac{4}{\pi(2+1/n)} = \frac{2}{\pi} \approx 0.64$$

Thus, the mean is more efficient than the median for all sample sizes for a normal population. For large samples, the mean requires only about 64% as many observations as the median to estimate μ with the same reliability. Using the Cramer–Rao inequality given by Eq. 8.60, one can confirm that $\bar{X}$ is an MVUE of the mean μ of a normal distribution.

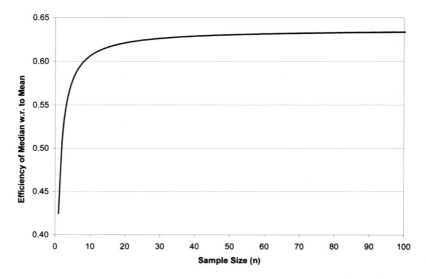

Figure 8-6 *Efficiency of median relative to mean.*

8.2.3 Mean Square Error

The bias measures the difference between the average value of an estimator and the quantity to be estimated. Unbiasedness may be a desirable property of an estimator, but there can be more than one unbiased estimator and sometimes a biased estimator may actually be superior. Another way to assess an estimator is to determine the estimator variance. The variance measures the spread or width of the estimator's distribution. Both the values of bias and variance contribute to the amount by which an estimator deviates from the quantity to be estimated. These two errors are often combined into the mean square error (MSE). Understanding that parameter a is fixed, and its estimator $\hat{a}$ is a random variable, the MSE is the expected value of the squared distance (error) between the two:

$$\text{MSE}(\hat{a}) = E\left[(\hat{a} - a)^2\right] \tag{8.61}$$

$$\text{MSE}(\hat{a}) = E\left[(\hat{a} - E[\hat{a}] + E[\hat{a}] - a)^2\right] = E\left[\{(\hat{a} - E[\hat{a}]) + (E[\hat{a}] - a)\}^2\right]$$

$$\text{MSE}(\hat{a}) = E\left[(\hat{a} - E[\hat{a}])^2\right] + E\left[(E[\hat{a}] - a)^2\right] + 2E\left[(\hat{a} - E[\hat{a}])(E[\hat{a}] - a)\right]$$

On simplifying, the third term cancels out and the remaining expression is

$$\text{MSE}(a) = \text{var}(a) + \left[\text{bias}(a)\right]^2 \tag{8.62}$$

Equation 8.62 shows that the MSE of parameter a is equal to the expected average squared deviation of the estimator from the true value. It can be computed as the bias squared plus the variance of the estimator. The MSE combines both bias and variance in a logical way and is therefore a convenient measure of how closely $\hat{a}$ approximates a.

Example 8.10 Let a_1 and a_2 be independent and identically distributed exponential random variables. Further, assume that a_1, a_2, and $b \times (a_1 + a_2)$ are all unbiased estimators of μ. Assume b to be constant such that $b = 0.5$. Determine which estimator will be the best estimator of μ.

Solution From Eq. 8.62, it is known that

$$MSE(a) = var(a) + [bias(a)]^2$$

Because all estimators are unbiased, bias(a_1) = bias(a_2) = bias$[0.5 \times (a_1 + a_2)]$ = 0. Therefore, MSE will be governed by the variance only and will be minimum for the variable having the smallest variance. Using the properties of the exponential distribution, one can write

$$var\,(a_1) = var\,(a_2) = \mu^2$$

$$var[b \times (a_1 + a_2)] = b^2[\,var\,(a_1) + var\,(a_2)\,] = 2b^2\mu^2$$

So,

$$var[b \times (a_1 + a_2)]_{b=0.5} = \mu^2/2$$

indicating that $b \times (a_1 + a_2)$ will be the best estimator of μ.

Example 8.11 Consider the previous example and determine what value of constant b minimizes the MSE of $b \times (a_1 + a_2)$.

Solution Let us represent the estimator $b \times (a_1 + a_2)$ by $\hat{a}$, so that

$$MSE(a) = var(a) + [bias(a)]^2, \quad Bias(a) = E[a] - a = b(\mu + \mu) - \mu = (2b - 1)\mu$$

As shown earlier var$[b \times (a_1 + a_2)] = 2b^2\mu^2$. So,

$$MSE(\hat{a}) = 2b^2\mu^2 + (2b - 1)^2\,\mu^2 = \left(6b^2 - 4b + 1\right)\mu^2$$

Now differentiating MSE with respect to b and equating it to zero, one determines the minimum of the MSE function:

$$\frac{d\left[MSE(\hat{a})\right]}{db} = 12b - 4 = 0$$

$$b = 1/3$$

8.2.4 Consistency

As already explained, the bias refers to the mean value of the estimator and the efficiency refers to the variance of the estimator. Now, we will discuss another property that refers to both the bias and the variance of the estimator. As shown in Eq. 8.62, MSE is a combination of bias and variance of an estimator. We term an estimate consistent if the MSE tends to be zero as the number of observations becomes large (i.e., $\lim_{n \to \infty} \text{MSE} \to 0$). Thus, a consistent estimate must be at least asymptotically unbiased. In other words, error in the estimator continuously decreases as the sample size increases.

Unbiased estimates whose errors never diminish as more data are collected do exist. Their variances remain nonzero no matter how much data are available. Inconsistent estimates may provide reasonable estimates when the amount of data is limited, but they have the counterintuitive property that the quality of the estimate does not improve as the number of observations increases. Although smaller MSE than a consistent estimate over a pertinent range of values of n may be appropriate in certain circumstances, consistent estimates are usually favored in practice.

8.2.5 Sufficiency

An estimate of a parameter a is termed sufficient if it uses all of the information that is contained in the sample and pertinent to the parameter estimation. More precisely, let a_1 and a_2 be two independent estimates of a. Estimate a_1 is considered a sufficient estimate if the joint probability distribution of a_1 and a_2 has the property

$$f(a_1, a_2) = f(a_1)f(a_2 \mid a_1) = f(a_1)K(x_1, x_2, \dots, x_n) \tag{8.63}$$

in which $f(a_1)$ is the distribution of a_1, $f(a_2 \mid a_1)$ is the conditional distribution of a_2 given a_1, and $K(x_1, x_2, \dots, x_n)$ is a function of x_is but not of a. If Eq. 8.63 holds, then a_2 does not produce any new information about a that is not already contained in a_1. In this case, a_1 is a sufficient estimate.

8.2.6 Standard Error

Another dimensionless performance measure used in hydrology is the normalized standard error (NSE), defined as

$$\text{NSE} = \sigma(\hat{a})/a \tag{8.64}$$

where $\sigma(.)$ denotes the standard deviation of a and is computed as

$$\sigma(\hat{a}) = \left[\frac{1}{n-1} \sum_{i=1}^{n} \left\{ \hat{a}_i - E(\hat{a}_i) \right\}^2 \right]^{1/2} \tag{8.65}$$

where the summations are over n estimates $\hat{a}$ of a. This measure is similar to the coefficient of variation.

8.2.7 Relative Mean Error

Another measure of error in assessing the goodness of fit of hydrologic models is the relative mean error (RME), defined as

$$\text{RME} = \frac{1}{n}\left(\sum_{i=1}^{n}\left[\frac{Q_0 - Q_c}{Q_0}\right]^2\right)^{0.5} \tag{8.66}$$

in which n is the sample size, Q_0 is the observed quantity of a given probability, and Q_c is the computed quantity of the same probability. Also used sometimes is the relative absolute error (RAE), defined as

$$\text{RAE} = \frac{1}{n}\sum_{i=1}^{n}\left|\frac{Q_0 - Q_c}{Q_c}\right| \tag{8.67}$$

8.2.8 Root Mean Square Error

The root mean square error (RMSE) is one of the most frequently employed performance measures and is defined as

$$\text{RMSE} = \sqrt{\text{MSE}} \tag{8.68}$$

Sometimes a normalized mean square error is used. Normalization is performed in two ways: (1) with respect to the mean of the true parameter a and (2) with respect to the standard deviation of parameter a. The corresponding expressions are given as

$$\text{NRMSE} = \frac{\text{RMSE}}{E[a]} \tag{8.69a}$$

$$\text{NRMSE} = \frac{\text{RMSE}}{\sigma(a)} \tag{8.69b}$$

In parameter estimation and model calibration, the smallest value of RMSE is preferred.

8.2.9 Robustness

Kuczera (1982a,b,c) defined a robust estimator as the one that is resistant and efficient over a wide range of fluctuations of population. Two criteria for resistant estimators are mini-max and minimum average RMSE. According to the mini-max criteria, the maximum RMSE for all population cases should be minimum. Thus, for a resistant estimator the average RMSE as well as the maximum RMSE should be minimum.

8.3 Interval Estimation of Parameters

For obtaining estimates of distribution parameters, such as the mean, the variance, the coefficient of skew, the covariance, and the correlation coefficient, the corresponding sample statistics are calculated from samples of observations and such estimates are called point estimates. The single point value obtained from the sample is peculiar to that particular sample used and another sample may yield a different point estimate. Sometimes even the addition of a single observation might measurably change the point estimate. Therefore, it is important to know the range of values for a sample statistic, which itself is random, as well as its probability distribution. For a given sample, one can determine with a specified probability of, say, 80%, 90%, or 95% the limits within which the distribution of parameters can be expected to lie. These limits are called *confidence limits* and the interval of uncertainty is called the *confidence interval*. Interval estimation does not increase the accuracy of estimation but allows for the quantification of uncertainty. An appreciation of this uncertainty is needed to avoid making unwarranted assertions.

8.3.1 Probability Distribution of the Sample Mean

The random variable X has a mean of μ and a variance of σ^2. Suppose that we have n independent observations of X. No assumption is made regarding the type of the distribution. The sample mean can now be written as

$$m_x = \frac{1}{n}X_1 + \frac{1}{n}X_2 + \frac{1}{n}X_3 + \dots + \frac{1}{n}X_n$$

(8.70)

The sample mean m_x is considered a random variable, for X_1, X_2,..., X_n are random variables. This can be seen by observing that any repetition of the n observations will result in different values for X_1, X_2, X_3,..., X_n. Therefore, X_1, X_2, X_3,..., X_n are regarded as n random variables. For each sample these variables will take on values in accord with their probability distributions. This means that m_x is the sum of n random variables X_i, each to be divided by n.

From the central limit theorem, one can first conclude that m_x is approximately normally distributed. As described in Chapter 5, the approximation is better when n is large. If X itself is already normally distributed, then the sample mean is normally distributed even if n is only 2. Second, taking the expectation of m_x in Eq. 8.70 one obtains

$$E(m_x) = \frac{1}{n}E(X_1) + \frac{1}{n}E(X_2) + \frac{1}{n}E(X_3) + \dots + \frac{1}{n}E(X_n)$$

(8.71)

The mean value of all the variables X_i is evidently μ_x. It follows that

$$E(m_x) = \mu_x$$

(8.72)

One can determine the variance of m_x. The terms in Eq. 8.70 are independent and for independent variables it is known that the variance of a sum is equal to the sum of the variances. Therefore,

$$\text{var}(m_x) = \frac{1}{n^2}\text{var}(X_1) + \frac{1}{n^2}\text{var}(X_2) + \frac{1}{n^2}\text{var}(X_3) + \ldots + \frac{1}{n^2}\text{var}(X_n) = \frac{\sigma^2}{n} \quad \text{(8.73)}$$

In summary, variable $m_x \sim N(\mu, \sigma/\sqrt{n})$; m_x is normally distributed with a mean equal to the mean of the variable and a standard deviation equal to the standard deviation of the variable divided by the square root of n. Equation 8.73 shows that the mean of a set of observations becomes less variable as the number of observations increases. Otherwise, m_x can be expected to approximately follow a normal distribution. In the limit the variance approaches zero and the sample mean approaches the mean of the distribution, μ. However, should X be normally distributed, then m_x is normally distributed, regardless of the sample size.

Example 8.12 At a building site, 45 samples of soil were taken and their analysis showed that the mean compressive strength was 35,000 kPa with a standard deviation of 600 kPa. Find the standard deviation of the mean. How many samples will be required to reduce this standard deviation by half?

Solution The standard deviation of the mean is

$$SD_{mean} = \sigma/\sqrt{n} = 600/\sqrt{45} = 89.44 \text{ kPa}$$

Clearly, if this value is to be halved, the number of required samples will have to be increased four times to $45 \times 4 = 180$. This might be quite an expensive proposition indeed.

To extend the discussion, one can also state that the variable $[(\mu - m)/(\sigma/\sqrt{n})]$ is $N(0,1)$. Invoking the properties of the normal distribution, one can state that, if a large number of samples are taken, about 95% of the time the value of the variable should lie within ± 1.96 standard deviation, or

$$P\left[-1.96 \leq \frac{\mu - m}{\sigma/\sqrt{n}} \leq 1.96\right] = 0.95$$

or

$$P\left[-1.96 \times \sigma/\sqrt{n} \leq \mu - m \leq 1.96 \times \sigma/\sqrt{n}\right] = 0.95$$

Using the given data one has

$$P[-1.96 \times 89.44 \leq 35000 - m \leq 1.96 \times 89.44] = 0.95$$

or

$$P[34{,}825 \leq m \leq 35{,}175] = 0.95$$

It is acknowledged here that the true mean is not known but the probability that the true mean lies in the range 34,825 to 35,175 is known to be 0.95. This leads to the interpretation of mean as a random variable instead of a fixed value. In other words, here we go from a point estimate of a parameter to an interval estimate. The width of the interval depends on three factors.

1. The width of the interval increases with increasing standard deviation of the data and vice versa, if all other things remain the same.
2. The width also depends on the probability, which was 0.95 in this example.
3. As this probability increases, the interval becomes wider, and vice versa.

Commonly used values of this probability are 0.99, 0.95, and 0.90. As seen in this example, reducing the width of the interval requires a larger number of samples.

Example 8.13 Consider measurement of river stage in a flood event. Assume that the fluctuations of water levels over a short span of time follow a normal distribution. Ten measurements of stage were taken and the mean stage was 295.384 m with a standard deviation of 0.15 m. Find the 95% confidence interval for the stage measurements.

Solution A 95% confidence interval implies a significance level = 0.05. The upper confidence limit is $u = \bar{x} + z_{\alpha/2} \times \sigma / \sqrt{n}$. From the tables of the standard normal distribution, $z_{\alpha/2} = z_{0.025} = 1.96$. Hence,

$$u = 295.384 + 1.96 \times 0.15/(10)^{0.5}$$

$$= 295.384 + 0.093 = 295.477 \text{ m}$$

and the lower limit is

$$l = 295.384 - 1.96 \times 0.15/(10)^{0.5}$$

$$= 295.384 - 0.093 = 295.291 \text{ m}$$

8.3.2 Probability Distribution of Sample Variance

Consider a normally distributed random variable X that has a mean of μ and a standard deviation of σ. The distribution of sample variance s^2 can be considered as follows. First, calculate the sample variance, assuming that parameter μ is known, as

$$s^2 = \frac{1}{n} \sum_{i=1}^{n} (X_i - \mu)^2 \tag{8.74}$$

If μ is not known then one can calculate the sample variance from the deviations from the sample mean m_x. Dividing each term inside the summation sign in Eq. 8.72 by σ_2, and bringing this factor outside the summation, one gets

$$s^2 = \frac{\sigma^2}{n} \sum_{i=1}^{n} \left[\frac{X_i - \mu}{\sigma} \right]^2 \qquad (8.75)$$

Each term in brackets within the summation sign in Eq. 8.75, Z_i, is a normally distributed random variable with a mean of zero and a standard deviation of one. The sum of squares of n such variables follows a distribution, known as the chi-square distribution with parameter n, denoted as $\chi^2(n)$:

$$\chi^2(n) = \sum_{i=1}^{n} Z_i^2 \text{ , where } Z \sim N(0,1), \quad Z = \frac{x - \mu}{\sigma} \qquad (8.76)$$

The chi-square distribution is a special case of the gamma distribution. (Recall that the gamma distribution arises from the sum of exponentially distributed variables.) PDF and CDF tables of the chi-square distribution are widely available. Parameter n in Eq. 8.76 is called the degrees of freedom. Since the probability of distribution of $c^2(n)$ is known, the probability distribution of s^2 is

$$s^2 = \frac{\sigma^2}{n} \chi^2(n) \qquad (8.77)$$

It can be shown that the mean of the chi-square distribution is equal to the degrees of freedom:

$$\text{var}(Z) = E(Z^2) - [E(Z)]^2$$

$$1 = E(Z^2) - 0$$

$$E(Z^2) = 1$$

$$E\left(Z_1^2 + Z_2^2 + Z_3^2 + \ldots + Z_n^2\right) = n$$

or

$$E(\chi^2) = v$$

and

$$\text{var}(\chi^2) = 2v$$

The chi-square distribution is additive. If $p = q + r$, then

$$\chi^2(p) = \chi^2(q) + \chi^2(r) \qquad (8.78)$$

Now consider the distribution of s^2 when this sample statistic is calculated from the sample as

$$s^2 = \frac{1}{n} \sum_{i=1}^{n} [X_i - m_x]^2 \qquad (8.79)$$

Since m_x is, in general, not the same as μ, division by σ^2 does not result in a standardized variable. One can write the departures as

$$\sum_{i=1}^{n} [X_i - \mu]^2 = \sum_{i=1}^{n} [(X_i - m) + (m - \mu)]^2$$

$$= \sum_{i=1}^{n} [X_i - m]^2 + \sum_{i=1}^{n} [m - \mu]^2 + \sum_{i=1}^{n} 2(X_i - m)(m - \mu)$$

$$= \sum_{i=1}^{n} (X_i - m)^2 + n(m - \mu)^2 + 2(m - \mu) \sum_{i=1}^{n} (X_i - m)$$

Because for each sample the sum of the departures from the sample mean m_x is equal to zero, one obtains

$$\sum_{i=1}^{n} (X_i - \mu)^2 = \sum_{i=1}^{n} (X_i - m)^2 + n(m - \mu)^2 \tag{8.80}$$

Dividing both sides of Eq. 8.74 by σ^2, one gets

$$\sum_{i=1}^{n} \left[\frac{X_i - \mu}{\sigma} \right]^2 = \sum_{i=1}^{n} \left[\frac{X_i - m}{\sigma} \right]^2 + \left[\frac{m - \mu}{\sigma / \sqrt{n}} \right]^2 \tag{8.81}$$

The term on the left side in Eq. 8.81 is the chi-square variable with n degrees of freedom. The second term on the right side is the chi-square variable with one degree of freedom, since the sample mean m_x is normally distributed with a mean of μ and a standard deviation of $\sigma / \sqrt{n}$. Noting that the chi-square distribution is additive, one can conclude that the first term on the right side is the chi-square variable with $n - 1$ degrees of freedom.

Returning to Eq. 8.79 for calculating the sample variance and dividing each term in the summation sign by σ^2, one obtains

$$s^2 = \frac{\sigma^2}{n} \sum_{i=1}^{n} \left(\frac{X_i - m}{\sigma} \right)^2$$

$$s^2 = \frac{\sigma^2}{n} \chi^2 (n - 1) \tag{8.82}$$

When s^2 is calculated from the sample mean, the effect is one of reducing the number of degrees of freedom of the chi-square variable by one. Figure 8-7 plots the chi-square distribution for selected degrees of freedom.

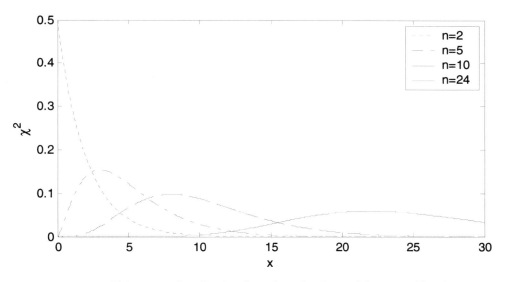

Figure 8-7 *Chi-square distribution for selected values of degrees of freedom.*

To determine the mean of the sample variance, take the expectation of Eq. 8.82 as

$$E(s^2) = \frac{\sigma^2}{n} E\left[\chi^2(n-1)\right] \tag{8.83}$$

It is known that the mean of the chi-square variable is equal to the number of degrees of freedom. Therefore,

$$E\left(s^2\right) = \frac{n-1}{n}\sigma^2 \tag{8.84}$$

Equation 8.82 shows that on average the variance of a sample from a distribution with a variance of σ^2 tends to be smaller than σ^2. This is commonly expressed by saying that Eq. 8.79 produces a biased estimate of the variance of σ^2. One can remove the bias in Eq. 8.79 by dividing by $(n-1)$ instead of n. Thus, an unbiased estimator of σ^2 is obtained as

$$s^2 = \frac{1}{n-1} \sum_{i=1}^{n} (X_i - m)^2 \tag{8.85}$$

The correction is also called small-sample correction. If Eq. 8.83 is used to calculate the sample variance then for the confidence limits one should use

$$s^2 = \frac{\sigma^2}{n-1} \chi^2(n-1) \tag{8.86}$$

The estimates of the $100(1 - \alpha)\%$ confidence interval for the variance can be constructed by

$$\frac{(n-1)s^2}{\chi^2_{\alpha/2}} \leq \sigma^2 \leq \frac{(n-1)s^2}{\chi^2_{(1-\alpha/2)}}$$

where $\chi^2_{\alpha/2}$ and $\chi^2_{(1-\alpha/2)}$ are the critical values of the chi-square distribution using $(n - 1)$ degree of freedom.

Example 8.14 To evaluate the variability of the compressive strength of concrete before use at a construction site, 25 test cylinders were prepared under "normal" control conditions. For each cylinder the compressive strength was measured. Thus 25 values of the compressive strength of the concrete were obtained. Then, the mean and standard deviation of the compressive strength were computed. These were, respectively, 34,000 kPa and 3,600 kPa². To check whether these 25 data points followed a normal distribution, they were plotted on a normal probability graph paper. It turned out that they reasonably represented a straight line. What are the 90% confidence limits of σ?

Solution To obtain 90% confidence limits, one must determine an upper limit and a lower limit such that there is a 5% probability that σ is larger than the upper limit and that there is a 5% probability that σ is smaller than the lower limit. Figure 8-7 shows the chi-square distribution for 24 degrees of freedom. It is now regarded that the unknown variance σ^2 is a random variable and the observed sample variance s^2 is fixed at $(3.6 \times 10^3)^2$ kPa². Here $100(1 - \alpha)\% = 90\%$; thus $\alpha = 0.1$ and $\alpha/2 = 0.05$. So we need to determine the $\chi^2_{0.05}$ and $\chi^2_{0.95}$ values corresponding to 24 degrees of freedom. From the χ^2 tables, one can read that the 95% and 5% confidence limits for $\chi^2(24)$ are 13.848 and 36.415, respectively. Corresponding to these, values of σ^2 are

$$\frac{(n-1)s^2}{\chi^2_{\alpha/2}} \leq \sigma^2 \leq \frac{(n-1)s^2}{\chi^2_{(1-\alpha/2)}}$$

$$\frac{(25-1)3600^2}{36.415} \leq \sigma^2 \leq \frac{(25-1)3600^2}{13.848}$$

$$8.54 \times 10^6 \leq \sigma^2 \leq 2.25 \times 10^7$$

Alternatively, the confidence bounds on the standard deviation are

$$2922.59 \leq \sigma \leq 4739.30$$

Hence, the limits of the standard deviation are 2,922.59 kPa² and 4,739.30 kPa².

8.3.3 Confidence Limits for the Mean

The sample mean is normally distributed if the variable is normally distributed or if the sample is large:

$$m_x \sim N\,(\mu, \frac{\sigma}{\sqrt{n}})$$ (8.87)

Equation 8.87 can be used to obtain confidence limits for μ provided that σ is known. One can write

$$\mu = m_x + \frac{\sigma}{\sqrt{n}} Z \text{ , where } Z \sim N(0,1)$$ (8.88)

Equation 8.88 shows that if the confidence limits of Z are known then those of μ can be calculated. The confidence limits for Z can be obtained from the table of the normal distribution. Then, substituting these in Eq. 8.88, one obtains the confidence limits for μ. In this calculation σ is not known and is replaced by the sample standard deviation s. Therefore, one defines a new variable T such that

$$\mu = m_x + \frac{s}{\sqrt{n}} T$$ (8.89)

which is analogous to Eq. 8.88, with the relationship between T and Z as

$$T = \frac{\sigma}{s} Z$$ (8.90)

It is known that

$$\frac{\sigma^2}{s^2} = \frac{n-1}{\chi^2(n-1)}$$

Therefore,

$$T = \frac{Z\sqrt{(n-1)}}{\sqrt{\chi^2(n-1)}}$$ (8.91)

The variable T follows Student's symmetrical t distribution with a zero mean, which resembles the standardized normal distribution. When parameter n, the number of degrees of freedom, becomes large, the T distribution approaches the normal distribution. Figure 8-8 shows the T distribution for selected values of degrees of freedom.

Example 8.15 Calculate the 90% confidence limits for the mean μ in Example 8.6 of the 25 test cylinders for which $m_x = 34{,}000$ kPa and $s = 3{,}600$ kPa.

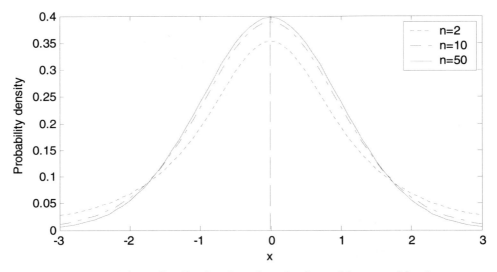

Figure 8-8 The T *distribution for selected values of degrees of freedom.*

Solution The relevant equation is

$$\mu = m_x + \frac{s}{\sqrt{n}} t_{n-1}$$

$$= 34{,}000 + (3{,}600/5)t_{24}$$

$$= 34{,}000 + 720 t_{24}$$

The 90% confidence limits for the t distribution with 24 degrees of freedom are ±1.711. Substitution in the preceding equation gives confidence limits as 32,768 kPa and 35,232 kPa.

Example 8.16 For the Sabarmati River data (Example 8.1), the mean and standard deviation were estimated as 664.29 m^3/s and 346.91 $(m^3/s)^2$ using 98 samples. Determine 95% confidence limits for the mean.

Solution As before, we write

$$\mu = m_x + \frac{s}{\sqrt{n}} t_{n-1}$$

or

$$\mu = 664.29 + 346.91/(98)^{0.5} \times t_{97}$$

For 98 degrees of freedom, the 95% confidence limits for the t distribution are ±1.96. Hence, $\mu = 664.29 \pm 346.91/(98)^{0.5} \times 1.96 = 733.61$ and 594.9 m^3/s.

8.4 Questions

8.1 Peak annual flow data of Buckhorn Creek observed at USGS station #02102192 near Cornith, North Carolina, are listed in Table Q8-1. Find the annual peak flow characteristics of the sample data and determine the candidate distribution(s) using the method of moments.

Table Q8-1

Year	1972	1973	1974	1975	1976	1977	1978	1979	1980	1981	1982	1983
Flow (cfs)	1530	3130	890	2150	891	1680	2820	1740	951	58	129	470

Year	1983	1984	1985	1986	1987	1988	1989	1990	1991	1992	1993	1994
Flow (cfs)	470	781	319	766	889	114	562	328	216	390	770	453

Year	1995	1996	1997	1998	1999	2000	2001	2002	2003	2004	2005	2006
Flow (cfs)	401	1940	480	1190	913	314	828	347	982	200	284	750

8.2 If one decides to fit an exponential distribution to the data given in Question 8.1, what will be its parameter based on the method of moments?

8.3 Using the maximum likelihood estimation procedure, find parameter α of the exponential distribution for the peak flow data of Buckhorn Creek given in Question 8.1.

8.4 The probability density function of a two-parameter exponential distribution is given as

$$f(x)\frac{1}{\alpha}\exp\left(-\frac{x-\varepsilon}{\alpha}\right), \varepsilon < x < \infty$$

Estimate the parameters of this distribution using the data of Question 8.1 and the method of maximum likelihood estimation.

8.5 Using the data of Question 8.1, find the parameters of a two-parameter exponential distribution using the PWM method.

8.6 Using the data of Question 8.1, find the parameters of the exponential distribution using the *L*-moment method.

8.7 Consider the data of Table Q8-7, in which x is the observed temperature value and y is the model simulated temperature value for a stream. The relationship between observed and model simulated temperatures can be represented by a linear relation of the form $y = a + bx$. Estimate parameters a and b using MOLS.

Table Q8-7

x	y	x	y	x	y	x	y
46.00	41.7	61	56.4	73.4	69.6	78.3	79.2
48.20	52.4	61.9	59.3	74.5	71.1	78.6	78.1
52.70	50.8	63.3	60.3	75	81.2	79	82.4
55.00	54	66.2	60.4	75.7	74.4	79.2	80.5
56.70	55.3	67.6	63.6	76.1	77.8	79.5	82.4
59.20	53.1	69.3	69.5	77	78.7	80.4	82
59.90	56.5	69.6	69.1	77	75.2	80.4	76
59.90	46.9	70.5	60.7	77	75.6	81.9	78.5
60.6	55.7	72.1	69.8	77.5	81.4	82	80.9
61	61.4	72.9	71.4	77.9	77.4	85.3	83.3

8.8 Solve Question 8.4 using the ordinary least-squares method.

8.9 Consider a model $y_t = b_1 + b_2 x_t + e_t$, where b_1 and b_2 are coefficients, x_t is an independent variable, and y_t is a dependent variable. Assuming that the model satisfies the classical assumptions, prove that the variance of the ordinary least-squares estimator declines to zero as the sample size increases.

8.10 A sample of 50 measurements of river stage were taken during various flood events. The actual mean flood level is unknown but the standard deviation can be assumed to be 3 ft. The computed value of the sample flood stage is 17.25 ft. Construct confidence interval estimates for the actual mean flood level for each of the following confidence levels: (a) 90%, (b) 95%, (c) 99%, and (d) 99.8%.

8.11 Solve Question 8.10 assuming the flood stage to be log-normally distributed.

8.12 For Buckhorn Creek (Question 8.1), the mean and standard deviation were estimated as 880 cfs and 757 cfs, respectively, from 34 samples. Determine the 95% confidence limits for the mean.

8.13 Table Q8-13 lists the daily dissolved oxygen (*DO*) concentration in a stream at a given station. Fit a most appropriate distribution for the daily *DO* concentration using a suitable method.

8.14 Using the daily *DO* concentration data of Question 8.13 construct confidence interval estimates for the mean daily *DO* concentration for each of the following confidence levels: (a) 90%, (b) 95%, (c) 99%, and (d) 99.8%.

Table Q8-13

Day	Observed DO (mg/L)	Day	Observed DO (mg/L)	Day	Observed DO (mg/L)	Day	Observed DO (mg/L)
1	3.3	9	7.7	17	8.4	25	9.4
2	6.6	10	7.8	18	8.5	26	9.6
3	6.8	11	7.8	19	8.5	27	9.6
4	6.8	12	7.9	20	8.9	28	10.2
5	7.5	13	8.1	21	9.1	29	10.2
6	7.5	14	8.2	22	9.2	30	10.4
7	7.7	15	8.1	23	9.2	31	10.5
8	7.7	16	8.2	24	9.3	32	8.2

8.15 For the daily *DO* concentration data of Question 8.13 determine the 95% confidence interval for the variance of the daily *DO* concentration.

8.16 For the peak annual flow of Buckhorn Creek given in Question 8.1 estimate parameters of the two-parameter gamma distribution function using the method of moments and maximum likelihood method and compare the two sets of parameter estimates.

Chapter 9

Entropy Theory and Its Applications in Risk Analysis

Environmental and water resource systems are inherently spatial and complex, and our understanding of these systems is less than complete. Many of the systems are either fully stochastic or part stochastic and part deterministic. Their stochastic nature can be attributed to randomness in one or more of the following components that constitute them: (1) system structure (geometry), (2) system dynamics, (3) forcing functions (sources and sinks), and (4) initial and boundary conditions. As a result, a stochastic description of these systems is needed, and entropy theory enables development of such a description.

Engineering decisions concerning environmental and water resource systems are frequently made with less than adequate information. Such decisions may often be based on experience, professional judgment, rules of thumb, crude analyses, safety factors, or probabilistic methods. Usually, decision making under uncertainty tends to be relatively conservative. Quite often, sufficient data are not available to describe the random behavior of such systems. Although probabilistic methods allow for a more explicit and quantitative accounting of uncertainty, their major difficulty occurs because of the lack of sufficient or complete data. Small sample sizes and limited information render estimation of probability distributions of system variables with conventional methods difficult. This problem can be alleviated by use of entropy theory, which enables determination of the least-biased probability distributions with limited knowledge and data. Where the shortage of data is widely rampant, as is normally the case in developing countries, entropy theory is particularly appealing.

Since the development of entropy theory by Shannon in the late 1940s and of the principle of maximum entropy by Jaynes in the late 1950s, there has been a proliferation in application of entropy. The real impetus to entropy-based modeling in environmental and water resources was however provided in the early 1970s, and a great variety of entropy-based applications have since been reported and new applications continue to unfold. The objective of this chapter is to briefly discuss entropy theory and demonstrate its usefulness for modeling and risk analysis in water resources and environmental systems.

9.1 History and Meaning of Entropy

Rudolph Clausius invented the term "entropy" in 1865 from the Greek meaning "transformation." He had noticed that a certain ratio was constant in reversible, or ideal, heat cycles. The ratio was heat exchanged to absolute temperature. Clausius decided that the conserved ratio must correspond to a real, physical quantity, and he named it entropy. For a closed system, entropy is the quantitative measure of the amount of thermal energy not available to do work. So it is a negative kind of quantity, the opposite of available energy. Obviously, in a closed system, entropy can never decrease. In a closed system, the available energy can never increase and its opposite, entropy (defined as unavailable energy), can never decrease. Further brooding reveals that there is nothing mysterious about this law. It is similar to saying that *things never organize themselves.*

Entropy can be considered as a measure of the degree of uncertainty or disorder, randomness, or lack of information about the microscopic configuration of particles of which a system is composed. A perfectly ordered system with a total number of quantum states equal to unity will have zero entropy, meaning a complete knowledge of the microscopic state of the system. Indirectly it also reflects the information content of space–time measurements. It has therefore been used as a measure of system diversity, system complexity, and system flexibility. Entropy is viewed in three different but related contexts and is hence typified by three forms: thermodynamical entropy, statistical entropy, and information-theoretical entropy. In environmental and water resources, the most frequently used form is the information-theoretical entropy. Before proceeding further, it will be instructive to briefly discuss the meaning of entropy.

Entropy originated in physics. It is an extensive property like mass, energy, volume, momentum, charge, or number of atoms of chemical species, but unlike these quantities, it does not obey a conservation law. Since entropy of a system is an extensive property, the total entropy of the system equals the sum of entropies of individual parts. The most probable distribution of energy in a system is the one that corresponds to the maximum entropy of the system. This occurs under the condition of dynamic equilibrium. During evolution toward an equilibrium or stationary state, the rate of entropy production per unit mass should be minimum, compatible with external constraints. This is the Prigogin principle. In thermodynamics, entropy is decomposed into two parts: (a) entropy exchanged between the system and its surroundings and (b) entropy produced in the system. According to the second law of thermodynamics, the entropy of a closed and isolated system always tends to increase. In hydraulics, entropy is a measure of the amount of irrecoverable flow energy expended by the hydraulic system to overcome friction. The system converts a portion of its mechanical energy to heat energy, which then is dissipated to the external environment. Thus, the process equation in hydraulics expressing energy (or head) loss originates indeed in the entropy concept. This conversion of energy is irreversible. Thus, this increase in entropy in a hydraulic system usually occurs as a result of this irreversible conversion of flow energy (mechanical) into heat through friction. This conversion represents the loss of energy and is nonrecoverable. The heat energy is conducted through the fluid and its bounding walls, if any; is lost to the atmosphere; and is eventually dissipated through space. The amount of energy so deployed is a function of the system geometry and the types of forces or energy affecting the system. Note that entropy is directly related to the total number of states available to the system. A perfectly ordered system with a total number of quantum states equal to unity corresponds to zero entropy and implies a complete knowledge of the microscopic state of the system.

Entropy has been employed in thermodynamics as a measure of the degree of ignorance about the true state of a system. For example, in a volume of gas, the greatest degree of order of the particles results when these particles are placed in a small space and are traveling with the same velocity. The thermodynamic entropy of such a system will be zero. However, as the particles begin to spread out in space and acquire different velocities, the disorder and thereby entropy increases. The thermodynamic entropy can be expressed as

$$H = -kN \sum_i \frac{N_i}{N} \ln \frac{N_i}{N} = -k_* \sum_i p_i \ln p_i \qquad \textbf{(9.1a)}$$

where H is the entropy of the system, p_i is the fraction of particles in energy state i, N is the total number of particles in the system, N_i is the number of particles in energy state i, and k_* is Boltzmann's constant.

If there were no energy loss, a hydraulic system would be orderly and organized. It is the energy loss and its causes that make the system disorderly and chaotic. Thus, entropy can be interpreted as a measure of the amount of chaos within a system. Algebraically, it is proportional to the logarithm of the probability of the state the system is in. The constant of proportionality is the Boltzmann constant and this defines the Boltzmann entropy or statistical entropy.

Shannon (1948) developed entropy theory for expression of information or uncertainty in communication. He expressed the average information conveyed per symbol j when the probability of the occurrence of symbol j in a message was p_j. Thus,

$$H=-kN\sum_j p_j \ln p_j =-k_* \sum_j p_j \ln p_j \qquad \text{(9.1b)}$$

where N is the total number of symbols in the message.

To further understand the informational aspect of entropy, we perform an experiment on a random variable X. There may be n possible outcomes $x_1, x_2,\ldots , x_n$, with probabilities $p_1, p_2,\ldots , p_n$; $P(X = x_1) = p_1$, $P(X = x_2) = p_2 ,\ldots , P(X = x_n) = p_n$. These outcomes can be described by

$$P(X)=(p_1,p_2,\ldots,p_n);\sum_i^n p_i =1;p_i \geq 0, \; i=1,2,\ldots,n \qquad \text{(9.1c)}$$

If this experiment is repeated, the same outcome is not likely, implying that there is uncertainty as to the outcome of the experiment. Based on one's knowledge about the outcomes, the uncertainty can be more or less. For example, the total number of outcomes is a piece of information and the number of those outcomes with nonzero probability is another piece of information. The probability distribution of the outcomes, if known, provides a certain amount of information. Shannon (1948) defined a quantitative measure of uncertainty associated with a probability distribution or the information content of the distribution in terms of entropy, $H(P)$ or $H(X)$, called Shannon entropy or informational entropy (with k_* taken as unity) as

$$H(X)=H(P)=-\sum_{i=1}^n p_i \ln p_i =E[-\ln p] \qquad \text{(9.2)}$$

If the random variable X is continuous then the Shannon entropy is expressed as

$$H(X)=-\int_0^\infty f(x)\ln[f(x)]dx=-\int \ln[f(x)]dF(x)=E[-\ln f(x)] \qquad \text{(9.3)}$$

where $f(x)$ is the probability density function of X, $F(x)$ is the cumulative probability distribution function of X, and $E[.]$ is the expectation of $[.]$.

Thus, entropy is a measure of the amount of uncertainty represented by the probability distribution and is a measure of the amount of chaos or of the lack of information about a system. If complete information is available, entropy $= 0$. Otherwise, it is greater than zero. The uncertainty can be quantified using entropy by taking into account all different kinds of available information. The Shannon entropy is the weighted Boltzmann entropy.

In frequency analysis, often the distribution function $f(x)$ is not known although some of its properties, such as lower and upper bounds, moments, etc., may be known. But these data are insufficient to define $f(x)$ uniquely and may delineate a set of feasible distributions. Each of these distributions contains a certain amount of uncertainty that can be expressed by employing the concept of entropy.

As an example, consider that weather at a given place on a given day has two possible outcomes: no rain R_0 with probability p_0 and rain R_1 with probability p_1. Based on an experiment, the following data are obtained:

	Scheme 1		Scheme 2	
	R_0	R_1	R_0	R_1
Probability	0.5	0.5	0.9	0.1

The entropy of the first scheme (by using Eq. 9.2) is

$$H(P) = -0.5 \ln(0.5) - 0.5 \ln(0.5) = 0.693$$

The entropy of the second scheme is

$$H(P) = -0.9 \ln(0.9) - 0.1 \ln(0.1) = 0.135$$

Clearly, the first scheme is about five-fold as uncertain as the second.

9.2 Principle of Maximum Entropy

To obtain an appropriate probability distribution for a given random variable, entropy should be maximized. By its nature, the entropy formula has its maximum value when all probabilities are equal. But this result seems reasonable only when we have no information. In practice, however, it is common that some information is available on the random variable. The chosen probability distribution should then be consistent with the given information. There can be more than one distribution consistent with the given information. From all such distributions, we should choose the distribution that has the highest entropy or uncertainty. To that end, Jaynes (1957) formulated the principle of maximum entropy (POME), a full account of which is presented in a treatise by Levine and Tribus (1979). According to POME, the minimally prejudiced assignment of probabilities is that which maximizes entropy subject to the given information; that is, POME takes into account all of the given information and at the same time avoids consideration of any information that is not given. The maximum-entropy distribution is maximally noncommittal with regard to missing information. It also has the property that no possibility is ignored; a positive weight is assigned to every situation that is not absolutely excluded by the given information.

The maximum entropy in the presence of some information will be less than the maximum entropy in the absence of that information. The difference between these two maximum entropies may be regarded as a measure of the bias resulting from the given information. Maximizing entropy amounts to minimizing this bias. For this reason, POME is said to give the minimally biased assignment of probabilities and POME may be called the principle of minimum bias or minimum prejudice.

If no information is available on the random variable, then all possible outcomes are equally likely, that is, $p_i = 1/n$, $i = 1, 2, 3, \ldots, n$. It can be shown that the Shannon entropy is maximum in this case and may indeed serve as an upper bound of entropy for all cases involving some information. In a more general case, let the information available about P or X be

$$p_i \geq 0, \quad \sum_{i=1}^{n} p_i = 1 \tag{9.4}$$

and

$$\sum_{i=1}^{n} g_r(x_i) p_i = a_r \quad r = 1, 2, \ldots, m \tag{9.5}$$

where m is the number of constraints, $m + 1 \leq n$, and g_r is the rth constraint. Equations 9.4 and 9.5 are not sufficient to determine P uniquely. Therefore, there can be many distributions that will satisfy Eq. 9.4 and Eq. 9.5. According to POME, there will be only one distribution that will correspond to the maximum value of entropy and this distribution can be determined using the method of Lagrange multipliers, which will have the following form:

$$p_i = \exp[-\lambda_0 - \lambda_1 g_1(x_i) - \lambda_2 g_2(x_i) \ldots - \lambda_m g_m(x_i)], \quad i = 1, 2, \ldots, n \tag{9.6}$$

where λ_i, $i = 0, 1, 2, \ldots, m$, are Lagrange multipliers, which are determined by using the information specified by Eq. 9.4 and Eq. 9.5.

According to the Shannon theory, entropy is an information measure; entropy defines a kind of measure on the space of probability distributions. Hence, the POME-based distribution is favored over those with less entropy among those that satisfy the given constraints. Intuitively, distributions of higher entropy represent more disorder, are smoother, are more probable, are less predictable, or assume less. The POME-based distribution is maximally noncommittal with regard to missing information and does not require invocation of ergodic hypotheses.

The concept of entropy provides a quantitative measure of uncertainty. To that end, consider a PDF $f(x)$ associated with a dimensionless random variable X. The dimensionless random variable may be constructed by dividing the observed quantities by its mean value (e.g., annual flood maxima divided by the mean annual flood). As usual, $f(x)$ is a possible function for every x in some interval (a, b) and is normalized to unity such that

$$\int_a^b f(x)dx = 1 \tag{9.7}$$

The most popular measure of entropy was first mathematically given by Shannon (1948) and has since been called the Shannon entropy functional (SEF), denoted as $I[f]$ or $I[x]$. It is a numerical measure of uncertainty associated with $f(x)$ in describing the random variable X and is defined as

$$I[f] = I[x] = -k\int_a^b f(x)\ln[f(x)/m(x)]dx \tag{9.8}$$

where $k > 0$ is an arbitrary constant or scale factor depending on the choice of measurement units, and $m(x)$ is an invariant measure function guaranteeing the invariance of $I[f]$ under any allowable change of variable and provides an origin of measurements of $I[f]$. Scale factor k can be absorbed into the base of the logarithm and $m(x)$ may be taken as unity so that Eq. 9.8 is often written as

$$I[f] = I[x] = -\int_a^b f(x)\ln[f(x)]dx; \qquad \int_a^b f(x)dx = 1 \tag{9.9}$$

We may think of $I[f]$ as the mean value of $-\ln[f(x)]$. Actually, $-I$ measures the strength and $+I$ measures the weakness. SEF allows us to choose the $f(x)$ that minimizes the uncertainty. Note that $f(x)$ is conditioned on the constraints used for its derivation. Singh (1988, 1998a, 1998b) has described the theory of entropy and has given expressions of SEF for a number of probability distributions.

Shannon (1948) showed that I is a unique function and the only one that satisfies the following properties:

1. It is a function of the probabilities $f_1, f_2, \ldots, f_n$, where n is the number of data points.
2. It follows an additive law, that is, $I[xy] = I[x] + I[y]$.
3. It monotonically increases with the number of outcomes when f_i are all equal.
4. It is consistent and continuous.

According to POME, "the minimally prejudiced assignment of probabilities is that which maximizes entropy subject to the given information." Mathematically, it can be stated as follows: Given m linearly independent constraints C in the form

$$C_i = \int_a^b y_i(x)f(x)dx, \quad i = 1, 2, \ldots, m \tag{9.10}$$

where $y_i(x)$ are some functions whose averages over $f(x)$ are specified, the maximum of I, subject to the conditions in Eq. 9.10, is given by the distribution

$$f(x) = \exp[-\lambda_0 - \sum_{i=1}^{m} \lambda_i y_i(x)] \tag{9.11}$$

where λ_i, $i = 0, 1, \ldots, m$, are Lagrange multipliers and can be determined from Eq. 9.10 and Eq. 9.11 along with the normalization condition in Eq. 9.9. An increase in the number of constraints leads to less uncertainty about the information concerning the system.

9.2.1 Entropy-based Parameter Estimation

The general procedure for deriving an entropy-based parameter estimation for a frequency distribution involves the following steps: (1) Define the given information in terms of constraints. (2) Maximize the entropy subject to the given information. (3) Relate the parameters to the given information. More specifically, let the available information be given by Eq. 9.10. Since POME specifies $f(x)$ by Eq. 9.11, inserting Eq. 9.11 in Eq. 9.9 yields

$$I[f] = \lambda_0 \int_a^b f(x)dx + \sum_{i=1}^{m} \lambda_i \int_a^b y_i(x)f(x)dx$$

or

$$I[f] = \lambda_0 + \sum_{i=1}^{m} \lambda_i C \tag{9.12}$$

In addition, the potential function or the zeroth Lagrange multiplier λ_0 is obtained by inserting Eq. 9.11 in Eq. 9.12 as

$$\int_a^b \exp[-\lambda_0 - \sum_{i=1}^{m} \lambda_i y_i] \, dx = 1 \tag{9.13}$$

resulting in

$$\lambda_0 = \ln \int_a^b \exp[-\sum_{i=1}^{m} \lambda_i y_i] \, dx \tag{9.14}$$

The Lagrange multipliers are related to the given information (or constraints) by

$$-\frac{\partial \lambda_0}{\partial \lambda_i} = C_i \tag{9.15}$$

It can also be shown that

$$\frac{\partial^2 \lambda_0}{\partial \lambda_i^2} = \text{var}[y_i(x)]; \quad \frac{\partial^2 \lambda_0}{\partial \lambda_i \partial \lambda_j} = \text{cov}[y_i(x), y_j(x)], \quad i \neq j \qquad (9.16)$$

With the Lagrange multipliers estimated from Eq. 9.15 and Eq. 9.16, the frequency distribution given by Eq. 9.11 is uniquely defined. It is implied that the distribution parameters are uniquely related to the Lagrange multipliers. Clearly, this procedure states that a frequency distribution is uniquely defined by specification of constraints and application of POME.

9.2.2 Parameter-Space Expansion Method

The parameter-space expansion method was developed by Singh and Rajagopal (1986). It employs an enlarged parameter space and maximizes entropy subject to both the parameters and the Lagrange multipliers. An important implication of this enlarged parameter space is that the method is applicable to virtually any distribution, expressed in direct form, having any number of parameters. For a continuous random variable X having a probability density function $f(x, \theta)$ with parameters θ, the SEF can be expressed as

$$I(f) = \int_{-\infty}^{\infty} f(x; \theta) \ln f(x, \theta) dx \qquad (9.17)$$

Parameters of this distribution, θ, can be estimated by maximizing $I(f)$. To apply the method, the constraints are first defined. Next, the POME formulation of the distribution is obtained in terms of the parameters by using the method of Lagrange multipliers. This formulation is used to define the SEF whose maximum is sought. If the probability distribution has n parameters, θ_i, $i = 1, 2, \ldots, n$, and the $(n - 1)$ Lagrange multipliers are λ_i, $i = 1, 2, \ldots, (n - 1)$, then the point where $I[f]$ is a maximum is a solution of $(2n - 1)$ equations:

$$\frac{\partial I[f]}{\partial \lambda_i} = 0, \quad i = 1, 2, \ldots, n-1 \qquad (9.18)$$

and

$$\frac{\partial I[f]}{\partial \theta_i} = 0, \quad i = 1, 2, \ldots, n \qquad (9.19)$$

Solution of Eq. 9.18 and Eq. 9.19 yields estimates of parameters of the distribution.

9.2.3 Principle of Minimum Cross Entropy

According to Laplace's principle of insufficient reason, all outcomes of an experiment should be considered equally likely unless there is information to the contrary. On the basis of intuition, experience, or theory, a random variable may have an a priori probability distribution. Then, the Shannon entropy is at a maximum when the probability distribution of the random variable is that one which is as close to the a priori distribution as possible. This is referred to as the principle of minimum cross entropy (POMCE), and under this principle the Bayesian entropy is minimized (Kullback and Leibler 1951). This is equivalent to maximizing the Shannon entropy.

To explain POMCE, let us suppose we guess a probability distribution for a random variable x as $Q = \{q_1, q_2, ..., q_n\}$ based on intuition, experience, or theory. This constitutes the prior information in the form of a prior distribution. To verify our guess we take some observations $X = (x_1, x_2, ..., x_n)$ and compute some moments of the distribution. To derive the distribution $P = \{p_1, p_2, ..., p_n\}$ of X we take all the given information and make the distribution as near to our intuition and experience as possible. Thus, POMCE is expressed as

$$D(P,Q) = \sum_{i=1}^{n} p_i \ln \frac{p_i}{q_i} \tag{9.20}$$

where the cross entropy D is minimized. If no a priori distribution is available and if according to Laplace's principle of insufficient reason Q is chosen to be a uniform distribution U, then Eq. 9.17 takes the form

$$D(P,U) = \sum_{i=1}^{n} p_i \ln \left[\frac{p_i}{1/n}\right] = \ln n \left(\sum_{i=1}^{n} p_i \ln p_i\right) \tag{9.21}$$

Hence, minimizing $D(P,U)$ is equivalent to maximizing the Shannon entropy. Because D is a convex function, its local minimum is its global minimum. Thus, a posterior distribution P is obtained by combining a prior Q with the specified constraints. The distribution P minimizes the cross (or relative) entropy with respect to Q, defined by Eq. 9.20, where the entropy of Q is defined as Eq. 9.2. Cross-entropy minimization results asymptotically from Bayes's theorem.

9.2.4 Joint Entropy, Conditional Entropy, and Transinformation

If there are two random variables X and Y with probability distributions $P(x) = \{p_1, p_2, ..., p_n\}$ and $Q(y) = \{q_1, q_2, ..., q_n\}$, which are independent, then the Shannon entropy of the joint distribution of X and Y is the sum of the entropies of the marginal distributions expressed as

$$H(P,Q) = H(X,Y) = H(P) + H(Q) = H(X) + H(Y) \tag{9.22}$$

If the two random variables are dependent then the Shannon entropy of the joint distribution is the sum of the marginal entropy of one variable and the conditional entropy of the other variable conditioned on the realization of the first. Expressed algebraically, this is

$$H(X,Y)=H(X)+H(Y|X) \tag{9.23}$$

where $H(Y|X)$ is the conditional entropy of Y conditioned on X. The conditional entropy can be defined as

$$H(X|Y)=-\sum_{i=1}^{n}\sum_{j=1}^{m}p(x_i,y_j)\ln\left[p\left(x_i|y_j\right)\right] \tag{9.24}$$

It is seen that if X and Y are independent then Eq. 9.23 reduces to Eq. 9.22. Furthermore, the joint entropy of dependent X and Y will be less than or equal to the joint entropy of independent X and Y, that is, $H(X, Y) \leq H(X) + H(Y)$. The difference between these two entropies defines *transinformation* $T(X, Y)$ or $T(P, Q)$ expressed as

$$T(X,Y)=H(X)+H(Y)-H(X,Y) \tag{9.25}$$

Transinformation represents the amount of information common to both X and Y. If X and Y are independent, then $T(X, Y) = 0$. Substitution of Eq. 9.23 in Eq. 9.25 yields

$$T(X,Y)=H(Y)-H(Y|X) \tag{9.26}$$

Equation 9.26 states that stochastic dependence reduces the entropy of Y.

9.3 Derivation of Parameters of the Normal Distribution Using Entropy

Frequency distributions that satisfy the given information are often needed. Entropy theory is ideally suited to that end. Indeed POME has been employed to derive a variety of distributions, some of which have found wide applications in environmental and water resources. Singh and Fiorentino (1992) and Singh (1998a) summarize many of these distributions. Let $p(x)$ be the probability distribution of X that is to be determined. The information on X is available in terms of constraints given by Eq. 9.2. Then, the entropy-based distribution is given by Eq. 9.6. Substitution of Eq. 9.5 in Eq. 9.4 yields

$$\exp(\lambda_0)=Z=\sum_{i=1}^{n}\exp[-\sum_{j=1}^{m}\lambda_j g_j(x_i)] \tag{9.27}$$

where Z is called the partition function and λ_0 is the zeroth Lagrange multiplier. The Lagrange parameters are obtained by differentiating Eq. 9.27 with respect to Lagrange multipliers:

$$\frac{\partial \lambda_0}{\partial \lambda_j} = -a_j = E[g_j], \quad j = 1, 2, 3, \ldots, m$$

$$\frac{\partial^2 \lambda_0}{\partial \lambda_j^2} = \mathrm{var}[g_j] \tag{9.28}$$

$$\frac{\partial^2 \lambda_0}{\partial \lambda_j \partial \lambda_k} = \mathrm{cov}[g_j, g_k]$$

$$\frac{\partial^3 \lambda_0}{\partial \lambda_j^3} = -\mu_3[g_j]$$

where $E[.]$ is the expectation, $\mathrm{var}[.]$ is the variance, $\mathrm{cov}[.]$ is the covariance, and μ_3 is the third moment about the centroid, all for g_j.

When there are no constraints, then POME yields a uniform distribution. As more constraints are introduced, the distribution becomes more peaked and possibly skewed. In this way, the entropy reduces from a maximum for the uniform distribution to zero when the system is fully deterministic.

The derivation of the normal distribution by the entropy method is described in the following.

9.3.1 Specification of Constraints

The probability density function of the normal distribution is

$$f(x) = \frac{1}{b\sqrt{2\pi}} \exp\left[-\frac{(x-a)^2}{2b^2} \right] \tag{9.29}$$

Taking the logarithm to the base e, one gets

$$\ln f(x) = -\ln \sqrt{2\pi} - \ln b - \frac{(x-a)^2}{2b^2}$$

$$= -\ln \sqrt{2\pi} - \ln b - \frac{x^2}{2b^2} - \frac{a^2}{2b^2} + \frac{2ax}{2b^2} \tag{9.30}$$

Multiplying Eq. 9.30 by $[-f(x)]$ and integrating between $-\infty$ to ∞, one gets

$$I(x) = -\int_{-\infty}^{\infty} f(x) \ln f(x) dx = [\ln \sqrt{2\pi} + \ln b + \frac{a^2}{2b^2}] \int_{-\infty}^{\infty} f(x) dx \tag{9.31}$$

$$+ \frac{1}{2b^2} \int_{-\infty}^{\infty} x^2 f(x) dx - \frac{a}{b^2} \int_{-\infty}^{\infty} x f(x) dx$$

From Eq. 9.31, the constraints appropriate for Eq. 9.29 can be written as

$$\int_{-\infty}^{\infty} f(x)dx = 1$$

$$\int_{-\infty}^{\infty} xf(x)dx = E[x] = \bar{x} \qquad (9.32)$$

$$\int_{-\infty}^{\infty} x^2 f(x)dx = E[x^2] = s_x^2 + \bar{x}^2$$

where $\bar{x}$ is the mean and s_x^2 is the variance of x.

9.3.2 Construction of the Zeroth Lagrange Multiplier

The least-biased probability density function $f(x)$ consistent with Eq. 9.32 and based on POME takes the form

$$f(x) = \exp(-\lambda_0 - \lambda_1 x - \lambda_2 x^2) \qquad (9.33)$$

where λ_0, λ_1, and λ_2 are Lagrange multipliers. Substitution of Eq. 9.33 in the normality condition in the first of Eq. 9.32 gives

$$\int_{-\infty}^{\infty} f(x)dx = \int_{-\infty}^{\infty} \exp(-\lambda_0 - \lambda_1 x - \lambda_2 x^2)dx = 1 \qquad (9.34)$$

Equation 9.34 can be simplified as

$$\exp(\lambda_0) = \int_{-\infty}^{\infty} \exp(-\lambda_1 x - \lambda_2 x^2)dx \qquad (9.35)$$

Equation 9.35 defines the partition function. Making the argument of the exponential as a square in Eq. 9.35, one gets

$$\exp(\lambda_0) = \int_{-\infty}^{\infty} \exp(-\lambda_1 x - \lambda_2 x^2 + \frac{\lambda_1^2}{4\lambda_2} - \frac{\lambda_1^2}{4\lambda_2})dx$$

$$= \exp\left(\frac{\lambda_1^2}{4\lambda_2}\right) \int_{-\infty}^{\infty} \exp - (x\sqrt{\lambda_2} + \frac{\lambda_1}{2\sqrt{\lambda_2}})^2 dx \qquad (9.36)$$

Now let

$$t = x\sqrt{\lambda_2} + \frac{\lambda_1}{2\sqrt{\lambda_2}} \qquad (9.37)$$

Then

$$\frac{dt}{dx} = \sqrt{\lambda_2} \tag{9.38}$$

Making use of Eqs. 9.37 and 9.38 in Eq. 9.36, we get

$$\exp(\lambda_0) = \frac{\exp\left(\dfrac{\lambda_1^2}{4\lambda_2}\right)}{\sqrt{\lambda_2}} \int_{-\infty}^{\infty} \exp(-t^2)dt = \frac{2\exp\left(\dfrac{\lambda_1^2}{4\lambda_2}\right)}{\sqrt{\lambda_2}} \int_{0}^{\infty} \exp(-t^2)dt \tag{9.39}$$

Consider the expression

$$\int_{0}^{\infty} \exp(-t^2)dt$$

Let $k = t^2$. Then $[dk/dt] = 2t$ and $t = k^{0.5}$. Hence, this expression can be simplified by making substitution for t to yield

$$\int_{0}^{\infty} \exp(-t^2)dt = \int_{0}^{\infty} \exp(-k)\frac{dk}{2k^{0.5}} = \frac{1}{2}\int_{0}^{\infty} k^{-0.5}\exp(-k)dk = \frac{1}{2}\int_{0}^{\infty} k^{[0.5-1]}\exp(-k)dk = \frac{\sqrt{\pi}}{2} \tag{9.40}$$

Substituting Eq. 9.40 in Eq. 9.39, one gets

$$\exp(\lambda_0) = \frac{2\exp\left(\dfrac{\lambda_1^2}{4\lambda_2}\right)}{\sqrt{\lambda_2}}\frac{\sqrt{\pi}}{2} = \exp\left(\dfrac{\lambda_1^2}{4\lambda_2}\right)\sqrt{\dfrac{\pi}{\lambda_2}} \tag{9.41}$$

Equation 9.41 is another definition of the partition function. The zeroth Lagrange multiplier λ_0 is given by Eq. 9.41 as

$$\lambda_0 = \frac{1}{2}\ln\pi - \frac{1}{2}\ln\lambda_2 + \frac{\lambda_1^2}{4\lambda_2} \tag{9.42}$$

One also obtains the zeroth Lagrange multiplier from Eq. 9.35 as

$$\lambda_0 = \ln\int_{-\infty}^{\infty} \exp(-\lambda_1 x - \lambda_2 x^2)dx \tag{9.43}$$

9.3.3 Relation Between Lagrange Multipliers and Constraints

Differentiating Eq. 9.43 with respect to λ_1 and λ_2, respectively, one obtains

$$\frac{\partial \lambda_0}{\partial \lambda_1} = -\frac{\int\limits_{-\infty}^{\infty} x \exp(-\lambda_1 x - \lambda_2 x^2) dx}{\int\limits_{-\infty}^{\infty} \exp(-\lambda_1 x - \lambda_2 x^2) dx} = -\int\limits_{-\infty}^{\infty} x \exp(-\lambda_0 - \lambda_1 x - \lambda_2 x^2) dx$$

$$= -\int\limits_{-\infty}^{\infty} x f(x) dx = -\overline{x}$$

(9.44)

$$\frac{\partial \lambda_0}{\partial \lambda_2} = -\frac{\int\limits_{-\infty}^{\infty} x^2 \exp(-\lambda_1 x - \lambda_2 x^2) dx}{\int\limits_{-\infty}^{\infty} \exp(-\lambda_1 x - \lambda_2 x^2) dx} = -\int\limits_{-\infty}^{\infty} x^2 \exp(-\lambda_0 - \lambda_1 x - \lambda_2 x^2) dx$$

$$= -\int\limits_{-\infty}^{\infty} x^2 f(x) dx = -(s_x^2 + \overline{x}^2)$$

(9.45)

Differentiating Eq. 9.42 with respect to λ_1 and λ_2, respectively, one obtains

$$\frac{\partial \lambda_0}{\partial \lambda_1} = \frac{2}{4} \frac{\lambda_1}{\lambda_2} = \frac{\lambda_1}{2\lambda_2}$$

(9.46)

$$\frac{\partial \lambda_0}{\partial \lambda_2} = -\frac{1}{2\lambda_2} - \frac{\lambda_1^2}{4\lambda_2^2}$$

(9.47)

Equating Eq. 9.44 to Eq. 9.46 and Eq. 9.45 to Eq. 9.47, one gets

$$\frac{\lambda_1}{2\lambda_2} = -\overline{x}$$

(9.48)

$$\frac{1}{2\lambda_2} + \frac{1}{4}\left(\frac{\lambda_1}{\lambda_2}\right)^2 = s_x^2 + \overline{x}^2$$

(9.49)

From Eq. 9.48, one gets

$$\lambda_1 = -2\lambda_2 \overline{x}$$

(9.50)

Substituting Eq. 9.50 in Eq. 9.49, one obtains

$$\frac{1}{2\lambda_2} + \frac{1}{4}\frac{4\lambda_2^2\bar{x}}{\lambda_2^2} = s_x^2 + \bar{x}^2 \Rightarrow \frac{1}{2\lambda_2} = s_x^2 \Rightarrow \lambda_2 = \frac{1}{2s_x^2} \tag{9.51}$$

Eliminating λ_2 in Eq. 9.48 then yields

$$\lambda_1 = -2\frac{1}{2s_x^2}\bar{x} = -\frac{\bar{x}}{s_x^2} \tag{9.52}$$

9.3.4 Relation Between Lagrange Multipliers and Parameters

Substitution of Eq. 9.42 in Eq. 9.33 yields

$$f(x) = [-\frac{1}{2}\ln\pi + \frac{1}{2}\ln\lambda_2 - \frac{\lambda_1^2}{4\lambda_2} - \lambda_1 x - \lambda_2 x^2]$$

$$= \exp[\ln(\pi)^{-0.5} + \ln(\lambda_2)^{0.5} - \frac{\lambda_1^2}{4\lambda_2} - \lambda_1 x - \lambda_2 x^2] \tag{9.53}$$

$$= (\pi)^{-0.5}(\lambda_2)^{0.5}\exp[-\frac{\lambda_1^2}{4\lambda_2} - \lambda_1 x - \lambda_2 x^2]$$

A comparison of Eq. 9.53 with Eq. 9.29 shows that

$$\lambda_1 = -a/b^2 \tag{9.54}$$

$$\lambda_2 = 1/(2b^2) \tag{9.55}$$

9.3.5 Relation Between Parameters and Constraints

The normal distribution has two parameters, a and b, which are related to the Lagrange multipliers by Eq. 9.54 and Eq. 9.55, which themselves are related to the constraints through Eq. 9.51 and Eq. 9.52 and in turn through the last two of Eq. 9.32. Eliminating the Lagrange multipliers between these two sets of equations, we obtain

$$a = \bar{x} \tag{9.56}$$

$$b = s_x \tag{9.57}$$

9.3.6 Distribution Entropy

Substitution of Eq. 9.56 and 9.57 in Eq. 9.31 yields

$$
\begin{aligned}
I(x) &= \left(\ln\sqrt{2\pi} + \ln s_x + \frac{\bar{x}^2}{2s_x^2} \right) \int_{-\infty}^{\infty} f(x)dx + \frac{1}{2s_x^2} \int_{-\infty}^{\infty} x^2 f(x)dx - \frac{\bar{x}}{s_x^2} \int_{-\infty}^{\infty} xf(x)dx \\
&= \left(\ln\sqrt{2\pi} + \ln s_x + \frac{\bar{x}^2}{2s_x^2} \right) + \frac{1}{2s_x^2}(\bar{x}^2 + s_x^2) - \frac{\bar{x}^2}{s_x^2} \\
&= \ln[s_x(2\pi\, e)^{0.5}]
\end{aligned}
\tag{9.58}
$$

9.3.7 Parameter-Space Expansion Method

The constraints for the parameter-space expansion method, following Singh and Rajagopal (1986), are given by the first of Eq. 9.32 and

$$
\int_{-\infty}^{\infty} \left(\frac{xa}{b^2} \right) f(x)dx = E\left(\frac{xa}{b^2} \right) = \frac{a}{b^2} E(x) = \frac{a\bar{x}}{b^2}
\tag{9.59}
$$

$$
\int_{-\infty}^{\infty} \left(\frac{x^2}{2b^2} \right) f(x)dx = E\left(\frac{x^2}{2b^2} \right) = \frac{s_x^2 + \bar{x}^2}{2b^2}
\tag{9.60}
$$

The PDF corresponding to POME and consistent with Eqs. 9.32, 9.59, and 9.60 takes the form

$$
f(x) = \exp[-\lambda_0 - \lambda_1 \frac{xa}{b^2} - \lambda_2 \frac{x^2}{2b^2}]
\tag{9.61}
$$

where λ_0, λ_1, and λ_2 are Lagrange multipliers. Insertion of Eq. 9.61 into Eq. 9.32 yields

$$
\exp(\lambda_0) = \int_{-\infty}^{\infty} \exp\left(-\lambda_1 \frac{xa}{b^2} - \lambda_2 \frac{x^2}{2b^2} \right) dx = \frac{b\sqrt{2\pi}}{\sqrt{\lambda_2}} \exp\left(\frac{a^2\lambda_1^2}{2\lambda_2 b^2} \right)
\tag{9.62}
$$

Equation 9.62 is the partition function. Taking the logarithm of Eq. 9.62 leads to the zeroth Lagrange multiplier, which can be expressed as

$$
\lambda_0 = \ln b + 0.5\ln(2\pi) - 0.5\ln\lambda_2 + \frac{a^2\lambda_1^2}{2\lambda_2 b^2}
\tag{9.63}
$$

The zeroth Lagrange multiplier is also obtained from Eq. 9.62 as

$$\lambda_0 = \ln \int_{-\infty}^{\infty} \exp[-\lambda_1 \frac{xa}{b^2} - \lambda_2(\frac{x^2}{2b^2})]dx \tag{9.64}$$

Introduction of Eq. 9.63 in Eq. 9.61 gives

$$f(x) = \frac{\sqrt{\lambda_2}}{b\sqrt{2\pi}} \exp\left[-(\frac{a^2\lambda_1}{2\lambda_2 b^2} + \frac{\lambda_1 xa}{b^2} + \frac{\lambda_2 x^2}{2b^2})\right] \tag{9.65}$$

A comparison of Eq. 9.65 with Eq. 9.29 shows that $\lambda_2 = 1$ and $\lambda_1 = -1$. Taking the logarithm of Eq. 9.65 and multiplying by $[-1]$, one gets

$$-\ln f(x) = -\frac{1}{2}\ln \lambda_2 + \ln b + \frac{1}{2}\ln(2\pi) + \frac{a^2\lambda_1^2}{2\lambda_2 b^2} + \frac{\lambda_1 xa}{b^2} + \frac{\lambda_2 x^2}{2b^2} \tag{9.66}$$

Multiplying Eq. 9.66 by $f(x)$ and integrating from minus infinity to positive infinity, we get the entropy function of the form

$$I(f) = -\frac{1}{2}\ln \lambda_2 + \ln b + \frac{1}{2}\ln(2\pi) + \frac{a^2\lambda_1^2}{2\lambda_2 b^2} + \frac{\lambda_1 a}{b^2}E[x] + \frac{\lambda_2}{2b^2}E[x^2] \tag{9.67}$$

9.3.8 Relation Between Distribution Parameters and Constraints

Taking partial derivatives of Eq. 9.67 with respect to λ_1, λ_2, a, and b individually, and then equating each derivative to zero, one obtains

$$\frac{\partial I}{\partial \lambda_1} = 0 = \frac{2a^2\lambda_1}{2\lambda_2 b^2} + \frac{a}{b^2}E[x] \tag{9.68}$$

$$\frac{\partial I}{\partial \lambda_2} = 0 = -\frac{1}{2\lambda_2} - \frac{a^2\lambda_1^2}{2\lambda_2^2 b^2} + \frac{1}{2b^2}E[x^2] \tag{9.69}$$

$$\frac{\partial I}{\partial a} = 0 = -\frac{2a\lambda_1^2}{2b\lambda_2} + \frac{\lambda_1}{b^2}E[x] \tag{9.70}$$

$$\frac{\partial I}{\partial b} = 0 = \frac{1}{b} - \frac{2a^2\lambda_1^2}{2\lambda_2 b^3} - \frac{2a\lambda_1}{b^3}E[x] - \frac{2\lambda_2}{2b^3}E[x^2] \tag{9.71}$$

Simplification of Eq. 9.68 through Eq. 9.71 results in

$$E[x] = a \tag{9.72}$$

$$E[x^2] = a^2 + b^2 \tag{9.73}$$

$$E[x] = a \tag{9.74}$$

$$E[x^2] = b^2 + a^2 \tag{9.75}$$

Equations 9.72 and 9.74 are the same, and so are Eqs. 9.73 and 9.75. Thus the parameter estimation equations are Eqs. 9.72 and 9.73.

9.4 Determination of Parameters of the Gamma Distribution

The two-parameter gamma distribution is commonly employed for synthesis of instantaneous or finite-period unit hydrographs (Dooge 1973) and also for flood frequency analysis (Haan 1977, Phien and Jivajirajah 1984, Yevjevich and Obseysekera 1984). By making two hydrologic postulates, Edson (1951) was perhaps the first to derive it for describing a unit hydrograph (UH). Using the theory of linear systems, Nash (1957, 1959, 1960) showed that the mathematical equation of the instantaneous unit hydrograph (IUH) of a basin represented by a cascade of equal linear reservoirs would be a gamma distribution. This also resulted as a special case of the general unit hydrograph theory developed by Dooge (1959). Using statistical and mathematical reasoning, Lienhard and associates (Lienhard 1964, Lienhard and Davis 1971, Lienhard and Meyer 1967) derived this distribution as a basis for describing the IUH. Thus, these investigators laid the foundation of a hydrophysical basis underlying the use of this distribution in synthesizing direct runoff. There has since been a plethora of studies employing this distribution in surface water hydrology (Gray 1961; Wu 1963; DeCoursey 1966; Dooge 1973; Gupta and Moin 1974; Gupta et al. 1974; Croley 1980; Aron and White 1982; Singh 1982a,b, 1988; Collins 1983).

If X has a gamma distribution then its PDF is given by

$$f(x) = \frac{1}{a\Gamma(b)} \left(\frac{x}{a} \right)^{b-1} e^{-x/a} \tag{9.76}$$

where $a > 0$ and $b > 0$ are parameters. The gamma distribution is a two-parameter distribution. Its CDF can be expressed as

$$F(x) = \int_0^\infty \frac{1}{a\Gamma(b)} \left(\frac{x}{a} \right)^{b-1} e^{-x/a} dx \tag{9.77}$$

If $y = x/a$ then Eq. 9.77 can be written as

$$f(y) = \frac{1}{\Gamma(b)} \int_0^y y^{b-1} \exp(-y) dy \qquad (9.78)$$

Abramowitz and Stegun (1965) express $F(y)$ as

$$F(y) = F(X^2|v) \qquad (9.79)$$

where $F(X^2|v)$ is the chi-square distribution with degrees of freedom as $v = 2b$ and $\chi^2 = 2y$. According to Kendall and Stuart (1965), for $v > 30$, the following variable follows a normal distribution with zero mean and variance equal to one:

$$u = \left[(\frac{\chi^2}{v})^{1/3} + \frac{2}{9v} - 1 \right] (\frac{9v}{2})^{1/2} \qquad (9.80)$$

This helps us to compute $F(x)$ for a given x by first computing $y = x/a$ and $\chi^2 = 2y$ and then inserting these values into Eq. 9.80 to obtain u. Given a value of u, $F(x)$ can be obtained from the normal distribution tables.

9.4.1 Ordinary Entropy Method

9.4.1.1 Specification of Constraints

Taking the logarithm of Eq. 9.76 to the base e, one gets

$$\ln f(x) = -\ln a\Gamma(b) + (b-1)\ln x - (b-1)\ln a - \frac{x}{a}$$

$$= -\ln a\Gamma(b) + (b-1)\ln a + (b-1)\ln x - [x/a] \qquad (9.81)$$

Multiplying Eq. 9.81 by $[-f(x)]$ and integrating between 0 and ∞, one obtains the function

$$I(f) = -\int_0^\infty f(x)\ln f(x) dx = [\ln a\Gamma(b) + (b-1)\ln a] \int_0^\infty f(x) dx$$

$$-(b-1)\int_0^\infty [\ln x] f(x) dx + \frac{1}{a} \int_0^\infty x f(x) dx \qquad (9.82)$$

From Eq. 9.82 the constraints appropriate for Eq. 9.76 can be written (Singh et al. 1985, 1986) as

$$\int_0^\infty f(x) = 1 \qquad (9.83)$$

$$\int_0^\infty x f(x) dx = \bar{x} \tag{9.84}$$

$$\int_0^\infty [\ln x] f(x) dx = E[\ln x] \tag{9.85}$$

9.4.1.2 Construction of the Zeroth Lagrange Multiplier

The least-biased PDF based on POME and consistent with Eq. 9.83 to Eq. 9.85 takes the form

$$f(x) = \exp[-\lambda_0 - \lambda_1 x - \lambda_2 \ln x] \tag{9.86}$$

where λ_0, λ_1, and λ_2 are Lagrange multipliers. Substitution of Eq. 9.86 in Eq. 9.83 yields

$$\int_0^\infty f(x) dx = \int_0^\infty \exp[-\lambda_0 - \lambda_1 x - \lambda_2 \ln x] dx = 1 \tag{9.87}$$

This leads to the partition function as

$$\exp(\lambda_0) = \int_0^\infty \exp[-\lambda_1 x - \lambda_2 \ln x] dx = \int_0^\infty \exp[-\lambda_1 x] \exp[-\lambda_2 \ln x] dx$$

$$= \int_0^\infty \exp[-\lambda_1 x] \exp[\ln x^{-\lambda_2}] dx \tag{9.88}$$

Let $\lambda_1 x = y$. Then $dx = dy / \lambda_1$. Therefore, Eq. 9.88 becomes

$$\exp(\lambda_0) = \int_0^\infty (\frac{y}{\lambda_1})^{-\lambda_2} \exp(-y) \frac{dy}{\lambda_1} = \frac{1}{\lambda_1^{1-\lambda_2}} \int_0^\infty y^{-\lambda_2} e^{-y} dy = \frac{1}{\lambda_1^{1-\lambda_2}} \Gamma(1-\lambda_2) \tag{9.89}$$

Thus, the zeroth Lagrange multipliers λ_0 is given by Eq. 9.89 as

$$\lambda_0 = (\lambda_2 - 1) \ln \lambda_1 + \ln \Gamma(1 - \lambda_2) \tag{9.90}$$

The zeroth Lagrange multiplier is also obtained from Eq. 9.88 as

$$\lambda_0 = \ln \int_0^\infty \exp[-\lambda_1 x - \lambda_2 \ln x] dx \tag{9.91}$$

9.4.1.3 Relation Between Lagrange Multipliers and Constraints

Differentiating Eq. 9.91 with respect to λ_1 and λ_2, respectively, produces

$$\frac{\partial \lambda}{\partial \lambda_1} = -\frac{\displaystyle\int_0^\infty x \exp[-\lambda_1 x - \lambda_2 \ln x] dx}{\displaystyle\int_0^\infty \exp[-\lambda_1 x - \lambda_2 \ln x] dx} = -\int_0^\infty x \exp[-\lambda_0 - \lambda_1 x - \lambda_2 \ln x] dx$$

$$= -\int_0^\infty x f(x) dx = -\overline{x} \tag{9.92}$$

$$\frac{\partial \lambda_0}{\partial \lambda_2} = -\frac{\displaystyle\int_0^\infty \ln x \exp[-\lambda_1 x - \lambda_2 \ln x] dx}{\displaystyle\int_0^\infty \exp[-\lambda_1 x - \lambda_2 \ln x] dx}$$

$$= -\int_0^\infty \ln x \exp[-\lambda_0 - \lambda_1 x - \lambda_2 \ln x] dx = -\int_0^\infty \ln x f(x) dx = -E[\ln x] \tag{9.93}$$

Also, differentiating Eq. 9.90 with respect to λ_1 and λ_2 gives

$$\frac{\partial \lambda_0}{\partial \lambda_1} = \frac{\lambda_2 - 1}{\lambda_1} \tag{9.94}$$

$$\frac{\partial \lambda_0}{\partial \lambda_2} = \ln \lambda_1 + \frac{\partial}{\partial \lambda_2} \Gamma(1 - \lambda_2) \tag{9.95}$$

Let $1 - \lambda_2 = k$. Then

$$\frac{\partial k}{\partial \lambda_2} = -1 \tag{9.96}$$

$$\frac{\partial \lambda_0}{\partial \lambda_2} = \ln \lambda_1 + \frac{\partial}{\partial k} \Gamma(k) \frac{\partial k}{\partial \lambda_2} = \ln \lambda_1 - \psi(k) \tag{9.97}$$

From Eq. 9.92 and Eq. 9.94 as well as Eqs. 9.93 and 9.96 and 9.97, one gets

$$\frac{\lambda_2 - 1}{\lambda_1} = -\overline{x} \; ; \; \overline{x} = \frac{k}{\lambda_1} \tag{9.98}$$

$$\Psi(k) - E[\ln x] = \ln \lambda_1 \tag{9.99}$$

From Eq. 9.98, $\lambda_1 = k / \bar{x}$, and substituting λ_1 in Eq. 9.99, one gets

$$E[\ln x] - \ln \bar{x} = \Psi(k) - \ln k \tag{9.100}$$

We can find the value of $k (= 1 - \lambda_2)$ from Eq. 9.100 and substitute it in Eq. 9.98 to get λ_1.

9.4.1.4 Relation Between Lagrange Multipliers and Parameters

Substituting Eq. 9.89 in Eq. 9.86 gives the entropy-based PDF as

$$\begin{aligned}
f(x) &= \exp[(1 - \lambda_2)\ln \lambda_1 - \ln \Gamma(1 - \lambda_2) - \lambda_1 x - \lambda_2 \ln x] \exp[\ln \lambda_1^{1-\lambda_2}] \\
&\quad \times \exp[\ln(\frac{1}{\Gamma(1-\lambda_2)})] \exp[-\lambda_1 x] \exp[\ln x^{-\lambda_2}] \\
&= \lambda_1^{1-\lambda_2} \frac{1}{\Gamma(1-\lambda_2)} \exp[-\lambda_1 x] x^{-\lambda_2}
\end{aligned} \tag{9.101}$$

If $\lambda_2 = 1 - k$ then

$$f(x) = \frac{\lambda_1^k}{\Gamma(k)} \exp[-\lambda_1 x] x^{k-1} \tag{9.102}$$

A comparison of Eq. 9.102 with Eq. 9.76 produces

$$\lambda_1 = 1/a \tag{9.103}$$

and

$$\lambda_2 = 1 - b \tag{9.104}$$

9.4.1.5 Relation Between Parameters and Constraints

The gamma distribution has two parameters, a and b, which are related to the Lagrange multipliers by Eq. 9.103 and Eq. 9.104, which themselves are related to the known constraints by Eq. 9.98 and Eq. 9.100. Eliminating the Lagrange multipliers between these two sets of equations, we get parameters directly in terms of the constraints as

$$ba = \bar{x} \tag{9.105}$$

$$\Psi(b) - \ln b = E[\ln x] - \ln \bar{x} \tag{9.106}$$

9.4.1.6 Distribution Entropy

Equation 9.82 gives the distribution entropy. Rewriting it, one gets

$$I(x) = -\int_0^\infty f(x)\ln f(x)dx$$

$$= [\ln a\Gamma(b) + (b-1)\ln a]\int_0^\infty f(x)dx - (b-1)\int_0^\infty \ln x f(x)dx + \frac{1}{a}\int_0^\infty x f(x)dx$$

$$= [\ln a\Gamma(b) + \ln a^{b-1}] - (b-1)E[\ln x] + \frac{\overline{x}}{a}$$

$$= \ln\left[a\Gamma(b)a^{b-1}\right] + \frac{\overline{x}}{a} - (b-1)E[\ln x]$$

$$= \ln\left[\Gamma(b)a^b\right] + \frac{\overline{x}}{a} - (b-1)E[\ln x] \tag{9.107}$$

9.4.2 Parameter-Space Expansion Method

9.4.2.1 Specification of Constraints

For this method, following Singh and Rajagopal (1986), one finds that the constraints are Eq. 9.83 and

$$\int_0^\infty \frac{x}{a} f(x)dx = E[\frac{x}{a}] \tag{9.108}$$

$$\int_0^\infty \ln(\frac{x}{a})^{b-1} f(x)dx = E[\ln(\frac{x}{a})^{b-1}] \tag{9.109}$$

9.4.2.2 Derivation of Entropy Function

The least-biased PDF corresponding to POME and consistent with Eqs. 9.83, 9.108, and 9.109 takes the form

$$f(x) = \exp[-\lambda_0 - \lambda_1(\frac{x}{a}) - \lambda_2 \ln(\frac{x}{a})^{b-1}] \tag{9.110}$$

where λ_0, λ_1, and λ_2 are Lagrange multipliers. Insertion of Eq. 9.110 into Eq. 9.83 yields the partition function

$$\exp(\lambda_0) = \int_0^\infty \exp[-\lambda_1(\frac{x}{a}) - \lambda_2 \ln(\frac{x}{a})^{b-1}]dx = a(\lambda_1)^{\lambda_2(b-1)-1}\Gamma[1 - \lambda_2(b-1)] \tag{9.111}$$

The zeroth Lagrange multiplier is given by Eq. 9.111 as

$$\lambda_0 = \ln a - [1 - \lambda_2(b-1)]\ln \lambda_1 + \ln \Gamma[1 - \lambda_2(b-1)] \qquad (9.112)$$

Also, from Eq. 9.112 one gets the zeroth Lagrange multiplier

$$\lambda_0 = \ln \int_0^\infty \exp[-\lambda_1(\frac{x}{a}) - \lambda_2 \ln(\frac{x}{a})^{b-1}]dx \qquad (9.113)$$

Introduction of Eq. 9.112 in Eq. 9.110 produces

$$f(x) = \frac{1}{a}(\lambda_1)^{1-\lambda_2(b-1)} \frac{1}{\Gamma[1-\lambda_2(b-1)]} \exp[-\lambda_1 \frac{x}{a} - \lambda_2 \ln(\frac{x}{a})^{b-1}] \qquad (9.114)$$

Comparison of Eq. 9.114 with Eq. 9.76 shows that $\lambda_1 = 1$ and $\lambda_2 = -1$. Taking the logarithm of Eq. 9.114 yields

$$\ln f(x) = -\ln a + [1 - \lambda_2(b-1)]\ln \lambda_1 - \ln \Gamma[1 - \lambda_2(b-1)] - \lambda_1 \frac{x}{a} - \lambda_2 \ln(\frac{x}{a})^{b-1} \qquad (9.115)$$

Multiplying Eq. 9.115 by $[-f(x)]$ and integrating from 0 to ∞ yields the entropy function of the gamma distribution. This can be written as

$$I(f) = \ln a - [1 - \lambda_2(b-1)]\ln \lambda_1 + \ln \Gamma[1 - \lambda_2(b-1)] + \lambda_1 E[\frac{x}{a}] + \lambda_2 E[\ln(\frac{x}{a})^{b-1}] \qquad (9.116)$$

9.4.3 Relation Between Parameters and Constraints

Taking partial derivatives of Eq. 9.116 with respect to λ_1, λ_2, a, and b separately and equating each derivative to zero, respectively, yields

$$\frac{\partial I}{\partial \lambda_1} = 0 = -[1 - \lambda_2(b-1)]\frac{1}{\lambda_1} + E(\frac{x}{a}) \qquad (9.117)$$

$$\frac{\partial I}{\partial \lambda_2} = 0 = +(b-1)\ln \lambda_1 - (b-1)\Psi(K) + E[\ln(\frac{x}{a})^{b-1}], \quad K = 1 - \lambda_2(b-1) \qquad (9.118)$$

$$\frac{\partial I}{\partial a} = 0 = +\frac{1}{a} - \frac{\lambda_1}{a}E[\frac{x}{a}] + \frac{(1-b)}{a}\lambda_2 \qquad (9.119)$$

$$\frac{\partial I}{\partial b} = 0 = +\lambda_2 \ln \lambda_1 + E[\ln(\frac{x}{a})^{b-1}]^{-\lambda_2}\Psi(K) \qquad (9.120)$$

Simplification of Eq. 9.117 to Eq. 9.120, respectively, gives

$$E(\frac{x}{a}) = b \tag{9.121}$$

$$E[\ln(\frac{x}{a})] = \Psi(k) \tag{9.122}$$

$$E(\frac{x}{a}) = b \tag{9.123}$$

Equation 9.121 is the same as Eq. 9.123. Therefore, Eq. 9.121 and Eq. 9.122 are the parameter estimation equations.

9.5 Application of Entropy Theory in Environmental and Water Resources

A historical perspective on entropy applications in environmental and water resources is given in Singh and Fiorentino (1992) and Singh (1998). Harmancio-glu and Singh (1998) discussed the use of entropy in water resources. Although entropy theory has been applied in recent years to a variety of problems in hydrology, its potential as a decision-making tool has not been fully exploited. What follows is a brief discussion highlighting this potential.

9.5.1 Information Content of Data

One frequently encounters a situation in which one wishes to exercise freedom of choice, evaluate uncertainty, or measure information gain or loss. The freedom of choice, uncertainty, disorder, information content, or information gain or loss has been variously measured by relative entropy, redundancy, and conditional and joint entropies by employing conditional and joint probabilities. As an example, in the analysis of empirical data, the variance has often been interpreted as a measure of uncertainty and as revealing gain or loss of information. However, entropy is another measure of dispersion and is an alternative to variance. This suggests that it is possible to determine the variance whenever it is possible to determine the entropy measure, but the reverse is not necessarily true. However, variance is not the appropriate measure if the sample size is small.

Since entropy is a measure of uncertainty or chaos, and variance is a measure of variability, the connection between them is of interest. In general, an explicit relation between entropy and variance does not exist. If two distributions have common variance, an entropy-based measure of affinity or closeness between the distributions can be defined (Mukherjee and Ratnaparkhi 1986). The affinity between two distributions is defined by the absolute difference between entropies of the two distributions, which can be shown to be the expectation of the likelihood ratio. This measure differs from Kullback's minimum distance information criterion. Likewise, a similarity function is defined as one minus the quotient of the affinity between any two distributions and the maximum value of affinity between the distributions. Thus, affinity (distance) is a monotonically decreasing function of similarity. The similarity factor can be used to cluster or group models.

To measure correlation or dependence between any two variables, an informational coefficient of correlation r_0 is defined as a function of transinformation T_0 as

$$r_0 = [1 - \exp(-2T_0)]^{0.5} \tag{9.124}$$

Transinformation given by Eq. 9.124 expresses the upper limit of common information between two variables and represents the level of dependence (or association) between the variables. It represents the upper limit of transferable information between the variables, and its measure is given by r_0. The ordinary correlation coefficient r measures the amount of information transferred between variables under specified assumptions, such as linearity and normality. An inference similar to that of the ordinary correlation coefficient r can be drawn by defining the amount (in percent) of transferred information by the ratio T/T_0, where T can be computed in terms of ordinary r.

9.5.2 Criteria for Model Selection

Usually there are more models than one needs and a choice has to be made as to which model to choose. Akaike (1973) formulated a criterion, called the Akaike information criterion (AIC), for selecting the best model from among several models as

$$AIC = -2 \log (\text{maximized likelihood}) + 2k \tag{9.125}$$

where k is the number of model parameters.

AIC provides a method of model identification and can be expressed as minus twice the logarithm of the maximum likelihood plus twice the number of parameters used to find the best model. The maximum likelihood and entropy are uniquely related. When there are several models, the model giving the minimum value of AIC should be selected. When the maximum likelihood is identical for two models, the model with the smaller number of parameters should be selected, for that will lead to smaller AIC and comply with the principle of parsimony.

9.5.3 Hypothesis Testing

Another important application of entropy theory is in testing of hypotheses (Tribus 1969). With use of Bayes's theorem in logarithmic form, an evidence function is defined for comparing two hypotheses. The evidence in favor of one hypothesis over its competitor is the difference between the respective entropies of the competition and the hypothesis under test. Defining *surprisal* as the negative of the logarithm of the probability, one can express the mean surprisal for a set of observations. Therefore, the evidence function for two hypotheses is obtained as the difference between the two values of the mean surprisal multiplied by the number of observations.

9.5.4 Risk Assessment

In common language, risk is the possibility of loss or injury and the degree of probability of such loss. Rational decision making requires a clear and quantitative way of expressing risk. In general, risk cannot be avoided and a choice has to be made among risks. There are different types of risk, such as business risk, social risk, economic risk, safety risk, investment risk, and occupational risk. To put risk in proper perspective, it is useful to clarify the distinction among risk, uncertainty, and hazard.

The notion of risk involves both uncertainty and some kind of loss or damage. Uncertainty reflects the variability of our state of knowledge or state of confidence in a prior evaluation. Thus, risk is the sum of uncertainty plus damage. Hazard is commonly defined as a source of danger and involves a scenario identification (e.g., failure of a dam) and a measure of the consequence of that scenario or a measure of the ensuing damage. Risk encompasses the likelihood of conversion of that source of danger into the actual delivery of loss, injury, or some form of damage. Thus, risk is the ratio of hazard to safeguards. By increasing safeguards, risk can be reduced but it can never be zero. Since awareness of risk reduces risk, awareness makes up part of the safeguards. Qualitatively, risk is subjective and is relative to the observer. Risk involves the probability of a scenario and the consequence resulting from the scenario happening. Thus, one can say that risk is probability and consequence. Kaplan and Garrick (1981) have analyzed risk by using entropy.

9.5.5 Safety Evaluation

Safety has two components: reliability and probabilistic risk assessment (PRA). In reliability, a safety margin or probability of failure is defined when maximum external loads are specified (i.e., design loads, such as design discharge). There can be three domains of safety:

1. Safe domain (no failure)
2. Potentially unsafe domain (failure is possible)
3. Unsafe (failure) domain (a failure is certain)

The safe and unsafe domains are separated by a limit state surface. A system will fail if the failure indicator reaches the limit state surface. If we consider a probability density function of failure at a value of the failure indicator then the cumulative probability of the failure indicator defines fragility.

In PRA, one considers the probability of failure from loads exceeding design loads. Through introduction of a hazard function, the probability density function of external loads is specified. Consider a hydraulic system with a set of random parameters. The system can fail in many ways and every failure mode can be described with a corresponding failure indicator involving a number of different failure modes, which is a function of the hazard parameters and the random parameters. For example, in case of an earth dam, depending on the failure mode, a failure indicator could be erosion at the bottom, reservoir water level, water leakage, displacement, etc. Hazard parameters could be extreme rainfall, reservoir level, peak discharge, depth of water at the dam top, etc. Structural random parameters could be strengths of materials, degree of riprap, degree of packing, internal friction, etc.

9.5.6 Reliability Analysis

Reliability of a system can be defined as the probability that the system will perform its intended function for at least a specified period of time under specified environmental conditions. Different measures of reliability are applied to different systems, depending on their objective. Indeed, the use of a particular system determines the kind of reliability measure that is most meaningful and most useful. As an example, the reliability measure of a dam is the probability of its survival during its expected life span. In contrast, the reliability measure associated with hydroelectric power plant components is the failure rate, since failure of a plant is of primary concern. Furthermore, at different times during its operating life a system may be required to have a different probability of successfully performing its required function under specified conditions. The term "failure" means that the system is not capable of performing its required function. We only consider the case where the system is either capable of performing its functions or not and exclude the case involving varying degrees of capability.

If the reliability is defined as the probability of success, that is, the system will perform its intended function for at least a defined period of time, then the reliability function can be computed directly from the knowledge of the failure time distribution. If the system is resurrected through repair and maintenance then the mean failure time is known as the mean (operating) time between failures. The mean time to failure is the expected time during which the system will perform successfully, also expressed as the expected life.

The rate at which failures occur in a time interval is the failure rate and is defined by the probability that a failure per unit time occurs in the interval, provided that a failure has not occurred prior to the beginning of the interval. The hazard rate (or hazard function) is defined by the limit of failure as the length of the time interval approaches zero. This implies the instantaneous failure rate.

9.5.7 Entropy Spectral Analysis for Flow Forecasting

Maximum entropy spectral analysis (MESA) was introduced by Burg (1975) and has several advantages over conventional spectral analysis methods. It has short and smooth spectra with high-degree resolutions (Fougere et al. 1976). The statistical characteristics used in stochastic model identification can also be estimated using MESA, thus permitting integration of spectral analysis and computations related to stochastic model development. Ulrych and Clayton (1976) reviewed principles of MESA and the closely related problem of autoregressive time series modeling. Shore (1979) presented a comprehensive discussion of minimum cross-entropy spectral analysis.

The relationship between spectrum $W(f)$ with frequency f of a stationary process $x(t)$ and entropy $H(f)$ can be expressed as

$$H(f) = \frac{1}{2}\ln(2w) + \frac{1}{4w}\int_{w}^{+w}\ln[W(f)]df \tag{9.126}$$

where w is the frequency band. Equation 9.126 is maximized subject to the constraint equations given as autocorrelations until log m:

$$\rho(n) = \int_{w}^{+w}W(f)\exp(i2\pi fn\Delta t)df, \ \ m \le n \le +m \tag{9.127}$$

where t is the sampling time interval and $i = (-1)^{1/2}$. Maximization of Eq. 9.127 is equivalent to maximizing

$$H(f) = \int_{w}^{+w}\ln[W(f)]df \tag{9.128}$$

which is known as the Burg entropy. The spectrum $W(f)$ can be expressed in terms of the Fourier series as

$$W(f) = \frac{1}{2w}\sum_{n=\infty}^{\infty}\rho(n)\exp[i2\pi nf\Delta t] \tag{9.129}$$

Substitution of Eq. 9.129 in Eq. 9.128 and maximization lead to MESA.

Jaynes (1982) has shown that MESA and other methods of spectral analysis such as Schuster, Blackman–Tukey, maximum likelihood, Bayesian, and autoregressive (AR, ARMA, or ARIMA) models are not in conflict, and that AR models are a special case of MESA. Krstanovic and Singh (1991a,b) employed MESA for long-term streamflow forecasting. Krstanovic and Singh (1993a,b) extended the MESA method to develop a real-time flood forecasting model. Padmanabhan and Rao (1986, 1988) applied MESA to analyze rainfall and river flow time series. Rao et al. (1984) compared a number of spectral analysis methods with

MESA and found MESA to be superior. Eilbert and Christensen (1983) analyzed annual hydrological forecasts for central California and found that dry years might be more predictable than wet years. Dalezios and Tyraskis (1989) employed MESA to analyze multiple precipitation time series.

9.5.8 Regional Precipitation Analysis and Forecasting

The Burg algorithm or MESA can be applied to identify and interpret multistation precipitation data sets and to explore spectral features that lead to a better understanding of rainfall structure in space and time (Dalezios and Tyraskis 1989). Then, multistation rainfall time series can be extrapolated to develop regional forecasting capabilities.

9.5.9 Grouping of River Flow Regimes

An objective grouping of flow regimes into regime types can be employed as a diagnostic tool for interpreting the results of climate models and flow sensitivity analyses. By minimizing an entropy-based objective function (such as minimum cross entropy) a hierarchical aggregation of monthly flow series into flow regime types can, therefore, be effectively performed, which will satisfy chosen discriminating criteria. Such an approach was developed by Krasovskaia (1997), who applied it to a regional river flow sample for Scandinavia for two different formulations of discriminating criteria.

9.5.10 Basin geomorphology

Entropy plays a fundamental role in characterization of landscape. Using the entropy theory for morphological analysis of river basin networks, Fiorentino et al. (1993) found the connection between entropy and the mean basin elevation to be linearly related to the basin entropy. Similarly, the relation between the fall in elevation from the source to the outlet of the main channel and the entropy of its drainage basin was found to be linear and so also was the case between the elevation of a node and the logarithm of its distance from the source. When a basin was ordered following the Horton–Strahler ordering scheme, a linear relation was found between the drainage basin entropy and the basin order. This relation can be characterized as a measure of the basin network complexity. The basin entropy was also found to be linearly related to the logarithm of the magnitude of the basin network. This relation led to a nonlinear relation between the network diameter and magnitude, where the exponent was found to be related to the fractal dimension of the drainage network.

9.5.11 Design of Hydrologic Networks

The purpose of measurement networks is to gather information in terms of data. Fundamental to evaluation of these networks is the ability to determine whether the networks are gathering the needed information optimally. Entropy theory is a natural tool to make that determination. Krstanovic and Singh (1992a,b) employed the theory for space and time evaluation of rainfall networks in Louisiana. The decision whether to keep or to eliminate a rain gauge was based entirely on reduction or gain of information at that gauge. Yang and Burn (1994) employed a measure of information flow, called directional information transfer index (DIT), between gauging stations in the network. The value of DIT varies from zero, where no information is transmitted and the stations are independent, to one, where no information is lost and the stations are fully dependent. Between two stations of one pair, the station with higher DIT value should be retained because of its greater capability of inferring information at the other side.

9.5.12 Reliability of Water Distribution Systems

Entropy-based measures have been developed for evaluation of reliability and redundancy of water distribution networks. These measures accurately reflect changes in network reliability. However, the redundancy of a network also depends on the ability of the network to respond to failure of one of its links. Awumah et al. (1990) applied them to evaluate reliability and redundancy of a range of network layouts and showed that the entropy-based redundancy measure was a good indicator of the relative performance implications of different levels of redundancy.

9.5.13 Subsurface Hydrology

In groundwater engineering, it is often true that limited measurements of aquifer and flow parameters, such as hydraulic conductivity, are available, and a large degree of uncertainty exists in the measured values of fundamental flow parameters. Woodbury and Ulrych (1993) used the principle of minimum relative entropy (POMRE) to determine these parameters. Barbe et al. (1994) applied POME to derive a probability distribution for piezometric head in one-dimensional steady groundwater flow in confined and unconfined aquifers, subject to the total probability law and conservation of mass. From a few measurements of transmissivity based on pumping tests and of piezometric head, Bos (1990) employed POME and Bayes's theorem to derive the probability distribution of transmissivity.

9.5.14 Hydraulics

Yang (1994) showed that the fundamental theories in hydrodynamics and hydraulics can be derived from variational approaches based on maximization of entropy, minimization of energy, or minimization of energy dissipation rate. Barbe et al. (1991) and Chiu and Murray (1992) applied POME to determine the probability distribution of velocity in nonuniform open-channel flow. The entropy-based velocity distribution fits experimental data very well and is of great practical value in hydraulic modeling.

9.5.15 Sediment Concentration and Discharge

Entropy theory has been successfully applied to derive velocity distributions in channels and rivers. The entropy-based velocity distribution can be employed to derive distribution of sediment concentration and computation of suspended sediment discharge. This approach has been followed by Chiu et al. (2000) and Choo (2000). Indeed, following this approach, one can also develop an efficient method of sediment discharge measurement.

9.5.16 Rating Curve

Moramarco and Singh (2001) employed the entropy theory to develop a method for reconstructing the discharge hydrograph at a river section where only water level is monitored and discharge is recorded at another upstream section. The method, which is based on the assumption that lateral inflows are negligible, has two parameters linked to remotely observed discharge and permits, without using a flood routing procedure and without the need of a rating curve at a local site, relating the local river stage with the hydraulic condition at a remote upstream section.

9.5.17 Water Quality Assessment

Environmental pollution can be perceived as a result of discharge of material and heat into the environment (water, air, and/or soil) through human activity of production and consumption. When a compound is added to pure water, the compound will dissolve and diffuse throughout water. The dissolution and diffusion imply an increase in the entropy of the solution (by virtue of its definition and the second law of thermodynamics) and an increase in the degree of pollution. This suggests that an increase in entropy implies water pollution. Water is extensively used in cooling, washing, disposal of waste material, and dissipation of waste heat. Water pollution can then be viewed as water initially containing a low value of entropy being eventually discharged with high value of entropy, which, in turn, increases the entropy of the environment. Thus, entropy can serve as a comprehensive index for assessment and control of pollution. To extend the argument further, the diversity of species of organisms in water or the diversity index (DI) is related to the degree of pollution. In general, the number of species decreases as the degree of pollution increases. The DI value can be calculated using entropy theory.

9.5.18 Design of Water Quality Networks

Entropy theory, when applied to water quality monitoring network design, yields promising results, especially in the selection of technical design features, such as monitoring sites, time frequencies, variables to be sampled, and sampling duration (Ozkul et al. 2000). Furthermore, it permits a quantitative assessment of efficiency and benefit/cost parameters (Harmancioglu et al. 1999).

Harmancioglu and Singh (1998) reviewed the advantages as well as the limitations of the entropy method as applied to the design of water quality monitoring networks. Given an observed change in water quality levels at a downstream location, the entropy-based formulation predicts the probabilities of each possible water quality level at each of the upstream stations.

9.5.19 Inadequate Data Situations

One of the main problems plaguing environmental and water resources development is the lack of data. Frequently, either the data are missing or are incomplete, or they are not of good quality, or the record is not of sufficient length. As a result, more often than not, the data themselves dictate the type of model to be used and not the availability of modeling technology. Many conventional models are not applicable when their data needs are not met. Furthermore, subjective information, such as professional experience, judgment, and practical or empirical rules, has played a significant role in hydrologic practice in many developing countries. Conventional models do not have the capability to accommodate such subjective information, although such information may be of good quality or high value. The potential for application of entropy theory is enormous in developing countries, for it maximizes the use of information contained in data, however little it may be, and it permits use of subjective information. Thus, in the face of limited data entropy theory results in a reliable solution of the problem at hand. Furthermore, it offers an objective avenue for drawing inferences as to the model results. In addition, entropy-based modeling is efficient, requiring relatively little computational effort, and is versatile in its applicability across many disciplines.

9.6 Closure

Entropy theory permits determination of the least-biased probability distribution of a random variable, subject to the available information. It suggests whether the available information is adequate and, if not, then additional information should be sought. In this way it brings the model, the modeler, and the decision maker closer together. As an objective measure of information or uncertainty, entropy theory allows us to communicate with nature, as illustrated by its application to the design of data acquisition systems, the design of environmental and hydrologic networks, and the assessment of reliability of these systems or networks. In a similar vein, it helps us to better understand physics or science of natural systems, such as landscape evolution, geomorphology, and hydrodynamics. A wide variety of seemingly disparate or dissimilar problems can be meaningfully solved with the use of entropy.

9.7 Questions

9.1 Take sample data of annual peak discharge from a gauging station on a river near your town. Fit the gamma distribution to the discharge data. Then using this distribution, determine the effect of sample size on the value of the Shannon entropy.

9.2 Use the same gamma distribution as in Question 9.1. Changing the parameter values determine the Shannon entropy and discuss the effect of parameter variation.

9.3 Determine the effect of discretization on the Shannon entropy.

9.4 Determine the constraints for the gamma distribution needed for estimation of its parameters using entropy. Then determine its parameters in terms of the constraints.

9.5 Determine the constraints for the Pearson type III distribution needed for estimation of its parameters using entropy. Then determine its parameters in terms of the constraints.

9.6 Determine the constraints for the Weibull distribution needed for estimation of its parameters using entropy. Then determine its parameters in terms of the constraints.

9.7 Determine the constraints for the Gumbel distribution needed for estimation of its parameters using entropy. Then determine its parameters in terms of the constraints.

9.8 Determine the constraints for the three-parameter log-normal distribution needed for estimation of its parameters using entropy. Then determine its parameters in terms of the constraints.

9.9 Determine the constraints for the logistic distribution needed for estimation of its parameters using entropy. Then determine its parameters in terms of the constraints.

9.10 Determine the constraints for the Pareto distribution needed for estimation of its parameters using entropy. Then determine its parameters in terms of the constraints.

9.11 Determine the constraints for the log-Pearson type III distribution needed for estimation of its parameters using entropy. Then determine its parameters in terms of the constraints.

9.12 Take monthly discharge data for several gauging stations along a river. Compute the marginal entropy of monthly discharge at each gauging station and plot it as a function of distance between gauging stations. What do conclude from this plot? Discuss it.

9.13   For the discharge data in Question 9.12, compute transinformation of monthly discharge. Are all gauging stations needed? What is the redundant information? What can be said about increasing or decreasing the number of gauging stations?

9.14   Consider a drainage basin that has a number of rainfall measuring stations. Obtain annual rainfall values for each gauging station. Using annual rainfall values, determine marginal entropy at each station and also compute transinformation among stations. Comment on the adequacy of the rain gauge network.

9.15   Consider the same basin and the rain gauge network as in Question 9.14. Now obtain monthly rainfall values and compute marginal entropy as well as transinformation. Comment on the adequacy of the rain gauge network. How does the adequacy change with reduced time interval? Which stations are necessary and which are not?

9.16   Can entropy be employed for designing a monitoring network? If yes, then how? Can entropy be employed for evaluating the adequacy of an existing network? If yes, how?

Part III

Uncertainty Analysis

Chapter 10

Error and Uncertainty Analysis

Uncertainty is defined as a measure of imperfect knowledge or probable error that can occur during the data collection process, modeling and analysis of engineering systems, and prediction of a random process. Engineering systems, such as wastewater treatment plants, soil remediation systems, water purification systems, flood control systems, and so on, are subject to uncertainty but decisions on their planning, design, operation, and management are often made without accounting for it. According to Chow (1979), uncertainty can be defined in simple language as the occurrence of events that are beyond human control. In water resources projects, there can be natural, model, parameter, data, computational, and operational uncertainties. Natural uncertainties, such as climatic, seismic, hydrologic, geologic, and structural, are associated with the intrinsic variability of the system. This implies that the performance indicators of the system will vary for different sets of equally likely input sequences. In this case, the system performance must be treated as a random variable. Model uncertainties arise when the model is unable to closely represent the true behavior of the system, because of inadequacy of model assumptions and hypotheses, less than accurate model parameters, and computational errors. The uncertainties that are associated with construction, maintenance, and management of the system are of operational type. The uncertainties in data arise from measurement errors, inadequate sampling, improper handling and retrieval, as well as unsatisfactory archiving and storage.

Uncertainty and risk are pervasive features and go hand in hand in many engineering systems. How to handle the risks often associated with uncertainty comprises one of the most difficult aspects of analysis, planning, and management of many civil engineering systems. Uncertainty is central to decision making and risk assessment problems. Questions of safety or reliability in environmental and water resources engineering arise principally because of the presence of uncertainty. However much one may like, uncertainties cannot be completely eliminated. At best, one can reduce them by better equipment, standard data-collection procedures, dense monitoring networks, and better models and maintenance. Uncertainty analysis is performed to determine the statistical properties of output as a function of statistical input parameters. This helps determine the contribution of each input parameter to the overall uncertainty of the model output and can be used to reduce the output uncertainty.

In most civil engineering–related projects, uncertainty analysis is the study of model output uncertainty as a function of a careful inventory of the different sources of uncertainty present in the model input parameters. Generally, the most frequent questions addressed by uncertainty analysis are the following: What is the prediction uncertainty resulting from all of the uncertainties in model inputs? How do uncertain inputs contribute to model prediction uncertainty? What input parameters need more data-collection effort? The objective of this chapter is to describe these issues, detailing different types, sources, and measures of uncertainty.

10.1 Types of Uncertainty

There are six sources of uncertainty in evaluating the reliability of environmental and water resources systems or in designing the systems based on reliability:

1. Natural uncertainties are associated with random temporal and spatial fluctuations inherent in natural processes (e.g., climatic variability, occurrence of hydrologic extremes, spatial variability of hydraulic conductivity of geologic formations, and occurrence of rainfall).

2. Model structure uncertainty reflects the inability of the simulation model or design technique to represent precisely the system's true behavior or process (e.g., use of Manning's formula for describing open-channel flow and use of a simple probability model to describe a random hydrologic process). This inability is caused by wrong assumptions employed for constructing the model.

3. Model parameter uncertainties reflect the variability in determining the parameters to be used in a model or design. For example, there can be uncertainty in Manning's roughness factor n in Manning's formula

$$Q = \frac{1}{n} A R^{2/3} S^{1/2}$$

or uncertainties in estimation of mean μ and standard deviation σ in the normal distribution

$$f(x) = \frac{1}{\sigma\sqrt{2\pi}} e^{-\frac{1}{2}\left(\frac{x-\mu}{\sigma}\right)^2}, \quad -\infty < x < \infty$$

Model parameter uncertainty occurs because an inadequate parameter estimation technique is used, inaccurate data are used for parameter estimation, or both.

4. Data uncertainties arise from (i) measurement inaccuracy and errors, (ii) inadequacy of the data gauging network, and (iii) data handling and transcription errors.

5. Computational uncertainties arise from truncation and rounding off errors in doing calculations.

6. Operational uncertainties are associated with construction, manufacturing, deterioration, maintenance, and other human factors that are not accounted for in the modeling or design procedure.

Uncertainty may also be classified into two categories:

1. Inherent or intrinsic—caused by randomness in nature.
2. Epistemic—caused by the lack of knowledge of the system or paucity of data.

10.1.1 Intrinsic Uncertainty

Environmental and water resources phenomena exhibit random variability and this variability is reflected when observations are made and samples are analyzed. For example, there is inherent randomness in the climatic system and it is impossible to precisely predict what the maximum rainfall would be in a given city in a given year, even if there is a long history of data. Similarly, there is no way to precisely predict the amount of sediment load that a given river will carry during a given week at a given location. Likewise, the maximum discharge of a river for a given year cannot be predicted in advance. Because randomness is an inherent part of nature, it is not possible to reduce the inherent uncertainty. There are local meteorological events, such as frost, hail, and gale (usually discussed by people with respect to possible climate change), that are difficult to predict. Following Pate-Cornell (1996), the intrinsic uncertainty can be divided into inherent uncertainty in time and inherent uncertainty in space.

10.1.1.1 Inherent Uncertainty in Time

A stochastic process expressed as a time series of a random variable exhibits uncertainty in time. For example, the time series of annual rainfall at a given station, the time series of the annual instantaneous maximum discharge of a river at a given station, the time series of annual 24-hour maximum rainfall at a gauging station, annual sediment yield of a watershed at its outlet, the time series of the annual 7-day minimum ozone level in a given city, the time series of annual 7-day low streamflow, and so on are examples of the inherent uncertainty in time.

10.1.1.2 Inherent Uncertainty in Space

Environmental and water resources systems exhibit uncertainty in space. For example, hydraulic conductivity in an aquifer varies from point to point in location as well as well in direction. Therefore, a space series of the hydraulic conductivity can be expressed along a given transect. This also applies to the hydraulic conductivity of soils. Likewise, the roughness of a river bed varies along the bed and can be expressed in space along the longitudinal direction. The river bed level varies longitudinally as well as transversely. Another example is atmospheric pollution in a large urban area, which varies at different points and in different directions and can be expressed as a function of space. Examples of this kind exhibiting spatial variability abound in environmental and hydraulic analyses.

10.1.2 Epistemic Uncertainty

Epistemic uncertainty is extrinsic and knowledge related. The knowledge relates to environmental and water resources engineering systems and their operation as well as to the collected data. Thus, this type of uncertainty is caused by a lack of understanding of the causes and effects occurring in a system. If a system is fully known, epistemic uncertainty may be caused by a lack of sufficient data. For example, using laboratory experimentation or computer simulation, it may be possible to construct a precise mathematical model for an environmental system but it will be impossible to determine the parameters for the vast range of conditions encountered in nature. Consider the case of flow in a channel. The flow dynamics is reasonably known but it is not possible to estimate the shear stress for the range of flow and morphologic conditions found in practice. Similarly, depth–duration–frequency curves can be constructed for a given area but these curves do not represent the true probabilistic nature of rainfall. Epistemic uncertainty changes with knowledge and can be reduced with increasing knowledge and a longer history of quality data. The knowledge can, in general, be increased by gathering data, research, experience, and expert advice. Following Pate-Cornell (1996), the epistemic uncertainty can be divided into statistical uncertainty and model uncertainty.

10.1.2.1 Statistical Uncertainty

Statistical uncertainty combines parameter uncertainty and distributional uncertainty (Vrijling and van Gelder 2000). These two types of uncertainties are not always independent and distinguishable. For example, the identification of a correct distribution model depends very much on the accuracy with which its parameters can be estimated. Consider the case of flood frequency analysis. There can be more than one frequency distribution that can fit instantaneous flood discharge data. The distribution that will better fit the data significantly depends on the accuracy with which the distribution parameters can be estimated. Vrijling and van Gelder (2000) proposed dividing the statistical uncertainty into statistical uncertainty in time and statistical uncertainty in space.

Parameter Uncertainty

Parameter uncertainty is caused by a lack of data, poor-quality data, or an inadequate method of parameter estimation. This type of uncertainty is widely prevalent in environmental and water resources analysis. Usually, the variance of parameter estimation is proportional to $1/N$, where N is the data length and the precision is proportional to $1/N^{0.5}$ (Burges and Lettenmaier 1982). This means that, to improve the precision of parameters by a factor of 2, the required data length will have to be four times as long. But the data themselves may have associated uncertainties; these could arise from measurement errors, inconsistency, errors during data recording, and inadequate representation of the variable owing to limited samples in spatial and temporal domains.

Distribution Uncertainty

It is not always clear which type of a probability distribution a particular environmental random variable follows. For example, the annual maximum instantaneous discharge of a river can be described by the log-Pearson type 3 distribution, the three-parameter log-normal distribution, the Pearson type 3 distribution, or the generalized extreme value distribution. In many cases it is difficult to discern the exact type of the distribution the annual instantaneous maximum discharge follows. Similar cases abound with a number of other environmental variables.

Statistical Uncertainty in Time

Consider, for example, the time series of the annual maximum instantaneous discharge of a river. In most cases the time series is short for determining the discharge of a long recurrence interval, say, 500 years. In this case then there is a scarcity of information. The same applies to droughts, minimum flows, and a host of other environmental variables. Although an estimate of a 500-year flood can be made using any of the available standard frequency analysis techniques, this estimate is subject to uncertainty and this is an example of the statistical uncertainty in time or statistical uncertainty of variations in time. This uncertainty can be reduced by lengthening the database.

Statistical Uncertainty in Space

Consider, for example, the spatial mapping of erosion in a large basin. There is very little information available to map spatial variability of erosion. Similarly, spatial mapping of hydraulic conductivity of an aquifer can be very difficult if there are not enough data available. Such examples abound in environmental analysis. Thus, the spatial mapping of an environmental variable is subject to uncertainty and this is an example of statistical uncertainty in space or statistical uncertainty of variations in space.

10.1.2.2 Model Uncertainty

Environmental and water resources models are imperfect because of uncertainties. Consider, for example, an air pollution model that describes the pollutant concentration in space and time in the atmosphere. The model is imperfect because there are many gaps in our knowledge about pollutant dispersion in the atmosphere or the model is simplified for practical reasons. In any case, the model is subject to uncertainty resulting from the way the model has been constructed.

10.2 Sources of Uncertainty: Types of Errors

Uncertainties in environmental analysis arise from (1) randomness of physical phenomena or (2) errors in data and modeling. The modeling and data errors are of two types: (1) systematic and (2) random. There is another source of error, called illegitimate error, that results from outright blunders and mistakes. Computational errors are of this type, but these can be avoided.

A random or stochastic phenomenon is characterized by the property that its repeated occurrences do not produce the same outcome. For example, when the same rainfall occurs at different times over a watershed, it produces different hydrographs. This means that the watershed response is a stochastic variable. In a laboratory experiment, when a measurement is repeatedly made, the measured values do not identically match; the deviations between the values are called random fluctuations. Essentially, these are random errors, which are also called statistical errors in common parlance. Roughly, random errors tend to be higher and lower than the true value about an equal number of times. Furthermore, estimators subject to only random errors tend to be consistent. Thus, random errors tend to exhibit a statistical regularity, not a deterministic one. Examples of such errors are numerous.

Systematic errors are characterized by their deterministic nature and are frequently constant. For example, a rain gauge located near a building tends to produce biased rainfall measurements. A gauge operator tends to operate the gauge in a biased manner. The efficiency of observation is often a source of systematic error. Changes in experimental conditions tend to produce systematic errors. For example, if the location of a rain gauge is changed, it will produce biased measurements.

10.3 Analysis of Errors

Errors are analyzed using a logical procedure, whether experimental or theoretical. In general, errors have definite limits. Random errors can be determined using statistical tools. Systematic errors can be evaluated experimentally, statistically, or by other means. Sometimes, some errors can be removed by using correction factors. For example, the U.S. National Weather Service pan is used to determine potential evaporation in an area by applying a correction factor to the pan evaporation measurement. There are many examples of this kind in environmental and water resources analyses.

There is another important but difficult aspect of data errors, relating to mistakes in the data and rejection of data. When data of natural phenomena are collected, anomalies are seemingly found. These anomalies and their causes should be carefully analyzed. The causes of these anomalies may be random or systematic. Under no circumstances should they be discarded, unless there is a very strong compelling reason to do so. For example, in frequency analysis of the annual instantaneous maximum discharge of a river, so-called outliers or inliers are usually encountered. They must be dealt with, not discarded. Most often, these anomalies are expected. If the normal probability law (or chi-square criterion) indicates that these anomalous values are expected, they must be retained. Consider another case. If the deviation of an anomalous value is too large and has a small chance of occurring, the Chauvenet criterion may be used as a guide for accepting or rejecting the value or more appropriately flagging the suspicious situation. According to this criterion, if the probability of the value deviating from the mean by the observed amount is $1/2N$ or less, where N = number of observations, then there is reason for suspicion or rejection.

Uncertainty reflects the degree of error in a measurement or result of a model application. When the collected experimental data or model outputs are subject to small errors, two terms—accuracy and precision—are often employed to characterize the uncertainty. The word accuracy is generally used to indicate the closeness of the agreement between an experimentally determined value of a quantity and its true value. An accurate result closely agrees with the actual value for that quantity. In other words, accuracy tells us how close a measurement is to an accepted standard. Precision describes how well repeated measurements agree with each other. In other words, precision tells us how close two or more measurements agree. It is worth mentioning here that precision does not necessarily indicate anything about the accuracy of the measurements. An experiment is considered good when it is both precise and accurate.

Let us consider an experiment. It is said to have high precision if it has small random error. It is said to have high accuracy if it has small systematic error. There are four possibilities for characterizing the obtained experimental data, as shown in Fig. 10-1:

(i) precise and accurate,

(ii) precise and inaccurate,

(iii) imprecise and accurate, and

(iv) imprecise and inaccurate.

Our objective is to reduce both systematic and random errors as much as possible. However, for economy of effort, one must try to strike a balance between these two sources of error, giving greater weight to the larger of the two.

Example 10.1 A watershed has a network of rain gauges for measurement of rainfall. It is ascertained that if a rain gauge measures rainfall within 2% to 5% of the true value, then the rain gauge is said to be accurate and precise. Based on an analysis of the rainfall observations, it has been determined that one of the rain gauges, called A, always measures rainfall about 10% to 15% away from the true value. There is another rain gauge, called B, which is found to measure rainfall within 2% of the true value. Another rain gauge, called C, is found to measure rainfall somewhat unpredictably, that is, sometimes 15% away from the true value, sometimes 20% away from the true value, and sometimes very close to the true value. Another rain gauge, called D, measures rainfall sometimes 5% higher, sometimes 5% lower, sometimes 2% higher, and sometimes 2% lower, but always within 5% of the true value. What can be said about the measurements of these rain gauges?

Solution The measurements of rain gauge A are precise but inaccurate, because rain gauge A has the repeatability quality. Rain gauge B is accurate and precise because it has the repeatability quality and its observations are close to the true values. Rain gauge C is inaccurate and imprecise because its measurements are neither repetitive nor accurate. Rain gauge D is imprecise but accurate because its observations are close to the true values but its characteristic is not repeatable. These characteristics are shown in Fig. 10-2.

10.3.1 Measures of Errors

The experimental data errors can be investigated using the criterion of repeatability of measurements. In observing natural phenomena, however, there is no way to repeat the measurement. In laboratory experimentation, if a measurement is repeated, say N times, and the measured values do not exactly match the "true" value, the differences can be analyzed. It should, however, be noted that the true value is never known and only an estimator can be obtained. If N becomes large, the arithmetic average of the measured values approaches a constant value and, if this is the case, the estimator approaches a constant value and

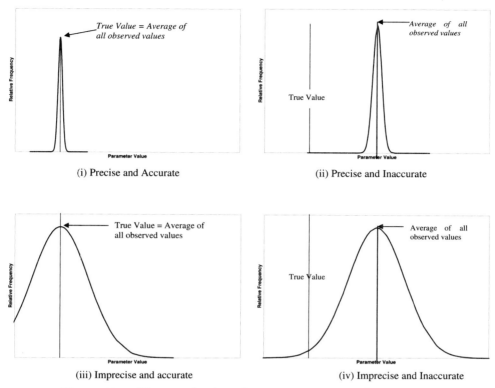

Figure 10-1 *Characterization of errors—accuracy and precision.*

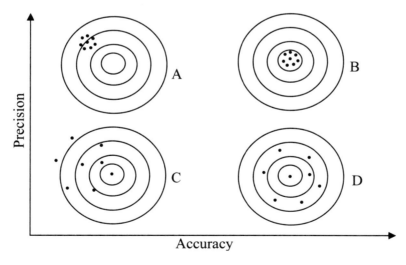

Figure 10-2 *Measurement of rainfall by four raingages. Gage A is precise, inaccurate; gage B is precise, accurate; gage C is imprecise, inaccurate; gage D is imprecise, accurate. The innermost circle indicates the true value.*

is qualified as a "consistent" estimator. Thus, consistency is one measure of experimental data error and is tied to the sample size. Ideally, the estimator should be consistent and without bias. However, the estimator, although consistent, may be biased if N is too small (i.e., the estimator may be either too large or too small). Thus, bias is another measure of experimental error that is tied to the sample size.

Not all statistics are unbiased estimators. Some are consistent estimators because they converge to the parent population as sample size increases but for a finite sample size they need correction to become an unbiased best estimate. For example, for N identical independent measurements, only the sample mean and sample variance are consistent statistics but only the sample mean is an unbiased estimator. To obtain an unbiased estimate of the variance, the sample variance is multiplied by the factor $N/(N-1)$ if the sample size is small.

10.3.2 Extraction of Information

Data are the source of information. They constitute the only medium of communication with nature. The purpose of data analysis is, therefore, to extract the maximum information. Statistical concepts used to analyze data are threefold:

1. Aggregate characteristics
2. Variation of individual values from aggregate properties
3. Frequency distribution of individual values

In the first case are mean (arithmetic, median, mode, harmonic, and geometric), deviation (mean and standard), variance, coefficient of variation, and higher order (such as skewness, kurtosis, etc.) or other types of moments (such as probability weighted, linear, and geometric). The moments and frequency distributions are interconnected. While computing these statistics, issues relating to rounding off and truncation errors have to be dealt with.

10.3.3 Analysis of Uncertainty

When analyzing uncertainty, it is assumed that the uncertainty can be measured quantitatively, at least in principle. For example, if an unbiased coin is tossed it is not known in advance whether the head or tail will turn up. But it is known that the event "head will turn up" has a probability of 1/2 or 50% each time the coin is tossed. Similarly, it is possible to assign a probability to the event that the peak flow in the Narmada River in any given year will exceed 3,000 m^3/s. One can also determine the probability that the compressive strength of concrete, manufactured in accordance with given specifications, will exceed 25,000 kN/m^2 or that the maximum number of trucks that have to wait at a toll station will exceed 50 on any given working day. The kind of uncertainty involved here is the uncertainty associated with the randomness of the event.

Investigating the uncertainty in the conclusions reached about uncertain events goes beyond the scope here. For example, one may calculate that the probability p of the peak flow in the Narmada River exceeding 3,000 m³/s in any given year is 2.5%. But there is an element of uncertainty in estimation of p, which depends on the length and quality of data. This raises the following question: What is the probability that p lies within a given range $p \pm \Delta p$? In each case, events can be analyzed such that reasonable people, using reasonable procedures, come up with reasonably close probability assessments. Here the objective is to rationally deal with uncertainty, not completely eliminate it.

To deal with uncertain events in the decision-making process, events that are certain to occur, or conclusions that are certainly true, must be fully taken into account. These can be given a weight of 1 (certain events). Impossible events, in contrast, are disregarded in decisions and these are given the weight of 0. Any in-between event is given a weight equal to the probability of its occurrence. Thus, the more likely an event is, the more weight it gets and the greater is its relative effect on the outcome or the decision. It is in this manner that the effective monetary value of the consequences of a decision was evaluated in Chapter 1.

Evaluation of safety and reliability requires information on uncertainty, which may be determined by the standard deviation or coefficient of variation. Questions of safety or reliability arise principally from the presence of uncertainty. Thus, an evaluation of the uncertainty is an essential part of the evaluation of engineering reliability. The uncertainty resulting from random variability in physical phenomena is described by a probability distribution function. For practical purposes, its description may be limited to (a) a central tendency and (b) its dispersion (e.g., standard deviation) or coefficient of variation.

To deal with uncertainty arising from prediction error (estimation error or statistical sampling error and imperfection of the prediction model), one normally employs the coefficient of variation or the standard deviation, which represents a measure of the random error. In effect, the random error is involved whenever there is a range of possible error. One source of random error is sampling error, which is a function of the sample size. The random sampling error can be expressed in terms of the coefficient of variation (CV) as $\Delta_1 = CV / \sqrt{N}$, where N = sample size.

Consider, for example, the mean annual rainfall for Baton Rouge to be 60.00 inches. Conceivably, this estimate of the true mean value would contain error. If the rainfall measurement experiment is repeated and other sets of data are obtained, the sample mean estimated from the other sets of data would most likely be different. The collection of all the sample means will also have a mean value, which may well be different from the individual sample mean values, and a corresponding standard deviation. Conceptually, the mean value of the collection of sample means may be assumed to be close to the true mean value (assuming that the estimator is unbiased). Then, the difference (or ratio) of the estimated sample mean (i.e., mean value of 60 inches) to the true mean is the systematic error, whereas the coefficient of variation or standard deviation of the collection of sample means represents a measure of the random error. In effect, random error is involved whenever there is a range of possible error. One source of random error is the error from sampling, which is a function of the sample size.

The systematic error is a bias in the prediction or estimation and can be corrected through a constant bias factor. The random error, called the standard error, requires statistical treatment. It represents the degree of dispersiveness of the range of possible errors. It may be represented by the standard deviation or coefficient of variation of the estimated mean value. An objective determination of the bias as well as the random error will require repeated data on the sample mean (or medians), which are hard to come by.

For a random phenomenon, prediction or estimation is usually confined to the determination of a central value (e.g., mean or median) and its associated standard deviation or coefficient of variation. The uncertainty associated with the error in the estimation of the degree of dispersion is of secondary importance, whereas the uncertainty resulting from error in the prediction of the central value is of first-order importance. To summarize, through methods of prediction one obtains

$$\overline{x} = \text{estimate of mean value}$$

$$\sigma_x = \text{estimate of the standard deviation}$$

For a set of observations, the mean value is

$$\overline{x} = \frac{1}{n} \sum_{i=1}^{n} x_i \tag{10.1}$$

and the variance is

$$\sigma_x^2 = \frac{1}{n-1} \sum_{i=1}^{n} (x_i - \overline{x})^2 \tag{10.2}$$

An assessment of the accuracy or inaccuracy of the prediction for the mean value is made to obtain e = bias correction for the error in the predicted mean value $\overline{x}$ and Δ = measure of the random error in $\overline{x}$. For quantification of uncertainty measures, we confine ourselves to the error of prediction or the error in the estimation of the respective mean values; that is, the systematic and random errors will refer to the bias and standard error, respectively, in the estimated mean value of a variable (or function of variables). It should be emphasized that the uncertainty measures are credible, for the validity of a calculated probability depends on credible assessments of the individual uncertain measures. Methods for evaluating uncertainty measures depend on the form of the available data and information.

The uncertainty associated with the inherent randomness is given by

$$CV = \frac{\sigma_x}{\overline{x}} \tag{10.3}$$

The mean value estimate may not be totally accurate relative to the true mean (especially for a small sample size n). The estimated mean value given here is unbiased as far as sampling is concerned; however, the random error of $\bar{x}$ is the standard error of $\bar{x}$, which is

$$\sigma_{\bar{x}} = \sigma_x / \sqrt{n} \tag{10.4}$$

The uncertainty associated with random sampling error is

$$\Delta_x = \sigma_{\bar{x}} / \bar{x} \tag{10.5}$$

This random error in $\bar{x}$ is limited to the sampling error only. There may, however, be other biases and random errors in $\bar{x}$, such as the effects of factors not included in the observational program.

Often, the information is expressed in terms of the lower and upper limits of a variable. Given the range of possible values of a random variable, the mean value of the variable and the underlying uncertainty may be evaluated by prescribing a suitable distribution within the range. For example, if a random variable X is assumed to be characterized by a uniform distribution with the lower and upper limits of x_l and x_u, respectively, then using Eq. 5.93 and Eq. 5.94 gives the mean, standard deviation, and CV of X as

$$\bar{x} = \frac{1}{2}(x_l + x_u), \quad \sigma_x = \frac{1}{2\sqrt{3}}(x_u - x_l) \tag{10.6}$$

$$CV = \frac{1}{\sqrt{3}}\left(\frac{x_u - x_l}{x_u + x_l}\right) \tag{10.7}$$

where the PDF of the variable is uniform between x_l and x_u, as shown in Fig. 10-3.

Alternatively, let the PDF be given by a symmetric triangular distribution with limits x_l and x_u, as shown in Fig. 10-4. By substituting $a = x_l$, $b = x_u$, and $c = (x_u + x_b)/2$ in Eq. 5.133 and Eq. 5.134, the corresponding CV would be

$$CV = \frac{1}{\sqrt{6}}\left(\frac{x_u - x_l}{x_u + x_l}\right) \tag{10.8}$$

With either the uniform or the symmetric triangular distribution, it is implicitly assumed that there is no bias within the prescribed range of values for X. However, if there is bias, a skewed distribution may be more appropriate. If the bias is judged to be toward the higher values within the specified range, then the upper triangular distribution as shown in Fig. 10-5 would be appropriate. In such a case, by substituting $a = x_l$ and $b = c = x_u$ in Eq. 5.133 and Eq. 5.134, the mean and CV of X can be determined as

$$\bar{x} = \frac{1}{3}(x_l + 2x_u) \tag{10.9}$$

$$CV = \frac{1}{\sqrt{2}}\left(\frac{x_u - x_l}{2x_u + x_l}\right) \tag{10.10}$$

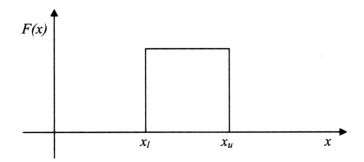

Figure 10-3 *Uniform probability density function between* x_l *and* x_u.

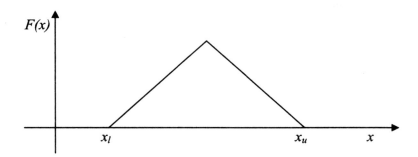

Figure 10-4 *Symmetric triangular probability density function between* x_l *and* x_u.

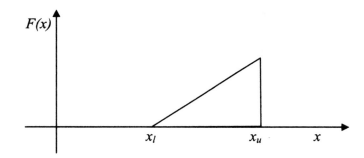

Figure 10-5 *Upper triangular probability density function between* x_l *and* x_u.

Conversely, if the bias is toward the lower range of values, the appropriate distribution may be a lower triangular distribution as shown in Fig. 10-6. Substituting $a = c = x_l$ and $b = x_u$ in Eq. 5.133 and Eq. 5.134 gives the corresponding mean and CV as

$$\bar{x} = \frac{1}{3}\left(2x_l + x_u\right) \tag{10.11}$$

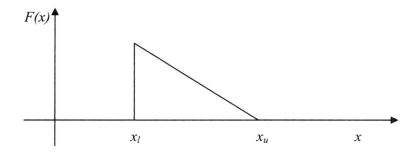

Figure 10-6 *Lower triangular probability density function between* x_l *and* x_u.

$$CV = \frac{1}{\sqrt{2}}\left(\frac{x_u - x_l}{x_u + 2x_l}\right) \tag{10.12}$$

Another distribution may be a normal distribution, as shown in Fig. 10-7, where the given limits may be assumed to cover $\pm 2\sigma$ from the mean value. In such cases, the mean value is

$$\overline{x} = \frac{1}{2}\left(x_u + x_l\right) \tag{10.13}$$

and the coefficient of variation is

$$CV = \frac{1}{2}\left(\frac{x_u - x_l}{x_u + x_l}\right) \tag{10.14}$$

The seemingly different types and sources of uncertainty can also be analyzed as follows. Let the true value of the variable be x and its prediction be given as $\hat{x}$. Let there be a correction factor λ to account for error in $\hat{x}$. Therefore, the true value x may be expressed as

$$x = \lambda\hat{x} \tag{10.15}$$

For random variable value X, the model $\hat{X}$ should be a random variable. The estimated mean value $\hat{x}$ and variance σ_x^2 (e.g., from a set of observations) are those of $\hat{x}$. Then, $CV = \sigma_x/\overline{x}$ represents the inherent variability. The necessary correction λ may also be considered a random variable, whose mean value "e" represents the mean correction for systematic error in the predicted mean value $\overline{x}$, whereas the CV of λ, Δ, represents the random error in the predicted mean value $\overline{x}$. If one assumes λ and $\hat{x}$ to be statistically independent, the mean value of X is

$$\mu_x = e\overline{x} \tag{10.16}$$

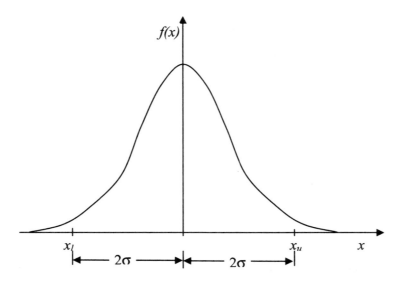

Figure 10-7 *Normal probability density function between* x_l *and* x_u.

The total uncertainty in the prediction of x then becomes

$$\Omega_x \cong \sqrt{CV_x^2 + \Delta_x^2}, \quad CV_x = \sigma_x / \bar{x} \tag{10.17}$$

The preceding analysis pertains to a single variable.

If Y is a function of several random variables $x_1, x_2, ..., x_n$, that is,

$$Y = g(x_1, x_2, ..., x_n) \tag{10.18}$$

the mean value and associated uncertainty of Y are of concern. A model (or function) $\hat{g}$ and a correction λ_g may be used, so

$$Y = \lambda_g \hat{g}(x_1, x_2, ..., x_n) \tag{10.19}$$

Thus, λ_g has a mean value of e_g and a CV of Δ_g. Using the first-order approximation gives the mean value of Y:

$$\mu_y \cong e_g \hat{g}\left(\mu_{x_1}, \mu_{x_2}, ..., \mu_{x_n}\right) \tag{10.20}$$

where e_g is the bias in $\hat{g}\left(\mu_{x1}, \mu_{x2}, ..., \mu_{xn}\right)$ and $\mu_{xi} = e_i \bar{x}_i$. Also, the total CV of Y is

$$\Omega_y^2 = \Delta_g^2 + \frac{1}{\mu_g^2} \sum_i \sum_j \rho_{ij} c_i c_j \sigma_{xi} \sigma_{xj} \tag{10.21}$$

in which $c_i = \partial g / \partial x_i$ evaluated at $\left(\mu_{x1}, \mu_{x2}, ..., \mu_{xn}\right)$ and ρ_{ij} = correlation coefficient between x_i and x_j.

Example 10.2 Consider the mean annual rainfall for Baton Rouge, which is given as 60 inches based on a sample of data. The mean rainfall estimated by the arithmetic mean method is about 5% to 10% higher than the true mean. Taking the sample standard deviation of 15 inches and the number of observations in the sample as 25, compute the total random error in the estimated mean value.

Solution The corresponding CV is $15/60 = 0.25$. Assume the random sampling error (expressed in terms of CV) would be

$$\Delta_1 = 0.25/(25)^{0.5} = 0.05$$

The systematic error or bias may arise from factors not accounted for in the prediction model that tends to consistently bias the estimate in one direction or the other. For example, the mean rainfall estimated by the arithmetic mean method may be about, say, 5% to 10% higher than the true mean, say, yielded by the isohyetal method. With this information, a realistic prediction of the mean rainfall requires a correction from 90% to 95% of the corresponding mean (arithmetic) rainfall. If a uniform PDF between this range of correction factors is assumed, then the systematic error in the estimated arithmetic mean rainfall of 60 inches will need to be corrected by a mean bias factor of

$$e = \frac{1}{2}(0.9 + 0.95) = 0.925$$

whereas the corresponding random error in the estimated mean value, expressed in terms of CV (see Eq. 10.7), is

$$\Delta_2 = \frac{1}{\sqrt{3}}\left(\frac{1.10 - 1.05}{1.10 + 1.05}\right) = \frac{1}{\sqrt{3}}\left(\frac{0.05}{2.15}\right) = 0.013$$

The total random error in the estimated mean value is, therefore,

$$\Delta = \sqrt{\Delta_1^2 + \Delta_2^2} = \sqrt{(0.05)^2 + (0.013)^2} = 0.052$$

Example 10.3 Consider a linear reservoir expressed as $S = KQ$, where S = storage (in m³), K = reservoir constant (the average travel or retention time), and Q = discharge (in m³/s). Assume K and Q are independent and their errors are independent and uncorrelated. Determine the CV of S.

Solution Since K and Q are independent and their errors are independent and uncorrelated, one can write

$$\sigma_S^2 = \left(\left.\frac{\partial f}{\partial K}\right|_p\right)^2 \sigma_K^2 + \left(\left.\frac{\partial f}{\partial Q}\right|_p\right)^2 \sigma_Q^2 \qquad \textbf{(10.22)}$$

where

$$\frac{\partial f}{\partial K}\bigg|_p = \overline{Q} \ , \ \frac{\partial f}{\partial Q}\bigg|_p = \overline{K}$$

Equation 10.22 can be expressed as

$$\sigma_S^2 = \overline{Q}^2 \sigma_K^2 + \overline{K}^2 \sigma_Q^2$$

Dividing both sides by $(\overline{S})^2 \left(= \overline{K}^2 \cdot \overline{Q}^2\right)$ one gets

$$\frac{\sigma_S^2}{\overline{S}^2} = \frac{\sigma_K^2}{\overline{K}^2} + \frac{\sigma_Q^2}{\overline{Q}^2}$$

Thus,

$$CV_S = \frac{\sigma_S}{\overline{S}} = \sqrt{\frac{\sigma_K^2}{\overline{K}^2} + \frac{\sigma_Q^2}{\overline{Q}^2}} = \sqrt{CV_K^2 + CV_Q^2} \qquad \textbf{(10.23)}$$

Example 10.4 The storage in a reservoir at the end of month $t+1$, S_{t+1}, can be computed using the continuity equation

$$S_{t+1} = S_t + I_t - R_t - E_t \qquad \textbf{(10.24)}$$

where I_t is inflow during month t, R_t is release during month t, and E_t is evaporation of water during month t. Of the variables on the right-hand side of Eq. 10.24, S_t is known at the beginning and the release to be made during the month (R_t) is assumed known. Variables I_t and E_t are assumed to be uncertain random variables, with mean values of 21.3 and 2.4 million m³, respectively, and standard deviations of 4.5 and 0.6 million m³, respectively. If the initial storage in the reservoir is 28.0 million m³ and the target release is 16.7 million m³, find the expected storage at the end of the month and its standard deviation.

Solution The expected value of the end-of-month storage (denoted by $E(S_{t+1})$) is

$$E(S_{t+1}) = S_t + E(I_t) - R_t - E(E_t) = 28.0 + 21.3 - 16.7 - 2.4 = 30.2 \text{ million m}^3 \quad \textbf{(10.25)}$$

The variance of the end-of-month storage is

$$\text{var}(S_{t+1}) = \text{var}(I_t) + \text{var}(E_t) = 4.5^2 + 0.6^2 = 20.61 (\text{million m}^3)^2 \qquad \textbf{(10.26)}$$

Hence,

$$SD(S_{t+1}) = 20.61^{0.5} = 4.54 \text{ million m}^3$$

Example 10.5 In the previous example, both reservoir inflow and evaporation depend upon the climate and hence may be correlated. From the analysis of historical data, this was found to be indeed the case and the correlation was –0.25. Determine the standard deviation of the expected storage at the end of the month.

Solution The variance of S_{t+1} can now be calculated as

$$\text{var}(S_{t+1}) = \text{var}(I_t) + \text{var}(E_t) - 2 \times \rho \times SD(I_t) \times SD(E_t) \tag{10.27}$$

$$= 4.5^2 + 0.6^2 - 2\times(-0.25) \times 4.5 \times 0.6 = 20.61 + 1.35 = 21.96$$

Hence,

$$SD(S_{t+1}) = 21.96^{0.5} = 4.67 \text{ million m}^3$$

The standard deviation is now increased. Note that the standard deviation depends upon the magnitude and sign of the correlation. In this example, if the correlation were positive, the standard deviation would have been less.

Example 10.6 Consider the rational method for computing peak discharge, $Q = CIA$, where Q = peak discharge in m^3/s, C = rational runoff coefficient (dimensionless), I = rainfall intensity in mm/hour, and A = drainage area in km^2. Assume that the variables C, I, and A are independent and that the errors in them are independent and uncorrelated. Express the relative error (standard deviation divided by mean) in the peak discharge as a function of errors in C, I, and A.

Solution Since C, I, and A are independent and their errors are independent and uncorrelated

$$\sigma_Q^2 = \left(\frac{\partial f}{\partial C}\Big|_p\right)^2 \sigma_C^2 + \left(\frac{\partial f}{\partial I}\Big|_p\right)^2 \sigma_I^2 + \left(\frac{\partial f}{\partial A}\Big|_p\right)^2 \sigma_A^2 \tag{10.28}$$

where

$$\frac{\partial f}{\partial C}\Big|_p = \bar{I}\,\bar{A}, \quad \frac{\partial f}{\partial I}\Big|_p = \bar{C}\,\bar{A}$$

and

$$\frac{\partial f}{\partial A}\Big|_p = \bar{C}\,\bar{I}$$

Equation 10.28 can be expressed as

$$\sigma_Q^2 = \bar{I}^2\bar{A}^2\sigma_C^2 + \bar{C}^2\bar{A}^2\sigma_I^2 + \bar{C}^2\bar{I}^2\sigma_A^2$$

But we have

$$\bar{Q} = \bar{C}\,\bar{I}\,\bar{A}$$

Thus,

$$\frac{\sigma_Q^2}{\overline{Q}^2} = \frac{\sigma_C^2}{\overline{C}^2} + \frac{\sigma_I^2}{\overline{I}^2} + \frac{\sigma_A^2}{\overline{A}^2}$$

$$E_Q = \sqrt{\frac{\sigma_Q^2}{\overline{Q}^2}} = \sqrt{\frac{\sigma_C^2}{\overline{C}^2} + \frac{\sigma_I^2}{\overline{I}^2} + \frac{\sigma_A^2}{\overline{A}^2}} = \sqrt{CV_Q^2} = \sqrt{CV_C^2 + CV_I^2 + CV_A^2} \qquad \textbf{(10.29)}$$

Example 10.7 The pollutant concentration C can be expressed as $C = Q_S/Q$, where Q is water discharge and Q_S is pollutant discharge. Express the error in C as a function of errors in Q_S and Q. Assume that Q_S and Q are independent and that the errors in them are independent and uncorrelated.

Solution Since Q_S and Q are independent and their errors are independent and uncorrelated

$$\sigma_C^2 = \left(\frac{\partial f}{\partial Q_S}\bigg|_p\right)^2 \sigma_{Q_s}^2 + \left(\frac{\partial f}{\partial Q}\bigg|_p\right)^2 \sigma_Q^2 \qquad \textbf{(10.30)}$$

where

$$\frac{\partial f}{\partial Q_s}\bigg|_p = \frac{1}{\overline{Q}} \text{ and } \frac{\partial f}{\partial Q}\bigg|_p = -\frac{\overline{Q_s}}{\overline{Q}^2}$$

Equation 10.30 can be expressed as

$$\sigma_C^2 = \frac{\sigma_{Q_s}^2}{\overline{Q}^2} + \frac{\overline{Q_s}^2 \sigma_Q^2}{\overline{Q}^4} \qquad \textbf{(10.31)}$$

But

$$\overline{C} = \frac{\overline{Q_s}}{\overline{Q}}$$

Thus,

$$\frac{\sigma_C^2}{\overline{C}^2} = \frac{\sigma_{Q_s}^2}{\overline{Q_s}^2} + \frac{\sigma_Q^2}{\overline{Q}^2} \text{ and } E_C = \sqrt{\frac{\sigma_C^2}{\overline{C}^2}} = \sqrt{\frac{\sigma_{Q_s}^2}{\overline{Q_s}^2} + \frac{\sigma_Q^2}{\overline{Q}^2}} \qquad \textbf{(10.32)}$$

Example 10.8 The cross-sectional area of a channel at a given location is measured by summing the cross-sectional areas of different segments. Assume the channel to be of trapezoidal form, which can be broken down into rectangular and triangular portions. Let the total area (A) be expressed as $A = x + y$, where x is the cross-sectional area of the rectangular part and y is the cross-sectional area of the two triangular parts. Express the error in A as a function of errors in x and y. Then, consider the following data: x (m²) = 100 ± 1.5 and y (m²) = 75 ± 0.5. Compute the error in the area.

Solution Since x and y are independent and their errors are independent and uncorrelated

$$\sigma_A^2 = \left(\frac{\partial f}{\partial x}\bigg|_p\right)^2 \sigma_x^2 + \left(\frac{\partial f}{\partial y}\bigg|_p\right)^2 \sigma_y^2 \qquad (10.33)$$

where

$$\frac{\partial f}{\partial x}\bigg|_p = 1 \, , \, \frac{\partial f}{\partial y}\bigg|_p = 1$$

Equation 10.33 can be expressed as

$$\sigma_A^2 = \sigma_x^2 + \sigma_y^2$$

But

$$\bar{A} = \bar{x} + \bar{y}$$

Thus,

$$\frac{\sigma_A^2}{\bar{A}^2} = \frac{\sigma_x^2}{(\bar{x} + \bar{y})^2} + \frac{\sigma_y^2}{(\bar{x} + \bar{y})^2}$$

and

$$E_A = \sqrt{\frac{\sigma_A^2}{\bar{A}^2}} = \sqrt{\frac{\sigma_x^2 + \sigma_y^2}{(\bar{x} + \bar{y})^2}}$$

Given $x = 100 \pm 1.5$ and $y = 75 \pm 0.5$, therefore,

$$E_A = \sqrt{\frac{\sigma_A^2}{\bar{A}^2}} = \sqrt{\frac{\sigma_x^2 + \sigma_y^2}{(\bar{x} + \bar{y})^2}} = \sqrt{\frac{1.5^2 + 0.5^2}{(100 + 75)^2}} = 9.035 \times 10^{-3}$$

Example 10.9 Daily lake evaporation is frequently estimated by measuring the lake levels at the beginning and at the end of the day (24 hours later). Consider that the lake level at the beginning of the day is 1,000 ± 10 cm and at the end of the day it is 998 ± 10 cm. Independently, the lake level is found to be measured with an accuracy of 1.0%. Compute the lake evaporation and determine its accuracy. Compare the accuracy of lake evaporation with the accuracy of the independent lake measurements.

Solution The depth of daily lake evaporation can be expressed as $z = x - y$, where x indicates the lake level at the beginning of the day ($x = 1,000 \pm 10$) and y indicates the lake level at the end of the day ($y = 998 \pm 10$). It is assumed that x and y are independent and their errors are independent and uncorrelated, so

$$\sigma_Z^2 = \left(\frac{\partial f}{\partial x}\Big|_p\right)^2 \sigma_x^2 + \left(\frac{\partial f}{\partial y}\Big|_p\right)^2 \sigma_y^2 \qquad (10.34)$$

where

$$\frac{\partial f}{\partial x}\Big|_p = 1 \text{ and } \frac{\partial f}{\partial y}\Big|_p = -1$$

Equation 10.34 can be expressed as

$$\sigma_Z^2 = \sigma_x^2 + \sigma_y^2$$

Thus,

$$\frac{\sigma_Z^2}{z^2} = \frac{\sigma_x^2}{(\bar{x} - \bar{y})^2} + \frac{\sigma_y^2}{(\bar{x} - \bar{y})^2}$$

and

$$E_Z = \sqrt{\frac{\sigma_Z^2}{z^2}} = \sqrt{\frac{\sigma_x^2 + \sigma_y^2}{(\bar{x} - \bar{y})^2}}$$

Since

$$E_x = \frac{\sigma_x}{\bar{x}} = E_y = \frac{\sigma_y}{\bar{y}} = 1.0\%$$

we have $\sigma_x = 0.01\bar{x}$ and $\sigma_y = 0.01\bar{y}$. Given $\bar{x} = 1,000$ and $\bar{y} = 998$, we have

$$E_Z = \sqrt{\frac{\sigma_Z^2}{z^2}} = \sqrt{\frac{\sigma_x^2 + \sigma_y^2}{(\bar{x} - \bar{y})^2}} = \sqrt{\frac{0.01^2(\bar{x}^2 + \bar{y}^2)}{(\bar{x} - \bar{y})^2}} = \sqrt{\frac{0.01^2(1000^2 + 998^2)}{(1000 - 998)^2}} = 7.06$$

$$z = 2 \pm (2 \times 7.06) = 2 \pm 14.12$$

Obviously, E_z is much larger than E_x or E_y.

Example 10.10 According to the Muskingum method for flow routing, $S_* = aI + (1 - a)Q$, where S_* = storage rate (storage/average time of travel), I = inflow rate, Q = outflow rate, and a = weighting factor. Assume the variables I and Q are independent. Using the least-square error criterion, determine the weighting factor in terms of the errors in I and Q. Derive the error in storage rate as a function of the errors in I and Q. Then express the storage rate in terms of the derived expression for a.

Solution Starting with the Muskingum equation $S_* = aI + (1-a)Q$, one can write

$$\sigma_S^2 = \left(\frac{\partial f}{\partial I}\bigg|_p\right)^2 \sigma_I^2 + \left(\frac{\partial f}{\partial Q}\bigg|_p\right)^2 \sigma_Q^2 \tag{10.35}$$

where

$$\frac{\partial f}{\partial I}\bigg|_p = a \text{ and } \frac{\partial f}{\partial Q}\bigg|_p = 1-a$$

Substituting these values in Eq. 10.35 gives

$$\sigma_S^2 = a^2\sigma_I^2 + (1-a)^2\sigma_Q^2$$

The least-square error criterion,

$$\frac{d\sigma_S^2}{da} = 2a\sigma_I^2 - 2(1-a)\sigma_Q^2 = 0$$

then gives

$$a = \frac{\sigma_Q^2}{\sigma_I^2 + \sigma_Q^2}$$

so

$$\sigma_S^2 = \left(\frac{\sigma_Q^2}{\sigma_I^2 + \sigma_Q^2}\right)^2 \sigma_I^2 + \left(\frac{\sigma_I^2}{\sigma_I^2 + \sigma_Q^2}\right)^2 \sigma_Q^2 = \frac{1}{\dfrac{1}{\sigma_I^2} + \dfrac{1}{\sigma_Q^2}}$$

and

$$S = \left(\frac{\sigma_Q^2}{\sigma_I^2 + \sigma_Q^2}\right)I + \left(\frac{\sigma_I^2}{\sigma_I^2 + \sigma_Q^2}\right)Q$$

10.3.4 Propagation of Errors

In most civil engineering projects we either use a mathematical model as an aid to analyze a given system or conduct experiments. In both cases we come across a variety of quantities, each with their own errors. Because of the uncertain nature of these quantities, these can be better represented by random variables.

Further, the system output or the overall result of an experiment is also uncertain and hence can be represented by a random variable as well. When input random variables of a given system, such as a model or an experiment associated with errors, are used in the calculation of overall system response, these errors propagate from input to output. Quite often we are interested in knowing how errors in mathematical models or instruments propagate throughout a given system so that we can estimate the magnitude of the error associated in the overall system response. Error propagation gives a valid estimate of the error involved in the mathematical result.

Let us consider a univariate relationship $Y = f(X)$ between a dependent variable Y and an input variable X. Further, let us consider that σ_x is the uncertainty (error) in x. How will the uncertainty associated with x be reflected in the uncertainty of y, denoted as σ_y? Figure 10-8 explains the various terms and their physical meanings in a graphical format. The functional relationship $f(x)$ can be written as

$$y = f(x) = f\left[\bar{x} + (x - \bar{x})\right] \tag{10.36}$$

Expanding $f(x)$ using Taylor's theorem gives

$$f(x) = f(\bar{x}) + \frac{dy}{dx}\bigg|_{x=\bar{x}} (x - \bar{x}) + \frac{1}{2!}\frac{d^2y}{dx^2}\bigg|_{x=\bar{x}} (x - \bar{x})^2 + \frac{1}{3!}\frac{d^3y}{dx^3}\bigg|_{x=\bar{x}} (x - \bar{x})^3 + \dots \tag{10.37}$$

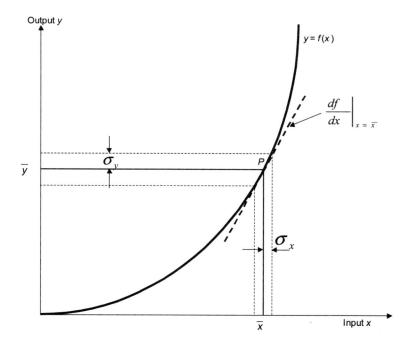

Figure 10-8 *Propagation of error.*

Now, truncating the series at the linear terms and solving for σ_y, we obtain

$$y = f(x) \approx f(\bar{x}) + \left.\frac{dy}{dx}\right|_{x=\bar{x}} (x - \bar{x}) \qquad (10.38)$$

Using the formula $\mathrm{var}[y] = E[y^2] - \bar{y}^2$ and Eq. 10.38, one can write

$$\sigma_y^2 = \left(\left.\frac{dy}{dx}\right|_{x=\bar{x}}\right)^2 \sigma_x^2 \text{ or } \sigma_y = \left.\frac{dy}{dx}\right|_{x=\bar{x}} \sigma_x \qquad (10.39)$$

In other words, uncertainty of the dependent variable Y can be determined by multiplying the uncertainty associated with the independent variable X ($= \sigma_x$) with the sensitivity of Y with respect to X at the operating point P ($= dy/dx|_{x=\bar{x}}$).

Equation 10.39 helps us understand how errors in mathematical models propagate throughout a given system. The error involved in the model output or in the overall result of the experiment is a function of (1) the error involved in the input random variable associated with a given system and (2) the functional form of a system. It is worth mentioning here that the estimate of system uncertainty obtained by Eq. 10.39 is not always accurate. The accuracy of Eq. 10.39 depends upon how accurate the approximate functional form (Eq. 10.38) is compared to the actual functional form (i.e., Eq. 10.37), which is directly related to the degree of nonlinearity in the functional form. If the functional form is linear, then the output uncertainty will only be governed by the uncertainty of the input variable.

If the dependent variable Y depends upon several input variables X_1, X_2, ..., X_n, we can write the truncated form of the Taylor series expansion up to the linear terms as

$$y = f(\bar{x}_1, \bar{x}_2, ..., \bar{x}_n) + \sum_{i=1}^{n} (x_i - \bar{x}_i)\left(\frac{\partial y}{\partial x_i}\right)_P \qquad (10.40)$$

where $P = (\bar{x}_1, \bar{x}_2, ..., \bar{x}_n)$ is a vector representing the expansion point. Thus, the variance of Y is

$$\sigma_y^2 = \left\{\left(\left.\frac{\partial f}{\partial x_1}\right|_P\right)^2 E[x_1 - \bar{x}]^2 + \left(\left.\frac{\partial f}{\partial x_2}\right|_P\right)^2 E[x_1 - \bar{x}]^2 + ...\right\}$$

$$+ \left\{2\left(\left.\frac{\partial f}{\partial x_1}\right|_P\right)\left(\left.\frac{\partial f}{\partial x_2}\right|_P\right) E[(x_1 - \bar{x}_1)(x_2 - \bar{x}_2)] + ...\right\} \qquad (10.41a)$$

Alternatively, Eq. 10.41a can be written as

$$\sigma_Y^2 = \sum_{i=1}^{n} \sum_{j=1}^{n} \left(\frac{\partial y}{\partial x_i} \right)_P \left(\frac{\partial y}{\partial x_j} \right)_P \text{cov}\left[x_i x_j \right] \qquad \text{(10.41b)}$$

For multivariate propagation of errors, two cases can be distinguished:

1. Input variables $x_1, x_2, ..., x_n$ are independent of each other (i.e., $\text{cov}(x_i, x_j) = 0$).
2. Input variables $x_1, x_2, ..., x_n$ are correlated to each other (i.e., $\text{cov}(x_i, x_j) \neq 0$).

We will discuss these two cases in what follows.

Case 1

If the basic variables are statistically independent, the expression for var(Y) becomes

$$\sigma_y^2 = \sum_{i=1}^{n} \left[\frac{\partial y}{\partial x_i}\bigg|_P \sigma_{x_i} \right]^2 \qquad \text{(10.42)}$$

If we want to estimate the relative uncertainty associated with operating point P, we normalize this equation with respect to $\bar{y} = f\left(\bar{x}_1, \bar{x}_2, ..., \bar{x}_n \right)$ as

$$\frac{\sigma_y}{\bar{y}} = \left\{ \sum_{i=1}^{n} \left[\frac{1}{\bar{y}} \frac{\partial y}{\partial x_i}\bigg|_P \sigma_{x_i} \right]^2 \right\}^{1/2} \qquad \text{(10.43)}$$

If we define the relative uncertainty associated with each independent variable x_i as

$$e_i = \frac{1}{\bar{y}} \frac{\partial y}{\partial x_i}\bigg|_P \sigma_{x_i} \qquad \text{(10.44)}$$

we can write the total relative uncertainty e_y as

$$e_y = \sqrt{e_1^2 + e_2^2 + ... + e_n^2} \qquad \text{(10.45)}$$

Thus when input variables are uncorrelated, the total relative error is the square root of the sum of squares of all individual relative errors.

Case 2

When input variables are correlated, one has to use Eq. 10.41. For the sake of simplicity, let us consider only two input variables. Rewriting Eq. 10.41 for the relationship $y = f(x_1, x_2)$ in which x_1 and x_2 are correlated gives

$$\sigma_y^2 = \left(\frac{\partial f}{\partial x_1}\bigg|_P\right)^2 \sigma_{x_1}^2 + \left(\frac{\partial f}{\partial x_2}\bigg|_P\right)^2 \sigma_{x_2}^2 + 2\left(\frac{\partial f}{\partial x_1}\bigg|_P\right)\left(\frac{\partial f}{\partial x_2}\bigg|_P\right)\text{cov}(x_1 x_2) \qquad (10.46)$$

Further, we know that

$$-\sqrt{\text{var}(x)\text{var}(y)} \leq \text{cov}(x_1 x_2) \leq \sqrt{\text{var}(x)\text{var}(y)} \qquad (10.47)$$

Now, using Eq. 10.46 and Eq. 10.47 one can conclude that when $\text{cov}(x_1, x_2)$ is negative, $\text{var}(y)$ is smaller than in the uncorrelated case, whereas, when $\text{cov}(x_1, x_2)$ is positive, $\text{var}(y)$ is larger than in the uncorrelated case.

Example 10.11 Consider the case of two independent variables x and y, with $z = f(x, y)$. Assume that the errors are independent.

Solution Using Eq. 10.42 gives the standard deviation of z as

$$\sigma_z = \sqrt{\left(\frac{\partial f}{\partial x}\right)^2 \sigma_x^2 + \left(\frac{\partial f}{\partial y}\right)^2 \sigma_y^2} \qquad (10.48)$$

If $z = xy$, then

$$\sigma_z = \sqrt{(\bar{y})^2 \sigma_x^2 + (\bar{x})^2 \sigma_y^2} \qquad (10.49)$$

Equation 10.49 is expressed more meaningfully in terms of the coefficient of variation (standard deviation divided by the mean) as

$$CV_z = \sqrt{CV_x^2 + CV_y^2} \qquad (10.50)$$

If $z = x/y$, then Eq. 10.50 also holds. If $z = x + y$ or $z = x - y$, then

$$\sigma_z = \sqrt{\sigma_x^2 + \sigma_y^2} \qquad (10.51)$$

Equation 10.51 shows why the methods of computation that are based only on the water balance equation are not preferred in environmental analysis. An example is the significant error obtained when evaporation from a lake or a watershed is computed based on water balance alone. The errors in the estimation of individual components will be accumulated in the estimate of evaporation, making it highly unreliable.

Another case is $z = x^m y^r$. In this case,

$$\varepsilon_z = \sqrt{m^2 \varepsilon_x^2 + r^2 \varepsilon_y^2} \qquad (10.52)$$

Equation 10.52 shows that the error multiplies in a nonlinear case. If $z = x^m$, then $CV_z = mCV_x$. If $z = cx$, where c is constant, then $CV_z = CV_x$.

It is now possible to determine the best value of a quantity x from two or more independent measurements whose errors may be different. Intuitively, the measurement with less error should carry more weight. However, how exactly the weighting should be done is not quite clear. To that end, the principle of least-square error may be invoked. Consider two independent measurements of X as x_1 and x_2, with their respective (plus or minus) errors as σ_1 and σ_2. It may be reasonable to assume an estimate of X as

$$\bar{x}_{12} = a\,x_1 + (1-a)x_2 \tag{10.53}$$

Equation 10.53 is similar to the Muskingum hypothesis used in flow routing. Then it can be shown that

$$x_{12} = \frac{\dfrac{x_1}{\sigma_1^2} + \dfrac{x_2}{\sigma_2^2}}{\dfrac{1}{\sigma_1^2} + \dfrac{1}{\sigma_2^2}} \tag{10.54}$$

$$\sigma_{12}^2 = [\frac{1}{\sigma_1^2 + \sigma_2^2}]^{-1} \tag{10.55}$$

Equations 10.54 and 10.55 can be generalized as

$$\bar{x} = \frac{\displaystyle\sum_{i=1}^{n}(\frac{x_i}{\sigma_i^2})}{\displaystyle\sum_{i=1}^{n}(\frac{1}{\sigma_i^2})} \tag{10.56}$$

and

$$\sigma_z = \left[\frac{1}{\displaystyle\sum_{i=1}^{n}\sigma_i^2}\right]^{-1/2} \tag{10.57}$$

Now consider the case when errors are correlated. Let $E = A/(A + B)$, where A and B are independent measurements, with their means and variances, respectively, denoted as $\bar{A}, \sigma_A^2$ and $\bar{B}, \sigma_B^2$. It can be shown that

$$\sigma_E^2 = \frac{(\bar{B})^2}{(\bar{A}+\bar{B})^2}\sigma_A^2 + \frac{(\bar{A})^2}{(\bar{A}+\bar{B})^2}\sigma_B^2 \qquad (10.58)$$

Equation 10.58 can also be derived by expressing $E = A/U$, where $U = A + B$, and then applying the Taylor series.

If $z = x + y$ or $z = x - y$, then it can be shown that

$$\sigma_z = \sqrt{\sigma_x^2 + \sigma_y^2 \pm 2\sigma_{xy}} \qquad (10.59)$$

Similarly, if $z = xy$, then

$$CV_z = \sqrt{CV_x^2 + CV_y^2 + 2\frac{\sigma_{xy}}{(\bar{x}\,\bar{y})}} \qquad (10.60)$$

If $z = x/y$, then

$$CV_z = \sqrt{CV_x^2 + CV_y^2 - 2\frac{\sigma_{xy}}{\bar{x}\,\bar{y}}} \qquad (10.61)$$

Equations 10.60 and 10.61 are similar, except for the sign of the covariance term.

If $z = x^m y^r$, then it can be shown that

$$CV_z = \sqrt{m^2 CV_x^2 + r^2 CV_y^2 + \frac{2mr\,\sigma_{xy}}{(\bar{x}\,\bar{y})}} \qquad (10.62)$$

Equations 10.61 and 10.62 contain covariance terms and should be calculated.

Example 10.12 Consider the case of independent and uncorrelated errors. The hydraulic radius R for a rectangular channel can be expressed as $R = bh/[b + 2h]$, where b = width and h = flow depth. Variables b and h can be considered independent. Derive the error in R as a function of errors in b and h. Assume that the means and standard deviations of b and h are known.

Solution Given that $R = bh/[b + 2h]$ and b and h are independent, we have

$$\partial R/\partial b = h/(b + 2h) - bh/(b + 2h)^2 = 2h^2/(b + 2h)^2$$

If the expression is evaluated at a point $p(b, \bar{h})$, then

$$(\partial R/\partial b)\,|_p = 2h^2/(b + 2h)^2$$

Similarly,

$$\partial R/\partial h = b/(b + 2h) - 2bh/(b + 2h)^2 = b^2/(b + 2h)^2$$

Hence,

$$(\partial R/\partial h)\,|_p = \bar{b}^2/(b + 2h)^2$$

Thus,

$$\sigma_R^2 = \left(\frac{\partial R}{\partial b}\bigg|_p\right)^2 \sigma_b^2 + \left(\frac{\partial R}{\partial h}\bigg|_p\right)^2 \sigma_h^2$$

or

$$\sigma_R = \sqrt{\left(\frac{2\bar{h}^2}{(\bar{b} + 2\bar{h})^2}\right)^2 \sigma_b^2 + \left(\frac{\bar{b}^2}{(\bar{b} + 2\bar{h})^2}\right)^2 \sigma_h^2}$$

Example 10.13 Consider the case described in Example 10.12. Let $R = A/B$, where $A = bh$ and $B = b + 2h$. Clearly, A and B are no longer independent. Derive the error in R as a function of errors in A and B. Compare this error in R with that derived in the previous example.

Solution Given that $R = A/B = bh/(b + 2h)$ and A and B are dependent, we have

$$\partial R/\partial A = 1/B, (\partial R/\partial A)\,|\,p = 1/B = 1/(b + 2h)$$

$$\partial R/\partial B = -A/B^2 (\partial R/\partial B)\,|\,p = -A/B^2 = -\bar{b}h/(b + 2h)^2$$

Since b and h are independent,

$$\sigma_A{}^2 = \bar{b}^2\sigma_h{}^2 + \bar{h}^2\sigma_b{}^2$$

and

$$\sigma_B{}^2 = \sigma_b{}^2 + 4\sigma_h{}^2$$

Thus,

$$\sigma_{AB}{}^2 = \overline{AB} - \bar{A}\bar{B}$$

$$= \overline{bh(b + 2h)} - \overline{bh}(\bar{b} + 2\bar{h})$$

$$= \overline{b^2h} + \overline{2bh^2} - \overline{b^2}\bar{h} - 2\bar{b}\,\overline{h^2}$$

$$= \overline{b^2h} + \overline{b^2}\bar{h} - \left(\overline{2bh^2} + 2\bar{b}\,\overline{h^2}\right)$$

$$= \sigma_b{}^2h + 2\bar{b}\sigma_h{}^2$$

Hence,

$$\sigma_R = \sqrt{\left(\left.\frac{\partial R}{\partial A}\right|_p\right)^2 \sigma_A^2 + \left(\left.\frac{\partial R}{\partial B}\right|_p\right)^2 \sigma_B^2 + 2\left(\left.\frac{\partial R}{\partial A}\right|_p\right)\left(\left.\frac{\partial R}{\partial B}\right|_p\right)\sigma_{AB}^2}$$

$$= \sqrt{\frac{\overline{b}^2\sigma_h^2 + \overline{h}^2\sigma_b^2}{(\overline{b}+2\overline{h})^2} + \frac{\overline{b}^2\overline{h}^2}{(\overline{b}+2\overline{h})^4}(\sigma_b^2 + 4\sigma_h^2) - \frac{2\overline{b}\overline{h}(\sigma_b^2\overline{h} + 2\overline{b}\sigma_h^2)}{(\overline{b}+2\overline{h})^3}}$$

$$= \frac{\sqrt{\left(4\overline{h}^4\sigma_b^2 + \overline{b}^4\sigma_h^2\right)}}{(\overline{b}+2\overline{h})^2}$$

Note that the expression for σ_R is the same for both cases. One reason is that, although A and B are considered dependent on each other, linear operations are involved in the computation of σ_R.

Example 10.14 Consider Manning's equation

$$V = \frac{1}{n}R^{2/3}S^{1/2}$$

where n = Manning's roughness factor, R = hydraulic radius, and S = slope. Express the error in V as a function of error in R, n, and S.

Solution Given Manning's equation, we have

$$\frac{\partial V}{\partial n} = -\frac{1}{n^2}R^{2/3}S^{1/2}$$

and

$$\left.\frac{\partial V}{\partial n}\right|_p = -\frac{1}{\overline{n}^2}\overline{R}^{2/3}\overline{S}^{1/2}$$

where $p = (n, \overline{R}, \overline{S})$.
    Similarly,

$$\frac{\partial V}{\partial R} = \frac{2}{3n}R^{-1/3}S^{1/2}$$

$$\left.\frac{\partial V}{\partial R}\right|_p = \frac{2}{3\overline{n}}\overline{R}^{-1/3}\overline{S}^{1/2}$$

and

$$\frac{\partial V}{\partial S} = \frac{1}{2n}R^{2/3}S^{-1/2}, \quad \left.\frac{\partial V}{\partial S}\right|_p = \frac{1}{2\overline{n}}\overline{R}^{2/3}\overline{S}^{-1/2}$$

and so

$$\sigma_V^2 = \left(\frac{\partial V}{\partial n}\bigg|_p\right)^2 \sigma_n^2 + \left(\frac{\partial V}{\partial R}\bigg|_p\right)^2 \sigma_R^2 + \left(\frac{\partial V}{\partial S}\bigg|_p\right)^2 \sigma_S^2$$

$$= \left(-\frac{1}{\overline{n}^2}\overline{R}^{2/3}\overline{S}^{1/2}\right)^2 \sigma_n^2 + \left(\frac{2}{3\overline{n}}\overline{R}^{-1/3}\overline{S}^{1/2}\right)^2 \sigma_R^2 + \left(\frac{1}{2\overline{n}}\overline{R}^{2/3}\overline{S}^{-1/2}\right)^2 \sigma_S^2$$

$$= \left(\frac{1}{\overline{n}}\overline{R}^{2/3}\overline{S}^{1/2}\right)^2 \left(\frac{\sigma_n^2}{\overline{n}^2} + \frac{4\sigma_R^2}{9\overline{R}^2} + \frac{\sigma_S^2}{4\overline{S}^2}\right)$$

$$\sigma_V = \left(\frac{1}{\overline{n}}\overline{R}^{2/3}\overline{S}^{1/2}\right)\sqrt{\frac{\sigma_n^2}{\overline{n}^2} + \frac{4\sigma_R^2}{9\overline{R}^2} + \frac{\sigma_S^2}{4\overline{S}^2}}$$

Example 10.15 Suppose the bottom width (x) and sides (y) of a rectangular channel have been measured independently. The measurements may be in error by ± 0.5 m. Thus, the bottom measurement is $x = (10 \pm 0.5)$ m and the side measurement is $y = (6 \pm 0.5)$ m. Find the best values of the area A and wetted perimeter P of the rectangular section and their standard deviations. Also, find the covariance if their correlation is 0.3.

Solution For a rectangular channel, we have

$$A = xy$$

$$P = x + 2y$$

Hence, the best values of these are

$$A = 10 \times 6 = 60 \text{ m}^2$$

$$P = 10 + (2 \times 6) = 22 \text{ m}$$

Now

$$\frac{\partial A}{\partial x} = y, \quad \frac{\partial A}{\partial x}\bigg|_{\overline{y}} = 6$$

$$\frac{\partial A}{\partial y} = x, \quad \frac{\partial A}{\partial y}\bigg|_{\overline{x}} = 10$$

and

$$\sigma_A^2 = \left(\frac{\partial A}{\partial x}\bigg|_{\overline{x}}\right)^2 \sigma_x^2 + \left(\frac{\partial A}{\partial y}\bigg|_{\overline{y}}\right)^2 \sigma_y^2$$

$$= 6^2 \times 0.5^2 + 10^2 \times 0.5^2 = 34$$

Hence, $\sigma_A = 5.83$ m^2.

Similarly,

$$\frac{\partial P}{\partial x} = 1, \quad \left.\frac{\partial P}{\partial x}\right|_{\bar{y}} = 1$$

$$\frac{\partial P}{\partial y} = 2, \quad \left.\frac{\partial P}{\partial y}\right|_{\bar{x}} = 2$$

$$\sigma_P^2 = \left(\left.\frac{\partial P}{\partial x}\right|_{\bar{x}}\right)^2 \sigma_x^2 + \left(\left.\frac{\partial P}{\partial y}\right|_{\bar{y}}\right)^2 \sigma_y^2$$
$$= 1 \times 0.5^2 + 2^2 \times 0.5^2 = 1.25$$

Hence, $\sigma_P = 1.128$ m and the covariance is

$$\sigma_{AP} = \rho\sigma_A \times \sigma_P = 0.3 \times 5.83 \times 1.128 = 1.955 \text{ m}^3$$

Example 10.16 The Universal Soil Loss Equation (USLE) is used to predict soil erosion. The USLE is given as $A = RKLSCP$, where A is the soil loss and $R, K, L, S, C,$ and P are input parameters. Assuming all the input variables are independent, estimate the uncertainty associated with the prediction of soil loss by erosion. Table E10-16a gives the mean and standard deviation of various parameters of the USLE.

Table E10-16a Mean and standard deviation of various parameters of the USLE.

Variable	Factor name	Mean value	Standard deviation
R	Rainfall Intensity	290 cm	72
K	Soil Erodibility	0.12	0.05
L	Slope Length Factor	1.15	0.05
S	Slope Gradient Factor	1.17	0.12
C	Cropping Practices	0.65	0.15
P	Erosion Control Practices	0.45	0.11

Solution Using Eq. 10.50, one can write

$$CV_A = \sqrt{CV_R^2 + CV_K^2 + CV_L^2 + CV_S^2 + CV_C^2 + CV_P^2}$$

The calculation is performed in Table E10-16b.

Table E10-16b *Calculation table for Example 10.16.*

Variable	Factor name	Mean value	Standard deviation	CV^2
R	Rainfall Intensity	290	72	0.06
K	Soil Erodibility	0.12	0.05	0.17
L	Slope Length Factor	1.15	0.05	0.00
S	Slope Gradient Factor	1.17	0.12	0.01
C	Cropping Practices	0.65	0.15	0.05
P	Erosion Control Practices	0.45	0.11	0.06

$$CV_A = \sqrt{CV_R^2 + CV_K^2 + CV_L^2 + CV_S^2 + CV_C^2 + CV_P^2} = 0.6$$

$$\overline{A} = \overline{RKLSCP} = 13.70 \; ; \; \sigma_A = \overline{A} \times CV_A = 13.70 \times 0.6 = 8.2$$

Example 10.17 By combining the rational method and Itensity-Duration-Frequency (IDF) curve for Kansas City, Missouri, the peak runoff Q is given as

$$Q = CA \frac{KF^x}{(T_c + b)^n}$$

in which $K = 1.74$, $x = 0.20$, $b = 0.20$, $n = 0.77$, and $F = 5$ are the fixed parameters, whereas the other parameters are uncertain with the characteristics given in Table E10-17.

Table E10-17 *Mean and coefficient of variation of parameters needed.*

Parameter	Mean	CV
C	0.45	0.33
T_c	0.37	0.62
A	12.00	0.10

Solution By substituting all the parameters that are constant, the peak runoff is given as

$$Q = CA \frac{KF^x}{(T_c + b)^n} = 2.41 CA (T_c + 0.2)^{-0.77}$$

Further, let us denote

$$Z = (T_c + 0.2)^{-0.77} = \beta^{-0.77}$$

where $\beta = T_c + 0.2$. Then

$$\text{mean}[\beta] = \text{mean}[T_c + 0.2] = \text{mean}[T_c] + 0.2 = 0.37 + 0.2 = 0.57$$

$$SD[\beta] = SD[T_c] = 0.37 \times 0.62 = 0.23$$

So,

$$CV_\beta = 0.23/0.57 = 0.41.$$

Further, from Eq. 9.52 we know that if $z = \beta^m$, then $CV_z = mCV_\beta$. Thus,

$$CV_z = (-0.77) \times 0.41 = -0.31$$

so that the peak runoff Q is given as

$$Q = 2.41CAZ$$

$$CV_Q = \sqrt{CV_C^2 + CV_A^2 + CV_Z^2} = \sqrt{0.33^2 + 0.10^2 + (-0.31)^2} = 0.46$$

This indicates that the peak runoff contains a significant amount of uncertainty. Substituting mean values of all parameters gives a mean peak runoff of 22.20 cfs.

10.4 Questions

10.1 Based on a sample of 20 years of data, the mean and standard deviation of the annual rainfall for Saint Tammany Parish, Louisiana, is 62.5 inches and 8.14 inches, respectively. The mean rainfall estimated by the arithmetic mean method is about 6.5% to 12.5% higher than the true mean. Estimate the overall random error in the estimated mean value.

10.2 Consider a linear reservoir expressed as $S = KQ$, where S = storage (in m^3), K = reservoir constant (in hours), and Q = discharge (in cfs). Further, the mean of K and Q are 20 hours and 50 cfs, respectively. Assume the coefficients of variation of K and Q to be 0.30 and 0.52, respectively.

    (a) Determine the mean storage and the uncertainty in the estimation of S assuming K and Q to be independent.

    (b) Determine the mean storage and the uncertainty in the estimation of S assuming K and Q to be correlated with a correlation coefficient of 0.52.

    (c) What is the magnitude of error involved if an engineer made an analysis by assuming K and Q as independent whereas the data show that both of these parameters were dependent?

10.3 The storage S_{t+1} in a reservoir at the end of month $t + 1$ is given as

$$S_{t+1} = S_t + I_t + R_t - O_t - E_t - L_t$$

where I_t is inflow, O_t is release, E_t is evaporation, R_t is rainfall, and L_t is seepage loss during month t. In this expression, S_t is known at the beginning and release O_t is to be made during the month t; thus neither S_t or O_t has any uncertainty. Variables I_t, E_t, R_t, and L_t are assumed to be uncertain random variables, with mean values of 21.3, 2.4, 3.2, and 1.2 million m³, respectively, and standard deviations of 4.5, 0.6, 0.75, and 0.45 million m³, respectively. If the initial storage in the reservoir is 30.0 million m³ and the target release is 18 million m³, find the expected storage at the end of the month and the amount of uncertainty associated with the storage.

10.4 In the previous example, both reservoir inflow and evaporation depend upon the climate and hence may be correlated. Assuming correlation coefficients of 0.10, 0.20, 0.30, 0.50, 0.60, and 0.80, determine the standard deviation of the expected storage at the end of the month.

10.5 Consider the rational method for computing peak discharge, $Q = CIA$, where Q = peak discharge in cfs, C = rational runoff coefficient (dimensionless), I = rainfall intensity in inches/hour, and A = drainage area in acres. The mean values of C, I, and A are 0.56, 1.25 inches/hour, and 92 acres, respectively. Assume the variables C, I, and A are independent and their CVs are 0.42, 0.28, and 0.07, respectively. Determine the mean and standard deviation of the peak discharge.

10.6 The outflow phosphorus load L_O from a lake is predicted from the following equation:

$$L_O = \frac{L_I}{(1 + KC_I T)}$$

where L_I = inflow phosphorus load (mg/m²), C_I = inflow phosphorus concentration (mg/m³), K = second-order phosphorus removal rate (1/day/mg/m³), and T = hydraulic residence time (days). For a lake, the mean values of L_I, K, C_I, and T are 1,200 mg/m² year, 0.0003 day⁻¹/mg/m³, 4,500 mg/m3, and 365 days and their coefficients of variation are 0.32, 0.23, 0.56, and 0.18, respectively. Determine the mean value of the outflow phosphorus load from the lake and its associated uncertainty.

10.7 The total maximum daily load (TMDL) represents the long-term average load consistent with a compliance rate of 50% and confidence level of 50%. To meet a specified lake target at the specified compliance rate (β) with a confidence level (α), the allocated long-term load (L_A) to a point source that discharges to a lake is determined as

$$L_A = (Q + KA)C \exp(-z_\beta S_V) \exp(-z_\alpha S_U)$$

where Q = long-term average lake outflow (m³/year), K = second-order phosphorus removal rate (1/day/mg/m³), C = lake phosphorus concentration (mg/m³), z_α and z_β = standard normal variate with upper probabilities α and β, S_U = model error CV for predicted lake phosphorus concentration, and S_V = year-to-year coefficient of variation of lake phosphorus concentration. Determine the expression of CV for the allocated long-term load L_A.

10.8   Mitchell (1948) suggested various equations for watershed lag time based on watershed characteristics such as drainage area (A), length (L), mean length (L_{ca}), and slope (S) for developing synthetic hydrographs in Illinois. These are as follows:

$$t = 3.85A^{0.35}L^{0.43}L_{ca}^{0.04}S^{-0.29}$$

$$t = 0.849A^{0.53}L^{0.26}L_{ca}^{-0.1}$$

$$t = 1.01A^{0.43}L^{0.12}L_{ca}^{0.20}$$

$$t = 1.05A^{0.6}$$

$$t = 1.17A^{0.59}$$

$$t = 0.537A^{0.70}$$

$$t = 6.64L_{ca}^{1.09}S^{-0.32}$$

$$t = 4.64A^{0.58}S^{-0.25}$$

Use the following data and determine the relationship that gives the lowest error in the lag time.

Parameter	A (square miles)	L (miles)	L_{ca} (miles)	S
Mean	296	25	13	2.0×10^{-3}
CV	0.17	0.12	0.12	0.32

10.9   For a given river reach, the Muskingum method for flow routing is represented as $S = K[0.25I + 0.75O]$, where S = storage rate, I = inflow rate, O = outflow rate, and K = storage constant. Assume the variables I and Q are independent. If mean values of K, I, and O are 22 hours, 80 cfs, and 65 cfs, respectively, with CV values of 0.30, 0.45, and 0.38, respectively, determine the CV and standard deviation of S.

10.10   Consider the case of independent and uncorrelated errors. The hydraulic radius R for a trapezoidal channel section is expressed as

$$R = \frac{y(b + my)}{b + 2y\sqrt{1 + m^2}}$$

where b = width, y = flow depth, and m is the side slope. Variables b, y, and m can be considered independent. Derive the error in R as a function of the errors in b, y, and m. Assume that the means and standard deviations of b, y, and m are known.

10.11 Consider Question 10.10 with the following data:

Parameter	b (ft)	y (ft)	m
Mean	50	10	2.5
CV	0.15	0.20	0.11

Determine the error in R if (a) all parameters are independent, (b) b and y are positively correlated with a correlation coefficient of 0.60, and (c) b and y are negatively correlated with a correlation coefficient –0.60.

10.12 Consider Manning's equation,

$$Q = \frac{1}{n} \frac{A^{5/3}}{P^{2/3}} S^{1/2}$$

where n = Manning's roughness factor, A = cross-sectional area, P = perimeter, and S = slope. Express the error in Q as a function of error in A, P, n, and S.

10.13 Based on the Soil Conservation Service method, the basin lag time is given as

$$t_L = \frac{L^{0.8} \left[(1000/CN) - 9 \right]^{0.7}}{1900 S^{0.5}}$$

where L = length along stream to basin divide in miles, CN = curve number, and S = % watershed slope. Assume for a given watershed that $L = 25$ miles, $CN = 75$, and $S = 0.2\%$. Determine the CV of watershed lag time.

10.14 Based on the Modified Rational Method (ISWM Design Manual for Development/Redevelopment 2004), the critical duration of the design storm is given as

$$T_d = \sqrt{\frac{2CAab}{Q_0}} - b$$

where T_d = critical storm duration (minutes), Q_0 = allowable release rate, C = developed condition Rational Method runoff coefficient, A = area (acres), and a and b are intensity-duration-frequency factors depending on the location and return period. The required storage volume for a detention basin is given as

$$V = 60\left[CAa - (2CabAQ_0)^{\frac{1}{2}} + \frac{Q_0(b-T_C)}{2}\right]\frac{P_{180}}{P_{T_d}}$$

where P_{180} = 3-hour (180-minute) storm depth, T_c = time of concentration, and P_{Td} = storm depth for the critical period.

Assuming that all parameters are independent, derive expressions for the coefficient of variation for T_d and V. Further, consider urban development near you and assume appropriate values for a small watershed and determine the mean and standard deviation of the required storage volume for the detention pond.

10.15 Based on Darcy's law, the flow rate through an aquifer is given as $Q = KIA$, where K = hydraulic conductivity (m/day), I = hydraulic gradient, and A = cross-sectional area (m²). Assume mean values of the various parameters as $I = 0.004$, $K = 50$ m/day, and $A = 1.0$ m. Further, assume CV values for I, K, and A of 0.11, 0.33, and 0.01, respectively. Determine the CV of Q.

10.16 Assuming complete and instantaneous mixing, the BOD L_0 of the mixture of streamwater and wastewater at the point of discharge is given as

$$L_0 = \frac{Q_w L_w + Q_r L_r}{Q_w + Q_r}$$

where Q_w = wastewater flow, Q_r = streamwater flow rate, L_w = ultimate BOD of wastewater, and L_r = ultimate BOD of streamwater. The following data about the river and wastewater are available:

Parameter	Q_w (m³/s)	Q_r (m³/s)	L_w (mg/L)	L_r (mg/L)
Mean	1.1	8.7	50	6.0
CV	0.20	0.42	0.12	0.34

Determine the mean and standard deviation of the flow and BOD at the point of discharge.

10.17 The BOD remaining at a point downstream of the point of discharge is given as $L(x) = L_0 \exp(-kx/v)$, where k = deoxygenation rate and v = stream velocity. Consider the data of Question 10.16 along with the mean values of k and v as 0.2 per day and 0.30 m/s, respectively, and their respective CV values of 0.15 and 0.30. Determine the mean and CV at a point 30,000 m downstream from the discharge.

10.18 A watershed has a network of snow-measuring devices for measurement of evaporation. It is ascertained that if a snow-measuring device measures within 5% to 10% of the true value, then it is said to be accurate and precise. Based on an analysis of the snow observations, it has been determined that one of the measuring devices, called A, always measures snowfall about 10% to 25% away from the true value. There is another measuring device, called B, that is found to measure snowfall within 5% away from the true value. Another measuring device, called C, is found to measure snowfall somewhat unpredictably (i.e., sometimes 10% away from the true value, sometimes 30% away from the true value, and sometimes very close to the true value). Another measuring device, called D, measures snowfall sometimes 5% higher, sometimes 5% lower, sometimes 2% higher, and sometimes 2% lower, but always within 5% away from the true value. What can be said about the measurements of these snow-measuring devices?

10.19 Consider the mean annual rainfall for Bowling Green, Kentucky, which is given as 50 inches based on a sample of data. The mean rainfall estimated by the arithmetic mean method is about 105% to 20% higher than the true mean. Taking the sample standard deviation of 10 inches and the number of observations in the sample as 30, compute the total random error in the estimated mean value.

10.20 Consider the relation for open channels as $Q = AV$, where A = cross-sectional area, V = cross-sectional average velocity, and Q = discharge (volume/time). Assume A and V are independent and their errors are independent and uncorrelated. Determine the CV of Q.

10.21 The Blaney–Criddle method for computing monthly evapotranspiration (ET) is $ET = KF$, where K is an empirically derived seasonal consumptive coefficient applicable to a particular crop and F is the sum of monthly consumptive use factors. Assuming that K and F are independent and uncorrelated, derive the CV of ET.

10.22 The energy balance for purposes of computing evaporation can be expressed as

$$R_n + E + G + H = 0$$

where R is the net radiation, E is evaporation, G is sensible heat flux from the bottom, and H is the sensible heat flux from air. Derive the relative error (standard deviation divided by the mean) of E.

10.23 Consider Darcy's equation for computing groundwater discharge: $Q = -KIA$, where Q = discharge in m^3/s, K = hydraulic conductivity in m/s, I = hydraulic gradient (dimensionless), and A = cross-sectional area of the aquifer in m^2. Assume that the variables K, I, and A are independent and that the errors in them are independent and uncorrelated.

Express the relative error (standard deviation divided by mean) in the discharge as a function of errors in K, I, and A.

10.24 The cross-sectional area of a channel at a given location is measured by summing the cross-sectional areas of different segments. Assume the channel is of compound form, which can be broken down into three rectangular portions. Let the total area (A) be expressed as $A = x + y$, where x is the cross-sectional area of the deeper rectangular part and y is the cross-sectional area of the two shallower rectangular parts, one on each side of the deeper part. Express the error in A as a function of errors in x and y. Then, consider the following data: x (m^2) = 200 ± 5 and y (m^2) = 50 ± 1.5. Compute the error in the area.

10.25 From the Thorthwaite method, monthly evapotranspiration (ET) can be computed as

$$ET = cT^a$$

where T is the mean monthly temperature, c is a coefficient, and a is an exponent. Derive the error in ET as function of the error in T.

10.26 Consider the case of a rectangular channel whose wetted perimeter can be $WP = b + 2h$, where b = width and h = flow depth. Variables b and h can be considered independent. Derive the error in WP as a function of the errors in b and h. Assume that the means and standard deviations of b and h are known.

10.27 Consider Chezy's equation, $V = CR^{1/2}S^{1/2}$, where C = Chezy's roughness factor, R = hydraulic radius, and S = slope. Express the error in V as a function of the errors in R, C, and S.

10.28 The rating curve for a river is usually defined as $Q = aA^b$, where Q is discharge (m^3/s), A is cross-sectional area (m^2), and a and b are constants. Derive the error in Q as a function of the error in A.

10.29 The base flow from a basin at any time t can be expressed as

$$Q_t = Q_0 K^t$$

where Q_t is the base flow at time t, Q_0 is the initial flow, and K is the recession constant (between 0 and 1, usually 0.85 to 0.99). Derive the error in Q_t as a function of the errors in Q_0 and K.

10.30 The time of concentration for a small watershed can be expressed as

$$T_c = CL_p^a S_p^b$$

where C is a coefficient, L_p is a length measure, and S_p is a slope measure. Compute the error as a function of the errors in L_p and S_p.

10.31 For a convex method of flow routing in a river, the outflow at the end of time interval $t + 1$, Q_{t+1}, can be computed using the continuity equation

$$Q_{t+1} = aQ_t + bI_t$$

where I_t is inflow at the beginning of the time interval t, Q_t is the outflow from the channel at the beginning of the time interval t, and a and b are constants. Variables I_t and Q_t are assumed to be uncertain random variables. Compute the error in Q_{t+1} as a function of the errors in I_t and Q_t.

Chapter 11

Monte Carlo Simulation

Simulation is the process of duplicating the behavior of an existing or proposed system. It consists of designing a model of the system and conducting experiments with this model either for better understanding of the functioning of the system or for evaluating various strategies for its management. The essence of simulation is to reproduce the behavior of the system in every important aspect to learn how the system will respond to conditions that may be imposed on it or that may occur in the future. Note that the model has to correctly reproduce those aspects of the system's response that are of interest. Also, the model should produce correct results for correct reasons.

Many problems require determination of the properties of the output of a system given the input and transfer function. When this transfer function is simple, the properties of the output can be obtained analytically. But when the transfer function is complex, the derivation of the properties of output may be difficult. For such systems, a possible way out is to prepare a model of the system, repeatedly subject it to input, observe the output, and analyze the output to infer its properties.

The main advantage of simulation models lies in their ability to closely describe reality. If a simulation model can be developed and is shown to represent the prototype system realistically, it can provide insight about how the real system might perform over time under varying conditions. Thus, proposed configurations of projects can be evaluated to judge whether their performance would be adequate or not before investments are made. In a similar manner, operating policies can be tested before they are implemented in actual control situations. Simulation is widely believed to be the most powerful tool to study complex systems.

These days most simulation experiments are conducted numerically using a computer. A mathematical model of the system is prepared and repeated runs of this model are taken. The results are available in the form of tables, graphs, or performance indices; these are analyzed to draw inferences about the adequacy of the model.

11.1 Basics of Monte Carlo Simulation

Design of real-world systems is generally based on observed historical data. For example, the observed streamflow data are used in sizing a reservoir, historical traffic data are used in design of highways, observed data are used in design of customer services, etc. However, it is frequently found that the historical records are not long enough and the observed pattern of data is not likely to repeat exactly. Besides, the performance of a system critically depends on the extreme values of input variables and the historical data may not contain the entire range of input variables. An important implication of nonavailability of data of adequate length is that one may not get a complete picture of the system performance and risks involved when historical data are used in analysis. Thus, for instance, the planner cannot determine the risks of a water supply system failing to meet the demand during its economic life because this requires a very large sample of data, which are not commonly available.

For many systems, some or all inputs are random, system parameters are random, initial conditions may be random, and boundary condition(s) may also be random in nature. The probabilistic properties of these are known. For analysis of such systems, simulation experiments may be conducted with a set of inputs that are synthetically (artificially) generated. The inputs are generated so as to preserve the statistical properties of the random variables. Each simulation experiment with a particular set of inputs gives an answer. When many such experiments are conducted with different sets of inputs, a set of answers is obtained. These answers are statistically analyzed to understand or predict the behavior of the system. This approach is known as Monte Carlo simulation (MCS). Thus, it is a technique to obtain statistical properties of the output of a system given the properties of inputs and the system. By using it, planners get better insight into the working of the system and can determine the risk of failure (e.g., chances of a reservoir running dry, pollution in a river basin exceeding the prescribed limits, or a customer service center failing to provide services within the promised time). Sometimes, MCS is defined as any simulation that involves the use of random numbers.

In Monte Carlo simulation, the inputs to the system are transformed into outputs by means of a mathematical model of the system. This model is developed such that the important features of the system are represented in sufficient detail. The main steps in Monte Carlo simulation are assembling inputs, preparing a model of the system, conducting experiments using the inputs and the model, and analyzing the output. Sometimes, a parameter of the system is systematically changed and the output is monitored in the changed circumstances to determine how sensitive it is to the changes in the properties of the system.

The main advantages of Monte Carlo simulation are that it permits detailed description of the system, its inputs, outputs, and parameters. All the critical parameters of the system can be included in its description. The other advantages include savings in time and expenses. It is important to remember that the synthetically generated data are no substitute for the observed data but this is a useful pragmatic tool that allows the analyst to extract detailed information from the available data. However, when using Monte Carlo simulation in practical risk and reliability analyses, a large amount of computation may be needed for generating random variables and these variables may be correlated. Because, these days, computing power is usually not a limitation, this consideration is not a serious limitation.

The generation of random numbers forms an important part of Monte Carlo simulation. In the early days, roulette wheels similar to those in use at Monte Carlo were used to generate random numbers, giving rise to the name of the technique. During the initial days of mathematical simulation, mechanical means were employed to generate random numbers. The techniques that were used to generate random numbers were drawing cards from a pack, drawing numbered balls from a vessel, reading numbers from a telephone directory, etc. Printed tables of random numbers were also in use for quite some time. The current approach is to use a computer-based routine to generate random numbers. This approach is discussed next.

11.2 Generation of Random Numbers

A number of arithmetic techniques are available for generating random numbers (e.g., the midsquare method, the congruence method, and composite generators). To illustrate the midsquare method, a four-digit integer (say 9603) is squared to obtain an integer of eight digits (in this case, 92217609). If necessary, zeros are appended to the left to make an eight-digit number. The middle four digits are picked as the next four-digit number and the random number is obtained by putting a decimal before these digits. Thus the first random number is 0.2176. Now, the number 2176 is squared to continue the process. The numbers are sequential because each new number is a function of its predecessor(s). Different procedures use different types of recursive equations to generate numbers.

Nearly all modern compilers have built-in routines to generate uniformly distributed random numbers between 0 and 1. The most popular random number generation method is the *congruence method*. To start the process, a number known as the "seed" is input to the equation, which gives a random number. This number is again input to the equation to generate another number and so on. When this process is repeated n times, n random numbers are obtained. The recursive equation commonly used to generate random numbers in the *linear congruential generator* (LCG) is

$$R_i = (aR_{i-1} + b) \text{ (modulo } d) \tag{11.1}$$

where R_i are integer variables and a, b, and d are positive integer constants that depend upon the properties of the computer. The word "modulo" denotes that the variable to the left of this word is divided by the variable to the right (in this case d) and the remainder is assigned the value R_i. The desired uniformly distributed random number is obtained as R_i/d. The initial value of the variable (R_0) in Eq. 11.1 is called the seed. The properties of the generated numbers depend on the values of constants a, b, and d, their relationships, and the computer used. The value of constant a needs to be sufficiently high; low values may not yield good results. Constants b and d should not have any common factors. The positive integers R_0, a, b, and d are chosen such that $d > 0$, $a < d$, and $b < d$.

In computer generation, the sequence of random numbers is repeated after a certain lag and it is desired that the length of this cycle should be as long as possible. This lag increases as d increases and therefore a large value of d should be chosen. Normally, d is set equal to the word length (the number of bits retained as a unit) of the computer; a typical value is $2^{31} - 1$. It is important to ensure that the length of the cycle is more than the numbers that are needed for the study.

Example 11.1 Generate 10 uniformly distributed random numbers using Eq. 11.1 with $a = 5$, $b = 3$, and $d = 7$. The seed R_0 can be assigned a value of 2.

Solution Equation 11.1 is rewritten as

$$R_i = (5 \times R_{i-1} + 3) \text{ (modulo } 7)$$

The results are given in Table E11-1.

It may be noted here that the numbers repeat after a cycle of 7. The length of this cycle is termed as the period of the generator, which is d here.

A drawback of using Eq. 11.1 is that it involves division, which requires more computer time than addition or subtraction. This can be avoided by making use of *integer overflow*. Integer overflow takes place when an attempt is made to store an integer that is larger than its word size. If the number of bits in a word of the computer is b then the largest integer that it can store is $2^b - 1$ (one bit is for sign). If a larger integer having y digits is given, only b digits are stored and the leftmost $y - b$ digits are lost. But how the integer overflow is implemented on a computer depends on its architecture and software. Improved versions of LCGs,

Table E11-1 Generation of uniformly distributed random numbers.

i	R_{i-1}	$(5 \times R_{i-1} + 3)$ (modulo 7)	U_i
1	2	$(5 \times 2 + 3) \bmod 7 = 13 \bmod 7$	0.857
2	6	$(5 \times 6 + 3) \bmod 7 = 33 \bmod 7$	0.714
3	5	$(5 \times 5 + 3) \bmod 7 = 28 \bmod 7$	0.000
4	0	$(5 \times 0 + 3) \bmod 7 = 3 \bmod 7$	0.429
5	3	$(5 \times 3 + 3) \bmod 7 = 18 \bmod 7$	0.571
6	4	$(5 \times 4 + 3) \bmod 7 = 23 \bmod 7$	0.286
7	2	$(5 \times 2 + 3) \bmod 7 = 13 \bmod 7$	0.857
8	6	$(5 \times 6 + 3) \bmod 7 = 33 \bmod 7$	0.714
9	5	$(5 \times 5 + 3) \bmod 7 = 28 \bmod 7$	0.000
10	0	$(5 \times 0 + 3) \bmod 7 = 3 \bmod 7$	0.429

known as prime modulus multiplicative LCGs, are widely used these days to generate random numbers.

Since these algorithms of random number generation are deterministic, the generated numbers can be duplicated again. Therefore, these numbers are not random in a strict sense and are called *pseudo random numbers*. A good random number generator should produce uniformly distributed numbers that do not have any correlation with each other, should be fast, should not require a large memory, and should be able to exactly reproduce a given stream of random numbers. It is also required that the algorithm be capable of generating several separate streams of random numbers.

After generation, the random numbers should be tested to ensure that they possess the desired statistical properties (i.e., the numbers are not serially correlated). The chi-square test is one such test that can be used to confirm that the numbers are uniformly distributed. Law and Kelton (1991) have discussed random number generation and tests in greater detail.

11.2.1 Transformation of Random Numbers

In the previous section, we have discussed a method to generate uniformly distributed random numbers. The input to the prototype system will have certain statistical properties and a certain probability distribution. This distribution for each input variable can be obtained from the analysis of historical data and the underlying physical process. The input random variables in Monte Carlo simulation should exhibit the same statistical properties and follow the same probability distribution. Therefore, the uniformly distributed random numbers must be converted to follow the desired probability distribution. The variables involved may either be continuous or discrete random variables.

11.3 Continuous Random Variates

Methods to generate random variates that follow a given distribution are described next. If the inverse form of a distribution can be easily expressed analytically, the inverse transformation is the simplest method. If an analytical expression for the inverse of the concerned distribution is not known, special algorithms are employed to efficiently generate numbers with such a distribution.

11.3.1 Inverse Transformation Method

In the inverse transformation method, first a uniformly distributed random number r_i in the range [0, 1] is generated. Now, let $F_Q(q)$ be the desired cumulative distribution function (CDF) of random variable Q, $0 \leq F_Q(q) \leq 1$. Therefore $F_Q^{-1}(r)$ can be defined for any value of r between 0 and 1. Note that $F_Q^{-1}(r)$ is the smallest q satisfying $F_Q(q) \geq r$. Then Q can be generated as

$$Q = F_Q^{-1}[r] \tag{11.2}$$

where F_Q^{-1} is the inverse of the cumulative distribution function of random variable Q. This method is known as the inverse transformation or inverse CDF method. The method is graphically illustrated in Fig. 11-1. It is useful when the inverse of the CDF of the random variable can be expressed analytically. This is a simple and computationally efficient method. However, it can be used for only those distributions whose inverse form can be easily expressed.

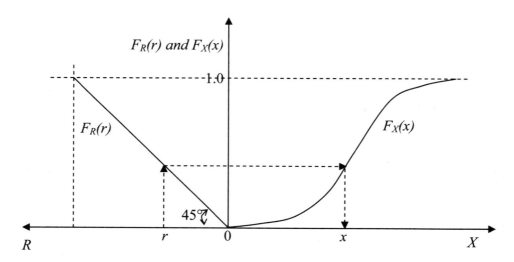

Figure 11-1 *Determination of random number x with desired distribution from uniformly distributed random number r.*

The inverse transformation method carries an intuitive appeal. In Fig. 11-2, the probability distribution function of X, $F(x)$, is plotted. The probability density function is nothing but the slope of this distribution curve. When the variates are generated, there should be relatively more variates in the zone where the peak of the density function lies or where the slope of the distribution function is more. The uniformly distributed random variables will be evenly spaced on the y axis (i.e., an interval $\Delta F(x)$ will have about the same number of variables irrespective of its location). Consequently, the corresponding intervals on the x axis, Δx_1 and Δx_2, will also have about the same number of variables. But Δx_1 is smaller than Δx_2. Therefore, the density of variates in Δx_1 will be higher than that in Δx_2. The following example demonstrates the use of this method to generate exponentially distributed random numbers.

Example 11.2 Generate random variates that follow an exponential distribution with parameter $\lambda = 2.3$.

Solution The cumulative distribution function of an exponential distribution is

$$F_X(x) = 1 - e^{-\lambda x}$$

Its inverse can be written as

$$x = F_X^{-1}[r] = -\ln(1-r)/\lambda \tag{11.3}$$

Since $(1-r)$ is uniformly distributed, this can be replaced by r, which is also uniformly distributed. Hence, exponentially distributed random variates with the desired property can be generated by

$$x = -\ln(r)/2.3$$

If the first uniformly distributed random number $r_1 = 0.89$, the corresponding variate x_1 will be

$$x_1 = -\ln(0.89)/2.3 = 0.05067$$

11.3.2 Composition Method

If a random variable has a composite probability distribution function, this property can be used to generate random variates by following the composition method. Let the probability of the variable $F_X(x)$ be

$$F_X(x) = \sum_{i=1}^{m} w_i F_X^i(x) \tag{11.4}$$

where w_i are the weights and $F_X^i(x)$ are the cumulative distribution functions. The weights should sum to unity. The requisite random number is generated in two stages. First, two uniformly distributed random numbers (u_1 and u_2) in the range [0, 1] are generated. The first number u_1 is used to select the appropriate CDF $F_x^i(x)$ for generation of the random number. The second number u_2 is used to determine the random variate according to the selected distribution.

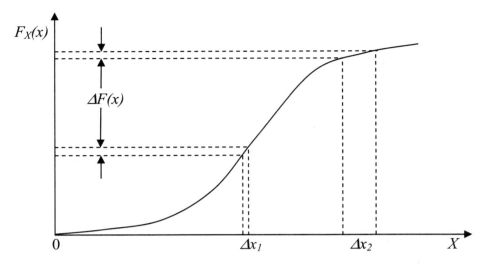

Figure 11-2 *Shape of the probability distribution function and density of generated variates.*

Example 11.3 Generate random variates that follow the following probability density function:

$$f_X(x) = 3/5 + x^3, 0 \le x \le 1 \tag{11.5}$$

Solution The given density function can be decomposed as

$$f_X(x) = 3/5 f_1(x) + 2/5 f_2(x)$$

where

$$f_1(x) = 1 \text{ and } F_1(x) = x, 0 \le x \le 1$$

$$f_2(x) = (5/2)x^3 \text{ and } F_2(x) = (5/8)x^4, 0 \le x \le 1$$

For the first function, $F_1(x) = x = u_2$ or variate $x = F_1^{-1}(u_2) = u_2$. For the second function, $F_2(x) = (5/8)x^4 = u_2$ or variate $x = F_2^{-1}(u_2) = (8u_2/5)^{0.25}$.

Note that here weights are 3/5 and 2/5 and they sum to unity. Now two uniformly distributed numbers are generated. Let these numbers be 0.538 and 0.181. Since $u_1 = 0.538$ is less than 3/5, u_2 is used to generate the variate by following $F_1(x)$. Hence,

$$\text{random variate } x_1 = F_1^{-1}(u_2) = 0.181$$

In the second attempt, let the uniformly distributed numbers be 0.722 and 0.361. Since u_1 is greater than 3/5, u_2 is used to generate the variate by following $F_2(x)$. Hence,

$$\text{random variate } x_2 = F_2^{-1}(u_2) = (8 \times 0.361/5)^{0.25} = 0.87$$

11.3.3 Function-Based Method

For some distributions, random variables can be expressed as functions of other random variables that can be easily generated. This property can be exploited for generation of random variables. For instance, the gamma distribution is nothing but the sum of exponential distributions. Thus, if a variable Z follows a gamma distribution with parameters (k, λ), then one can write

$$Z_i = X_{1i} + X_{2i} + X_{3i} + \ldots + X_{ki} \tag{11.6}$$

where $X_{1i}, X_{2i}, \ldots + X_{ki}$ are k independent exponentially distributed random variables with parameter λ. The procedure to generate a gamma-distributed random variate by following this method consists of generating random variates that follow exponential distributions having parameter λ. Now, k such variates are added to obtain one gamma-distributed variate.

Example 11.4 Generate gamma-distributed random variates with parameters $(4, 0.8)$ using the function-based method.

Solution The inverse transform method can be used to generate exponentially distributed random variates. We first generate a uniformly distributed random number u in the range $[0, 1]$. Using it, a random variate that follows the exponential distribution with parameter λ can be obtained by

$$x = -\ln(u)/\lambda \tag{11.7}$$

Let the first four uniformly distributed random numbers $u_1, \ldots, u_4$ be 0.5371, 0.1814, 0.6193, and 0.1319. The corresponding exponentially distributed random variates with parameter 0.8 (see Example 11.2) will be 0.777, 2.134, 0.599, and 2.532. Therefore, the gamma-distributed variate with parameter $(4, 0.8)$ will be

$$g = \frac{1}{\lambda} \sum_{i=1}^{4} \ln u_i = 0.777 + 2.134 + 0.599 + 2.532 = 6.042 \tag{11.8}$$

Recall that the sum of two independent gamma-distributed variates with parameters (λ, k_1) and (λ, k_2) is a gamma-distributed variate with parameters (λ, k_1+k_2).

Example 11.5 Generate normally distributed random numbers with parameters $(3.9, 1.6)$ using the function-based method.

Solution If u_1 and u_2 are two independent uniformly distributed random numbers in the range $[0, 1]$ then a pair of independent normally distributed random variates (m_x, σ_x^2) can be obtained by

$$x_1 = m_x + \sigma_x \sqrt{-2\ln u_1} \times \cos(2\pi u_2) \tag{11.9}$$

$$x_2 = m_x + \sigma_x \sqrt{-2\ln u_1} \times \sin(2\pi u_2) \tag{11.10}$$

This method was developed by Box and Muller (1958).

Let two uniformly distributed random numbers be 0.3465 and 0.8552. Hence,

$$x_1 = 3.9 + 1.6 \sqrt{-2 \times \ln(0.3465)} \times \cos 2\pi \times 0.8552 = 5.325$$

$$x_2 = 3.9 + 1.6 \sqrt{-2 \times \ln(0.3465)} \times \sin 2\pi \times 0.8552 = 2.057$$

Further, one can also generate log-normally distributed random variates by the transformation

$$x_1 = e^{5.325} = 205.4084$$

$$x_2 = e^{2.057} = 7.822$$

The mean of these log-normally distributed variates will be $m_{x(ln)} = \exp(m_x + \sigma_x^2/2)$ and the variance will be $\sigma_{x(ln)}^2 = \exp(2m_x + \sigma_x^2)[\exp(\sigma_x^2) - 1]$. If the aim is to generate log-normally distributed random variates with mean $m_{x(ln)}$ and variance $\sigma_{x(ln)}^2$ then these can be obtained as follows. First, generate normally distributed random variates with mean and variance as

$$m_x = \ln(m_{x(ln)}^2 / \sqrt{\sigma_{x(ln)}^2 + m_{x(ln)}^2}) \tag{11.11}$$

$$\sigma_x^2 = \ln[(m_{x(ln)}^2 + \sigma_{x(ln)}^2)/m_{x(ln)}^2] \tag{11.12}$$

and then transform them to the log-normal domain.

A common method for generating normally distributed random variates based on the central limit theorem is to use

$$x = \sum_{i=1}^{n} u_i - \frac{n}{2}$$

where u is a uniformly distributed random variable in [0,1]; commonly $n = 12$ is adopted. Of course, the method gives approximate numbers and hence its use is discouraged. Another simple method to generate normally distributed random variates with mean zero and standard deviation unity makes use of the approximation of the lambda distribution:

$$x = 4.91[u^{0.14} - (1 - u)^{0.14}] \tag{11.13}$$

The accuracy of this approximation is 0.0032 for $|x| < 2$ and 0.0038 for $2 < |x| < 3$, and the probability of x being outside $|x| > 4.91$ is less than 10^{-6} (Salas 1993).

Another approximation (Salas 1993) for the normal distribution based on polynomial equations is

$$x(u) = t + \frac{p_0 + p_1 t + p_2 t^2 + p_3 t^3 + p_4 t^4}{q_0 + q_1 t + q_2 t^2 + q_3 t^3 + q_4 t^4}, \quad 0.5 \le u < 1 \tag{11.14}$$

$$x(u) = -x(1 - u), \quad\quad\quad\quad 0 < u < 0.5$$

where $t = [-2 \ln (1 - u)]^{0.5}$, and the coefficients are $p_0 = -0.322232431088$, $p_1 = -1$, $p_2 = -0.342242088547$, $p_3 = -0.0204231210245$, $p_4 = -0.0000453642210148$, $q_0 = 0.099348462606$, $q_1 = 0.588581570495$, $q_2 = 0.531103462366$, $q_3 = 0.103537752285$, and $q_4 = 0.0038560700634$. Equation 11.14 can also be used to generate normally distributed numbers from the us.

11.4 Discrete Random Variates

The problem of generating discrete random variates is to transform a uniformly distributed random number to the desired discrete mass function. If u is the uniformly distributed random number then one way to obtain the corresponding random variate x_j that follows the desired CDF is

$$F_X(x_{j-1}) < u \le F_X(x_j) \tag{11.15}$$

It can be seen that this process is basically an inverse transformation method but it additionally requires a numerical search to arrive at the desired variate. However, this method may not be the most efficient approach to generate a discrete random variate.

Example 11.6 Generate random variates that follow a binomial distribution with parameters $(5.0, 0.4)$.

Solution The CDF of the binomial distribution is given by

$$F_X(j) = \sum_{i=0}^{j} \binom{5}{i}(0.4)^i (0.6)^{5-i}, \quad j = 0, 1, ..., 5 \tag{11.16}$$

The CDF is tabulated as follows.

i	0	1	2	3	4	5
$F_X(i)$	0.0778	0.337	0.6826	0.913	0.9898	1.000

Now a uniformly distributed random number is generated. Let it be $u = 0.5$. Since $F_X(1) < u \le F_X(2)$, the requisite variate $x = 2$. Under certain circumstances, the binomial distribution can be approximated by the normal distribution and this property can be used in data generation (see Chapter 5).

Example 11.7 Generate random variates that follow a Poisson distribution with parameter $\lambda = 4$.

Solution The CDF of the Poisson distribution is given by

$$F_X(x) = \sum_{i=0}^{x} \frac{\lambda^i e^{-\lambda}}{i!}$$ (11.17)

For $i = 0$,

$$F_X(x) = 4^0 \times 2.71828^{-4}/0! = 1 \times 0.0183/1 = 0.0183$$

The CDF is tabulated as follows:

i	0	1	2	3	4	5
$F_X(i)$	0.018	0.091	0.237	0.432	0.627	0.783

Now a uniformly distributed random number is generated. Let it be $u = 0.6$. Since $F_X(3) < u \le F_X(4)$, the requisite number $x = 3$. For the Poisson distribution, when parameter λ is large ($\lambda > 10$), the distribution can be approximated by the normal distribution, $N(\lambda - 0.5, \sqrt{\lambda})$ (Rubinstein 1981). This property can be utilized to generate Poisson-distributed random numbers when λ is large.

A general method that can be used to generate any discrete random variate having a finite range of values is the alias method. Although somewhat complex, this is a general and efficient method. The method has been described by Kronmal and Peterson (1979).

11.5 Jointly Distributed Random Variates

Consider a set of n random variables $X_1, X_2, \ldots, X_n$. These variables may be independent or dependent. If the random variables are independent, the joint probability distribution can be obtained by multiplying their marginal probability distribution functions:

$$f_{X_1, X_2, \ldots, X_n}(x_1, \ldots, x_n) = \prod_{i=1}^{n} f_{X_i}(x_i)$$ (11.18)

Here $f_{X_i}(x_i)$ is the marginal probability distribution function of X_i.

If the random variables are dependent, then their joint probability distribution function becomes

$$f_{X_1, X_2, \ldots, X_n}(x_1, \ldots, x_n) = f_{X_1}(x_1) f_{X_2}(x_2 \mid x_1) \ldots f_{X_n}(x_n \mid x_1, \ldots, x_{n-1})$$ (11.19)

where $f_{X_i}(x_i \mid x_1, \ldots, x_{i-1})$ is the conditional PDF of X_i, given $X_1 = x_1, X_2 = x_2, \ldots, X_{i-1} = x_{i-1}$. The joint cumulative distribution is

$$F_{X_1, X_2, \ldots, X_n}(x_1, \ldots, x_n) = F_{X_1}(x_1) F_{X_2}(x_2 \mid x_1) \ldots F_{X_n}(x_n \mid x_1, \ldots, x_{n-1})$$ (11.20)

Equation 11.20 provides a basis to generate jointly distributed random variables. Assume that uniformly distributed random numbers have been generated. Now, a value of variable X_1 may be generated by the inverse transformation using Eq. 11.2. With this value of X_1 and the inverse form of the conditional cumulative distribution function $F_{X2}(x_2 | x_1)$, the value of random variable X_2 can be obtained from

$$x_2 = F_{X_2}^{-1}(u_2 | x_1) \tag{11.21}$$

Proceeding in this way, one can obtain the values of X_n as

$$x_n = F_{X_n}^{-1}(u_n | x_1, \dots x_{n-1}) \tag{11.22}$$

Example 11.8 Generate numbers that follow a bivariate normal distribution with parameters (3.9, 1.6) and (4.5, 2.2). The coefficient of correlation (ρ) between the numbers is 0.65.

Solution Let the means of the two distributions be given by m_x and m_y and standard deviations by σ_x and σ_y. Their joint probability density function can be written as

$$f_{X,Y}(x,y) = f_{Y|X}(y | x) f_X(x) \tag{11.23}$$

We first generate a normally distributed number x. For this purpose, a uniformly distributed number u is generated. Let this be 0.791. Treating this as the probability, we can read the corresponding value of the standard normal variate (z) from the table of normal distribution and it turns out to be 0.81. Hence, the corresponding normally distributed number is

$$x = m_x + z \times \sigma_x = 3.9 + 0.81 \times 1.6 = 5.196$$

Knowing x, we can compute the conditional mean of the second number Y by

$$E(Y|x) = m_Y + \rho \times (\sigma_Y/\sigma_X) \times (x - m_x) \tag{11.24}$$

$$= 4.5 + 0.65 \times (2.2/1.6) \times (5.196 - 3.9) = 5.6583$$

The conditional standard deviation is given by

$$\sigma_{Y|x} = \sigma_Y \sqrt{1 - \rho^2} = 2.2\sqrt{1 - 0.65^2} = 1.672 \tag{11.25}$$

We generate another uniformly distributed number, which turns out to be 0.43. Treating this as probability gives the value of z as -0.18. Hence, the value of Y will be

$$y = 5.6583 - 0.18 \times 1.672 = 4.3573$$

Thus, the first pair of generated numbers is (5.196, 4.3573).

Abramowitz and Stegun (1965) have given an approximate formula to calculate standard normal variates given the probability of exceedance without use of a table, with an error of approximation less than 4.5×10^{-4}. Let P represent the nonexceedance probability and Q the probability of exceedance: $Q = 1 - P$. Compute W by

$$W = \sqrt{-2 \times \ln Q} \qquad (11.26)$$

The corresponding standard normal variate z is

$$z = W - \frac{2.515517 + 0.802853\ W + 0.010328\ W^2}{1 + 1.432788\ W + 0.189269\ W^2 + 0.001308\ W^3} + \varepsilon(Q) \qquad (11.27)$$

For example, if $P = 0.791$, $Q = 1 - 0.791 = 0.209$ and $W = 1.7694$ from Eq. 11.26. From Eq. 11.27, $z = 0.8097$. If P is 0.43, $Q = 0.57$ and $W = 1.0603$ from Eq. 11.26. From Eq. 11.27, $z = -0.176$.

11.6 Simulation of Systems with Random Inputs

During the past few decades, Monte Carlo techniques have been applied to a wide range of problems in water resources, such as estimation of mean areal precipitation (Shih and Hamrick 1975), surface water hydrology (e.g., Labadie et al. 1987), and groundwater problems (e.g., Jones 1990). In this section, several real-life examples are presented to illustrate the Monte Carlo simulation technique and its strengths.

11.6.1 Monte Carlo Simulation of Reservoir Design

Monte Carlo simulation enables the analyst to study and quantify the influence of variability in system input and demands. The influence of input variability in system design is demonstrated through the following example.

Example 11.9 The purpose of this example is to illustrate the strength of Monte Carlo simulation for sizing of a storage reservoir. Assume that, at the site of interest, the statistical characteristics of annual streamflows, which are log-normally distributed, are known. The inflows have mean $= 660 \times 10^6$ m^3, standard deviation $= 175 \times 10^6$ m^3, lag–1 autocorrelation coefficient $\rho = 0.2$, and skewness $\gamma = 0.99$. The generated inflows, quite naturally, should preserve these statistical properties. Here, we adopt a three-parameter log-normal distribution to model the flows x.

Let a be the lower bound of the flows. Accordingly, $(x - a)$ will be log-normally distributed, or $y = \ln (x - a)$ will be normally distributed. For the random variate x, the mean μ_x, standard deviation σ_x, lag–1 correlation coefficient ρ_x, and skewness γ_x are related to statistical properties of y (Matalas 1967) by

$$\mu_x = a + \exp(\sigma_y^2 / 2 + \mu_y) \tag{11.28}$$

$$\sigma_x^2 = \exp\{2(\sigma_y^2 + \mu_y)\} - \exp(\sigma_y^2 + 2\mu_y) \tag{11.29}$$

$$\gamma_x = \frac{\exp(3\sigma_y^2) - 3\exp(\sigma_y^2) + 2}{\{\exp(\sigma_y^2) - 1\}^{1.5}} \tag{11.30}$$

$$\rho_x = \frac{\exp(\sigma_y^2 \rho_y) - 1}{\exp(\sigma_y^2) - 1} \tag{11.31}$$

Knowing the values of μ_x, σ_x, ρ_x, and γ_x for the historic data, we can find the values of a, μ_y, σ_y, and ρ_y for y. First, using Eq. 11.30 we have

$$\frac{\exp(3\sigma_y^2) - 3\exp(\sigma_y^2) + 2}{\{\exp(\sigma_y^2) - 1\}^{1.5}} = 0.99$$

the solution of which is

$$\sigma_y^2 = 0.096659 \text{ or } \sigma_y = 0.3109$$

Now, from Eq. 11.29, we get

$$\exp\{2(0.096659 + \mu_y)\} - \exp(0.096659 + 2\mu_y) = 175^2 = 30625$$

The solution of this equation gives $\mu_y = 6.26$. Next, we have from Eq. 11.31

$$\frac{\exp(0.096659\rho_y) - 1}{\exp(0.096659) - 1} = 0.2$$

This gives $\rho_x = 0.2079$. Finally, from Eq. 11.28 we obtain

$$a + \exp(0.096659/2 + 6.26) = 660$$

which yields $a = 111 \times 10^6$ m^3.

Now by using these properties, flows in the log domain are generated by

$$(y_{i+1} - \mu_y) = \rho_y (y_i - \mu_y) + \sigma_y t_i (1 - \rho_y^2)^{0.5} \tag{11.32}$$

where t_i are normally distributed random variates with mean zero and standard deviation of unity. The synthetic flows in real domain are obtained by the transformation

$$x_i = \exp(y_i) + a \tag{11.33}$$

This procedure was used to generate 500 traces of inflows, each 100 years long, which is the useful life of a typical storage reservoir.

A simple procedure to determine the required storage capacity is the sequent peak algorithm (see Jain and Singh 2003). In the present case, the storage capacity was obtained for a draft equal to 0.7 times the average inflows. These estimates are shown in Fig. 11-3. The mean storage was 125×10^6 m^3, and the standard deviation of storage values was 46×10^6 m^3. It is interesting to note that the maximum and minimum values of storage were 363×10^6 and 13×10^6 m^3. Clearly, deciding on the size of the reservoir after such an analysis will be a much better decision than just using the available (and usually small sample) data.

11.6.2 Monte Carlo Simulation of Reservoir Operation

The operation policies for a reservoir are commonly developed by using either optimization techniques or simulation. Sometimes a combination of these two techniques is also used. An important question that these techniques cannot answer is the influence of the variability of inflows and demands on the planning, design, and performance of the system. It has been pointed out that the deterministic models are optimistic because they overestimate benefits and underestimate costs (Loucks et al. 1981). Monte Carlo simulation enables the analyst to study and quantify the influence of variabilities in system input and demands.

Example 11.10 Consider a multipurpose reservoir that is being planned to serve irrigation and municipal water supply demands. The demands for these purposes are known. The reservoir is operated by following the linear operation policy shown in Fig. 11-4. According to this policy, if, in a particular period, the amount of water available in storage is less than the target demand, all the available water is released. If the available water is more than the target demand but less than the target demand plus the available storage capacity, a release equal to the target demand is made and the excess water is stored in the reservoir. If, even after making releases equal to the target demands, there is no space to store the excess water, all the water in excess of the maximum storage capacity is released.

Let A_w represent the available water and T the target demand. Mathematically, the policy can be expressed as

$$\text{if } A_w \leq T, \text{ release} = A_w$$

$$\text{if } T < A_w \leq S_{max} + T, \text{ release} = T \qquad \textbf{(11.34)}$$

$$\text{if } A_w > S_{max} + T, \text{ release} = A_w - S_{max}$$

The reservoir will be empty in the first case and full in the third.

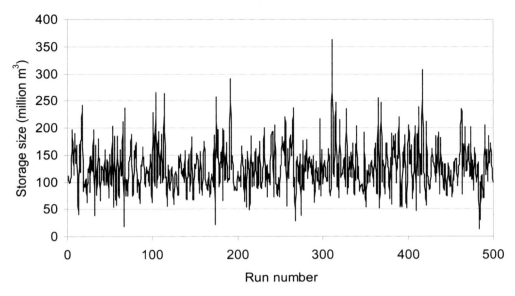

Figure 11-3 *Variation of storage values for the various simulation runs.*

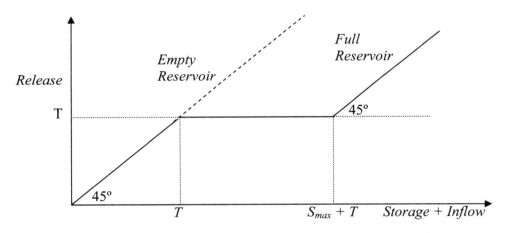

Figure 11-4 *Operation policy for the reservoir in Example 11.10.*

In this example, the influence of variability in inflow will be studied. As in most planning studies, annual data will be used. The inflows to the reservoir under examination are normally distributed with mean = 660×10^6 m^3 and standard deviation = 350×10^6 m^3. The annual inflows have a small positive autocorrelation $\rho = 0.3$.

The maximum possible storage capacity of the reservoir at the site of interest is 750×10^6 m^3. The depth of water evaporated from the reservoir every year depends on the climate and the amount of water available in storage. Here, for simplicity, it is assumed that 4% of the available water is lost from evaporation. The annual water demand from the reservoir is 700×10^6 m³.

Solution Inflows to the reservoir are needed to simulate its operation. In Monte Carlo simulation, synthetically generated inflows are used. A number of techniques for generation of inflow data are available. In the present case, the following model (ASCE 1996) was used to generate annual flows whose statistical properties are the same as those of the observed flows:

$$x_{j+1} = \bar{x} + r(x_j - \bar{x}) + s(1 - r^2)N_j \tag{11.35}$$

where x_j is the flow for year j, $\bar{x}$ is the mean of the inflows, r is the correlation coefficient, and s is the standard deviation. Further, the N_js are standard normal deviates that can be generated by using techniques explained previously. Equation 11.35 can be employed to generate an annual streamflow sequence.

One drawback of the model for inflow generation used here is that it ignores skewness of inflows. To preserve the skewness of the historic data, the numbers N_j in Eq. 11.35 should be transformed as follows (ASCE 1996):

$$W = \frac{2}{g}\left[1 + \frac{gN_j}{6} - \frac{g^2}{36}\right]^3 - \frac{2}{g} \tag{11.36}$$

where g represents the coefficient of skewness of historical data. Naturally, the generated inflows will be different if the transformation given by Eq. 11.36 is followed.

By using this model, the inflow data for 1,000 years were generated. The operation was simulated for 1,000 years by following the standard linear operating policy given by Eq. 11.34 and assuming an initial reservoir content of 500×10^6 m³. As a result, the reservoir releases and end-of-year storage for these 1,000 years are produced. From this information, the probability of different ranges of release can be computed, as shown in Table E11-10a.

Table E11-10a *Probability of different release ranges and associated benefits.*

Release range (10^6 m³)	< 400	400–600	600–800	800–1,000	>1,000
Benefits	2.0	6.0	10.0	7.0	5.0
Frequency	0	40	899	46	15
Probability	0	0.040	0.899	0.046	0.015

Let the annual benefits from the release in different ranges (b_i) be as given in the second row of this table. Hence, the expected annual benefit (B) from the reservoir can be computed as

$$B = \sum b_i R_i$$

where R_i is the probability of release in range i. Therefore,

$$B = 2.0 \times 0.0 + 6.0 \times 0.040 + 10.0 \times 0.899 + 7.0 \times 0.046 + 5.0 \times 0.015 = 9.627$$

Once a model for data generation and system operation has been formulated, its power can be exploited to evaluate the sensitivity of various parameters of the system, such as the inflow properties or the level of demands. One can take repeated runs of the system to analyze how the expected benefits will change if the correlation of inflows were different or if the magnitude of target demands changed (assuming that the benefit for different flows remains constant, although this can also be studied if enough data were available).

Table E11-10b gives the results of some sensitivity runs. In this table, the term base run refers to the run with the given system parameters. In the sensitivity runs, one parameter is changed at a time and the release probability and expected benefits are computed.

Several inferences can be derived from Table E11-10b by comparing the results of base run with other runs. Thus, one sees that, as the correlation of inflows increases, the benefits from the operation increase (the correlation for the base run being 0.3). Moreover, when the skewness of historical flows is preserved, the benefits are less. Table E11-10b also shows that, as the target demand is increased, the benefits decrease because the reservoir can no longer supply enough water owing to limited storage and inflows. Note that the mean of the inflows is 660×10^6 m^3, whereas the volume of demands was increased from 700×10^6 to 900×10^6 m^3. When the target demand was 700×10^6 m^3, the probability of release being in the range 600–800 was 0.899 but when the demand was raised to 900×10^6 m^3, the probability of release being in the range 800–1000 was only 0.827, indicating that the frequency of failures had increased. Notice also that when the reservoir storage capacity was reduced from 750×10^6 to 600×10^6 m^3, the benefits were reduced from 9.627 to 9.302 units. In this case also, the probability of release being in the range 600–800 decreased from 0.899 to 0.808, indicating a larger number of failures.

Table E11-10b Results of sensitivity runs: reservoir operation example.

Release range (10^6 m^3)	<400	400–600	600–800	800–1,000	>1,000	Expected benefits
Benefits	2.0	6.0	10.0	7.0	5.0	
Base run	0	0.040	0.899	0.046	0.015	9.627
Inflow correlation = 0.5	0	0.020	0.949	0.028	0.003	9.821
Inflow correlation = 0.1	0	0.054	0.866	0.058	0.022	9.5
Skewness considered using Eq. 11.30	0	0.015	0.841	0.091	0.053	9.402
Target demand = 900	0	0.040	0.118	0.827	0.015	7.284
Reservoir capacity = 600	0	0.040	0.808	0.111	0.041	9.302

Performing a sensitivity analysis is one way to use the power of Monte Carlo simulation. After the model has been developed, it is easy to change the key parameters and study the impact of the change. However, to make full use of this power, it is essential that the runs be carefully planned and that physical realism not be lost.

11.6.3 Estimation of Parameters of a Probability Distribution

As an example, consider the generalized Pareto (GP) distribution, which has been applied to a number of areas, encompassing socioeconomic processes, physical and biological phenomena, reliability analysis, and analyses of environmental extremes. This distribution has been discussed in Chapter 5; Singh (1998b) provides a detailed description of this distribution. The three-parameter generalized Pareto distribution (GPD3) is a flexible distribution and is, therefore, better suited to model those environmental and hydrologic processes that have heavy-tailed distributions.

The GPD3 distribution has the CDF

$$F(x) = \begin{cases} 1 - \left(1 - a\dfrac{x-c}{b}\right)^{\frac{1}{a}}, & a \neq 0 \\ 1 - \exp\left(-\dfrac{x-c}{b}\right), & a = 0 \end{cases} \tag{11.37}$$

and the PDF

$$f(x) = \begin{cases} \dfrac{1}{b}\left(1 - a\dfrac{x-c}{b}\right)^{\frac{1}{a}-1}, & a \neq 0 \\ \dfrac{1}{b}\exp\left(\dfrac{x-c}{b}\right), & a = 0 \end{cases} \tag{11.38}$$

The range of x is $c \leq x \leq \infty$ for $a \leq 0$ and $c \leq x \leq b/a + c$ for $a \geq 0$.

In environmental and water resources engineering, the most commonly used methods of parameter estimation are the methods of moments, maximum likelihood, and probability-weighted moments. These have been discussed in Chapter 8. These methods and their variants were considered in this example.

Specifically, four methods of parameter estimation were compared for the GPD3 distribution: two methods of moments, probability-weighted moments, and maximum likelihood estimation. The parameter estimators resulting from these methods are briefly outlined.

The regular moment estimators of the GPD3 distribution were derived by Hosking and Wallis (1987). It is important to note that $E[1 - a(x - c)/b]^r = 1/(1 + ar)$ if $[1 + ra] > 0$. Thus, the rth moment of X exists only if $a \geq [1/r]$. Provided that these moments exist, the moment estimators are

$$\bar{x} = c + \frac{b}{1+a} \tag{11.39}$$

$$s^2 = \frac{b^2}{(1+a)^2(1+2a)} \tag{11.40}$$

$$G = \frac{2(1-a)(1+2a)^{1/2}}{1+3a} \tag{11.41}$$

where $\bar{x}$, s^2, and G are, respectively, the mean, the variance, and the skewness coefficient. Equations 11.39 to 11.41 constitute a system of equations that are solved for obtaining parameters a, b, and c, given the value of $\bar{x}$, s^2, and G.

11.6.3.1 Modified Moment Estimators

The regular method of moments is valid only when $a > -1/3$, which limits its practical application. Second, its estimation of shape parameter (a) depends on the sample skewness alone, which in reality could be grossly different from the population skewness. Following Quandt (1966), a modified version of this method (MM1) involves restructuring Eq. 11.41 in terms of some known property of the data. In this modification, Eq. 11.37 is replaced by $E[F(x_1)] = F(x_1)$, which yields

$$1 - a\frac{x_1 - c}{b} = \left(\frac{n}{n+1}\right)^a \tag{11.42}$$

Eliminating b and c by substitution of Eq. 11.40 and Eq. 11.41 into Eq. 11.42, one gets an expression for the estimation of a:

$$\bar{x} - x_1 = \left[\frac{1}{1+a} - \frac{1}{a} + \frac{1}{a}\left(\frac{n}{n+1}\right)^a\right]S(1+a)(1+2a)^{1/2}, \quad a \neq 0 \tag{11.43}$$

With the value of a obtained from the solution of Eq. 11.43, estimates of b and c are obtained from the solution of Eq. 11.40 and Eq. 11.41.

11.6.3.2 Probability-Weighted Moment Estimators

The probability-weighted moments (PWM) estimators for GPD3 are given (Hosking and Wallis 1987) as

$$a = \frac{W_0 - 8W_1 + 9W_2}{-W_0 + 4W_1 - 3W_2} \tag{11.44}$$

$$b = -\frac{(W_0 - 2W_1)(W_0 - 3W_2)(-4W_1 + 6W_2)}{(-W_0 + 4W_1 - 3W_2)^2} \tag{11.45}$$

$$c = \frac{2W_0W_1 - 6W_0W_2 + 6W_1W_2}{-W_0 + 4W_1 - 3W_2} \qquad (11.46)$$

where the rth probability-weighted moment, W_r, is given by

$$W_r = \frac{1}{r+1}\left(c + \frac{b}{a}\right) - \frac{b}{a}\frac{1}{(a+r+1)}, r = 0,1,2,\dots \qquad (11.47)$$

Thus, Eq. 11.44 to Eq. 11.46 yield parameter estimates in terms of PWM. For a finite sample size, consistent moment estimates ($\overline{W}_r$) can be computed as

$$\overline{W}_r = \frac{1}{n}\sum_i^n (1 - F_{i:n})^r x_{i:n} \qquad (11.48)$$

where $x_{1:n} \le x_{2:n} \le \dots \le x_{n:n}$ is the ordered sample and $F_{i:n}=(i-0.35)/n$ (Landwehr et al. 1979b).

11.6.3.3 Maximum Likelihood Estimators

The maximum likelihood (ML) equations of Eq. 11.38 are given (Moharram et al. 1993) as

$$\sum_{i=1}^n \frac{(x_i - c)/b}{1 - a(x_1 - c)/b} = \frac{n}{1 - a} \qquad (11.49)$$

$$\sum_{i=1}^n \ln[1 - a(x_i - c)/b] = -na \qquad (11.50)$$

A maximum likelihood estimator cannot be obtained for c, since the maximum likelihood function (L) is unbounded with respect to c. However, since c is the lower bound of the random variable X, one may maximize L subject to the constraint $c \le x_1$, the lowest sample value. Clearly, L is maximum with respect to c when $c = x_1$. Thus, the regular maximum likelihood estimators (RMLE) are given as $c = x_1$ and the values of a and b are given by Eq. 11.49 and Eq. 11.50. This causes a large-scale failure of the algorithm, particularly for sample sizes < 20. By ignoring the smallest observation x_1, however, the pseudo-MLE (PMLE) can be obtained as follows: First assume an initial value of ($c < x_1$). Then estimate a and b. Thereafter, re-estimate c, and then repeat this procedure until the parameters no longer change.

Example 11.11 Compare the various parameter estimation methods for the three-parameter generalized Pareto distribution.

Solution To compare the various methods of parameter estimation, error-free data were generated by Monte Carlo simulations using an experimental design commonly employed in environmental and water resources engineering.

The inverse of Eq. 11.37 is

$$x(F) = c + \frac{b}{a}\left[1 - (1 - F)^a\right], \quad a \neq 0 \tag{11.51}$$

$$x = c - b\ln(1 - F), \quad a = 0 \tag{11.52}$$

where $X(P)$ denotes the quantile of the cumulative probability P or nonexceedance probability $1 - P$. To assess the performance of the parameter estimation methods outlined here, Monte Carlo sampling experiments were conducted using Eq. 11.51 and Eq. 11.52.

In engineering practice, observed samples may be frequently available, for which the first three moments (mean, variance, and skewness) are computed. Thus, parameter estimates and quantiles for the commonly encountered peak characteristics data are frequently characterized by using the coefficients of variation and skewness. Because this information is readily derivable from a given data set, a potential candidate for estimating the parameters and quantiles of GPD3 will be the one that performs the best in the expected observed ranges of the coefficient of variation and skewness. Thus, the selection of the population parameter ranges is very important for any simulation study. To that end, the following considerations were made:

1. For a given set of data, the distribution parameters are not known in advance; usually known quantities are the sample coefficients of variation and skewness, if they exist. Therefore, if one were somehow able to classify the best estimators with reference to these readily knowable data characteristics, that would be preferable from a practical standpoint as compared to classifying the best estimators in terms of an *a priori* unknown parameter.

2. Not all the estimators perform well in all population ranges, so a range where all the estimators can be applied has to be selected. Fortunately, in real life most commonly encountered data lie within the range considered in this study.

3. Because the sample skewness in GPD may correspond to more than one variance, parameter estimation based on skewness alone is misleading. To avoid this folly, evaluation of the estimators is based on the parameters.

Keeping the above considerations in mind, we investigated parameters using a factorial experiment within a space spanned by $\{a_i, b_j, c_k\}$, where $\{a_i = -0.1, -0.05, 0.0, 0.05, 0.1\}$, $\{b_j = 0.25, 0.50\}$, and $\{c_i = 0.5, 1.0\}$. Twenty GPD3 population cases, listed in Table E11-11, were considered. The ranges of data characteristics were also computed so that the results of this study could be related to the commonly used data statistics. For each population case, 20,000 samples of size 10, 20, 50, 100, 200, and 500 were generated, and then parameters and quantiles were estimated.

11.6.3.4 Performance Indices

Although standardized forms of bias, standard error, and root mean square error are commonly used in engineering practice (Kuczera 1982a,b), the performance of parameter estimators was evaluated by using their well-defined statistics. These performance indices were defined, respectively, as

$$\text{BIAS} = E(\theta) - \theta \qquad\qquad (11.53)$$

$$\text{SE} = S(\theta) \qquad\qquad (11.54)$$

$$\text{RMSE} = E\left[(\theta - \theta)^2\right]^{0.5} \qquad\qquad (11.55)$$

Table E11-11 *GPD3 population cases considered in the sampling experiment.*

Population parameters			Population statistics		
a	b	c	Mean	Coefficient of variation	Skewness
−0.10	0.25	0.5	0.78	0.40	2.81
		1.0	1.28	0.24	
	0.50	0.5	1.06	0.59	
		1.0	1.56	0.40	
−0.05	0.25	0.5	0.76	0.36	2.34
		1.0	1.26	0.22	
	0.50	0.5	1.03	0.54	
		1.0	1.53	0.36	
0.00	0.25	0.5	0.75	0.33	2.00
		1.0	1.25	0.20	
	0.50	0.5	1.00	0.50	
		1.0	1.50	0.33	
0.05	0.25	0.5	0.74	0.31	1.73
		1.0	1.24	0.18	
	0.50	0.5	0.98	0.47	
		1.0	1.48	0.31	
0.10	0.25	0.5	0.73	0.29	1.52
		1.0	1.23	0.17	
	0.50	0.5	0.95	0.43	
		1.0	1.45	0.29	

where $\hat{\theta}$ is an estimate of θ (parameter or quantile), $E[\cdot]$ denotes the statistical expectation, and $S[\cdot]$ denotes the standard deviation of the respective random variable. If n is the total number of random samples then the mean and standard deviation were calculated as

$$E\left[\hat{\theta}\right] = \frac{1}{n}\sum_{i=1}^{n}\hat{\theta}_i \qquad (11.56)$$

$$S\left[\hat{\theta}\right] = \left\{\frac{1}{n-1}\sum_{i=1}^{n}\left[\hat{\theta}_1 - E(\hat{\theta})\right]^2\right\}^{0.5} \qquad (11.57)$$

The root mean square error can also be expressed as

$$\text{RMSE} = \left[\frac{n-1}{n}\text{SE} + \text{BIAS}^2\right]^{0.5} \qquad (11.58)$$

These indices were used to measure the variability of parameter and quantile estimates for each simulation. Although they were used to determine the overall "best" parameter estimation method, our interest lies in the bias and variability of estimates of quantiles in the extreme tails of the distribution (non-exceedance probability P = 0.9, 0.99, 0.999) when the estimates are based on small samples ($n \leq 50$). Owing to the limited number of random number of samples (20,000 here) used, the results are not expected to reproduce the true values of BIAS, SE, RMSE, and $E[\hat{\theta}]$. Nevertheless, they provide a means to compare the performance of estimation methods used. The computed values of BIAS and RMSE in quantiles are tabulated as ratios $\hat{X}(F)/X(F)$ rather than for the estimator $\hat{X}(F)$ itself.

11.6.3.5 Robustness

Kuczera (1982a,b) defined a robust estimator as the one that is resistant and efficient over a wide range of population fluctuations. If an estimator performs steadily without undue deterioration in RMSE and bias, then it can be expected to perform better than other competing estimators under population conditions different from those on which conclusions were based. Two criteria to identify a resistant estimator are mini-max and minimum average RMSE (Kuczera 1982b). Based on this mini-max criterion, the preferred estimator is the one whose maximum RMSE for all population cases is minimum. The minimum average criterion is to select the estimator whose RMSE average over the test cases is minimum.

11.6.3.6 Results and Discussion

The performance of a parameter estimator depends on the following:

1. Sample size n
2. Population parameters
3. The distribution parameter
4. The probability of exceedence or quantile

The results for the five cases are summarized in Tables 11-1 through 11-4.

11.6.3.7 Bias in Parameter Estimates

Table 11-1 displays the results of bias in parameters. For shape parameter (a), for small sample sizes ($n \leq 50$), RME and MM1 exhibited less bias, particularly for increasing population values of a, for smaller sample sizes ($n \leq 20$). The moment-based methods tended to estimate a without responding to changes in b and c. MLE did not do well for small sample sizes but improved consistently whereas its bias increased with increasing population values of a. PWM exhibited less bias for $a > 0$. From the bias results of all 20 populations, we concluded that for increasing sample sizes ($n \geq 50$) PWM and MLE would be acceptable.

For scale parameter (b), RME and MM1 responded linearly to the change in population values of b. MM1 performed better for all sample sizes and population cases. For larger sample sizes ($n \geq 100$), all the estimators, except RME, which had a higher bias for $a < 0$, showed comparable absolute bias. For the location parameter (c), all the methods, except MLE, showed a negative bias. The reason for the positive bias of MLE is rooted in the solution procedure adopted for MLE. Another important finding in this regard is that the bias in c was dictated only by the population values of a and b.

11.6.3.8 RMSE in Parameter Estimates

Table 11-2 summarizes the root mean square error in the results of parameter estimates. MM1 performed well only for c, but for other parameters it showed a pattern similar to that of RME. PWM demonstrated a consistent improvement for all population cases as sample size increased. MLE did not do well when $n \leq 20$; however, as the sample size increased, it outperformed other methods. With increasing population a, MLE exhibited a consistent improvement in RMSE for both a and b. RME and MM1 did not respond to changes in populations c and b while estimating parameter a, as was indicated by the bias results earlier.

11.6.3.9 Bias in Quantile Estimates

The bias results of quantile estimation by GPD3 are summarized in Table 11-3. In general, for all nonexceedance probabilities, among the three moment-based methods RME performed better for lower values of P in terms of bias in quantile

Table 11-1 Bias in the parameters of GDP3 for spaces spanned by a = (−0.1, −0.05, 0.0, 0.05, 0.1), b = 0.50, c = 1.0.

Size	Method	Bias(a) (b = 0.50 c = 0.50)					Bias(b) (b = 0.50 c = 0.50)					Bias(c) (b = 0.50 c = 0.50)				
	a	−0.100	−0.050	0.000	0.050	0.100	−0.100	−0.050	0.000	0.050	0.100	−0.100	−0.050	0.000	0.050	0.100
10	RME	0.324	0.296	0.263	0.232	0.199	0.276	0.236	0.201	0.170	0.141	−0.085	−0.070	−0.055	−0.046	−0.037
	MM1	0.266	0.252	0.237	0.220	0.209	0.161	0.148	0.141	0.132	0.124	−0.012	−0.010	−0.009	−0.009	−0.009
	PWM	0.188	0.175	0.159	0.140	0.125	0.140	0.131	0.124	0.114	0.106	−0.041	−0.040	−0.037	−0.036	−0.036
	MLE	0.259	0.267	0.275	0.275	0.289	0.168	0.180	0.194	0.189	0.204	0.051	0.050	0.051	0.050	0.049
20	RME	0.230	0.206	0.181	0.158	0.140	0.199	0.167	0.142	0.115	0.097	−0.069	−0.054	−0.043	−0.032	−0.024
	MM1	0.139	0.127	0.113	0.102	0.097	0.076	0.067	0.062	0.054	0.051	−0.003	−0.003	−0.002	−0.002	−0.002
	PWM	0.091	0.083	0.071	0.062	0.054	0.065	0.060	0.057	0.051	0.046	−0.018	−0.018	−0.017	−0.016	−0.015
	MLE	0.134	0.138	0.142	0.142	0.150	0.090	0.097	0.105	0.102	0.110	0.025	0.025	0.025	0.025	0.025
50	RME	0.136	0.112	0.094	0.077	0.064	0.124	0.096	0.077	0.059	0.046	−0.050	−0.036	−0.026	−0.018	−0.013
	MM1	0.065	0.052	0.046	0.040	0.036	0.034	0.026	0.025	0.021	0.018	−0.001	0.000	0.000	0.000	0.000
	PWM	0.038	0.030	0.027	0.022	0.018	0.026	0.021	0.022	0.019	0.016	−0.007	−0.007	−0.006	−0.006	−0.005
	MLE	0.056	0.058	0.059	0.059	0.062	0.031	0.033	0.036	0.035	0.037	0.010	0.010	0.010	0.010	0.010
100	RME	0.090	0.070	0.056	0.043	0.035	0.086	0.063	0.046	0.034	0.026	−0.038	−0.025	−0.017	−0.011	−0.007
	MM1	0.035	0.028	0.023	0.020	0.018	0.018	0.014	0.012	0.010	0.009	0.000	0.000	0.000	0.000	0.000
	PWM	0.018	0.015	0.013	0.010	0.009	0.012	0.011	0.010	0.009	0.008	−0.003	−0.003	−0.003	−0.003	−0.003
	MLE	0.027	0.027	0.028	0.028	0.030	0.013	0.013	0.014	0.014	0.015	0.005	0.005	0.005	0.005	0.005
200	RME	0.057	0.044	0.032	0.025	0.017	0.057	0.040	0.027	0.020	0.013	−0.026	−0.016	−0.010	−0.006	−0.004
	MM1	0.019	0.015	0.012	0.010	0.008	0.010	0.008	0.006	0.005	0.004	0.000	0.000	0.000	0.000	0.000
	PWM	0.009	0.008	0.007	0.006	0.003	0.006	0.006	0.005	0.005	0.004	−0.002	−0.002	−0.001	−0.001	−0.001
	MLE	0.012	0.012	0.013	0.013	0.013	0.007	0.007	0.008	0.007	0.008	0.002	0.002	0.002	0.002	0.002
500	RME	0.032	0.023	0.015	0.010	0.008	0.033	0.021	0.014	0.008	0.006	−0.016	−0.009	−0.005	−0.003	−0.002
	MM1	0.008	0.007	0.004	0.004	0.004	0.004	0.004	0.002	0.002	0.002	0.000	0.000	0.000	0.000	0.000
	PWM	0.003	0.004	0.002	0.002	0.002	0.002	0.003	0.002	0.002	0.002	−0.001	−0.001	−0.001	−0.001	0.000
	MLE	0.004	0.004	0.004	0.004	0.005	0.002	0.002	0.003	0.003	0.003	0.001	0.001	0.001	0.001	0.001

Table 11-2 RMSE in the parameters of GPD3 for space spanned by a = (0.1, 0.05, 0.0, 0.05, 0.1), b = 0.50, c = 1.0.

Size	Method	RMSE(a) (b = 0.50 c = 0.50)					RMSE(b) (b = 0.50 c = 0.50)					RMSE(c) (b = 0.50 c = 0.50)				
	a	-0.100	-0.050	0.000	0.050	0.100	-0.100	-0.050	0.000	0.050	0.100	-0.100	-0.050	0.000	0.050	0.100
10	RME	0.408	0.389	0.364	0.344	0.324	0.435	0.393	0.353	0.323	0.297	0.151	0.128	0.112	0.099	0.100
	MM1	0.489	0.495	0.496	0.502	0.513	0.391	0.388	0.385	0.388	0.388	0.061	0.059	0.061	0.060	0.059
	PWM	0.386	0.381	0.372	0.361	0.359	0.322	0.312	0.303	0.291	0.284	0.078	0.075	0.074	0.071	0.070
	MLE	0.456	0.410	0.390	0.398	0.386	0.379	0.322	0.306	0.313	0.303	0.072	0.071	0.072	0.070	0.069
20	RME	0.300	0.285	0.269	0.257	0.249	0.298	0.265	0.244	0.222	0.208	0.113	0.093	0.079	0.068	0.062
	MM1	0.270	0.271	0.271	0.269	0.279	0.204	0.198	0.197	0.194	0.198	0.027	0.027	0.026	0.026	0.027
	PWM	0.263	0.258	0.252	0.246	0.243	0.208	0.199	0.196	0.188	0.185	0.049	0.047	0.046	0.045	0.044
	MLE	0.305	0.274	0.261	0.266	0.258	0.247	0.210	0.200	0.204	0.198	0.036	0.035	0.035	0.035	0.035
50	RME	0.187	0.172	0.162	0.153	0.149	0.182	0.158	0.143	0.131	0.123	0.075	0.060	0.050	0.044	0.040
	MM1	0.154	0.150	0.148	0.147	0.150	0.112	0.109	0.107	0.105	0.106	0.010	0.010	0.010	0.010	0.010
	PWM	0.164	0.160	0.156	0.153	0.152	0.123	0.119	0.116	0.113	0.112	0.029	0.027	0.027	0.026	0.026
	MLE	0.182	0.164	0.156	0.159	0.154	0.139	0.118	0.113	0.115	0.111	0.014	0.014	0.014	0.014	0.014
100	RME	0.133	0.122	0.114	0.108	0.103	0.127	0.112	0.099	0.091	0.086	0.056	0.044	0.036	0.032	0.029
	MM1	0.106	0.103	0.101	0.100	0.101	0.075	0.074	0.072	0.072	0.072	0.005	0.005	0.005	0.005	0.005
	PWM	0.115	0.112	0.110	0.108	0.107	0.083	0.083	0.081	0.079	0.078	0.020	0.019	0.019	0.018	0.018
	MLE	0.113	0.102	0.097	0.099	0.096	0.083	0.071	0.067	0.069	0.067	0.007	0.007	0.007	0.007	0.007
200	RME	0.099	0.091	0.083	0.077	0.072	0.092	0.081	0.071	0.065	0.061	0.040	0.032	0.027	0.024	0.021
	MM1	0.078	0.074	0.071	0.069	0.070	0.054	0.052	0.051	0.050	0.050	0.050	0.002	0.002	0.002	0.002
	PWM	0.083	0.080	0.078	0.076	0.076	0.060	0.058	0.056	0.056	0.055	0.014	0.013	0.013	0.013	0.012
	MLE	0.079	0.071	0.067	0.069	0.067	0.058	0.046	0.046	0.047	0.046	0.004	0.004	0.004	0.004	0.004
500	RME	0.068	0.062	0.055	0.050	0.047	0.063	0.055	0.047	0.043	0.039	0.028	0.022	0.018	0.016	0.014
	MM1	0.050	0.047	0.044	0.043	0.043	0.034	0.033	0.032	0.031	0.031	0.001	0.001	0.001	0.001	0.001
	PWM	0.052	0.050	0.049	0.048	0.048	0.037	0.036	0.036	0.035	0.035	0.009	0.008	0.008	0.008	0.008
	MLE	0.047	0.042	0.040	0.041	0.040	0.034	0.029	0.027	0.028	0.027	0.001	0.001	0.001	0.001	0.001

Table 11-3 Bias in the quantiles of GPD3 for space spanned by a = (−0.1, −0.05, 0.0, 0.05, 0.1), b = 0.50, c = 1.0.

Size	Method	0.900 (b = 0.50 c = 0.50)					0.950 (b = 0.50 c = 0.50)					0.999 (b = 0.50 c = 0.50)				
X(F)		1.795	1.720	1.651	1.587	1.528	2.246	2.116	1.998	1.891	1.794	5.476	4.625	3.953	3.421	2.994
a		−0.100	−0.050	0.000	0.050	0.100	−0.100	−0.050	0.000	0.050	0.100	−0.100	−0.050	0.000	0.050	0.100
10	RME	−0.018	−0.034	−0.045	−0.053	−0.071	−0.073	−0.081	−0.084	−0.086	−0.097	−0.342	−0.302	−0.260	−0.219	−0.189
	MM1	−0.035	−0.038	−0.036	−0.037	−0.039	−0.077	−0.075	−0.067	−0.063	−0.061	−0.201	−0.161	−0.124	−0.086	−0.058
	PWM	−0.007	−0.008	−0.003	−0.001	0.000	−0.029	−0.026	−0.018	−0.011	−0.007	0.027	0.041	0.058	0.077	0.090
	MLE	−0.004	−0.006	−0.002	−0.001	0.000	0.045	0.026	−0.024	0.023	0.016	0.055	0.081	0.094	0.156	0.196
20	RME	0.017	0.007	0.003	−0.005	−0.011	−0.025	−0.029	−0.027	−0.030	−0.032	−0.250	−0.210	−0.167	−0.135	−0.109
	MM1	−0.020	−0.020	−0.016	−0.018	−0.018	−0.047	−0.043	−0.035	−0.033	−0.031	−0.128	−0.097	−0.063	−0.046	−0.032
	PWM	−0.009	−0.007	−0.002	−0.003	−0.002	−0.020	−0.016	−0.008	−0.007	−0.005	0.053	0.057	0.069	0.066	0.066
	MLE	−0.008	−0.007	−0.003	−0.004	−0.003	0.006	−0.020	−0.015	0.013	0.010	0.057	0.068	0.034	−0.008	−0.044
50	RME	0.020	0.012	0.007	0.002	−0.002	−0.006	−0.008	−0.009	−0.011	−0.012	−0.163	−0.123	−0.093	−0.069	−0.052
	MM1	−0.010	−0.009	−0.007	−0.007	−0.008	−0.024	−0.020	−0.015	−0.014	−0.014	−0.069	−0.041	−0.027	−0.017	−0.011
	PWM	−0.005	−0.004	−0.001	−0.001	−0.002	−0.010	−0.007	−0.003	−0.002	−0.002	0.030	0.039	0.037	0.037	0.035
	MLE	−0.005	−0.005	−0.001	−0.002	−0.003	−0.005	−0.008	−0.005	−0.004	0.004	0.043	0.031	0.019	−0.030	−0.042
100	RME	0.017	0.010	0.005	0.001	0.000	0.001	−0.003	−0.005	−0.006	−0.006	−0.111	−0.080	−0.058	−0.039	−0.028
	MM1	−0.005	−0.005	−0.004	−0.004	−0.004	−0.013	−0.010	−0.009	−0.007	−0.007	−0.037	−0.022	−0.014	−0.008	−0.005
	PWM	−0.003	−0.002	−0.001	−0.001	−0.001	−0.005	−0.003	−0.002	−0.001	−0.001	0.021	0.022	0.020	0.020	0.019
	MLE	−0.003	−0.002	−0.001	−0.002	−0.001	−0.005	−0.006	−0.005	−0.003	−0.002	0.028	0.007	−0.013	−0.024	−0.034
200	RME	0.013	0.007	0.002	0.001	0.000	0.003	−0.001	−0.004	−0.003	−0.003	−0.071	−0.050	−0.035	−0.023	−0.012
	MM1	−0.003	−0.002	−0.003	−0.002	−0.002	−0.007	−0.005	−0.005	−0.004	−0.003	−0.018	−0.012	−0.008	−0.005	−0.001
	PWM	−0.001	−0.001	−0.001	0.000	0.000	−0.002	−0.001	−0.002	−0.001	0.000	0.012	0.010	0.009	0.010	0.012
	MLE	−0.001	−0.001	−0.002	0.000	0.000	−0.002	−0.003	−0.005	−0.001	0.000	0.012	0.004	−0.005	−0.012	−0.016
500	RME	0.009	0.004	0.002	0.000	0.000	0.003	0.000	−0.001	−0.002	−0.001	−0.041	−0.026	−0.016	−0.010	−0.006
	MM1	−0.001	−0.001	−0.001	−0.001	−0.001	−0.003	−0.002	−0.002	−0.002	−0.001	−0.008	−0.005	−0.002	−0.002	−0.001
	PWM	0.000	0.000	0.000	−0.001	0.000	−0.001	0.000	0.000	−0.001	0.000	0.005	0.004	0.006	0.004	0.004
	MLE	0.000	0.000	−0.002	−0.001	0.000	−0.001	0.000	0.000	−0.001	0.000	0.005	0.003	−0.003	−0.001	−0.004

Table 11-4 RMSE in the parameters of GPD3 for space spanned by a = (−0.1,−0.05, 0.0, 0.05, 0.1), b = 0.50, c = 1.0.

Size	Method	0.900 (b = 0.50 c = 0.50)					0.950 (b = 0.50 c = 0.50)					0.999 (b = 0.50 c = 0.50)				
X(F)		1.795	1.720	1.651	1.587	1.528	2.246	2.116	1.998	1.891	1.794	5.476	4.625	3.953	3.421	2.994
a		−0.100	−0.050	0.000	0.050	0.100	−0.100	−0.050	0.000	0.050	0.100	−0.100	−0.050	0.000	0.050	0.100
10	RME	0.289	0.262	0.236	0.217	0.204	0.316	0.287	0.258	0.236	0.219	0.394	0.383	0.366	0.350	0.333
	MM1	0.254	0.239	0.220	0.204	0.190	0.287	0.269	0.247	0.228	0.212	0.549	0.532	0.508	0.484	0.459
	PWM	0.252	0.237	0.219	0.204	0.190	0.296	0.277	0.254	0.234	0.216	0.963	0.878	0.796	0.715	0.650
	MLE	0.267	0.249	0.228	0.208	0.190	0.405	0.360	0.325	0.312	0.262	1.338	1.186	1.043	0.901	0.780
20	RME	0.210	0.186	0.169	0.151	0.138	0.237	0.210	0.189	0.169	0.153	0.361	0.343	0.322	0.298	0.275
	MM1	0.182	0.168	0.157	0.145	0.135	0.210	0.194	0.180	0.164	0.152	0.465	0.444	0.415	0.380	0.348
	PWM	0.181	0.168	0.159	0.145	0.135	0.216	0.199	0.186	0.169	0.155	0.738	0.664	0.586	0.513	0.449
	MLE	0.188	0.17	0.159	0.145	0.135	0.270	0.247	0.212	0.177	0.150	0.959	0.844	0.733	0.605	0.503
50	RME	0.132	0.118	0.105	0.096	0.089	0.153	0.137	0.121	0.108	0.098	0.292	0.275	0.252	0.227	0.204
	MM1	0.115	0.108	0.099	0.092	0.086	0.136	0.126	0.115	0.106	0.098	0.347	0.325	0.294	0.264	0.238
	PWM	0.115	0.109	0.100	0.093	0.087	0.140	0.130	0.118	0.109	0.100	0.462	0.416	0.363	0.318	0.280
	MLE	0.115	0.109	0.100	0.093	0.087	0.155	0.137	0.116	0.098	0.080	0.522	0.453	0.382	0.322	0.263
100	RME	0.093	0.083	0.075	0.068	0.063	0.110	0.097	0.086	0.076	0.069	0.245	0.223	0.200	0.177	0.155
	MM1	0.081	0.076	0.070	0.065	0.060	0.098	0.091	0.083	0.075	0.069	0.275	0.248	0.220	0.193	0.172
	PWM	0.082	0.077	0.071	0.066	0.062	0.101	0.093	0.085	0.077	0.071	0.328	0.290	0.253	0.221	0.195
	MLE	0.082	0.077	0.071	0.065	0.060	0.103	0.092	0.080	0.068	0.056	0.345	0.290	0.238	0.195	0.162
200	RME	0.066	0.058	0.053	0.048	0.044	0.078	0.068	0.060	0.054	0.049	0.201	0.179	0.156	0.134	0.115
	MM1	0.058	0.054	0.050	0.046	0.043	0.072	0.065	0.059	0.054	0.048	0.213	0.185	0.160	0.139	0.123
	PWM	0.059	0.055	0.051	0.047	0.043	0.072	0.066	0.060	0.055	0.050	0.230	0.202	0.177	0.155	0.136
	MLE	0.058	0.054	0.049	0.045	0.041	0.072	0.064	0.055	0.047	0.038	0.226	0.196	0.163	0.127	0.098
500	RME	0.041	0.037	0.034	0.030	0.028	0.048	0.043	0.038	0.034	0.031	0.144	0.127	0.106	0.090	0.075
	MM1	0.037	0.034	0.032	0.029	0.027	0.045	0.041	0.037	0.034	0.031	0.140	0.120	0.102	0.088	0.078
	PWM	0.037	0.034	0.032	0.030	0.027	0.046	0.042	0.038	0.034	0.031	0.143	0.126	0.111	0.097	0.085
	MLE	0.036	0.033	0.030	0.027	0.023	0.044	0.040	0.035	0.029	0.023	0.139	0.121	0.098	0.075	0.058

when $a > 0$ and sample size $n \leq 20$. As sample size increased ($n \geq 50$), MM1 performed comparatively better than did RME in all ranges. RME and MM1 tended to further underestimate the quantiles with increasing population c for a given value of b for $P \leq 0.995$. However, for $P = 0.999$ the trend was reversed. On the whole, with increasing b the absolute bias of these methods increased for sample sizes $n \geq 10$. PWM and MLE outperformed the other methods in terms of the absolute value of the bias for all sample sizes and quantile ranges. PWM and MLE responded positively in terms of bias to both b and c when one was varied keeping the other constant.

11.6.3.10 RMSE in Quantile Estimates

Table 11-4 gives RMSE in quantile estimates for $P = 0.9, 0.95,$ and 0.999 for selected five population cases. For probability of nonexceedance $P \leq 0.90$, PWM exhibited the least RMSE for small samples ($n > 20$) when $a \leq 0$ for $b = 0.25$ and $c = 0.50$. As populations b and c increased, MM1 showed better results in terms of RMSE for all quantiles and sample ranges. The RME performance was better for small sample sizes when $P \geq 0.90$. For small sample sizes, MLE did poorly but as the sample size increased, both PWM and MLE exhibited a consistent improvement. MLE did perform best only when $n \geq 50$ with $a > 0$. RMSE of the quantiles responded positively to the increase in population b, whereas the opposite was seen for the case of a population c increase.

11.6.3.11 Robustness Evaluation

The relative robustness of different methods of parameter and quantile estimation can be judged from Table 11-1 and 11-2. In terms of parameter bias, PWM performed in a superior manner for most of the data ranges, whereas MLE performed better for large sample sizes. In terms of parameter RMSE, RME did well with small sample sizes. For the bias in quantiles, PWM and MLE performed better. For RMSE in the quantiles, MM1 did better in most cases. On the whole, PWM showed the most consistent behavior. Thus, PWM would be the preferred method.

11.6.3.12 Summary

An evaluation of the relative performance of four methods for estimating parameters and quantiles of the three-parameter GPD was performed by using Monte Carlo simulation. The generation of a large number of sample data and their analysis enabled a comparison of the various methods of parameter estimation. No single method was found to be preferable to another for all population cases considered. On the whole, the PWM method was found to perform in a consistent fashion. When a clear choice of a particular method is in doubt, PWM can be the most reliable and should be the preferred method.

11.6.4 Robustness of a Frequency Distribution

This example demonstrates the application of Monte Carlo technique to examine the robustness of a frequency distribution and to determine whether the estimates are biased.

Example 11.12 Annual maximum flow data for a river in South America for 10 years are arranged in descending order in Table E11-12a. The results of a detailed study showed that the GEV-PWM distribution fits the data for the region. The following regional formula was developed for the region:

$$Q(T) = (u + \alpha Y_T)aA^b \tag{11.59}$$

where A is the catchment area (in km^2) and

$$Y_T = \left\{1 - \left[-\ln(1 - 1/T)\right]^K\right\}/K \tag{11.60}$$

The regional parameters of the GEV distribution were computed using the PWM estimation method. These are $K = -0.247$, $u = 0.448$, $\alpha = 0.493$, $a = 20.91$, and $b = 0.46$. Estimate the bias in the regional formulas using Monte Carlo simulations.

Solution The CDF of the GEV distribution is

$$F(z) = \exp\left\{-\left[1 - K\left(\frac{z - u}{\alpha}\right)\right]^{\frac{1}{K}}\right\}$$

Table E11-12a Annual maximum flow data.

Rank	Q (m³/s)
1	5,111.1
2	4,352.0
3	4,089.0
4	3,228.3
5	3,014.0
6	2,999.6
7	2,927.8
8	2,489.4
9	2,424.3
10	2,339.4

and the PWM estimators are related to data statistics by the following relations:

$$K = 7.859\ C + 2.9554\ C^2$$

$$C = 2/(T3 + 3) - \ln(2)/\ln(3)$$

$$\alpha = \frac{L2 \times K}{\Gamma(1 + K) \times (1 - 2^{-K})}$$

$$u = L1 + \frac{\alpha}{K}[\Gamma(1 + k) - 1]$$

$$T3 = L3/L1$$

where $L1$, $L2$, and $L3$ are the L-moments.

From the data, the mean of annual floods is

$$x_m = 3,297 \text{ cumec}$$

We generate a random number as given in the second column of Table E11-12b. If x_T is the flood for T-year return period, we compute x_T/x_m by

$$x_T / x_m = u + \alpha\left\{1 - \left[-\ln(F)\right]^K\right\}/K$$

Table E11-12b

	F	x_T/x_m	x_T	x_T in descending order	B1	B2	B3
1	0.440575	0.54848	1808.338	8437.501	843.75	843.75	843.75
2	0.101219	0.078529	258.9088	4413.256	392.29	343.25	294.22
3	0.28496	0.338954	1117.531	3671.264	285.54	214.16	152.97
4	0.798873	1.338567	4413.256	2515.626	167.71	104.82	59.90
5	0.947568	2.559145	8437.501	1808.338	100.46	50.23	21.53
6	0.280841	0.333588	1099.839	1404.637	62.43	23.41	6.69
7	0.351404	0.426035	1404.637	1388.346	46.28	11.57	1.65
8	0.347675	0.421094	1388.346	1117.531	24.83	3.10	0.00
9	0.73204	1.113516	3671.264	1099.839	12.22	0.00	0.00
10	0.575503	0.763005	2515.626	258.9088	0.00	0.00	0.00
			Average	2611.525	1935.51	1594.29	1380.70

The values of x_T/x_m are stored in the third column and the corresponding values x_T are stored in the fourth column with $x_m = 3,297$ cumec. In the fifth column, the values of the fourth columns are arranged in descending order. The values of $B1$, $B2$, and $B3$ are listed in the next three columns.

Now, the *L*-moments are computed by

$$\lambda_1 = B0 = 2{,}611.525$$

$$\lambda_2 = 2B1 - B0 = 2 \times 1{,}935.51 - 2{,}611.525 = 1{,}259.505$$

$$\lambda_3 = 6B2 - 6B1 + B0 = 6 \times 1{,}594.29 - 6 \times 1{,}935.51 + 2{,}611.525 = 564.2006$$

$$\lambda_4 = 20B3 - 30B2 + 12B1 - B0 = 399.8626$$

Next, we determine the *L*-moment ratios:

$$L\text{-}CV = \lambda_2/\lambda_1 = 0.48229$$

$$L\text{-}SK = \lambda_3/\lambda_2 = 0.44795$$

$$L\text{-}KR = \lambda_4/\lambda_3 = 0.70872$$

The GEV parameters using these *L*-moments can be computed as follows:

$$c = 2/(L\text{-}SK + 3) = -0.050876$$

$$k = 7.859c + 2.985c^2 = -0.3922182$$

$$\alpha = \frac{\lambda_2 k}{\Gamma(1+k)}(1 - 2^{-k}) = 0.411066$$

$$u = 1 + \frac{\alpha}{k}[\Gamma(1+k) - 1] = 0.504225$$

Using these computed parameters, we compute floods for various return periods as shown in Table E11-12c.

In a similar manner, two more replications of the procedure were made. The results for the second replication are shown in Table E11-12d.

Now, the *L*-moments are

$$\lambda_1 = 2{,}816.911, \; \lambda_2 = 1{,}178.135, \; \lambda_3 = 180.8436, \; \lambda_4 = 198.4997$$

and the *L*-moment ratios are

$$L\text{-}CV = 0.41824$$

$$L\text{-}SK = 0.15350$$

$$L\text{-}KR = 1.09763$$

The GEV parameters using these *L*-moments can be computed as follows:

$$c = 0.003286, \; k = -0.0258, \; \alpha = 0.588463, \; u = 0.635062$$

The floods for various return periods for this replication are shown in Table E11-12e.

For the third replication the data are listed in Table E11-12f.

Table E11-12c

T	F	$(\ln F)^k$	$u + \alpha/k[1 - (\ln F)^k]$	X(T)
50	0.98	4.619749	4.297679	11221.24
100	0.99	6.074979	5.822743	15203.18
200	0.995	7.980631	7.819844	20417.61
500	0.998	11.43841	11.44355	29879.12
1000	0.999	15.01464	15.1914	39664.74
10000	0.9999	37.05028	38.28449	99960.81

Table E11-12d

	F	x_T/x_m	x_T	x_T in descending order	B1	B2	B3
1	0.637087	0.882038	2908.079	6870.646	687.06	687.06	687.06
2	0.680785	0.979472	3229.32	4507.997	400.71	350.62	300.53
3	0.199327	0.225736	744.2525	4267.184	331.89	248.92	177.80
4	0.805957	1.367303	4507.997	3229.32	215.29	134.56	76.89
5	0.512151	0.656233	2163.601	2908.079	161.56	80.78	34.62
6	0.047668	0.03175	104.681	2163.601	96.16	36.06	10.30
7	0.356743	0.43313	1428.029	1945.318	64.84	16.21	2.32
8	0.787364	1.294263	4267.184	1428.029	31.73	3.97	0.00
9	0.915208	2.083908	6870.646	744.2525	8.27	0.00	0.00
10	0.469038	0.590027	1945.318	104.681	0.00	0.00	0.00
			Sum	2611.525	1997.52	1558.18	1289.52

Table E11-12e

T	F	$(\ln F)^k$	$u + \alpha/k[1 - (\ln F)^k]$	X(T)
50	0.98	1.105912	3.043800	8571.342
100	0.99	1.126014	3.501944	9861.475
200	0.995	1.146406	3.966698	11170.22
500	0.998	1.173876	4.592755	12933.2
1000	0.999	1.195073	5.075847	14293.59
10000	0.9999	1.268234	6.743243	18988.97

Table E11-12f

	F	x_T/x_m	x_T	x_T in descending order	B1	B2	B2
1	0.551562	0.721099	2377.464	5496.240	549.62	549.62	549.62
2	0.096165	0.069667	229.6935	2981.103	264.99	231.86	198.74
3	0.647532	0.904186	2981.103	2956.833	229.98	172.48	123.20
4	0.620144	0.847434	2793.988	2793.988	186.27	116.42	66.52
5	0.644095	0.896825	2956.833	2377.464	132.08	66.04	28.30
6	0.864891	1.667043	5496.24	1890.559	84.02	31.51	9.00
7	0.30977	0.371299	1224.171	1249.217	41.64	10.41	1.49
8	0.457785	0.573418	1890.559	1224.171	27.20	3.40	0.00
9	0.117616	0.105994	349.4614	349.4614	3.88	0.00	0.00
10	0.315583	0.378895	1249.217	229.6935	0.00	0.00	0.00
			Sum	2154.873	1519.69	1181.75	976.88

Now, the *L*-moments are

$$\lambda_1 = 2{,}154.873, \ \lambda_2 = 884.4992, \ \lambda_3 = 127.2336, \ \lambda_4 = 166.6144$$

and the *L*-moment ratios are

$$L\text{-}CV = 0.41046$$

$$L\text{-}SK = 0.14385$$

$$L\text{-}KR = 1.30952$$

The GEV parameters using these *L*-moments can be computed as follows:

$$c = 0.005233, \ k = -0.041, \ \alpha = 0.569528, \ u = 0.652727$$

The floods for various return periods for this replication are shown in Table E11-12g.

Now we analyze the observed data (Table E11-12h).

Here we get

$$x_m = 3{,}297.49$$

with *L*-moments

$$\lambda_1 = 3{,}297.49, \ \lambda_2 = 526.1456, \ \lambda_3 = 152.575, \ \lambda_4 = 57.02214$$

and *L*-moment ratios

$$L\text{-}CV = 0.15956$$

$$L\text{-}SK = 0.28999$$

$$L\text{-}KR = 0.37373$$

Table E11-12g

T	F	$(\ln F)^k$	$u + \alpha/k[1 - (\ln F)^k]$	X(T)
50	0.98	1.173487	3.059658	6590.504
100	0.99	1.207565	3.532600	7609.221
200	0.995	1.242504	4.017475	8653.642
500	0.998	1.290149	4.678705	10077.93
1000	0.999	1.327367	5.195218	11190.5
10000	0.9999	1.458811	7.019405	15119.8

Table E11-12h

SN	X	B1	B2	B3
1	5111.1	511.11	511.11	511.11
2	4352	386.84	338.49	290.13
3	4089	318.03	238.53	170.38
4	3228.3	215.22	134.51	76.86
5	3014	167.44	83.72	35.88
6	2999.6	133.32	49.99	14.28
7	2927.8	97.59	24.40	3.49
8	2489.4	55.32	6.92	0.00
9	2424.3	26.94	0.00	0.00
10	2339.4	0.00	0.00	0.00
Sum	32974.9	1911.82	1387.67	1102.13

The GEV parameters using these *L*-moments can be computed as follows:

$$c = -0.023024, \; k = -0.179, \; \alpha = 0.189, \; u = 0.8501$$

The floods for various return periods for this replication are shown in Table E11-12i.

The floods for the various return periods computed using the observed and generated data are summarized in Table E11-12j.

From this table, one can conclude the following:

1. Bias increases with return period of the flood.
2. All estimates using the generated data are overestimated.

Here, results from only a few replications have been shown. When 10,000 replications were made, the bias for T = 1,000 years was nearly 11%. So the GEV distribution can be considered to be a robust model for the region.

11.7 Monte Carlo Integration

When integration involves problems in one or two dimensions and the integrands are well behaved (say, there is no discontinuity), conventional numerical integration methods, such as the trapezoidal rule or Simpson's rule, are commonly employed. But the accuracy of conventional numerical integration deteriorates rapidly as the dimension of integration increases. For integrations involving multiple dimensions, the Monte Carlo method is a suitable numerical integration technique. Let the integration problem be

$$F = \int_a^b f(x)\,dx \qquad (11.61)$$

Table E11-12i

T	F	$(\ln F)^k$	$u + \alpha/k[1 - (\ln F)^k]$	$X(T)$
50	0.98	2.010628	1.914808	6313.123
100	0.99	2.278294	2.196797	7242.84
200	0.995	2.580415	2.515086	8292.239
500	0.998	3.04115	3.000475	9892.567
1000	0.999	3.443191	3.42403	11289.03
10000	0.9999	5.199913	5.274758	17390.88

Table E11-12j

T	\multicolumn{4}{c}{$X(T)$ from replication number}	$X(T)$ from obs. data	Bias
	1	2	3	Mean		
50	11221.24	8571.342	6590.504	8794.362	6313.123	39.30288
100	15203.18	9861.475	7609.221	10891.29	7242.84	50.37322
200	20417.61	11170.22	8653.642	13413.83	8292.239	61.76362
500	29879.12	12933.2	10077.93	17630.08	9892.567	78.21543
1000	39664.74	14293.59	11190.5	21716.28	11289.03	92.36621
10000	99960.81	18988.97	15119.8	44689.86	17390.88	156.973

where $a \leq x \leq b$. Monte Carlo integration using the *sample-mean method* is based on the premise that the integral Eq. 11.61 can be expressed as

$$F = \int_a^b \left[\frac{f(x)}{h_X(x)} \right] h_x(x)\, dx \qquad (11.62)$$

where the PDF, $h_x(x) \geq 0$, is defined over the interval $a \leq x \leq b$. The transformed integral given by Eq. 11.62 is equivalent to the computation of expectation of the term inside the square brackets; that is,

$$F = E_X \left[\frac{f(x)}{f_X(x)} \right] \qquad (11.63)$$

where X is a random variable whose PDF $h_X(x)$ is defined over $a \leq x \leq b$. Now F can be computed by the Monte Carlo method as

$$\hat{F} = \frac{1}{N} \sum_{i=1}^{N} \left[\frac{f(x_i)}{h_X(x_i)} \right] \qquad (11.64)$$

Here, x_i is the ith random variate generated according to $f_x(x)$ and N is the number of random variates generated. Computations using the sample-mean Monte Carlo integration method are carried out in the following steps:

1. Select $h_X(x)$ defined over the region of the integral from which N random variables are generated.
2. Compute $f(x_i)/h_X(x_i)$ for $i = 1, 2, \ldots, N$.
3. Calculate the sample average based on Eq. 11.64 as the estimate of F.

11.8 Variance Reduction Techniques

Monte Carlo simulation involves sampling and therefore the results obtained will entail sampling errors. These errors decrease with increasing sampling size, but increasing sample size to achieve higher precision generally means an increase in computer time to generate random numbers and for data processing. Several variance reduction techniques have been developed to obtain high precision in the Monte Carlo simulation results without having to substantially increase the required sample size. Some of these techniques are:

1. Antithetic variates (Hammersley and Morton 1956)
2. Correlated sampling (Rubinstein 1981)
3. Latin hypercube sampling (Pebesma and Heuvelink 1999)
4. Importance sampling
5. Stratified sampling (Cochran 1972)
6. Control variates

These are described in what follows.

11.8.1 Antithetic Variates Technique

Assume that two unbiased estimators α_1 and α_2 of a parameter α have been computed. The two estimators can be averaged to obtain another estimator of α as

$$\alpha_{avg} = \frac{1}{2}(\alpha_1 + \alpha_2) \tag{11.65}$$

The new estimator is also unbiased and its variance is

$$\text{var}(\alpha_{\text{avg}}) = \frac{1}{4}\left[\text{var}(\alpha_1) + \text{var}(\alpha_2) + 2\text{cov}(\alpha_1, \alpha_2)\right] \tag{11.66}$$

In Monte Carlo simulation, estimators α_1 and α_2 depend on random variates that are generated. Of course, these variates are related to the standard uniform random variates used to generate random variates. Thus, α_1 and α_2 are functions of the two standard uniform random variables U_1 and U_2. It is clear from Eq. 11.66 that the variance of α_{avg} can be reduced if one can generate random variates that yield strongly negative correlations between α_1 and α_2.

A negative value of $\text{cov}[\alpha_1 (U_1), \alpha_2(U_2)]$ can be obtained by generating U_1 and U_2, which are negatively correlated. A simple approach that produces negatively correlated uniform random variates and demands minimal computation is to set $U_1 = 1 - U_2$.

11.8.2 Correlated Sampling Technique

Many times the basic objective of Monte Carlo simulation is to evaluate the sensitivity of system performance or to determine the difference in the performance of the system under several configurations and designs. Correlated sampling techniques are especially effective in such situations. Let a design involve a vector of N random variables $X = (X_1, X_2, ..., X_N)$. These variables could be correlated, having a joint PDF $f_X(x)$; these could also be independent, each having a PDF $f_i(x_i)$, $i = 1, 2, ..., N$.

As an example, consider two designs A and B of a system; the performance of the system is denoted by Z. Hence the performance of the system under design A will be

$$Z_A = g\,(A,\, X) \tag{11.67}$$

where $A = (a_1, a_2, ..., a_m)$ is a set of parameters for design A. Similarly, the performance of another design B will be

$$Z_B = g\,(B,\, X) \tag{11.68}$$

where $B = (b_1, b_2, ..., b_m)$ is a set of parameters for design B. Now, the difference in performances for the two designs is

$$Z = Z_A - Z_B \tag{11.69}$$

Since both Z_A and Z_B involve the same random numbers, these may be highly correlated. Hence, the method of correlated sampling may be used effectively to estimate statistical properties of Z. Let the mean value of these be $\overline{Z}$. The variance of $\overline{Z}$ is given by

$$\text{var}(\overline{Z}) = \text{var}(\overline{Z}_A) + \text{var}(\overline{Z}_B) - 2\text{cov}(\overline{Z}_A, \overline{Z}_B) \tag{11.70}$$

If Z_A and Z_B are positively correlated, $\text{cov}(\overline{Z}_A, \overline{Z}_B) > 0$ and the variance of $\overline{Z}$ in Eq. 11.70 will be less than the sum of their individual variances:

$$\text{var}(\overline{Z}) < \text{var}(\overline{Z}_A) + \text{var}(\overline{Z}_B) \tag{11.71}$$

Higher reduction in the variance of $\overline{Z}$ can be achieved by increasing correlation between the random numbers. The random numbers z_{Aj} and z_{Bj} will be positively correlated if they are generated as follows (Ang and Tang 1984):

$$z_{Aj} = g\left[A, F_{X_1}^{-1}(u_1), F_{X_2}^{-1}(u_2), \cdots, F_{Xn}^{-1}(u_n)\right] \tag{11.72}$$

$$z_{Bj} = g\left[B, F_{X_1}^{-1}(u_1), F_{X_2}^{-1}(u_2), \cdots, F_{Xn}^{-1}(u_n)\right] \tag{11.73}$$

where $(u_1, u_2, \ldots, u_n)$ is an independent set of uniformly distributed random numbers.

11.8.3 Latin Hypercube Simulation

Latin hypercube simulation (LHS) is a stratified sampling approach that allows efficient estimation of the statistics of output. In LHS the probability distribution of each basic variable is subdivided into N ranges, each with a probability of occurrence equal to $1/N$. Random values of the basic variable are simulated such that each range is sampled only once. The order of the selection of the ranges is randomized and the model is executed N times with the random combination of basic variables from each range for each basic variable. The output statistics, distributions, and correlations may then be approximated from the sample of N output values. The stratified sampling procedure of LHS converges more quickly than other stratified sampling procedures. As with Monte Carlo simulation, accuracy is a function of the number of samples. If N equals twice the number of suspected important parameters, it might provide a good balance of accuracy and economy for models with a large number of parameters.

For correlated basic variables, there are several procedures to generate a stratified sample. One of the popular methods is the exact sampling method. For the case of two dependent basic variables, the marginal distribution of one of the basic variables can be stratified and sampled as in LHS, and the paired value of the codependent basic variable can be randomly sampled from the conditional distribution using the value of the first basic variable. These pairs remain together in the random assignment to model runs. For N-dependent basic variables, the conditional sampling can be extended, or the joint distribution can be sampled randomly in a stratified manner. The N-tuples remain together in the random assignment to model runs.

Several computer packages containing routines for Monte Carlo and LHS methods are reported in the literature. The U.S. Environmental Protection Agency has approved some computer packages for LHS and these are available through EPA's Exposure Models Library. Monte Carlo and LHS have been applied to model groundwater contamination and reliability assessment of civil engineering structures.

11.8.4 Importance Sampling Technique

Rather than spreading the sampling points uniformly, the importance sampling technique picks up more sampling points in that part of the domain that is most important for the study. Rubinstein (1981) has demonstrated that the errors associated with the sample mean method (see Section 11.7) can be considerably reduced if $h_X(x)$ is chosen such that it has a shape similar to that of $f(x)$. However, there may be problems in drawing samples from $h_X(x)$, especially when $|f(x)|$ is not well behaved. Thus, while implementing this technique, there is a trade-off between the desired error reduction and the difficulties of sampling from a specific zone of the domain.

11.8.5 Stratified Sampling Technique

Stratified sampling is a well-established area in statistical sampling (Cochran 1972). In many ways, it is similar to importance sampling. The basic difference between the two techniques is that the stratified sampling technique takes more observations at regions that are more "important," whereas importance sampling chooses an optimal PDF. Variance reduction by the stratified sampling technique is achieved by taking more samples at important subregions. In this approach, the population consisting of N samples is subdivided in M units of on-overlapping zones, which are called strata (see Fig. 11-5). Samples are drawn from each stratum, the drawing of samples being independent in different strata.

There are many advantages of dividing the population area into strata:

1. Data with desired precision can be drawn from a specific stratum.
2. Stratification may be desirable from administrative or logistic considerations.
3. Different sampling approaches may be necessary for different strata. For instance, if an Environmental Impact Assessment (EIA) survey is to be carried for industries in an area, the aspects that are important in metal industries and in textile industries will be different.
4. Stratification helps divide a heterogeneous population into groups of more homogenous population. The advantage of a homogenous population is that, even with a small sample, reasonably reliable estimates of parameters can be obtained.

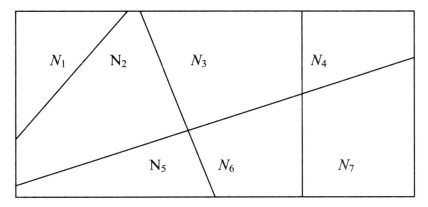

Figure 11-5 *Stratified sampling: population space of N samples divided into 7 strata.*

The theory of stratified sampling examines the issues of how to divide the population into strata, how many strata should there be, and how the domains of the strata should be determined. Cochran (1972) has described the technique in greater detail.

11.8.6 Control Variates

Sometimes, accuracy of estimation can be improved by the use of an indirect estimator. Let $\overline{Z}$ be a direct estimator and $\overline{Y}$ an indirect estimator. Let

$$\overline{Y} = \overline{Z} - \alpha(X - \mu_X) \tag{11.74}$$

where X is a random variable with known mean μ_X and α is a coefficient; X is correlated with $\overline{Z}$. Variate X is called a controlled variate for $\overline{Z}$. X may represent the performance function of a very simple model of the prototype that allows an analytical determination of μ_X. We have

$$E(\overline{Y}) = E(\overline{Z}) - \alpha[E(X) - \mu_X] = E(\overline{Z}) \tag{11.75}$$

Clearly, if $(\overline{Z})$ is an unbiased estimator then $(\overline{Y})$ is also an unbiased estimator. The variance of $(\overline{Z})$ can be obtained by

$$\operatorname{var}(\overline{Y}) = \operatorname{var}(\overline{Z}) - \alpha^2 \operatorname{var}(X) - 2\alpha \operatorname{cov}(\overline{Z}, X) \tag{11.76}$$

Therefore, the variance of $\overline{Y}$ will be less than the variance of $\overline{Z}$ if

$$2\alpha \operatorname{cov}(\overline{Z}, X) > \alpha^2 \operatorname{var}(X) \quad \text{or} \quad 2\operatorname{cov}(\overline{Z}, X) > \alpha \operatorname{var}(X) \tag{11.77}$$

In that case, the indirect estimator $\overline{Y}$ is more accurate than the direct estimator $\overline{Z}$. The value of α can be selected to obtain the maximum reduction in variance.

11.9 Reducing Parameter Uncertainty

Parameters of mathematical hydrologic models are estimated by using the available data of the real system. Many times, there may be large uncertainties in parameter values if enough data are not available. Two nonparametric statistical methods—namely the jackknife and bootstrap methods, which make no assumption about normality—are especially useful in such cases. Although these methods require large computations, the calculations are tractable. These methods are discussed next.

11.9.1 Jackknife Method

The jackknife, introduced in the late 1950s, is an attempt to answer an important statistical question: Having computed an estimate of some quantity of interest, what accuracy can be attached to the estimate? Standard deviation is a commonly used expression for the "accuracy" of an estimate. The jackknife method can be employed to compute the standard deviation of parameter estimates as follows:

1. Determine the parameter of interest, α_s, from the data of N samples.

2. For each observation $i = 1, \ldots, N$, compute α_s^i using the data of $(N - 1)$ observations; the ith observation is ignored while computing α_s^i. As a result of this step, N values of α_s are obtained.

3. The accuracy of α_s can be determined by

$$\alpha^J = \sqrt{\frac{N-1}{N} \sum_{i+1}^{N} (\alpha_s^i - \alpha_s)^2} \tag{11.78}$$

where α^J is the jackknife standard deviation of the estimate α_s.

Example 11.13 Twenty students were asked to measure the water level of Narmada River at a gauging site. The measured values (in meters) are 290.940, 290.870, 291.010, 290.950, 291.070, 291.110, 291.090, 290.640, 290.680, 290.750, 290.890, 290.580, 290.750, 291.120, 290.630, 290.890, 290.610, 290.870, 291.060, and 291.070. Determine the accuracy of the mean of these measurements using the jackknife method.

Solution There are 20 different values of the water stage of Narmada River. The mean μ of these 20 measurements of river stage is 290.879 m. Following step 2 of the jackknife method, 20 values of mean river stage were computed by ignoring one observation at a time. These mean values μ_i (in meters) are 290.928, 290.985, 291.030, 291.086, 291.132, 291.183, 291.236, 291.313, 291.363, 291.412, 291.457, 291.526, 291.570, 291.603, 291.682, 291.721, 291.788, 291.827, 291.869, and 291.922.

Applying Eq. 11.78 with $N = 20$, we obtain

$$\mu^J = \sqrt{\frac{20-1}{20}} \times 7.97 = 2.7516$$

11.9.2 Bootstrap Method

The bootstrap method was developed by Efron (1977). It can be used to determine the accuracy of any estimate determined from sample data. The computational steps of this method are as follows:

1. Let there be N data items x_i, $i = 1, ..., N$, and let G be their empirical distribution. The probability of occurrence of each datum is $1/N$.

2. Generate N new random data items x_i^*, $i = 1, ..., N$. Each new data item is a replacement for one of the N original numbers. This new set of items is called the bootstrap sample.

3. Estimate the value of the desired parameter α for the bootstrap sample x_i^*, $i = 1, ..., N$.

4. Repeat steps 2 and 3 a large number of times (say, M). Each time a new bootstrap sample is used and parameter α is estimated. Finally, we will have M independent sets of bootstrap statistics α_s^*.

5. The variance of α_s can be calculated as

$$\text{var}(\alpha_s^*) = \frac{1}{M} \sum_{m=1}^{M} \left(\alpha_s^{*(m)} - \bar{\alpha}_s^*\right)^2 \tag{11.79}$$

where $\overline{\alpha_s^*}$ is the mean of α_s^*.

The jackknife and bootstrap methods can be used to compute the variance of any statistic (e.g., mean, standard deviation, and skewness) that are determined from the sample data.

11.10 Uncertainty and Sensitivity Analysis Using Monte Carlo Simulation

The discussion on model uncertainty in the preceding chapters shows that by sensitivity analysis the relative importance of model parameters (and variables) can be assessed by perturbing one parameter (or variable) at a time about a selected value in parameter space and determining the sensitivity of model output to such a perturbation in the form of a sensitivity coefficient. This type of sensitivity analysis is designated as local sensitivity analysis (LSA). The sensitivity coefficient can be expressed in several ways. One way to express it is as follows. Denoting the model output by Y and the jth parameter by x_j, we can express the sensitivity coefficient s_j for the jth parameter as

$$s_j = \left(\frac{dY}{dx_j}\right)\Big|_{x_0} = \frac{\Delta Y}{\Delta x_j} = \frac{Y(x_{j0} + \Delta x_j) - Y(x_{j0})}{\Delta x_j} \tag{11.80}$$

The sensitivity coefficient can also be expressed in dimensionless form as

$$s_j(\%) = \left(\frac{dY/Y(x_{j0})}{dx_j/x_{j0}}\right)\Big|_{x_0} = \frac{\Delta Y}{\Delta x_j}\left(\frac{x_{j0}}{Y(x_{jo})}\right) = \frac{Y(x_{j0} + \Delta x_j) - Y(x_{j0})}{\Delta x_j}\left(\frac{x_{j0}}{Y(x_{j0})}\right) \tag{11.81}$$

Local sensitivity analysis may be able to decipher sources that significantly contribute to model uncertainty. For example, a high sensitivity parameter may have less influence on the model uncertainty than a parameter that is much less sensitive but is more uncertain. Melching (2001) has reasoned that to evaluate the contribution of each model parameter and input variable to the overall uncertainty of model output, uncertainty analysis must integrate the effects of sensitivity as well as uncertainty.

In many practical cases, model sensitivity varies from one region of parameter space to another. In such cases, LSA yields limited information and therefore global sensitivity analysis (GSA) should be employed. GSA permits evaluation of the model response over the parameter space (i.e., the pattern of change in model output to changes in parameters over the entire range). To perform GSA using MCS-based schemes, one generates sets (say, I) of model parameters (say J), $I > J$, following their statistical characteristics in a defined parameter range. Then, each generated parameter set is used in the model to produce output. Thus one would obtain I sets of model outputs, that is,

$$(x_{1,i}, x_{2,i}, ..., x_{J,i}) \rightarrow \text{Model} \rightarrow y_i, i = 1, 2, ..., I$$

where $x_{j,i}$ is the jth parameter generated in the ith parameter set, and y_i is the corresponding ith model output. Now one can use regression and correlation analysis for I outputs and I sets of J parameters to determine the relative importance of each of the J parameters and then define sensitivity and uncertainty indicators.

The accuracy of MCS-based GSA depends on the sample size I for the number of parameters J. Using the LHS scheme, McKay (1988) suggested I to be greater than or equal to twice J, whereas Iman and Helton (1985) as well as Manache (2001) found that $I = 4J/3$ would be adequate. The importance of each model parameter can be determined by using the coefficient of correlation (or the Pearson product moment correlation coefficient) indicating the strength of the linear relationship between model output and parameters. It is possible that there is a monotonic nonlinear relationship between model output and parameters. Then Spearman's rank correlation coefficient can be computed. To that end, the data generated for each parameter and model output are ranked in either ascending order or descending order and then the rank correlation coefficient is computed as

$$r_{R(y), R(x_j)} = \frac{\sum\limits_{i=1}^{I} [R(x_{j,i}) - \frac{I+1}{2}][R(y_i) - \frac{M+1}{2}]}{I(I^2 - 1)/12} = 1 - \frac{6 \sum\limits_{i=1}^{I} [R(x_{j,i}) - R(y_i)]^2}{M(M^2 - 1)/12} \quad \text{(11.82)}$$

where $j = 1, 2, ..., J$, r is the rank correlation coefficient, $R(y_i)$ is the rank of the i-generated model output value of y, and $R(x_{i,j})$ is the ith generated value of parameter x_j. If the rank correlation coefficient is higher than the Pearson product correlation coefficient then a nonlinear relationship exists between model output and parameter values. In this case the sensitivity coefficient changes with parameter values.

In practice a second-order regression equation suffices to relate the model output to model parameters:

$$y = a_0 + \sum_{j=1}^{J} b_j x_j + \sum_{j=1}^{J} c_j x_j^2 + \sum_{j=1}^{J-1} \sum_{m=j+1}^{J} d_{jm} x_j x_m + \varepsilon \quad \text{(11.83)}$$

where a, b, c, and d are regression coefficients; ε is the error term denoting the deviation between model response and regression relation-produced output. For purposes of discussion of GSA, only the linear term of Eq. 11.83 is retained:

$$y = a_0 + \sum_{j=1}^{J} b_j x_j + \varepsilon \quad \text{(11.84)}$$

Equation 11.84 shows that the global sensitivity coefficient associated with model parameters b_j is the same as the regression coefficient $dy/dx_j = b_j$, reflecting the average sensitivity of the model response to a unit change in the model parameter.

For comparing model sensitivity to parameters having different units, it is better to standardize model parameters and model output as

$$x_* = \frac{(x - \bar{x})}{s_x}, \quad y_* = \frac{(y - \bar{y})}{s_y}$$

Equation 11.84 can now be expressed as

$$y_* = \sum_{j=1}^{J} b_{*j} x_{*j} \quad \text{(11.85)}$$

where b_* is the standardized regression coefficient and is related to the regression coefficient b_j as

$$b_{*j} = \frac{s_j}{s_y} b_j \tag{11.86}$$

Now the contribution of each model parameter to the total model output variability can be expressed as

$$s_y^2 = \sum \sum (b_j s_j)(b_m b_m) r_{x_j, x_m} + s_\varepsilon^2 \tag{11.87}$$

Here r_{x_j, x_m} is the correlation coefficient between model parameter x_j and x_m, and s_ε^2 is the mean square error. If model parameters are independent, each term in the summation $b_j^2 s_j^2$ represents the regression sum of squares, indicating the contribution of parameter x_j to the overall model uncertainty. If model parameters are not independent, then correlation among parameters might increase or decrease the total model output variability; this, of course, depends on the algebraic sign of the sensitivity coefficient and of the correlation coefficient. In such cases partial correlation coefficients can be employed to incorporate the influence of other correlated parameters.

11.11 Questions

11.1   Generate 20 uniformly distributed random numbers using Eq. 11.1 with $a = 4$, $b = 2$, and $d = 5$. Take an appropriate value of seed R_0.

11.2   Generate values of a random variable that follows an exponential distribution with parameter $\lambda = 5.0$.

11.3   Generate values of a random variable that follows a symmetric triangular distribution.

11.4   Generate values of a random variable that follows an asymmetric (to the left) triangular distribution.

11.5   Generate values of a random variable that follows an asymmetric (to the right) triangular distribution.

11.6   Generate a random variable that follows the probability density function

$$f_X(x) = 3/4 + x^3, 0 \le x \le 1$$

11.7   Generate gamma-distributed random variates with parameters (5, 0.5) using the function-based method.

11.8   Generate normally distributed random numbers with parameters $(5, 2.0)$ using the function-based method.

11.9   Generate values of a random variate that follows a binomial distribution with parameters $(10.0, 0.5)$.

11.10   Generate values of a random variate that follows a Poisson distribution with parameter $\lambda = 8$.

11.11   Generate numbers that follow a bivariate normal distribution with parameters $(4, 1.5)$ and $(6.0, 2.5)$. The coefficient of correlation (ρ) between the numbers is 0.5.

11.12   Twenty values of the water level of Narmada River were measured at a gauging site. The measured values (in meters) are 291.20, 291.50, 291.50, 290.05, 292.01, 291.20, 291.10, 290.58, 290.78, 290.85, 290.90, 290.35, 290.65, 291.08, 290.43, 290.78, 290.82, 290.95, 291.35, and 291.29. Determine the accuracy of the mean of these measurements using the jack-knife method.

Chapter 12

Stochastic Processes

A stochastic (or random) process represents a family of functions. Consider, for example, a time series of river discharge at a gauging station, as shown in Fig. 12-1. It is observed that discharge X varies in an erratic manner, such that it is hard to represent it algebraically. Successive observations of daily discharge, however, exhibit a strong dependence or memory in time. The dependence is measured by serial correlation and suggests that a family of random variables may describe the discharge time series, one for every day. The probability density function of each random variable may or may not be the same for each day and its parameters may be correlated with those at other days. Thus, $X(t)$ is a stochastic process. For a specific day, $X(t = t^1)$ is a random variable at time $t = t^1$. Of course, a stochastic process can also be represented as $X(t, \lambda)$, where λ represents possible outcomes. For a specific outcome $\lambda = \lambda_1$, $X(t, \lambda_1)$ is a simple time function. For a specific time t^1, $X(t_1, \lambda)$ is a random variable whose outcomes are the intersection of $X(t, \lambda)$ and the time $t = t^1$. For simplicity, we represent a stochastic process by $X(t)$. Thus a stochastic process is a random variable for any given value of time t and can be considered as the collection of all possible records of variation of the observed quantity (e.g., discharge) in time. The collection is called the *ensemble*. In most practical applications, there is only one observation of the time variation of the phenomenon under consideration, and this particular record is called a *realization of the stochastic process*, as shown in Fig. 12-1.

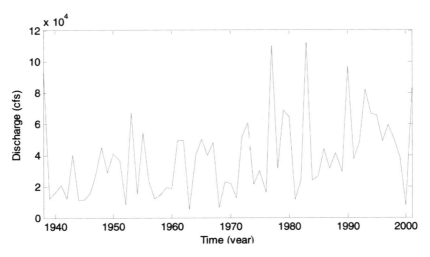

Figure 12-1 *A single realization of a stochastic process.*

Since $X(t)$ is a random variable at each time, $X(t)$ will have a probability density function (PDF), denoted as $f_X(x, t)$, and a cumulative distribution function (CDF) defined as $F_X(x,t) = P(X(t) \leq x)$. Note that both PDF and CDF are functions of time and related as usual:

$$f_X(x,\ t) = \frac{\partial F_X(x,t)}{\partial x}$$

The function $f_X(x, t)$ is also called the first-order density of $X(t)$.

For two assigned times $t_1, t_2, X(t_1)$ and $X(t_2)$ are random variables. Their joint distribution depends on the values of t_1 and t_2 and can be written as

$$F_X(x_1,x_2;t_1,t_2) = P(X(t_1) \leq x_1, X(t_2) \leq x_2) \tag{12.1}$$

and

$$f_X(x_1,x_2;t_1,t_2) = \frac{\partial^2 F_X(x_1,x_2;t_1,t_2)}{\partial x_1 \partial x_2} \tag{12.2}$$

Here $F_X(x_1, x_2; t_1, t_2)$ is also called the second-order distribution of $X(t)$, and $f_X(x_1, x_2; t_1, t_2)$ is called the second-order probability density function of $X(t)$.

12.1 Mean and Variance

The mean of a stochastic process $X(t)$ is defined as

$$\mu_X(t) = E[X(t)] = \int_{-\infty}^{\infty} x f_X(x, t) dx \tag{12.3}$$

where $\mu_X(t)$ is the time-dependent mean of the process $X(t)$. The variance of the process is expressed as

$$\sigma_X^2(t) = E[X(t) - \mu(t)]^2 = \int_{-\infty}^{\infty} (x - \mu(t))^2 f_X(x,t)dx \tag{12.4}$$

12.2 Covariance and Correlation

A complete probabilistic description of a stochastic process entails consideration of the interrelationship among random variables resulting from the stochastic process at different times. This can be expressed by using the covariance between $X(t_1) = X_1$ and $X(t_2) = X_2$:

$$cov(X(t_1), X(t_2)) = E\{[X(t_1) - \mu_X(t_1)][X(t_2) - \mu_X(t_2)]\} \tag{12.5}$$

If $t_1 = t_2$, Eq. 12.5 yields the variance as a function of time. cov $[X(t_1), X(t_2)]$ is the autocovariance of the random process $X(t)$ at times t_1 and t_2. Equation 12.5 can be expressed as

$$\begin{aligned} cov[X(t_1), X(t_2)] &= E[X(t_1)X(t_2) - \mu_X(t_2)X(t_1) - \mu_X(t_1)X(t_2) + \mu_X(t_1)\mu_X(t_2)] \\ &= E[X(t_1)X(t_2)] - m_X(t_1)m_X(t_2) \\ &= R_X(t_1, t_2) - \mu_X(t_1)\mu_X(t_2) \end{aligned} \tag{12.6}$$

where $R_X(t_1, t_2)$ is the autocorrelation of $X(t)$ at t_1 and t_2 and is indeed the joint moment of random variables $X(t_1)$ and $X(t_2)$:

$$R_X(t_1, t_2) = E[X(t_1)X(t_2)] = \int_{-\infty}^{\infty} \int_{-\infty}^{\infty} x_1 x_2 f_X(x_1, x_2; t_1, t_2)dx_1 dx_2 \tag{12.7}$$

For $t_1 = t_2 = t$, Eq. 12.6 can be written as:

$$R_X(t,t) = \int_{-\infty}^{\infty} x^2 f(x,t)dx$$

The covariance is an even function of time lag $\tau = t_1 - t_2$, that is,

$$cov(\tau) = cov(-\tau) \tag{12.8}$$

For $t_1 = t_2 = t$, Eq. 12.6 yields

$$cov[X(t), X(t)] = R_X(t,t) - \mu_X^2(t) = \sigma_X^2(t) \tag{12.9}$$

which is the variance of X as a function of t.

The autocorrelation coefficient ρ of two random variables is defined as the ratio of the covariance to the product of their standard deviations. Hence $\rho_X(t_1, t_2)$ is written as

$$\rho_X(t_1, t_2) = \frac{\text{cov}(t_1, t_2)}{\sigma_X(t_1)\sigma_X(t_2)} \tag{12.10}$$

For $t_1 = t_2 = t$, $\rho_X(t, t) = 1$. The covariance reduces to the variance:

$$\sigma_X^2(t) = E[X(t) - \mu(t)]^2 = \int_{-\infty}^{\infty} [x - \mu(t)]^2 f_X(x, t)dx$$

Example 12.1 Consider a stochastic process expressed as $X(t) = at$, where a is a random variable and t is time. Determine the mean, autocovariance function, autocorrelation function, variance, and autocorrelation coefficient.

Solution Applying Eq. 12.3 for the mean gives

$$E[X(t)] = \mu_X(t) = E[at] = t\mu_a$$

where μ_a is the mean of a. Applying Eq. 12.7 for the autocorrelation function gives

$$R(t_1, t_2) = E[(at_1)(at_2)] = E[a^2 t_1 t_2] = t_1 t_2 E[a^2]$$

The autocovariance function is found by applying Eq. 12.6:

$$\begin{aligned} \text{cov}(t_1, t_2) &= R_X(t_1, t_2) - \mu_X(t_1)\mu_X(t_2) \\ &= t_1 t_2 E(a^2) - t_1 t_2 [E(a)]^2 \\ &= t_1 t_2 \{E(a^2) - [E(a)]^2\} = t_1 t_2 \, \text{var}(a) = t_1 t_2 \sigma_a^2 \end{aligned}$$

where σ_a^2 is the variance of a. Applying Eq. 12.9 gives the variance:

$$\sigma_x^2(t) = t^2 E(a^2) - t^2 [E(a)]^2 = t^2 \sigma_a^2$$

Finally, applying Eq. 12.10 yields the correlation coefficient:

$$\rho_X(t_1, t_2) = \frac{\text{cov}(t_1, t_2)}{\sigma_X(t_1)\sigma_X(t_2)} = \frac{t_1 t_2 \sigma_a^2}{\sqrt{t_1^2 \sigma_a^2}\sqrt{t_2^2 \sigma_a^2}} = \frac{t_1 t_2 \sigma_a^2}{t_1 t_2 \sigma_a^2} = 1$$

Example 12.2 Consider a stochastic process expressed as $Q(t, K) = Q_0 \exp(-kt)$, in which Q_0 and K are independent random variables and k is a specific value of K. Let $K \sim N(\mu_K, \sigma_K^2)$ and $Q_0 \sim N(\mu_Q, \sigma_Q^2)$. Determine the mean, variance, autocorrelation function, covariance function, and correlation coefficient of $Q(t)$. This equation is used to describe a streamflow recession process.

Solution Applying Eq. 12.3 for the mean gives

$$E[Q(t)] = E[Q_0 e^{-kt}] = E[Q_0]E[e^{-kt}]$$

Here $E[e^{-kt}]$ can be evaluated by noting that K is a normally distributed random variable. Therefore,

$$E\left(e^{-kt}\right) = \int_{-\infty}^{\infty} e^{-kt} f_K(k)dk = \frac{1}{\sqrt{2\pi}\,\sigma_K} \int_{-\infty}^{\infty} e^{-kt} \exp\left[-\frac{(k-\mu_K)^2}{2\sigma_K^2}\right]dk$$

$$= \exp\left(\frac{\sigma_K^2 t^2}{2} - \mu_K t\right)$$

where $f_K(k)$ is the PDF of K. The term within the integral sign can be evaluated as follows: First, consider

$$\exp\left[-\frac{(k^2 + \mu_K^2 - 2k\mu_K + 2\sigma_K^2 kt)}{2\sigma_K^2}\right]$$

$$= \exp\left\{-\frac{\left[k-(\mu_K - \sigma_K^2 t)\right]^2 + 2\mu_K \sigma_K^2 t - \sigma_K^4 t^2}{2\sigma_K^2}\right\}$$

$$= \exp\left\{-\frac{\left[k-(\mu_K - \sigma_K^2 t)\right]^2}{2\sigma_K^2}\right\}\exp\left(\frac{\sigma_K^2 t^2}{2} - \mu_K t\right)$$

Therefore, the integral term becomes

$$\int_{-\infty}^{\infty} \exp\left[-\frac{k-(\mu_K - \sigma_K^2 t)^2}{2\sigma_K^2}\right]\exp\left(\frac{\sigma_K^2 t^2}{2} - \mu_K t\right)dk$$

$$= \exp\left(\frac{\sigma_K^2 t^2}{2} - \mu_K t\right)\int_{-\infty}^{\infty} \exp\left\{-\frac{\left[k-(\mu_K - \sigma_K^2 t)\right]^2}{2\sigma_K^2}\right\}dk$$

Hence,

$$E\left(e^{-kt}\right) = \exp\left(\frac{\sigma_K^2 t^2}{2} - \mu_K t\right)\frac{1}{\sqrt{2\pi}} \int_{-\infty}^{\infty} \exp\left\{-\frac{\left[k-(\mu_K - \sigma_K^2)\right]^2}{2\sigma_K^2}\right\}dk$$

This yields the mean of Q as

$$E[Q(t)] = E\left(Q_0\right)\exp\left(\frac{\sigma_K^2 t^2}{2} - \mu_K t\right) \text{ or } \mu_Q(t) = \mu_{Q_0} \exp\left(\frac{\sigma_K^2 t^2}{2} - \mu_K t\right)$$

From Eq. 12.7 for the autocorrelation function, one gets

$$R(t_1,t_2) = E\left[(Q_0 e^{-kt_1})(Q_0 e^{-kt_2})\right] = E\left(Q_0^2\right) E\left(e^{-kt_1-kt_2}\right)$$

$$= E\left(Q_0^2\right) \exp\left[\frac{\sigma_K^2(t_1^2+t_2^2)}{2} - \mu_K(t_1+t_2)\right]$$

Equation 12.6 gives the autocovariance function:

$$\text{cov}(t_1,t_2) = E\left[Q_0^2\right] \exp\left[\frac{\sigma_K^2(t_1^2+t_2^2)}{2} - \mu_K(t_1+t_2)\right]$$

$$- [E(Q_0)]^2 \exp\left[\frac{\sigma_K^2(t_1^2+t_2^2)}{2} - \mu_K(t_1+t_2)\right]$$

$$= \left\{E\left(Q_0^2\right) - [E(Q_0)]^2\right\} \exp\left[\frac{\sigma_K^2(t_1^2+t_2^2)}{2} - \mu_K(t_1+t_2)\right]$$

Referring to Eq. 12.9 for the variance $(t_1 = t_2 = t)$, one obtains

$$\sigma_Q^2 = E\left[Q_0^2\right] \exp\left[\sigma_K^2 t^2 - 2\mu_K t\right] - [E(Q_0)]^2 \exp\left[\sigma_K^2 t^2 - 2\mu_K t\right]$$

$$= \exp\left[\sigma_K^2 t^2 - 2\mu_K t\right] - \left\{E(Q_0)^2 - [E(Q_0)]^2\right\}$$

$$= \sigma_{Q_0}^2 \exp\left[\sigma_K^2 t^2 - 2\mu_K t\right] = \sigma_{Q_0}^2 \frac{E(Q)}{E(Q_0)}$$

where $\sigma_{Q_0}^2$ is the variance of Q_0.

Application of Eq. 12.10 yields the correlation coefficient:

$$\rho_Q(t_1,t_2) = \frac{E[Q_0^2]\exp[\frac{\sigma_K^2(t_1^2+t_2^2)}{2} - \mu_K(t_1+t_2)] - [E(Q_0)]^2 \exp[\frac{\sigma_K^2(t_1^2+t_2^2)}{2} - \mu_K(t_1+t_2)]}{\sigma_{Q_0}^2 \sqrt{\exp\left[2t_1^2\sigma_K^2 - 2\mu_K t_1\right]\exp\left[2t_2^2\sigma_K^2 - 2\mu_K t_2\right]}}$$

Example 12.3 Let the head in an aquifer be described by

$$H = H_0 - \frac{(H_0 - H_L)}{L} x$$

where H_0 is the head at $x = 0$, H_L is the head at $x = L$ (the length of the aquifer), and x is the distance coordinate. This is the solution of the one-dimensional Laplace equation for a homogenous isotropic confined aquifer. This equation can be represented as $H = H_0 + Bx$, where $B = -(H_0 - H_L)/L$ and H_0 is a stochastic process in x (space), with H_0 and B independent random variables. Determine the mean, variance, autocorrelation function, autocovariance function, and autocorrelation coefficient.

Solution Applying Eq. 12.3 for the mean gives

$$E[H(x)] = E(H_0 + Bx) = \mu_{H_0} + x\mu_B$$

where μ_{H_0} and μ_B are the mean of H_0 and the mean of B, respectively. Using Eq. 12.7 for the autocorrelation function yields

$$R(x_1, x_2) = E\left[(H_0 + Bx_1)(H_0 + Bx_2)\right]$$
$$= E\left(H_0^2\right) + x_1 E(H_0)E(B) + x_2 E(H_0)E(B) + x_1 x_2 E\left(B^2\right)$$
$$= E\left(H_0^2\right) + (x_1 + x_2)E(H_0)E(B) + x_1 x_2 E\left(B^2\right)$$

The autocovariance function is given by Eq. 12.6 as

$$\text{cov}(x_1, x_2) = R(x_1, x_2) - \mu_H(x_1)\mu_H(x_2)$$
$$= E\left(H_0^2\right) + (x_1 + x_2)E(H_0)E(B) + x_1 x_2 E\left(B^2\right) - (\mu_{H_0} + x_1\mu_B)$$
$$(\mu_{H_0} + x_2\mu_B)$$
$$= E\left(H_0^2\right) + (x_1 + x_2)\mu_{H_0}\mu_B + x_1 x_2 E\left(B^2\right) - \mu_{H_0}^2 - x_2\mu_{H_0}\mu_B$$
$$- x_1\mu_{H_0}\mu_B - x_1 x_2\mu_B^2$$
$$= \sigma_{H_0}^2 + x_1 x_2 E\left(B^2\right) - x_1 x_2\mu_B^2$$

where $\sigma_{H_0}^2$ is the variance of H_o.
Applying Eq. 12.9 gives the variance of H:

$$\sigma_H^2(x) = R(x, x) - \mu_H(x)\mu_H(x) = \sigma_{H_0}^2 + x^2 E\left(B^2\right) - x^2\mu_B^2 = \sigma_{H_0}^2 + x^2\sigma_B^2$$

where σ_B^2 is the variance of B.
The autocorrelation coefficient is given as

$$\rho_H(x_1, x_2) = \frac{\sigma_{H_0}^2 + x_1 x_2 E\left(B^2\right) - x_1 x_2\mu_B^2}{\sqrt{\sigma_{H_0}^2 + x_1^2\sigma_B^2}\sqrt{\sigma_{H_0}^2 + x_2^2\sigma_B^2}}$$

Example 12.4 Let the potential rate of infiltration be described by the Horton model: $f(t) = f_0 + (f_0 - f_c)\exp(-Kt)$, where f_0 is the initial infiltration rate, f_c is the final (or steady) infiltration rate, and K is a parameter. The Horton model can be simply written as $f(t) = f_0 + B\exp(-Kt)$, where $B = (f_0 - f_e)$. Let $f(t)$ be a stochastic process with f_0, B, and K as independent random variables. K is normally distributed. Determine the mean, variance, autocorrelation function, covariance function, and autocorrelation coefficient.

Solution Applying Eq. 12.3 for the mean gives

$$E[f(t)] = E[f_0 + B\exp(-Kt)] = E(f_0) + E(B)E[\exp(-Kt)]$$

$$= \mu_{f_0} + \mu_B \int_{-\infty}^{\infty} \exp(-Kt) \frac{1}{\sigma_K \sqrt{2\pi}} \exp\left[-\frac{(K-\mu_K)^2}{2\sigma_K^2}\right] dK$$

$$= \mu_{f_0} + \mu_B \exp\left(\frac{\sigma_K^2 t^2}{2} - \mu_K t\right)$$

where μ_{f_0} is the mean of f_0, μ_B is the mean of B, μ_K is the mean of K, and σ_K^2 is the variance of K. Applying Eq. 12.7 yields the autocorrelation function:

$$R(t_1, t_2) = E\{[f_0 + B\exp(-Kt_1)][f_0 + B\exp(-Kt_2)]\}$$

$$= E(f_0^2) + E(f_0)E(B)E[\exp(-Kt_2)] + E(f_0)E(B)E[\exp(-Kt_1)]$$
$$+ E(B^2)E(\exp(-Kt_1))E[\exp(-Kt_2)]$$

$$= E(f_0^2) + \mu_{f_0}\mu_B \exp\left(\frac{\sigma_K^2 t_2^2}{2} - \mu_K t_2\right) + \mu_{f_0}\mu_B \exp\left(\frac{\sigma_K^2 t_1^2}{2} - \mu_K t_1\right)$$
$$+ E(B^2)\exp\left[\frac{\sigma_K^2}{2}(t_1^2 + t_1^2) - \mu_K(t_1 + t_2)\right]$$

Using Eq. 12.6 for the autocovariance function, one obtains

$$\text{cov}(t_1, t_2) = R(t_1, t_2) - \mu_f(t_1)\mu_f(t_2)$$

$$= R(t_1, t_2) - \left[\mu_{f_0} + \mu_B \exp\left(\frac{\sigma_K^2 t_1^2}{2} - \mu_K t_1\right)\right]\left[\mu_{f_0} + \mu_B \exp\left(\frac{\sigma_K^2 t_2^2}{2} - \mu_K t_2\right)\right]$$

$$= R(t_1, t_2) - (\mu_{f_0})^2 - \mu_{f_0}\mu_B \exp\left(\frac{\sigma_K^2 t_2^2}{2} - \mu_K t_2\right) - \mu_{f_0}\mu_B \exp\left(\frac{\sigma_K^2 t_1^2}{2} - \mu_K t_1\right)$$
$$- \mu_B^2 \exp\left[\frac{\sigma_K^2}{2}(t_1^2 + t_2^2) - \mu_K(t_1 + t_2)\right]$$

$$= E\left(f_0^2\right) + \mu_{f_0}\mu_B \exp\left(\frac{\sigma_K^2}{2}t_2^2 - \mu_K t_2\right) + \mu_{f_0}\mu_B \exp\left(\frac{\sigma_K^2}{2}t_1^2 - \mu_K t_1\right)$$

$$+ E\left(B^2\right)\exp\left[\frac{\sigma_K^2}{2}(t_1^2 + t_2^2) - \mu_K(t_1 + t_2)\right] - (\mu_{f_0})^2 - \mu_{f_0}\mu_B \exp\left(\frac{\sigma_K^2 t_2^2}{2} - \mu_K t_2\right)$$

$$- \mu_{f_0}\mu_B \exp\left(\frac{\sigma_K^2 t_1^2}{2} - \mu_K t_1\right) - (\mu_B)^2 \exp\left[\frac{\sigma_K^2}{2}(t_1^2 + t_2^2) - \mu_K(t_1 + t_2)\right]$$

$$= \sigma_{f_0}^2 + \sigma_B^2 \exp\left[\frac{\sigma_K^2}{2}(t_1^2 + t_2^2) - \mu_K(t_1 + t_2)\right]$$

From Eq. 12.9, the variance of $f(t)$ is

$$\sigma_f^2(t) = \sigma_{f_0}^2 + \sigma_B^2 \exp[\sigma_K^2 t^2 - 2\mu_K t]$$

From Eq. 12.10 the correlation function is obtained as

$$\rho_f(t_1, t_2) = \frac{\sigma_{f_0}^2 + \sigma_B^2 \exp\left[\frac{\sigma_K^2}{2}\left(t_1^2 + t_2^2\right) - \mu_K(t_1 + t_2)\right]}{\sqrt{\sigma_{f_0}^2 + \sigma_B^2 \exp\left(\sigma_K^2 t_1^2 - 2\mu_K t_1\right)}\sqrt{\sigma_{f_0}^2 + \sigma_B^2 \exp\left(\sigma_K^2 t_2^2 - 2\mu_K t_2\right)}}$$

12.3 Stationarity

A stochastic process is considered stationary if its probabilistic descriptions (e.g., statistics) are independent of a shift in time. This means that joint distributions would be invariant with a shift of the time origin. Two processes $X(t)$ and $X(t + \tau)$ have the same statistics for any τ, that is, $f_X(x,t) = f_X(x)$ is independent of t; $\mu_X(t) = \mu_X$ is constant; $f_X(x_1, x_2; t_1, t_2) = f_X(x_1, x_2; \tau)$, $\tau = t_1 - t_2$, depends on the time difference τ; $R_X(t_1, t_2) = R(\tau)$ depends on the time difference; $\text{cov}[x(t_1), x(t_2)] = \text{cov}(x; \tau)$ depends on the time difference; and $\rho_X(t1, t2) = \rho_X(\tau)$ also depends on the time difference. If $t_1 = t_2$, $\tau = 0$, then $R(t, t)$ gives the variance of the process.

The stationarity property may also be extended to n-dimensional vectors. For example, when the joint distribution of n-dimensional random vectors $\{X_1(t), X_2(t), ..., X_n(t)\}$ and $\{X_1(t + \tau), X_2(t + \tau), ..., X_n(t + \tau)\}$ have the same statistical characteristics (e.g., mean, variance, etc.) for all τ, the stochastic process $X(t)$ is stationary. If a stochastic process does not satisfy this condition, the process is called an *evolutionary stochastic* or *nonstationary process*.

The concept of stationarity alludes to a similar structure of variability at different times (i.e., some kind of repetition is implied in the process). This is an important property in that statistical interpretations can be based on the analysis of a single realization. If the joint distribution is invariant, irrespective of time, the stochastic process is strictly stationary. When the mean of $X(t)$ and $X(t + \tau)$ converge to the population mean, the process is said to be stationary in the mean or first-order stationary. A weakly stationary stochastic process is defined with use of the autocovariance function:

$$\text{cov}[X(t), X(t + \tau)] = R(\tau) \tag{12.11}$$

which depends only on the time difference τ. Such a process is a weakly stationary stochastic process. Its mean value function $E[X(t)]$ is constant, independent of t, and its autocovariance function depends only on τ for all t (i.e., second-order stationary). Many times "stationary" implies second-order stationary. The condition of a constant mean can be relaxed in many environmental processes. Thus, with known mean, the second-order stationarity condition can be applied to the resulting zero-mean process.

If further higher-order (third, fourth,...) moments of the series are independent of time but depend on τ and converge to the higher-order population moments as a large number of samples are drawn, the series is said to be higher order or strictly stationary. Usually a hydrologic series is tested only up to second-order stationarity properties.

Consider an example of a single realization of a stationary process as shown in Fig. 12-1. The covariance function can be determined by taking the average of the lagged product of the departure of X from its constant mean, and repeating the process for all lags τ. For a simple realization, the covariance is determined by the time average of the lagged product; this is done for all possible lags. Physically, the covariance function exhibits the degree of correlation between the processes at adjacent points in time. For a continuous process, one can expect a high correlation at points close to each other (e.g., daily streamflow), but the correlation decreases as the lag between points increases (e.g., monthly streamflow).

A common covariance function has a negative exponential form:

$$\text{cov}(t) = s^2 \exp[-|t|/l] , \quad \tau \geq 0 \tag{12.12a}$$

where λ is a parameter, called integral scale, and can be expressed as

$$\lambda = \int_0^\infty \rho(\tau)d\tau \ , \quad \rho(\tau) = \frac{\text{cov}(\tau)}{\sigma^2} \tag{12.12b}$$

This covariance function is plotted in Fig. 12-2.

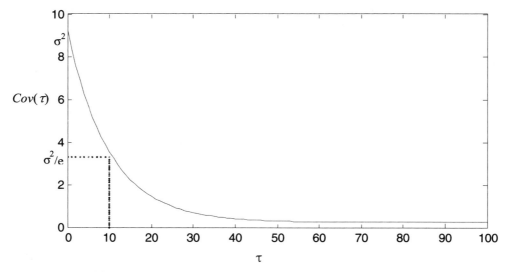

Figure 12-2 *Negative exponential covariance function.*

12.4 Correlogram

The autocorrelation coefficient (Eq. 12.10) can be expressed in terms of τ, the lag or separation time, as

$$\rho_X(t, t+\tau) = \frac{\text{cov}(t, t+\tau)}{\sigma_X^2(t)} \tag{12.13}$$

If $\tau = 0$, $\rho_X(t, t) = \rho_X(t = 0) = 1$. As τ increases, $\rho_X(\tau)$ decreases and vice versa. Hence ρ_X is a measure of linear dependence. A decline in its value with increasing τ suggests a decrease in the memory of the process. A graph of ρ_X with τ is referred to as the correlogram. The pace at which $\rho_X(\tau)$ decreases as τ increases is a measure of serial dependence of the stochastic process. When a correlogram exhibits a periodic increase and decrease with lag, this indicates that a deterministic component, such as seasonal variability, is present in the process. Fig. 12-3 shows a correlogram of a stochastic process where the autocorrelation function smoothly decreases with increasing τ; and Fig. 12-4 shows a correlogram of a stochastic process having periodicity.

In environmental and water resources, observations are often made at time intervals such as, say, 12 hours, 24 hours, 1 week, 1 month, or 1 year. In such cases, a sample autocorrelation coefficient is estimated. This measures the linear statistical dependence of consecutive observations. For such observations, τ can be represented as $\tau = k\Delta t$, $k = 0, 1, 2,\ldots$. For a fixed time interval $t = 1$ Eq. 12.3 can be estimated as

The first part in Eq. 12.17 yields that X will have zero mean, and the second part shows that increments of Z at two different frequencies are uncorrelated (i.e., the Z process will have orthogonal increments). If $\omega_1 = \omega_2 = \omega$, then

$$E[dZ(\omega)dZ^*(\omega)] = d\Phi(\omega) = S(\omega)d\omega \tag{12.18}$$

where (ω) is the integrated spectrum, and $S(\omega)$ is the spectral density function or simply the spectrum.

If Eq. 12.16 is viewed as a Fourier transform then dZ can be considered as the random amplitude. Equation 12.18 shows that the spectrum is proportional to the square of the random amplitude per frequency increment. This means that the spectrum must be non-negative at all frequencies.

For a stationary stochastic process, the covariance function can be expressed as the inverse of the Fourier transform of the spectrum:

$$\text{cov}(\tau) \text{ or } R(\tau) = \int_{-\infty}^{\infty} e^{i\omega t} S(\omega)d\omega \tag{12.19a}$$

or

$$R(\tau) = \int_{-\infty}^{\infty} S(\omega)\cos(\omega\tau)d\omega \tag{12.19b}$$

Likewise, the spectral density is the Fourier transform of the autocovariance function:

$$S(\omega) = \frac{1}{2\pi} \int_{-\infty}^{\infty} e^{-i\omega\tau} R(\tau)d\tau \tag{12.20a}$$

or

$$S(\omega) = \frac{1}{2\pi} \int_{-\infty}^{\infty} R(\tau)\cos(\omega\tau)d\tau \tag{12.20b}$$

The spectrum is an even function: $S(\omega) = S(-\omega)$.
If the process is not zero mean, then

$$\text{cov}(\tau) = \int_{-\infty}^{\infty} S(\omega)e^{i\omega\tau}d\omega = \int_{-\infty}^{\infty} S(\omega)\cos(\omega\tau)d\omega \tag{12.21}$$

and

$$S(\omega) = \frac{1}{2\pi} \int_{-\infty}^{\infty} \text{cov}(\tau)e^{-\omega\tau}d\tau = \frac{1}{2\pi} \int_{-\infty}^{\infty} \text{cov}(\tau)\cos(\omega\tau)d\tau \tag{12.22}$$

If the process is covariance stationary, one may normalize the spectral density function by dividing it by the variance σ_X^2. This means that the covariance function can be replaced in Eq. 12.22 by the autocorrelation coefficient:

$$S_*(\omega) = \frac{1}{2\pi} \int_{-\infty}^{\infty} \rho_X(\tau) e^{-\omega\tau} d\tau = \frac{1}{2\pi} \int_{-\infty}^{\infty} \rho_X(\tau) \cos(\omega\tau) d\tau \qquad (12.23)$$

where $S_*(\omega)$ is the normalized spectral density function. Likewise, Eq. 12.21 can be recast as

$$\rho_X(\tau) = \int_{-\infty}^{\infty} S_*(\omega) e^{i\omega\tau} d\omega = \int_{-\infty}^{\infty} S_*(\omega) \cos(\omega\tau) d\omega \text{ s} \qquad (12.24)$$

Equation 12.24 indicates that area under the normalized spectral density function is unity since $\rho(0) = \cos(0) = 1$.

If $\tau = 0$, $\text{cov}(0) = \sigma^2 = $ variance, then Eq. 12.19 reduces to

$$R(0) = \sigma^2 = \int_{-\infty}^{\infty} S(\omega) d\omega \qquad (12.25)$$

Interpreted physically, the spectrum represents a distribution of variance over frequency. When divided by the variance, the spectrum is analogous to a probability density function. This explains the designation of spectral density function. When divided by the variance, the integrated spectrum is analogous to the cumulative probability distribution function. The spectral density function helps determine the frequencies that dominate the variance. A graph of the spectral density function shows predominant frequencies relative to less dominant frequencies. Thus, the spectral density is a function of frequency in cycles per unit of time, frequency in radians per unit of time, or period in units of time. These three quantities are related as

$$\omega = \frac{2\pi}{T} = 2\pi f \qquad (12.26)$$

where T is period in units of time and f is frequency in cycles per unit of time. Therefore,

$$S(\omega) = 2\pi S(f) \qquad (12.27)$$

Equations 12.19 to 12.25 can also be expressed in terms of f.

Example 12.5 Consider the exponential covariance function given by Eq. 12.11. Express the relation between the covariance and spectrum.

Solution Substituting Eq. 12.11 in Eq. 12.20 gives

$$S(\omega) = \frac{\sigma^2}{2\pi} \int\limits_{-\infty}^{\infty} \exp[-i\omega\tau]\exp[-|\tau|/\lambda]d\tau$$

$$= \frac{\sigma^2}{2\pi}\left[\int\limits_{0}^{\infty} e^{-i\omega\tau - \tau/\lambda}d\tau + \int\limits_{-\infty}^{0} e^{-i\omega\tau + \tau/\lambda}d\tau\right] = \frac{\sigma^2\lambda}{\pi(1+\lambda^2\omega^2)} \qquad \textbf{(12.28)}$$

Equation 12.28 is graphed in Fig. 12-5 with $\sigma = 3$ and $\lambda = 1$ as an example, which shows the maximum spectral amplitude at zero frequency. The zero intercept for the spectrum is proportional to the integral scale.

In civil engineering most of the observations are taken at discrete time intervals, not continuously. Quite often, these observations are taken at a constant time interval t. Then the oscillation with the highest frequency is defined as

$$f_{\max} = \frac{\Delta t}{2} \qquad \textbf{(12.29)}$$

where $f_{\max}$ is the maximum frequency, also called the Nyquist frequency.

For a sample of observations the spectral density function can be computed from the sample autocorrelation coefficient and integration of Eq. 12.20:

$$S(f) = [\rho_X(0) + 2\sum_{k=1}^{M-1} \rho_X(k)\cos(2\pi k f \Delta t) + \rho_X(M)\cos(2\pi M f \Delta t)]\Delta t \qquad \textbf{(12.30)}$$

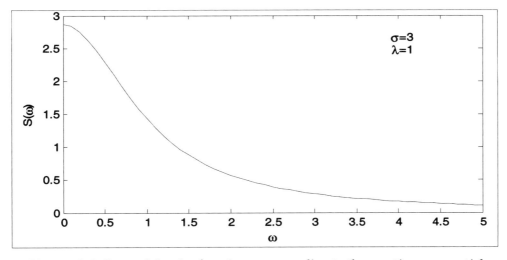

Figure 12-5 *Spectral density function corresponding to the negative exponential covariance function.*

where M is the maximum lag, which should be a small portion of N data points, for example, $M \leq 0.25N$. Equation 12.30 should be used for frequencies:

$$f = \frac{kf_{max}}{M} \tag{12.31}$$

Now consider two zero-mean random processes $X(t)$ and $Y(t)$. The covariance function of X and Y can be expressed as

$$\text{cov}[X(t+\tau), Y(\tau)] = E[X(t+\tau)Y(t)] = R_{XY}(\tau) \tag{12.32}$$

In terms of the cross–spectral density function S_{XY}, one has

$$R_{XY}(\tau) = \int_{-\infty}^{\infty} \exp(i\omega\tau) S_{xy}(\omega) d\omega \tag{12.33}$$

or

$$S_{XY}(\omega) = \frac{1}{2\pi} \int_{-\infty}^{\infty} e^{-i\omega\tau} R_{XY}(\tau) d\tau \tag{12.34}$$

The cross-covariance and cross–spectral density yield information on how two processes X and Y are related in time. The function S_{XY} is complex because R_{XY} is neither an even nor an odd function. The cross-correlation function satisfies the following relation:

$$R_{XY}(\tau) = E[X(t_*)Y(t_* - \tau)] = E[X(t_* + \tau)Y((t_* + \tau) - \tau)] = R_{YX}(-\tau) \tag{12.35}$$

Here R_{XY} is real because X and Y are real. Thus, the real and imaginary parts of S_{XY} are

$$S_{XY}(\omega) = C_{XY}(\omega) - iQ_{XY}(\omega) = A_{XY}(\omega)\exp\left[-i\theta_{XY}(\omega)\right] \tag{12.36}$$

where C_{XY} is the cospectrum, a measure of in-phase covariance; Q_{XY} is the quadspectrum, a measure of out-of-phase covariance; and θ_{XY} is the phase spectrum expressed as

$$\theta_{XY} = \tan^{-1}(Q_{XY}/C_{XY}) \tag{12.37}$$

The cospectrum is expressed as

$$C_{XY}(\omega) = \int_{-\infty}^{\infty} R_{XY}(\tau)\cos(\omega\tau) d\tau \tag{12.38}$$

and the quadspectrum as

$$Q_{XY}(\omega) = \int_{-\infty}^{\infty} R_{XY}(\tau)\sin(\omega\tau) d\tau \tag{12.39}$$

12.6 Stochastic Processes

Many environmental and water resources processes, such as rainfall, discharge in a river, reservoir water level, and pollutant concentration in a river, are stochastic processes. Time series analysis techniques are frequently used to study these processes. For a stochastic process represented as $X(t)$, $t \in T$, if T is an infinite sequence, then the process becomes $X(m)$, $X(m-1)$, ..., or $X(m-1)$, $X(m)$, ..., or $X(0)$, $X(1)$, ... and is a discrete-parameter process. Any process whose parameter set is finite or enumerable is called a discrete-parameter process. However, if T is an interval, then the process is a continuous-parameter family and is called a continuous-parameter process. A process with a nonenumerable set is a continuous process. Note that $x(t)$ in practice is the observation of X at time t, and T is the time range involved.

12.6.1 Counting Process

In environmental and water resources, the frequency of occurrence of random events and the time interval between successive events are of considerable interest. Consider a time interval $t = 0$ to t in which the total number of occurrences of an event is represented by $N(t)$. Then $N(t)$ is called a counting process of the series of events, as shown in Fig. 12-6. This process is an integer-valued continuous-time stochastic process.

The time interval between successive occurrences of events $T_1 = t_1$, $T_2 = t_2 - t_1$, etc. is defined as the interarrival time. If the interarrival times are independent, identically distributed random variables with distribution F, then the process is called a renewal process (or renewal counting process). Specifically, if the interarrival times follow an exponential distribution, the stochastic process is called a Poisson process.

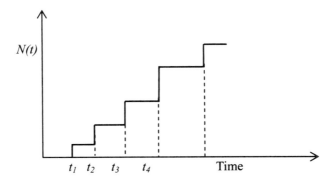

Figure 12-6 *A counting process.*

12.6.2 Poisson Process

A counting process is a Poisson process with mean rate (or intensity) υ if $N(t)$ has stationary independent increment, $N(0) = 0$, and the number in any interval of length $[N(t + \tau) - N(t)]$, is Poisson distributed with mean $\upsilon\tau$. The Poisson increment process is covariance stationary.

Example 12.6 Consider Examples 4.18 and 4.19. Suppose that the occurrences of drought events may be considered as a Poisson process with rate $\upsilon = 1.79$, and the interarrival time has the exponential distribution with parameter $\lambda = 0.124$. Then, find $P(N(s) = 5)$. (Hint: $N(s) = 5$ means by time s, a total of five droughts have occurred but the sixth has not, which can be seen from Fig. 12-7.)

Solution To solve this problem, we need to know that $N(s) = 5$ if $T_4 \leq s \leq T_5$. Then, let $T_4 = t$ and $T_5 > s$. We have the time interval $t_6 = T_5 - T_4 = T_5 - t_4 > s - t$, which is independent of T_4 according to the properties of the Poisson process. With this information in hand, we can solve this problem as

$$P(N(s) = 5) = \int_0^s P(T_4 = t)P(T_5 > s \mid T_4 = t)dt = \int_0^s P(T_4 = t)P(t_5 > s - t)dt \qquad (12.40)$$

Consider T_4 is the summation of independent interarrival times (i.e., $T_s = t_1 + t_2 + t_3 + t_4 + t_5$ with $t_1 = T_1$; $t_2 = T_2 - T_1$; etc.). Since each interarrival time has an exponential distribution, the summation of these interarrival times has a gamma distribution (i.e., $T_s \sim \text{gamma}(5, 0.124)$). Thus, Eq. 12.40 can be expressed as

$$P(N(s) = 5) = \int_0^s \lambda e^{-\lambda t} \frac{(\lambda t)^{5-1}}{(5-1)!} e^{-\lambda(s-t)}dt = \frac{\lambda^5}{(5-1)!} e^{-\lambda s} \int_0^s t^{5-1}dt$$
$$= e^{-\lambda s} \frac{(\lambda s)^5}{5!} = e^{-0.124s} \frac{(0.124s)^5}{5!}$$

12.6.3 Bernoulli Process

A counting process X, that is, $X = \{x_1, x_2, \ldots, x_n\}$, is called a Bernoulli process if the x_is are independent identically distributed Bernoulli trials. In other words, each trial can only have one of two outcomes: success (1) or failure (0); rain (1) or dry (0); hot (1) or cold (0), etc., and with $P(x_i = 1) = p$.

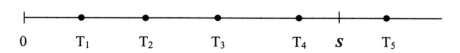

Figure 12-7 *Schematic of Poisson process*

Example 12.7 Consider rainy (1) or dry (2) weather conditions in any given day in summer. Suppose the probability of a rainy day is 0.2. Determine the probability of 4 rainy days occurring in 20 days.

Solution In this problem, the weather condition denoted as X has only two possible outcomes: rain (1) or no rain (0). Also suppose that in a given day whether it rains or not does not depend on any previous weather condition. Thus the weather condition X can be considered as a Bernoulli process. Then each x_i is Bernoulli distributed with parameter $P = 0.3$.

The probability of 4 rainy days occurring in 20 days can be computed as follows: This can be expressed as the probability of 4 successes (rainy days) in a total of 20 trials (days). If we treat weather conditions (X) as independent identically distributed Bernoulli trials, then the probability of 4 rainy days in a total of 20 days is binomial distributed as $B(20, 0.3)$, so

$$P(4 \text{ rainy days in 20 days}) = \binom{20}{4} \cdot 0.3^4 (1 - 0.3)^{16} = 0.13$$

12.6.4 Gaussian Process

A stochastic process $X(t)$ is said to be a normal or Gaussian process if the random variable $X(t)$ for any given time t is normally distributed with covariance function $CX(t_1, t_2)$. The assumption of the process being ergodic is implied here. A stochastic process $\{X(t), t \in T\}$ is called Gaussian if the joint distribution of every finite set of the $X_i s$ is Gaussian, which is represented as

$$f_X(x) = \frac{1}{(2\pi)^{N/2} |\Sigma_X|^{1/2}} e^{-\left[\frac{1}{2}(x-\mu_X)^T \Sigma_X^{-1}(x-\mu_X)\right]}, x \in \mathbb{R}^n \tag{12.41}$$

where

$$\mu_X = \begin{pmatrix} \mu_X(t_1) \\ \mu_X(t_2) \\ \vdots \\ \mu_X(t_1) \end{pmatrix}, \quad \Sigma_X = \begin{vmatrix} C_X(t_1, t_1) & C_X(t_1, t_2) & \dots C_X(t_1, t_N) \\ \vdots & \vdots & \vdots & \vdots \\ C_X(t_N, t_1) & C_X(t_N, t_2) & \dots C_X(t_N, t_N) \end{vmatrix}$$

Example 12.8 Suppose the sequence $X = \{x_1, x_2, \dots, x_n\}$ with each $x_i \sim N(\mu_{x_i}, \sigma_{x_i}^2)$, and let $x_1, \dots, x_n$ be independent. Prove that $Z = X_1 + X_2 + \dots + X_n$ is also a Gaussian process.

Solution The moment-generating function of the normal distribution is expressed as

$$mgf_X(t) = \exp(\mu t + \frac{\sigma^2 t^2}{2})$$

Then

$$mgf_Z(t) = E(e^{tZ}) = E[e^{t(X_1 + \ldots + X_n)}] = E[e^{tX_1}]E[e^{tX_2}]\ldots E[e^{tX_n}]$$

$$= \exp(\mu_{x_1}t + \frac{\sigma_{x_1}^2 t^2}{2})\exp(\mu_{x_2}t + \frac{\sigma_{x_2}^2 t^2}{2})\ldots\exp(\mu_{x_n}t + \frac{\sigma_{x_n}^2 t^2}{2})$$

$$= \exp\left[t\left(\mu_{x_1} + \mu_{x_2} + \ldots + \mu_{x_n}\right) + \frac{\left(\sigma_{x_1}^2 + \sigma_{x_2}^2 + \ldots + \sigma_{x_n}^2\right)t^2}{2}\right]$$

Thus Z is also a Gaussian process.

12.6.5 Markov Process

A Markov process is a process satisfying the following condition: For any integer $n \geq 1$, if $t_1 < \ldots < t_n$ are parameter values, the conditional probabilities x_{t_n} relative to $x_{t_1}, \ldots, x_{t_{n-1}}$ are the same as those relative to $x_{t_{n-1}}$ in the sense that, for each λ,

$$P\left\{x_{t_n}(w) \leq \lambda \mid x_{t_1}, \ldots, x_{t_{n-1}}\right\} = P\left\{x_{t_n}(w) \leq \lambda \mid x_{t_{n-1}}\right\} \qquad (12.42)$$

Thus the probability of transition from $x_{t_{n-1}}$ at t_{n-1} to x_{t_n} at t_n can be expressed only in terms of the state at t_{n-1} and information about previous states is not needed. The conditional probabilities are also termed as transition probabilities. The set of these probabilities are expressed through a transition matrix.

Example 12.9 Consider the weather type in Baton Rouge in summer. Let the probability(rain) = 0.2 and probability(no rain) = 0.8. Then if we know today is a rainy day, is tomorrow a dry day, given only today's weather type? Is this process a stochastic process? If it is not, can it be made a Markov process?

It is clear that it is not guaranteed that tomorrow's weather type only depends on today's weather type; the Markov property does not hold in this case, so it is not a Markov process. But one can find a way to make this non-Markov process a Markov process. If we have the weather type information of yesterday, then the combination of weather type of yesterday and today can be considered as a state. The probability of tomorrow being a rain day is given in Table E12-9.

Table E12-9 *Weather type structure.*

Yesterday	Today	Probability of rain day tomorrow
Rain	Rain	0.6
Dry	Rain	0.5
Rain	Dry	0.3
Dry	Dry	0.2

Solution

Thus the state can be defined as

1: (R, R), 2: (D, R), 3: (R, D), 4: (D, D)

Now it is Markov process with the transition matrix T given as

$$T = \begin{vmatrix} 0.6 & 0 & 0.4 & 0 \\ 0.5 & 0 & 0.5 & 0 \\ 0 & 0.3 & 0 & 0.7 \\ 0 & 0.2 & 0 & 0.8 \end{vmatrix}$$

12.6.6 Wiener–Levy Process

The Weiner–Levy process is a stochastic process and is known as the Brownian motion process. It describes the random movement of an extremely small particle immersed in a liquid or a gas. It satisfies the following conditions: (1) X_t has stationary-independent increments and $X_t \sim N(0, \sigma^2 t)$, which shows that the variance of the Wiener–Levy process increases linearly with time t. (2) $E[x_t] = 0$ for all time, and $X_t(0) = 0$ with probability 1. (3) Every independent increment is normally distributed (i.e., say, $s < t$, $X_t - X_s \sim N[0, \sigma^2(t-s)]$ in which σ is constant.

Example 12.10 Let X_t be a Weiner–Levy process. What is the joint probability density function of (X_s, X_t) where $s < t$.

Solution Since X_t is a Weiner–Levy process, then $X_t \sim N(0, \sigma^2 t)$, and $X_s \sim N(0, \sigma^2 s)$. From probability theory, we have

$$f_{(X_s, X_t)}(x, y) = f_{X_s}(x) f_{X_t|X_s}(y \mid x)$$

Considering property (3) of the Wiener–Levy process, we have that

$$X_t \mid X_s \sim N\left[x, \sigma^2(t-s)\right]$$

which can be proved as follows:

$$F_{X_t|X_s=x}(y) = P\left[X_t \leq y \mid X_s = x\right]$$
$$= P\left[X_t - X_s \leq y - x \mid X_s = x\right]$$
$$= P\left[X_t - X_s \leq y - x\right] \quad \text{(the increment is independent of the former state } X_s)$$
$$= \Phi\left(\frac{y-x}{\sigma\sqrt{t-s}}\right) \quad \text{(property 3)}$$

Thus

$$f_{(X_s, X_t)}(x, y) = f_{X_s}(x)f_{X_t|X_s}(y \mid x)$$

$$= \frac{1}{2\pi\sigma\sqrt{s(t-s)}} \exp\left[-\frac{x^2}{2\sigma^2 s} - \frac{(y-x)^2}{2\sigma^2(t-s)}\right]$$

12.6.7 Shot-Noise Process

A stochastic process is a shot-noise process if it is defined by a sequence of impulses applied to a system at random time τ_k. It can be expressed in the form

$$X(t) = \sum_{k=1}^{N(t)} A_k \omega(t, \tau_k) \tag{12.43}$$

where $\omega(t, \tau_k)$ is the response of a system at time t resulting from impulse A_k at time τ_k, A_k is a set of independent identically distributed random variables, and $N(t)$ is a counting process with interarrival time τ_k, often taken as a Poisson process.

12.7 Time Series Analysis

In time series analysis, ergodicity is invoked and time-average statistics are therefore assumed to represent ensemble values. When trend and periodic compounds have been removed from the time series, the ergodicity assumption is nearly valid.

12.7.1 Trend

12.7.1.1 Concept of Time Trend

When a time series is plotted, the series values may, on average, increase or decrease. This increasing or decreasing tendency defines a trend. The trend may be the result of low-frequency oscillation, depending on the time scale of observation. If the time scale of observation is large, trends may be identified as seasonal or periodic components.

Consider a stochastic process

$$X(t) = Y(t) + Z(t) \tag{12.44}$$

where $Y(t)$ is the deterministic trend and $Z(t)$ is the random component. The deterministic trend can be observed from the solution of the governing deterministic differential equation of the system. The deterministic trend normally accounts for the largest portion of the total magnitude of a stochastic process. This is a result of inherent determinism in the process.

The component $Z(t)$ accounts for the randomness inherent in the process, errors in the model hypothesis and its parameter estimation, and errors in the data. Thus, $Z(t)$ is modeled statistically. The magnitude of its variance $\sigma_Z^2(t)$ is a measure of its relative importance to $X(t)$. The larger the value of $\sigma_Z^2(t)$, the more important the uncertainty in $X(t)$. If $Y(t)$ is determined and represented by a function, say a polynomial, then the fitted values are subtracted from $X(t)$, and then the random component $Z(t)$ is analyzed by using the serial correlogram for identification and removal of periodic components, if any.

12.7.1.2 Time Trend Removal

The deterministic trend $Y(t)$ can be removed from the observed time series. The widely used technique is to fit a low-order polynomial to the data using the least square method discussed in Chapter 8. Now, considering this special problem, we let the deterministic trend $Y(t)$ be represented by a Kth-order polynomial ($K \le 3$; Bendat and Piersol 1980) as

$$Y(j) = \sum_{i=0}^{K} a_i (j\Delta t)^i, j = 1, 2, 3, ..., n \qquad (12.45)$$

and then the least square fit is represented as

$$A = \sum_{i=0}^{n} [x(i) - y(i)]^2 = \sum_{i=0}^{n} \left[x(i) - \sum_{j=0}^{K} a_j (i\Delta t)^j \right]^2 \qquad (12.46)$$

Taking the partial derivative of A with respect to a_j and setting it equal to zero yields

$$\sum_{j=0}^{K} a_j \sum_{i=1}^{n} (i\Delta t)^{k+m} = \sum_{i=1}^{n} x(i)(i\Delta t)^m, \quad m = 0, 1, 2, ..., K \qquad (12.47)$$

in Eqs. 12.44 to 12.46, n is sample size, t is the sampling interval, a_j is regression coefficient.

To this end, the deterministic component, also called the trend component $Y(t)$, can be determined and removed from the time series.

Example 12.11 Consider the monthly discharge in July from the Colorado River near the Grand Canyon, Arizona. Determine the deterministic component (time trend) $Y(t)$ and random component $Z(t)$ of the data.

Solution The monthly discharge data at the Colorado River near the Grand Canyon, Arizona, are given in Table E12-11.

Let $K = 1$. Then, for this problem, Eq. 12.45 to Eq. 12.47 can be written as

$$Y(i) = a_0 + a_1 t(i) \text{, where } t = 1 \text{ and } i = 1, \ldots, 82 \qquad \textbf{(12.48)}$$

$$Q = \sum_{i=1}^{N} \left[x(i) - \left(a_0 + a_1 t(i) \right) \right]^2 \qquad \textbf{(12.49)}$$

$$\begin{cases} \dfrac{\partial Q}{\partial a_0} = a_0 N + a_1 \sum_{i=1}^{N} t(i) - \sum_{i=1}^{N} x(i) = 0 \\ \dfrac{\partial Q}{\partial a_1} = a_0 \sum_{i=1}^{N} t(i) + a_1 \sum_{i=1}^{N} \left[t(i) \right]^2 - \sum_{i=1}^{N} \left[x(i) t(i) \right] = 0 \end{cases} \qquad \textbf{(12.50)}$$

Table E12-11 *Monthly discharge data in July at the Colorado River near the Grand Canyon.*

Year	Discharge (cfs)	Year	Discharge (cfs)	Year	Discharge (cfs)	Year	Discharge (cfs)
1923	37840	1945	28160	1967	11270	1989	13580
1924	17060	1946	12760	1968	14060	1990	12970
1925	24190	1947	31750	1969	16160	1991	15150
1926	23230	1948	16410	1970	13250	1992	14290
1927	41100	1949	34600	1971	15170	1993	14520
1928	25260	1950	22790	1972	14170	1994	13880
1929	34410	1951	22720	1973	10910	1995	18310
1930	18790	1952	25860	1974	20080	1996	16480
1931	8195	1953	15939	1975	20260	1997	22020
1932	33610	1954	10860	1976	13120	1998	20640
1933	19200	1955	10050	1977	14440	1999	18660
1934	2380	1956	9722	1978	11340	2000	8703
1935	24620	1957	65590	1979	13950	2001	13460
1936	17000	1958	11110	1980	25400	2002	15079
1937	22230	1959	12939	1981	13700	2003	15240
1938	28520	1960	11030	1982	13430	2004	15409
1939	7611	1961	6780	1983	55550		
1940	7040	1962	29620	1984	35400		
1941	28510	1963	1755	1985	28290		
1942	21870	1964	1368	1986	21470		
1943	23730	1965	11780	1987	18380		
1944	30150	1966	11350	1988	11890		

From Eq. 12.50 one can obtain

$$a_0 = \frac{N\sum_{i=1}^{N}x(i)\sum_{i=1}^{N}[t(i)]^2 - \sum_{i=1}^{N}x(i)\left[\sum_{i=1}^{N}t(i)\right]^2 - N\sum_{i=1}^{N}[x(i)t(i)] - \sum_{i=1}^{N}x(i)\sum_{i=1}^{N}t(i)}{N^2\sum_{i=1}^{N}[t(i)]^2 - N\left[\sum_{i=1}^{N}t(i)\right]^2} = 2.165 \times 10^5$$

$$a_1 = \frac{N\sum_{i=1}^{N}[t(i)x(i)] - \sum_{i=1}^{N}x(i)\sum_{i=1}^{N}t(i)}{N\sum_{i=1}^{N}[t(i)]^2 - \left[\sum_{i=1}^{N}t(i)\right]^2} = -100.6$$

From this analysis, we see that there is a decreasing trend existing in the discharge time series considered. Figure 12-8a shows the discharge time series and the corresponding time trend $Y(t)$. Figure 12-8b shows the discharge time series after the time trend removal.

12.7.2 Periodicity

Cyclic, periodic, or seasonal fluctuations in time series are other deterministic components. These components are detected by periodic oscillations in the correlogram or by frequencies of oscillations in the spectral density function. A time series containing a trend and a periodic component can be represented as

$$X(t) = V(t) + W(t) + Z(t) \tag{12.51}$$

where $W(t)$ is the periodic component. Periodicity is usually modeled using harmonic functions. The fitted periodic values are then subtracted from $X(t)$ and the random component $Z(t)$ is analyzed further using correlogram and statistical techniques.

Note that in Eqs. 12.44 and 12.48 deterministic components $Y(t)$ and $W(t)$ are added to $Z(t)$ or $X(t)$ is partitioned into $Y(t)$, $W(t)$, and $Z(t)$. This linear addition or subtraction will not be valid if the differential equation for the periodic component or the trend is nonlinear or their coefficients are random.

12.7.3 Random Component

If trends and periodicity have been properly analyzed and detected and subtracted from $X(t)$, the random component $Z(t)$ should be a zero-mean process. The next step in the analysis of $Z(t)$ is to determine the correlogram and the spectral density function. This is needed to reveal hidden periodic components, dominant frequencies, and most importantly serial dependence in time. If the correlogram shows that the series is uncorrelated for lags greater than 1, then a white-noise process can be employed to represent $Z(t)$. If the correlogram is an exponentially

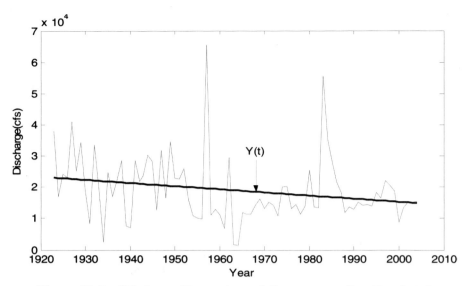

Figure 12-8a *Discharge time series and the corresponding time trend.*

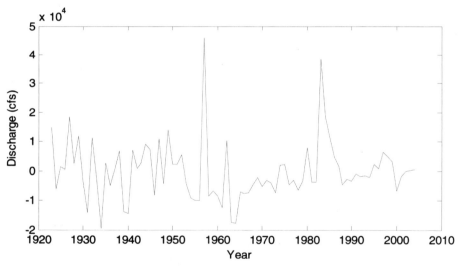

Figure 12-8b *Discharge time series after time trend removal.*

decaying curve or the spectral density shows a smooth combination of several frequencies, then a colored-noise process may be employed to represent $Z(t)$. If the autocorrelation function shows serial dependence up to a certain lag, then an autoregressive process may be a suitable representation of $Z(t)$.

12.8 Questions

12.1 The potential rate of infiltration, $f(t)$ at time t, in a soil is described by the Kostiakov equation as

$$f(t) = at^{-0.5}$$

where a is a parameter. Assuming a as a random variable, determine the mean, autocovariance function, autocorrelation function, variance, and autocorrelation coefficient of f.

12.2 For a linear watershed, the discharge $Q(t)$ at time t during depletion can be described as

$$Q(t) = Q_0 K^t$$

where Q_0 is the discharge at the start of depletion and K is the depletion constant. Assuming K as a random variable, determine the mean, autocovariance function, autocorrelation function, variance, and autocorrelation coefficient of Q. [Hint: Take logarithmic transformation of the equation and then do the derivation for logarithmic quantities.]

12.3 If kinematic wave theory is used for modeling overland flow on a plain then the depth hydrograph during the rising limb is described as

$$h(t) = qt$$

where q is rainfall intensity and t is time. Assuming q as a random variable, determine the mean, autocovariance function, autocorrelation function, variance, and autocorrelation coefficient of h.

12.4 If kinematic wave theory is used for modeling overland flow on a plain then the discharge hydrograph during the equilibrium state is described as

$$Q(t) = qx$$

where q is rainfall intensity and x is distance. Assuming q as a random variable, determine the mean, autocovariance function, autocorrelation function, variance, and autocorrelation coefficient of Q.

12.5 Consider a stochastic process expressed as

$$f(t) = f_c + st^{-0.5}$$

where f_c and s are independent normally distributed random variables and t is time. This is Philip's equation for describing the potential rate of infiltration in soil. Determine the mean, variance, autocorrelation function, covariance function, and correlation coefficient of $f(t)$.

12.6 For ice and snow melt conditions the base-flow recession at time t can be adequately represented as

$$Q(t) = at^{-n} + b$$

where $Q(t)$ is discharge at time t, a and b are parameters, and n is an exponent. Assuming a and b as random variables, determine the mean, variance, autocorrelation function, covariance function, and correlation coefficient of $Q(t)$.

12.7 Consider rainy (1) or dry (2) weather conditions on any given day in summer. Suppose the probability of a rainy day is 0.1. Determine the probability of 5 rainy days occurring in 15 days. Now suppose the probability of a rainy day is 0.3; determine the probability of 5 rainy days occurring in 15 days.

12.8 In the Muskingum method of flow routing in a river reach, the storage $S(t)$ in the reach at any time t can be expressed as

$$S(t) = aI(t) + bQ(t)$$

where a and b are constant parameters, $I(t)$ is inflow to the reach at time t, and Q is outflow from the reach at time t. Assume that I and Q are normally distributed. Show whether S is normally distributed or not.

12.9 Assume that in summer in New Orleans the probability(rain) on any day is 0.15 and probability(no rain) = 0.85. If it rains today, will it rain or be dry tomorrow, given only today's weather type? What type of a stochastic process is this?

12.10 Consider monthly discharge for January from the Amite River near Denham Springs, Louisiana. Determine the deterministic component (time trend) $Y(t)$ and random component $Z(t)$ of the discharge data.

Chapter 13

Stochastic Differential Equations

In civil and environmental engineering, the vast majority of problems until recently were formulated as deterministic problems, with the assumption that system parameters or variables were known. This was particularly the case in structural engineering, geotechnical engineering, and engineering mechanics. The deterministic formulations are simple to solve and require less data. In recent years there has been a growing realization that, for a variety of reasons, these deterministic representations may be inadequate. For instance, if some characteristic of a system does not remain constant and keeps changing, it may have to be treated as a random variable. The boundary conditions of a system may be subject to random fluctuations. It is also possible that the only information on some parameters of the system available resides in terms of their statistical properties. In addition, measurement errors in system variables are common because of either limitations of measurement techniques or simplifying assumptions. Furthermore, variables may have large spatial and temporal variability, whereas the measurements are made at discrete points, far from each other. It is possible that the combined effect of all these factors on the system behavior is not significant, in which case the system may be treated as deterministic and its behavior may be described deterministically without significant errors. If, however, the impact of the random fluctuations is large, it may be necessary to represent the behavior of the system using a stochastic differential equation (SDE). These equations abound in engineering and scientific applications. This chapter provides a brief discussion of such equations and their application.

13.1 Ordinary Differential Equations

Differential equations result from the application of physical, chemical, biological, and other governing laws to a system. The fundamental laws commonly used to describe environmental and water resources systems are the laws of conservation of mass, momentum, and energy and laws of thermodynamics. When these laws are applied to a system or process undergoing changes in temporal and spatial domains, one obtains equations that contain derivations. An equation that involves derivatives is known as a differential equation (DE) (e.g., $dy/dx = ax + b$, where x and y are variables and a and b are parameters).

A DE has some independent variable(s) and some dependent variable(s). The independent variables are usually space–time coordinates. The system being described by the DE may be subject to an input forcing function. The coefficients of the DE may be constants or functions of independent and/or dependent variables. The coefficients of the DE represent some physical quantities related to the system and these can be constants, some function of dependent variables, and/or some function of independent variables.

An equation that contains total derivatives with respect to one variable is called an ordinary differential equation (ODE); when partial derivatives with respect to more than one independent variable are involved, one gets partial differential equations (PDEs). The differential equations (both ODE and PDE) are dealt with in detail in many standard mathematical texts, such as Ayres (1952).

If a DE describes a system in a physical region, such as flow in a channel, the solution may require the value of the system variables at the boundary of the physical region and the problem is known as a *boundary value problem*. The equations that involve the element of time may require knowledge of initial conditions or the value of system variables at time $t = 0$. Such problems are known as *initial value problems*.

A differential equation may be solved analytically; that is, it may be possible to obtain a relation describing a dependent variable as a function of an independent variable. Note that, in many cases, an analytical solution may be quite difficult to obtain and, in some cases, it may not exist at all. Another way of solving a differential equation is numerically, by discretizing the various differential terms in the space–time domains. The solution describes the dependent variable as a function of the independent variable and no derivatives are involved. Upon substitution, the solution should satisfy the DE. Various software packages are widely available these days to numerically solve a DE. There are myriad numerical techniques available to solve differential equations.

13.2 Partial Differential Equations

Partial differential equations are equations that contain an unknown function of two or more variables and its partial derivatives with respect to these variables. These equations must involve at least two independent variables. If x and y are the independent variables and z is a dependent variable then

$$x\frac{\partial z}{\partial x} + y\frac{\partial z}{\partial y} = z \tag{13.1}$$

is a PDE. The order of a PDE is the order of the highest derivative. Thus Eq. 13.1 involves partial differentials of first order and hence it will be classified as a PDE of order one. Similarly, the equation

$$\frac{\partial^2 z}{\partial x^2} + \frac{\partial^2 z}{\partial x \partial y} + \frac{\partial^2 z}{\partial y^2} = 0 \tag{13.2}$$

is a PDE of second order. A PDE can be a linear or a nonlinear equation. The equation

$$\frac{\partial z}{\partial x} + y\frac{\partial z}{\partial y} = z \tag{13.3}$$

is a linear equation, since the coefficients of the derivatives do not depend on the dependent variable. If the coefficients associated with the derivatives of the dependent variable are functions of the dependent variable, it is a nonlinear PDE. For example,

$$z\frac{\partial^2 z}{\partial x^2} + \frac{\partial^2 z}{\partial y^2} = 1 \tag{13.4}$$

and

$$\frac{\partial z}{\partial x} + \ln z\frac{\partial z}{\partial y} = 0 \tag{13.5}$$

are examples of nonlinear PDEs.

13.3 Stochastic Differential Equations

There are many types of SDEs. For example, an SDE may be an ODE or a PDE, or it may be linear or nonlinear. For purposes of understanding it is a good idea to classify SDEs.

An SDE whose left-hand side is linear in both dependent and independent variables is known as a linear equation of first order. For example,

$$\frac{dy}{dx} + yP = Q \tag{13.6}$$

is a linear equation of first order, where y is a stochastic dependent variable, Q is a forcing function, P is a sink function, and x is an independent variable. Functions P and Q may be stochastic. Another example of a linear equation is

$$\frac{dy}{dx} + xy = 5x \tag{13.7}$$

In contrast, the equation

$$\frac{dy}{dx} + xy^2 = 5x \tag{13.8}$$

is not a linear equation.

A given system may receive input at time $t = 0$ that is not deterministic. In ordinary differential equations, the initial condition(s) may be random variables. In a partial differential equation, it may be specified as a random process. Depending upon the properties of the system, the uncertain input may be further propagated or it may be dissipated. Given sufficient data, the problem is to find the probability distribution of the system output. However, at times, enough information may not be available to determine the complete distribution and one may have to be content with only the first few moments of it. In all of these situations SDEs arise. Depending on the way randomness is considered, stochastic differential equations can also be classified as (i) differential equations with random initial conditions, (ii) differential equations with random forcing functions, (iii) differential equations with random boundary conditions, (iv) differential equations with random coefficients, (v) differential equations with random geometrical domains, and (vi) differential equations that combine two or more of these conditions. A solution of an SDE is a stochastic process that satisfies it. Because the dependent variable is stochastic, the concepts of mean square continuity, stochastic differentiation, and stochastic integration are invoked. These concepts define the continuity, differentiation, and integration of a stochastic process.

13.4 Fundamental Concepts in Solving SDEs

Since the solution of an SDE is in terms of stochastic variable(s), the concepts such as continuity, differentiation, and integration are modified and defined to take stochasticity into account. These concepts are introduced in what follows. The concept of mean square continuity is useful in the study of stochastic processes. A process $X(t)$ is said to be continuous in mean square sense if it satisfies the condition

$$E[X(t + \tau) - X(t)]^2 \to 0 \text{ as } \tau \to 0 \qquad (13.9)$$

where $\tau > 0$ is the time lag or delay. Expanding Eq. 13.9, one obtains

$$E[X(t + \tau) - X(t)]^2 = E[X(t + \tau)^2 - E[X(t+\tau)X(t)] - E[X(t)X(t+\tau)] + E[X(t)^2] \qquad (13.10)$$

The right-hand side of this equation approaches zero as $\tau \to 0$. Clearly, the process is continuous if $E[X(t_1)X(t_2)]$ is continuous along the time axis. This implies that

$$E[X(t + \tau)] \to E[X(t)] \text{ as } \tau \to 0 \qquad (13.11)$$

A related concept in differentiation is of mean square derivative of a stochastic process. A process has mean square derivative at t if the following limit is satisfied in the mean square sense:

$$\lim_{\tau \to 0} E\left[\frac{X(t+\tau) - X(t)}{\tau} - \frac{dX(t)}{dt}\right]^2 = 0 \qquad (13.12)$$

13.4.1 Stochastic Differentiation

The mean square derivative is useful because its properties can be represented in terms of the second-order properties of the stochastic process, that is, the covariance function. A stationary stochastic process $X(t)$ is differentiable in the mean square sense if its autocorrelation function $R(\tau)$ is differentiable up to the second order. The derivative of the expected value of $X(t)$ is equal to the expected value of the derivative of $X(t)$. This property can be generalized to an nth derivative if it exists. If $X(t)$ is nonstationary then it is differentiable in the mean square if the second-order partial derivative of its autocorrelation function $R(t_1, t_2)$ with respect to t_1 and t_2 [i.e., $\partial^2 R(t_1,t_2)/\partial t_1 \partial t_2$], exists at $t_1 = t_2$. Similarly, a stochastic process $X(t)$ is nth-order differentiable if $\partial^{2n} R(t_1,t_2)/\partial t_1^n \partial t_2^n$ exists at $t_1 = t_2$.

13.4.2 Stochastic Integration

A mean-square integral of a stochastic process involves the limit of the sum in the mean square sense. Thus, a stochastic process $X(t)$ is integrable if

$$\int_a^b X(t)dt = \lim_{\Delta t_i \to 0} \sum_i X(t_i)\Delta t_i \qquad (13.13)$$

exists. Similar to stochastic differentiation, $X(t)$ is integrable over the interval (a, b) if the double integral of the autocorrelation function is bounded:

$$\int_a^b \int_a^b |R_X(t_1, t_2)| \, dt_1 \, dt_2 < \infty \qquad (13.14)$$

The condition for the existence of a mean-square derivative can also be considered from the spectral representation, which is an integral in the mean square sense. The derivative of $X(t)$ can be expressed as

$$W(t) = \frac{dX}{dt} = \int_{-\infty}^{\infty} e^{i\omega t} i\omega \, dZ_X(\omega) = \int_{-\infty}^{\infty} e^{i\omega t} \, dZ_W(\omega) \qquad (13.15)$$

Here $X(t)$ is considered to be a zero-mean stochastic process, ω represents angular frequency, and $dZ_W(\omega)$ represents the Fourier amplitude of the stochastic process. If the derivative $W(t)$ is stationary, then

$$dZ_W = i\omega \, dZ_X \qquad (13.16)$$

The spectrum of the derivative can then be expressed as

$$S_{WW}(\omega) = \omega^2 S_{XX}(\omega) \qquad (13.17)$$

The covariance function of the derivative W can be expressed as

$$R_{WW}(\tau) = \int_{-\infty}^{\infty} e^{i\omega\tau} S_{WW}(\omega) d\omega = \int_{-\infty}^{\infty} e^{i\omega\tau} \omega^2 S_{XX}(\omega) \qquad (13.18a)$$

$$d\omega = -\frac{d^2 R_{XX}}{d\tau^2} \qquad (13.18b)$$

For $\tau = 0$, the variance of the derivative follows:

$$\sigma_W{}^2 = \int_{-\infty}^{\infty} \omega^2 S_{XX}(\omega) d\omega = -\frac{d^2 R_{XX}}{d\tau^2}\Big|_{\tau=0} < \infty \qquad (13.19)$$

Since

$$W(t) = \frac{dX(t)}{dt}$$

the covariance functions of X and W are related as

$$\frac{d^2 R_{XX}(\tau)}{d\tau^2} = -R_{WW}(\tau) \qquad (13.20)$$

by Eq. 13.18. However, Eq. 13.17 shows a simple algebraic relation between their spectra:

$$S_{XX}(\omega) = \frac{S_{WW}(\omega)}{\omega^2} \qquad (13.21)$$

Thus, for X to be stationary, its variance must be finite:

$$\sigma_X^2 = \int_{-\infty}^{\infty} \frac{S_{WW}(\omega)}{\omega^2} d\omega < \infty \tag{13.22}$$

Using Eq. 13.13, one can write

$$E[|\int_a^b X(t)dt|]^2 = \int_a^b \int_a^b R_X(t_1, t_2)dt_1 dt_2 \tag{13.23}$$

The order of integration and expectation is interchangeable. To illustrate it, consider a process $Y(t)$ as

$$Y(t) = \int_0^t X(s)ds \tag{13.24}$$

Then

$$E[Y(t)] = \mu_Y(t) = E[\int_0^t X(s)ds] = \int_0^t E[X(s)]ds = \int_0^t \mu_X(s)ds \tag{13.25}$$

The property can be extended to obtain the correlation function as follows:

$$
\begin{aligned}
E[Y(t_1)Y(t_2)] = R_Y(t_1, t_2) &= E[(\int_0^{t_1} X(t)dt)(\int_0^{t_2} X(s)ds)] \\
&= E[\int_0^{t_1}\int_0^{t_2} X(t)X(s)dtds] \\
&= \int_0^{t_1}\int_0^{t_2} E[X(t)X(s)]dtds = \int_0^{t_1}\int_0^{t_2} R_X(t, s)dtds
\end{aligned}
\tag{13.26}
$$

In a similar manner, the autocovariance of a stochastic integral can be determined. Thus,

$$
\begin{aligned}
\text{cov}_Y(t_1, t_2) &= R_Y(t_1, t_2) - \mu_Y(t_1)\mu_Y(t_2) \\
&= \int_0^{t_1}\int_0^{t_2} R_X(t, s)dtds - \int_0^{t_1}\int_0^{t_2} \mu_X(t)\mu_X(s)dtds
\end{aligned}
\tag{13.27}
$$

If $t_1 = t_2 = t$ is inserted in Eq. 13.27, the result is the variance of the stochastic integral:

$$
\begin{aligned}
\sigma_Y^2(t) &= \text{cov}_Y(t, t) = R_Y(t, t) - \mu_Y^2(t) \\
&= \int_0^t\int_0^t R_X(z, s)dz\,ds - \int_0^t\int_0^t \mu_X(z)\mu_X(s)dz\,ds
\end{aligned}
\tag{13.28}
$$

$$\rho_Y(t_1,t_2) = \frac{\int_0^{t_1}\int_0^{t_2} R_X(t,s)\,dt\,ds - \int_0^{t_1}\int_0^{t_2}\mu_X(t)\mu_X(s)\,dt\,ds}{\sqrt{\int_0^{t_1}\int_0^{t_1} R_X(z,s)\,dz\,ds - \int_0^{t_1}\int_0^{t_1}\mu_X(z)\mu_X(s)\,dz\,ds}} \tag{13.29}$$
$$\times \frac{1}{\int_0^{t_2}\int_0^{t_2} R_X(z,s)\,dz\,ds - \int_0^{t_2}\int_0^{t_2}\mu_X(t)\mu_X(s)\,dz\,ds}$$

Now, let us look at some examples. The first example treats outflow as a random function.

Example 13.1 The water level in a lake in India during the summer months of no rainfall is governed by the following differential equation:

$$K\frac{\partial H}{\partial t} = -Q$$

where Q is discharge from the lake through an outlet and K is a parameter. The water level in the lake is expressed as

$$H(t) = H_0(t) + h(t)$$

where H_0 is a deterministic function of time and $h(t)$ is a random process with $E[h(t)] = 0$ and with the autocorrelation function represented as $R_h(t_1,t_2) = C_V^2(K)\exp[-a|t_2 - t_1|]$, where a is the correlation time parameter and $C_V^2(K)$ is the coefficient of variation of K. Determine the mean, the autocorrelation function, and the covariance function of discharge.

Solution The differential equation is expressed as

$$Q(t) = -K\left\{\frac{\partial}{\partial t}[H_0 + h(t)]\right\}$$

Taking the expectation gives

$$E[Q(t)] = -E\{K\frac{\partial}{\partial t}[H_0 + h(t)]\} = -K\frac{\partial H_0}{\partial t} - KE\left[\frac{\partial}{\partial t}h(t)\right]$$

Assuming the stochastic process mean-square continuity, we can extend Eq. 13.11 and Eq. 13.12 to the derivative of the stochastic process as

$$E\left[X'(t)\right] = \frac{dE[X(t)]}{dt}$$

We then have

$$KE\left[\frac{\partial}{\partial t}h(t)\right] = K\frac{dE[h(t)]}{dt} = 0$$

Thus, we get

$$E[Q(t)] = -E\{K\frac{\partial}{\partial t}[H_0 + h(t)]\} = -K\frac{\partial H_0}{\partial t}$$

The autocorrelation function of discharge can be expressed as

$$E[Q(t_1)Q(t_2)] = R_Q(t_1,t_2) = E\{[-K\frac{\partial H_0(t_1)}{\partial t_1} - K\frac{\partial h(t_1)}{\partial t_1}][-K\frac{\partial H_0(t_2)}{\partial t_2} - K\frac{\partial h(t_2)}{\partial t_2}]\}$$

$$= E[K^2\frac{\partial H_0(t_1)}{\partial t_1}\frac{\partial H_0(t_2)}{\partial t_2} + K^2\frac{\partial H_0(t_1)}{\partial t_1}\frac{\partial h(t_2)}{\partial t_2} + K^2\frac{\partial H_0(t_2)}{\partial t_2}\frac{\partial h(t_1)}{\partial t_1} + K^2\frac{\partial h(t_1)}{\partial t_1}\frac{\partial h(t_2)}{\partial t_2}]$$

$$= K^2 E[\frac{\partial H_0(t_1)}{\partial t_1}\frac{\partial H_0(t_2)}{\partial t_2}] + K^2\frac{\partial^2}{\partial t_1 \partial t_2}E\left[h(t_1)h(t_2)\right]$$

$$= K^2\frac{\partial^2}{\partial t_1 \partial t_2}[R_{H_0}(t_1,t_2)] + K^2\frac{\partial^2}{\partial t_1 \partial t_2}[R_{h_0}(t_1,t_2)]$$

$$= K^2\frac{\partial^2}{\partial t_1 \partial t_2}[R_{H_0}(t_1,t_2)] + K^2\frac{\partial^2}{\partial t_1 \partial t_2}[C_V^2\exp(-a|t_2-t_1|)]$$

$$= K^2\frac{\partial^2}{\partial t_1 \partial t_2}[R_{H_0}(t_1,t_2)] - a^2 K^2 C_V^2(K)\exp(-a|t_2-t_1|)$$

The variance is obtained by inserting $t_1 = t_2 = t$, which yields

$$\sigma_Q^2(t) = R_Q(t,t) - \mu_Q^2(t) = K^2\left[\frac{\partial H_0(t)}{\partial t}\right]^2 + K^2 a^2 C_V^2 - K^2\left(\frac{\partial H_0}{\partial t}\right)^2 = K^2 a^2 C_V^2$$

The case of random initial condition is exemplified next.

Example 13.2 A linear differential equation

$$\frac{dS}{dt} = -Q, \quad S = KQ$$

is frequently used for stream base-flow recession. Here S is the storage in a watershed at time t, Q is discharge, and K is the residence time. The initial condition is the following: At $t = 0$, $Q(0) = Q_0$. It is assumed that Q_0 is a random variable with mean μ_{Q_0} and variance $\sigma_{Q_0}^2$. Determine the solution of the differential equation and the mean μ and the variance, covariance, and autocorrelation function of discharge Q.

Solution The differential equation can be recast as

$$\frac{dQ}{dt} + \frac{1}{K}Q = 0$$

Its solution is

$$Q(t) = Q_0 \exp(-t/K)$$

Here Q_0 is a random variable. Therefore $Q(t)$ is a stochastic process comprising a family of exponential recessions with random initial value Q_0.
The mean of $Q(t)$ is

$$E[Q(t)] = E[Q_0 e^{-t/K}] = E[Q_0]E[e^{-t/K}] = \mu_{Q_0} e^{-t/K}$$

The autocorrelation function of $Q(t)$ is obtained as

$$E[Q(t_1)Q(t_2)] = R_Q(t_1, t_2) = E[(Q_0 e^{-t_1/K})(Q_0 e^{-t_2/K})] = E[Q_0^2]e^{-(t_1+t_2)/K}$$
$$= (\sigma_{Q_0}^2 + \mu_{Q_0}^2)\exp[-(t_1 + t_2)/K]$$

The covariance function of $Q(t)$ is given as

$$\text{cov}(t_1, t_2) = R_Q(t_1, t_2) - \mu_Q(t_1)\mu_Q(t_2)$$
$$= (\sigma_{Q_0}^2 + \mu_{Q_0}^2)\exp[-(t_1 + t_2)/K] - \mu_{Q_0}^2 \exp[-(t_1 + t_2)/K]$$
$$= \sigma_{Q_0}^2 \exp[-(t_1 + t_2)/K]$$

The variance of $Q(t)$ is obtained by setting $t_1 = t_2$, which gives

$$\sigma_Q^2 = R_Q(t,t) - \mu_Q^2(t) = \sigma_{Q_0}^2 \exp[-2t/K]$$

The correlation coefficient of $Q(t)$ is

$$\rho_Q(t_1, t_2) = \frac{\sigma_{Q_0}^2 \exp[-(t_1 + t_2)/K]}{\sigma_{Q_0}^2 \sqrt{\exp(-2t_1/K)}\sqrt{\exp(-2t_2/K)}} = 1$$

Taking K as 24 hours, mean Q_0 as 10 m^3/s^2, and variance as 68 m^3/s^2, we can plot $E[Q(t)]$ and the variance, covariance, and autocorrelation function of $Q(t)$. The results are presented in Fig. 13-1.
Now consider the case where the input is random.

Example 13.3 A surface runoff hydrograph from an area represented by a linear reservoir can be described mathematically as

$$\frac{dQ(t)}{dt} + AQ(t) = AP(t) = P_*(t), \quad Q(0) = 0$$

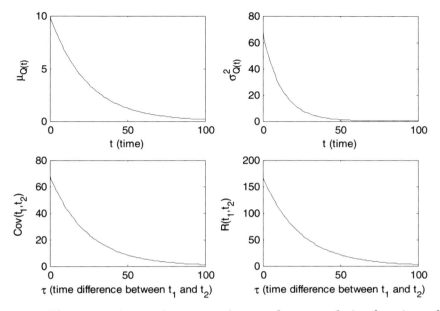

Figure 13-1 *The expectation, variance, covariance and autocorrelation functions of* Q(t).

where $P(t)$ is rainfall intensity, Q is unit surface runoff hydrograph, A is a reservoir coefficient, and t is time. It is assumed that A is a random variable uniformly distributed as

$$f(A) = \frac{1}{A_u - A_L}$$

with A_L = lower limit of A and A_u = upper limit, and that $P(t)$ is a stochastic process expressed as $P_* = A^2 \exp(-At)$. Determine the mean, variance, covariance, autocorrelation function, and the coefficient of correlation of Q. This problem is discussed by Lin and Wang (1996).

Solution The solution of the differential equation is

$$Q(t) = \int_0^t P_*(s)\exp[-A(t-s)]ds = \int_0^t A^2 \exp(-As)\exp[-A(t-s)]ds$$

$$= A^2 t \exp(-At)$$

The stochastic process $Q(t)$ now has an explicit solution. Its mean is expressed as

$$E[Q(t)] = \mu_Q(t) = E[A^2 t \exp(-At)] = \int_{A_L}^{A_u} A^2 t \exp(-At)\frac{1}{(A_u - A_L)}dA$$

$$= \frac{\left[(A_L t + 1)^2 + 1\right]\exp(-A_L t) - \left[(A_u t + 1)^2 + 1\right]\exp(-A_u t)}{t^2 (A_u - A_L)}$$

$$R_Q(t_1, t_2) = E[Q(t_1)Q(t_2)] = E[A^4 t_1 t_2 \exp(-At_1 - At_2)]$$

$$= \int_{A_L}^{A_u} A^4 t_1 t_2 \exp(-At_1 - At_2) \frac{1}{A_u - A_L} dA$$

$$= \frac{-B_1 \exp[-A_u(t_1 + t_2)] + B_2 \exp[-A_L(t_1 + t_2)]}{t_1^4 t_2^4 (A_u - A_L)}$$

Letting $t' = t_1 + t_2$, we get

$$B_1 = A_u^4 t'^4 + 4A_u^3 t'^3 + 12A_u^2 t'^2 + 24A_u t' + 24$$

$$B_2 = A_L^4 t'^4 + 4A_L^3 t'^3 + 12A_L^2 t'^2 + 24A_L t' + 24$$

The covariance is obtained as

$$\text{cov}(t_1, t_2) = R_Q(t_1, t_2) - \mu_Q(t_1)\mu_Q(t_2)$$

$$= \frac{-B_1 \exp[-A_u(t_1 + t_2)] + B_2 \exp[-A_L(t_1 + t_2)]}{t_1^4 t_2^4 (A_u - A_L)} -$$

$$\frac{[C_1 \exp(-A_L t_1) - C_2 \exp(-A_u t_1)][C_3 \exp(-A_L t_2) - C_4 \exp(-A_u t_2)]}{t_1^2 t_2^2 (A_u - A_L)^2}$$

$$C_1 = 2 + 2A_L t_1 + A_L^2 t_1^2; \; C_2 = 2 + 2A_u t_1 + A_u^2 t_1^2$$

$$C_3 = 2 + 2A_L t_2 + A_L^2 t_2^2; \; C_4 = 2 + 2A_u t_2 + A_u^2 t_2^2$$

The variance is obtained by setting $t_1 = t_2$ as

$$\sigma_Q^2 = R_Q(t,t) - \mu_Q^2(t)$$

$$= \frac{1}{A_u - A_L}\left[\frac{B_3 \exp(-2A_L t) - B_4 \exp(-2A_u t)}{4t^3}\right] - \frac{[B_5 \exp(-A_L t) - B_6 \exp(-A_u t)]^2}{(A_u - A_L)^2 t^4}$$

$$B_3 = 3 + 6A_L t + 6A_L^2 t^2 + 4A_L^3 t^3 + 2A_L^4 t^4$$

$$B_4 = 3 + 6A_u t + 6A_u^2 t^2 + 4A_u^3 t^3 + 2A_u^4 t^4$$

$$B_5 = A_L^2 t^2 + 2A_L t + 2$$

$$B_6 = A_u^2 t^2 + 2A_u t + 2$$

The correlation of Q is

$$\rho_Q(t_1, t_1) = \frac{\text{cov}[t_1, t_2]}{\sigma_Q^2(t_1)\sigma_Q^2(t_2)} = 1$$

Taking the lower and upper limits of A as 0.10 and 0.5 hour^{-1}, respectively, we can plot the mean, variance, covariance function, and autocorrelation function of $Q(t)$. The results are given in .com.

The next two examples deal with the cases where a parameter in the IUH is a random variable.

Example 13.4 A watershed is represented by a cascade of n equal reservoirs, each with reservoir coefficient k considered as a random variable. This representation is referred to as the Nash cascade and is popularly used for modeling surface runoff. The IUH of this cascade is

$$h_n = \frac{k}{(n-1)!}(kt)^{n-1}e^{-kt}$$

Because k is a random variable, h_n is the stochastic IUH of the n-reservoir cascade. Determine the mean, variance, and the first three moments of h_n (or IUH). Assume that k has a normal distribution with mean μ_k and variance σ_k^2. This watershed problem was discussed by Lin and Wang (1996). Take the mean of k as 2.14, σ_k as 0.25, and $n = 3$. Plot the computed functions of $h_n(t)$.

Solution The mth moment of h_n can be expressed as

$$E[h_n^m(t)] = E[\frac{k}{(n-1)!}(tk)^{n-1}e^{-kt}]^m = \int_{-\infty}^{\infty}[\frac{k}{(n-1)!}(kt)^{n-1}e^{-kt}]^m f_k(k)dk$$

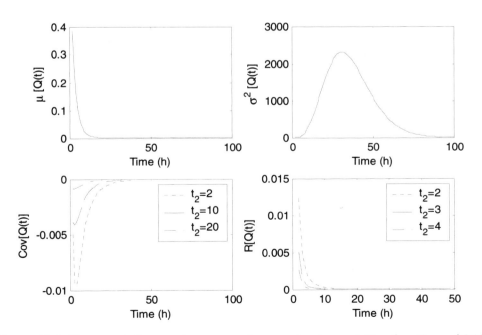

Figure 13-2 *The expectation, variance, covariance and autocorrelation functions of* Q(t).

Here

$$f_k(k) = \frac{1}{\sqrt{2\pi}\sigma_k}\exp[-\frac{(k-\mu_K)^2}{2\sigma_k^2}]$$

Then

$$E[h_n^m(t)] = E[\frac{k}{(n-1)!}(tk)^{n-1}e^{-kt}]^m = \int_{-\infty}^{\infty}[\frac{k}{(n-1)!}(kt)^{n-1}e^{-kt}]^m\frac{1}{\sigma_k\sqrt{2\pi}}\exp\left[-\frac{(k-\mu_k)^2}{2\sigma_k^2}\right]dk$$

For $m = 1$,

$$E[h_n(t)] = [\frac{\mu_k}{(n-1)!}(\mu_k t)^{n-1}e^{-\mu_k t}][\frac{\exp(\sigma_k^2 t^2/2)M_n}{\mu_k^n}]$$

where M_n is the nth moment of the normally distributed random variable whose mean is ($\mu_k - \sigma_k^2 t$) and variance is σ_k^2 (i.e., $N(\mu_k - \sigma_k^2 t, \sigma_k^2)$). The first three moments of this variable are

$$M_1 = \mu_k - \sigma_k^2 t \ , n = 1$$

$$M_2 = \sigma_k^2 + (\mu_k - \sigma_k^2 t)^2 \ , n = 2$$

$$M_3 = 3(\mu_k - \sigma_k^2 t)\sigma_k^2 + (\mu_k - \sigma_k^2 t)^3 \ , n = 3$$

The variance of $h_n(t)$ can be obtained as

$$\mathrm{var}[h_n(t)] = E[h_n^2(t)] - E[h_n(t)]^2$$

$$= \left[\frac{\mu_k}{(n-1)!}(\mu_k t)^{n-1}e^{-\mu_k t}\right]^2\left[\frac{\exp(\sigma_k^2 t^2)}{\mu_k^{2n}}\right]\left[\exp(\sigma_k^2 t^2)M_{2n}^* - M_n^2)\right]$$

where M_{2n}^* denotes the moments of order $2n$ of the random variable $\tilde{N}(\mu_k - \sigma_k^2 t, \sigma_k^2)$. The moments M_2^*, M_4^*, and M_6^* are

$$M_2^* = \sigma_k^2 + (\mu_k - 2\sigma_k^2 t)^2 \ , n = 1$$

$$M_4^* = 3\sigma_k^4 + 6(\mu_k - 2\sigma_k^2 t)^2\sigma_k^2 + (\mu_k - 2\sigma_k^2 t)^4 \ , n = 2$$

$$M_6^* = 9\sigma_k^6 + 36(\mu_k - 2\sigma_k^2 t)^2\sigma_k^4 + 14(\mu_k - 2\sigma_k^2 t)^4\sigma_k^2 + (\mu_k - 2\sigma_k^2 t)^6 \ , n = 3$$

The computed mean and standard deviation of function $h_n(t)$ are given in Fig. 13-3.

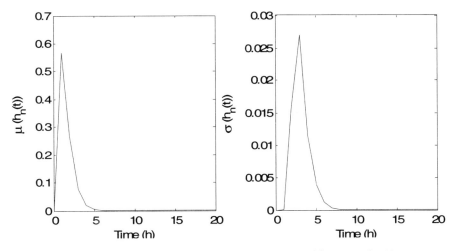

Figure 13-3 *Mean and standard deviation of function* $h_n(t)$.

Example 13.5 In Example 13.4, k was assumed normally distributed. Now, assume that k is gamma distributed:

$$f_k(k) = \frac{\lambda}{\Gamma(\alpha)}(\lambda k)^{n-1}\exp(-\lambda k)$$

Determine the mean and variance of the IUH in Example 13.4. Plot the function $f_k(k)$ and the mean and standard deviations of function $h_n(t)$.

Solution Following the procedure used in Example 13.4, we find the mean of $h_n(t)$ from

$$E\big[h_n(t)\big] = \frac{t^{n-1}}{(n-1)!}(\frac{\lambda}{\lambda+t})^{\alpha}M_n$$

where M_n is the nth moment of the random variable distributed as $G(\alpha, \lambda+t)$ and can be expressed as

$$M_n = (\lambda+t)^{-n}\prod_{i=1}^{n}(\alpha+i-1)$$

The variance of $h_n(t)$ is

$$\mathrm{var}[h_n(t)] = \left[\frac{t^{n-1}}{(n-1)!}\right]^2\left[(\frac{\lambda}{\lambda+\lambda t})^{\alpha}M_{2n}^* - (\frac{\lambda}{\alpha+t})^{2\alpha}M_n^2\right]$$

where M_{2n}^* is the moment of order $2n$ of the gamma-distributed random variable $G(\alpha, \lambda + 2t)$:

$$M_{2n}^* = (\lambda + 2t)^{-2n} \prod_{i=1}^{n} (\alpha + 2i - 1)$$

Taking $\alpha = 0.25$, $\lambda = 2.1$, and $n = 3$, we can plot the mean and standard deviation of the IUH (see Fig. 13-4). It is worth noting that not all gamma-distributed random coefficients can be studied by this method, because the variance calculated in this manner might be smaller than zero; thus only a narrow range of parameters α and λ may be applicable.

The next two examples deal with reservoirs. Example 13.6 focuses on a cascade of linear reservoirs with a random parameter. Example 13.7 relates to the topic of a lumped linear reservoir.

Example 13.6 In Example 13.4, the Nash cascade of equal linear reservoirs, each with parameter k, can be recast as

$$k\frac{dQ_i}{dt} + Q_i = Q_{i-1} , I = 1, 2,..., n$$

where n is the number of reservoirs, $Q_i(0) = 0$, $i = 1,2,...,n$, Q_i is the outflow from the ith reservoir, and t is time. If the cascade is subject to a unit impulse of input upstream (i.e., at the first reservoir), then the cascade response from this input will be the IUH, denoted as $h_n(t)$. Assuming n as 2 and k to be represented as $k = \bar{k} + k_*$, where $\bar{k}$ = mean value of k and k_* = zero-mean Gaussian random variable, determine the mean and variance of the IUH and plot these functions. This situation was addressed by Sarino and Serrano (1990). Take $\bar{k} = 10$ hours and $\sigma_{k_*} = 5$ hours. Plot the functions of the IUH.

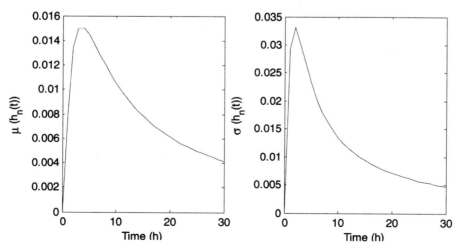

Figure 13-4 Mean and standard deviation of function $h_n(t)$.

Solution Consider the IUH (or outflow) of the first reservoir expressed as

$$k\frac{dh_1}{dt} + h_1 = \delta(t)$$

Replacing k by $\bar{k} + k_*$, one obtains

$$(\bar{k} + k_*)\frac{dh_1}{dt} + h_1 = \delta(t)$$

or

$$\bar{h}\frac{dh_1}{dt} + h_1 = \delta(t) - k_*\frac{dh_1}{dt}$$

or

$$\frac{dh_1}{dt} + \frac{h_1}{\bar{k}} = \frac{\delta(t)}{\bar{k}} - \frac{k_*}{\bar{k}}\frac{dh_1}{dt} , \quad h_1(0) = 0$$

To determine the IUH, consider the right side as $I_1(t)$. Therefore,

$$\frac{dh_1}{dt} + \frac{h_1}{\bar{k}} = I_1(t)$$

The unit impulse response of this equation is $e^{-t/\bar{k}}$, if $I_1(t)$ is represented by a unit impulse or delta function. Therefore, the IUH of the first reservoir is obtained by convoluting the IUH with $I_1(t)$ as

$$h_1(t) = e^{-t/\bar{k}}h_1(0) + \frac{1}{\bar{k}}\int_0^t e^{-(t-s)/\bar{k}}\delta(s)ds - \frac{k_*}{\bar{k}}\int_0^t e^{-(t-s)/\bar{k}}\frac{dh_1(s)}{ds}ds$$

Its solution is

$$h_1(t) = \frac{e^{-t/\bar{k}}}{\bar{k}} - \frac{k_*}{\bar{k}}e^{-t/\bar{k}}\int_0^t e^{s/\bar{k}}\frac{dh_1(s)}{ds}ds$$

Note that h_1 whose selection is sought appears on the right side as well. Sarino and Serrano (1990) approximated h_1 on the right side as a series:

$$h_1 = \sum_{i=1}^{4} h_{*i}$$

Thus, $h_1(t)$ can be written as

$$h_1(t) = \frac{e^{-t/\bar{k}}}{\bar{k}} - \frac{k_*}{\bar{k}}e^{-t/\bar{k}}\int_0^t e^{s/\bar{k}}\frac{d}{ds}\sum_{i=1}^{4}h_{*i(\lambda)}ds = \frac{e^{-t/\bar{k}}}{\bar{k}} - \frac{k_*}{\bar{k}}e^{-t/\bar{k}}\sum_{i=1}^{4}\int_0^t e^{s/\bar{k}}\frac{d}{ds}h_{*i}(s)ds$$

Now we solve the second term on the right side term by term. The first term h_{*1} is taken as the previous term, which then is obtained from $h_1(t)$:

$$h_{*1}(s) = \frac{e^{-t/\bar{k}}}{\bar{k}}$$

The second term now becomes

$$h_{*2}(s) = -\frac{k_*}{\bar{k}} e^{-s/\bar{k}} \int_0^s e^{\xi/\bar{k}} \frac{d}{d\xi}(e^{-\xi/\bar{k}}) d\xi = \frac{k_*}{\bar{k}^3} s e^{-s/\bar{k}}$$

Now the third term becomes

$$h_{*3}(s) = -\frac{k_*}{\bar{k}} e^{-s/\bar{k}} \int_0^s e^{\xi/\bar{k}} \frac{d}{d\xi}(\frac{k_*}{\bar{k}} \xi e^{-\xi/\bar{k}}) d\xi = \frac{k_*^2}{\bar{k}^4} e^{-s/\bar{k}} + \frac{k_*^2}{2\bar{k}} s^2 e^{-s/\bar{k}}$$

and the fourth term becomes

$$h_{*4}(s) = -\frac{k_*}{\bar{k}} e^{-s/\bar{k}} \int_0^s e^{\xi/\bar{k}} \frac{d}{d\xi}(h_{*3}(\xi)) d\xi = e^{-s/\bar{k}}(\frac{k_*^3}{\bar{k}^6} s^2 + \frac{k_*^3}{\bar{k}^5} s + \frac{k_*^3}{6\bar{k}^7} s^3)$$

Summing h_{*1}, h_{*2}, h_{*3}, and h_{*4}, we obtain the IUH of the first reservoir:

$$h_1(t) = e^{-t/\bar{k}}(\frac{1}{\bar{k}} + \frac{k_* t}{\bar{k}^3} - \frac{k_*^2 t}{\bar{k}^4} + \frac{k_*^2 t^2}{2\bar{k}^5} + \frac{k_*^3 t}{\bar{k}^5} - \frac{k_*^3 t^2}{\bar{k}^6} - \frac{k_*^3 t^3}{6\bar{k}^7})$$

Now we consider the outflow from the second reservoir, which will be the IUH of the two reservoirs in series. The input to the second reservoir is the outflow from the first reservoir, which is given by $h_1(t)$ as already calculated. Following the previous procedure and avoiding algebraic details, we find $h_2(t)$ to be

$$h_2(t) = e^{-t/\bar{k}}(\frac{t}{\bar{k}^2} - \frac{k_* t}{\bar{k}^4} + \frac{k_* t^2}{\bar{k}^4} + \frac{k_*^2 t}{\bar{k}^4} - \frac{2k_*^2 t^2}{\bar{k}^5} - \frac{k_*^3 t}{\bar{k}^5} + \frac{k_*^2 t^3}{2\bar{k}^6} + \frac{3k_*^3 t^2}{2\bar{k}^6} + \frac{k_*^2 t^4}{6\bar{k}^8} - \frac{9k_*^3 t^3}{6\bar{k}^7})$$

The mean of $h_2(t)$ is derived as

$$E[h_2(t)] = e^{-t/\bar{k}}[\frac{t}{\bar{k}^2} - (\frac{t}{\bar{k}^4} - \frac{t^2}{\bar{k}^4}) \int_{-\infty}^{\infty} k_* \frac{e^{-k_*^2/2\sigma_{k_*}^2}}{\sqrt{2\pi}\sigma_{k_*}} dk + (\frac{t}{\bar{k}^4} - \frac{2t^2}{\bar{k}^5} + \frac{t^3}{2\bar{k}^6}) \int_{-\infty}^{\infty} k_*^2 \frac{e^{-k_*^2/2\sigma_{k_*}^2}}{\sqrt{2\pi}\sigma_{k_*}} dk_*$$

$$+ (-\frac{t}{\bar{k}^5} + \frac{3t}{2\bar{k}^6} + \frac{t^4}{6\bar{k}^8} - \frac{9t^3}{6\bar{k}^7}) \int_{-\infty}^{\infty} k_*^3 \frac{e^{-k_*^2/2\sigma_{k_*}^2}}{\sqrt{2\pi}\sigma_{k_*}} dk_*]$$

Terms within integrals are moments of first, second, and third orders of k_* about the origin. The first-order moment is zero. Neglecting the third-order moments, one obtains

$$E[h_2(t)] = \sigma^2_{k_*} e^{-t/\bar{k}} \left[\frac{t}{\sigma^2_{k_*} \bar{k}^2} + \frac{t}{\bar{k}^4} - \frac{2t^2}{\bar{k}^5} + \frac{t^3}{2\bar{k}^6} \right]$$

Similarly, the variance of the IUH is obtained as

$$\mathrm{var}[h_2(t)] = E[h_2(t)]^2 - (E[h_2(t)])^2$$

Avoiding algebraic details, one gets

$$\mathrm{var}[h_2(t)] = \sigma^2_{k_*} e^{-2t/\bar{k}} \left(\frac{t^2}{\bar{k}^6} - \frac{2t^3}{3\bar{k}^7} + \frac{t^4}{3\bar{k}^8} \right)$$

The mean and covariance function are plotted in Fig. 13-5.

Example 13.7 Consider a phreatic aquifer, as shown in Fig. 13-6, represented by a lumped parameter linear reservoir as

$$q = a(h - H)$$

with its water balance given as

$$s\frac{dh}{dt} + a(h - H) = R$$

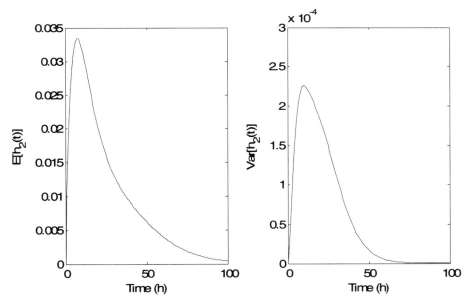

Figure 13-5 *Mean and covariance function of* $h_2(t)$.

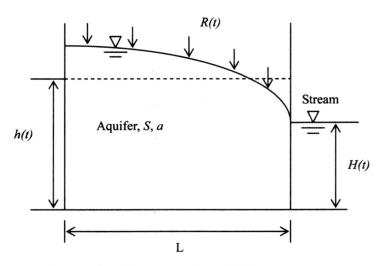

Figure 13-6 *Phreatic aquifer with linear reservoirs.*

where q is the outflow per unit area, s is the average storage coefficient or specific yield, R is the recharge rate, h is the average thickness of the saturated zone, H is the elevation of water level in an adjacent water body, a is an outflow constant, and t is time. Treating s and a as constant, and H, h, and R as stationary zero-mean random processes, determine the spectral solution of the equation. This example is discussed by Gelhar (1974). Take $a = 0.25$ and $s = 0.2$, and plot the ratios $S_{hh}(\omega)/S_{RR}(\omega)$.

Solution Using the stochastic Fourier–Stieltjes integral, we can express the random functions as

$$h(t) = \int_{-\infty}^{\infty} e^{i\omega t} dZ_h(\omega)$$

$$H(t) = \int_{-\infty}^{\infty} e^{i\omega t} dZ_H(\omega)$$

$$R(t) = \int_{-\infty}^{\infty} e^{i\omega t} dZ_R(\omega)$$

where dZ_h, dZ_H, and dZ_R are, respectively, the Fourier amplitudes of h, H, and R; and ∞ is the angular frequency. The Fourier amplitudes satisfy the usual properties as discussed in the preceding chapter. Substituting these into the differential equation yields

$$dZ_h(\omega) = \frac{a dZ_H(\omega) + dZ_R(\omega)}{(a + is\omega)}$$

Recall that, for a random process with independent increments, one gets

$$E[dZ_X(\omega_1)dZ_X^*(\omega_2)] = 0 \, , \ \omega_1 \neq \omega_2$$

$$E[dZ_X(\omega)dZ_X^*(\omega)] = S_{XX}(\omega)d\omega$$

where dZ_X^* is the complex conjugate of dZ_X, and S_{XX} is the spectral density function of X. In this specific problem, X is taken as $X = h, H$, or R. This leads to the Fourier transform relation between the autocovariance $R_{XX}(\tau)$ and spectral density function $S_{XX}(\omega)$ of X:

$$R_{XX}(\tau) = \int_{-\infty}^{\infty} e^{i\omega\tau} S_{XX}(\omega)d\omega = E[X(t+\tau)X^*(\tau)]$$

$$S_{XX}(\omega) = \frac{1}{2\pi} \int_{-\infty}^{\infty} e^{-i\omega\tau} R_{XX}(\tau)d\tau$$

where τ is the lag between $X(t)$ and $X(t + \tau)$.

Inserting dZ_h in the spectral density function expression, one obtains

$$E[dZ_h(\omega)dZ_h^*(\omega)] = E[(\frac{a}{a+i\omega s}dZ_H + \frac{1}{a+i\omega s}dZ_R)(\frac{a}{a-i\omega s}dZ_H^* + \frac{1}{a-i\omega s}dZ_R^*)]$$

$$S_{hh(\omega)}d\omega = E[\frac{a^2}{a^2+\omega^2 s^2}dZ_H dZ_H^*] + E[\frac{a}{a^2+\omega^2 s^2}dZ_H dZ_R^*]$$

$$+ E[\frac{a}{a^2+\omega^2 s^2}dZ_R dZ_H^*] + E[\frac{1}{a^2+\omega^2 s^2}dZ_R dZ_R^*]$$

$$= \frac{a^2}{a^2+\omega^2 s^2}S_{HH}(\omega)d\omega + \frac{a}{a^2+\omega^2 s^2}S_{HR}d\omega + \frac{a}{a^2+\omega^2 s^2}S_{RH}d\omega$$

$$+ \frac{1}{a^2+\omega^2 s^2}S_{RR}d\omega$$

$$S_{hh} = \frac{1}{a^2+\omega^2 s^2}[a^2 S_{HH} + a S_{HR} + a S_{RH} + S_{RR}]$$

where S_{HR} is the cross-spectrum, which is related to the cross-correlation function as

$$S_{HR}(\omega) = \frac{1}{2\pi} \int_{-\infty}^{\infty} e^{-i\omega\tau} R_{HR}(\tau)d\tau = C_{HR}(\omega) - iQ_{HR}(\omega)$$

The real part $C_{HR}(\omega)$ is the cospectrum and the imaginary part $Q_{HR}(\omega)$ is the quadspectrum. The cross-correlation function R_{HR} is expressed as

$$R_{HR}(\tau) = \frac{1}{2\pi} \int\limits_{-\infty}^{\infty} e^{i\omega\tau} S_{HR}(\omega) d\omega = E[H(t+\tau)R^*(\tau)]$$

Therefore, $S_{HR} = S_{RH}$. Hence,

$$S_{hh} = \frac{1}{a^2 + \omega^2 s^2} [a^2 S_{HH} + 2aS_{HR} + S_{RR}]$$

This shows how the spectral density functions of recharge R and input H, along with their cospectrum, determine the spectral density function of h.

When $\tau = 0$, the mean square fluctuation $E[h^2]$ is

$$E[h^2] = \int\limits_{-\infty}^{\infty} S_{hh}(\omega) d\omega$$

where S_{hh} is as already specified.

Likewise, the input–output cross-spectra can be obtained as follows:

$$E[dZ_R(\omega_1)dZ_h^*(\omega_2)] = E[dZ_R(\omega_1)(\frac{a}{a - i\omega s} dZ_H^* + \frac{1}{a - i\omega s} dZ_R^*)]$$

$$S_{Rh} = (aS_{RH} + S_{RR})(a + i\omega s)/(a^2 + \omega^2 s^2)$$

$$E[dZ_H(\omega_1)dZ_h^*(\omega_2)] = E[dZ_H(\omega_1)(\frac{a}{a - i\omega} dZ_H^* + \frac{1}{a - i\omega} dZ_R^*)]$$

$$S_{Hh} = (aS_{HH} + S_{HR})(a + i\omega s)/(a^2 + \omega^2 s^2)$$

One can draw inferences in the response of the linear reservoir using these spectra. For example, if $H = 0$, then

$$S_{Hh} = (aS_{HH} + S_{HR})(a + i\omega s)/(a^2 + \omega^2 s^2)$$

The term $1/(a^2 + \omega^2 s^2)$ is the square of the modulus of the frequency response. Here $(a^2 + \omega^2 s^2) = S_{RR}/S_{hh}$. These are also referred to as transfer functions, for they provide amplitude attenuation between input and output function in the frequency domain. That is, high-frequency variations in recharge R will be attenuated in the output h.

If H is constant, then the spectral representation of the linear reservoir is

$$S_{RR} = a^2 S_{hh}$$

The ratio $S_{hh}(\omega)/S_{RR}(\omega)$ is plotted in Fig. 13-7.

The next example considers the case where the input is random.

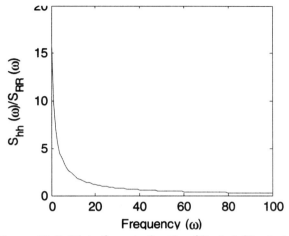

Figure 13-7 *Plot of spectral ratio of* $S_{hh}(\omega)/S_{RR}(\omega)$.

Example 13.8 Consider a drainage system as shown in Fig. 13-8, where the drain spacing is $2L$ and the maximum water table height above the drain is $M(t)$. One-dimensional flow to parallel drains can be described by the Dupuit approximation as

$$s\frac{\partial h}{\partial t} = T\frac{\partial^2 h}{\partial^2 t} + R$$

where the variables are defined in the preceding example. This drainage problem has been discussed by Duffy et al. (1984). The Darcy equation can be written as

$$q = \frac{T}{L}\frac{\partial h}{\partial x}\Big|_{x=0}$$

where q is the aquifer outflow per unit area. Determine the spectral solution of $h(t)$. It is assumed that $h(t)$ and $R(t)$ are stochastic processes.

Solution To reduce these processes to zero-mean processes, let

$$h = y + E[h] , \quad q = Q + E[q] , \quad R = P + E[R]$$

Substituting into the differential equation gives

$$s\frac{\partial}{\partial t}(y + \bar{h}) = T\frac{\partial^2}{\partial x^2}(y + \bar{h}) + P + \bar{R}$$

where the bar denotes the mean value. This yields

$$s\frac{\partial y}{\partial t} = T\frac{\partial^2 y}{\partial x^2} + P + [T\frac{\partial^2 \bar{h}}{\partial x^2} + \bar{R}]$$

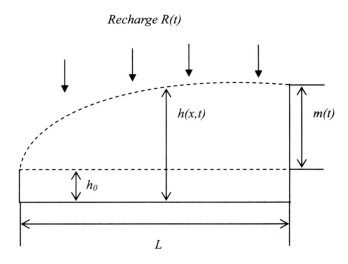

Figure 13-8 *Description of drainage system.*

This equation has two parts: a mean part and a fluctuating part. The mean part is

$$T\frac{\partial^2 \overline{h}}{\partial x^2} = -\overline{R} \ , \ \ h = h_0 \ , \ \ x = 0 \ ; \ \ h = h_0 \ , \ \ x = L$$

which has the solution

$$\overline{h} = h_0 + \overline{R}x(2T)^{-1}(2L - x) = \overline{M}(x)$$

This describes the steady-state water table distribution between drains. The stochastic partial differential equation becomes

$$s\frac{\partial y}{\partial t} = T\frac{\partial^2 y}{\partial x^2} + P \ , \ \ x = 0 \ , \ \ y = 0 \ , \ \ x = L \ , \ \ \partial y / \partial x = 0$$

In terms of the Fourier amplitudes, this equation reduces to

$$i\omega s dZ_y(\omega, x) = T\frac{d^2}{dx^2}[dZ_y(\omega, x)] + dZ_p(\omega)$$

$$x = 0 \ , \ dZ_y = 0 \ ; \ x = L \ , \ \frac{d}{dx}(dZ_y) = 0$$

Using the properties of the Fourier amplitudes gives the spectral solution:

$$S_{RR}(\omega) = \frac{s^2 \omega^2}{F(\omega, x)} S_{yy}(\omega)$$

where

$$F(\omega, X) = \left[g \times g^* - (g + g^*) + 1 \right]$$

$$g = \cosh[b(x - L)] / \cosh(bL)$$

$$b = (1 + i)(\tfrac{1}{2} |\omega| sT^{-1})^{1/2}$$

S_{RR} = recharge spectrum, and S_{yy} = water table spectrum. The asterisk means the complex conjugate.

Perturbation can also be applied to the Darcy equation for drain-flow spectral response. Applying the Stieltjes integral to Darcy's equation, one has

$$dZ_Q(\omega, x)|_{x=0} = \frac{T}{L} \frac{d}{dx} \left[dZ_y(\omega, x) \right]\Big|_{x=0} = \frac{T}{L} \frac{dZ_y(\omega, x)}{(f-1)} \frac{df}{dx} \Big|_{x=0}$$

Following the same procedure, one obtains

$$S_{QQ}(\omega) = S_{yy}(\omega, x)\omega sT^{-1}[G(\omega, L) / F(\omega, x)]$$

where

$$G(\omega, L) = \tanh(bL) * \tanh(b^* L)$$

Using $L = 200$ m, $T = 200$ m^2/day, $s = 0.25$, $\overline{R} = 25$ cm, and $h_0 = 20$ m, we plot the spectral ratios in Fig. 13-9.

Example 13.9 Consider an unconfined aquifer extending infinitely in the horizontal plane. The aquifer receives recharge R from rainfall over a circular area with radius a as shown in Fig. 13-10. The recharge relation is expressed simply as

$$\alpha = \frac{R}{P}$$

where α is the rainfall recharge coefficient, R is the amount of groundwater recharge, and P is the amount of rainfall. The governing equation in radial coordinates for groundwater flow is obtained from the Dupuit–Forchheimer theory as

$$C \frac{\partial H_1}{\partial t} = \frac{1}{r} \frac{\partial}{\partial r} (Tr \frac{\partial H_1}{\partial r}) + R, \ 0 \le r \le a$$

$$C \frac{\partial H_2}{\partial t} = \frac{1}{r} \frac{\partial}{\partial r} (Tr \frac{\partial H_2}{\partial r}), \ a \le r \le \infty$$

where C is the storage coefficient, T is the transmissivity, H_1 and H_2 are the hydraulic heads at locations 1 and 2, R is the groundwater recharge, r is the radial distance from the center of recharge, and t is time. The initial and boundary conditions are

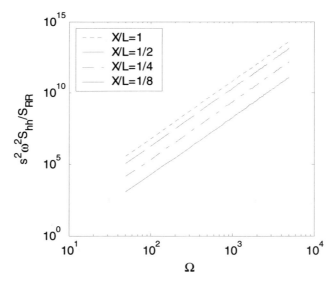

Figure 13-9 *Plot of spectral ratio of* $s^2\omega^2 S_{hh}(\omega)/S_{RR}(\omega)$

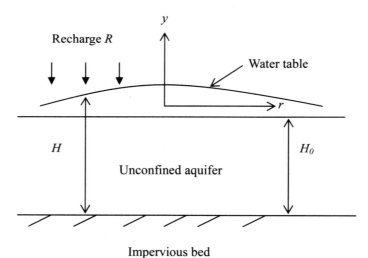

Figure 13-10 *Unconfined aquifer, with H as hydraulic head.*

$$H_1(r, 0) = H_0$$

$$\frac{\partial H_1}{\partial r}\Big|_{r=0} = 0 , \; H_1(a, t) = H_2(a, t)$$

$$\frac{\partial H_1}{\partial r}\Big|_{r=a} = \frac{\partial H_1}{\partial r}\Big|_{r=a} , \; H_2(\infty, t) = H_0 = \text{constant head}$$

$$H_2(r, 0) = H_0$$

Assuming recharge as a stationary random process, determine the spectral density functions of recharge and the water table or hydraulic head.

Solution To derive the spectral density functions, let the hydraulic head and recharge be represented as the sum of mean and perturbation:

$$H = \overline{H} + h , \quad R = \overline{R} + q$$

where $\overline{H}$ is the mean of H, h is the perturbation of H around its mean, $\overline{R}$ is the mean of R, and q is the perturbation of R around its mean. The perturbations are such that $E[h] = 0$ and $E[q] = 0$. Both h and q are therefore zero-mean stationary stochastic processes. Inserting these representations in the governing equation, one obtains

$$C\frac{\partial}{\partial t}(H + h_1) = \frac{1}{r}\frac{\partial}{\partial r}\left[rT\frac{\partial}{\partial r}(H + h_1)\right] + \overline{R} + q$$

$$C\frac{\partial}{\partial t}(H + h_2) = \frac{1}{r}\frac{\partial}{\partial r}(rT\frac{\partial}{\partial r}(H + h_2))$$

Removing the average terms and neglecting the products of perturbation quantities, one gets

$$C\frac{\partial h_1}{\partial t} = \frac{1}{r}\frac{\partial}{\partial r}(rT\frac{\partial h_1}{\partial r}) + q , \quad 0 \le r \le a$$

$$C\frac{\partial h_2}{\partial t} = \frac{1}{r}\frac{\partial}{\partial r}(rT\frac{\partial h_2}{\partial r}) , \quad a \le r \le \infty$$

In a similar manner, the initial and boundary conditions reduce to

$$h_1(r, 0) = 0 , \quad h_2(r, 0) = 0 , \quad h_2(\infty, t) = 0$$

$$h_1(a, t) = h_2(a, t)$$

$$\frac{\partial h_1}{\partial r}\Big|_{r=0} = 0 , \quad \frac{\partial h_1}{\partial r}\Big|_{r=a} = \frac{\partial h_2}{\partial r}\Big|_{r=a}$$

For spectral representation of h and q, one can write the Fourier–Stieltjes integrals as

$$h(r, t) = \int\limits_{-\infty}^{\infty} e^{i\omega t} dZ_h(r, \omega)$$

$$q(t) = \int_{-\infty}^{\infty} e^{i\omega t} dZ_q(\omega)$$

where ∞ represents the angular frequency, dZ_h is the Fourier amplitude corresponding to h, and dZ_q is the Fourier amplitude corresponding to q. Both satisfy the following properties:

$$E[dZ_h(\omega, r)] = 0 , \quad E[dZ_h(\omega_1, r)dZ_h^*(\omega_2, r)] = 0 , \quad \omega_1 \neq \omega_2$$

$$E[dZ_q(\omega)] = 0 , \quad E[dZ_q(\omega_1)dZ_q^*(\omega_2)] = 0 , \quad \omega_1 \neq \omega_2$$

$$E[dZ_h(\omega)dZ_h^*(\omega)] = S_h(\omega)d\omega$$

$$E[dZ_q(\omega)dZ_q^*(\omega)] = S_q(\omega)d\omega$$

where Z_h^* is the complex conjugate of dZ_h, Z_q^* is the complex conjugate of Z_q, $S_h(\omega)$ is the spectral density function of h, and $S_q(\omega)$ is the spectral density function of q.

Introducing the Fourier–Stieltjes integrals into the governing equations and boundary conditions gives

$$C\frac{\partial}{\partial t}[\int_{-\infty}^{\infty} e^{i\omega t} dZ_{h_1}(r, \omega)] = \frac{1}{r}\frac{\partial}{\partial r}(rT\frac{\partial}{\partial r}[\int_{-\infty}^{\infty} e^{i\omega t} dZ_{h_1}(r, \omega)] + \int_{-\infty}^{\infty} e^{i\omega t} dZ_q(\omega))$$

which becomes

$$\frac{\partial^2}{\partial r^2} dZ_{h_1} + \frac{1}{r}\frac{\partial}{\partial r} dZ_{h_1} - iC\frac{\omega}{T} dZ_{h_1} = -\frac{dZ_q}{T}$$

Similarly, the second equation is simplified as

$$C\frac{\partial}{\partial t}[\int_{-\infty}^{\infty} e^{i\omega t} dZ_{h_2}(r, \omega)] = \frac{1}{r}\frac{\partial}{\partial r}(rT\frac{\partial}{\partial r}[\int_{-\infty}^{\infty} e^{i\omega t} dZ_{h_2}(r, \omega)]) , \quad 0 \leq r \leq a$$

which reduces to

$$\frac{\partial^2}{\partial r^2} dZ_{h_2} + \frac{1}{r}\frac{\partial}{\partial r} dZ_{h_2} - iC\frac{\omega}{T} dZ_{h_2} = 0 , \quad 0 \leq r \leq \infty$$

Likewise the boundary conditions become

$$\frac{\partial}{\partial r} dZ_{h_1} |_{r=0} = 0 , \quad dZ_{h_1} |_{r=a} = dZ_{h_2} |_{r=a}$$

$$\frac{\partial}{\partial r} dZ_{h_1} \big|_{r=a} = \frac{\partial}{\partial r} dZ_{h_2} \big|_{r=a}, \quad dZ_{h_1} \big|_{r \to \infty} = 0$$

These two equations can now be solved. Let

$$dZ_{h_1} = M(r) - i\frac{dZ_q}{C\omega}$$

Inserting this in the first equation, one obtains

$$\frac{\partial^2}{\partial r^2} M + \frac{1}{r}\frac{\partial M}{\partial r} = iC\frac{\omega}{T}M$$

The solution of this equation (Carslaw and Jaeger 1959) is

$$M = PI_0\left[\left(\frac{iCw}{T}\right)^{1/2} r\right] + QK_0\left[\left(\frac{iCw}{T}\right)^{1/2} r\right]$$

where I_0 is the first-kind modified Bessel function of zeroth order and K_0 is the second-kind modified Bessel function of zeroth order. Then,

$$dZ_{h_1} = \frac{dZ_q}{Cw} f_{h_1}$$

Likewise, the solution of the second equation is

$$dZ_{h_2} = \frac{dZ_q}{Cw} f_{h_2}$$

where

$$f_{h_1}(r, \omega) = \frac{K_1(\sqrt{\frac{C\omega}{T}}e^{i\pi/4}a)I_0(\sqrt{\frac{C\omega}{T}}e^{i\pi/4}r)}{K_0(\sqrt{\frac{C\omega}{T}}e^{i\pi/4}a)I_1(\sqrt{\frac{C\omega}{T}}e^{i\pi/4}a) + K_1(\sqrt{\frac{C\omega}{T}}e^{i\pi/4}a)I_0(\sqrt{\frac{C\omega}{T}}e^{i\pi/4}a)} - 1$$

$$f_{h_2}(r, \omega) = \frac{K_0(\sqrt{\frac{C\omega}{T}}e^{i\pi/4}r)I_1(\sqrt{\frac{C\omega}{T}}e^{i\pi/4}a)}{K_0(\sqrt{\frac{C\omega}{T}}e^{i\pi/4}a)I_1(\sqrt{\frac{C\omega}{T}}e^{i\pi/4}a) + K_1(\sqrt{\frac{C\omega}{T}}e^{i\pi/4}a)I_0(\sqrt{\frac{C\omega}{T}}e^{i\pi/4}a)} - 1$$

where I_0 and I_1 are the first-kind modified Bessel functions of zeroth and first order, respectively, and K_0 and K_1 are the second-kind modified Bessel functions of zeroth and first orders, respectively.

The spectral density functions now follow:

$$S_{h_1} = \frac{S_q}{C^2 \omega^2} f_{h_1}(r, \omega) f_{h_1}^*(r, \omega) , \; 0 \leq r \leq a$$

$$S_{h_2} = \frac{S_q}{C^2 \omega^2} f_{h_2}(r, \omega) f_{h_2}^*(r, \omega) , \; a \leq r \leq \infty$$

where S_{h_1} is the spectral density function of h_1, S_{h_2} is the spectral density function of h_2, and S_q is the spectral density function of q.

The spectral relation between rainfall and recharge is

$$S_q = \alpha^2 S_P$$

Using this relation, one can derive the spectral relation between rainfall and hydraulic head:

$$\frac{S_P}{S_{h_1}} f_{h_1}(r, \omega) f_{h_1}^*(r, \omega) = \frac{C^2}{\alpha^2} \omega^2 , \; 0 \leq r \leq a$$

$$\frac{S_P}{S_{h_2}} f_{h_2}(r, \omega) f_{h_2}^*(r, \omega) = \frac{C^2}{\alpha^2} \omega^2 , \; a \leq r \leq \infty$$

Plots of S_p / S_{h_1} and S_p / S_{h_2} as functions of ∞ (in cycles/minute) are given in Fig. 13-11 for the case where a = 50 m, K = 0.004 m/minute, C = 0.0015, aquifer thickness = 150 m, and α = 0.5. Figure 13-11(a) shows S_p / S_{h_2} for r = 1, 20, and 50 m and Fig. 13-11(b) shows S_p / S_{h_2} for r = 50, 100, and 150 m.

Example 13.10 Consider one-dimensional flow in a phreatic aquifer described as

$$\frac{\partial h}{\partial t} = K \frac{\partial^2 h}{\partial x^2} , \; h(x, 0) = 0 , \; h(0, t) = g(t) , \; h(x, t) = 0 \text{ as } x \to \infty$$

where h is the hydraulic head (or piezometric height), t is time, x is the space coordinate, and

$$K = \frac{kH}{m}$$

in which k is the hydraulic conductivity, m is porosity, and H is the aquifer thickness. The IUH or impulse response function (IRF) for $g(t) = \delta(t)$ for the partial differential equation is given as (Venetis 1970; Singh 1989):

$$u(x, t) = (\frac{4Kt}{x^2})^{-3/2} \exp(-\frac{x^2}{4Kt}) \frac{x^2}{4K}$$

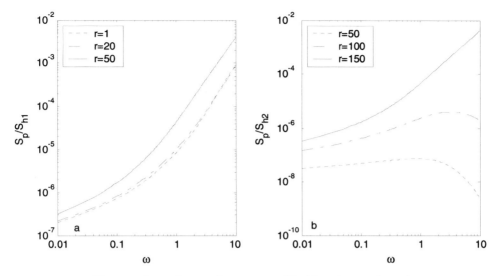

Figure 13-11 *Spectral response of* S_p/S_{h1}. *(a: r a; b: r a).*

Determine the autocovariance function of h assuming it to be a zero-mean random process. Also assume the input $h(0, t)$ to be independent and its covariance defined by the Dirac delta function $\delta(\tau)$ as $C_h(0, \tau) = E[h(0, t)h(0, t + \tau)] = \delta(\tau)$.

Solution Recall that

$$h(x, t) = \int_0^t u(x, t - s)h(0, s)\, ds$$

where $u(x, t)$ is the IUH and $h(0, t)$ is the input at $x = 0$. The IUH is due to $h(0,t) = \delta(0,t)$.

The autocovariance function of $h(x, t)$, $C_h(\tau)$, is expressed as

$$C_h(x, \tau) = E[h(x, t)h(x, t + \tau)]$$

$$= \int_0^\infty \int_0^\infty u(x, z)u(x, y)E[h(0, t - y)h(0, t + \tau - z)]dy\, dz$$

$$= \int_0^\infty u(x, y)u(x, y + \tau)dy$$

The correlation function of $h(x, t)$ is defined as

$$\rho_h(x, \tau) = \frac{C_h(x, \tau)}{C_h(x, 0)}$$

Integration of ρ_h over $\tau > 0$ yields the time scale T, which measures the persistence or memory of the water table or hydraulic head. Substituting the IUH into the expression for $C_h(x,\tau)$, one gets

$$C_h(x,\tau) = (\frac{2x^2 k}{K})^2 \int_0^\infty [\frac{4Kt}{x^2}(\frac{4Kt}{x^2} + \tau)]^{-3/2} \exp[-\frac{8Kt + x^2\tau}{4Kt(4kt + x^2\tau)}]\frac{x^2}{4Kt} dt$$

For $\tau = 0$, $C_h(x,0)$ is

$$C_h(x,0) = (\frac{2x^2 k}{K})^2 \int_0^\infty [\frac{4Kt}{x^2}]^{-3} \exp[-\frac{x^2}{2Kt}]dt = (\frac{2x^2 k}{K})^2 \int_0^\infty \frac{1}{t^3} \exp[-\frac{x^2}{2Kt}]dt$$

$$\approx \frac{2x^4 k^2}{K^2}$$

The correlation function can be described as

$$\rho_h(x,\tau) = 4\int_0^\infty [\frac{4Kt}{x^2}(\frac{4Kt}{x^2} + \tau)]^{-3/2} \exp[-\frac{8kt + x^2\tau}{4Kt(4Kt + x^2\tau)}]\frac{4K}{x^2} dt$$

Its integration yields the correlation time:

$$T_h = \frac{\pi x^2}{2K}$$

Take $x = 1,000$ m, $m = 0.35$, $H = 50$ m, and $k = 5 \times 10^{-4}$ m/s. The correlation function of h is plotted in Fig. 13-12.

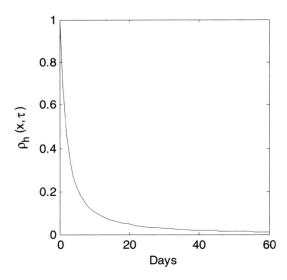

Figure 13-12 Correlation function of h.

Example 13.11 A linearized form of the Saint-Venant equation when applied to flow in a river can be expressed, after neglecting inertia terms, as

$$\frac{\partial^2 q}{\partial x^2} = \frac{3}{A_0^2} \frac{Q_0^2}{C^2 H_0^2} \frac{\partial q}{\partial x} + \frac{2}{A_0^2} \frac{Q_0}{C^2 H_0^2} \frac{\partial q}{\partial t}$$

where Q_0 is the mean river flow, q is the fluctuation of the flow, H_0 is the mean flow depth, A_0 is the mean cross-sectional area, C is Chezy's roughness coefficient, x is the distance along the channel, and t is time. The IUH of this equation is (Dooge 1973; Singh 1989):

$$u(x, t) = \frac{x}{\sqrt{4\pi \dfrac{A_0 C^2 H_0^2}{2Q_0}}} \frac{1}{t^{3/2}} \exp[-\frac{(x - \dfrac{3}{2}\dfrac{Q_0}{A_0} t)^2}{\dfrac{4 A_0 C^2 H_0^2}{2Q_0}}]$$

For many river reaches the IUH can be expressed as a gamma function:

$$u(n, t) = \frac{1}{k\Gamma(n)} (\frac{t}{k})^{n-1} e^{-t/k}$$

where n is the number of subreaches arranged in series to represent the reach, k is the travel time of each subreach, and $u(n, t)$ is the IUH. Determine the covariance function of flow and its time scale.

Solution In the equation

$$C_q(n, \tau) = \int_0^\infty (\frac{1}{k\Gamma(n)})^2 (\frac{s}{k})^{n-1} e^{-s/k} (\frac{s+\tau}{k})^{n-1} e^{-(s+\tau)/k} ds$$

we expand the term $(s + \tau/k)^{n-1}$ using the binomial series. This yields

$$C_q(n, \tau) = \frac{1}{\Gamma(n)} e^{-\tau/k} \sum_{r=0}^{n-1} (\frac{\tau}{k}) \frac{\Gamma(2n-r-1)}{\Gamma(n-r)\Gamma(r+1)} (\frac{1}{2})^{2n-r-1}$$

For $\tau = 0$,

$$C_q(n, 0) = \frac{\Gamma(2n-1)}{\Gamma^2(n)} 2^{-2n+1}$$

The correlation function, thus, becomes

$$\rho_q(n, \tau) = \frac{C_q(n, \tau)}{C_y(n, 0)} = e^{-\tau/k} \sum_{r=0}^{n-1} (\binom{n-1}{r})(\frac{\tau}{K})^r \frac{\Gamma(2n-r-1)}{\Gamma(2n-1)} 2^r$$

Integration over $\tau > 0$ leads to the time scale of the flow process:

$$T_q = \int_0^\infty \rho_q(n, \tau)d\tau = k \sum_{r=0}^{n-1} \binom{n-1}{r} 2^n \frac{\Gamma(2n-r-1)}{\Gamma(2n-1)} \cdot \Gamma(r+1)$$

The instantaneous hydrographs are plotted in Fig. 13-13 for a river reach of length = 20 km; $Q = 500$ m^3/s; $n = 2, 3, 4,$ and 5; and $k = 0.2, 0.3,$ and 0.5 for time in days.

Example 13.12 Consider flood routing in a channel reach using the Muskingum method, expressed as

$$\frac{dV}{dt} = I - Q$$

$$V = K[xI + (1-x)Q]$$

where V is the storage in the channel at time t, I is rate of flow to the channel, Q is the outflow from the channel, x is a weighting factor, and K is the average travel time of the reach or average residence time.

Combining the two equations, one gets

$$K(1-X)\frac{dQ}{dt} + Q = I - Kx\frac{dI}{dt}$$

Assuming I and Q as zero-mean random processes, and treating x and K as constants, determine the spectral solution of Q as well as V.

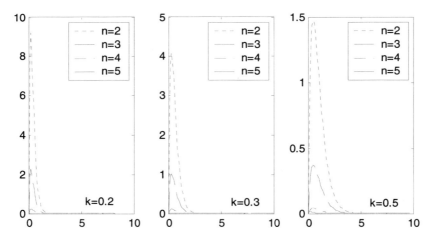

Figure 13-13 *Instantaneous hydrograph with different n and k.*

Solution By using the Fourier–Stieltjes integral, the Muskingum differential equation can be written as

$$dZ_Q = \frac{(1 - xKi\omega)}{1 + K(1 - x)i\omega} dZ_I$$

Then we have

$$E[dZ_Q \, dZ_Q^*] = S_{QQ}(\omega) = E[\frac{(1 - xKi\omega)dZ_I}{1 + K(1 - x)i\omega} \cdot \frac{(1 + xKi\omega)dZ_I^*}{1 - K(1 - x)i\omega}]$$

This yields

$$S_{QQ} = \frac{1 + x^2 K^2 \omega^2}{1 + \omega^2 K^2 (1 - x)^2} S_{II}$$

or

$$S_{QQ}/S_{II} = \frac{1 + x^2 K^2 \omega^2}{1 + \omega^2 K^2 (1 - x)^2}$$

This shows the spectral density function of outflow from the channel reach as a function of the spectral density function of inflow to the reach. Usually $x < 0.5$; therefore, high-frequency variations in the inflow hydrograph will be attenuated in the outflow.

For the Muskingum storage–discharge relation, one obtains

$$S_{VV}(\omega) = E[(Kx \, dZ_I + K(1 - x)dZ_Q)(xK \, dZ_I^* + K(1 - x)dZ_Q^*)]$$
$$= K^2 x^2 S_{II} + K^2 (1 - x)^2 S_{QQ} + 2K^2 x(1 - x)S_{IQ}]$$

where S_{IQ} is the inflow–outflow spectrum or cross-spectrum of I and Q. This expresses the spectral density function of storage in the channel in terms of the spectral density function of inflow and the cross-spectrum of I and Q. One can draw inferences on the Muskingum hypothesis. For example, the contribution to the spectrum of storage is the highest by the channel outflow, the second highest by the combined inflow and outflow, and the lowest by the channel inflow. A plot of S_{QQ}/S_{II} for $K = 24$ hours and $x = 0.10, 0.25, 0.4$ is given in Fig. 13-14.

Example 13.13 Lake or reservoir remediation requires an evaluation of the residence time. The residence time provides an estimate of the time required to clean the reservoir, the time for which aquatic organisms will be exposed to contaminants, the effects of incidental releases of pollutants on ecosystems, the time to simulate and control lake water quality, and so on. A simple differential equation governing the time variation of chemical concentration $C(t)$ in a reservoir is expressed as

$$\frac{dC(t)}{dt} = -\frac{q}{V}C(t) , \quad C(0) = C_0$$

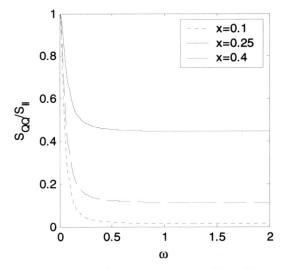

Figure 13-14 Spectral response of S_{QQ}/S_{II}.

where q is the discharge leaving the reservoir, V is the volume of the reservoir, C_0 is the initial concentration, and t is time. Assuming that q is a stochastic process, determine the expected value of C and the residence time. Compare the residence time derived with the residence time assuming q to be a constant or a mean value. For simplicity, assume C to be the excess concentration above the background level. This problem is discussed by Maran (2002).

Solution When q is constant, the solution of the differential equation is

$$C(t) = C_0 \exp(-\frac{q}{V}t) = C_0 \exp(-kt)$$

where $k = q/V$ is the rate constant for advection.

When q is considered as a stochastic process, the residence time T_r is then defined as

$$T_r = \frac{1}{K} = \frac{V}{q}$$

Thus T_r is the residence time from advection. In addition, T_r is a stochastic process and so is k. Since q is not constant in time, the concentration usually does not exhibit a simple exponential decay. If $q(t)$ were known explicitly then the differential equation could be solved explicitly and T_r can be derived. However, $q(t)$ is not known. Therefore, the equation for time variation of C is

$$\frac{dC(t)}{dt} = -k(t)C(t)$$

where $k(t)$ is a random function. Assume $k(t)$ to be stationary. That is, its expected value and all successive moments are independent of time:

$$\mu_k = E[k(t)] = \frac{E[q(t)]}{V} = \frac{\mu_q}{V}$$

$$\sigma_k^2 = E[k(t) - \mu_k]^2 = \frac{\sigma_q^2}{V^2}$$

$$r_k(\tau) = \frac{1}{\sigma_k^2} E[(k(t) - \mu_k)(k(t - \tau) - \mu_k)] = r_q(\tau)$$

$$p_k(k) = V p_q(kV)$$

where μ_k, σ_k^2, r_k, and p_k are, respectively, the mean, variance, autocorrelation function, and PDF of k; E is expectation; and μ_q, σ_q^2, r_q, and p_q are, respectively, the mean, variance, autocorrelation function, and the PDF of q.

Treating C_0 as a constant, we now determine the expectation of $C(t)$ from the statistical characterization of q using the given differential equation. A general solution is not tractable so an approximate solution is derived. To that end, a discharge autocorrelation time τ_{ac} is defined. This defines the time lag such that the autocorrelation is zero or negligible for $\tau \geq \tau_{ac}$. Two cases can be considered: (1) τ_{ac} is small and (2) τ_{ac} is large. The first case is considered here.

The random function $k(t)$ can be expressed as a sum of two components:

$$k(t) = \mu_k + \sigma_k k_1(t)$$

where $k_1(t)$ is a dimensionless random function having zero mean and unit variance and autocorrelation equal to r_k or r_q. Here σ_K signifies the magnitude of the fluctuations. The differential equation can now be written as

$$\frac{dC(t)}{dt} = -(\mu_k + \sigma_k k_1)C$$

To solve this equation, we consider a change of variable:

$$C(t) = \exp(-\mu_k t) y(t)$$

The differential equation in $y(t)$ becomes

$$\frac{dy(t)}{dt} = -\sigma_k k_1 y \;, \;\; y(0) = y_0$$

Its solution is

$$y(t) = y_0 \exp[-\sigma_k \int_0^t k_1(s) ds]$$

The expected value of $y(t)$ can be determined by using the cumulant expansion presented by Kubo (1962). For any random variable X, cumulants are defined as

$$E[\exp(ax)] = \exp[\sum_{n=1}^{\infty} \frac{1}{n!} a^n C_X^n]$$

where C_X^n is the nth term in the cumulant expansion of $X = \langle\langle x^n \rangle\rangle$. The double angle ($\langle\langle \cdot \rangle\rangle$) notation implies cumulant. Cumulants and moments are related (Singh 1988). For the first four cumulants,

$$C_X = \langle\langle x \rangle\rangle = E[X] = \mu_X$$

$$C_{X^2} = \langle\langle x^2 \rangle\rangle = E(X^2) - [E(X)]^2 = \sigma_X^2$$

$$C_{X^3} = \langle\langle x^3 \rangle\rangle = E(X^3) - 3E(X^2)E(X) + 2[E(X)]^3$$

$$C_{X^4} = \langle\langle x^4 \rangle\rangle = E(X^4) - 4E(X^3)E(X) - 3\left[E(X^2)\right]^2 + 12E(X^2) + [E(X)]^2 - 6[E(X)]^4$$

Note that, for a random variable with zero mean, the first three cumulants are equal to the first three moments. Furthermore $C_{X_1, X_2, ..., X_n} = 0$ if one of the random variables, $X_1, X_2, ..., X_n$, is independent of the others. For example, $\langle\langle k_1(t)k_1(t-\tau) \rangle\rangle$ vanishes if $\tau > \tau_{ac}$.

Making use of the cumulant relations, one can express the expected value of $y(t)$ as

$$\frac{E[y(t)]}{y_0} = E\left\{\exp\left[-\sigma_k \int_0^t k_1(s)ds\right]\right\}$$

$$= \exp\left[\sum_{n=2}^{\infty} \frac{(-\sigma_k)^n}{n!} \int_0^t dt_1 ... \int_0^t dt_n C_{K_1}(t_1)C_{k_1}(t_2)...C_{k_1}(t_n)\right]$$

$$= \exp\left[\sum_{n=2}^{\infty} (-\sigma_k)^n\right] \int_0^t dt_1 \int_0^{t_1} dt_2 ... \int_0^{t_{n-1}} dt_n ... \langle\langle k_1(t_1)k_2(t_2)...k_1(t_n) \rangle\rangle$$

The nth term in the summations yields a contribution of order $\sigma_k t(\sigma_k \tau_{ac})^{n-1}$, where $\sigma_k \tau_{ac}$ is an expansion parameter. To get an idea of the approximation, the zero-order approximation is 1, the first-order term is 0 because the expected value of k_1 vanishes, and the second-order term is of the order $\sigma_k^2 \tau_{ac} t$.

Using the second-order approximation only, one obtains

$$\frac{E[y(t)]}{y_0} = \exp[\sigma_k^2 \int_0^t dt_1 \int_0^{t_1} dt_2 C_{k_1 k_2}(t_1, t_2)]$$

$$= \exp[\sigma_k^2 \int_0^t dt_1 \int_0^{t_1} d\tau\, r_k(\tau)] = \exp[\sigma_k^2 \int_0^t \Gamma(\tau)d\tau]$$

where $\Gamma(t) = \int_0^t r_k(\tau)d\tau$. This is the solution of the following differential equation:

$$\frac{d[E(y)]}{dt} = \sigma_k^2 \Gamma(t) E(y)$$

or

$$\frac{dE(C)}{dt} = -[\mu_k - \sigma_k^2 \Gamma(t)] E[C]$$

This solution presents the variability of the average concentration, provided $\sigma_k \tau_{ac} \ll 1$. If $t \gg \tau_{ac}$, $\Gamma(t) \approx \tau_{ac}$ and the decay rate of the average concentration is decreased by a quantity proportional to the variance of fluctuations and to their autocorrelation time. This differential equation is valid when μ_k dominates or when $\sigma_k / \mu_k \ll 1$ even if $\sigma_k \tau_{ac}$ is not smaller than one.

Neglecting terms of higher order than $\sigma_k \tau_{ac}$, we can write

$$\frac{dE(C)}{dt} = -[\mu_k - \sigma_k^2 \tau_{ac}] E[C]$$

Its solution is

$$E(C) = C_0 \exp[-(\mu_k - \sigma_k^2 \tau_{ac})t]$$

The effect of discharge fluctuation is one of normalization of the mean discharge with an additional term proportional to the variance and the autocorrelation time of the discharge series. In this case, the effective rate constant of advection is given as $k_{eff} = \mu_k - \sigma_k^2 \tau_{ac}$, which is reduced by the amount $\sigma_k^2 \tau_{ac}$.

Now the residence time is computed. For the autocorrelation time of the discharge time series being much smaller than the residence time in the zero-order approximation, the residence time T_r is obtained as

$$T_r^{(1)} = \frac{1}{\mu_k - \sigma_k^2 \tau_{ac}} = \frac{1}{\mu_k} \frac{1}{[1 - \frac{\sigma_k^2 \tau_{ac}}{\mu_k}]} \cong T_r^{(0)}(1 + \frac{\sigma_k^2 \tau_{ac}}{\mu_k})$$

where $T_r^{(0)}$ is the residence time obtained from the mean discharge (zero-order approximation, equal to $1/\mu_k$), and terms of second or higher order in $\sigma_k^2 \tau_{ac}$ have been neglected. One can write

$$\frac{T_r^{(1)} - T_r^{(0)}}{T_r^{(0)}} = \frac{\sigma_q^2 \tau_{ac}}{E[q]V}$$

This specifies the increase in the residence time owing to discharge fluctuations.

Now the case of large autocorrelation time is considered. In this case k is a random variable, not a random function. For initial concentration C_0 and rate constant k, let the solution of the concentration equation be denoted as $f(t, k)$. Then the PDF of the concentration at time t, p_C, can be computed from the PDF of k, p_K, as

$$P_C(C, t) = \int_0^\infty \delta[C - f(t, k)] p_k(k) dk$$

where δ is the Dirac delta function. The probability that the concentration at time t is between C and $C + dC$ is equal to $p_C(C, t)dC$. Using the solution of C, one obtains

$$p_C(C, k) = V \int_0^\infty p_q(kV) \delta(C - C_0 e^{-kt}) dk$$

Using the change of variable $s = C_0 \exp(-kt)$ gives

$$p_C(C, t) = \frac{V}{t} \int_0^\infty p_q(\frac{V}{t} \ln \frac{C_0}{s}) \delta(C - s) \frac{ds}{s} = \frac{V}{Ct} p_q(\frac{V}{t} \ln \frac{C_0}{C})$$

Now the expected value of concentration is

$$\mu_C(t) = E[C(t)] = \int_0^\infty C p_C(C, t) dC$$

and with change of variable from C to q, one obtains

$$\mu_C(t) = C_0 \int_0^\infty e^{-qt/V} p_q(q) dq$$

which is the Laplace transform of the discharge PDF.

Similarly, the variance of $C(t)$ is

$$\sigma_C^2(t) = \mu_C(2t)\mu_C(0) - [\mu_C(t)]^2$$

The asymptotic behavior of the mean concentration, that is, as $t \to 0$ (t small compared with $T_r^{(0)}$), can be obtained by expanding the exponential to the first order:

$$\frac{\mu_C(t)}{C_0} = \int_0^\infty (1 - \frac{qt}{V}) p_q(q) dq = 1 - \frac{t}{T_r^{(0)}}$$

which is the same as the asymptotic behavior of the deterministic equation.

However, if $t \to \infty$, only a small region near the origin is significant. Therefore, p_q can be expanded around zero as

$$\frac{\mu_C}{C_0} = \int_0^\infty e^{-qt/V}[p_q(0) + p_q'(0)q + \frac{1}{2}p_q''(0)q^2 + \ldots]dq$$

$$= \frac{V}{t}[p_q(0) + p_q'(0)\frac{V}{t} + \frac{1}{2}p_q''(0)(\frac{V}{t})^2 + \ldots]$$

This shows that the expected value of concentration decreases with a power of $1/t$.

The residence time in this case can be computed as

$$T_r = \frac{1}{C_0}\int_0^\infty \mu_C(t)dt = V\int_0^\infty \frac{p_q(q)}{q}dq = VE[\frac{1}{q}]$$

The probability that the residence time is between T and $T + dT$ can be obtained from the discharge PDF as

$$p_T(T) = p_q(V/T)\frac{V}{T^2}dT$$

Using this expression one can calculate the mean of T, as T_r above. Similarly, the variance of T is

$$\sigma_T^2 = V^2 \sigma_{q^*}^2 \ , \quad q^* = 1/q$$

Example 13.14 Consider a lake with a residence time based on the mean discharge as 50 hours. Assume the lake outflow is characterized by a log-normal probability distribution with a coefficient of variation of 0.5 and a long autocorrelation time. The initial concentration is $C_0 = 1.5$ mg/L. Compute the PDFs of the residence time and concentration and the expected value of the residence time.

Solution The PDF of $k \ (= q/V)$ is

$$p_k(k) = \frac{1}{s_k k \sqrt{2\pi}} \exp[-\frac{1}{2}(\frac{\ln k - m_k}{s_k})^2]$$

Parameters m_k and s_k are related to CV and $T_r^{(0)}$ by

$$CV^2 = \exp(s_k^2) - 1, \ m_k = -\ln T_r^{(0)} - \frac{1}{2}s_k^2$$

where $\exp(m_k)$ is the geometric mean of k, which is the geometric mean discharge divided by volume. The plot of p_k is shown in Fig. 13-15.

The PDF of the residence time is computed by using the relation between p_T and p_q. This is also a log-normal PDF with parameters $s_T = s_K$ and $m_T = -m_k$. The coefficient of variation CV of T equals the CV of q, and the geometric mean of T equals the ratio of the reservoir volume and the geometric mean of discharge. The PDF of T is shown in Fig. 13-16.

The expected value of residence time is computed as

$$E[T] = \frac{V}{E[q]}(1 + CV^2) = T_r^{(0)}(1 + CV^2) = 50 \times (1 + 0.5^2) = 62.5 \text{ hours}$$

The geometric mean of T is

$$T_g = \frac{E[T]}{\sqrt{1 + CV_q^2}} = T_r^{(0)}\sqrt{1 + CV_q^2} = 50 \times \sqrt{1 + 0.5^2} = 55.9 \text{ hours}$$

Now the PDF of concentration can be written as

$$p_C(C, t) = \frac{1}{C \ln(\frac{C_0}{C})S_q\sqrt{2\pi}} \exp[-\frac{1}{2}(\frac{\ln\ln(C_0/C) - m(t)}{S_q})^2]$$

where $m(t) = \ln(t/T_g)$.

Figure 13-17 plots the PDFs of C at various times. The expected value of C is plotted in Fig. 13-18.

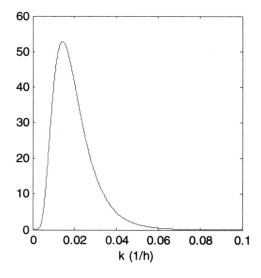

Figure 13-15 *Probability density function of* k.

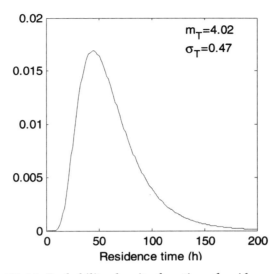

Figure 13-16 *Probability density function of residence time* (T).

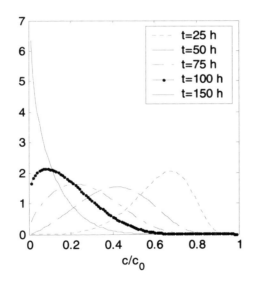

Figure 13-17 *Probability density functions for concentration at various times.*

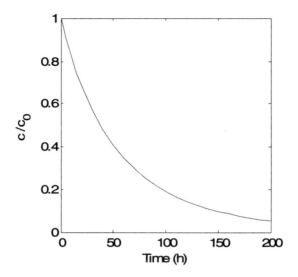

Figure 13-18 *The expected value of concentration versus time.*

13.5 Questions

13.1 A linear differential equation

$$\frac{dS}{dt} = -Q \, , \; S = KQ$$

is frequently used for stream base-flow recession. Here S is the storage in a watershed at time t, Q is discharge, and K is the residence time. The initial condition is the following: At $t = 0$, $Q(0) = Q_0$. It is assumed that K is a random variable with mean μ_K and variance σ_K^2. Determine the solution of the differential equation and the mean μ and the variance, covariance, autocorrelation function of discharge Q.

13.2 Consider the differential equation in Question 13.1. Assume that both K and Q_0 are normally distributed random variables. Determine the solution of the differential equation and the mean μ and the variance, covariance, autocorrelation function of discharge Q.

13.3 The surface runoff hydrograph from an area represented by a linear reservoir can be described mathematically as

$$K\frac{dQ(t)}{dt} + Q(t) = P \, , \; Q(0) = 0$$

where $P(t)$ is rainfall intensity, Q is unit surface runoff hydrograph, K is a reservoir coefficient, and t is time. It is assumed that P is a uniformly distributed random variable. Determine the mean, variance, covariance, autocorrelation function, and the coefficient of correlation of Q.

13.4 A watershed is represented by a cascade of n equal reservoirs each with reservoir coefficient k considered as a random variable. This representation is referred to as the Nash cascade and is popularly used for modeling surface runoff. The IUH of this cascade is

$$h_n = \frac{k}{(n-1)!}(kt)^{n-1}e^{-kt}$$

where h_n is the stochastic instantaneous unit hydrograph of the n-reservoir cascade and k is an exponentially distributed random variable. Determine the mean, the variance, and the first three moments of h_n (or IUH).

13.5 For the IUH in Question 13.4, if k is assumed to have a uniform distribution then determine the mean and variance of the IUH.

13.6 Consider the Muskingum method of flow routing in a river reach. The governing equation can be expressed as

$$K(1-x)\frac{dQ}{dt}+Q=I+xK\frac{dI}{dt}$$

where I is inflow to the reach, Q is the outflow from the reach, x is a weighting factor, and K is the storage delay time. Considering x and K as constant and I as a stationary zero-mean random process, determine the spectral solution of the equation.

13.7 The rate of infiltration, $f(t)$, at time t in a soil column can be described by (1) the continuity equation

$$\frac{dS(t)}{dt}=f_s-f(t)$$

where $S(t)$ is potential water storage space available in the soil column at time t and f_s is the seepage rate, assumed to be constant, and (2) a relation between $S(t)$ and $f(t)$. One such simple relation is

$$S_0-S(t)=\frac{a}{f(t)}$$

where a is a parameter and S_0 is the initial value of S. Assuming a to be a normally distributed random variable, derive the spectral solution of the equation. Assume that S and f are stochastic processes.

13.8   For the situation in Question 13.7, make the additional assumption that f_s is also a random variable. Derive the spectral solution of the infiltration equation.

13.9   Solve Question 13.7 if a is uniformally distributed.

Part IV

Risk and Reliability Analysis

Chapter 14

Reliability Analysis and Estimation

Engineering projects are always subject to a possibility of failure in achieving their intended objectives. Failure of a system can be defined as an event in which the system fails to function with respect to its desired objectives. There are two types of failure in engineering projects: (1) performance failure and (2) structural failure. Performance failure is said to take place when the system is unable to perform as per the expectation and hence undesirable consequences occur, although the structure of the system has not been altered. For example, flood control structures may not be able to protect an area from extreme floods, water supply systems may not deliver enough water, a canal may not convey enough water for irrigation of agricultural crops, or storm sewer systems may fail to convey excessive urban runoff. Structural failure involves damage or change of the structure as the result of the system load exceeding the system capacity (resistance) and hence hindering the ability of the structure to serve its intended objectives. Examples include the washout of a dam because of overtopping, the breach of a levee because of erosion, buckling of a beam, breaking of a bridge, or failure of a pump. There are various definitions of risk for different purposes, including the probability of failure, the reciprocal of the expected length of time before failure, the expected cost of failure, and the actual cost of failure.

Reliability, the complement of risk, is defined as the probability of nonfailure. Two types of application of reliability analysis to engineering problems are:

1. Evaluation or review of the reliability or safety of an existing or predetermined system
2. Design of a new system on a reliability basis

Evaluating reliability is one of the most important problems in a multitude of civil engineering projects, particularly those having catastrophic consequences in the event of structural failure, such as failure of a dam. Reliability engineering means considering tolerances in design parameters, uncertainties in environment, uncertainties in application (e.g., usage scenarios), and variations in manufacturing.

The design and analysis of any civil engineering project are subjected to uncertainty because of inherent uncertainty in natural systems; a lack of understanding of the causes and effects as well as interactions in various physical, chemical, and biological processes occurring in natural systems; and insufficient data. As a result of these uncertainties, the performance of a project is uncertain. A reliable assessment of the performance of any water resources project requires an assessment of the validity of predicted loads (such as discharges and pollutant loads) and capacities (e.g., ability to perform under a given load without any harm). Typically, the loads are assessed by using models having a number of parameters that can be determined with varying degrees of certainty. These parameters are best represented as random variables. Consequently, the model response, being a function of random variables, is best represented as a random variable. For reliable design and analysis of a project, it becomes necessary to address the uncertain nature of model outputs. Reliability, risk, and uncertainty analyses are therefore becoming increasingly important in modeling and designing water resources infrastructure and decision support systems. In some cases, performing an uncertainty analysis is mandatory, particularly when critical decisions involve potentially high levels of risk. Quantification of the underlying reliability is central to each of these systems.

14.1 Approaches to Reliability Evaluation

There are two approaches considered for safety evaluation of engineering systems. In the traditional approach one uses conservative assumptions in the underlying process by defining a worst-case scenario, whereas the probabilistic approach involves a probabilistic assessment of the performance of the underlying system in all possible conditions. Before discussing these approaches any further we need to define the terms most frequently used in the literature on reliability analysis in civil engineering. These terms include resistance, strength, capacity, load, input, and demand.

14.1.1 Traditional Approach

Traditionally, evaluations of safety and adequacy of engineering systems were expressed in terms of safety margins and safety factors to compensate for uncertainties in loading and material properties and inaccuracies in geometry and theory. In this approach point design evaluations are performed by assuming precise and fixed values of design parameters in a specific environment, and then the steady-state and/or transient performance of that design is predicted. The use of precisely defined single values in analysis—known as the deterministic approach or point design evaluation—represents not what an engineer needs to accomplish, but rather what is convenient to numerically solve, assuming inputs that are known precisely. Specifically, point design evaluation is merely a subprocess of what an engineer must do to produce a useful and efficient design. Sizing, selecting, and locating components and coping with uncertainties and variations comprise the real tasks. Point design simulations by themselves cannot produce effective designs; they can only verify deterministic instances of them. The safety factor in this approach accounts for the condition of the future, the engineer's judgment, and the degree of conservatism incorporated into the parameter values.

In the traditional approach of safety analysis, the worst-case scenarios are considered to determine the load and capacity of a system and tolerances stacked up in terms of safety factors and margins, as shown in Fig. 14-1. In most cases, such safety margins and factors are seldom based on any mathematical rigor or true knowledge of the underlying risk and results in an overdesign. This leads to designs that are heavier and costlier than they need to be, and in some cases such designs do not even result in greater safety or reliability.

As an example of the traditional approach, let us consider the total maximum daily load (TMDL) process. TMDL is a written plan established through analysis to ensure that a water body will attain and maintain water quality standards including consideration of existing pollutant loads and reasonably foreseeable increases in pollutant loads. The TMDL process is an essential element of the water quality–based approach to watershed management. A water body's allowable pollutant load contains waste load allocations for point sources, load allocations for nonpoint sources, a margin of safety sufficient to account for uncertainty and lack of knowledge, and an allowance for future growth. The allowable pollutant load must ensure that the water body will attain and maintain water quality standards regardless of seasonal variations or design flow conditions and in consideration of reasonably foreseeable increases in pollutant loads. We can illustrate this by showing how the allowable pollutant load is the total of these components:

$$TMDL = WLA + LA + MOS \tag{14.1}$$

where TMDL = allowable pollutant load governed by the assimilative pollutant capacity of a water body, WLA = waste load allocation for point sources, LA = load allocation for nonpoint sources, and MOS = margin of safety, which is

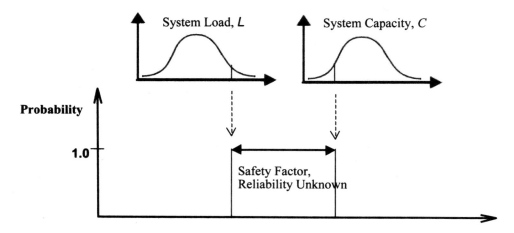

Figure 14-1 Traditional approach of safety analysis considered in engineering system.

established by leaving a portion of the assimilative capacity unallocated or by use of conservative analytical assumptions to account for the uncertainties in establishing the TMDL (e.g., derivation of numerical targets, modeling assumptions, or effectiveness of proposed management actions, etc.). Rewriting Eq. 14.1, we can express the margin of safety as

$$MOS = TMDL - (WLA + LA) \qquad (14.2a)$$

and the corresponding factor of safety (FOS) or safety factor can be expressed as

$$FOS = capacity/total \ load \qquad (14.2b)$$

In Eq. 14.2a, (WLA + LA) is the total pollutant load (*L*).

Example 14.1 For a given watershed the point-source load WLA is 65 lb/day and the non-point-source load is 1,258 lb/day. The TMDL capacity at the outlet of this watershed is determined to be 1,678 lb/day. Determine the margin of safety and factor of safety.

Solution By using Eqs. 14.2a and 14.2b the margin of safety is calculated as

$$MOS = capacity - load = 1,678 - (65 + 1,258) = 355 \ lb/day$$

The factor of safety is given as

$$FOS = capacity/total \ load = 1,678/(65 + 1,258) = 1.27$$

As an alternative to stacking up worst-case margins and uncertainties, the engineer could combine these factors statistically to yield information about the degree of confidence ("reliability") in a particular point design. In other words, the engineer could generate not just a single performance predictions but also a distribution of performance predictions with associated probabilities of occurrence, as discussed in the following section.

14.1.2 Probabilistic Approach

The probabilistic design approach is a logical extension of the traditional safety method. Probabilistic calculation techniques are more laborious and complicated than deterministic ones, but they correspond better with the aim of producing sophisticated designs and yield insight into actual risks. The "safety coefficient" used in deterministic practice actually says little about safety and nothing about reliability. The same value of safety factor may mean something completely different, depending on the mechanism.

In a probabilistic approach both the system capacity and loading can take on a wide range of values by explicitly incorporating uncertainty in system parameters. The parameter uncertainty can be quantified through statistical analysis (as described in Chapters 3 and 4) of existing data or judgmentally assigned. Even if judgmentally assigned, the probabilistic results will be more meaningful than a deterministic analysis because the engineer provides a measure of the uncertainty of his or her judgment in each parameter. This can be interpreted that the load and the capacity have fixed but unknown values. A sufficiently low probability of failure is obtained if the safety margin is at least not negative for a physically possible but unlikely combination of the capacity and the load. By using this approach, the probability that a design will achieve its required performance (i.e., the reliability) can be calculated, providing an assessment of risk or confidence in the design and quantifying the amount of overdesign or underdesign. A robust design means factoring reliability into the development of the design itself: designing for a target reliability and thereby avoiding either costly overdesign or dangerous underdesign in the first place. Such an approach eliminates a deterministic stackup of tolerances, worst-case scenarios, safety factors, and margins that have been the traditional approaches for treating uncertainties. In other words, in the probabilistic reliability analysis, an engineer could generate not just single performance predictions but also a distribution of performance predictions with associated probabilities of occurrence, as shown graphically in Fig. 14-2.

Probabilistic reliability analysis is a technique for identifying, characterizing, quantifying, and evaluating the probability of a pre-identified hazard. It is widely used by private and government agencies to support regulatory and resource allocation decisions. In most hydrologic, hydraulic, and environmental engineering projects, empirically developed or theoretically derived mathematical models are used to evaluate a system's performance. These models involve several uncertain parameters that are difficult to accurately quantify. An accurate reliability assessment of such models would help the designer build more reliable systems and aid the operator in making better maintenance and scheduling decisions.

The reliability of a system can be most realistically measured in terms of probability. The failure of a system can be considered as an event in which the demand, or loading L, on the system exceeds the capacity, or resistance R, of the system so that the system fails to perform satisfactorily for its intended use. The

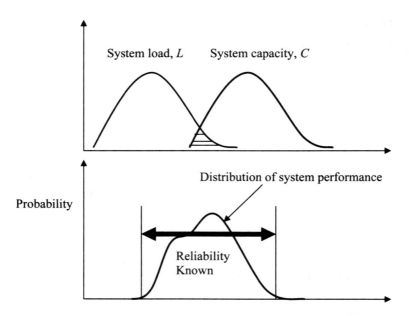

Figure 14-2 *Probabilistic approach of safety analysis of engineering systems.*

objective of reliability analysis is to ensure that the probability of the event $(C < L)$ throughout the specified useful life is acceptably small. The risk P_f, defined as the probability of failure, can be expressed as (Ang and Tang 1984, Yen et al. 1986)

$$P_f = P(L > C) \qquad (14.3)$$

where P denotes the probability function. Equation 14.3 can be rewritten in terms of the performance function Z as

$$P_f = P(Z < 0) \qquad (14.4)$$

where Z is defined alternatively as

$$Z = C - L \qquad (14.5)$$

$$Z = \frac{C}{L} - 1 \qquad (14.6)$$

$$Z = \ln\left(\frac{C}{L}\right) \qquad (14.7)$$

The reliability $\Re$ of the system can be written as

$$\Re = P(Z > 0) = 1 - P_f \qquad (14.8)$$

In general, from Eq. 14.3 risk can be expressed as

$$P_f = \int_a^b \int_d^l f_{C,L}(c,l)\, dc\, dl \tag{14.9}$$

where $f_{C,L}(c,l)$ is the joint probability density function of C and L; d and l are the lower and upper bounds of C; and a and b are the lower and upper bounds of L, respectively. The capacity C and load L are random variables that, in general, are the resultant of many uncertain variables of the system under consideration, such as weather parameters; location of the water table; temperature; flow quantities, such as runoff, peak discharge, and volume; contaminant concentration in soil, water, and air; minimum dissolved oxygen in a stream; material characteristics; and process-specific variables of an engineering system under consideration, to name only a few. Therefore, a generic performance function Z can be written as

$$Z = g(X_1, X_2, X_3, \ldots, X_n) \tag{14.10}$$

The corresponding reliability $\Re$ of the system can be written as

$$\Re = P(Z > 0) = 1 - P_f = 1 - \int \int \ldots \int f_{X_1, X_2, \ldots, X_n}(x_1, x_2, \ldots, x_n)\, dx_1\, dx_2 \ldots dx_n \tag{14.11}$$

14.1.3 Traditional Versus Probabilistic Approaches

In many civil engineering systems, component dimensions, environmental factors, material properties, and external loads are design variables. These variables may be characterized with statistical modes. The deterministic approach seeks out and defines a worst case or an extreme value to meet in the design. The probabilistic approach utilizes the statistical characterization and attempts to provide a desired reliability in the design. The deterministic approach introduces conservatism by specifying a factor of safety to cover unknowns. The probabilistic approach depends on the statistical characterization of a variable to determine its magnitude and probability. The amount of data (how well the variable is defined) influences its extreme values. Application of a factor of safety to cover unknowns has a history of success. The danger in this approach is that the factor of safety may be too large, or in some cases too small. Because it has worked in the past is no guarantee that it will suffice in the future. The whole approach of worst-case extremes can lead to compounding and inefficiency. To select a factor of safety solely on the basis of "it worked in the past" should be examined.

14.1.4 Reliability Measures

The reliability index β is a measure of the reliability of an engineering system that reflects both the mechanics of the problem and the uncertainty in the input variables. This index was developed in structural engineering to provide a measure of comparative reliability without having to calculate an exact value of the probability of failure. The reliability index is defined in terms of the expected value and standard deviation of the performance function, and it permits comparison of reliability among different structures or modes of performance without having to calculate absolute probability values. Figure 14-3 depicts a graphical interpretation of the reliability index. The calculation of the reliability index requires two pieces of information:

1. The performance function $Z = 0$ must be defined and its expected value $\mu(Z)$ and standard deviation $\sigma(Z)$ must be evaluated.

    If Z is defined by Eq. 14.5 ($Z = C - L$), the boundary separating safe ($Z > 0$) and unsafe regions ($Z < 0$) is called the *limit state* ($Z = 0$). Definitions of these regions are presented in Fig. 14-3.

2. The assumption that the distribution of Z is normally distributed holds good.

Based on these requirements, the probability of failure P_f is given as

$$P_f = P(Z<0) = \int_{-\infty}^{0} f_Z(z)dZ = \int_{-\infty}^{0} \frac{1}{\sigma(Z)\sqrt{2\pi}} exp\left[-\frac{1}{2}\left(\frac{z-\mu(Z)}{\sigma(Z)}\right)^2\right] \qquad \textbf{(14.12)}$$

As shown in Fig. 14-3, the probability of failure is the area of the PDF of Z below 0. Thus, by substituting $Z = 0$ in Eq. 14.12 and writing $\mu(Z)/\sigma(Z) = \beta$, Eq. 14.12 becomes

$$P_f = \int_{-\infty}^{0} \frac{1}{\sigma(Z)\sqrt{2\pi}} exp\left[-\frac{1}{2}(-\beta)^2\right] = \Phi(-\beta) = 1 - \Phi(\beta) \qquad \textbf{(14.13)}$$

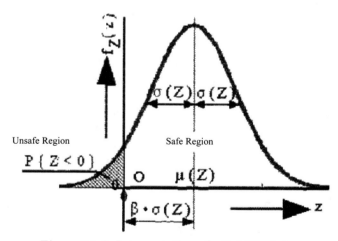

Figure 14-3 *Interpretation of reliability index.*

The reliability index is a number given by the number of standard deviations of the performance functions by which the expected condition exceeds the limit state. In other words, the reliability index is a ratio of the expected value and standard deviation of Z. The reliability index is also seen to be the reciprocal of the coefficient of variation of Z [i.e., $\beta = \mu(Z)/\sigma(Z) = 1/CV(Z)$].

Further, using Eq. 14.13, we can define the reliability index as

$$\beta = -\Phi^{-1}\left(P_f\right) \tag{14.14}$$

in which $\Phi^{-1}(P_f)$ is the inverse of the standard normal probability distribution function. Typical values of β lie between 1 and 4, corresponding to probabilities of failure ranging from on the order of 15% to 0.003%, as shown in Table 14-1. The relationship between β and P_f is unique. In particular, the probability of failure decreases with increasing values of β. The choice between using β or P_f as a measure of design risk is a matter of convenience.

The reliability index concept has gained considerable popularity. However, reliability index values are not absolute measures of probability. The assessment of reliability of engineering systems is made entirely by comparing the calculated reliability index with that found to be adequate on the basis of previous experience with the engineering systems under consideration. Engineering systems and performance modes with higher indices are considered more reliable than those with lower indices.

Although the probability of failure appears to be more physically meaningful, it can be cumbersome to use when the value becomes very small, and it carries the negative connotation of failure. The reliability index is a more convenient number to report, although it must be appreciated that a change in β cannot be readily correlated to a change in P_f because their relationship is highly nonlinear. However, it is useful to note that the probability of failure is approximately divided by half when β increases by 0.2 for β values lying between 2.8 and 3.6, as shown in Table 14-1.

To further clarify the concept of limit state, the graphical schematic in Fig. 14-4 depicts the limit state function in the L and C coordinate system.

Further, defining reduced variables as

$$U_L = \frac{L - \mu(L)}{\sigma(L)} \text{ and } U_C = \frac{C - \mu(C)}{\sigma(C)} \tag{14.15}$$

and substituting values of L and C in Eq. 14.5 gives the performance function in the limit state as

$$Z = C - L = 0, \text{ i.e., } \sigma(C)U_C - \sigma(L)U_L + \mu(C) - \mu(L) = 0 \tag{14.16}$$

Now, plotting the performance function (Eq. 14.16) in the reduced coordinate system in Fig. 14-5, we see that the straight line generated by this expression is displayed at a distance equal to the reliability index β from the origin. The shortest

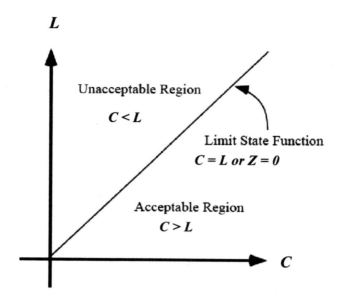

Figure 14-4 *Limit state function.*

Table 14-1 *Relationship between reliability index (β) and probability of failure (P_f).*

Reliability index β	1	1.2	1.4	1.6
Probability of failure $P_f = \Phi(-\beta)$	0.159	0.115	0.0808	0.0548

Reliability index β	1.8	2	2.2	2.4
Probability of failure $P_f = \Phi(-\beta)$	0.0359	0.0228	0.0139	8.20×10^{-3}

Reliability index β	2.6	2.8	3	3.2
Probability of failure $P_f = \Phi(-\beta)$	4.66×10^{-3}	2.56×10^{-3}	1.35×10^{-3}	6.87×10^{-4}

Reliability index β	3.4	3.6	3.8	4
Probability of failure $P_f = \Phi(-\beta)$	3.37×10^{-4}	1.59×10^{-4}	7.23×10^{-5}	3.16×10^{-5}

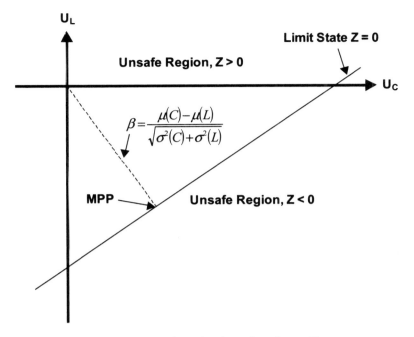

Figure 14-5 *Limit state function in reduced coordinate system.*

distance of this line from the origin is equal to the length of the perpendicular drawn from the origin, which is given as

$$\beta = \frac{\mu(C) - \mu(L)}{\sqrt{\sigma^2(C) + \sigma^2(L)}}$$ (14.17)

If the C and L are log-normally distributed, then the performance function $Z = \ln(C) - \ln(L)$. The mean of the performance function is

$$\mu_Z = \mu_{\ln(C)} - \mu_{\ln(L)}$$ (14.18)

and the variance is

$$\sigma_Z^2 = \sigma_{\ln(C)}^2 + \sigma_{\ln(L)}^2 - 2\,\mathrm{cov}[\ln(C), \ln(L)]$$ (14.19)

For statistically independent C and L,

$$\sigma_Z^2 = \sigma_{\ln(C)}^2 + \sigma_{\ln(L)}^2$$ (14.20)

Using the relations

$$\mu_{\ln(X)} = \ln \frac{\bar{X}}{\sqrt{1 + CV^2(X)}}$$

and

$$\sigma_{\ln(X)} = \sqrt{\ln\left[1 + CV^2(X)\right]}$$

for the log-normal distribution gives the reliability index $\beta = \mu_{\ln M}/\sigma_{\ln M}$ as

$$\beta = \frac{\mu_{\ln(C)} - \mu_{\ln(L)}}{\sqrt{\sigma^2_{\ln(C)} + \sigma^2_{\ln(L)}}} = \frac{\ln\left(\frac{\mu_C}{\mu_L}\sqrt{\frac{1+CV_L^2}{1+CV_C^2}}\right)}{\sqrt{\ln[(1+CV_C^2)(1+CV_L^2)]}} \tag{14.21}$$

Example 14.2 It is well known that the TMDL process is inherently uncertain and a deterministic approach to determining the factor of safety and margin of safety may not be justified. Assume that the magnitudes of uncertainty (represented by the coefficient of variation, CV) associated with TMDL, WLA, and LA are 0.26, 0.18, and 0.33, respectively. Using the values of Example 14.1 as the mean values for point and non-point-source loads and the TMDL capacity, determine the reliability index and the corresponding probability of failure. Assume TMDL, WLA, and LA are independent and normally distributed.

Solution Equation 14.2a and Eq. 14.2b define the performance function as the conventional MOS:

$$Z = \text{TMDL} - (\text{WLA} + \text{LA})$$

The mean of Z is calculated as

$$\mu_Z = \mu_{\text{TMDL}} - (\mu_{\text{WLA}} + \mu_{\text{LA}}) = 1{,}678 - (65 + 1{,}258) = 355 \text{ lb/day}$$

We know that $\sigma = \mu \times CV$ and

$$\sigma_z^2 = \sigma_{\text{TMDL}}^2 + (\sigma_{\text{WLA}}^2 + \sigma_{\text{LA}}^2)$$

$$= (1{,}678 \times 0.26)^2 + (65 \times 0.18)^2 + (1{,}258 \times 0.33)^2$$

$$= 362{,}818.35$$

Thus, the standard deviation of Z is

$$\sigma_z = 602.34$$

So the reliability index is then

$$\beta = \mu_Z / \sigma_z = 355/602 = 0.59$$

and the probability of failure becomes

$$P_f = 1 - \Phi(\beta) = 0.28$$

Alternatively, one can also use Eq. 14.17 directly as

$$\beta = \frac{\mu(TMDL) - [\mu(WLA) + \mu(LA)]}{\sqrt{\sigma^2(TMDL) + \sigma^2(WLA) + \sigma^2(LA)}} = \frac{355}{602} = 0.59$$

Therefore, there is a 28% chance that the TMDL will be violated.

Examples 14.1 and 14.2 give us some idea about the difference between deterministic and probabilistic approaches.

Example 14.3 In a multipurpose storage reservoir, storage is allocated for various purposes. The average volume of space for flood control is about 10 million m³ with a standard deviation of 1.5 million m³. Analysis of historical data shows that the mean of the volume of the largest flood in a given year is 7.5 million m³ with a standard deviation of 2.4 million m³. Find the probability of not being able to contain the largest flood.

Solution Assume that the load and resistance are normally distributed. Then using Eq. 14.13 gives the failure probability as

$$P_f = 1 - \Phi\left(\frac{\mu_C - \mu_L}{\sqrt{\sigma_C^2 + \sigma_L^2}}\right) = 1 - \Phi\left(\frac{10 - 7.5}{\sqrt{1.5^2 + 2.4^2}}\right)$$
$$= 1 - \Phi(0.8834) = 1 - 0.81 = 0.19$$

Thus, there is a 19% risk that the available storage will be inadequate to contain the incoming flood.

Example 14.4 Using the data of Example 14.3, compute the risk of not being able to contain the flood if the load and resistance follow the log-normal distribution.

Solution We first calculate the coefficient of variation for load and resistance:

$$CV_L = 2.4/7.5 = 0.32$$

$$CV_R = 1.5/10 = 0.15$$

Using Eq. 14.21 gives the probability of failure or risk as

$$P_f = 1 - \Phi\left[\frac{\ln\left(\frac{\mu_C}{\mu_L}\sqrt{\frac{1 + CV_L^2}{1 + CV_C^2}}\right)}{\sqrt{\ln[(1 + CV_C^2)(1 + CV_L^2)]}}\right] = 1 - \Phi\left[\frac{\ln\left(\frac{10}{7.5}\sqrt{\frac{1 + 0.32^2}{1 + 0.15^2}}\right)}{\sqrt{\ln[(1 + 0.15^2)(1 + 0.32^2)]}}\right]$$

$$= 1 - \Phi(0.325/0.367) = 1 - \Phi(0.886) \approx 0.19$$

It may be noted from the results that the difference in the failure probability is negligible when the variables are assumed to follow the log-normal distribution rather than the normal distribution.

Example 14.5 A company has been granted a license to discharge waste into a stream. Under the licensing arrangement, the company must comply with certain conditions, one of which concerns the concentration of pollutant in the stream at a monitoring point 100 meters downstream from the outfall. Specifically, there should be a chance of less than 1% of the pollutant concentration exceeding 10 mg/L during any one month. The stream has been monitored daily, since the company began operations, and the data suggest that the monthly maximum concentration is approximately normally distributed, with a mean of 6.3 mg/L and a standard deviation of 2.1 mg/L. Does it appear that the company is complying with the condition of its license?

Solution Let C represent the maximum monthly concentration of the concerned pollutant. We are given $\mu_C = 6.3$ mg/L and $\sigma_C = 2.1$ mg/L. Then use the following steps:

Step 1: Define the performance function as $Z = 10 - C$.

Step 2: Determine the mean and standard deviation of Z.

The mean value of Z can be determined by taking expectation of Z, $E[Z] = \mu_Z = 10 - E[C] = 10 - 6.3 = 3.7$, and the variance of Z will remain the same as that of C. Thus, $\sigma_Z = 2.1$ mg/L.

Step 3: Now, the reliability index is

$$\beta = \frac{\mu(Z)}{\sigma(Z)} = 1.762$$

The probability of failure or violating the maximum pollutant concentration is given as

$$P_f = \Phi(-1.762) = 1 - \Phi(1.762) = 1 - 0.961 = 0.039 = 3.9\%$$

Thus, it appears that the company is not complying with its license conditions.

14.2 Reliability Analysis Methods

Ideally, a probability distribution function should be obtained to do a complete assessment of reliability analysis of a given system. As shown in Eq. 14.11, this requires determination of the joint probability distribution function for all the significant sources of uncertainty affecting the output of the system under consideration. However, the determination of probability distributions for the basic variables can be quite difficult and involves several assumptions. Further, the multivariate combination and integration of the input variable distributions is a daunting task. In real-life problems, the aggregation of uncertainties in the basic variables of a model into measures of overall system reliability are done in an approximate manner. Several methods that have been used in water resources and environmental engineering have been discussed.

14.2.1 Direct Integration

The direct integration method is based on the direct integration of the joint probability density function of the basic random variables involved in design. For direct integration, Eq. 14.11 and Eq. 14.9 are integrated over the probability density function of the random variables. Equation 14.11 can be simplified for some specified cases, because in practical engineering problems it is difficult to estimate the joint probability distribution function of the performance function. Further, one can employ standard integration methods, such as analytical integration and advanced numerical integration methods.

Example 14.6 The drinking water demand of a city follows a normal distribution with a mean of 700 m^3 per day and a standard deviation of 125 m^3. (a) If the water distribution network can provide a constant rate of supply of 900 m^3 per day, determine the risk of failure on a typical day. (b) What is the risk if the standard error of the estimated water distribution network capacity is 100 m^3?

Solution

(a) Here demand follows a normal distribution. The risk is the probability of load exceeding 900. This case is depicted graphically in Fig. 14-6. Mathematically, it can be written as

$$R = Pr(\text{variable demand} > \text{constant capacity})$$
$$= Pr(D > C) = Pr(D > 900)$$
$$= 1 - Pr(D \leq 900)$$

where $Pr(\ )$ stands for probability.

$$R = \int_{900}^{\infty} \frac{1}{125\sqrt{2\pi}} \exp\left(-\frac{(D-700)^2}{2 * 125^2}\right) dD$$

Using the standard normal variate for evaluating the integral gives

$$R = \int_{1.6}^{\infty} \frac{1}{\sqrt{2\pi}} \exp\left(-\frac{u^2}{2}\right) du = 1 - \int_{-\infty}^{1.6} \frac{1}{\sqrt{2\pi}} \exp\left(-\frac{u^2}{2}\right) du$$
$$= 1 - 0.9453 = 0.0547$$

Thus, there is a 5.47% chance that the demand on a day will exceed the capacity of the system.

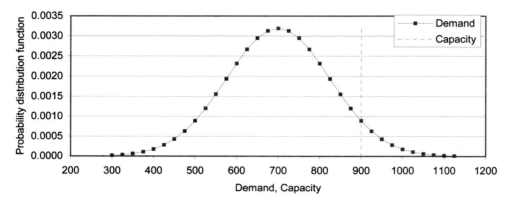

Figure 14-6 *Demand and capacity of a city water distribution network.*

(b) At this stage, the capacity is also a normally distributed variable with a mean of 900 and standard deviation of 100:

$$f(C) = \frac{1}{100\sqrt{2\pi}} \exp\left(-\frac{(C-900)^2}{2*100^2}\right) dC$$

Note that here D and C are considered as independent of each other. However, in real life there are situations when a significant correlation exists between D and C. Figure 14-7 depicts this case. In this case, the risk of failure is computed by using Eq. 14.9 as

$$R = \int_{-\infty}^{\infty}\left[\int_{C}^{\infty}\frac{1}{125\sqrt{2\pi}}\exp\left(-\frac{(D-700)^2}{2\times125^2}\right)dD\right]\frac{1}{100\sqrt{2\pi}}\exp\left(-\frac{(C-900)^2}{2\times100^2}\right)dC = 0.1001$$

Hence, the risk of failure is about 10% when the capacity is assumed to follow a normal distribution. This is about twice the value obtained when the capacity was assumed to be constant.

The integral here can also be numerically integrated to determine risk. To that end, it is convenient to write it in the following form:

$$R = \int_{-\infty}^{\infty}\left[1-\int_{\infty}^{C}\frac{1}{125\sqrt{2\pi}}\exp\left(-\frac{(D-700)^2}{2\times125^2}\right)dD\right]\frac{1}{100\sqrt{2\pi}}\exp\left(-\frac{(C-900)^2}{2\times100^2}\right)dC$$

A small discretization of the dummy variable is necessary to obtain correct results. The limits of integration are decided based on experience; a value of (mean ± 6 times the standard deviation) would be adequate in most cases.

If D and C are correlated, their joint distribution needs to be found to evaluate the integral. If both these variables follow a normal distribution, the joint probability distribution is given by

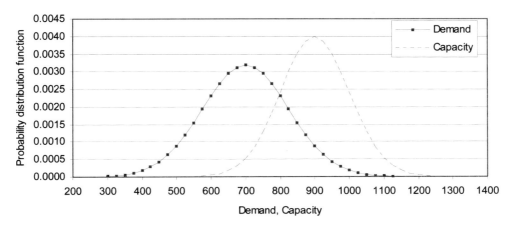

Figure 14-7 *Demand and capacity of example water distribution network.*

$$f(c,d) = \frac{1}{2\pi\sigma_c\sigma_d\sqrt{1-\rho^2}} \exp\left(-\frac{1}{2(1-\rho^2)}\left[\frac{(c-\mu_c)^2}{\sigma_c^2} + \frac{(d-\mu_d)^2}{\sigma_d^2} - 2\rho\frac{(c-\mu_c)}{\sigma_c}\frac{(d-\mu_d)}{\sigma_d}\right]\right)$$

where ρ is the correlation between C and D. This equation can be obtained numerically to obtain the value of $f(c, d)$ if the mean, standard deviation, and correlation coefficient are known.

Example 14.7 Let the statistical parameters of water demand and capacity of the supply system be the same as in Example 14.6 and assume these are correlated with $\rho = 0.2$. Assuming their joint distribution to be normal, determine the probability $P(C \leq 900, D \geq 900)$. Also determine this probability if $\rho = 0.5$.

Solution To obtain correct results by numerical integration, it is necessary to use a small increment for the variables. In the present case, a value of 1.0 may be adopted for both the variables. The results yield

$$P(C \leq 900, D \geq 900, \rho = 0.2) = 0.0186$$

$$P(C \leq 900, D \geq 900, \rho = 0.5) = 0.0069$$

14.2.2 First-Order Approximation Method

The first-order approximation (FOA) method can be used to estimate the amount of uncertainty, or scatter, of a dependent variable owing to uncertainty about the independent variables included in a functional relationship. Benjamin and Cornell (1970) have described the FOA technique in detail.

To present the general methodology, consider an output random variable Y, which is a function of n random variables. Mathematically, Y can be expressed as

$$Y = g(\underline{X}) \tag{14.22}$$

where $\underline{X} = (X_1, X_2, ..., X_n)$, a vector containing n random variables. In FOA, a Taylor series expansion of the model output is truncated after the first-order term:

$$Y = g\left(\underline{X}_e\right) + \sum_{i=1}^{n}\left(X_i - X_{ie}\right)\left(\frac{\partial g}{\partial X_i}\right)_{X_e} \tag{14.23}$$

where $\underline{X}_e = (X_{1e}, X_{2e}, ..., X_{ne})$, a vector representing the expansion points. In FOA applications to water resources and environmental engineering, the expansion point is commonly the mean value of the basic variables. Thus, the expected value and variance of Y are

$$E[Y] \approx g\left(\overline{\underline{X}}\right) \tag{14.24}$$

$$\mathrm{var}(Y) = \sigma_Y^2 \approx \sum_{i=1}^{n}\sum_{j=1}^{n}\left(\frac{\partial g}{\partial X_i}\right)_{\overline{X}_i}\left(\frac{\partial g}{\partial X_j}\right)_{\overline{X}_j} E\left[\left(X_i - \overline{X}_i\right)\left(X_i - \overline{X}_j\right)\right] \tag{14.25}$$

where σ_Y is the standard deviation of Y and $\overline{X} = \left(\overline{X}_1, \overline{X}_2, ... \overline{X}_n\right)$ is a vector of mean values of the input basic variables. If the basic variables are statistically independent, the expression for var(Y) becomes

$$\mathrm{var}(Y) = \sigma_Y^2 \approx \sum_{i=1}^{n}\left[\left(\frac{\partial g}{\partial X_i}\right)_{\overline{X}_i} \sigma_{X_i}\right]^2 \tag{14.26}$$

To avoid the inconvenience of differentiating the performance function, it is possible to derive simplified expressions for some commonly used functional forms, such as multiplicative forms, additive forms, and their combined forms.

A multiplicative-type model is frequently encountered in hydrologic studies (e.g., daily streamflow; peak runoff; annual floods; annual, monthly, and daily rainfall; soil loss; and sediment transport). In hydraulics many equations are of multiplicative type. Examples are flow over control structures, such as weirs, spillways, overfalls, and sluices; channel control equations, such as Manning's equation; and pipe flow resistance equations such as Hazen–Williams and Darcy–Weisbach equations. In environmental engineering, many of the equations for predicting water quality and pollution used in risk assessment are of multiplicative type. In the multiplicative form the output random variable Y is expressed as the multiplication of n power functions:

$$Y = C_0 X_1^{r_1} X_2^{r_2} ... X_n^{r_n} = C_0\prod_{i=1}^{n} X_i^{r_i} \tag{14.27}$$

where C_0 and r_i are constants and the X_is are independent stochastic input random variables. For the multiplicative form of Eq. 14.27, the approximate mean of the model output, $\hat{\mu}_Y$, can be written as

$$\hat{\mu}_Y = C_0\prod_{i=1}^{n} \mu_{X_i}^{r} \tag{14.28}$$

where μ_{Xi} = mean of X_i. By using Eq. 14.26, the approximate variance of the multiplicative $\hat{\sigma}_Y^2$ can be approximated as

$$\hat{\sigma}_Y^2 = C_0^2 \prod_{i=1}^{n} \mu_{X_i}^{2r_i} \sum_{i=1}^{n} r_i^2 CV_{X_i}^2 \tag{14.29}$$

where r_i is the exponent of ith power function and

$$CV_{X_i} = \sigma_{X_i} / \mu_{X_i} = \text{coefficient of variation of } X_i$$

The approximate coefficient of variation of Y, $\hat{C}V_Y$, can be evaluated as

$$\hat{C}V_Y = \left(\frac{C_0^2 \prod_{i=1}^{n} \mu_{X_i}^{2r_i} \sum_{i=1}^{n} r_i^2 CV_{X_i}^2}{C_0^2 \prod_{i=1}^{n} \mu_{X_i}^{2r_i}} \right)^{0.5} = \left(\sum_{i=1}^{n} r_i^2 CV_{X_i}^2 \right)^{0.5} \tag{14.30}$$

The additive form is obtained when two or more power functions are added. This form is often encountered in reliability analysis of engineering systems. The general additive form is written as

$$Y = C_1 X_1^{r_1} + C_2 X_2^{r_2} + \ldots + C_n X_n^{r_n} = \sum_{i=1}^{n} C_i X_i^{r_i} \tag{14.31}$$

The mean of Y is then estimated as

$$\hat{\mu}_Y = \sum_{i=1}^{n} C_i \mu_{X_i}^{r_i} \tag{14.32}$$

Similarly, the variance of the additive model can be approximated by

$$\hat{\sigma}_Y^2 = \sum_{i=1}^{n} C_i^2 r_i^2 \mu_{X_i}^{2r_i} CV_{X_i}^2 \tag{14.33}$$

So $\hat{C}V_Y$ can be evaluated by

$$\hat{C}V_Y = \frac{\sqrt{\sum_{i=1}^{n} C_i^2 r_i^2 \mu_{X_i}^{2r_i} CV_{X_i}^2}}{\sum_{i=1}^{n} C_i \mu_{X_i}^{r_i}} \tag{14.34}$$

The other functional form is a combination of multiplicative and additive forms. This form is obtained when two or more multiplicative forms having common power function(s) are added. The general form can be represented as

$$Y = C_0 X_1^{r_1} X_2^{r_2} \dots X_m^{r_m} \left(C_1 X_{m+1}^{r_{m+1}} + C_2 X_{m+2}^{r_{m+2}} + \dots + C_n X_{m+n}^{r_{m+n}} \right)$$

$$= C_0 \prod_{i=1}^{m} X_i^{r_i} \sum_{j=1}^{n} C_{j+m} X_{j+m}^{r_{j+m}} \qquad (14.35)$$

For evaluating the mean and variance of combined forms, such as Eq. 14.35, the mean and variance of the additive part must be determined first by using Eq. 14.32 and Eq. 14.33. Next, Eqs. 14.28, 14.29, and 14.30 are used to determine the mean, variance, and CV of Y by treating the combined form as a multiplicative form and assuming the additive part as a multiplicative component with known mean and variance.

Then the mean and variance of the performance function Z are determined. To estimate the reliability of the system, $\Re$, typically one assumes that Z is normally distributed. Taking $P_Z(z)$ to be a normal distribution with its parameters $E[Z]$ and σ_Z determined by FOA, one can determine the risk and reliability of a given system using the concept of reliability index β as discussed in the preceding section.

The great advantage of FOA is its simplicity: It requires knowledge of only the first two statistical moments of the basic variables and simple sensitivity calculations about selected central values. FOA is an approximate method that may suffice for many applications, but the method does have several theoretical and conceptual shortcomings. The main weakness of the FOA method is that it is assumed that a single linearization of the system performance function at the central values of the basic variables is representative of the statistical properties of system performance over the complete range of basic input variables. The accuracy of the estimates is influenced in part by the degree of nonlinearity in the functional relationship and the importance of higher-order terms, which are truncated in the Taylor series expansion. In applying FOA in reliability analyses, it is generally assumed that the performance function is normally distributed, which is seldom true. Therefore, any attempt to characterize the tails of the actual distribution based on an assumption of normality is likely to result in an inexact answer.

Example 14.8 Solve Example 14.2 using FOA by defining the performance function in two ways: (1) as the conventional MOS and (2) as the FOS.

Solution

Case 1: The solution for the conventional MOS method is given in Example 14.2.

Case 2: The conventional factor of safety is defined as FOS = capacity/load; thus the performance function is

$$Z = \frac{\text{TMDL}}{\text{WLA+LA}} - 1$$

Using Eq. 14.24 gives the mean of the performance function as

$$\mu(Z) = \frac{\mu(\text{TMDL})}{\mu(\text{WLA}) + \mu(\text{LA})} - 1 = \frac{1678}{(65 + 1258)} - 1 = 0.27$$

Using Eq. 14.26 gives the variance of the performance function:

$$\sigma_Z^2 = \left[\frac{1}{\mu(\text{WLA}) + \mu(\text{LA})}\right]^2 \left\{\sigma_{\text{TMDL}}^2 + \left[\frac{\mu(\text{TMDL})}{\mu(\text{WLA}) + \mu(\text{LA})}\right]^2 (\sigma_{\text{WLA}}^2 + \sigma_{\text{LA}}^2)\right\}$$

$$= \left[\frac{1}{65 + 1258}\right]^2 \left\{436.28^2 + \left[\frac{1678}{65 + 1258}\right]^2 (11.7^2 + 415.14^2)\right\} = 0.267$$

So,

$$\sigma_Z = 0.517$$

The reliability index is

$$\beta = 0.27 / 0.517 = 0.52$$

and the probability of failure is

$$P_f = 1 - \Phi(0.52) = 0.30$$

that is, 30%.

The values of reliability index and failure probability obtained in case 1 and case 2 show that risk estimates are different for the two mechanically equivalent formulations of the performance function. This indicates that the probability of failure or the reliability of a system depends upon the type of formulation of the objective function. This is a serious problem and is generally known as the lack of invariance.

Example 14.9 In a stream the concentration C of a pollutant is generally given in the form of a power function as $C = aQ^b$, where Q is the streamflow and a and b are some constants. Let us consider the following performance function:

$$Z = C_{\max} - aQ^b$$

where $C_{\max}$ is the maximum allowable pollutant concentration. If $C_{\max} = 10$ mg/L, $a = 2.10 \times 10^{-8}$, $b = 3$, $\mu_Q = 800$ cfs, and $CV_Q = 0.33$, determine the probability of violating the allowable stream standard.

Solution Substituting values of a, b, and $C_{\max}$ we can write the performance function as

$$Z = C_{\max} - C = 10 - (2.10 \times 10^{-8})Q^3$$

First, let us calculate the mean and standard deviation of C using FOA. Using Eq. 14.24 gives the mean of C:

$$\mu_C = (2.10 \times 10^{-8}) \times (800)^3 = 10.75$$

Using Eq. 14.26 gives the standard deviation of C:

$$\sigma_C = (2.10 \times 10^{-8}) \times (3) \times (800)^{3-1} \times (0.33 \times 800) = 10.64$$

Now,

$$E[Z] = E[C_{max}] - E[C] = 10 - 10.75 = -0.75$$

Further, $\text{var}(Z) = \text{var}(C)$ because $\text{var}(C_{max}) = 0$; thus

$$\sigma_Z = 10.64$$

Therefore, the reliability index is

$$\beta = \mu_Z / \sigma_Z = -0.75/10.64 = -0.071$$

The probability of failure is then

$$P_f = 1 - \Phi(-0.071) = 0.53$$

that is, 53%.

Example 14.10 In drinking water distribution systems, maintaining a chlorine residual provides protection against contamination from leaks, regrowth of microbial contamination, cross connections, and other breakdowns. Most network modeling packages assume that chlorine decay follows first-order kinetics. The chlorine concentration C (mg/L) at time t is given by the following equation:

$$C = C_0 \exp(-kt)$$

where C_0 is the initial chlorine concentration (mg/L) and k is the overall decay constant (L/hour). Assuming mean and coefficient of variation of k to be 0.14 L/hour and 0.33, respectively, determine the reliability that a location having a travel time of 20 hours will be having at least 0.2 mg/L of residual chlorine. Assume $C_0 = 4$ mg/L.

Solution The water is safe when the concentration of free residual chlorine is greater than or equal to 0.2 mg/L. The performance function Z can be defined as

$$Z = C - C_{min} = C_0 e^{-kt} - C_{min}$$

with

$$\sigma_k = 0.14 \times 0.33 = 0.046$$

$$\mu_Z = C_0 e^{-\mu_k t} - C_{min} = 4 \times \exp(-0.14 \times 20) - 0.2 = 0.04$$

and

$$\text{var}[Z] = \left(\frac{dC}{dk}\right)^2_{\mu_k} \sigma_k^2 = \left(-C_0 t e^{-kt}\right)^2_{\mu_k} \sigma_k^2 = \left(t^2 C^2 e^{-2kt} \sigma_k^2\right)_{\mu_k}$$
$$= (20 \times 4 \times \exp(-0.14 \times 20) \times 0.046)^2 = 0.051$$

Thus,

$$\sigma_Z = 0.225$$

Therefore, the reliability index is

$$\beta = \mu_Z / \sigma_Z = 0.04/0.225 = 0.19$$

and the probability of failure is

$$P_f = 1 - \Phi(0.19) = 0.42$$

that is, 42%. Therefore the reliability is only 58%.

Example 14.11 Alluvial streambeds downstream of an outlet facility may be seriously scoured under jet action. The degree of scour depends on the characteristics of the jet leaving the outlet facility, the depth of tailwater, and properties of the bed material, which are uncertain in nature. Using the following relationship for the terminal scour depth, find the reliability of the outlet structure against the scouring induced by vertical jets downstream of the outlet facility:

$$d_s = \lambda \frac{110 b^{0.862} u^{0.891} Dg^{1.128}}{y^{0.431} W_f^{2.01}}$$

Assume that the foundation depth $d_f = 12$ m. The other parameters are given in Table E14-11. The situation is depicted in Fig. 14-8.

Solution The objective function is defined as

$$Z = d_f - d_s$$

Table E14-11 *Scour parameters.*

Parameter	Mean	CV	PDF
b (m)	0.30	0.01	Log-normal
u (m/s)	7.0	0.20	Symmetrical triangular
Dg (m)	0.005	0.05	Uniform
y (m)	4.0	0.20	Symmetrical triangular
W_f (m/s)	0.30	0.20	Symmetrical triangular
λ (m)	1.0	0.19	Normal

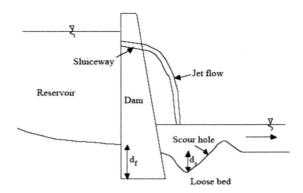

Figure 14-8 *Definition sketch for scouring parameters.*

where d_f is foundation depth and d_s is the scouring depth. The dam is safe when the depth of scour is less than the foundation depth. By using Eq. 14.28, the mean of d_s is determined to be 3.46. Thus, $\mu_Z = 12 - 3.46 = 8.54$. Now, using Eq. 14.30, we can determine the coefficient of variation of d_s as

$$\hat{C}V_{d_s} = \left(\sum_{i=1}^{n} r_i^2 CV_{X_i}^2 \right)^{0.5}$$

$$= \left[\begin{array}{l} (0.862 \times 0.01)^2 + (0.891 \times 0.2)^2 + (1.128 \times 0.05)^2 \\ + (-0.431 \times 0.2)^2 + (-2.01 \times 0.2)^2 + (1 \times 0.19)^2 \end{array} \right]^{0.5}$$

$$= 0.49$$

Thus,

$$\sigma_{ds} = 3.46 \times 0.49 = 1.7$$

Now

$$\text{var}(Z) = \text{var}(d_f) + \text{var}(d_s)$$

but $\text{var}(d_f) = 0$, so we have

$$\text{var}(Z) = \text{var}(d_s)$$

and thus

$$\sigma_Z = 1.7$$

So, the reliability index is

$$\beta = \mu_Z / \sigma_Z = 8.54 / 1.7 = 5.03$$

and the probability of failure is

$$P_f = 1 - \Phi(5.03) = 2.48 \times 10^{-7}$$

14.2.3 Corrected FOA Method

In this section, the properties of statistical expectation of a random variable are used to derive moments of various model forms. When Y is defined by a multiplicative form such as Eq. 15.27 with strictly independent input parameters, X_is, the mean of Y, μ_Y, can be written as

$$\mu_Y = E[Y] = C_0 \prod_{i=1}^{n} E\left[X_i^{r_i}\right] = C_0 \prod_{i=1}^{n} \mu_{Y_i} \tag{14.36}$$

where $E[\ ]$ is an expectation operator and μ_{Y_i} is the mean of the ith power function,

$$\mu_{Y_i} = E\left[X_i^{r_i}\right] \tag{14.37}$$

The coefficient of variation of Y, CV_Y, can be written as

$$CV_Y = \left[\prod_{i=1}^{n}\left(CV_{Y_i}^2 + 1\right) - 1\right]^{0.5} \tag{14.38}$$

The variance of Y, σ_Y^2, can be written as

$$\sigma_Y^2 = \left(\mu_Y CV_Y\right)^2 \tag{14.39}$$

Equation 14.38 shows that the output uncertainty of a multiplicative model is governed by the most uncertain component function.

Using the additive form (Eq. 14.31) gives the mean of Z:

$$\mu_Y = \sum_{i=1}^{n} C_i \mu_{Y_i} \tag{14.40}$$

Similarly, the variance of Y, σ_Y^2, can be written as

$$\sigma_Y^2 = \sum_{i=1}^{n} C_i^2 \sigma_{Y_i}^2 \tag{14.41}$$

Equation 14.41 shows that the magnitude of C_i is as important as uncertainty in the component function (σ_{Y_i}).

For evaluating the mean and variance of combined forms of Z, such as Eq. 14.35, the mean and variance of the additive part must be determined first. Then μ_Z, CV_Z, and σ_Z^2 are determined by treating the combined form as a multiplicative form and considering the additive part as a multiplicative component with known mean and variance.

Knowledge of the relative error corresponding to FOA estimates ($\hat{\mu}_Y$ and $\hat{\sigma}_Y^2$) can be used to correct them to obtain their exact values. Consider a power function

$$Y = f(X) = cX^r \tag{14.42}$$

where c and r are constants. The FOA estimate for the mean $\hat{\mu}_Y$ is

$$\hat{\mu}_Y = c\mu_X^r \qquad (14.43)$$

The FOA estimate for the variance of Y, $\hat{\sigma}_Y^2$, is

$$\hat{\sigma}_Y^2 = c^2 r^2 \mu_X^{2(r-1)} \sigma_X^2 = c^2 r^2 \mu_X^{2r} CV_X^2 . \qquad (14.44)$$

These estimates for μ_Y and σ_Y contain errors. The exact value of any moment can be computed as

$$\text{Exact value} = \frac{\text{FOA estimate}}{1 - \text{E}(.)} \qquad (14.45)$$

where $\text{E}(.)$ is the relative error in a moment estimated using FOA. Analytical relationships for $\text{E}(.)$ in FOA estimates for means and variances of component functions were developed (Tyagi 2000) for generic power and exponential functions for five common distributions. These analytical expressions can be used as a guide for judging the suitability of FOA by determining the relative errors in the most sensitive parameters. Further, when the relative error is more than the acceptable error, these analytical relationships enable one to correct the FOA estimates for means and variances of model components to their true values. Using these corrected values of means and variances of model components, one can determine the exact values of mean and variance of an overall model output. Table 14-2 and Table 14-3 present the developed expressions for $\text{E}(\hat{\mu}_Y)$ and $\text{E}(\hat{\sigma}_Y^2)$ for a generic power function $(Y = cX^r)$. The correction factors for the normal distribution have been presented graphically in Fig. 14-9 and Fig. 14-10.

In Table 14-2 and Table 14-3, to avoid the singularity at $r = -1$, r should be taken as -0.9999, and to avoid the singularity at $r = -2$, r should be taken as -1.9999.

Similarly, Table 14-4 and Table 14-5 present the developed expressions for $\text{E}(\hat{\mu}_Y)$ and $\text{E}(\hat{\sigma}_Y^2)$ for a generic exponential function $(Y = be^{cX})$.

Example 14.12 Solve Example 14.8 using the corrected FOA method.

Solution

Case 1: As the performance function $Z = TMDL - (WLA + LA)$ is linear, there is no error involved in the FOA application. So, both the reliability index β and probability of failure will remain the same as calculated in Example 14.8.

Case 2: In this case the performance function is nonlinear:

$$Z = \frac{TMDL}{WLA + LA} - 1 = \frac{TMDL}{TL} - 1 = TMDL \times TL^{-1} - 1$$

where $TL = WLA + LA$. First, the mean and variance of TL are determined:

$$E[TL] = E[WLA] + E[LA] = 65 + 1258 = 1323$$

Table 14-2 *Generalized relative error in the FOA predicted mean of a power function*

Distribution	Relative error in FOA predicted mean, $E(\hat{\mu}_Y)$
Uniform	$$1-\frac{2\sqrt{3}(r+1)CV_X}{\left[\left(1+CV_X\sqrt{3}\right)^{(r+1)}-\left(1-CV_X\sqrt{3}\right)^{(r+1)}\right]}$$
Symmetrical triangular	$$1-\frac{6(r+1)(r+2)CV_X^2}{\left[\left(1+CV_X\sqrt{6}\right)^{(r+2)}+\left(1-CV_X\sqrt{6}\right)^{(r+2)}-2\right]}$$
Normal	$$1-\left(1+CV_X^2\right)^{\frac{1}{2}r(1-r)}$$
Gamma	$$1-\frac{CV_X^{-2r}\Gamma\left(CV_X^{-2}\right)}{\Gamma\left[CV_X^{-2}\left(1+rCV_X^2\right)\right]}$$
Exponential	$$1-\frac{1}{\Gamma(r+1)}$$

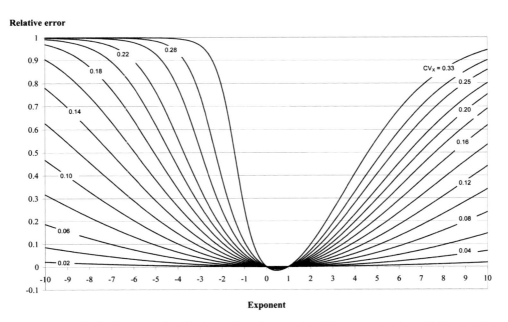

Figure 14-9 *Relative error in FOA predicted mean for* CV_x *ranging from 0.01 to 0.33, where X is normally distributed.*

Table 14-3 *Generalized relative error in the FOA predicted variance of a power function.*

Distribution	Relative error in FOA predicted variance, $E\left(\hat{\sigma}_Y^2\right)$
Uniform	$1-\dfrac{12\ (2r+1)\ r^2(r+1)^2 CV_X^4}{\left\{2\sqrt{3}CV_X(r+1)^2\left[\left(1+CV_X\sqrt{3}\right)^{2r+1}-\left(1-CV_X\sqrt{3}\right)^{2r+1}\right]-(2r+1)\left[\left(1+CV_X\sqrt{3}\right)^{r+1}-\left(1-CV_X\sqrt{3}\right)^{r+1}\right]^2\right\}}$
Symmetrical triangular	$1-\dfrac{36\ (2r+1)\ r^2(r+1)^2(r+2)^2 CV_X^6}{\left\{3(r+1)(r+2)^2 CV_X^2\left[\left(1+CV_X\sqrt{6}\right)^{2r+2}+\left(1-CV_X\sqrt{6}\right)^{2r+2}-2\right]-(2r+1)\left[\left(1+CV_X\sqrt{6}\right)^{r+2}+\left(1-CV_X\sqrt{6}\right)^{r+2}-2\right]^2\right\}}$
Normal	$1-\dfrac{r^2 CV_X^2\left(CV_X^2+1\right)^r}{\left(CV_X^2+1\right)^{r^2}\left[\left(CV_X^2+1\right)^{r^2}-1\right]}$
Gamma	$1-\dfrac{r^2 CV_X^{2(1-2r)}\left[\Gamma\left(CV_X^{-2}\right)\right]^2}{\Gamma\left[CV_X^{-2}\left(1+2rCV_X^2\right)\right]\Gamma\left(CV_X^{-2}\right)-\left\{\Gamma\left[CV_X^{-2}\left(1+rCV_X^2\right)\right]\right\}^2}$
Exponential	$1-\dfrac{r^2}{\left[\Gamma(2r+1)-\Gamma^2(r+1)\right]}$

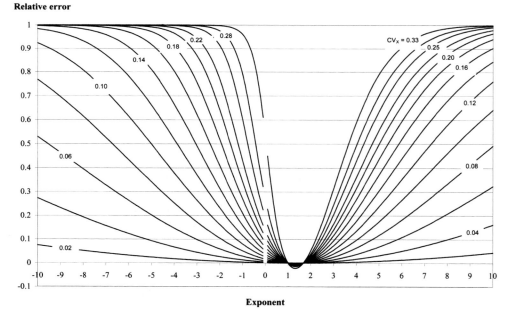

Figure 14-10 *Relative error in FOA predicted variance for* CV_x *ranging from 0.01 to 0.33, where X is normally distributed.*

Table 14-4 *Generalized relative error in FOA predicted mean of an exponential function.*

Distribution	Relative error in FOA predicted variance, $E(\hat{\mu}_Y)$
Uniform	$1 - \dfrac{2\sqrt{3}c\mu_x CV_x e^{\sqrt{3}c\mu_x CV_x}}{\left(e^{2\sqrt{3}c\mu_x CV_x} - 1\right)}$
Symmetrical triangular	$1 - \dfrac{6c^2\mu_x^2 CV_x^2 e^{c\mu_x CV_x \sqrt{6}}}{\left(e^{c\mu_x CV_x \sqrt{6}} - 1\right)^2}$
Normal	$1 - \exp\left[-\dfrac{1}{2}c^2\mu_x^2 CV_x^2\right]$
Gamma	$1 - \left(1 - c\mu_x CV_x^2\right)^{\frac{1}{CV_x^2}} \exp(c\mu_x)$
Exponential	$1 - (1 - c\mu_x)\exp(c\mu_x)$

$$\text{var[TL]} = \text{var[WLA]} + \text{var[LA]} = (65 \times 0.18)^2 + (1258 \times 0.33)^2 = 415.30^2$$

Now the corrected FOA is used to correct both the FOA estimated mean and the variance of TL^{-1}. From Figs. 14-9 and 14-10, the corrections in the FOA mean and variance are determined corresponding to $CV = 0.31$ and exponent $r = -1$. The correction factors are 0.25 and 0.96, respectively. So the correct mean and variance of the power function TL^{-1} are 1.01×10^{-3} and 1.41×10^{-6}, respectively.

After determining the corrected mean and variance of the individual power function, Eq. 14.37 and Eq. 14.38 are used to determine the mean and variance of the performance function Z. So,

$$\mu_Z = 1.69 - 1.0 = 0.69$$

$$\sigma_Z = 2.1$$

Assuming Z to be normally distributed gives the reliability index:

$$\beta = 0.69/2.1 = 0.33$$

The probability of failure is

$$P_f = 1 - \Phi(0.33) = 0.37$$

that is, 37%.

It is worth mentioning that the distribution of Z may not be a normal distribution [as evidenced by a very high CV (=1/0.33 = 3) value] and hence the

Table 14-5 *Generalized relative error in FOA predicted variance of an exponential function.*

Distribution	Relative error in FOA predicted variance, $E\left(\hat{\sigma}_Y^2\right)$
Uniform	$1-\dfrac{12c^4\mu_x^4 CV_x^4 e^{2\sqrt{3}c\mu_x CV_x}}{\left(e^{2\sqrt{3}c\mu_x CV_x}-1\right)\left[\left(\sqrt{3}c\mu_x CV_x-1\right)e^{2\sqrt{3}c\mu_x CV_x}+\sqrt{3}c\mu_x CV_x+1\right]}$
Symmetrical triangular	$1-\dfrac{72c^6\mu_x^6 CV_x^6 e^{2\sqrt{6}c\mu_x CV_x}}{\left(e^{\sqrt{6}c\mu_x CV_x}-1\right)^2\left[\left(3c^2\mu_x^2 CV_x^2-2\right)\left(e^{2\sqrt{6}c\mu_x CV_x}+1\right)+2e^{\sqrt{6}c\mu_x CV_x}\left(3c^2\mu_x^2 CV_x^2+2\right)\right]}$
Normal	$1-\dfrac{c^2\sigma_x^2}{\exp\left(c^2\sigma_x^2\right)\left[\exp\left(c^2\sigma_x^2\right)-1\right]}$
Gamma	$1-\dfrac{c^2\mu_x^2 CV_x^2\exp(2c\mu_x)}{\left(1-2c\mu_x CV_x^2\right)^{-\frac{1}{CV_x^2}}-\left(1-c\mu_x CV_x^2\right)^{-\frac{2}{CV_x^2}}}$
Exponential	$1-\dfrac{c^2\mu_x^2\exp(2c\mu_x)}{\left(1-2c\mu_x\right)^{-1}-\left(1-c\mu_x\right)^{-1}}$

P_f estimate may contain error; the correct P_f should have been 28%. The corrected FOA method is particularly useful when only mean and variance of a model output is needed (e.g., in uncertainty analysis). It is advised that one should determine higher moments and find a suitable distribution for Z. Using this distribution along with its parameters one can calculate the correct value of P_f. This will be further discussed in another approach, called the generic expectation function approach.

Example 14.13 Solve Example 14.9 using the corrected FOA method assuming Q is characterized by (1) a normal distribution, (2) a log-normal distribution, and (3) a gamma distribution.

Solution Substituting values of *a*, *b*, and C_{max} one can write the performance function as

$$Z = C_{max} - C = 8 - (2.10 \times 10^{-8})Q^3$$

First, let us calculate the mean and standard deviation of C using the corrected FOA method for the simple power function Q^3. The calculation is presented in the tabular form in Table E14-13. Since the table only considers the power function Q^3, the mean of the power function $C = b \times Q^3$ is determined as

$$E[C] = 2.10 \times 10^{-8} \times E[Q^3]$$

Table E14-14a Calculations using corrected FOA method.

Characteristics of parameter k				**Characteristics of the function $C = C_0 \exp(-kt)$**					
Distribution	Mean μ_k	CV_k	SD σ_k	FOA Mean	Relative error in FOA mean*	Correct mean	FOA variance	Relative error in variance*	Correct variance
(1)	(2)	(3)	(4)	(5)	(6)	(7)	(8)	(9)	(10)
Gamma	0.14	0.33	0.046	0.06	0.30	0.098	0.051	0.38	0.04
Normal	0.14	0.33	0.046	0.06	0.3475	0.091	0.051	0.73	0.01

*Entries in columns (6) and (9) are calculated by using formulas presented in Tables 14-3 and 14-4.

Table E14-14b Calculations of performance function Z.

Z	FOA	Corrected FOA	
		$K = \text{Gamma}(0.14, 0.046^2)$	$k = \text{Normal}(0.14, 0.046^2)$
Mean	0.043	0.15	0.17
Variance	0.051	0.08	0.19
SD	0.22	0.29	0.43
CV	5.20	1.94	2.51
b	0.19	0.52	0.40
Failure probability	0.42	0.30	0.34

Notes: $E[Z] = E[C - C_{min}] = \mu_k - C_{min}$; $\text{var}(Z) = \text{var}(C)$; reliability index $\beta = \mu_Z / \sigma_Z$; the probability of failure $P_f = 1 - \Phi(\beta)$.

Solution The objective function is defined as $Z = d_f - d_s$ where d_f (=12 m) is the foundation depth and d_s is the scouring depth defined as

$$d_s = \lambda \frac{110b^{0.862} u^{0.891} Dg^{1.128}}{y^{0.431} W_f^{2.01}}$$

First, we determine the correct mean and variance of d_s without considering the constant $C = 110$. Calculations are performed in Table E14-15.
Now using Eq. 14.39 we have

$$\mu_{d_s} = E[d_s] = C_0 \prod_{i=1}^{n} E\left[X_i^{r_i}\right] = C_0 \prod_{i=1}^{n} \mu_{Y_i}$$
$$= 110 \times 0.35 \times 5.65 \times 0.0025 \times 0.56 \times 12.87 \times 1$$
$$= 4.01$$

Table E14-15 *Calculation process using corrected FOA method.*

Function	Exponent	Distribution	Mean	CV	SD
$b^{0.862}$	0.862	Normal	0.3	0.01	0.003
$u^{0.891}$	0.891	Symmetrical triangular	7	0.20	1.40
$Dg^{1.128}$	1.128	Uniform	0.005	0.05	0.0003
$y^{-0.431}$	−0.431	Symmetrical triangular	4	0.20	0.80
$W_f^{-2.01}$	−2.01	Symmetrical triangular	0.3	0.20	0.06
1	1	Normal	1	0.19	0.19

FOA mean	Rel. errror	Corr. mean	FOA var.	Rel. error	Corr. var	CV^2
0.35	0.00	0.35	0.0051	7.50×10^{-5}	0.0051	7.43×10^{-5}
5.66	0.00	5.65	1.02	0.00	1.02	0.03
0.0025	0.0002	0.0025	2.05×10^{-8}	−0.0002	2.05×10^{-8}	0.0032
0.55	0.01	0.56	0.00	0.14	2.62×10^{-3}	0.01
11.246	0.13	12.874	20.44	0.44	36.57	0.22

The coefficient of variation of d_s can be determined using Eq. 14.41 as

$$CV_{d_s} = \left[\prod_{i=1}^{n} \left(CV_{Y_i}^2 + 1 \right) - 1 \right]^{0.5}$$

$$= \left[\begin{array}{c} \left(1 + 7.43 \times 10^{-05}\right)\left(1 + 0.03\right)\left(1 + 0.0032\right) \\ \times \left(1 + 0.01\right)\left(1 + 0.22\right)\left(1 + 0.04\right) - 1 \end{array} \right]^{0.5}$$

$$= 0.57$$

Thus the standard deviation of d_s is

$$\sigma_{d_s} = 4.01 \times 0.57 = 2.27$$

Now, one can determine the mean and standard deviation of Z:

$$E[Z] = E[d_f - d_s] = 12 - E[d_s]$$

$$= 12 - 4.01 = 7.99$$

Because $\text{var}(Z) = \text{var}(d_s)$,

$$\sigma_Z = \sigma_{d_s} = 2.27$$

The reliability index is

$$\beta = \mu_Z/\sigma_Z = 7.99/2.27 = 3.52$$

and the probability of failure is

$$P_f = 1 - \Phi(3.52) = 2.17 \times 10^{-4}$$

14.2.4 Response Surface Method

The response surface (RS) method is very similar to the FOA method. Whereas the FOA method deals directly with the performance function, the RS method involves approximating the original, complicated system performance function with a simpler, more computationally tractable system model. This approximation typically takes the form of a first- or second-order polynomial:

$$Y = g(\underline{X}) \approx G(\underline{X}) \approx a_0 + a_1 X_1 + ... + a_n X_n + a_{n+1} X_1^2 + ... + a_{2n} X_n^2 + a_{2n+1} X_1 X_2 + ... \quad \textbf{(14.46)}$$

where $G(\underline{X})$ is the approximate function representing the original function $g(\underline{X})$. Determination of the constants is accomplished through a linear regression about some nominal value, typically the mean. Given the new performance function, the analysis proceeds in exactly the same manner as the FOA method. This method has not been used much in the area of water resources and environmental engineering.

14.2.5 Monte Carlo Simulation

In Monte Carlo simulation (MCS), probability distributions are assumed for the uncertain input variables for the system being studied. Random values of each of the uncertain variables are generated according to their respective probability distributions and the model describing the system is executed. By repeating the random generation of variable values and model execution steps many times, statistics and an empirical probability distribution of the model output can be determined. The accuracy of the statistics and probability distribution obtained from MCS is a function of the number of simulations performed and the adequacy of the assumed parameter distributions. The MCS method has been described in detail in Chapter 11.

Monte Carlo simulation is an art; it requires judgment on the part of the modeler to create theoretical input sample distributions that are representative of the populations and to estimate the number of trials needed to generate the input and output density functions. There is no strictly defined answer to either of these questions.

A key problem in applying the MCS method is estimating the necessary sample size. One empirical test to determine the adequacy of the sample size consists of iterating the sample program with increasingly greater sample sizes and estimating the convergence rate of the sample mean value toward the population mean. The error in the estimation of the population mean is inversely proportional to the square root of the number of trials. To improve the estimate by a factor of 2, the sample size must increase by a factor of 4. If the sample size is n, the standard deviation of the mean is $1/\sqrt{n}$ times the standard deviation of the population. This indicates that the sample size must be large. As the sample size increases, the precision of the empirical percentile estimates of a model output improves. However, the rate of convergence to the true distribution decreases as the size of the sample increases.

The requirement of generating very large samples poses a serious problem. The MCS method often entails sample sizes that are in the range of 5,000 to 20,000 members. Generally, the number of required samples increases with the variances and the coefficient of skewness of the input distributions.

Another simulation technique similar to MCS is Latin hypercube sampling (LHS) in which a stratified sampling approach is used. In LHS the probability distribution of each basic variable is subdivided into nonoverlapping intervals (say, m) each with equal probability $(1/m)$. Random values of the basic variables are simulated such that each range is sampled only once. The order of the selection of the ranges is randomized and the model is executed m times with the random combination of basic variables from each range for each basic variable. The output statistics and distributions may then be approximated from the sample of m output values. It has been shown that the stratified sampling procedure of LHS converges more quickly than an equidistribution sampling employed in MCS. Except for reducing the computation effort to some extent, LHS has the same problems that are associated with MCS.

14.2.6 Second-Order Approximation Method

In the second-order approximation (SOA) method, a Taylor series expansion of a model is truncated after the second-order term. Consider a model represented by Eq. 14.22. The second-order Taylor series expansion of Y is given as

$$Y = g\left(\underline{X}_e\right) + \sum_{i=1}^{n}\left(X_i - X_{ie}\right)\left(\frac{\partial g}{\partial X_i}\right)_{X_e} + \frac{1}{2}\sum_{i=1}^{n}\left(X_i - X_{ie}\right)^2\left(\frac{\partial^2 g}{\partial X_i^2}\right)_{X_e} \tag{14.47}$$

In SOA, the expansion point is commonly the mean value of the basic variables. By considering all input variables to be statistically independent and taking the expectation of Eq. 14.47, the expected value Y is given as

$$E[Y] \approx g\left(\underline{\overline{X}}\right) + \frac{1}{2}\sum_{i=1}^{n}\left(\frac{\partial^2 g}{\partial X_i^2}\right)\text{var}\left(X_i\right) \tag{14.48}$$

The variance of Y is given as

$$\text{var}(Y) = \sigma_Y^2 \approx \sum_{i=1}^{n} \left(\frac{\partial g}{\partial X_i} \right)_{\overline{X}_i}^2 \text{var}(X_i) - \frac{1}{4} \sum_{i=1}^{n} \left(\frac{\partial^2 g}{\partial X_i^2} \right)_{\overline{X}_i}^2 \text{var}^2(X_i) + \sum_{i=1}^{n} \left(\frac{\partial g}{\partial X_i} \right)_{\overline{X}_i}^2 \left(\frac{\partial^2 g}{\partial X_i^2} \right)_{\overline{X}_i}^2 E\left[(X_i - \overline{X}_i)^3 \right]$$

$$+ \frac{1}{4} \sum_{i=1}^{n} \left(\frac{\partial^2 g}{\partial X_i^2} \right)_{\overline{X}_i}^2 E\left[(X_i - \overline{X}_i)^4 \right] + \sum_{i=1}^{n-1} \sum_{j=i+1}^{n} \left(\frac{\partial^2 g}{\partial X_i^2} \right) \left(\frac{\partial^2 g}{\partial X_j^2} \right) \text{var}(X_i) \text{var}(X_j)$$

$$\tag{14.49}$$

The SOA method has been used only for evaluating the mean of the model output. It has not been used much for variance evaluation because of the involvement of complicated calculations in approximating the model output variance.

14.2.7 First-Order Reliability Method

The first-order reliability method (FORM) is characterized by an iterative linear approximation to the performance function. Fundamentally, this method can be considered as an extension to the FOA method and is also known as the advanced first-order approximation (AFOA) method, which was developed to address some of the technical difficulties of FOA. One of the major problems with the FOA technique was the lack of invariance of the solution relative to the formulation of the performance function. Simple algebraic changes in the problem formulation can lead to significant changes in assessing the propagation of uncertainty. Hasofer and Lind (1974) presented a methodology that specifically addressed this issue by requiring expansion about a unique point in the feasible solution space. It should be mentioned that Fruedenthal (1956) also proposed a method suggesting similar restrictions on the expansion point.

Hasofer and Lind (1974) proposed taking the Taylor series expansion at a likely point on the failure surface of the performance function. Rackwitz (1976) implemented the ideas of Hasofer and Lind. The failure surface is defined by the equation $Z = 0$. The perpendicular drawn on the failure surface from the origin cuts the failure surface at a point called the failure point. The distance of the failure point from the origin is a measure of reliability. The expected value and variance of Z can be obtained by first solving $Z = 0$ to find the failure point $\underline{X}^*$ and then expanding Z about $\underline{X}^*$ using a Taylor series expansion as

$$E[Z] \approx \sum_{i=1}^{n} \left(\frac{\partial Z}{\partial X_i} \right)_{X_i^*} (\overline{X}_i - X_i^*) \tag{14.50}$$

$$\text{var}(Z) = \sigma_Z^2 \approx \sum_{i=1}^{n} \sum_{j=1}^{n} \left(\frac{\partial Z}{\partial X_i} \right)_{X_i^*} \left(\frac{\partial Z}{\partial X_j} \right)_{X_j^*} E\left[(X_i - X_i^*)(X_j - X_j^*) \right] \tag{14.51}$$

where σ_Z is the standard deviation of Z. For the case of statistically independent basic variables, var(Z) is rewritten as

$$\text{var}(Z) = \sigma_Z^2 \approx \sum_{i=1}^{n} \left(\frac{\partial Z}{\partial X_i} \right)_{X_i^*}^{2} \sigma_{X_i}^2 \qquad (14.52)$$

As described earlier, knowing the reliability index β [$\beta = \mu(Z)/\sigma(Z)$], one can determine the probability of failure P_f and reliability $\Re$. For models having a linear failure surface and all the basic variables normally distributed, the estimates of P_f and $\Re$ are exact.

For most modeling problems, it is very unlikely that all basic input variables will be normally distributed. Rackwitz (1976) proposed a transformation technique in which the values of the CDF and PDF of the non-normal distributions are the same as those of the equivalent normal distributions at the failure point $\underline{X}^*$, also known as the most probable point (MPP; see Fig. 14-5). Consider an input random variable X_i for which the PDF and CDF are given as $p_{Xi}(x_i)$ and $P_{Xi}(x_i)$, respectively. Equating the cumulative probabilities at the failure point we have

$$\Phi\left(\frac{x_i^* - \mu_{X_i}^N}{\sigma_{X_i}^N} \right) = P_{X_i}\left(x_i^* \right) \qquad (14.53)$$

where $\mu_{X_i}^N$ and $\sigma_{X_i}^N$ are the mean value and standard deviation of the equivalent normal distribution for X_i; $P_X\left(x_i^* \right)$ is the original CDF of X_i; and $\Phi(.)$ is the CDF of the standard normal distribution. Using Eq. 14.53, one can write the mean of the equivalent normal distribution as

$$\mu_{X_i}^N = x_i^* - \sigma_{X_i}^N \Phi^{-1}\left[P_{X_i}\left(x_i^* \right) \right] \qquad (14.54)$$

Now equating the corresponding PDF ordinates at x_i^* gives

$$\frac{1}{\sigma_{X_i}^N} \phi\left(\frac{x_i^* - \mu_{X_i}^N}{\sigma_{X_i}^N} \right) = p_{Xi}\left(x_i^* \right) \qquad (14.55)$$

where $\phi(.)$ is the PDF of the standard normal distribution. Combining Eq. 14.54 and Eq. 14.55, one obtains the standard deviation of the equivalent normal distribution as

$$\sigma_{X_i}^N = \frac{\phi\left\{ \Phi^{-1}\left[P_{X_i}\left(x_i^* \right) \right] \right\}}{p_{Xi}\left(x_i^* \right)} \qquad (14.56)$$

The key to FORM is the determination of the failure point for the Taylor series expansion. Shinozuka (1983) has shown that for FORM the reliability index β is the shortest distance in the standardized space between the system mean state and the failure surface. Thus, if the failure point is determined correctly, it represents the most likely combination of input variable values that produce the critical target level. Ang and Tang (1984) and Haldar and Mahadevan (2000) present detailed mathematical treatment and interpretation of FORM. The Hasofer and Lind approach can be summarized as follows:

1. Formulate the performance function or the limit state in terms of the original design space, that is, $X = \{X_1, X_2, X_3, ...,X_n\}$ [the X space as shown in Fig. 14-11(a)].
2. Define the independent and standardized normal vector $U = \{U_1, U_2, U_3, ...,U_n\}$ by transforming the input variables into an equivalent normal distribution.
3. Transform the performance function into the standard normal space [the U space as shown in Fig. 14-11(b)].
4. Search for the minimum β.
5. Determine the probability of failure or reliability of the system corresponding to the obtained reliability index β.

Figure 14-11 illustrates the concept of reliability index and MPP search for a two-variable case in the standard normal space. After completing steps 1 and 2, one focuses on the transformed performance function curve [i.e., $G(U_1, U_2) = 0$]. Next, among the various possible β values, the minimum β is sought. The corresponding point is called the MPP. The process of determining the minimum β value can be mathematically expressed as follows:

$$\text{minimize } \beta = \sqrt{U_1^2 + U_2^2}$$

$$\text{subjected to } G(U_1, U_2) = 0 \tag{14.57}$$

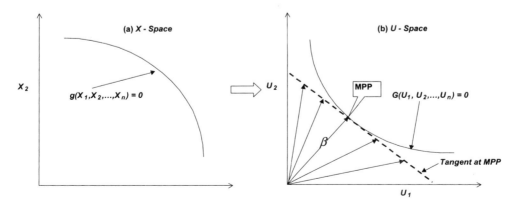

Figure 14-11 *Transformation of input variables, nonlinear limit state function, MPP, and reliability index in FORM.*

For a more general system consisting of n input variables, Eq. 14.57 can be written as

$$\text{minimize } \beta = \sqrt{\sum_{i=1}^{n} U_i^2} = \sqrt{U^T U}$$

$$\text{subjected to } G(U_1, U_2, \ldots, U_n) = 0 \tag{14.58}$$

Therefore, the difficulty then lies in determining the minimum distance for a general nonlinear function. This is essentially a nonlinear, constrained optimization problem. Thus, determination of β requires application of a constrained nonlinear optimization, such as the generalized reduced-gradient algorithm used by Cheng (1982), the Lagrange multiplier approach used by Shinozuka (1983), or the iterative optimization method suggested by Rackwitz (1976).

14.2.7.1 Rackwitz's Numerical Algorithm

As mentioned earlier, Rackwitz's approach deals with the variables in the reduced space (i.e., the U space as shown in Fig. 14-11). Further, it requires partial derivatives of the performance function with respect to each of the reduced random variables followed by extensive iterative calculations until convergence is reached.

Consider a general function $Z = g(X_1, X_2, \ldots, X_n)$. A typical stepwise approach suggested by Rackwitz for determining the reliability index can be given as follows:

Step 1. Formulate the limit state in terms of uncorrelated and independent normally distributed standardized random variables. If we assume X_i to be a normally distributed random variable, the standardized random variable U_i can be written as $U_i = (X_i - \mu_i)/\sigma_i$.

Step 2. Assume an initial value for each input random variable. Let us represent these values as X_i^* s in the X space and U_i^* in the reduced U space. Typically, the value for the first iteration is taken as the mean of the variables.

Step 3. Evaluate the partial derivatives

$$\partial g / \partial U_i = (\partial g / \partial X_i) \times (\partial X_i / \partial U_i) = \sigma_i (\partial g / \partial X_i)$$

at the assumed points in step 2.

Step 4. Determine the direction cosines as

$$\alpha_i = \frac{(\partial g / \partial U_i)_{U*}}{\sqrt{\sum_{i=1}^{n} (\partial g / \partial U_i)_{U*}^2}} \tag{14.59}$$

Step 5. Formulate each reduced variable in terms of the reliability index as $U_i = -\alpha_i \beta$ or in the X space as $X_i = \mu_i - \alpha_i \beta \sigma_i$.

Step 6. Substitute values of U_i or X_i in the limit state function $g(U_i) = 0$ or $g(X_i) = 0$ and solve for β.

Step 7. Using the result from step 6 as a new starting point repeat steps 2 through 6 until the starting point in step 2 and resulting solution in step 6 converge to the same solution.

14.2.7.2 Lagrange Multiplier Method

Using the Lagrange multiplier method, the objective function and the constraint can be combined into a single function L. Since minimizing β also means minimizing β^2, Eq. 14.58 can be written as

$$\text{minimize } L = \sum_{i=1}^{n} \left(\frac{X_i - \mu_i}{\sigma_i} \right)^2 + \lambda g\left(X_1, X_2, \ldots, X_n\right) \tag{14.60}$$

The function L is called the Lagrangean function and the parameter λ the Lagrange multiplier. Taking the partial derivatives of L with respect to X_i and λ and equating to zero one gets the following equations:

$$\frac{\partial L}{\partial X_i} = 0 \tag{14.61}$$

$$\frac{\partial L}{\partial \lambda} = 0 \tag{14.62}$$

Solving Eq. 14.61 and Eq. 14.62, one gets the coordinates of the MPP and can evaluate the reliability index as

$$\beta = \sqrt{\sum_{i=1}^{n} \left(\frac{X_i - \mu_i}{\sigma_i} \right)^2}$$

Example 14.16 Solve case 2 of Example 14.8 using the Rackwitz and Lagrange multiplier methods.

Solution The performance function is given as

$$Z = \frac{\text{TMDL}}{\text{WLA} + \text{LA}} - 1$$

Rackwitz's Method

Step 1: In this example all the variables are assumed to be independent and normally distributed. The first step is to transform the random variables into the reduced space as

$$U_1 = (\text{TMDL} - \mu_{\text{TMDL}})/\sigma_{\text{TMDL}} \; ; \; U_2 = (\text{WLA} - \mu_{\text{WLA}})/\sigma_{\text{WLA}}$$

and

$$U_3 = (\text{LA} - \mu_{LA})/\sigma_{LA}$$

where the mean and standard deviations are given in Table E14-16a.

The first iteration proceeds as follows:

Step 2: Assume initial values for each input random variable, $U_1^* = U_2^* = U_3^* = 0$; that is, $\text{TMDL}^* = 1678$, $\text{WLA}^* = 65$, and $\text{LA}^* = 1258$.

Step 3: Evaluate the partial derivatives:

$$\partial g / \partial U_1 = \sigma_{\text{TMDL}} \frac{1}{(\text{WLA+LA})}$$

$$\partial g / \partial U_2 = \sigma_{\text{WLA}} \frac{-\text{TMDL}}{(\text{WLA+LA})^2}$$

$$\partial g / \partial U_3 = \sigma_{LA} \frac{-\text{TMDL}}{(\text{WLA+LA})^2}$$

For the assumed MPP, the values of the partial derivatives are

$$\partial g / \partial U_1 \big|_{U_1^*} = 436 \times \frac{1}{(65 + 1258)} = 0.33$$

$$\partial g / \partial U_2 = 12 \times \frac{-1678}{(65 + 1258)^2} = -0.01$$

$$\partial g / \partial U_3 = 415 \times \frac{-1678}{(65 + 1258)^2} = -0.40$$

Table E14-16a

Variable	Standard deviation	Mean
TMDL	436	1678
WLA	12	65
LA	415	1258

Step 4. Determine the direction cosines as

$$\alpha_1 = \frac{(\partial g/\partial U_1)_{U_*}}{\sqrt{\sum_{i=1}^{3}(\partial g/\partial U_i)^2_{U_*}}} = \frac{0.33}{\sqrt{(0.33)^2 + (-0.01)^2 + (-0.40)^2}} = 0.64$$

$$\alpha_2 = \frac{(\partial g/\partial U_2)_{U_*}}{\sqrt{\sum_{i=1}^{3}(\partial g/\partial U_i)^2_{U_*}}} = \frac{-0.01}{\sqrt{(0.33)^2 + (-0.01)^2 + (-0.40)^2}} = -0.02$$

$$\alpha_3 = \frac{(\partial g/\partial U_3)_{U_*}}{\sqrt{\sum_{i=1}^{3}(\partial g/\partial U_i)^2_{U_*}}} = \frac{-0.40}{\sqrt{(0.33)^2 + (-0.01)^2 + (-0.40)^2}} = -0.77$$

Step 5. Formulate each reduced variable in terms of the reliability index as $U_i = -\alpha_i \beta$ or in X space as $X_i = \mu_i - \alpha_i \beta \sigma_i$. Thus the coordinates of the failure point are

$$U_1 = -\alpha_1\beta = -0.64\beta \text{ or } \text{TMDL} = 1678 - 0.64 \times 436\beta = 1678 - 278.29\beta$$

$$U_2 = -\alpha_2\beta = 0.02\beta \text{ or } \text{WLA} = 65 + 0.02 \times 12\beta = 65 + 0.24\beta$$

$$U_3 = -\alpha_3\beta = 0.77\beta \text{ or } \text{LA} = 1258 + 0.77 \times 415\beta = 1258 + 319.55\beta$$

Step 6. Substitute values of U_i or X_i in the limit state function $g(U_i) = 0$ or $G(X_i) = 0$ and solve for β:

$$Z = \frac{\text{TMDL}}{\text{WLA+LA}} - 1 = \frac{1678 - 278.29\beta}{(65 + 0.24\beta) + (1258 + 319.55\beta)} - 1 = \frac{1678 - 278.29\beta}{1323 + 319.79\beta} - 1 = 0$$

and so

$$\beta = \frac{1678 - 1323}{319.79 + 278.84} = 0.593$$

Thus the revised failure point is

$$\text{TMDL} = 1678 - 278.29 \times 0.593 = 1513$$

$$\text{WLA} = 65 + 0.25 \times 0.593 = 65.15$$

$$\text{LA} = 1258 + 319.55 \times 0.593 = 1447.49$$

This completes the first iteration.

Step 7. Using the result from step 6 as a new starting point steps 2 to 6 are repeated until the starting points in step 2 and resulting solution in step 6 converge to the same solution. The new iterations are presented in Table E14-16b. Note that the solution converges in the fourth iteration. Another point to be noted in this example is that the reliability index β remains the same in all iterations, which is not a necessity.

Lagrange Multiplier Method

The Lagrangean function for this problem is

$$L = \sum_{i=1}^{n} \left(\frac{X_i - \mu_i}{\sigma_i} \right)^2 + \lambda \left(\frac{X_1}{X_2 + X_3} - 1 \right) \tag{14.63}$$

in which X_1, X_2, and X_3 are TMDL, WLA, and LA and are the mean and standard deviation of the respective X_i. Taking the partial derivatives of L and setting them equal to zero, one obtains the following four nonlinear equations:

$$2(X_1 - \mu_1) + \frac{\lambda}{X_2 + X_3} \sigma_1^2 = 0 \tag{14.64}$$

$$2(X_2 - \mu_2) - \frac{\lambda X_1}{(X_2 + X_3)^2} \sigma_2^2 = 0 \tag{14.65}$$

Table E14-16b

Iteration	Variable, X_i	Assumed failure point	Standard deviation X_i	Mean X_i	$\partial g / \partial U_i$	α_1	New X_i
2nd	TMDL	1513.98	436	1678	0.289	0.724	1491.91
	WLA	65.15	12	65	−0.008	0.019	65.13
	LA	1446.35	415	1258	−0.275	0.69	1426.77
				$\beta = 0.589$			
3rd	TMDL	1491.91	436	1678	0.292	0.724	1491.76
	WLA	65.13	12	65	−0.008	0.019	65.13
	LA	1426.77	415	1258	−0.278	0.689	1426.63
				$\beta = 0.589$			
4th	TMDL	1491.76	436	1678	0.292	0.724	1491.76
	WLA	65.13	12	65	−0.008	0.019	65.13
	LA	1426.63	415	1258	−0.278	0.689	1426.63
				$\beta = 0.589, P_f = 1 - \Phi(0.59) = 0.28$			

$$2(X_3 - \mu_3) - \frac{\lambda X_1}{(X_2 + X_3)^2} \sigma_3^2 = 0 \qquad (14.66)$$

$$\frac{X_1}{X_2 + X_3} - 1 = 0 \qquad (14.67)$$

Defining the problem in tabular form in an Excel spreadsheet and invoking Excel's optimization module SOLVER by equating Eq. 14.64 to zero, changing X_i, and setting Eqs. 14.65, 14.66, and 14.67 equal to zero, we obtain the solution shown in Table E14-16c.

Table E14-16c

Item	TMDL	WLA	LA	λ
Mean	1678	65	1258	
CV	0.26	0.18	0.33	
Standard deviation	436	12	415	
Assumed solution, X_i	1491.76	65.13	1426.63	2.92
$u_i^2 = \left[(X_i - \mu_i)/\sigma_i\right]^2$	0.1822	0.0001	0.1650	
Equation	14.64	14.65	14.66	14.67
Partial derivative	0.00	0.00	0.00	0.00
$\beta = \sqrt{\sum_{i=1}^{3}\left(\dfrac{X_i - \mu_i}{\sigma_i}\right)^2} = 0.589$				
$Z = 0.000$				
$P_f = 0.278$				

14.2.7.3 Ellipsoid Approach

Low (1996) and Low and Tang (1997) proposed the ellipsoid approach for calculating the reliability index in which an ellipsoid just touches the limit state surface in the original space of the variables. In matrix notation one can express the formulation of the reliability index given by Hasofer and Lind (1974) in matrix form (Ditlevsen 1981, Veneziano 1974) as

$$\beta = \min_{X \in F} \sqrt{(\underline{X} - \underline{\mu})^T \underline{C}(\underline{X} - \underline{\mu})} \qquad (14.68)$$

where $\underline{X}$ represents the vector of the random variables, $\underline{\mu}$ are the corresponding mean values, $\underline{C}$ is the covariance matrix, F is the failure region, and T indicates the transpose of a matrix. Substituting $\beta = 1$ in Eq. 14.68, we obtain the following ellipse:

$$\left(\underline{X} - \underline{\mu}\right)^T \underline{C} \left(\underline{X} - \underline{\mu}\right) = 1 \tag{14.69}$$

To explain the ellipsoid approach, let us consider a simple system

$$Z = \left(25.50 - 1.41X_1 + 0.039X_1^2\right) - X_2$$

containing only two random variables X_1 and X_2. For this case Eq. 14.69 reduces to

$$\begin{bmatrix} (X_1 - \mu_1) & (X_2 - \mu_2) \end{bmatrix} \begin{bmatrix} \sigma_1^2 & \rho\sigma_1\sigma_2 \\ \rho\sigma_1\sigma_2 & \sigma_2^2 \end{bmatrix}^{-1} \begin{bmatrix} (X_1 - \mu_1) \\ (X_2 - \mu_2) \end{bmatrix} = 1 \tag{14.70}$$

Simplifying Eq. 14.70, one obtains

$$\frac{(X_1 - \mu_1)^2}{(1 - \rho^2)\sigma_1^2} + \frac{(X_2 - \mu_2)^2}{(1 - \rho^2)\sigma_2^2} - \frac{2\rho(X_1 - \mu_1)(X_2 - \mu_2)}{(1 - \rho^2)\sigma_1\sigma_2} = 1 \tag{14.71}$$

Equation 14.71 is the 1σ dispersion ellipse plotted for various values of ρ assuming $\mu_1 = \mu_2 = 9$, $\sigma_1 = 3$, and $\sigma_1 = 2$. Figure 14-11 also shows the critical ellipse for $\rho = 0.7$, which just touches the limit state line $\left(25.50 - 1.41X_1 + 0.039X_1^2\right) - X_2 = 0$. The critical ellipse is the 1σ dispersion ellipse corresponding to $\rho = 0.7$ expanded by β times so that it becomes tangent to the limit state line.

Low (1996) has explained the meaning of the reliability index as follows. Suppose the 1σ scatter ellipse gradually expands without changing its original aspect ratio. The equation of the ellipse at any time is then obtained by substituting $k\sigma_1$ for σ_1 and $k\sigma_2$ for σ_2 in Eq. 14.71. The resulting equation is

$$\frac{(X_1 - \mu_1)^2}{(1 - \rho^2)\sigma_1^2} + \frac{(X_2 - \mu_2)^2}{(1 - \rho^2)\sigma_2^2} - \frac{2\rho(X_1 - \mu_1)(X_2 - \mu_2)}{(1 - \rho^2)\sigma_1\sigma_2} = k^2 \tag{14.72}$$

When the expanding ellipse just touches the failure surface (or limit state surface), the value of k is the reliability index β, as shown in Fig. 14-12. The corresponding point of contact on the failure surface is the MPP having the coordinate vector $\underline{X}^*$. Therefore, the reliability index

$$\beta = \sqrt{\frac{\left(X_1^* - \mu_1\right)^2}{(1 - \rho^2)\sigma_1^2} + \frac{\left(X_2^* - \mu_2\right)^2}{(1 - \rho^2)\sigma_2^2} - \frac{2\rho\left(X_1^* - \mu_1\right)\left(X_2^* - \mu_2\right)}{(1 - \rho^2)\sigma_1\sigma_2}} \tag{14.73}$$

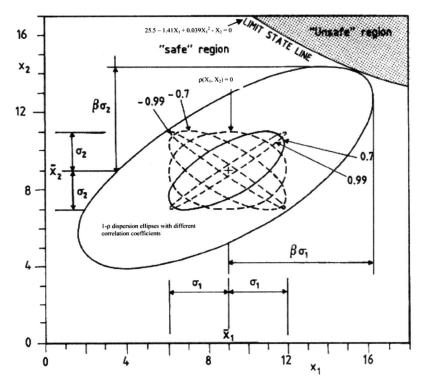

$$25.5 - 1.41X_1 + 0.039X_1^2 - X_2 = 0$$

Figure 14-12 *1–σ dispersion ellipses for various values of ρ and the critical ellipse for =0.7.*

The search for the most critical point (X_1^*, X_2^*) on the failure surface and subsequent evaluation of β index using Eq. 14.73 can be formulated into an optimization problem as follows (Low 1996):

$$\text{minimize } \beta = \sqrt{\frac{(X_1 - \mu_1)^2}{(1-\rho^2)\sigma_1^2} + \frac{(X_2 - \mu_2)^2}{(1-\rho^2)\sigma_2^2} - \frac{2\rho(X_1 - \mu_1)(X_2 - \mu_2)}{(1-\rho^2)\sigma_1\sigma_2}} \quad \textbf{(14.74a)}$$

$$\text{subject to } Z(X_1, X_2) = 0 \quad \textbf{(14.74b)}$$

where $Z(X_1, X_2) = 0$ is the equation describing the limit state function. A similar method can be used for systems involving more than two random variables, in which case the ellipses will become ellipsoids or hyper-ellipsoids.

Example 14.17 Solve Example 14.16 using the ellipsoid approach.

Solution Because the variables are uncorrelated, the problem of determining the reliability index, Eq. 14.74, can be written as

$$\text{minimize } \beta = \sqrt{\frac{(X_1 - \mu_1)^2}{\sigma_1^2} + \frac{(X_2 - \mu_2)^2}{\sigma_2^2} + \frac{(X_3 - \mu_3)^2}{\sigma_3^2}} \text{ subject to } Z(X_1, X_2, X_3) = 0$$

The input variables to the problem are shown in Table E14-17. Invoking the optimization module SOLVER in Excel by setting β = minimum, changing assumed X_i values, and subjecting the problem to the constraint $Z = 0$, one can perform the calculations listed in Table E14-17.

Example 14.18 A culvert has been designed for a carrying capacity of 30 cfs. Based on the rational formula, the 5-year flow at the culvert is given as $Q = 2.41 \times C \times A \times (T_c + 0.2)^{-0.77}$. The data are shown in Table E14-18a. Determine the probability of failure of the culvert using FORM assuming all the variables are independent and normally distributed.

Solution The objective function becomes

$$Z = 30 - Q = 30 - 2.41 \times C \times A \times (T_c + 0.2)^{-0.77}$$

Because the variables are uncorrelated, by using the ellipsoid method the problem for determining the reliability index, Eq. 14.74, can be written as

$$\text{minimize } \beta = \sqrt{\frac{(X_1 - \mu_1)^2}{\sigma_1^2} + \frac{(X_2 - \mu_2)^2}{\sigma_2^2} + \frac{(X_3 - \mu_3)^2}{\sigma_3^2}}$$

subject to $Z(X_1, X_2, X_3) = 0$, where $X_1 = A$, $X_2 = C$, and $X_3 = T_c$.

Defining the problem as shown in Table E14-18b and invoking SOLVER in Excel by setting β = minimum, changing assumed X_i values, and subjecting them to the constraint $Z = 0$, one can perform the calculations listed in Table E14-18b.

Table E14-17

Item	TMDL	WLA	LA
Mean	1678	65	1258
CV	0.26	0.18	0.33
Standard deviation	436.28	11.7	415.14
Assumed solution, X_i	1491.76	65.13	1426.63
$U_i^2 = \left[(X_i - \mu_i)/\sigma_i\right]^2$	0.18	0.001	0.16
Nonlinear optimization			
$\min \beta = \sqrt{\sum_{i=1}^{3}\left(\dfrac{X_i - \mu_i}{\sigma_i}\right)^2}$ = 0.59, after minimization			
subjected to $Z = 0$			
$P_f = 1 - \text{normal}(0.59) = 0.28$			

Table E14-18a

Function	Mean	SD
A	12.00	1.20
C	0.45	0.15
T_c	0.37	0.23

Table E14-18b

Item	C	A	T_C
Mean	0.45	12.00	0.37
CV	0.33	0.10	0.62
Standard deviation	0.15	1.20	0.23
Assumed solution, X_i	0.52	12.19	0.21
$U_i^2 = \left[(X_i - \mu_i)/\sigma_i\right]^2$	0.21	0.02	0.46
Nonlinear optimization			
$\min \ \beta = \sqrt{\sum_{i=1}^{3}\left(\dfrac{X_i - \mu_i}{\sigma_i}\right)^2} = 0.83$, after minimization			
subjected to $Z = 0$			
$P_f = 1 - \Phi(0.83) = 0.20$			

Example 14.19 Solve Example 14.18 using FORM if C and T_c are correlated with $\rho(C, T_c) = 0.75$.

Solution This problem can be solved by formulating the problem in the matrix notation:

$$\beta = \min_{X \in F} \sqrt{\left(\underline{X} - \underline{\mu}\right)^T \underline{C}\left(\underline{X} - \underline{\mu}\right)}$$

subjected to $Z = 0$

To explain how this calculation is performed, let us assume a starting point $A = 10$, $C = 1$, and $T_c = 0.5$. The corresponding $\left(\underline{X} - \underline{\mu}\right)$ matrix can be written as

$$\left(\underline{X} - \underline{\mu}\right) = \begin{pmatrix} 10 - 12 \\ 1 - 0.45 \\ 0.5 - 0.37 \end{pmatrix} = \begin{pmatrix} -2 \\ 0.55 \\ 0.13 \end{pmatrix}$$

Thus the $(\underline{X} - \underline{\mu})^T$ matrix becomes

$$(\underline{X} - \underline{\mu})^T = (-2 \quad 0.55 \quad 0.13)$$

The covariance matrix C_o can be written as

$$C_o = \begin{pmatrix} 1.44 & 0 & 0 \\ 0 & 0.02 & 0.03 \\ 0 & 0.03 & 0.05 \end{pmatrix}$$

and its inverse is

$$C_o = \begin{pmatrix} 0.69 & 0 & 0 \\ 0 & 103.65 & -50.79 \\ 0 & -50.79 & 44.24 \end{pmatrix}$$

Now, $C_o(\underline{X} - \underline{\mu})$ is determined as

$$\underline{C_o}(\underline{X} - \underline{\mu}) = \begin{pmatrix} -1.39 \\ 50.40 \\ -22.18 \end{pmatrix}$$

and $(\underline{X} - \underline{\mu})^T C_o(\underline{X} - \underline{\mu})$ is determined by matrix multiplication. This value comes out to be 27.62. At this point the objective function $Z = -1.72$. So this point is not located on the limit state surface. Now this process of matrix manipulation is performed with the constraint $Z = 0$ using SOLVER. The mean, standard deviation (SD), and covariance matrix are entered as given in Table E14-19. Then some X values are assumed as the starting point. The matrices $(X - \mu)$, $(X - \mu)^T$, and C_o are determined using Excel spreadsheet formulas in terms of cells. Then formulas for matrix multiplication $C_o(X - \mu)$ and $(X - \mu)^T C_o(X - \mu)$ are fed into the program. The reliability index is defined as $\beta = \sqrt{(\underline{X} - \underline{\mu})^T C_o(\underline{X} - \underline{\mu})}$. Now, SOLVER is invoked to minimize β by changing X values subject to $Z = 0$. Table E14-19 shows the solution obtained, which gives $\beta = 1.47$. Using this value, one gets a probability of failure of 0.07 or 7%.

Example 14.20 Solve Example 14.19 using FORM if C, T_c, and A are characterized by log-normal, gamma, and triangular distributions, respectively. Assume $\rho = 0.75$ between C and T_c.

Solution This problem is the same as given in Example 14.19, except for the type of distributions used to characterize C, T_c, and A. The FORM approach for determining the reliability index given by Hasofer and Lind assumes that all the ran-

Table E14-19

Variable X_i	X_i value	Mean $X_i(\mu)$	SD $X_i(\sigma)$	Covariance matrix, C		
A	12.46	12.00	1.20	1.44	0	0
C	0.42	0.45	0.15	0	0.02	0.03
T_c	0.12	0.37	0.23	0	0.03	0.05

$(X-\mu)^T$			Inverse of covariance matrix, C^{-1}			$(X-\mu)$
0.46	−0.03	−0.25	0.69	0.00	0.00	0.46
			0.00	103.65	−50.79	−0.03
			0.00	−50.79	44.24	−0.25

$C^{-1}(X-\mu)$	$(X-\mu)^T C^{-1}(X-\mu)$	b	P_f	Performance function, Z
0.32				
9.37	2.16	1.47	0.07	-8.15×10^{-8}
−9.34				

dom variables are normally distributed. Therefore, all the non-normal variables need to be transformed into their equivalent normal variables. This can be accomplished by using Eq. 14.54 and Eq. 14.56. As evident from these equations, the parameters of the equivalent normal distribution, $\mu_{X_i}^N$ and $\sigma_{X_i}^N$, are functions of the assumed failure point X_i^*; these are used in calculations to get a better estimate of the failure point; the transformation process is performed in each iteration step of optimization. Once the parameters of equivalent normal distributions for all the variables in each iteration are determined, the problem becomes the same as given in Example 14.19. A stepwise solution is given as follows:

Step 1: Assume a starting value for x_i^*; generally mean values are taken as the starting point (see Table E14-20a).

Table E14-20a

Variable	Assumed x value, x_i^*
A	11.43
C	0.45
T_c	0.37

Step 2: Determine parameters of all the non-normal distributions. For a symmetric triangular distribution the parameters are

$$\hat{a} = \mu_X \left(1 - \sqrt{6}CV_X\right) \quad \text{Par. 1}$$

$$\hat{b} = \mu_X \left(1 + \sqrt{6}CV_X\right) \quad \text{Par. 2}$$

For the log-normal distribution, the parameters are

$$\hat{\mu}_Y = \frac{1}{2}\ln\left(\frac{\mu_X^2}{1+CV_X^2}\right) \quad \text{Par. 1}$$

$$\hat{\sigma}_Y = \sqrt{\ln\left(1+CV_X^2\right)} \quad \text{Par. 2}$$

For the gamma distribution, the parameters are

$$\hat{\alpha} = \frac{1}{CV_X^2} \quad \text{Par. 1}$$

$$\hat{\beta} = \mu_X CV_X^2 \quad \text{Par. 2}$$

The values of estimated parameters using these equations are given in columns 5 and 6 of Table E14-20b.

Step 3: Determine the values of the cumulative distribution function $[P_{X_i}(x_i^*)]$ and the probability density function $[p_{Xi}(x_i^*)]$ at the assumed point x_i^* (columns 7 and 8 of Table E14-20b).

Step 4: Using the calculated values of the cumulative distribution function $[P_{X_i}(x_i^*)]$ and the probability density function $[p_{Xi}(x_i^*)]$ at the assumed point x_i^*, determine $\Phi^{-1}\left[P_{X_i}(x_i^*)\right]$, in other words, the normal standard variate (Z value), corresponding to the cumulative probability $P_{X_i}(x_i^*)$ (column 9). Now using this Z value, one can easily determine the density of the standard normal distribution $\phi\left\{\Phi^{-1}\left[P_{X_i}(x_i^*)\right]\right\}$ (column 10).

Step 5: Determine the standard deviation of the equivalent normal distribution using Eq. 14.53 as

$$\sigma_{X_i}^N = \frac{\phi\left\{\Phi^{-1}\left[P_{X_i}(x_i^*)\right]\right\}}{p_{Xi}(x_i^*)} = \frac{\text{Column } 10}{\text{Column } 8}$$

Now using the values of standard deviation $\sigma_{X_i}^N$ and $\Phi^{-1}\left[P_{X_i}(x_i^*)\right]$, determine values of $\mu_{X_i}^N$ by using the following relationship (Eq. 14.51):

$$\mu_{X_i}^N = x_i^* - \sigma_{X_i}^N \Phi^{-1}\left[P_{X_i}(x_i^*)\right] \quad \text{(column 11 in Table E14-20b).}$$

Step 6: After transforming the non-normal distributions, perform an optimization by using the values of mean and standard deviation of equivalent normal distributions. Values of $\mu_{X_i}^N$ and $\sigma_{X_i}^N$ are calculated in the Excel spreadsheet using a dynamic link between the value of x_i^* and the optimization given in Table E14-20c so that as the value of x_i^* changes, the values of

Table E14-20b

Variable (1)	Distribution (2)	Mean (3)	CV (4)	Par. 1 (5)	Par. 2 (6)
A	Triangular	11.43	0.1	8.63	14.23
C	Log-normal	0.45	0.33	−0.85	0.32
T_c	Gamma	0.37	0.61	2.65	0.14

$P_{X_i}(x_i^*)$ (7)	$p_{Xi}(x_i^*)$ (8)	$\Phi^{-1}\left[P_{X_i}(x_i^*)\right]$ (9)	$\phi\left\{\Phi^{-1}\left[P_{X_i}(x_i^*)\right]\right\}$ (10)	$\mu_{X_i}^N$ (11)	$\sigma_{X_i}^N$ (12)
0.48	0.35	−0.04	0.4	11.43	1.13
0.42	1.21	−0.21	0.39	0.47	0.32
0.03	0.94	−1.9	0.07	0.2	0.07

Table E14-20c

Variable X_i	X_i value[a]	Mean[b] $\mu_{X_i}^N$	SD[c] $\sigma_{X_i}^N$	Covariance matrix, C		
A	11.430	11.431	1.20	1.44	0	0
C	0.450	0.398	0.322	0	0.02	0.03
T_c	0.370	0.323	0.230	0	0.03	0.05

$(X-\mu)^T$			Inverse of covariance matrix, C^{-1}			$(X-\mu)$
−0.001	0.052	0.047	0.69	0.00	0.00	−0.001
			0.00	103.65	−50.79	0.052
			0.00	−50.79	44.24	0.047

$C^{-1}(X-\mu)$

0.00
2.95
−0.53

$(X-\mu)^T C^{-1}(X-\mu)$	b	P_f	Performance function, Z
0.13	0.36	0.36	10.89

a. As assumed in step 1.
b. From column 11.
c. From column 12.

$\mu_{X_i}^N$ and $\sigma_{X_i}^N$ are automatically updated and used in the optimization. The optimization table is the same as used in Example 14.19.

FORM is quite accurate because it is able to overcome model nonlinearity problems, and no additional assumption about the distribution type of the performance function is required. It is still an approximation method because the performance function is approximated by a linear function at the design point, and accuracy problems may arise when the performance function is strongly nonlinear (Cawlfield and Wu 1993, Zhao and Ono 1999). Another disadvantage of FORM is that determination of the linearization point is generally not easy, depending upon the nature and complexity of the system for which the reliability, risk, or uncertainty analysis is being studied (Melching and Anmangandla 1992). Further, the magnitude of acceptable convergence may affect the accuracy of the reliability estimates. In some cases, the magnitude of the convergence error may not be reduced after a certain level.

14.2.8 Second-Order Reliability Method

The second-order reliability method (SORM) has been used extensively in structural reliability analyses. It has been established as an attempt to improve the accuracy of FORM. SORM involves approximating the limit state surface function at the design point by a second-order surface, and the failure probability is given as the probability content outside the second-order surface. Several researchers compared the reliability estimates of various engineering systems based on FORM and SORM and reported that their results were in good agreement when the limit state surface at the design point in the standard normal space is nearly flat. However, when the limit state function contains highly nonlinear terms, or when the input random variables have an accentuated non-normal character, SORM tends to produce more accurate results than does FORM. But computational requirements of SORM are much higher than those of FORM. For all practical purposes the FORM-based reliability estimates are considered ᵊfficient. For further details of SORM, see Haldar and Mahadevan (2000).

9 Generic Expectation Function Method

ᵃr from the preceding discussion that reliability and risk analysis of a given ⁻equires knowledge of the distribution of its performance function. In ᵗhis distribution cannot be determined exactly because of insufficient ᵗr mathematical complexity involved in the distribution derivation pro- ᵥ cases, the true form of the performance function or output distribu- ᵗuired. A very good estimate of system reliability can be obtained if ᵗ model output are known correctly. As far as the distribution of the concerned, several forms of distributions can be assumed. Knowl- ⁻ order moments of a model output helps in identifying the can- ₅ for the model output and provides more flexibility to include

Table 14-6 *Generic expectation functions for some commonly used probability density functions.*

Name	Generic expectation function, $E[X^r]$
Uniform	$E[X^r] = \int_{-\infty}^{\infty} \frac{1}{(b-a)} X^r dX = \frac{\mu_X^r}{2\sqrt{3}(r+1)CV_X}\left[\left(1+CV_X\sqrt{3}\right)^{r+1} - \left(1-CV_X\sqrt{3}\right)^{r+1}\right]$
Symmetrical triangular	$E[X^r] = \frac{\mu_X^r}{6(r+1)(r+2)CV_X^2}\left[\left(1+CV_X\sqrt{6}\right)^{r+2} + \left(1-CV_X\sqrt{6}\right)^{r+2} - 2\right]$
Unsymmetrical triangular	$E[X^r] = \frac{2\left[(b-c)a^{r+2} + (c-a)b^{r+2} + (a-b)c^{r+2}\right]}{(r+1)(r+2)(b-c)(c-a)(b-a)}$
Log-normal	$E[X^r] = \mu_X^r\left(1+CV_X^2\right)^{\frac{r(r-1)}{2}}$
Gamma	$E[X^r] = \mu_X^r CV_X^{2r} \exp\left\{\ln\left[\Gamma\left(CV_X^{-2} + r\right)\right] - \ln\left[\Gamma\left(CV_X^{-2}\right)\right]\right\}$
Exponential	$E[X^r] = \mu_X^r \Gamma(r+1)$
Normal	$E[X^r] = \mu_X^r\left[1 + \frac{r(r-1)}{2}CV_X^2 + \dots + \frac{r(r-1)(r-2)\dots(r-n+1)}{2^{n/2}(n/2)!}CV_X^n + \dots\right]$ $E[X^r] = \mu_X^r \sum_{n=0}^{r/2}\binom{r}{2n}CV_X^{2n}E[z^{2n}]$, when r is even $E[X^r] = \mu_X^r \sum_{n=0}^{(r-1)/2}\binom{r}{2n}CV_X^{2n}E[z^{2n}]$, when r is odd

14.2.10 Point Estimation Methods

Many times the PDF of a random variable is not available. Therefore the uncertainty of the variable is expressed in terms of its statistical moments. To that end, point estimation methods are frequently employed. These methods are computationally straightforward and can be employed for determining statistical moments of any order of a function involving several variables correlated or uncorrelated. A short discussion of these methods is given here. The point estimation (PE) method was originally proposed by Rosenblueth (1975) to deal with symmetric, correlated, stochastic input parameters. The method was later extended to the case involving asymmetric random variables (Rosenblueth 1981).

Table 14-7 *Generic expectation functions for some commonly used probability density functions.*

Name	Generic expectation function, $E\left[be^{cx}\right]$
Uniform	$$\frac{b^r}{2\sqrt{3}rc\mu_x CV_x}\left[e^{rc\mu_x\left(1+CV_x\sqrt{3}\right)}-e^{rc\mu_x\left(1-CV_x\sqrt{3}\right)}\right]$$
Symmetrical triangular	$$\frac{b^r}{6r^2c^2\mu_x^2 CV_x^2}\left[e^{\frac{1}{2}rc\mu_x\left(1+CV_x\sqrt{6}\right)}-e^{\frac{1}{2}rc\mu_x\left(1-CV_x\sqrt{6}\right)}\right]$$
Unsymmetrical triangular	$$\frac{2b^r\left[(\alpha-\beta)\exp(rc\omega)+(\beta-\omega)\exp(rc\alpha)+(\omega-\alpha)\exp(rc\beta)\right]}{r^2c^2(\beta-\alpha)(\omega-\alpha)(\beta-\omega)}$$ where α, β, and ω are the min, max, and mod of the random variable X.
Normal	$$b^r\exp\left(rc\mu_x+\frac{1}{2}r^2c^2\mu_X^2 CV_X^2\right)$$
Gamma	$$b^r\left(1-cr\mu_x CV_x^2\right)^{-\frac{1}{CV_x^2}}$$
Exponential	$$\frac{b^r}{\left(1-cr\mu_x\right)}$$

Point estimate methods are procedures where probability distributions for continuous random variables are modeled by discrete "equivalent" distributions having two or more values. The elements of these discrete distributions (or *point estimates*) have specific values with defined probabilities such that the first three moments of the discrete distribution match those of the continuous random variable. With only a few values over which to integrate, the moments of the performance function are easily obtained. First we summarize the PE method developed by Rosenblueth (1981), which is applicable to both symmetric and nonsymmetric and to correlated and uncorrelated random input variables.

14.2.10.1 Rosenblueth Method

Consider a variable y as a function of variables X_i, $i= 1, 2, 3, ..., n$; $Y = f(X_1, X_2, X_3, ..., X_n)$. The Rosenblueth method bases the probability distribution of y on the first three moments of independent variables X_i, $i= 1, 2, 3, ..., n$. The probability distribution of each independent variable is approximated by concentrating the entire probability mass at two points, X_{j-} and X_{j+}, each having a specific weight, P_- and P_+, on the distribution. Let us consider a continuous random variable X_i with its mean, standard deviation, and skew represented by $\mu(X_i)$, $\sigma(X_i)$, and $\gamma(X_i)$, respectively, as shown in Fig. 14-13. The random variable X_i is represented

Table 14-8 Results by using generic expectation function method.

t (hours)	Mean and variance of C using FOA			Error in FOA mean and variance using corrected FOA		First four moments of C using the generic expectation method				Statistical characteristics of C using the first four moments about the origin				
	FOA mean	FOA var	FOA CV	Error mean	Error var	1	2	3	4	Mean	Var	CV	Skew	Kurtosis
(1)	(2)	(3)	(4)	(5)	(6)	(7)	(8)	(9)	(10)	(11)	(12)	(13)	(14)	(15)
1	3.48	0.03	0.05	0.00	−0.03	3.48	12.14	42.44	148.65	3.48	0.03	0.05	−0.51	3.31
2	3.02	0.08	0.09	0.00	−0.05	3.04	9.29	28.65	88.98	3.04	0.07	0.09	−0.37	3.06
3	2.63	0.13	0.14	0.01	−0.06	2.65	7.16	19.65	54.73	2.65	0.12	0.13	−0.24	2.89
4	2.28	0.18	0.18	0.02	−0.07	2.32	5.56	13.68	34.49	2.32	0.17	0.18	−0.12	2.78
5	1.99	0.21	0.23	0.03	−0.07	2.04	4.35	9.66	22.23	2.04	0.20	0.22	0.00	2.73
6	1.73	0.23	0.28	0.04	−0.07	1.79	3.42	6.90	14.61	1.79	0.21	0.26	0.12	2.72
7	1.50	0.24	0.32	0.05	−0.06	1.58	2.71	4.99	9.78	1.58	0.22	0.30	0.23	2.76
8	1.31	0.23	0.37	0.06	−0.04	1.39	2.16	3.65	6.66	1.39	0.22	0.34	0.34	2.84
9	1.13	0.22	0.42	0.08	−0.02	1.23	1.73	2.70	4.61	1.23	0.22	0.38	0.45	2.95
10	0.99	0.21	0.46	0.09	0.00	1.09	1.39	2.01	3.23	1.09	0.21	0.42	0.56	3.10
11	0.86	0.19	0.51	0.11	0.03	0.96	1.12	1.52	2.30	0.96	0.20	0.46	0.66	3.28
12	0.75	0.17	0.55	0.13	0.06	0.86	0.91	1.15	1.65	0.86	0.18	0.50	0.77	3.49
13	0.65	0.15	0.60	0.15	0.10	0.76	0.75	0.88	1.20	0.76	0.17	0.54	0.87	3.74
14	0.56	0.13	0.65	0.17	0.13	0.68	0.61	0.68	0.88	0.68	0.15	0.58	0.98	4.03
15	0.49	0.12	0.69	0.19	0.17	0.60	0.50	0.53	0.66	0.60	0.14	0.62	1.08	4.35
16	0.43	0.10	0.74	0.21	0.21	0.54	0.42	0.41	0.49	0.54	0.13	0.66	1.19	4.71
17	0.37	0.08	0.79	0.23	0.25	0.48	0.35	0.33	0.37	0.48	0.11	0.70	1.29	5.11
18	0.32	0.07	0.83	0.25	0.30	0.43	0.29	0.26	0.28	0.43	0.10	0.74	1.40	5.55
19	0.28	0.06	0.88	0.28	0.34	0.39	0.24	0.21	0.22	0.39	0.09	0.78	1.51	6.03
20	0.24	0.05	0.92	0.30	0.38	0.35	0.20	0.16	0.17	0.35	0.08	0.82	1.62	6.57
21	0.21	0.04	0.97	0.32	0.42	0.31	0.17	0.13	0.13	0.31	0.07	0.86	1.73	7.15
22	0.18	0.03	1.02	0.35	0.46	0.28	0.14	0.11	0.10	0.28	0.06	0.91	1.85	7.79
23	0.16	0.03	1.06	0.37	0.50	0.25	0.12	0.09	0.08	0.25	0.06	0.95	1.97	8.49
24	0.14	0.02	1.11	0.39	0.54	0.23	0.10	0.07	0.06	0.23	0.05	0.99	2.09	9.24

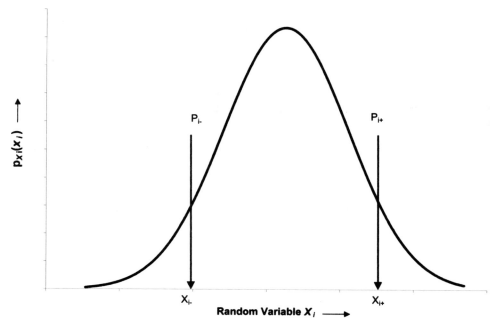

Figure 14-13 *Point estimate method.*

by two point estimates, X_{i-} and X_{i+}, with probability concentrations P_{i-} and P_{i+}, respectively. Because the two point estimates and their probability concentrations form an equivalent probability distribution for the random variable, the two P values must sum to unity. The two point estimates and probability concentrations are chosen to match three moments of the random variable. These two probability masses (P_{i-} and P_{i+}) are located x'_{i-} and x'_{i+} standard deviations above and below the mean:

$$X_{i+} = \mu(X_i) + x'_{i+}\sigma(X_i) \tag{14.85}$$

$$X_{i-} = \mu(X_i) + x'_{i-}\sigma(X_i) \tag{14.86}$$

$$x'_{i+} = \frac{\gamma(X_i)}{2} + \sqrt{1 + \left(\frac{\gamma(X_i)}{2}\right)^2} \tag{14.87}$$

$$x_{i-} = x'_{i+} - \gamma(X_i) \tag{14.88}$$

These equations result in $2N$ point locations for N random variables. For an input variable that is symmetric about the mean the two probability masses are located one standard deviation above and below the mean. The probability masses for each of the two points are calculated as

$$P_{i+} = \frac{x'_{i-}}{x'_{i+} + x'_{i-}} \tag{14.89}$$

$$P_{i-} = 1 - P_{i+} \tag{14.90}$$

Now, once the two probability masses and their locations for each random variable are determined, there will be 2^N points on the performance function at which the values of the performance function need to be determined. For example, for a bivariate performance function $Y(X_1, X_2)$, four point evaluations need to be done as follows:

$$\begin{aligned}
Y_{++} &= Y\left[\mu(X_1) + x'_{1+}\sigma(X_1), \mu(X_2) + x'_{2+}\sigma(X_2)\right] \\
Y_{-+} &= Y\left[\mu(X_1) - x'_{1-}\sigma(X_1), \mu(X_2) + x'_{2+}\sigma(X_2)\right] \\
Y_{+-} &= Y\left[\mu(X_1) + x'_{1+}\sigma(X_1), \mu(X_2) - x'_{2-}\sigma(X_2)\right] \\
Y_{--} &= Y\left[\mu(X_1) - x'_{1-}\sigma(X_1), \mu(X_2) - x'_{2-}\sigma(X_2)\right]
\end{aligned} \tag{14.91}$$

Now corresponding to each value of the performance functions, its probability mass represented by $f_{(\delta_1, \delta_2, \ldots, \delta_n)}$, in which δ_1 is a sign indicator that can only be + or −, will be calculated using the following relationship:

$$f_{(\delta_1, \delta_2, \ldots, \delta_n)} = \prod_{i=1}^{N} P_{i,\delta_i} + \sum_{i=1}^{N-1}\left(\sum_{j=i+1}^{N} \delta_i \delta_j a_{ij}\right) \tag{14.92}$$

where a_{ij} is defined as

$$a_{ij} = \frac{\left(\dfrac{\rho_{ij}}{2^N}\right)}{\sqrt{\displaystyle\prod_{i=1}^{N}\left[1 + \left(\dfrac{\gamma_i}{2}\right)^2\right]}} \tag{14.93}$$

After evaluating all the point locations of the performance function and their corresponding probability masses, one can approximate the rth moment of the performance function about the origin as

$$E\left[Y^r\right] = \sum P_{(\delta_1, \delta_2, \ldots, \delta_n)} Y^r_{(\delta_1, \delta_2, \ldots, \delta_n)} \tag{14.94}$$

From this information, the mean and the variance of the performance function are determined as

$$\mu_Y = E[Y] \tag{14.95}$$

$$\mathrm{var}(Y) = E\left[Y^2\right] - (\mu_Y)^2 \tag{14.96}$$

Now, one can determine the reliability index and the corresponding reliability and probability of failure.

As is seen, this method requires 2^N model evaluations to estimate a single statistical moment of the model output. For a complex model with a large number of parameters, Rosenblueth's PE method is computationally intensive and may sometimes be impractical. Further, a reliability analysis requires knowledge of higher-order moments to approximate the distribution of the output random variable. This makes the method even more computationally extensive. Thus, although Rosenblueth's method is quite efficient for problems with a small number of uncertain basic variables, its computational requirements are similar to those of MCS for a model having a large number of parameters. For example, a model having between 10 and 15 parameters will require 1,024 to 32,768 model evaluations.

Harr (1989) modified Rosenblueth's method to reduce its computational requirements from 2^N to $2N$ for an N input parameter model by using the first two moments of the random variables. This method does not provide the flexibility to incorporate known higher-order moments of input random variables. Chang et al. (1995) showed that the estimated uncertainty feature of model output could be inaccurate if the skewness of a random variable is not accounted for.

Example 14.22 Solve Example 14.13 using the PE method.

Solution The performance function is given as $Y = C_{max} - C = 8 - (2.10 \times 10^{-8})Q^3$. In case 1, Q is normally distributed and thus the skew is zero. In case 2, Q is defined by the gamma distribution giving a skew of $2CV$, and when Q is defined by the log-normal distribution the skew is $3CV + CV^3$.

The calculations are summarized in Table E14-22.

Example 14.23 Consider the following performance function:

$$Y = X_2 - \left(0.04X_1^2 - 1.41X_1 + 25.5\right)$$

Assume that the means of X_1 and X_2 are 9 and 20 and that their standard deviations are 3 and 2, respectively. Determine the reliability index and failure probability using the PE method for the following cases:

(a) Both X_1 and X_2 are independent and normally distributed.

(b) Both X_1 and X_2 are dependent and normally distributed with covariance of 4.2.

(c) X_1 is characterized by a gamma distribution and X_2 is log-normally distributed with covariance of 4.2.

Solution

(a) See Table 14-23a.

(b) See Table 14.23b.

(c) See Table 14.23c.

Table E14-22

Item and formula	Case 1	Case 2	Case 3
	X = normal	X = gamma	X = log-normal
Known information			
$\mu(Q)$	800	800	800
$CV(Q)$	0.33	0.33	0.33
$\sigma(Q)$	264	264	264
$\gamma(Q)$	0	0.66	1.03
Calculated PE method parameters			
$x'_+ = \dfrac{\gamma(Q)}{2} + \sqrt{1 + \left(\dfrac{\gamma(Q)}{2}\right)^2}$	1	1.4	1.7
$x'_- = x'_+ - \gamma(Q)$	1	0.8	0.7
$Q_+ = \mu(Q) + x'_+\sigma(Q)$	1064	1178.5	1261.6
$Q_- = \mu(Q) - x'_-\sigma(Q)$	536	595.8	609.3
$P_+ = \dfrac{x'_-}{x'_+ + x'_-}$	0.50	0.35	0.29
$P_- = 1 - P_+$	0.50	0.65	0.71
Determination of moments of objective function and reliability index			
$Y_+ = 10 - 2.1 \times 10^{-8} Q_+^3$	−15.30	−24.37	−32.17
$Y_- = 10 - 2.1 \times 10^{-8} Q_-^3$	6.77	5.56	5.25
$E[Y] = P_{i+}Y_+ + P_{i-}Y_-$	−4.26	−4.93	−5.69
$E[Y^2] = P_{i+}Y_+^2 + P_{i-}Y_-^2$	139.87	228.19	322.02
$\text{var}[Y] = E[Y^2] - \{E[Y]\}^2$	121.68	203.89	289.65
$\sigma(Y)$	11.03	14.28	17.02
$\beta = \mu(Y)/\sigma(Y)$	−0.39	−0.35	−0.33
$P_f = 1 - \Phi(\beta)$	0.65	0.64	0.63

Table E14-23a

Item	X_1 = normal	X_2 = normal	Item	Value
$\mu(X_i)$	9.00	20.00	$\text{cov}(X_1, X_2)$	0
$CV(X_i)$	0.33	0.10	$\rho(X_1, X_2)$	0.00
$\sigma(X_i)$	3.00	2.00	a_{12}	0.00
$\gamma(X_i)$	0.00	0.00	$P_{++} = P_{1+}P_{2+} + a_{12}$	0.25
$x'_{i+} = \dfrac{\gamma(X_i)}{2} + \sqrt{1 + \left(\dfrac{\gamma(X_i)}{2}\right)^2}$	1.00	1.00	$P_{--} = P_{1-}P_{2-} + a_{12}$	0.25
$x'_{i-} = x'_{i+} - \gamma(X_i)$	1.00	1.00	$P_{+-} = P_{1+}P_{2-} - a_{12}$	0.25
$X_{i+} = \mu(X_i) + x'_{i+}\sigma(X_i)$	12.00	22.00	$P_{-+} = P_{1-}P_{2+} - a_{12}$	0.25
$X_{i-} = \mu(X_i) - x'_{i-}\sigma(X_i)$	6.00	18.00	$Y_{++} = Y\left[\mu(X_1) + x'_{1+}\sigma(X_1), \mu(X_2) + x'_{2+}\sigma(X_2)\right]$	7.66
$P_{i+} = \dfrac{x'_{i-}}{x'_{i+} + x'_{i-}}$	0.50	0.50	$Y_{--} = Y\left[\mu(X_1) - x'_{1-}\sigma(X_1), \mu(X_2) - x'_{2-}\sigma(X_2)\right]$	−0.48
$P_{i-} = 1 - P_{i+}$	0.50	0.50	$Y_{+-} = Y\left[\mu(X_1) + x'_{1+}\sigma(X_1), \mu(X_2) - x'_{2-}\sigma(X_2)\right]$	3.66
			$Y_{-+} = Y\left[\mu(X_1) - x'_{1-}\sigma(X_1), \mu(X_2) + x'_{2+}\sigma(X_2)\right]$	3.52
$E[Y] = P_{++}Y_{++} + P_{--}Y_{--} + P_{+-}Y_{+-} + P_{-+}Y_{-+}$				3.59
$E[Y^2] = P_{++}Y^2_{++} + P_{--}Y^2_{--} + P_{+-}Y^2_{+-} + P_{-+}Y^2_{-+}$				21.17
$\text{var}[Y] = E[Y^2] - \{E[Y]\}^2$				8.28
$\sigma(Y)$				2.87
$\beta = \mu(Y)/\sigma(Y)$				1.24
$P_f = 1 - \Phi(\beta)$				0.11

Table E14-23b

Item	X_1 = normal	X_2 = normal	Item	Value
$\mu(X_i)$	9.00	20.00	$\text{cov}(X_1,X_2)$	4.2
$CV(X_i)$	0.33	0.10	$\rho(X_1,X_2)$	0.70
$\sigma(X_i)$	3.00	2.00	a_{12}	0.18
$\gamma(X_i)$	0.00	0.00	$P_{++} = P_{1+}P_{2+} + a_{12}$	0.43
$x'_{i+} = \dfrac{\gamma(X_i)}{2} + \sqrt{1 + \left(\dfrac{\gamma(X_i)}{2}\right)^2}$	1.00	1.00	$P_{--} = P_{1-}P_{2-} + a_{12}$	0.43
$x'_{i-} = x'_{i+} - \gamma(X_i)$	1.00	1.00	$P_{+-} = P_{1+}P_{2-} - a_{12}$	0.08
$X_{i+} = \mu(X_i) + x'_{i+}\sigma(X_i)$	12.00	22.00	$P_{-+} = P_{1-}P_{2+} - a_{12}$	0.08
$X_{i-} = \mu(X_i) - x'_{i-}\sigma(X_i)$	6.00	18.00	$Y_{++} = Y\left[\mu(X_1)+\sigma(X_1),\mu(X_2)+\sigma(X_2)\right]$	7.66
$P_{i+} = \dfrac{x'_{i-}}{x'_{i+} + x'_{i-}}$	0.50	0.50	$Y_{--} = Y\left[\mu(X_1)-\sigma(X_1),\mu(X_2)-\sigma(X_2)\right]$	−0.48
$P_{i-} = 1 - P_{i+}$	0.50	0.50	$Y_{+-} = Y\left[\mu(X_1)+\sigma(X_1),\mu(X_2)-\sigma(X_2)\right]$	3.66
			$Y_{-+} = Y\left[\mu(X_1)-\sigma(X_1),\mu(X_2)+\sigma(X_2)\right]$	3.52
$E[Y] = P_{++}Y_{++} + P_{--}Y_{--} + P_{+-}Y_{+-} + P_{-+}Y_{-+}$				3.59
$E[Y^2] = P_{++}Y_{++}^2 + P_{--}Y_{--}^2 + P_{+-}Y_{+-}^2 + P_{-+}Y_{-+}^2$				26.97
$\text{var}[Y] = E[Y^2] - \{E[Y]\}^2$				14.08
$\sigma(Y)$				3.75
$\beta = \mu(Y)/\sigma(Y)$				0.96
$P_f = 1 - \Phi(\beta)$				0.17

Table E14-23c

Item	$X_1 =$ gamma	$X_2 =$ log-normal	Item	Value
$\mu(X_i)$	9.00	20.00	$\text{cov}(X_1,X_2)$	4.2
$CV(X_i)$	0.33	0.10	$\rho(X_1,X_2)$	0.70
$\sigma(X_i)$	3.00	2.00	a_{12}	0.18
$\gamma(X_i)$	0.67	0.30	$P_{++} = P_{1+}P_{2+} + a_{12}$	0.32
$x'_{i+} = \dfrac{\gamma(X_i)}{2} + \sqrt{1+\left(\dfrac{\gamma(X_i)}{2}\right)^2}$	1.39	1.16	$P_{--} = P_{1-}P_{2-} + a_{12}$	0.55
$x_{i-} = x'_{i+} - \gamma(X_i)$	0.72	0.86	$P_{+-} = P_{1+}P_{2+} - a_{12}$	0.02
$X_{i+} = \mu(X_i) + x'_{i+}\sigma(X_i)$	13.16	22.32	$P_{-+} = P_{1-}P_{2+} - a_{12}$	0.11
$X_{i-} = \mu(X_i) - x'_{i-}\sigma(X_i)$	6.84	18.28	$Y_{++} = Y[\mu(X_1)+\sigma(X_1),\mu(X_2)+\sigma(X_2)]$	8.4
$P_{i+} = \dfrac{x'_{i-}}{x'_{i+} + x'_{i-}}$	0.34	0.43	$Y_{--} = Y[\mu(X_1)-\sigma(X_1),\mu(X_2)-\sigma(X_2)]$	0.55
$P_{i-} = 1 - P_{i+}$	0.66	0.57	$Y_{+-} = Y[\mu(X_1)+\sigma(X_1),\mu(X_2)-\sigma(X_2)]$	4.4
			$Y_{-+} = Y[\mu(X_1)-\sigma(X_1),\mu(X_2)+\sigma(X_2)]$	4.6
$E[Y] = P_{++}Y_{++} + P_{--}Y_{--} + P_{+-}Y_{+-} + P_{-+}Y_{-+}$				3.59
$E[Y^2] = P_{++}Y_{++}^2 + P_{--}Y_{--}^2 + P_{+-}Y_{+-}^2 + P_{-+}Y_{-+}^2$				25.69
$\text{var}[Y] = E[Y^2] - \{E[Y]\}^2$				12.81
$\sigma(Y)$				3.57
$\beta = \mu(Y)/\sigma(Y)$				1.00
$P_f = 1 - \Phi(\beta)$				0.15

Table E14-24b

Item	X_1	X_2	X_3
$\mu(X_i)$	40	50	1000
$CV(X_i)$	0.125	0.05	0.20
$\sigma(X_i)$	5	2.5	200
$\gamma(X_i)$	0.2	0.7	-0.66
$x'_{i+} = \dfrac{\gamma(X_i)}{2} + \sqrt{1 + \left(\dfrac{\gamma(X_i)}{2}\right)^2}$	1.10	1.41	0.72
$x'_{i-} = x'_{i-} - \gamma(X_i)$	0.90	0.71	1.38
$x_{i+} = \mu(X_i) + x'_{i+}\sigma(X_i)$	45.52	53.52	1200.00
$x_{i-} = \mu(X_i) - x'_{i-}\sigma(X_i)$	35.48	48.23	800.00
$P_{i+} = \dfrac{x'_{i-}}{x'_{i+} + x'_{i-}}$	0.45	0.33	0.66
$P_{i-} = 1 - P_{i+}$	0.55	0.67	0.34

Table E14-24c

Item	$Y = f(x_1, x_2, x_3)$
Y_{+++}	1236.66
Y_{++-}	1636.66
Y_{+-+}	995.50
Y_{+--}	1395.50
Y_{-++}	698.76
Y_{-+-}	1098.76
Y_{--+}	510.83
Y_{---}	910.83

14.2.10.2 Harr's Method

In Harr's method (Harr 1989) one assumes that the entire probability mass distribution of an independent variable x_i is distributed between two points, x_{i-} and x_{i+}. The mth moment of the probability distribution of Y is calculated as

$$E[Y^m] = \frac{\sum_{i=1}^{n} \lambda_i y_i^m}{\sum_{i=1}^{n} \lambda_i} \tag{14.97}$$

Table E14-24d

Performance function		Probability		$Y_{ijk}P_{ijk}$	$Y_{ijk}^2P_{ijk}$
Point notation	Value	Notation	Value		
Y_{+++}	1236.66	P_{+++}	0.199	246.5	304848.1
Y_{++-}	1636.66	P_{++-}	0.041	66.5	108773.4
Y_{+-+}	995.50	P_{+-+}	0.208	206.9	205956.1
Y_{+--}	1395.50	P_{+--}	0.002	3.5	4838.6
Y_{-++}	698.76	P_{-++}	0.021	14.4	10029.9
Y_{-+-}	1098.76	P_{-+-}	0.074	81.7	89750.4
Y_{--+}	510.83	P_{--+}	0.229	117.0	59754.7
Y_{---}	910.83	P_{---}	0.226	205.7	187390.9
		Sum	1.00	942.07	971341.97

Table E14-24e

Item	Value
$E[Y]=P_{+++}Y_{+++}+P_{++-}Y_{++-}+P_{+-+}Y_{+-+}+P_{+--}Y_{+--}+P_{-++}Y_{-++}+P_{-+-}Y_{-+-}$ $+P_{--+}Y_{--+}+P_{---}Y_{---}$	942.07
$E[Y^2]=P_{+++}Y_{+++}^2+P_{++-}Y_{++-}^2+P_{+-+}Y_{+-+}^2+P_{+--}Y_{+--}^2+P_{-++}Y_{-++}^2+P_{-+-}Y_{-+-}^2$ $+P_{--+}Y_{--+}^2+P_{---}Y_{---}^2$	971341.97
$var[Y]=E[Y^2]-\{E[Y]\}^2$	83841.09
$\sigma(Y)$	289.55
$\beta=\mu(Y)/\sigma(Y)$	3.25
$P_f=1-\Phi(\beta)$	5.70×10^{-4}

where y_i is the mean of y_{i+} and y_{i-}; $y_i=(y_{i+}+y_{i-})/2=[f(x_{i+})+f(x_{i-})]/2$; and λ_i are the eigenvalues obtained as the correlation matrix ρ of variables decomposed using the orthogonal transformation method into an eigenvector matrix $(w_1, w_2, w_3,..., w_n)$, W, its transpose W^T, and a diagonal matrix $\overline{\Delta}$ containing the eigenvalues $\lambda_1, \lambda_2,..., \lambda_n$:

$$\rho = W\lambda W^T \qquad (14.98)$$

where superscript T denotes the transpose of the matrix. The uncorrelated standardized coordinates of the vectors of the n random variables x_+ and x_- are generated as

$$x_{i-} = \bar{\mu} - \sqrt{n}\sqrt{\bar{D}}\,w_i \text{ and } x_{i+} = \bar{\mu} + \sqrt{n}\sqrt{\bar{D}}\,w_i \tag{14.99}$$

where $\bar{\mu}$ is the vector of the expected values of the random variables, $X_1, X_2, \ldots, X_n, \bar{D}$ is the diagonal matrix of the variance of the random variables, and w_i is the eigenvector associated with the eigenvalue λ_i.

Chang et al. (1995) modified Harr's method by evaluating y as

$$E[Y^m] = \frac{\sum\limits_{i=1}^{n} y_i^m}{n} \tag{14.100}$$

in which y_i is calculated as before.

The weighting factor for each independent variable x_i is considered for the modified uncorrelated standardized coordinates in the eigenspace as

$$x_{i-} = \bar{\mu} - \sqrt{\bar{D}}\,\overline{W}\sqrt{\bar{\Delta}}(\sqrt{n}e_i) \text{ and } x_{i+} = \bar{\mu} + \sqrt{\bar{D}}\,\overline{W}\sqrt{\bar{\Delta}}(\sqrt{n}e_i) \tag{14.101}$$

where $\overline{W}$ is the eigenvector matrix, $\bar{\Delta}$ is the diagonal matrix of the eigenvalues, and $\bar{e}_i$ is a unit vector with ith element equal to 1 and 0 everywhere else.

The Harr method is computationally more efficient than the Rosenblueth method because it reduces the computational runs from 2^N to $2N$ and uses only the first and second moments of each stochastic variable.

14.2.10.3 Li's Method

In Li's method (Li 1992) one assumes that the entire probability mass of a random variable is concentrated at three points, x_{i-}, x_+, and μ, having, respectively, the probability values of p_-, p_+, and p_0. The probability distribution of y is obtained from the first four moments of independent variables. The mth moment of the probability distribution of y is calculated as

$$E[Y^m] = (1 - \frac{3n}{2} + \frac{\eta}{2} + \sum_{i=1}^{n} p_{i0}) + \sum_{i=1}^{n}[(p_{i+} - \eta_i + 1)y_{i+}^m + p_{i-}y_{i-1}^m)] + \sum_{j}^{n}\sum_{i<j}^{n} y_{ij}^m \eta_{ij} \tag{14.102}$$

where η is the sum of all the η_i, η_i is the sum of all η_{ij} with respect to i, and $\eta_{ij} = \rho_{ij}/(x'_{i+}x'_{j+})$. Note that $\eta_{ii} = 1$. The points x_{i-}, x_{i+}, and μ are computed as

$$x'_{i+1} = \mu_i + x'_{j+}\sigma,$$
$$x_{i-} = \mu_i + x'_{i-}\sigma,$$
$$x'_{i+} = \frac{\gamma_1 + \sqrt{4k_i - 3\gamma_i^2}}{2},$$
$$x'_{i-} = \frac{\gamma_i - \sqrt{4k_i - 3\gamma_i^2}}{2},$$
$$x_{i0} = \mu$$

where k is the coefficient of kurtosis. The weight of each point is given as

$$p_{i+} = \frac{1}{x'_{i+}(x'_{i+} - x'_{i-})}, \; p_{i-} = \frac{1}{x'_{i-}(x'_{i-} - x'_{i+})}, p_{i0} = 1 - p_{i+} - p_- \qquad (14.103)$$

This method is efficient and accurate. Thus the reliability of the system is $R = 1 - P_f = 0.999$ (i.e., 99.9%).

14.2.10.4 Modified Rosenblueth's Method

Tsai and Franceschini (2003) modified the Rosenblueth method for cases involving more than three stochastic variables. The mth moment of the probability distribution of y is calculated as

$$E[Y^m] = [(1-n) + \frac{\eta - N}{2}]\bar{y}^m + \sum_{i=1}^{N}[(p_{i+} - \eta_i + 1)y_{i+}^n] + \sum_{i=1}^{N-1}\sum_{j=i+1}^{N} y_{ij}^m \eta_{ij} \qquad (14.104)$$

This modification preserves the capabilities of the original Rosenblueth method and is an improvement at the same time.

The discrepancy between observed and computed y can be expressed as

$$E(Y^m) - \bar{Y}^m = \frac{\eta - 3N}{2}\bar{y}^m + \sum_{i=1}^{N}[(p_{i+} - \eta_i + 1)y_{i+}^m + p_{i-}y_{i-}^m] + \sum_{i=1}^{N-1}\sum_{j=i+1}^{N} y_{ij}^m \eta_{ij} \quad (14.105)$$

Thus the reliability of the system is $R = 1 - P_f = 0.999$ (i.e., 99.9%).

14.2.10.5 Characteristics of Point Estimation Methods

The various point estimation methods can be compared based on the moments to be used, intensity of computation, and the capability to deal with variables. These are summarized in Table 14-9.

Table 14-9

Characteristics	Rosenblueth's method	Harr's method	Modified Harr's method	Li's method	Modified Rosenblueth's method
Moments needed	3	2	2	4	3
Intensity of computation	2^N	$2N$	$2N$	$(N^2+3N+2)/2$	$(N^2+3N+2)/2$
Capability to consider correlated variables	yes	yes	yes	yes	yes
Capability to consider asymmetric variables	yes	yes	yes	yes	yes

14.2.11 Transform Methods

Tung (1990) used the Mellin transform to calculate higher-order moments of a model output. Application of the Mellin transform, however, is cumbersome, and it cannot be universally applied. As pointed out by Tung, the Mellin transform may not be analytic under certain combinations of distribution and functional forms. In particular, problems may arise when a functional relationship consists of input variable(s) with negative exponent(s). When component functions of a given model have forms other than power functions, it cannot be applied. Further, no formulation was suggested to obtain the moments of a model output having nonstandard normally distributed input variable(s).

14.3 Reliability Analysis of Dynamic Systems

So far we have discussed various methods dealing with estimation of static systems (i.e., case in which reliability of a system is time independent). But reliability is not always independent of time; rather it is highly time dependent and hence time (e.g., the length or amount of use) can be used as a surrogate to determine the reliability of a system. For example, reliability of many civil engineering systems, such as a city's water distribution system, sanitary sewer system, or combined sewer system, reduces with time owing to wear and tear and other reasons.

14.3.1 Time to Failure

The time to failure is used as an indicator of reliability of a system or its components in early studies. In this concept, reliability is defined as the probability that a system would perform adequately for at least a specified period of time and under specified operating conditions. Failure designates the inability of a system to perform its intended function. Of course, all systems fail eventually. (As a sage once said: Time is a great teacher, but it kills all its pupils.) However, from an engineering viewpoint, it is the survival time before failure that determines whether the system was successful or not. Thus, the performance period can be designated as the time to failure. The performance period t is a random variable that has a probability density function $f(t)$ and a cumulative distribution function $F(t)$. Therefore, the reliability function $R(t)$ can be defined as

$$R(t) = \int_t^\infty f(t)dt = 1 - F(t) \tag{14.106}$$

The expected (or mean) time to failure (MTTF) of a system, or its expected life, is the expected value of time during which the system will be reliable (or operate successfully), that is,

$$\text{MTTF} = E(t) = \mu_t = \int_0^\infty tf(t)dt \tag{14.107}$$

This is the expected life, which can also be computed as

$$E[t] = \int_0^\infty R(t)dt \tag{14.108}$$

Equation 14.108 can be derived as follows. Differentiating Eq. 14.107 with respect to time, one obtains

$$\frac{dR(t)}{dt} = R'(t) = -f(t) \tag{14.109}$$

Substituting Eq. 14.109 into Eq. 14.107, we have

$$E(t) = -\int_0^\infty tR'(t)dt \tag{14.110}$$

Integrating by parts gives

$$E(t) = -tR(t)\Big|_0^\infty + \int_0^\infty R(t)dt \tag{14.111}$$

Since all systems must fail eventually, $R(t)$ approaches zero faster than t approaches infinity. Hence,

$$E(t) = \int_0^\infty R(t)dt$$

which is the same as Eq. 14.108.

The variance of the time to failure (or life), σ_t^2, can be computed as

$$\sigma_t^2 = \text{var}[t] = \int_0^\infty (t - \mu_t)^2 f(t) dt \tag{14.112}$$

The variance of the time to failure can also be expressed as

$$\sigma_t^2 = \text{var}(t) = \int_0^\infty 2tR(t)dt - \left[\int_0^\infty R(t)dt\right]^2 \tag{14.113}$$

This can be shown as follows. Substituting Eq. 14.109 into Eq. 14.112, we have

$$\sigma_t^2 = -\int_0^\infty (t - \mu_t)^2 R'(t)dt \tag{14.114}$$

$$= -\int_0^\infty (t^2 - 2\mu_t^t t + \mu_t^2)R'(t)dt \tag{14.115}$$

Integrating by parts gives

$$\sigma_t^2 = -\int_0^\infty t^2 R'(t)dt + 2\mu_t \int_0^\infty tR'(t)dt - \mu_t^2 \int_0^\infty R'(t)dt$$

$$= -\int_0^\infty t^2 dR(t) + 2\mu_t \int_0^\infty -tf(t)d(t) + \mu_t^2 = -t^2 R(t)\big|_0^\infty + \int_0^\infty 2tR(t)dt - 2\mu_t^2 + \mu_t^2$$

$$= \int_0^\infty 2tR(t)dt - \mu_t^2 = \int_0^\infty 2tR(t)dt - \left[\int_0^\infty R(t)dt\right]^2$$

which is the same as Eq. 14.113.

In general,

$$E[t^k] = \int_0^\infty t^k f(t)dt = \int_0^\infty kt^{k-1}R(t)dt \tag{14.116}$$

Example 14.25 A pump that is used to withdraw groundwater in an agricultural area is found to fail about twice per year. Assuming that the mean time to failure follows an exponential distribution, find the reliability of the pump for a period of 200 days and the mean time to failure.

Solution Since the pump fails about two times per year, for an exponential distribution parameter λ is the rate of failure per day:

$$\lambda = 2/365 = 0.0055 \text{ per day}$$

From Eq. 14.94, the reliability of the pump $R(t)$ is

$$R(t) = \int_t^\infty \lambda \exp(-\lambda t)dt = \exp(-\lambda t) = \exp(-0.0055 \times 200) = 0.33$$

Also,

$$\text{MTTF} = \int_0^\infty t 0.0055 \exp(-0.0055t)dt = 1/0.0055 = 182 \text{ days}$$

14.3.2 Hazard Function

Another aspect of reliability is the assessment of failure time. In general, the failure rate of a system can be time variant as the system ages. Given that a system has survived up to a time t_1, the probability that it will fail in the next time interval Δt is the conditional probability:

$$P\left[\, t_1 < t \le t_1 + \Delta t \,|\, t > t_1 \,\right] = \frac{P\left[t_1 \le t \le t_1 + \Delta t\right]}{P\left[t > t_1\right]} \tag{14.117}$$

Dividing both sides by Δt and taking the limiting case as $\Delta t \to 0$, we obtain

$$h(t) = \lim_{\Delta t \to 0} \frac{R(t) - R(t + \Delta t)}{\Delta t R(t)} = \frac{-R'(t)}{R(t)} = \frac{f(t)}{R(t)} \tag{14.118}$$

which defines the hazard function $h(t)$. Thus, the hazard function is seen to be the rate of change of the conditional probability of failure, given that the system has survived to time t. Conceptually, the hazard function is the failure rate: If $h(t)$ increases in time, the rate of failure increases.

Integrating both sides of Eq. 14.118 and assuming that the system is perfect initially, $R(0) = 1$, we obtain

$$R(t) = \exp\left[-\int_0^t h(t)dt\right] \tag{14.119}$$

and directly from Eq. 14.118,

$$f(t) = h(t)\exp\left[-\int_0^t h(t)dt\right] \tag{14.120}$$

To obtain Eq. 14.119, Eq. 14.120 is written as

$$h(t) = \frac{dF/dt}{1 - F(t)} \tag{14.121}$$

Therefore,

$$\frac{dF}{1 - F(t)} = h(t)dt$$

On integration we get

$$\ln[1 - F(t)] = -\int_0^t h(t)dt$$

$$R(t) = \exp\left[-\int_0^t h(t)dt\right]$$

which is the same as Eq. 14.119.

14.3.3 Bathtub Distribution for Hazard Function

All products, systems, assemblies, components, and parts exhibit different failure rates over their service lives. Although the shape of the curve varies, most exhibit a low failure rate during most of their useful lives and higher failure rates at the beginning and end of their useful lives. For many systems, the hazard function can be modeled as a bathtub distribution, as shown in Fig. 14-14. In this distribution three regions can be identified:

1. Warranty failure
2. Chance failure
3. Aging failure

The first region represents the break-in or debugging period and is commonly covered by the manufacturer's warranty (for example, in commercial products such as a pump, TV, car, etc.). In this initial period of operation, there is a high failure rate (or hazard); these failures can be due to such factors as manufacturing errors, imperfections, nonadherence to standards, poor quality control, and human factors. In this region, both the hazard function and the failure rate decrease with time. Strict quality control and inspection during construction and manufacture can reduce the high initial failure rate.

The second region represents the period during which the hazard function represents the likelihood of failure during the service life of the system. This period represents the occurrence of chance or random failures. During this period, the rate of failure is small and fairly constant. The third period represents the region in which the hazard level begins to increase again as the result of aging. This is the time when the economic life of the system or component comes to an end and a decision needs to be made concerning replacement, life extension, or repair.

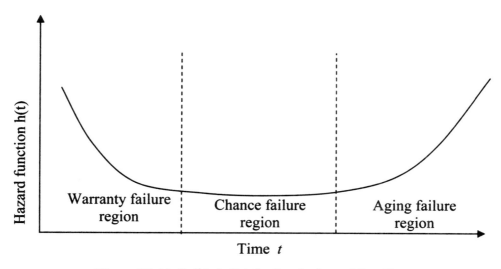

Figure 14-14 *Bathtub distribution for hazard function.*

It is desirable to select a probability distribution for the hazard function on the basis of shape characteristics that appear to model the expected performance. For the second region or the middle portion of Fig. 14-14, the failure rate is fairly constant,

$$h(t) = \lambda = \text{constant} \tag{14.122}$$

Substituting into Eq. 14.120, we get

$$f(t) = \lambda e^{-\lambda t} \tag{14.123}$$

which is the exponential probability distribution. The reliability function $R(t)$ is

$$R(t) = \int_t^\infty \lambda e^{-\lambda t} dt = e^{-\lambda t} \tag{14.124}$$

The mean time to failure is

$$\text{MTTF} = E[t] = \int_0^\infty e^{-\lambda t} dt = \frac{1}{\lambda} \tag{14.125}$$

which is the expected value of the exponential distribution and the reciprocal of the hazard rate. The implication is that the greater the hazard or failure rate, the shorter will be the expected time to failure.

Example 14.26 A survey of highway pavement in a part of Louisiana identified 20 sections with fairly similar constituent thicknesses, properties, and traffic loadings. Indications were that expensive rehabilitation was required, on average, 10 years after the sections were constructed and accepted. What is the reliability of the 20 sections, as a whole and 10 years and 20 years after acceptance?

Solution If expensive rehabilitation denotes failure, MTTF = 10 years, and the failure rate λ is 0.10 years. Therefore,

$$R \text{ (after 10 years)} = e^{-\lambda t} = e^{-(0.10)10} = 0.37 = 37\%$$

$$R \text{ (after 20 years)} = e^{-(0.10)20} = 0.135 = 14\%$$

14.3.4 Reliable Life

Another measure of reliability is the reliable life, t_R. This corresponds to the time required for the reliability to decrease to a specified level. For a constant failure rate, $h(t) = \lambda$, we can write from Eq. 14.124

$$R = e^{-\lambda t_R}$$

$$\ln R = -\lambda t_R \text{ or } t_R = -\frac{1}{\lambda} \ln R \qquad (14.126)$$

For a constant failure rate of 0.10 per year ($\lambda = 0.10$), the reliable life for a reliability of 10% is

$$t_{R=10\%} = -10 \times \ln(0.10) = 23 \text{ years}$$

To determine the time until only one of the 20 pavements in the previous example is reliable, $R = 1/20 = 0.05$, we calculate

$$t_{R=5\%} = -10 \times \ln(0.05) = 30 \text{ years}$$

The expected time for half of the pavements to show massive distress in this example, $R = 0.5$, would be

$$t_{R=50\%} = -10 \times \ln(0.50) = 7 \text{ years}$$

Now recall the Poisson distribution:

$$f(x) = \frac{(\lambda t)^x e^{-\lambda t}}{x!} \qquad (14.127)$$

where λ = mean occurrence rate, x = random variable, and t = time interval. The Poisson distribution models the probability of occurrence of events during a time interval t, where the mean occurrence rate is λ. Then $\mu = \lambda t$.

Suppose that the probability of occurrence of a storm with a rainfall amount that is capable of causing severe damage to a system follows a Poisson distribution. Here λ is the rate at which the killer storm is expected to occur. Within the

Example 14.28 Suppose that a system is only 75% reliable (or adequate) one year after it has begun to wear out. Assuming the failure rate to increase linearly with time, estimate the reliability of the system for the next 2 years.

Solution Equation 14.130 can be graphed as shown in Fig. 14-16. For $t = 1$ and $R = 75\%$, from Fig. 14-16, $k = 0.58$.

The time required to reach other estimates of reliability are then easily scaled. In a sense the remaining reliability can be thought of as the percent worth of the system. For example, from Fig. 14-16, after 3 years of deterioration without maintenance, one can estimate that the system would be worth approximately 8% of its value after the initiation of the wearing-out mode of distress.

14.4 Reliability Analysis of a Multicomponent System

The reliability of a multicomponent system depends on the reliability of its components. For reliability analysis, statistical distributions are fitted to data of failures. In this analysis, selection of an appropriate probability distribution is crucial.

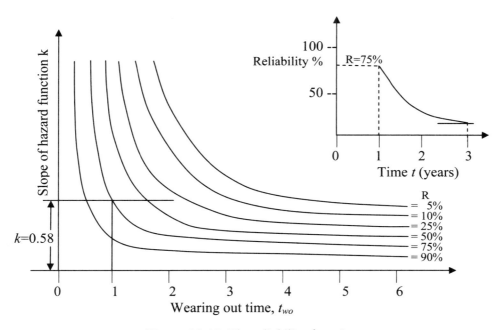

Figure 14-16 *The reliability function.*

14.4.1 Events and Fault-Tree Analysis

All failures begin with an adverse initiating event, which usually is the failure of a component of the system. This event may or may not trigger other events and may initiate a sequence of adverse events with progressively more damaging repercussions. Many such initial events are ignored because their influence appears to be limited or "local." But the final outcome depends upon how wide-spread is the influence of this initial event, the health of the related components, and their loads. Numerous examples can be recalled from everyday life where a catastrophe resulted because of the failure to notice an initial event, failure to visualize the consequences, or failure to initiate ameliorative action in time. A reliability analysis attempts to identify all these scenarios, determine the chances of their occurrence, and ascertain how these influence the safety of the system.

An event tree for a given frequency specifies a range of possible outcomes, so that the event frequency of a particular outcome is given by the product of the initial frequency with all the probabilities at each of the intervening steps. Such a chain is referred to as an accident sequence. It shows how a failure may propagate through a complex system. Figure 14-17 shows a simple event-tree diagram for failure of an earthen dam. Since this tree has been drawn for illustrative purposes only, each event is shown to result in two follow-up events. But this is not a limitation of the method and a particular event may result in more than two events. Further, the event tree for a real-life case will usually be quite larger than that shown here. Another important aspect of this tree is that when computing the probability of occurrence of a branch, the events are assumed to be independent. In Fig. 14-17, the consequences are expressed in qualitative terms; these may also be expressed in monetary terms, in terms of area influence, in terms of generation of energy, etc.

High rainfall	High water level in the dam	Small boil on the dam	Probability of occurrence	Consequences
Safe: $1 - P(E)$	Safe: $1 - P(E_1)$	Initiating event $P(E_2)$	$P(E_2)* \{1-P(E_1)\}$ $\{1-P(E)\}$	Small local damage
Unsafe: $P(E)$			$P(E_2)* \{1-P(E_1)\}$ $P(E)$	Severe damage to dam
Safe: $1- P(E)$	Unsafe: $P(E_1)$		$P(E_2)* P(E_1) \{1- P(E)\}$	Large but local damage
Unsafe: $P(E)$			$P(E)*P(E_1)P(E_2)$	Failure of dam

Figure 14-17 *Event tree for failure of an earthen dam.*

After the event tree is prepared, follow-up steps are necessary to attend to the branch that has high associated risk. The risk can be minimized by replacing the component that is perceived to initiate, contribute, or quicken failure.

A fault tree gives a reverse representation of the process, working back from a particular event (known as the top event) through all the chains of events that are precursors of the top event. There can be more than one top event. But for a top event to take place, some other events, known as lower-level events, must take place. At the bottom are the events known as basic events; these cannot be decomposed further and the failure probabilities for these events need to be known. The key components of a fault tree are thus event specifications and logic gates (with jargon heavily borrowed from the electronics and communication field). A fault tree for failure of an earthen dam is shown in Fig. 14-18. At the first level, the major causes of failure of such a dam are identified and the next level lists the various causes that may result in overtopping of the dam.

The main outcome of fault-tree analysis is the probability of occurrence of the top event. This probability is stated in terms of OR (union) or AND (intersection) of the basic events. Knowing the probability of the basic events, one can compute the probability of the top event. This works well for small fault trees, but for large problems the computations become complex. An efficient method, known as the *cut set* approach, is employed to increase efficiency. A cut set is a set of basic events whose joint occurrence causes the top event to take place. A minimal cut set of the system comprises the set of components that, when they fail, cause failure of the system. If any component of this system works, the remaining components in this set will collectively no longer cause the failure of the system. Thus, in a cut set, the nonoccurrence of any basic event will lead to the nonoccurrence of the top event. A complex system, such as a large water distribution network, can be subdivided into a number of cut sets working in parallel and any one of these can result in the occurrence of the top event. Thus, the system is a union (joined by an OR switch) of the cut sets.

Clearly, the application of these techniques requires an extensive database on the occurrence of events and failures. Such databases obviate the necessity of assuming some value based on subjective judgment. Although many databases exist, one has to be cautious in pooling data from different sources since the purpose, categorization, etc. are not likely to be the same across the databases.

Example 14.29 A water distribution system is shown in Fig. 14-19. The failure probabilities of various pipes are as follows: Pipe 1 (P_1) = 0.004, Pipe 2 (P_2) = 0.003, Pipe 3 (P_3) = 0.002, and Pipe 4 (P_4) = 0.005. Draw the fault-tree diagram for the system and determine the failure probability of no supply from source to outlet.

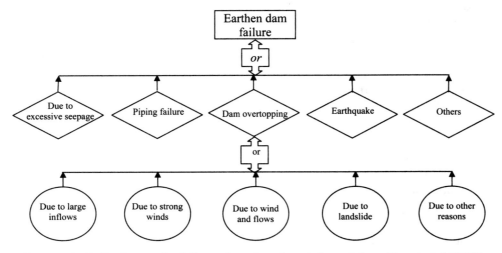

Figure 14-18 *Fault-tree for failure of earthen dam (adapted from Yen et al. [1986]).*

Figure 14-19 *Example water distribution network.*

Solution The fault-tree diagram of the system is drawn in Fig. 14-20 using AND and OR gates. The probability of no supply from source to outlet is

$P_{\text{no supply}}$ = P[failure of Pipe 1 AND failure of Pipe 2]

   OR [failure of Pipe 3 AND failure of Pipe 4]

$= [P_1 \cap P_2] \cup [P_3 \cap P_4]$

$= (P_1 \times P_2) + (P_3 \times P_4) - (P_1 \times P_2) \times (P_3 \times P_4)$

$= (0.004 \times 0.003) + (0.002 \times 0.005) - (0.004 \times 0.003) \times (0.002 \times 0.005)$

$= 12 \times 10^{-6} + 10 \times 10^{-6} - 12 \times 10^{-6} \times 10^{-6}$

$\approx 22 \times 10^{-6}$

Evidently, this is a very low probability, partly because of redundancy in the network.

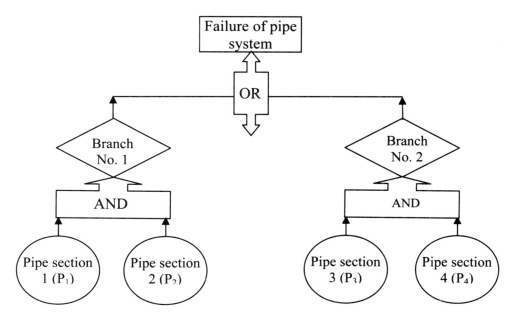

Figure 14-20 *Fault-tree diagram of water distribution system of Fig. 14-19.*

14.5 Reliability Programming

An optimization problem in which some or all of the data are random is termed a stochastic programming problem. For such a program, we need to define the concepts of feasible and optimal solutions that will account for the random nature of the problem. There are various approaches to solve reliability programming problems. Here, the discussion focuses on the chance constrained programming method. In this method, the concept of decision rule is important.

A decision rule is a function that maps a random variable into a decision. There are two types of decision rules to determine the optimal values of the decision variables X_j. In a nonzero-order decision rule, the values of the decision variables are based upon values of random variables that are observed during the time horizon; the values of the decision variables for stage t are specified as explicit functions of the outcomes of the random variables for the stages $j = 1, \ldots, t - 1$. In other words, we wait for the values of the random elements to become known before determining X_j. In a special subclass of the general nonzero-order rules, the so-called zero-order decision rules, the values of all decision variables are determined before the actual values of the random elements become known. Of course, one must decide in advance how the knowledge of the sample values of the random elements will be used. For example, consider regulation of spillway gates of a reservoir to control flooding in downstream areas. If this regulation is decided after the actual value of random inflows to the reservoir are observed, this will be termed as a nonzero-order rule. If the spillway gate opening is given by $g = f(x)$ and $f()$ is determined *a priori*, this will be termed as a zero-order rule.

Thus, the main difference between the two types of rules is that, according to the nonzero-order decision rule, the exact value of a decision variable with respect to stage t can only be computed after the outcomes of all random variables concerning the preceding $t - 1$ stages have been observed whereas, according to the zero-order rule, the value of the decision variable is exactly known at the beginning of stage t.

With respect to the random nature of the variables, when the unknown values of the decision variables are assumed to be deterministic, the decision rule is called a nonrandomized decision rule. Since the random variations in the parameters of a problem induce random variations in the optimal values of decision variables X_j, we can have a chance mechanism to determine the optimal values of X_j. The rules governing such a mechanism are called randomized decision rules. In these rules, X_j are treated as random variables and consequently we have to find their probability distributions. In the example just cited, the reservoir inflows are treated as random variables whose probability distribution must be determined to develop and implement the decision rule.

Reliability constraints are frequently imposed on the system under consideration so as to ensure a certain level of reliability regarding its performance.

14.5.1 Chance Constraints

In many real-life problems, some of the inputs that influence the decision variables may be random. Therefore, the constraints defining the limits of associated variables should also specify the percentage of time that these limits can be exceeded, if any. Chance-constrained models typically have the constraints that limit the permissible range of decision variables. Constraints that explicitly do this are termed chance constraints.

A chance constraint to ensure that some variable $\hat{x}$ is not greater than the value of a random variable X at least some fraction α of the time can be written as

$$\text{prob}\{ \hat{x} \leq X\} \geq \alpha \text{ or prob}\{ \hat{x} \geq X\} \leq (1 - \alpha) \tag{14.133}$$

where prob denotes probability. Similarly, if some variable $\bar{x}$ is to be no less than the value of a random variable X at least some fraction α of the time, the relevant chance constraint becomes

$$\text{prob}\{\bar{x} \geq X\} \geq \alpha \text{ or prob}\{\bar{x} \leq X\} \leq (1 - \alpha) \tag{14.134}$$

The constraint given by Eq. 14.133 can be satisfied if we can ensure that the variable $\hat{x}$ is less than or equal to that value $x^{(\alpha)}$ of the random variable X that is exceeded a fraction α of the time:

$$\hat{x} \leq x^{(\alpha)} \tag{14.135}$$

Similarly, to ensure that Eq. 14.134 is satisfied, it is sufficient to ensure that the variable $\bar{x}$ is greater than or equal to that particular value $x^{(1-\alpha)}$ of the random variable X that is exceeded a fraction $(1 - \alpha)$ of the time:

$$\bar{x} \geq x^{(1-\alpha)} \tag{14.136}$$

Here α and $(1 - \alpha)$ denote the probabilities of exceedance. Figure 14-21 illustrates the interpretation of $x^{(\alpha)}$ and $x^{(1-\alpha)}$. Basically, Eq. 14.135 and Eq. 14.136 are the deterministic equivalents of chance constraints defined by Eq. 14.133 and Eq. 14.134. We are able to define these deterministic equivalents because the probability distribution of random variable X is assumed to be known, and hence the particular values $x^{(\alpha)}$ and $x^{(1-\alpha)}$ of that random variable can be computed.

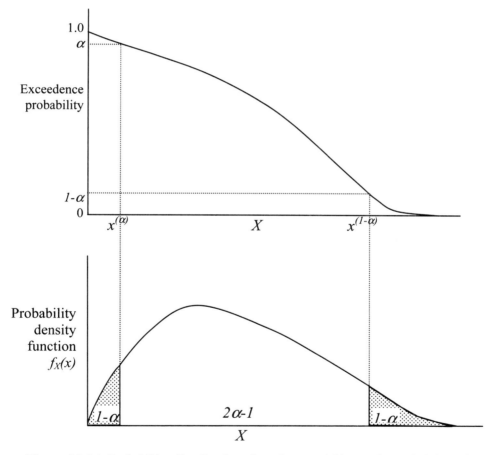

Figure 14-21 *Probability distribution of random variable X. The probability of exceedence of* $x^{(\alpha)}$ *is* α.

14.5.2 Reservoir Design Using Chance-Constrained Optimization

Let us illustrate the application of chance-constrained optimization to real-life cases with the help of an example of a reservoir. The future inflows into a reservoir are not deterministic in nature; rather these are stochastic. Hence the inflow to a reservoir in a particular period of time can be predicted only with some probability. Chance-constrained models for reservoir management problems have two sets of constraints. One set limits the permissible range of reservoir storage volumes, and the other set limits the permissible range of reservoir releases. These constraints defining the limits of storage volumes and releases should also specify the probability with which these limits can be exceeded.

Consider a reservoir that is to be built and operated for the purpose of (a) supplying water for irrigation, domestic, and industrial needs, and hydropower generation; (b) flood control; and (c) recreation. It is desired that the reservoir be as small as possible, thus reducing the cost of dam construction, while meeting the stated objectives.

In reservoir design and operation problems, for each period t in a year, the inflows I_t, the initial storage volumes S_t, and the releases R_t are random. Observed historical records can be analyzed to estimate the probability distribution of inflows, but the probability distributions of random variables S_t and R_t depend on the operating policy. Since the optimal operating policy is unknown, the distributions of initial storage volumes S_t and the releases R_t are unknown. Thus it is not possible to derive deterministic equivalents similar to Eq. 14.135 and Eq. 14.136 of chance constraints limiting the range of storage volumes or releases unless the unknown distributions of S_t and R_t are functions of the known distributions of inflow I_t. One way to overcome this problem in a linear programming (LP) application is through the use of what is known as a linear decision rule (LDR). Basically, LDRs define S_t and R_t in terms of I_t. The linear decision rule for reservoir design and management was proposed by ReVelle et al. (1969).

The LDR permits the use of an LP algorithm for solving reservoir management problems. Mathematically, it is indeed an advantage since it considerably reduces the number of possible operating policies that need to be examined. Consider the following LDR, which defines the initial reservoir storage volume in period $t + 1$:

$$S_{t+1} = \lambda_t I_t + b_t \quad \forall\ t \tag{14.137}$$

Here, λ_t is a known coefficient ($0 \le \lambda_t \le 1$) and b_t is an unknown, unrestricted, deterministic variable defined for each within-year period t (Loucks and Dorfman 1975). This is termed as the LDR parameter for the tth month of the year and is to be determined. Another form of LDR, which expresses release in terms of storage, is

$$R_t = S_t - b_t \tag{14.138}$$

Assume that, over many years, the initial storage volumes S_t, in each within-year period t, were to be within certain lower $\hat{s}_t$ and upper $\bar{s}_t$ limits at least some fraction α_t of the time:

$$\text{prob}\{\, \hat{s}_t \le S_t \} \ge \alpha_t \quad \forall\, t \tag{14.139}$$

$$\text{prob}\{\, \bar{s}_t \ge S_t \} \ge \alpha_t \quad \forall\, t \tag{14.140}$$

Similarly, let the reservoir releases R_t be within the range $\hat{r}_t$ to $\bar{r}_t$ at least some fraction β_t of the time:

$$\text{prob}\{\, \hat{r}_t \le R_t \} \ge \beta_t \quad \forall t \tag{14.141}$$

$$\text{prob}\{\, \bar{r}_t \ge R_t \} \ge \beta_t \quad \forall t \tag{14.142}$$

Because the probability distributions of all S_t and R_t are unknown, they must be replaced by a function of random variables whose distributions are known before deterministic equivalents can be defined.

Substituting Eqs. 14.138 into 14.139 and 14.140 permits the definition of deterministic equivalents, since the distributions of the random inflow variables I_t are known. Similarly, deterministic equivalents of Eq. 14.141 and Eq. 14.142 can also be defined by using the continuity equation:

$$R_t = S_t + I_t - S_{t+1} , t = 1, 2, \ldots, T; T + 1 = 1 \tag{14.143}$$

and the linear decision rule given by Eq. 14.137 or Eq. 14.138. It is then possible to determine the optimal values of the decision variables.

In a reservoir design problem, the objective function is to minimize the capacity of the reservoir (C). It can be expressed mathematically as

$$\text{minimize } C \tag{14.144}$$

The objective function is subject to a number of constraints as discussed next.

Freeboard Constraint

To provide flood control, the storage S_{t+1} at the end of period t should be such that the freeboard volume $C - S_{t+1}$ is at least v_t with reliability (say) 90%. Here, v_t is the flood storage capacity required at the end of the tth month of the year. Mathematically, the constraint can be written in deterministic form as

$$C - S_{t+1} \ge v_t \tag{14.145}$$

The reservoir mass balance equation using an explicit statement of chance constraints is

$$S_{t+1} = S_t + I_t^{0.90} - R_t \tag{14.146}$$

where $I_t^{0.90}$ is the flow for month t that is available 90% of the time (see $x^{(\alpha)}$ in Fig. 14-21). Putting Eq. 14.138 in Eq. 14.146, we have

$$S_{t+1} = S_t + I_t^{0.90} - S_t + b_t = I_t^{0.90} + b_t \qquad (14.147)$$

Putting Eq. 14.147 in Eq. 14.145, we have

$$C - b_t - I_t^{0.90} \geq v_t \qquad (14.148)$$

or

$$C - b_t \geq v_t + I_t^{0.90}, \ t = 1, 2, 3, \ldots, n \qquad (14.149)$$

Since $I_t^{0.90}$ is exceeded, on average, 10% of the time, this constraint should hold 90% of the time. If monthly data are being used, the value of n will be 12.

Minimum Storage Requirement Constraint

The minimum storage in the reservoir, S_{t+1}, at the end of period t should be at least S_{min} with (say) 90% reliability. Here 90% reliability can be achieved by considering inflow that is available 90% of the time or not available only 10% of the time, $I_t^{0.10}$. Mathematically, the constraint can be written in deterministic form as

$$S_t \geq S_{min} \qquad (14.150)$$

We express the minimum storage to be maintained as a fraction a_m of the reservoir capacity. Putting Eq. 14.147 in Eq. 14.150, we have

$$b_t + I_t^{0.10} \geq a_m C \qquad (14.151)$$

or

$$a_m C - b_t \leq I_t^{0.10}, \ t = 1, 2, 3, \ldots, n \qquad (14.152)$$

Minimum Water Supply Requirement Constraint

The release at any time (R_t) should exceed the minimum committed release value (q_t) with a 90% reliability. Here 90% reliability can be achieved by considering inflow $I_t^{0.10}$. Mathematically, the constraint can be written in deterministic form as

$$R_t \geq q_t \qquad (14.153)$$

Equation 14.157 implies that

$$S_t = b_{t-1} + I_{t-1}^{0.10} \qquad (14.154)$$

Putting Eq. 14.157 and Eq. 14.154 in Eq. 14.146, we have

$$R_t = b_{t-1} - b_t + I_{t-1}^{0.10} \qquad (14.155)$$

Chapter 15

Risk Analysis and Management

Risk is an inherent part of life and engineering decisions. The word "risk" seems to have been derived from Spanish or Portuguese. Originally, it referred to sailing into uncharted waters and it had therefore an orientation in space. Another illustration of spatial orientation would be individuals or governments or banks lending money for projects. With progression of time, its connotation assumed a time dimension. For example, in water resources and environmental engineering risk may entail calculation of the probable adverse consequences of environmental and water resources projects to be built for people in the project area. Normally, one would want to minimize the risk of undesirable consequences or outcomes of a decision. In most cases it is not possible to completely eliminate risk; however, one can mitigate it. Before initiating a discussion on risk, the frequently used terms are defined first.

15.1 Basic Definitions

15.1.1 Risk

Risks are possibilities that human activities or natural events lead to consequences that affect what humans value. The definition of risk varies with the purpose—it also changes with discipline. In general terms, risk can be defined as the potential loss resulting from the convolution of hazard and vulnerability. Mathematically, risk may be expressed as the probability of surpassing a determined level of economic, social, or environmental consequence at a certain site and during a certain period of time. Convolution is a mathematical term that refers to concomitance and mutual conditioning of hazard and vulnerability. In other words, a system cannot experience a risk if it is not exposed to a hazard and is vulnerable. Hazard and vulnerability are mutually conditioning situations and neither can exist on its own. Altering one or two of the components of risk alters the overall system risk as well. However, in many cases it is not possible to modify hazard to reduce risk; one has to reduce the vulnerability of the system as a measure of prevention or mitigation, a process also known as risk reduction.

Most commonly, in civil engineering risk has been defined as the probability of a system failure, the reciprocal of the expected length of time before a system failure takes place, or some measure of the cost of failure. The Royal Society (1983) defined risk as the probability that a particular adverse event (an event whose occurrence produces harm, such as a 100-year flood, a category-5 hurricane, a 50-ft storm surge, or a 300-mile/hour tornado) occurs during a stated period of time. Thus, the concept of risk combines a probabilistic measure of the occurrence of the adverse event with a measure of the consequences of the occurrence of that event. The occurrence may include the amount or intensity, starting time, or duration.

An important point to note is that risk is not viewed in a positive sense. Consider, for example, that a person receives information that he is likely to receive an award of either 1 million dollars or 10 million dollars. Although the person is not certain as to the amount he will receive, it is safe to say that he will not be under risk.

Risk involves uncertainty as well as loss or damage. For example, the risk of flooding involves the probability of occurrence of the flood as well as the damage that might result from the flood event. Therefore,

$$Risk = Hazard\ uncertainty + Consequence\ owing\ to\ the\ system's$$
$$vulnerability\ to\ hazard\ (damage\ or\ loss) \tag{15.1}$$

Sometimes risk is defined as the probability times consequences. A drawback of this definition is that it equates risk of a high-probability, low-damage event with that of a low-probability, high-damage event. Clearly, in real life these two events may not amount to the same risk.

The converse of loss is benefit, which is defined in terms of gain or improvement for a human being, a society, a nation, the human population, or the planet. Expected benefit includes an estimate of the probability of achieving the gain. Gain and loss are often measured in economic terms but in real life they can also be in nontangible terms.

Example 15.1 Most people are always interested in weather. What is the risk from the coldest weather on record next winter in Baton Rouge, Louisiana?

Solution Weather plays an important role in our daily lives. We want to know whether the coming winter in Baton Rouge will be the coldest on record and its consequences if it is. Thus, risk (assuming cold weather is harmful) in this case is the probability of occurrence of the coldest winter next year in Baton Rouge and the ensuing consequences in terms of crop damage, bursting of pipes, traffic accidents, higher heating bills, and so on.

15.1.2 Conditionality of Risk

All risks are conditional and the conditions are often implied by the context and are not explicitly stated. For example, the risk of death from a flood is relatively small in the United States, but its value will significantly differ from one place to the other and from one country to the other, depending on the climate, flood protection schemes, warning issued, communication, rescue operations, people's perception, etc. In contrast, the risk of death from a flood of the same magnitude can be relatively large in, say, Bangladesh. The tsunami of December 26, 2004, caused an unimaginable loss of human life in Asian countries. If the same tsunami were to occur in the United States, it is almost certain that the loss of human life would be little because of advanced warning systems, better infrastructure, and the receptivity of Americans to warning. In light of what happened in Louisiana in general and New Orleans in particular when Hurricane Katrina hit, the level of preparedness in the United States has come under question.

15.1.3 Hazard

Hazard is a situation or occurrence of an event that could, in particular circumstances, lead to harm. Hazard can be considered as a latent danger or an external risk factor of an exposed system. This can be mathematically expressed as the probability of occurrence of an event of certain intensity at a specific site and during a determined period of exposure. Thus, hazard is a source of risk and risk includes the chances of conversion of that source into actual loss. For example, it is not advisable to drive when road conditions are icy and not favorable for driving. It is not advisable to go to a beach when there is a warning of tsunami. It is hazardous to cross a river under spate by swimming, because the chances of drowning are relatively significant even for an expert swimmer. But if the swimmer attempts to cross the river by a motorboat, equipped with a powerful engine, rugged body, and life jackets, etc., the risk of drowning is considerably smaller. Thus safeguards help reduce risk. Mathematically, one can write (Kaplan and Garrick 1981)

$$Risk = Hazard/Safeguards \tag{15.2}$$

Equation 15.2 uses division rather than subtraction. As safeguards tend to zero, even a small hazard can lead to a high value of risk; as safeguards increase, the risk becomes smaller. In day-to-day life this equation is seen to work. For example, implementation of strict traffic rules in many countries has reduced the risk of accidents to a small number, whereas in some countries road traffic is considered as highly risky. After the tragic experience of Hurricane Katrina in 2005 in New Orleans and along the Mississippi Gulf Coast, people in western Louisiana and Texas heeded the warnings about Hurricane Rita and the result was relatively little loss of human life.

Harm is defined as the loss to a human life (or human population) or plant life and animals as a consequence of damage, where damage is the inherent quality of loss suffered by an entity (physical or biological or social). Identification of hazard comprises an important part of risk assessment. The techniques of hazard identification include safety audit, hazard survey, hazard indices, and hazard and operability studies. Many countries have standardized and documented procedures of hazard identification.

Detriment is a numerical measure of the expected harm or loss associated with an adverse event. It is generally the integrated product of risk and harm and is often expressed in terms of costs measured by a monetary currency (say, dollars), or loss in expected years of life, or loss of productivity. A determination of detriment is often needed for cost–benefit or risk–benefit analysis. It is a numerical way of comparing different events associated with the same hazard or the combined effects of events from different hazards.

Example 15.2 Consider a detention pond for local flood control in an urban area. What could the risk be from this detention structure?

Solution The detention dam may be overtopped and breached. As a result, the dam breach may cause harm to people in the urban area. The risk would be the probability of specified damage or harm in a given period.

For water control structures, hazard from failure depends on the size of the structure. Therefore, decisions about the recommended design load (or design flood) are based upon the size of the structure and its hazard potential. The recommendations of the U.S. Army Corps of Engineers regarding selection of spillway design flood for a dam are given in Table 15-1. A typical classification of reservoirs according to size and the hydraulic head is given in Table 15-2. The hazard potential classification of reservoirs is given in Table 15-3.

15.1.4 Disaster

The term disaster has the connotation of an event capable of inflicting damage or causing danger to human and/or animal life and/or property. A disaster can be anthropogenic or natural. Hurricanes, typhoons, cyclones, earthquakes, tsunamis, lightning, and land subsidence are examples of natural disasters. Anthropogenic disasters include dam breaching, levee failure, chemical spills, nuclear

Table 15-1 *Recommendations regarding selection of design flood.*

Hazard	Size	Design flood
Low	Small	50 to 100 years
	Intermediate	100 years to 0.5 PMF
	Large	0.5 PMF to PMF
Significant	Small	100 years to 0.5 PMF
	Intermediate	0.5 PMF to PMF
	Large	PMF
Large	Small	0.5 PMF to PMF
	Intermediate	PMF
	Large	PMF

Note: PMF = probable maximum flood.

Table 15-2 *Size classification of reservoirs.*

Type of Dam	Storage capacity (million m³)	Hydraulic head (m)	Inflow design flood
Small	0.5–10	7.5–12	100 year
Intermediate	10–60	12–30	SPF
Large	> 60	> 30	PMF

Note: SPF = standard project flood.

Table 15-3 *Hazard potential classification of reservoirs.*

Category	Loss of life	Economic loss
Low	None expected	Minimal
Significant	Few	Appreciable
High	More than a few	Excessive

explosions, bomb blasts, bridge collapses, train accidents, and so on. Disasters have occurred from time immemorial and will continue to occur. We cannot eliminate disasters but we can mitigate their impact.

15.1.5 Characterization of Risk and Concept of Risk Triplet

Characterization of risk helps establish its specific context. To characterize risk, two basic elements are necessary:

1. Probability of occurrence of a hazard
2. Extent of damage, which is governed by the vulnerability of a given system to exposed hazard.

Risk assessment evaluates and integrates these two components. To represent risk, the concept of triplet definition is widely applied. This can be parameterized by asking three questions:

1. What is the harmful event occurring or what can go wrong during a given event?
2. What is the probability of the event occurring?
3. What are the consequences if the event occurred?

These three questions are answered by preparing a list of scenarios in which the answers to the questions are arranged as a triplet (Kaplan and Garrick 1981):

$$R = [s_i, p_i, x_i], i = 1, 2, ..., N \qquad (15.3)$$

where s_i is the scenario or hazardous event identification, p_i is the probability of that event or scenario, x_i is a measure of loss and represents the consequences of that event or scenario, and R is the risk. The scenario list can be conveniently arranged in the form of a table. Such a table of scenarios for the failure of a dam is shown in Table 15-4. In risk analysis, we try to determine how the events will turn out in the future if certain actions are initiated (or not initiated).

Table 15-4 *Scenarios (illustrative list) for dam failure.*

Scenario or event	Probability	Consequences
Structural failure		
Dam overtopping	0.001	Failure of main dam
Piping	0.002	Excessive erosion
Failure of spillway	0.0004	May lead to failure of main dam
Sloughing on dam slopes	0.0009	Localized damage to dam body
Earthquake of magnitude > 8 on Richter scale occurs	0.0001	Extensive damage to the dam and the appurtenant works
Performance failure		
Flood bigger than design flood enters reservoir	0.0002	Large flows cause damage in downstream areas
Water inflows are very small	0.0001	Reservoir cannot meet the intended objectives
Power plant turbine fails	0.00015	Less electricity is generated
Others	0.00003	Unknown

If this table contains all the possible scenarios, it is the estimation of risk. Of course, the list of scenarios in a real-life case can be quite lengthy. Kaplan and Garrick (1981) suggest that a category "others," encompassing all the scenarios that have not been thought of, may be added to the list for the sake of completion. Of course, the problem of assigning a probability to this category remains to be tackled. Logically, the probability for the event in this category will be very small because the event has not happened; otherwise it would have been included in the list.

The triplet-based definition of risk and previous discussion suggest that hazard can be defined as a subset of risk—a set of doublets:

$$H = [s_i, x_i], i = 1, 2, \ldots, N \tag{15.4}$$

If the consequences are arranged in order of increasing severity of damage

$$x_1 \leq x_2 \leq \ldots \leq x_N$$

then the second column of Table 15-2 can be accumulated and a smooth curve between x and p can be plotted, as shown in Fig. 15-1. Such curves are termed risk curves.

15.2 Risk and Its Classification

Risk can be classified into three categories:

1. Risks for which statistics of identified casualties are available
2. Risks for which there may be some evidence, but where the connection between suspected cause and damage cannot be established
3. Estimates of probabilities of events that have not yet occurred.

Additionally, there are risks that are unforeseen.

All systems have a probability of failure no matter how small it is and the complete avoidance of all risk of calamitous failure is not possible. The objective is to reduce the probability to an acceptable level of individual and societal risk. An engineering approach to quantify risk begins with a physical appreciation of possible failure mechanisms or modes and their analysis. This entails quantification of the reliability of the components and examination of the systematic failure to establish the overall vulnerability of the complete system, based on experience, and verified by analysis, testing, and inspection.

An examination of past events helps with an understanding of failure modes. Consider, for example, the case of river levees for flood control. In light of the potential for major consequences involved in project failures, it is not advisable to wait for disasters to occur such that a body of case histories can be built to form a basis for policy decision making. Therefore, an anticipatory approach based on judgment and experience is required, and such an approach can be developed using risk estimation through methods based on a systematic decomposition of a complex system into its component subsystems and the use of predictive techniques and modeling. Thereafter failure mechanisms can be analyzed and risk estimated by pooling together models of individual subsystems. This method requires a wide range of data on past failures and knowledge about the various processes that could occur. The results of such a method are subject to substantial

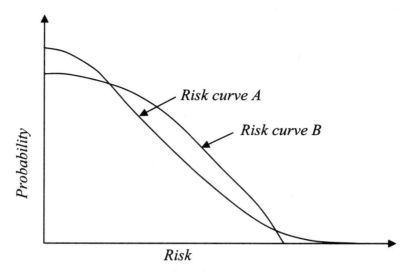

Figure 15-1 *Smooth risk curves.*

uncertainties owing to inadequacies in the data and insufficient accuracy of scientific knowledge.

There are other more traditional methods in widespread use that are essentially deterministic in nature. A deterministic method can be illustrated by the employment of a factor of safety, which can be defined as the ratio of design strength to design stress. A multitude of factors affect both design demand and design capacity. In practice, therefore, there will be a distribution of demand and capacity, both having mean values with a spread about those means. If the mean demand is smaller than the mean capacity, then it can be shown that failure will occur where the upper end of the demand distribution encounters the lower end of the capacity distribution. This leads to definitions of safety factors and safety margins in probabilistic terms.

Although a deterministic approach incorporates the concept of the variability of demand and capacity, it implies that there is a level of probability of failure that is acceptable for design purposes, and that level can be quantified. In contrast, the probabilistic approach includes the low-probability events in the overall assessment. By necessity, sufficient data must be available.

In terms of decision making, the deterministic approach incorporates implicit value judgments as to an acceptable standard of practice and is derived from an extension of past practice and experience, which may be inadequate to deal with rapidly changing technology. In contrast, the probabilistic approach describes hazard in terms of risk of failure and its associated consequences. Thus, it enables making an acceptable decision based on a design process and making needed judgments.

The risk of failure and its consequences are significantly influenced by management. Effective management and auditing of safety involves many of the principles of total quality management (TQM). These ensure the maintenance of safe practices laid down in the safety guidelines. Management of safety also involves the training of staff to observe, record, and report; not to panic; to systematically react to the onset of a potential disaster; and to organize evacuation and rescue procedures in the event of a disaster. The absence or lack of adequate management and auditing of safety may exacerbate a major disaster. This is more or less what happened when Hurricane Katrina hit New Orleans, Louisiana.

15.2.1 Aspects of Risk

Risk has two aspects—negative and positive—and they emanate from the early beginnings of modern industrial society. Traditional cultures did not have a concept of risk because they did not need one. The concept of risk is widely used in a society that is future oriented, which sees the future to be managed. In earlier times or even today in many countries, the ideas of fate, faith, luck, destiny, the will of God, etc. are prevalent. We now tend to substitute risk in their place. Acceptance of risk may also be viewed as a condition of excitement, thrill, and adventure. For example, some people get pleasure from the risks of rock climbing, mountain hiking, gambling, skiing, surfing, canoeing, driving fast, ballooning, the plunge of a fairground rollercoaster, etc. One can argue then that a positive embrace of risk is the very source of energy that creates wealth in a modern society. This perhaps is the basis of the common phrase "No risk, no gain."

Risk serves as a dynamic force that compels a society to carve its future rather than to leave it to religion, faith, luck, destiny, God, tradition, or vagaries of nature. For example, a modern capitalistic society carves its future using risk and this is reflected through the calculation of expected profit and loss or expected monetary value in future. We wish to minimize a multitude of risks, such as those related to human health, environmental pollution, disease epidemic, drug addiction, social violence, wildfires, and flooding. This indeed is the basis of insurance policies. Insurance is the baseline against which people are prepared to take risks. It is the basis of security where fate is replaced by an active engagement with the future. Insurance is only conceivable where there is belief in a humanly engineered future. Although insurance provides security, it may be actually parasitic upon risk and people's attitudes toward it. Indeed, risk is traded off in exchange for payment. The trading and offloading of risk is actually the backbone of a capitalist economy. Thus, the idea of risk is involved in modernity. Risk is supposed to be a way of regulating the future, or normalizing it and bringing it under control. Our efforts to control the future tend to compel us to look for different ways of relating to uncertainty.

15.2.2 Types of Risk

The notion of risk is inseparable from the concepts of probability and uncertainty. Risk is not the same as hazard or danger. It relates to hazards that are assessed in terms of future possibilities. Two types of risks can be distinguished: external risk and manufactured risk. External risk is the risk that stems from outside, from the fixities of tradition or nature. Manufactured risk is the risk that is created by our actions and can occur in a situation that we have very little experience of confronting. For example, environmental risks of global warming or climate change are manufactured risks, influenced by intensifying globalization. This is the risk created by the very impact of our developing knowledge upon the world. Flooding risk from land use change, such as urbanization, can be categorized as a manufactured risk.

From the earliest days of human civilization up to the threshold of modern times, risks were primarily due to external sources (natural): floods, famines, plagues, earthquakes, tsunamis, etc. Recently, the focus has shifted from what nature does to us to what we have done to nature. This marks the transition from the predominance of external risk to that of manufactured risk. Much of what used to be natural is not completely natural any more. Therefore, natural phenomena, such as floods, droughts, diseases, extreme weather, and land subsidence, are not entirely natural; rather, they are being influenced by human activities, as suggested by their unusual features.

As a manufactured risk expands, there is a new riskiness to risk. The very idea of risk is tied to the possibility of calculation. However, in many cases we simply do not know what the level of risk is and we could not know for sure until it is too late. In these circumstances, there are two extremes to characterizing risk. On the one hand, if the risk is real, there must be an explicit statement to that effect and the risk must be emphasized even at the cost of scaremongering. On the other hand, if the risk turns out to be minimal, there will indeed be accusations of scaremongering. Furthermore, if the risk is not emphasized and it turns out to be significant, there will then be accusations of cover-up. This is illustrated in Fig. 15-2.

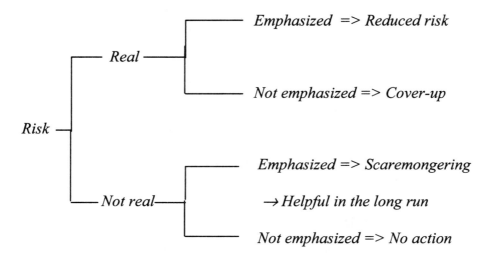

Figure 15-2 *Characterization of risk.*

In many cases of manufactured risk, it is difficult to say when there is scaremongering and when there is not. In the hurricane season of 2004, it was assessed that the New Orleans area might get hit by a hurricane named Ivan. Therefore, local and state governments decided to evacuate people before the arrival of Ivan. There was massive migration of people from New Orleans to areas 100 or more kilometers away. Roads were jammed and people took hours to travel even short distances. Fortunately, it turned out that Hurricane Ivan spared New Orleans, but it did hit parts of Florida and Alabama. A couple of days after people had returned to New Orleans, they started complaining about the false warning and unnecessary evacuation. This was a typical case of deciding between two options: (1) evacuating and following a safer route or (2) taking a chance and not evacuating.

As science and technology become more and more integral parts of our life, there must be a more engaged relationship between science and technology and people. Scientific findings can no longer be accepted on their face value, especially in situations of manufactured risk, if only because scientific findings are sometimes conflicting. Thus, one way to cope with the rise of manufactured risk is to employ the precautionary principle. It presupposes action about issues even though there is insufficient scientific evidence about them. Sometimes this principle is not helpful in coping with risk and responsibility. Sometimes we need to be bold rather than cautious in supporting innovation or other forms of change.

Risk involves a number of unknowns, and we need to reduce these unknowns or manage them. This leads to the concept of risk management: the balance of risk and danger. Merely taking a negative attitude toward risk is not scientific nor healthy. Risk needs to be managed: Active risk-taking is a core element of a dynamic economy and innovative society.

15.2.3 Acceptable Risk

One of the dilemmas in risk assessment is defining "acceptable risk." The acceptability depends on the context in which to assess risk and the availability of financial resources. Furthermore, it also depends on whether one is taking an individual view or a much broader view. However, it may be added here that the notion of an acceptable probability is highly subjective and it is difficult to arrive at a rational number by using some objective criteria. Consider a design that can give protection against an event x up to a certain magnitude x_d. The design is acceptable if the exceedance probability, $\mathrm{Prob}(x \geq x_d) = 1 - \mathrm{Prob}(x_d)$, is smaller than the acceptable value P_{Acc} (Plate 2002). The acceptable value of risk for a person reflects his or her preference; Vrijling et al. (1995) suggested that it can be written as

$$P_{\mathrm{Acc}} = \frac{\beta_i \times 10^{-4} / \mathrm{year}}{v_{ij}} \tag{15.5}$$

where β_i can range from 0.01 for a high risk for an action that gives no benefit to the person to 10 for a risky activity that brings high satisfaction to the person; and v_{ij} is the vulnerability of an individual to an event x_i. For a nation, the acceptable probability, according to Vrijling et al. (1995), becomes

$$P_{Acc} = \frac{10^{-3}/\text{year}}{n^2} \text{ for } n \geq 10 \text{ casualties} \qquad (15.6)$$

A problem with the idea of acceptable risk is that acceptability or otherwise of a risk can only be expressed along with the associated costs and benefits. Given an option, one may not be willing to accept any risk at all. At a given time, a risk may be taken only if some benefits are associated with it and these cannot be obtained in another way unless a higher risk is taken. Logically, a decision maker will choose the optimum mixture of risk, cost, and benefit and might be willing to take a higher risk only if it is associated with either less cost or more benefit.

The second point to highlight here is the implicit assumption that risks are linearly comparable, but this is not true. For instance, one cannot say in Fig. 15-1 whether risk is higher in case of the curve A or curve B. A way out is to somehow reduce the risk curves to single numbers by defining a utility function, which depends on x, and then integrate to determine the expected utility.

15.3 Risk Criteria

Natural hazards, such as hurricanes, floods, droughts, earthquakes, tornadoes, winds, snow avalanches, traffic accidents, tsunamis, lightning, and others, have always been a source of concern to planners, designers, operators, and managers of engineering systems. Because of their destructive nature, they are assigned a major role in design considerations. There are many phenomena for which it is difficult to assess the degree of risk. By increasing the factor of safety, one can reduce the risk but there is a natural reluctance to pay for the often exorbitant cost that is associated with the safety factor. Even with an increase in safety, there will always be some risk because it is not possible to build a system that is so strong that it can withstand all conceivable disasters. Thus, in reality risk cannot be eliminated entirely, but it can be reduced to an acceptable level.

A *risk criterion* is a qualitative and quantitative assessment of the acceptable standard of risk with which to compare the assessed risk. A risk criterion is employed to balance the risk of loss against the cost of increase in safety. There are several risk criteria that are employed in engineering. Some of the simple criteria include (Borgman 1963)

1. Return period
2. Encounter probability
3. Distribution of waiting time

4. Distribution of total damage
5. Probability of zero damage
6. Mean total damage

The choice of the most appropriate criterion is a matter of engineering judgment. These criteria emphasize different risk aspects and can be derived from mathematical risk models.

15.3.1 Dimensionality of Risk

Risk cannot be perceived simply as a one-dimensional objective concept, such as the product of the probabilities and consequences of any event. Risk perception is inherently multidimensional and personalistic, with a particular risk or hazard meaning different things to different people and different things in different contexts. Given the conditional nature of all risks, assessments of risks are derived from social and institutional assumptions and processes; that is, risk is socially constructed.

15.3.2 Risk Quantification

Statistical estimation of a risk may involve developing probabilities of future events, based on a statistical analysis of past historical events. It also involves the appraisal of the significance of a given quantitative (or, when acceptable, qualitative) measure of risk.

15.4 Risk Assessment and Management

Risk assessment is a systematic, analytical method used to determine the probability of adverse effects, whereas *risk management* is a systematic process of making decisions to accept a known or assumed risk and/or the implementation of actions to reduce the harmful consequences or probability of occurrence. Risk management is also concerned with the mitigation of those risks derived from unavoidable hazards through the optimum specification of warning and safety devices and risk control procedures, such as contingency plans and emergency actions. Hazard identification, risk analysis, risk criteria, and risk acceptability generally define risk management. Although risk evaluation and management is largely a technical exercise, economic and social aspects (often remaining in the background) are always important and influence the selection of critical parameters. Assessment and management do overlap but are separate tasks. There is need to bring together natural science expertise and knowledge about human behavior and the operation of human institutions.

The steps of risk management are depicted in Fig. 15-3. These days, most large organizations integrate risk management in overall decision making. There is considerable diversity of opinion as to the identification, measurement, and regulation of risk. Risk management may entail

1. The degree of anticipation to be adopted
2. The extent of blame (orientation) in management systems
3. The contribution of quantitative assessment techniques
4. The feasibility of institutional design
5. The cost of risk reduction
6. The desirable level of participation
7. The regulatory budget

15.4.1 Hazard Identification and Management

Environmental management and civil engineering systems are always subject to hazards that may arise in the form of natural inputs; the possibility of sabotage is also increasing with time. A reservoir may be subject to a big flood and a sky-scraper may be hit by a tornado. The cause of hazard may be an event that takes place suddenly and is difficult to predict in advance, such as an earthquake or a tornado. The hazard-producing event (e.g., a flood) may take some time to develop and can be forecasted so that a timely management action can be initiated. Events such as droughts (also termed a "creeping disaster") take a long time to develop but even then their occurrence cannot be avoided and the only possible way to tackle them is preparedness.

Hazard identification is the first step in risk estimation and involves the human element—cultural, social, organizational, group, and individual—which is frequently a contributing cause of disasters. It depends on knowledge, experience, forecasting, engineering judgment, and imagination. The key is to include all hazards, however unlikely these may be.

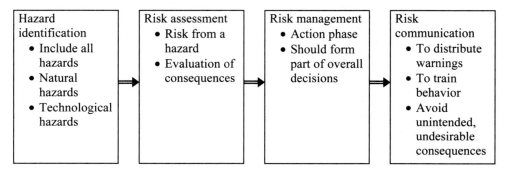

Figure 15-3 Steps in risk management.

Hazard identification and reliability and failure analysis
pilation of possible failures, each with a set of parameters for c
eling. Hazard management depends on the types of hazards to

1. Natural hazards (emanating from the physical environme
2. Technological hazards (from human-created technology)
3. Social hazards (from within human society)

Note that hazard identification does not quantify risk. Quantificat.   ⌐⌐ is
carried out by reliability and failure analysis followed by consequence modeling.

The events that produce hazards are characterized by their magnitude and
an event is termed as hazardous when the magnitude crosses over a threat-
producing threshold. This concept can be appreciated by noting that a perennial
river always carries some flow but a flood is said to occur when the flow crosses
over the channel section to inflict damage to the adjoining areas. In statistical
analysis, the term *return period* is frequently used to denote the magnitude of an
extreme event. As an example, flood frequency analysis based on the log-
Pearson type 3 distribution was performed by using 51 years of streamflow data
(1949–2000) from U.S.G.S. Station #08053000 on the Elm Fork Trinity River near
Lewisville, Texas, to estimate flood magnitudes for various return periods as
presented in Table 15-5.

Hazard maps are prepared to convey the susceptibility of an area or place.
For example, earthquake zone maps indicate the chances of occurrence of earthquakes of various intensities. Likewise, flood-plain zoning maps indicate the
area adjoining a river that is likely to be submerged by floods of various return
periods. Thus three-dimensional maps with video animation for easy interpretation can be prepared to indicate likely submergence areas. The Federal Emergency Management Agency (FEMA) publishes flood hazard information for
various counties subjected to flooding. As an example, a Flood Insurance Rate
Map, presented in Fig. 15-4, helps identify areas subjected to flooding in St. Clair
County, Illinois.

Table 15-5 *Estimated streamflow during flood.*

Return period, T (years)	Exceedance probability, 1/T	Flood magnitude, Q (cfs)
50	0.02	154,254
100	0.01	204,693
200	0.005	265,672
500	0.002	382,028

15.4.1.1 Environmental Hazard

An *environmental hazard* is an event, or continuing process, that, if realized, will
lead to circumstances having the potential to degrade, directly or indirectly, the

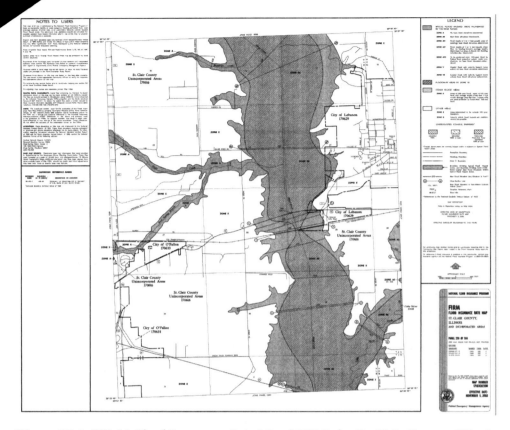

Figure 15-4 *FEMA Flood Insurance Rate Map (FIRM) for St. Clair County, Illinois.*

quality of the environment in the short or longer term. Risks affecting the environment should not be confused with risks caused by environmental effects, either natural or anthropogenic.

15.4.2 Expression of Risk

There are considerable difficulties in determining and quantifying the public and political understanding of risk, and especially in defining an acceptable social risk. One of the most widely used expressions is individual risk, which may be in terms of mortality rate or harm caused to an individual. Another measure may be to relate the detriment to any one of a variety of measures of activity rather than to a simple unit of time. Loss-of-life expectancy is another expression of risk. Frequency against consequence, a graph of frequency of events versus the consequence, can be used to estimate risk.

15.4.2.1 Individual Risk

Individual risk is the frequency at which an individual may be expected to sustain a given level of harm from the realization of a specified hazard.

15.4.2.2 Societal Risk

Societal risk is the relation between frequency of occurrence and the number of people in a given society suffering from a specified level of harm from the realization of a specified hazard. Public attitudes favoring less risk are compatible with the desire to develop better technology and greater regulation of technology.

15.4.2.3 Economic Risk

Economic risk is related to financial losses that represent the commercial consequences of a hazard. The financial loss associated with a product or system from potential hazards may be caused by loss of production or damage or be the result of other financial consequences. Risk to human safety can also have economic consequences. The loss can be either partial or total, and it can be temporary or permanent. The loss can be financial or related to safety. Costs can include those associated with capital, operations, maintenance, and/or life cycles.

15.4.2.4 Environmental Risk

Environmental risk is a measure of potential threats to the environment; it combines the probability that events will cause or lead to degradation of the environment and severity of that degradation. A common application of risk assessment methods is to evaluate human health and ecological impacts of chemical releases into the environment. Information gathered from environmental data collection, monitoring, or modeling is incorporated into models of human and ecosystem exposure, and conclusions on the likelihood of adverse effects are determined. As such, risk assessment can be an important tool for making decisions with environmental consequences. Almost always, when the results from environmental risk assessment are used, they are incorporated into the decision-making process along with economic, societal, technological, and political consequences of a proposed action. The two most widely used examples of environmental risk assessment are (1) human health risk assessment, to examine the effects of an agent on humans, and (2) ecological risk assessment, to examine the effects of an agent on ecosystems.

15.4.2.5 Human Health Risk Assessment

Human health risk assessment has received increased attention because of the recognition of both the potential threat to human health from hazardous substances and the potential for releases into the environment. Recognizing the extent of the hazardous waste problem and the role of risk assessment, the U.S.

Environmental Protection Agency has developed assessment procedures that are used for a variety of purposes. Risk assessment is used for designating substances as hazardous and establishing minimum quantities for reporting releases when they would present substantial danger. In addition, risk assessment is used to evaluate the relative dangers of various sites to establish priorities for response actions and for developing, evaluating, and selecting appropriate response actions at the contaminated site. For example, risk assessment is used to evaluate threats to public health posed by a Superfund site.

The risk assessment is carried out in four steps (USEPA 1989). The first step is hazard identification, in which chemicals of concern are selected based on their toxicity, mobility, spatial distribution, and concentration. In the second step, exposure assessment, all possible exposure pathways (e.g., inhalation, ingestion, and dermal) are identified. In the third step, intake doses of the pre-identified contaminants absorbed through the various exposure routes are estimated. The final step is risk characterization, in which the magnitude of risk is calculated. Quantitative uncertainty analysis is necessary when screening level calculations indicate a potential problem, remediation may result in high costs, or it is necessary to establish the relative importance of contaminants and exposure pathways.

After the exposure point concentrations of specific chemicals through relevant pathways are estimated, it is necessary to estimate the amount of a substance taken by a person. For calculating the chronic daily intake, the generic equation for intake dose is given as

$$CDI = \frac{C \times CR \times EF \times ED}{BW \times AT} \tag{15.7}$$

where CDI is the chronic daily intake (mg/kg/day), C is the chemical concentration, contacted over the exposure period (mg/L), CR is the contact rate, the amount of contaminated medium contacted per unit time (L/day), EF is the exposure frequency (day/year), ED is the exposure duration (years), BW is the body weight, and AT is the averaging time.

The risk from a carcinogenic chemical is calculated as

$$\text{Risk} = CDI \times SF \tag{15.8}$$

where CDI is the chronic daily intake and SF is the carcinogen slope factor.

The risk from a noncarcinogenic chemical is calculated as

$$HI = CDI/RfD \tag{15.9}$$

where HI is the hazard index, CDI is the chronic daily intake, and RfD is the reference dose.

Example 15.3 To demonstrate application of this method of human health risk quantification, let us consider risk assessment associated with the ingestion of contaminated soils. Ingestion of soils contaminated by high-molecular-weight contaminants, such as polychlorinated biphenyl (PCB), is a potential source of human exposure to toxicants. The following equation (USEPA 1990) is used to estimate the probability of excess lifetime cancer R_c from the ingestion of contaminated soil:

$$R_c = \frac{CS \times IR \times CF \times FI \times EF \times ED}{BW \times AT} SF \tag{15.10}$$

where CS is the chemical concentration in the soil (mg/kg), CF is a conversion factor (10^{-6} kg/mg), IR is the ingestion rate (mg soil/day), FI is the fraction ingested from contaminated sources (nondimensional), EF is the exposure frequency (days/year), ED is the exposure duration (years), BW is the body weight (kg), AT is the averaging time (period over which exposure is averaged in days), and SF is the slope factor or cancer potency factor ((kg-day)/mg).

Solution Using the FOA method described in Chapter 14, we find that the mean of a function $y = x_1 x_2 ... x_n$ is

$$\bar{y} = \bar{x}_1 \bar{x}_2 ... \bar{x}_n \tag{15.11}$$

Applying Eq. 15.10 in the context of Eq. 15.7 allows us to determine the expected value of the excess lifetime cancer R_c as

$$R_c = \frac{155 \times 100 \times 10^{-6} \times 0.909 \times 17.4 \times 13 \times 2.25}{15.6 \times 70 \times 365} = 1.8 \times 10^{-5}$$

It has been customary to represent the environmental risk in terms of its expected value. Let us assume that there is another pathway (e.g., through groundwater or inhalation) of being exposed to the contaminant. Now, one has to determine the human health risk through this pathway too and compare it with the other to set the priorities of mitigation. But there is a shortcoming in the approach that allows one to distinguish between two pathways on the basis of their expected values of risk. If the expected values of both the risks are equal, one cannot decide which pathway is riskier as there is a state of ambiguity. Similar situations may arise when risks at multiple sites are compared to identify the site having the highest risk. For this reason, it is necessary to determine the uncertainty involved in calculating the values of risks. As a device for assessing the quality of risk, one can use the coefficient of variation, which is given as (Eq. 10.50, Chapter 10)

$$CV_y = \sqrt{CV_1^2 + CV_2^2 + ... + CV_n^2} \tag{15.12}$$

Using Eq. 15.12 and the CV values of various parameters given in Table E15-3 gives the coefficient of variation of excess lifetime cancer R_c from the ingestion of contaminated soil as 2.57. Because $CV_y \gg 1$, there is considerable uncertainty in the R_c value and thus its use as a representative risk is not justified.

Table E15-3 *Statistical data for Example 15.3.*

Parameter	Symbol	Parameter values	
		Mean	CV
Contaminant concentration (mg/kg)	CS	155	0.39
Ingestion rate (mg/day)	IR	100	1.26
Fraction ingested	FI	0.909	0.03
Exposure frequency (days/year)	EF	17.4	1.0
Exposure duration (years)	ED	13.0	1.0
Body weight (kg)	BW	15.6	0.23
Averaging time (years)	AT	70.0	0.19
Slope factor (kg-day/mg)	SF	2.25	1.66

15.4.2.6 Ecological Risk Assessment

Ecological risk is defined as the conditional probability or likelihood of an adverse ecological event occurring, along with an evaluation of its consequences. An adverse ecological event might include local species extinction, population change, change in community structure, change in growth and reproduction, individual loss, ecosystem stability, and physiological processes such as photosynthesis, energy flow, and nutrient cycling. An ecological risk assessment is a qualitative or quantitative appraisal of the actual or potential impacts of stressors (i.e., contaminants) on plants and animals at a site, other than humans and domesticated species. It determines whether living organisms and/or their environment have been adversely affected, or may be affected in the future owing to existing conditions. It uses information from scientific studies, surveys, and site characteristics to estimate ecological risk. Ecological risk exists when a stressor (contaminant) is in contact with any part of the ecosystem long enough and at a level that is able to cause an adverse effect. Unlike human health risk assessments, ecological risk assessments usually address risk at the population, community, or ecosystem level. An ecological stressor is something (e.g., a chemical compound) that has the potential to cause an adverse effect.

15.4.3 Assessment of Risk

After the hazard-producing events have been identified and their data have been collected, an assessment of risk can be made. In many cases, unwanted situations arise if a variable exceeds or is below a critical value. Flood is a typical example of the former and drought of the latter. But there may be instances when the critical value itself is not a constant. For example, consider estimating the probability of exceedance of a critical level of pollution in a river. Besides the rate of entry of pollutants, the water quality also depends upon discharge and the self-purifying capacity of the river. A data bank on the occurrence of events and failures is needed for estimation of engineering risk and many such data banks are available.

Example 15.4 The town of Risky is located on the banks of Floody River. The river flow follows an extreme value type I distribution with a mean of 256.8 m^3/s and a standard deviation of 78.2 m^3/s. The rating curve of Floody River at a gauging site near Risky is given by

$$Q = 97.03 \times (h - 214.5)0.64 \qquad\qquad (15.13)$$

The Noah Shipping Company wants to construct an office on a plot of land near the river at an elevation of 223.0 m. Find the risk of flooding at this plot every year. What will be the risk if the ground elevation is raised 0.75 m by filling?

Solution We first need to calculate the parameters of the EVI distribution as described in Chapter 5. Here $m_Q = 256.8$ m^3/s and $\sigma_Q = 78.2$ m^3/s. Therefore,

$$\alpha = \frac{1.282}{78.2} = 0.0164$$

$$u = 256.8 - \frac{0.5772}{0.0164} = 221.605$$

From the rating curve expressed by Eq. 15.13, the flow at stage 223 m will be 381.7 m^3/s. Hence the probability that the flow in a given year will exceed a value $q = 381.7$ m^3/s will be

$$P[Q \geq q] = 1 - F_Q(q) = 1 - \exp - \exp[-0.0164(q - 221.605)]\}$$

The probability that this flow q is exceeded is

$$P[Q \geq 381.7] = 1 - \exp\{-\exp[-0.0164(381.7 - 221.605)]\}$$
$$= 0.0695$$

Thus there is about 7% risk that the Noah Shipping Company office will be under water in a given year. If this risk is perceived to be high and yet the company wants to build office at this very site, it can consider raising the plinth level by soil filling. If soil is filled so that the ground elevation is raised by 0.75 m, the new elevation will be 223.75 m. The corresponding discharge (calculated by assuming that the same rating curve remains valid) is 403 m^3/s and for this discharge, the risk of flooding is about 5% each year. Clearly, the risk of flooding has been reduced by about 2% by raising the plinth level by 0.75 m.

The methodology illustrated in this example is employed to construct floodplain zoning maps. If a hazard-producing event leads to failure of a structure or its component, risk assessment includes the consequences of this failure also.

A criticism of risk assessment is that any numerical estimation may be highly uncertain because of its dependence on future human actions and developments in the area. For example, consider a flood control dam that is constructed in an area subject to frequent flooding. As a result of protection provided by the dam, the area may witness rapid growth in industrial and construction activities. If the dam fails 20 years after construction, the damage might be much more than what would have occurred if the dam had not been constructed in the first place.

15.4.4 Risk Mitigation

Risk mitigation is the action phase of risk management wherein the best strategy for risk mitigation is decided. The occurrence of hazards generated by human actions can be largely controlled through robust rules and enforcement but the same is not true for natural hazards. The latter have to be understood and dealt with so that their undesired consequences are minimized. For example, one cannot stop the occurrence of hurricanes. All that can be done is to set up a system for their forecasting and take steps to minimize the damage to life and property. For risk mitigation to be effective, it must be part of an overall decision-making process and the procedures should be reviewed and modified on a routine basis in light of new information.

15.4.5 Safety and Safety Management

Safety is a measure of the freedom from unacceptable risks of personal harm. The objective of applying organizational and management principles is to achieve optimum safety with high confidence. This encompasses planning, organizing, controlling, coordinating all contributory development, and operational activities.

15.4.6 Communication of Risk

To distribute timely warnings during an emergency, to change beliefs and behavior, or to avoid unintended consequences constitutes communication of risk. Such communication provides information about the existence and nature of a threat and the seriousness of risk and details the steps that can be taken to mitigate its effects. Successful communication requires desire and interest by both the giver and receiver of information. Generally, if the action required to mitigate risk is likely to cost money or change habits, people are likely to reject new information, rationalize why it is not applicable to them, find fault with the information or its source, or otherwise create a way to avoid dealing with risk (HEC 1990). People, in general, have erroneous notions about risk. For example, the common thinking is that after a large flood has occurred, the probability of another similar flood is very small. In reality, however, such probability has not at all diminished. Common ways of communicating risk are through mass media, public meetings, and written material.

15.4.7 Institutional Design

For public management of risk, there are three territorial levels: supranational, national, and subnational. Each is subject to a complex set of underlying rules.

15.5 Risk Modeling

Fundamental to risk modeling is the assessment of uncertainties. These cover a broad range of types. Some are quantifiable, whereas others are not.

15.5.1 Statistical Independence–Based Risk Model

Consider a hazard time series where at each integer value of time, say, 1, 2, 3, …, along the time axis t, hazard events occur. The hazard time series may be wave heights, flood discharges, temperatures, and so on. Each hazard event has an intensity, denoted by X, which measures the degree of danger the event unleashes. It is assumed that X is a random variable with the distribution function $F(x)$ defined as the probability that X will be less than or equal to x (a specific value of X) for one of the selected future events. In mathematical notation, $F(x) = P(X \leq x)$, where $P(.)$ denotes the probability that the event within parenthesis will not occur. It is assumed that the intensities of hazard events are statistically independent and have the same distribution $F(x)$. For statistical independence, the time scale for measuring X may be in years. If the flow discharge of a river at a given gauging station, for example, represents the time series, then X may represent the instantaneous maximum annual flood discharge. At the end of the first year ($t = 1$), the intensity of the hazard event would be the maximum discharge for that year. Similarly, if the time series is represented by the wave height at a particular location, then the largest wave height during a year would be one value of X for that year.

15.5.2 Return Period and Waiting Time

In hydraulic design, the return period concept is commonly used, because of its apparent simplicity. However, it is prone to misinterpretation and misuse. The return period $T(x)$ is defined as

$$T(x) = \frac{1}{1 - \text{Prob}(X \leq x)} = \frac{1}{\text{Prob}(X \geq x)} \tag{15.14}$$

where $T(x)$ represents the average time between hazard events having intensities equal to or exceeding x. It does not mean that the event will certainly take place. The events occurring in the time interval $n - 1 < t \leq n$ are plotted at $t = n$, not at their actual time of occurrence in the interval. $T(x)$ will be slightly greater than the real average time between the events with intensities x. The magnitude of the difference will depend on the time scale used. For example, the return period for the time measured in years will be different from the return period measured in decades.

Consider a random variable $W(x)$ that denotes the time interval between two successive exceedances of x. Then, $W(x)$ can be referred to as the waiting time. If an exceedance of x has just occurred, one can compute the probability that the next exceedance will occur n time units away. Because of the assumption of statistical independence, what occurs during one time unit has no effect on the probabilities of future occurrences. This means that the same result would be obtained if any integer on the time axis was selected as the initial point, irrespective of whether an exceedance has just previously occurred. If $W = n$, this means that there must have been $n - 1$ hazard events without an exceedance of x (probability of each $= q$) followed by an exceedance (probability $= p$). Here $p = 1 - F(x)$ and $q = F(x) = 1 - p$. Thus,

$$P(N = n) = q^{n-1} p = [F(x)]^{n-1} [1 - F(x)], \quad n = 1, 2, \ldots$$

$$= q^{n-1} (1 - q) = q^{n-1} - q^n \tag{15.15}$$

The expected or theoretical average of n can be expressed as

$$T = 1\, P(N = 1) + 2\, P(N = 2) + 3\, P(N = 3) + \ldots \tag{15.16}$$

or

$$T = E[N] = \sum_{n=1}^{\infty} n\, P(N = n) = \sum_{n=1}^{\infty} n q^{n-1} - \sum_{n=1}^{\infty} n q^n \tag{15.17}$$

$$= [1 + 2q + 3q^2 + 4q^3 + \ldots] - q - 2q^2 - 3q^3 - \ldots$$

$$= 1 + q + q^2 + q^3 + \ldots = \frac{1}{1-q} = \frac{1}{1-F(x)} \tag{15.18}$$

This equation can also be derived by noting that

$$\begin{aligned} T &= p(1 + 2q + 3q^2 + 4q^3 + \ldots) \\ &= p\frac{d}{dq}(1 + q + q^2 + q^3 + \ldots) \\ &= p\frac{d}{dq}(\frac{1}{1-q}) = \frac{p}{(1-q)^2} = \frac{p}{p^2} = \frac{1}{p} = T = \frac{1}{1-F(x)} \end{aligned} \tag{15.19}$$

The quantity T, as used in the hydrologic literature, is the average return period. Thus, we state that, on average, a flood above a level x will occur once every T years. It should be noted that the distribution function of N is a geometric progression:

$$P(N \le n) = p(1 + q + q^2 + q^3 + q^4 + \ldots + q^{n-1}) = p\frac{1-q^n}{1-q} = 1 - q^n \tag{15.20}$$

Since $q = F(x) = 1 - \dfrac{1}{T}$, Eq. 15.20 can be expressed in terms of the return period as

$$P(N \le n) = 1 - \left(1 - \frac{1}{T}\right)^T \tag{15.21}$$

Next, our interest is in computing var [N]. By definition, we have

$$\text{var }[N] = E\,[N]^2 - [E\,(N)]^2\,,\,0 < p < 1 \tag{15.22}$$

Now,

$$E[N]^2 = \sum_{n=1}^{\infty} n^2\,P(N=n) = \sum_{n=1}^{\infty} n^2\,q^{n-1} - \sum_{n=1}^{\infty} n^2\,q^n$$

$$= [1 + 2^2 q + 3^2 q^2 + 4^2 q^3 + \dots\,] - q - 2^2 q^2 - 3^2 q^3 - \dots$$

$$= 1 + 3q + 5q^2 + 7q^3 + 9q^4 + 11q^5 + \dots = \frac{1+q}{(1-q)^2}$$

Hence,

$$\text{var}[N] = \frac{1+q}{(1-q)^2} - \frac{1}{(1-q)^2} = \frac{q}{(1-q)^2} = \frac{q}{p^2} \tag{15.23}$$

Example 15.5 Given that $q = 0.9$ and 0.99, find the variance of the return period.

Solution When $q = 0.9$, $p = 1 - 0.9 = 0.1$. Hence, $T = 10$ and

$$\text{var}[T] = \frac{q}{p^2} = \frac{0.9}{0.01} = \frac{9}{0.1} = 90$$

In the second case, if $q = 0.99$, then $p = 0.01$ and $T = 100$, and so

$$\text{var}[T] = \frac{0.99}{0.0001} = \frac{99}{0.01} = 9900$$

These examples show that $E[N]$ is not a good measure. Therefore, it is better to calculate the probability of no exceedance within a given period, say, n; that is, we want

$$P(N > n) = \alpha$$

$$= \sum_{k=n+1}^{\infty} P(N=k) = \sum_{k=n+1}^{\infty} q^{k-1}(1-q)$$

$$= \sum_{k=n+1}^{\infty} q^{k-1} - \sum_{k=n+1}^{\infty} q^k \tag{15.24}$$

$$= (q^n + q^{n+1} + \dots) - (q^{n+1} + q^{n+2} + \dots)$$

This yields

$$\alpha = q^n = [F(x)]^n \tag{15.25}$$

This equation has three variables: α, n, and x. Recall that

$$T = \frac{1}{1-q} \quad \text{or} \quad q = 1 - \frac{1}{T}$$

$$\alpha = \left(1 - \frac{1}{T}\right)^n = \left(1 - \frac{1}{T}\right)^T \quad \text{for} \quad n = T \tag{15.26}$$

$$\lim_{T \to \infty} \left(1 - \frac{1}{T}\right)^T = e^{-1} \cong 0.368 \tag{15.27}$$

or P (at least one exceedance within a large return period) $= 1 - 0.368 = 0.632$.

Example 15.6 Let the return period of a flood be $T = 10$ years. Find the probability of at least one exceedance within the return period.

Solution Given $T = 10$, then $\alpha = (1 - 0.1)^{10} = 0.9^{10} = 0.356$. Therefore, for $T = 10$, $P(A) = 1 - 0.356 = 0.644$.

15.5.3 Relation to Risk

Let p be the probability that a value of the random variable X will be equal to or greater than x. Here $q = 1 - p$ or $p = 1 - q$. The probability that x will occur in the next year by definition is $p = 1/T$. The probability that x will not occur in the next year is

$$q = 1 - p = 1 - \frac{1}{T}$$

The probability that x will be equaled or exceeded in any n successive years is given by

$$\left(1 - \frac{1}{T}\right)^n$$

The probability that x will occur for the first time in n years is

$$q^{n-1} p = \left(1 - \frac{1}{T}\right)^{n-1} \frac{1}{T}$$

The probability that x will occur at least once in the next n years is the sum of the probabilities of its occurrence in the first, second, ... to nth years and is therefore

$$p + pq + pq^2 + \ldots + pq^{n-1}$$

Thus, the probability that the event will occur only once is

$$R=1-q^n=1-\left(1-\frac{1}{T}\right)^n$$

The probability R is called risk. This can also be obtained directly from the probability of nonexceedance in n years as

$$R=1-\left(1-\frac{1}{T}\right)^n \tag{15.28}$$

This equation can be used to calculate the probability that x will occur within its return period:

$$P_T=1-\left(1-\frac{1}{T}\right)^T \tag{15.29}$$

For large T, as already shown,

$$P_T=1-e^{-1}=0.63$$

This indicates that the probability that x will occur within its return period is about 64%. Thus a dam designed to withstand a flood with a 25-year return period has a 64% chance that this design flood will be exceeded before the end of the first 25-year period.

For design purposes it might be desirable to specify some probability that the undesirable event would occur within the design period and calculate the required return period. If R is the risk that the event will occur within the design period then

$$R=1-q^n=1-\left(1-\frac{1}{T}\right)^n$$

Thus, one can compute the values of the design return period T corresponding to a number of values of the risk R and the design period n.

The probability R is also called the encounter probability. Suppose a dam is built for a postulated life of L time units (say, years). The probability that an event with intensity x will occur during the life of the dam is the encounter probability, $E(x)$, and is a measure of risk. The probability of no exceedance during L time units is $[F(x)]^L$. Hence, the probability of one or more exceedances is

$$E = 1 - [F(x)]^L \tag{15.30}$$

The relationship between E and T is given by

$$E=1-\left[1-\frac{1}{T}\right]^L \tag{15.31}$$

Equations 15.30 and 15.31 have the same appearance because there are one or more exceedances of x in time L if $N \leq L$. Hence, $E = P(N \leq L)$. Table 15-7 shows values of the encounter probability for various values of the estimated life L and return periods T. Table 15-6 shows the return periods for various values of the encounter probability and estimated life. A comparison of the waiting time and the encounter probability brings out several interesting properties. For example, a dam with a 50-year life has a better than even chance of encountering 50-year floods during its life. Indeed the probability is 0.636. Thus, depending on the amount of risk one is willing to take, a much higher return period flood will have to be used for a 50-year dam. As an example, for a 10% risk, a 475-year flood will have to be used.

Table 15-6 *Return periods T_1 for estimated life L and encounter probability $E_1 [= 1 - (1 - 1/T)^L]$.*

L	E_1								
	0.02	0.05	0.10	0.15	0.20	0.30	0.40	0.50	0.70
1	50	20	10	7	5	3	3	2	1
2	99	39	19	13	9	6	4	3	2
3	149	59	29	19	14	9	6	5	3
4	198	78	38	25	18	12	8	6	4
5	248	98	48	31	23	15	10	8	5
6	297	117	57	37	27	17	12	9	6
7	347	137	67	44	32	20	14	11	6
8	396	156	76	50	36	23	16	12	7
9	446	176	86	56	41	26	18	13	8
10	495	195	95	62	45	29	20	15	9
12	594	234	114	74	54	34	24	18	10
14	693	273	133	87	63	40	28	21	12
16	792	312	152	99	72	45	32	24	14
18	892	351	171	111	81	51	36	26	15
20	990	390	190	124	90	57	40	29	17
25	1238	488	238	154	113	71	49	37	21
30	1485	585	285	185	135	85	59	44	25
35	1733	683	333	216	157	99	69	51	30
40	1981	780	380	247	180	113	79	58	34
45	2228	878	428	277	202	127	89	65	38
50	2475	975	475	308	225	141	98	73	42

Example 15.7 Suppose a dam is designed with a projected life of 25 years. The designer wants to take only a 10% chance that the dam will be overtopped within this period. What return period flood should be used?

Solution Given $n = 25$, $R = 10\%$, one gets $T = 238$ years. This is the return period of the flood one should use in design. A useful approximation for the previous expression for R is

$$T = n \left(\frac{1}{R} - \frac{1}{2} \right)$$

This is a good approximation if $n \geq 10$ and $R \leq 0.5$.

To give an idea of the magnitude of the error involved, consider the problem of estimating the probability of occurrence of some value of the annual maximum flood. For the samples of sizes likely to be used in environmental and water resources engineering, the 95% confidence interval for an estimated probability is surprisingly large. For a value with an observed return period of 10 years (i.e., probability 0.10 from a sample of 30), one can state, with 95% confidence, that the true probability lies between 0.02 and 0.27, corresponding to return periods of 50 and 3.7 years, respectively. In this case, the results are not very meaningful.

15.6 Decision Making Under Uncertainty

As discussed in Chapter 1, it is not possible to deal with every kind of uncertainty. Only the kind of uncertainty that can be measured quantitatively, at least in principle, can be considered. To account for uncertainty in decision making, ideally one must consider the most general type of decision situation in which all available options have been identified, all differences in possible consequences have been determined and quantified, and all probabilities have been assigned to all possible outcomes of each decision. In the real world, however, one will have to make do with a less complete specification of the decision situation. Let us consider the following two examples.

1. An inspector must monitor the quality control of a production process. She knows from experience that, under properly controlled conditions, the probability of a defect is p percent per item. The inspector cannot check all items that are produced, so she takes samples at regular intervals, checks them, and takes action if a sample appears to be significantly inferior to what she expects. The inspector knows that a substandard sample may simply be due to chance; it may also be due to a breakdown in quality control. What is the proper decision when faced with a particular substandard sample?

Table 15-7 Encounter probabilities E_1 [$= 1 - (1 - 1/T)^L$] for estimated life L and return period T_1.

L	\(T_1\)																	
	5	10	15	20	25	30	40	50	60	80	100	120	160	200	250	300	400	500
1	0.200	0.100	0.067	0.050	0.040	0.033	0.025	0.020	0.017	0.012	0.010	0.008	0.006	0.005	0.004	0.003	0.002	0.002
2	0.360	0.190	0.129	0.098	0.078	0.066	0.049	0.040	0.033	0.025	0.020	0.017	0.012	0.010	0.008	0.007	0.005	0.004
3	0.488	0.271	0.187	0.143	0.115	0.097	0.073	0.059	0.049	0.037	0.030	0.025	0.019	0.015	0.012	0.010	0.007	0.006
4	0.590	0.344	0.241	0.185	0.151	0.127	0.096	0.078	0.065	0.049	0.039	0.033	0.025	0.020	0.016	0.013	0.010	0.008
5	0.672	0.410	0.292	0.226	0.185	0.156	0.119	0.096	0.081	0.061	0.049	0.041	0.031	0.025	0.020	0.017	0.012	0.010
6	0.738	0.469	0.339	0.265	0.217	0.184	0.141	0.114	0.096	0.073	0.059	0.049	0.037	0.030	0.024	0.020	0.015	0.012
7	0.790	0.522	0.383	0.302	0.249	0.211	0.162	0.132	0.111	0.084	0.068	0.057	0.043	0.034	0.028	0.023	0.017	0.014
8	0.832	0.570	0.424	0.337	0.279	0.238	0.183	0.149	0.126	0.096	0.077	0.065	0.049	0.039	0.032	0.026	0.020	0.016
9	0.866	0.613	0.463	0.370	0.307	0.263	0.204	0.166	0.140	0.107	0.086	0.073	0.055	0.044	0.035	0.033	0.025	0.020
10	0.893	0.651	0.498	0.401	0.335	0.288	0.224	0.183	0.155	0.118	0.096	0.080	0.061	0.049	0.039	0.032	0.025	0.020
12	0.931	0.718	0.563	0.460	0.387	0.334	0.262	0.215	0.183	0.140	0.114	0.096	0.072	0.058	0.047	0.039	0.030	0.024
14	0.956	0.771	0.619	0.512	0.435	0.378	0.298	0.246	0.210	0.161	0.131	0.111	0.084	0.068	0.055	0.046	0.034	0.028
16	0.972	0.815	0.668	0.560	0.480	0.419	0.333	0.276	0.236	0.182	0.149	0.125	0.095	0.077	0.062	0.052	0.039	0.032
18	0.982	0.850	0.711	0.603	0.520	0.457	0.366	0.305	0.261	0.203	0.165	0.140	0.107	0.086	0.070	0.058	0.044	0.035
20	0.988	0.878	0.748	0.642	0.558	0.492	0.397	0.332	0.285	0.222	0.182	0.154	0.118	0.095	0.077	0.065	0.049	0.039
25	0.996	0.928	0.822	0.723	0.640	0.572	0.469	0.397	0.343	0.270	0.222	0.189	0.145	0.118	0.095	0.080	0.061	0.049
30	0.999	0.958	0.874	0.785	0.706	0.638	0.532	0.455	0.396	0.314	0.260	0.222	0.171	0.140	0.113	0.095	0.072	0.058
35	0.999+	0.976	0.911	0.834	0.760	0.695	0.588	0.507	0.445	0.356	0.297	0.254	0.197	0.161	0.013	0.110	0.084	0.068
40	0.999+	0.985	0.937	0.871	0.805	0.742	0.637	0.554	0.489	0.395	0.331	0.284	0.222	0.182	0.148	0.125	0.095	0.077
45	0.999+	0.991	0.955	0.901	0.841	0.782	0.680	0.597	0.531	0.432	0.364	0.314	0.246	0.202	0.165	0.140	0.107	0.086
50	0.999+	0.955	0.968	0.923	0.870	0.816	0.718	0.636	0.568	0.467	0.395	0.342	0.269	0.222	0.182	0.154	0.118	0.095

2. A new concrete-casting procedure is being tested in a factory. When compared with the present procedure, it appears to result in fewer defective casts. But in a limited number of comparisons, the difference may be due to chance. Should management be advised to change to the new procedure considering that the results are, to a degree, uncertain and the cost of the changeover considerable?

These two decision situations and many others like them have in common that there are only two options. The inspector might take an action, or she might decide not to take any action. A changeover in the concrete-casting procedure may or may not be initiated. In each case, there are two possible but unknown "states of nature." A breakdown in control has or has not occurred; the new procedure is or is not better. Assumptions regarding the state of nature are commonly called hypotheses. One can call the hypothesis of no breakdown and no improvement as hypothesis I, or the null hypothesis. The alternative is called the alternative hypothesis or hypothesis II. The decisions corresponding to these hypotheses are also indicated by I or II. Either decision may be right or wrong with greater or smaller probability and both desirable and undesirable consequences can be associated with the outcomes "right" or "wrong." One can therefore construct a decision tree for this kind of a decision situation, as shown in Fig. 15-5.

The problem with this kind of a decision tree is that usually only the probabilities of p/I and q/I can readily be determined, since they are conditional upon a known situation: normal quality control conditions or proven performance of the current concrete-casting technique. But what is the probability of events when a new factor, such as breakdown in control or a brand new concrete casting procedure, enters in the picture? It is sometimes possible to tighten up the alternative hypothesis so that the conditional probabilities p/II and q/II can be estimated. Often, however, this cannot be done reliably. Then, one cannot determine the effective monetary value (EMV) or effective utility value (EUV) and neither can one determine the decision rule that optimizes EMV or EUV. An added problem is often the difficulty in quantifying the consequences in these decision problems..

In situations like this, one must accept or reject hypothesis I on the basis of probabilities p/I and q/I only. For example, the inspector does not wish to raise a fuss about quality control unless she is fairly certain that something is wrong. She may therefore decide against action (decision I) unless the probability of being right (p/I) is high, say, about 90%. This means that she will be wrong not more than 10% of the times she decides to act (on average). Similarly, one would not wish to embark on a costly changeover process in a factory unless the probability of being wrong was quite small. That probability might be set at 1% or less.

Although there is a degree of arbitrariness in the probability one accepts of wrongly acting on the null hypothesis, or wrongly taking decision I, such a probability limit is, of course, not set without taking the consequences into consideration. If they are not quantified, the decision rule becomes judgmental. In the statistical literature the error of wrongly rejecting the null hypothesis is usually called a type I

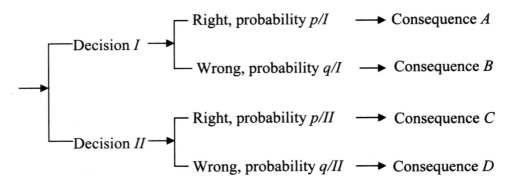

Figure 15-5 Decision tree.

error. Its probability is usually designated by symbol α. A type II error occurs when the null hypothesis is wrongly accepted. Its probability is usually designated by the symbol β. In the decision tree of Fig. 15-5, $q/I = \alpha$ and $q/II = \beta$. We now illustrate these decision problems with a number of examples.

Example 15.8 A steel manufacturing plant produces steel reinforcing bars in accordance with given specifications. An inspector, charged with monitoring the quality control, is informed that under normal operating conditions the tensile strength of a given type of bar is normally distributed with a mean of 275 MPa and a standard deviation of 20 MPa. The inspector is to test bars at regular intervals and report a possible breakdown in quality control if a test bar shows a tensile strength of less than 250 MPa. What is the probability of a false alarm?

Solution The null hypothesis is that there is no breakdown in quality control and that therefore the tensile strength X is assumed to be normally distributed with a mean of 275 MPa and a standard deviation of 20 MPa. The probability that X will be smaller than 250 MPa is equal to the probability that z will be smaller than -1.25 (i.e., $z = (250 - 275)/20 = -1.25$), which is 0.1057 (i.e., Prob($z \le -1.25$) = $\phi^{-1}(-1.25)$ = 0.1057). The probability of a type I error is, therefore, 10.57%. Figure 15-6 shows the PDF of X on the assumption of hypothesis I. This PDF is a conditional distribution. The so-called acceptance region and the critical region for decision I have been indicated as well as the probability of committing a type I error.

It should be noted that there appeared to be no concern about abnormally high tensile strength values in the problem just described and that led to a so-called one-sided test. One could argue, however, that high tensile strength values are often associated with a lack of ductility and are undesirable in reinforcing steel for that reason. This might lead to the requirement that a report is to be made when the test strength falls outside the interval between 250 and 300 MPa. Then one has a two-sided test and the probability of a type I error would rise to 21.14%.

Furthermore, the problem as formulated here does not permit any conclusion about the probability of committing a type II error, namely, falsely

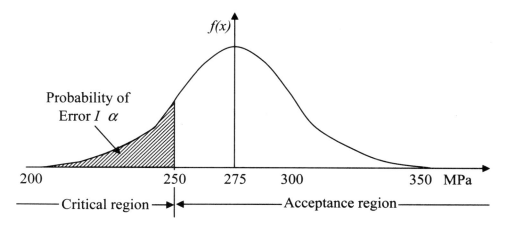

Figure 15-6 *Probability density function of* X *on the assumption of hypothesis* I.

concluding that there has been no breakdown in quality control. Even if it were assumed that X remained normally distributed, one could assign neither a mean nor a standard deviation to the distribution without additional information.

Let us now assume that the inspector is asked not simply to report but to take immediate action if the probability was less than 10% that a too high or too low test value was due to chance only. The inspector must now define the acceptance region for the null hypothesis of no breakdown in quality control. This requires the definition of two limits, x_1 and x_2, such that there is a 5% probability that X is smaller than x_1 and a 5% probability that X is larger than x_2. This decision rule is frequently expressed in terms of confidence or significance. It is said that there is a 90% confidence that decision I is correct. Alternatively, one can say that hypothesis I is rejected (if indeed it is) at the 10% significance level. The determination of the acceptance region is elementary. The limiting values of Z for the 90% confidence interval are ±1.645. The corresponding values for X are 242.1 and 307.9 MPa. Figure 15-7 shows the PDF and the acceptance region.

Obviously, without additional information very little can be said about the probability of a type II error. However, knowledge of the production process and the effect of factors that may get out of control may provide the inspector with some clues. It may well be, for example, that whatever goes wrong is not likely to change the type of distribution and that the change can be expected to manifest itself in the mean of the tensile strength rather than in the standard deviation. The null hypothesis $X \sim N(275, 20)$ may then have to be compared with the alternative hypothesis $X \sim N(\mu, 20)$, where $\mu \neq 275$.

The probability that X lies between the two limits x_1 and x_2, which define the acceptance region, is the probability of the type II error if $\mu \neq 275$. This probability β can readily be calculated if μ is given. Since it is not, all one can do is express β as a function of μ. This function, known as the *operating characteristic curve*, is shown in Fig. 15-8a. The maximum value of β is evidently 0.95 for a value of μ equal to 275 MPa.

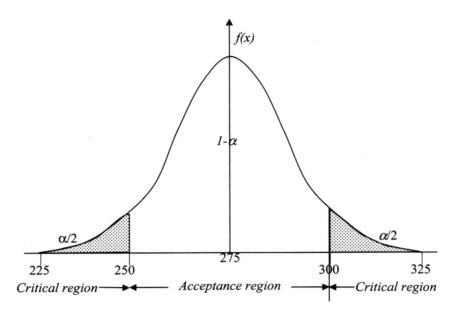

Figure 15-7 Probability density function of X and the acceptance region.

The operating characteristic curve (Fig. 15-8a,b) shows that for larger deviations from the normal μ of 275 MPa, the probability of a type II error becomes progressively smaller. It remains substantial, however, even for very significant deviations from the normal value. This, of course, is because a single test $(N = 1)$ is not a good indicator of the possibility of a breakdown in quality control.

One can reduce the probability of the type II error by narrowing the acceptance region. But that would be at the expense of increasing the type I error. The only way to obtain better control over the production process—that is, to reduce the type II error without increasing the type I error—is by increasing the number of samples, N. Let us assume that the inspector takes $N = 25$ samples from each shift and determines the mean tensile strength. In the null hypothesis, the sample mean is normally distributed with a mean of 275 MPa and a standard deviation of $20 / \sqrt{25} = 4$ MPa. The acceptance region for the sample mean is now $275 \pm 1.645(4) = 275 \pm 6.58$ MPa. The 90% confidence limits for μ are again twice as far from the mean as the limits of the acceptance region, which means 275 ± 13.16 MPa, as shown in Fig. 15-9.

To see a clearer picture of how the number of samples reduces the type II error without increasing the level of type I error, operating characteristic curves were plotted for both $N = 1$ and $N = 25$ in Fig. 15-10. One sees that, for a deviation of 20 MPa in the mean value, there is negligible type II error when 25 samples are examined, but this error may be as high as 82% when only one sample is examined in the quality control process.

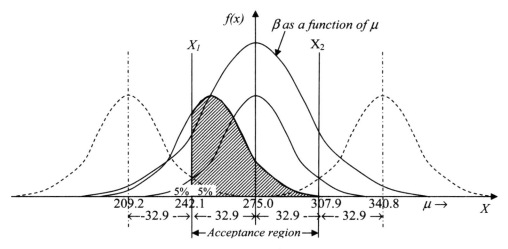

Figure 15-8a *Operating characteristics curve.*

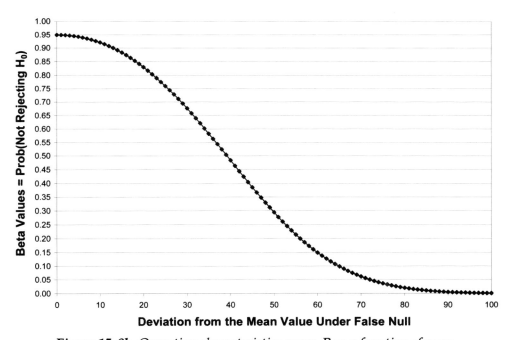

Figure 15-8b *Operating characteristics curve,* B *as a function of mean.*

Example 15.9 A firm uses a wastewater treatment procedure, referred to as method A. In the past, this method has resulted in 37 days out of 228 days when the outfall quality exceeded the prescribed limits. The firm is informed of an alternative method, Method B. Method B is in use at another place, which has produced 28 days of exceeding waste limits out of 337 days. Method B looks superior, but before considering a changeover, the firm wants to be certain that the difference is not due to chance. How should the firm assess the situation?

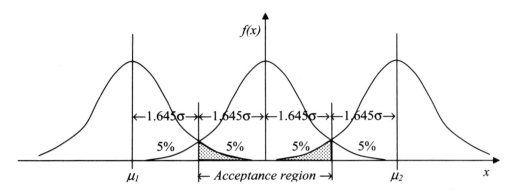

Figure 15-9 *Probability density function and the acceptance region.*

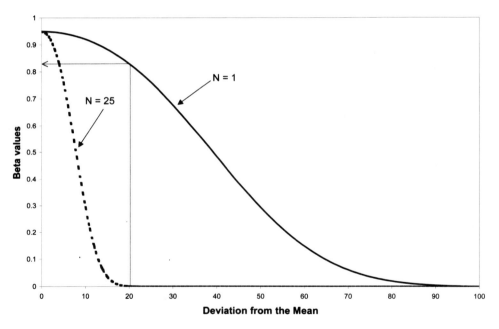

Figure 15-10 *Effect of sample size on reducing the type II error without increasing the type I error.*

Solution The firm argues that there is bound to be a difference in the proportion of water quality violations even if the two methods are equivalent. This leads us to adopt as null hypothesis the assumption that the two methods do not differ in their probabilities of producing violations of waste norms. This hypothesis should not be rejected unless the probability of a type I error is less than, say, 5%.

First, the random variable for which the acceptance region and the critical regions are to be defined is to be identified. One can begin by defining as X_{228} the number of defectives in a sample of 228, and as X_{337} the number of defectives in a sample of 337. One is interested, however, in the difference between X_{337} and X_{228}, which is denoted by Y. Can we determine the probability distribution of Y, assuming the null hypothesis to be true?

If the null hypothesis were true, then one can pool the two samples and state that one had 65 days of exceeding waste limits out of 565 days. The probability of violation per single sample is then $p = 65/565 = 0.115$. Now, X_{337} and X_{228} follow binomial distributions as

$$X_{337} \sim B(337, 0.115)$$

$$X_{228} \sim B(228, 0.115)$$

For X_{337}, the variance is

$$npq = 337 \times (0.115) \times (0.885) = 34.30$$

For X_{228}, the variance is

$$npq = 228 \times (0.115) \times (0.885) = 23.20$$

In both cases, the value of npq is well over 9; hence the approximation is justified. One can write

$$X`_{337} \sim N(337 \times 0.115, \sqrt{34.3}) = N(38.76, 5.86)$$

$$X`_{227} \sim N(228 \times 0.115, \sqrt{23.2}) = N(26.22, 4.82)$$

One can now determine the distribution of $Y` = X`_{337} - X`_{228}$. The mean of $Y`$ is equal to the difference between the means in the $X`$ distributions. Because the two samples are independent, the variance of $Y`$ is equal to the sum of the variances of the $X`$ distributions. Therefore,

$$Y` \sim N[12.54, (34.30 + 23.20)^{1/2}]$$

$$\sim N(12.54, 7.58)$$

The acceptance region is bounded by the $Y`$ value corresponding to $Z = -1.645$ or $Y` = 0.07$. Actually, Y was observed to be $28 - 37 = -9$. It follows that Y lies well within the critical region, and the null hypothesis should be rejected, as shown in Fig. 15-11. The Z value corresponding to the observed value of Y is equal to $(-9 - 12.54)/7.58 = -2.84$. The probability of getting a deviation this large or larger is only 0.23%. The null hypothesis would therefore have to be rejected at the 1% level.

It is not always possible to use the normal approximation. Consider the following example.

Example 15.10 Experience shows that, under the prevalent conditions of control, a production process will show 5% defective items. An inspector is told that he must take action if the probability of a breakdown in control exceeds 95%. Periodically, he takes a sample of 50 items. What minimum number of defectives in this sample should he regard as reason for action?

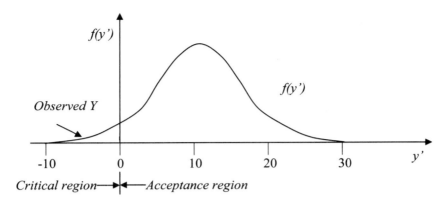

Figure 15-11 *Acceptance and critical regions.*

Solution The random variable for which the acceptance region must be determined is evidently the number of defective items in a sample of 50. Calling this variable X, one can write

$$X \sim B(50, 0.05)$$

if the null hypothesis "no breakdown in control" holds. The variance is now $npq = 50(0.05)(0.95) = 2.375$, which is substantially smaller than 9. The normal distribution is, therefore, not an acceptable approximation. In such cases, the Poisson distribution often makes a good approximation. A rule in this regard is that the Poisson distribution is acceptable if $n \geq 20$ and $p \leq 0.05$. This is the case here. The Poisson distribution has one parameter, the average number of defectives in a sample of 50, which is 2.5:

$$X^{\hat{}} \sim P(2.5)$$

The value of $X^{\hat{}}$ that has a 5% probability of being exceeded can be found in standard tables. The critical number is 6, as shown in Fig. 15-12.

Example 15.11 An experiment is carried out to determine the effect of the loading rate on the measured compressive strength of concrete test cylinders. To this end, 54 test cylinders were loaded to failure at a slow speed and 36 were tested at a fast speed. The results of both tests were found to be normally distributed. The first run gave a mean strength of 28 MPa and the second gave a mean strength of 30.5 MPa. The standard deviation was not significantly different for the two runs and was taken to be 4.0 MPa. Is the difference significant at the 95% confidence level?

Solution For the null hypothesis, one can assume that there is no difference in the strength measurement. The random variable for which the acceptance region must be determined is the difference in the average strength D for samples of size 54 (m_{54}) and 36 (m_{36}):

$$D = m_{36} - m_{54}$$

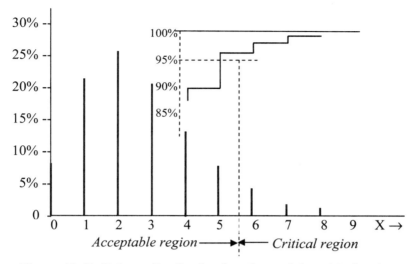

Figure 15-12 Poisson distribution function and the critical region.

Since m_{54} and m_{36} are both normally distributed, the difference is also normally distributed. On the basis of the null hypothesis, there is no difference in the means of m_{54} and m_{36}, and their variances are, respectively, 16/54 and 16/36. One can therefore write

$$D \sim N[0,(\frac{16}{54} + \frac{16}{36})^{0.5}]$$

or

$$D \sim N(0,0.86)$$

To determine the acceptance region, consider that there is little reason to expect the faster test to result in lower strength values. The experiment therefore is regarded as a one-sided test. The acceptance region is then determined by $Z < 1.645$ or $D < 1.41$. It has been observed that $D = 2.5$; this point lies in the critical region and the null hypothesis must therefore be rejected. The difference in measured strength is significant.

Notice that in this example the two samples were assumed to be independent; otherwise one could not add the variances to obtain the variance of the difference. This condition is not always met in practice.

Suppose one would not wish to set up a special testing program for the determination of the effect of the loading rate. Instead, one could, whenever a cylinder had to be tested, produce two cylinders from the same batch. One would be tested at the standard slow rate, but the other at the fast rate. The samples would then be paired. One would expect a correlation between sample items since the effect of different aggregate, different water/cement ratio, different cement content, etc. would be reflected in each of the pairs in the same way. The variance of the difference is then reduced by twice the covariance:

$$\text{var}(X - Y) = \text{var}(X) + \text{var}(Y) - \text{cov}(X,Y) \tag{15.32}$$

The correct procedure in this case is to determine the differences between the paired items first. Let this difference be denoted by

$$Y = X_1 - X_2$$

One then has n values of Y and can determine the mean m_y, which can be assumed to be normally distributed. Assume that n is large so that the variances of Y can be determined reliably from the sample. On the basis of the null hypothesis, there is no difference in the real mean strength, so μ_y is zero. The standard deviation of m_y will be $s_y/n^{0.5}$. One can then determine the acceptance region of m_y and see if the observed value of m_y falls within it.

15.6.1 t Test and F Test

Another complication arises if the true variance is not known but must be estimated from a relatively small sample. It is known that under such circumstances, the mean of a sample of n follows a t distribution with $N - 1$ degrees of freedom. If N is larger than 30, this need not be a problem, since the t distribution for large samples is practically identical with the normal distribution. But if the samples are smaller, then a slightly different procedure must be followed.

Example 15.12 A builder wants to determine whether two kinds of cement, cement A and cement B, produce different tensile strengths in mortar. The builder manufactures six mortar briquettes with each cement and determines the tensile strength for each briquette. The results are as follows:

cement A: 4,600, 4,710, 4,820, 4,670, 4,760, 4,478 MPa

cement B: 4,400, 4,450, 4,700, 4,400, 4,170, 4,100 MPa

Is there a significant difference in the tensile strengths at the 95% confidence level?

Solution Routine calculations show the following data for the sample statistics:

$$m_a = 4673 \text{ MPa} s_a = 121 \text{ MPa}$$

$$m_b = 4370 \text{ MPa} s_b = 214 \text{ MPa}$$

The problem is, in principle, quite similar to the one discussed earlier. One must judge whether the observed difference in sample means m_a and m_b is significant. The first difference, however, is that in the previous problem the standard deviation was given. Here, one has to determine the standard deviation from the samples, and the two observed standard deviations are disconcertingly different. The first step, therefore, should be to determine whether it is reasonable to assume that the two cements do lead to the same standard deviation and that the observed difference is due only to chance. One might anticipate the results and assume that both s_a and s_b come from the same distribution and their difference is due to chance.

This does not solve the problem of what to take for the standard deviation. The proper procedure is to calculate the pooled variance of the two samples by adding the sums of the squares of the deviations from the sample means and dividing that sum by the total number of degrees of freedom, $(N_1 - 1) + (N_2 - 1)$. Since s_a and s_b have already been calculated, one can simply multiply each by $N - 1$, adding the results and dividing by $(N_1 - 1) + (N_2 - 1)$. In this particular case, where the sample sizes are equal, the procedure is the same as taking the average of the variances. The result is that the pooled variance is 30,313 and the pooled standard deviation is 174 MPa.

The second difference from the previous problem is that the sample sizes are small and the standard deviation is determined from the samples. The variance of the mean in a sample of 6 is calculated from $s_m^2 = 30,313/6 = 5,052$, and the variance of the difference between the two means of samples of 6 is equal to twice this number or 10,104. This makes the standard deviation of the difference equal to 100.52 MPa. But this is a sample standard deviation, not a true standard deviation. Under these circumstances, one must use the t distribution instead of the normal distribution. In this case the t distribution has 10 degrees of freedom $(N_1 + N_2 - 2)$:

$$D = 0 + 100.52\, T_{10}$$

At the 95% confidence level, the acceptance region for the t distribution with 10 degrees of freedom is ± 2.228. This gives an acceptance region for D between +214 and –224 MPa. The observed difference was $4,673 - 4,370 = 303$ MPa. This is well outside the acceptance region, so the null hypothesis of "no difference" must be rejected.

One can now consider here only the relatively simple problem of judging the null hypothesis that the variances from two samples are chance observations of the same random variable and that the difference is therefore not significant. It has been seen that the sample variance is a random variable, the distribution of which is related to the χ^2 distribution:

$$s^2 = \frac{\sigma^2}{N-1}\chi^2(N-1) \tag{15.33}$$

Suppose now that one has two samples for which the sample variances have been calculated. One wants to test the null hypothesis that the sample variances are estimates of the same parameter σ^2. One then defines a variable F, which is the ratio of the two sample variances:

$$F = \frac{s_1^2}{s_2^2} \tag{15.34}$$

Substitution of Eq. 15.28 into this equation results in

$$F = \frac{\chi^2(N-1)(N-2)}{\chi^2(m-1)(m-2)} \tag{15.35}$$

where N and m are sample sizes. The terms $(N-1)$ and $(m-1)$ are the degrees of freedom in Eq. 15.35.

The F distribution is a well-known distribution for which tabulated values are readily available. Tables giving values of the cumulative distribution for a range of degrees of freedom and several levels of significance are widely available. It should be noted that to calculate F, one must put the largest variance in the numerator and the smallest in the denominator. This is because the F test has been set up as a one-sided test in which the alternative to the null hypothesis is the alternative hypothesis that s_1 is larger than s_2.

One can demonstrate the use of the F test by examining s_a^2 and s_b^2 in the previous example where these variances had been obtained from two samples of 6 items. There

$$s_a^2 = 14{,}626 \text{ and } s_b^2 = 46{,}000$$

Then one can calculate F to be $46{,}000/14{,}626 = 3.15$.

For 5 degrees of freedom in the numerator and the denominator, one obtains the following critical values for F:

$$\text{for } \alpha = 10\%, F_{crit} = 3.45$$

$$\text{for } \alpha = 5\%, F_{crit} = 5.05$$

$$\text{for } \alpha = 1\%, F_{crit} = 10.97$$

It can be seen that the null hypothesis cannot be ruled out even at the 90% confidence level. Note that the result obtained here is largely negative. The small samples make it impossible to rule out the null hypothesis that the variances are obtained from the same random variable. That does not mean, however, that the results should inspire confidence in the correctness of the null hypothesis.

15.7 Questions

15.1 There is always a great deal of interest in weather. What is the risk from the coldest weather on record occurring next winter in Houston, Texas?

15.2 Consider a small dam pond for local flood control in an urban area. What could the risk be from the failure of this detention structure?

15.3 What is the health risk from air pollution?

15.4 Consider instantaneous peak discharge data for a number of years for a river near the town in which you live. Assume that the peak discharge data follow a two-parameter log-normal distribution. The rating curve at the nearest gauging site is also known. A private company wants to construct a chemical plant near the river. Compute the risk that the plant will not be flooded by a 100-year flood. Also, what will be the reduction in flooding risk if the plant were elevated one meter above the ground?

15.5   Find the variance of the return period for different values of nonexceedance probabilities.

15.6   What is the probability that a 500-year flood will occur at least once in 100 years?

15.7   Suppose a dam is designed with a projected life of 100 years. The designer wants to take only a 5% chance that the dam will be overtopped within this period. What return period flood should be used?

15.8   A chemical plant has installed a pollution abatement device, referred to as Method A. This method has resulted in 40 days out of 300 days when the poor air quality exceeded the prescribed limits. For fear of avoiding penalties, the plant operator considers an alternative method, Method B, which has produced 30 days of exceeding waste limits out of 300 days when used at another place. Method B looks superior, but before considering a changeover, the plant operator wants to be certain that the difference is not due to chance. How should the plant operator assess the situation?

15.9   At a cement manufacturing plant, the quality of cement is to be inspected. It is found that 5% of the cement bags do not meet the prescribed quality standard. The production process must be amended if the probability of defective cement bags exceeds 5%. For cement testing, a sample of 30 bags is used. What minimum number of bags in this sample should be regarded as reason for action? For simplicity, the normal approximation can be employed here.

Chapter 16

Reliability Analysis of Water Distribution Networks

A water distribution network (WDN) is designed and constructed to supply water to the user in accordance with the demand (or load) at sufficient pressure. Knossos, near Heraklion, the modern capital city of Crete, was a large and developed city of Europe during the Neolithic Age (ca. 5700–2800 B.C.). Its inhabitants, numbering tens of thousands, were supplied water through an elaborate network of tubular conduits (Mays 2000). These days, a WDN is considered to be a key infrastructure requirement of a modern city and a measure of the standard of living of the society.

The users of a municipal water supply system are spread all over the city area, which may be of the order of tens of square kilometers. Thus, water demand has a spatial distribution. The demand of water also changes with season, day of the week, and time of day. As a city expands, water demand increases. Occasionally, there may be high demand of water to meet unusual events, such as fires, organization of special events, etc. To meet the needs of the users, water is pumped from a source, such as a river, a reservoir, a lake, or an aquifer. Depending upon the quality of water received, treatment is provided to the raw water. Treated water may be stored in huge tanks before it is supplied to the users. Tanks are necessary so that the pumps can operate at their peak efficiency. Demand is variable: The WDN must be able to provide water with enough pressure, to provide water when a pump or treatment plant is nonoperational, to meet a sudden large demand, and to suppress hydraulic transients.

In a large WDN, pumping stations, treatment plants, and overhead storage tanks are located at many places. These are connected to the demand centers and among themselves through a network of (underground) pipes of various diameters and materials. The pipes join together at nodes. Valves are installed in the network at various places to control the flow of water and its pressure. Major components of a WDN are pumps, storage tanks, pipes, and valves. Typically, a pump gets water from a source. At times, a pipe may break, a pump may stop working, a valve may leak, or a treatment plant may have to be shut down for repair and maintenance.

A reliable WDN meets the demand placed upon it without undue failures. In a WDN, failure is said to take place when either the pressure or flow or both drop below a specified value at one or more nodes. Thus, the reliability of a network is the probability that it can satisfactorily meet the demand. A WDN has two types of reliabilities: mechanical and hydraulic. Mechanical reliability refers to the satisfactory operation of various components, such as, pumps, pipes, and valves in the network. Hydraulic reliability measures the performance of a WDN in meeting its demands in terms of the quantity of water at desired pressure and at required time. Since water is essential for life, WDNs are designed for high reliability, among other things by providing loops and valves to control flow. This chapter focuses on the hydraulic reliability of a WDN.

16.1 Relevant Principles of Hydraulics

Before the concept of reliability of WDNs can be addressed, it is pertinent to discuss relevant basic principles of flow in closed conduits. The texts by Jeppson (1977) and Mays (2000) provide a detailed treatment of WDNs, their components, and their design. For the purpose of mathematical analysis, users of water are grouped together to form nodes of the network and these nodes are connected by pipes, which are represented as links. Here, we consider only the main pipes; the smaller pipes that connect consumers with the network are ignored.

The first principle in dealing with pipe flows is the continuity of matter. According to the principle of continuity for an incompressible fluid, the sum of volumes of water entering a junction (ΣV_{in}) equals the flow leaving the junction (ΣV_{out}) over a given time, that is,

$$\Sigma V_{in} = \Sigma V_{out} \tag{16.1}$$

Further, according to the principle of conservation of energy, the total energy of flow at two cross sections will be the same if there is no energy loss. The total energy in terms of the head of water is expressed by the Bernoulli equation. For flow between two cross sections 1 and 2, we have

$$z_1 + \frac{p_1}{\gamma} + \frac{v_1^2}{2g} = z_2 + \frac{p_2}{\gamma} + \frac{v_2^2}{2g} + h_{L1-2} \qquad (16.2)$$

where z_1 is the elevation head, p_1 is the pressure head, γ is the unit weight of water, v_1 is the velocity of flow; subscript 1 denotes that the variables refer to cross section 1; and h_{L1-2} denotes head loss between cross sections 1 and 2. The terms in Eq. 16.2 are explained in Fig. 16-1. The hydraulic grade line (HGL) is a line that is p/γ above the center line of the pipe. If a piezometer is attached to the pipe, water will rise up to the HGL. The energy grade line (EGL) is $v^2/2g$ above the HGL.

The loss of energy in a pipe network takes place because of the roughness of pipes, turbulence, and viscous stress. Some energy is also lost in contractions, expansions, bends, joints, and valves. This loss is termed as minor loss. Generally, the minor head loss is proportional to $v^2/2g$.

Example 16.1 The discharge passing through a horizontal pipe of diameter = 60mm is 0.005 m^3/s. The pressures at the upstream and downstream sections are 15 and 11 kPa, respectively. What is the head loss in the pipe?

Solution Since the diameter of the pipe is constant, $v^2/2g$ will be the same at both ends. Because the pipe is horizontal, the head loss will be

$$h_L = \frac{p_1}{\gamma} - \frac{p_2}{\gamma} = \frac{(15-11)\text{kPa}}{9.89\text{kN/m}^3} = 0.404 \text{ m}$$

Example 16.2 For the pipe of Example 16.1, let the elevation of the upstream end be 2 m higher than that of the downstream end. Further, at the downstream end, the pipe diameter is 50 mm (whereas it is 60 mm at the upstream end). Find the head loss between the two sections.

Solution The velocity of flow at section 1 will be

$$v_1 = Q/A_1 = 0.005/[\pi(60 \times 10^{-3}/2)^2] = 1.77 \text{ m/s}$$

At section 2,

$$v_2 = Q/A_2 = 0.005/[\pi(50 \times 10^{-3}/2)^2] = 2.55 \text{ m/s}$$

Energy Eq. 16.2 gives

$$2 + \frac{15}{9.89} + \frac{1.77^2}{2 \times 9.81} = 0 + \frac{11}{9.89} + \frac{2.55^2}{2 \times 9.81} + h_{L,1-2}$$

so

$$h_{L1-2} = 2.232 \text{ m}$$

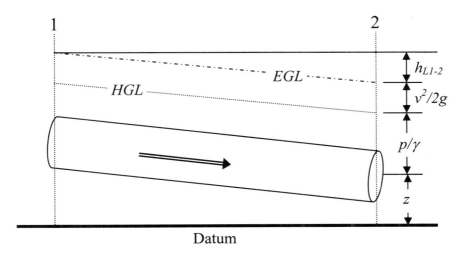

Figure 16-1 *Energy terms for pipe flow.*

16.1.1 Pipe Flow Equations

In a WDN, the velocity of flow in pipes is usually computed by using the Hazen–Williams equation

$$v = 0.849 \, C \, R^{0.63} \, S_f^{0.54} \qquad (16.3)$$

where v is the flow velocity (in meters per second), R is the hydraulic radius (in meters), S_f is the friction slope (in meters per meter), and C is the Hazen–Williams roughness coefficient, which depends upon the pipe properties. The hydraulic radius is the ratio of cross-sectional area and wetted perimeter. For a circular pipe, $R = A/P = \pi r^2 / 2\pi r = r/2$, where r is the radius of the pipe.

For a smooth plastic pipe, a typical value of the coefficient C may be about 150 (metric units). The head loss from friction (in meters) per 1,000 m pipe length can be computed by

$$h_L = \left(\frac{151Q}{CD^{2.63}} \right)^{1.85} = KQ^{1.85} \qquad (16.4)$$

where D is the diameter of the pipe (in meters) and K is the pipe coefficient.

Another commonly used equation for head loss in pipes is the Darcy–Weisbach equation

$$h_L = f \frac{L}{D} \frac{V^2}{2g} = KQ^2 \qquad (16.5)$$

where L is pipe length (in meters) and f is the Darcy–Weisbach friction factor, which depends, among other things, on the relative roughness of the pipe and the Reynolds number (Re). The Reynolds number is the ratio of inertial forces to viscous forces (vD/v, where v is the kinematic viscosity). For laminar flow (Re < 2,100),

$$f = 64/\text{Re} \qquad (16.6)$$

For turbulent flow (Re > 2000), the roughness is obtained from the Karman and Prandtl equations:

$$\text{smooth pipe: } \frac{1}{\sqrt{f}} = 2\log_{10}(\text{Re}\sqrt{f}) - 0.8 \quad \text{for Re} > 3000 \qquad (16.7)$$

$$\text{rough pipe: } \frac{1}{\sqrt{f}} = 1.14 - 2\log_{10}(\frac{k_s}{D}) \qquad (16.8)$$

Knowing the relative roughness and the Reynolds number, one can read the friction factor for a pipe from the Moody diagram. The relative roughness equals k_s/D. Here, k_s is the equivalent sand roughness, which is the resistance character-istics produced by a pipe of the same diameter, internally coated with sand par-ticles having diameter k_s. From a computational viewpoint, it is convenient to use the equation proposed by Swamee and Jain (1976) to compute f:

$$f = \frac{0.25}{\left[\log\left(\dfrac{k_s}{3.7D} + \dfrac{5.74}{\text{Re}^{0.9}}\right)\right]^2} \qquad (16.9)$$

For flow in open channels, the velocity can be computed using Manning's equation

$$v = \frac{1}{n_m} R^{2/3} S_0^{1/2} \qquad (16.10)$$

where S_0 is the slope of the channel bed and n_m is a coefficient, known as Man-ning's roughness coefficient, whose values depend upon the properties of the channel cross section; higher values represent a "rough" cross section. Barnes (1962) has tabulated values of n for natural channels. For a concrete channel, a typical value of n may be 0.013, and n for a straight earthen section may be 0.02. For a steel pipe, flowing partially full, n is about 0.012; it is 0.014 for a cast-iron pipe.

Another popular equation to compute the velocity in a channel is Chezy's equation

$$v = C\,(RS_0)^{0.5} \qquad (16.11)$$

where C is the Chezy's resistance coefficient. It is related with Manning's n_m by

$$C = \frac{1}{n_m} R^{1/6} \qquad (16.12)$$

Example 16.3 A 500-m-long pipe with a diameter of 0.5 m carries water at a velocity of 4 m/s. Determine the head loss in the pipe if the relative roughness is $k_s = 0.0005$ m. Assume the kinematic viscosity is $\nu = 1.0 \times 10^{-6}$ m^2/s.

Solution The Reynolds number is

$$Re = 4 \times 0.5 / (1.0 \times 10^{-6}) = 2 \times 10^6$$

and the relative roughness is $0.0005/0.5 = 0.001$. The value of f can be obtained from the Moody diagram or from Eq. 16.7 or Eq. 16.9. Equation 16.7 gives

$$f = \frac{1}{[1.14 - 2\log_{10}(0.001)]^2} = 0.0196$$

and from Eq. 16.9, one obtains

$$f = \frac{0.25}{\left[\log\left(\dfrac{0.005}{3.7 \times 0.5} + \dfrac{5.74}{(2 \times 10^6)^{0.9}}\right)\right]^2} = 0.0198$$

Now, the head loss can be computed by using Eq. 16.4:

$$h_L = 0.0198 \times 500 \times 4^2 / (0.5 \times 2 \times 9.81) = 16.14 \text{ m}$$

16.1.2 Pipe Networks

In a WDN, pipes may be connected in series, in parallel, or in branches. For the purpose of solution, the subnetworks are represented by an equivalent pipe. A network and a pipe are equivalent when both carry the same discharge for the same head loss. When n pipes are joined in a series (see Fig. 16-2), the total head loss of the equivalent pipe (h_{Le}) with the pipe coefficient (K_e) is the sum of head losses (h_L) of individual pipes with pipe coefficient K:

$$h_{Le} = \sum_{i=1}^{n} h_{Li} \tag{16.13}$$

and

$$K_e = \sum_{i=1}^{n} K_i \tag{16.14}$$

When pipes are connected in parallel as shown in Fig. 16-3, the head loss in each pipe between the junctions will be the same. Thus

$$h_{L1} = h_{L2} = h_{L3} = \ldots$$

$$Q = Q_1 + Q_2 \tag{16.15}$$

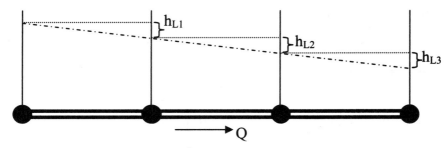

Figure 16-2 *A network of pipes in series.*

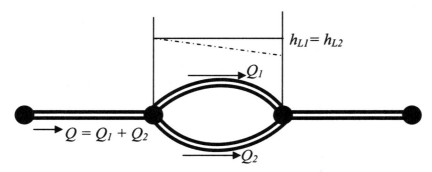

Figure 16-3 *A network of pipes in parallel.*

Substituting the value of Q from Eq. 16.4 or Eq. 16.5, one obtains

$$\left(\frac{h_L}{K_e}\right)^{1/n} = \left(\frac{h_L}{K_1}\right)^{1/n} + \left(\frac{h_L}{K_2}\right)^{1/n} \tag{16.16}$$

or

$$\left(\frac{1}{K_e}\right)^{1/n} = \left(\frac{1}{K_1}\right)^{1/n} + \left(\frac{1}{K_2}\right)^{1/n} \tag{16.17}$$

When the Hazen–Williams equation is used, n will be 1.85; it will be 2 when the Darcy–Weisbach equation is used.

Often, the division of flows in parallel pipes is also required. To that end, we have

$$h_{L1} = h_{L2}$$

so

$$f_1 \frac{L_1}{D_1} \frac{v_1^2}{2g} = f_2 \frac{L_2}{D_2} \frac{v_2^2}{2g}$$

or

$$\frac{v_1}{v_2} = \sqrt{\left(\frac{f_2}{f_1} \frac{L_2}{L_1} \frac{D_2}{D_1}\right)} \tag{16.18}$$

Example 16.4 Two tanks are connected through two pipes as shown in Fig. 16-4. The flow of water from the upper tank to the lower one is at $0.04 \text{ m}^3/\text{s}$ and the Darcy–Weisbach friction factor is 0.03. Find the elevation of water in the lower tank if the elevation of water in the upper tank is 100 m.

Solution The water velocities in the pipes are

$$v_1 = Q/A_1 = 0.04/[3.14 \times 0.10^2] = 1.274 \text{ m/s}$$

$$v_2 = Q/A_2 = 0.04/[3.14 \times 0.15^2] = 0.566 \text{ m/s}$$

The energy equation can be written as

$$z_2 = z_1 - h_{L2} - h_{L2}$$
$$= 100 - 0.03 \frac{400}{0.20} \frac{1.274^2}{2 \times 9.81} - 0.03 \frac{600}{0.30} \frac{0.566^2}{2 \times 9.81} = 94.06 \text{m}$$

Example 16.5 Two tanks are connected through pipes as shown in Fig. 16-5. The length of each pipe is 100 m. The diameter of pipes 1 and 2 is 40 cm and that of pipe 3 is 30 cm. The elevation of water in the first tank is 100 m and it is 90 m in the second tank. If the Darcy–Weisbach friction factor is 0.02 for all the pipes, find the velocity of water in the pipes and the discharge through pipe 1.

Solution Since pipe 2 and pipe 3 are in parallel connection, the head loss in them will be equal. Therefore, from the Darcy–Weisbach equation, one has

$$f \frac{L_3}{D_3} \frac{v_3^2}{2g} = f \frac{L_2}{D_2} \frac{v_2^2}{2g}$$

or

$$f \frac{100}{0.3} \frac{v_3^2}{2 \times 9.81} = f \frac{100}{0.4} \frac{v_2^2}{2 \times 9.81}$$

which upon simplifying gives

$$v_2 = 1.155 v_3$$

The discharge through pipe 1 will be the same as the sum of discharges through pipes 2 and 3. Hence,

$$A_1 v_1 = A_2 v_2 + A_3 v_3$$

so

$$3.14 \times (40^2/4)v_1 = 3.14 \times (40^2/4)v_2 + 3.14 \times (30^2/4)v_3$$

$$1{,}600 v_1 = 1{,}600 \times v_2 + 900 \times v_3$$

$$= 1{,}600 \times (1.155) \times v_3 + 900 \times v_3$$

which gives

$$v_1 = 1.7175 v_3$$

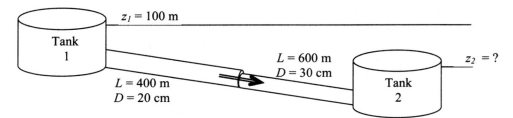

Figure 16-4 *Two tanks connected by pipes in series.*

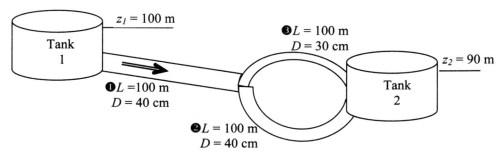

Figure 16-5 *Two tanks connected by pipes in parallel.*

The total head loss through the system is 10 m. This permits us to write

$$0.02\frac{100}{0.4}\frac{v_1^2}{2\times9.81}+0.02\frac{100}{0.3}\frac{v_3^2}{2\times9.81}=10$$

Substituting the value of v_1 and solving yields $v_3 = 3.027$ m/s. Hence,

$$v_1 = 5.199 \text{ m/s}$$

and

$$Q_1 = 3.14 \times (0.40^2/4) \times 5.199 = 0.653 \text{ m}^3/\text{s}$$

16.2 Analysis of a WDN

A typical WDN consists of pipes connected in series, in parallel, and in branches. In addition, there are valves, meters, etc. that result in head loss. To simplify the analysis, the constituents are conceptually combined together to form *equivalent pipe*. The movement of water through a WDN with a known demand must satisfy the laws of conservation of mass and energy. Two approaches are commonly used for steady-state analysis of a WDN. The equations that express the conservation of energy in terms of head are known as loop equations. In another type of formulation, the mass balance is expressed in terms of head at the junction nodes and the resulting equations are known as node equations. These two formulations are discussed in what follows.

16.2.1 Loop Formulation

A pipe network consists of a number of pipes joining together. The end points of the pipes are junction nodes or fixed head nodes. Primary loops in a network include all closed pipe circuits within the network. The number of pipes (P), the number of junction nodes (j), the number of primary loops (n), and the number of fixed head nodes (f) are related by

$$P = j + n + f - 1 \tag{16.19}$$

The mass balance (Eq. 16.1) for each junction node can be written as

$$\Sigma Q_{\text{in},\,j} - \Sigma Q_{\text{out},j} = Q_{e,j} \tag{16.20}$$

where $Q_{e,j}$ is the external inflow and demand at node j. The energy conservation equation for each primary loop can be written as

$$\Sigma h_L = \Sigma E_p \tag{16.21}$$

where h_L is the energy loss in each pipe and E_p is the energy imparted to the flow by the pumps. If no energy is imparted, the right-hand side will be zero.

16.2.2 Hardy–Cross Method

This method developed by Cross (1936) is the oldest systematic method and is frequently used to solve WDN hydraulic design problems. The computation of the loop method begins with a set of assumed flow rates that satisfy the continuity equation. Application of head-loss equations gives nonzero residual head and the discharge corrections are found using the Hardy–Cross formula.

In the Hazen–Williams, Manning, or Darcy–Weisbach methods, the head loss h (in meters) in a pipe carrying discharge at Q (m^3/s) is given by (see Eq. 16.4)

$$h = k\,Q^n \tag{16.22}$$

where k is a constant and n is an exponent. If the Hazen–Williams equation is used, $n = 1.85$.

In the absence of the knowledge of discharge Q flowing through a pipe, let the assumed discharge be Q_1. We can write

$$Q = Q_1 + \Delta \tag{16.23}$$

where Δ is the error in the assumed discharge. Substituting Q in Eq. 16.22 yields

$$kQ^n = k(Q_1 + \Delta)^n = k[Q_1^n + n\,Q_1^{n-1}\,\Delta + \ldots]$$

If Δ is small compared to Q, the higher order terms can be neglected and this yields

$$kQ^n = k[Q_1^n + n\,Q_1^{n-1}\,\Delta] \tag{16.24}$$

For a pipe loop, the sum of head losses for all pipes must be zero. This yields

$$\Sigma k Q^n = 0$$

or

$$\Sigma k [Q_1^n + n\, Q_1^{n-1}\, \Delta] = 0$$

and so

$$\Delta = -\frac{\Sigma k Q_1^n}{\left| n \Sigma k Q_1^{n-1}\right|} = -\frac{\Sigma h}{\left| n (\Sigma h / Q)\right|} \tag{16.25}$$

Note that the term in the numerator has a sign whereas, in the denominator, the absolute value of the terms is taken. Eq. 16.25 is used in the Hardy–Cross method to get the value of correction that is applied to the assumed flow through a pipe to obtain a better value. The steps of the Hardy–Cross method are as follows:

1. Assume a distribution of flow in the network that should satisfy the continuity equation. Ensure that, at a junction, the sum of flows entering must be equal to the sum of flows leaving.
2. Determine the head loss in each pipe. By convention, clockwise flows are given a positive sign and counterclockwise flows are given a negative sign.
3. Compute head loss for each loop.
4. Determine correction term using Eq. 16.25. If the largest of the corrections is smaller than a predetermined limit, stop. Otherwise, compute corrected flows and go to step 2.

The computations of this method can be easily programmed on a computer.

Example 16.6 A simple WDN is shown in Fig. 16-6. The network consists of two loops. The diameters and lengths of the various pipes are shown in the diagram. Water enters the network at node A and the direction of flows through the various pipes is shown with the help of arrows. The demands at various nodes are shown in the diagram. Assume the roughness coefficient is $C = 100$ for all pipes. Compute the flow in each pipe by the Hardy–Cross method.

Solution To begin calculations, assume a flow in each pipe such that the demands are met and mass balance is maintained. Now compute the head loss for each pipe. The computations are carried out iteratively. As an example, the head loss h_L for pipe AB is computed by the use of Eq. 16.4 with the Hazen–Williams roughness coefficient $C = 100$:

$$h_{LAB} = \left(\frac{151 \times 0.9}{100 \times 0.55^{2.63}}\right)^{1.85} = 40.4233\text{m}$$

The solution is given in Table E16-6a.

Therefore, the correction to the flows for the top loop is 0.0485 m³/s and for the bottom loop, it is –0.0298 m³/s. For pipe BE, which is common to both loops, the net correction will be 0.0485 – 0.0298 = 0.0187 m³/s. With the corrected flows, we proceed to the second iteration (Table E16-6b).

At this stage, the largest of the corrections is quite small and the discharges obtained after this correction can be considered to be close to the true values. Generally, the convergence is rapid in the Hardy–Cross method even if the initial guess is not good.

16.2.3 Node Formulation: Nonlinear

In this formulation, the analysis is carried out in terms of the unknown total head (H) at each junction node. By using the continuity equation, the discharge in a pipe that connects nodes i and j can be written as

$$Q_{ij} = [(h_i - h_j)/K_{ij}]^{1/1.85} \tag{16.26}$$

where h_i and h_j are the heads at nodes i and j, respectively, and K_{ij} is the pipe coefficient for the connecting pipe. The main advantage of the node formulation is that it has fewer equations, but the equations are nonlinear, which means they are difficult to solve by hand. Figure 16-7 shows a node that receives flow from node i and two pipes carrying water to nodes $i + 1$ and $i + 2$. Also, the discharge Q_{out} leaves the network at this node.

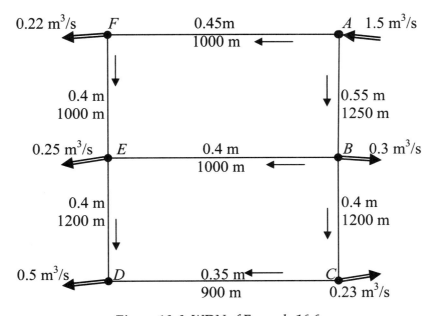

Figure 16-6 *WDN of Example 16.6.*

Table E16-6a First iteration.

Pipe	Q (m³/s)	Diameter (m)	Length (m)	h_L (m)	h_L/Q [m/(m³/s)]
Top loop *ABEF*					
AB	0.9	0.55	1250	40.4233	44.9147
BE	0.2	0.4	1000	9.4230	47.1151
FE	−0.38	0.4	1000	−30.8947	81.3019
AF	−0.6	0.45	1000	−40.5491	67.5819
			Sum	−21.5976	240.9136
		Correction $\Delta_1 = -(-21.5976)/(1.85 \times 240.9136) = 0.0485$			
Bottom loop *BCDE*					
BC	0.43	0.4	1200	40.7640	101.9100
CD	0.17	0.35	1000	13.3590	78.5825
DE	−0.33	0.4	1200	−28.5573	86.5371
EB	−0.2	0.4	900	−8.4807	42.4036
			Sum	17.0850	309.4332
		Correction $\Delta_2 = -(-17.085)/(1.85 \times 309.4332) = 0.0298$			

Table E16-6b Second iteration.

Pipe	Q (m³/s)	Diameter (m)	Length (m)	h_L (m)	h_L/Q [m/(m³/s)]
Top loop *ABEF*					
AB	0.9485	0.55	1250	44.5453	46.9639
BE	0.2187	0.4	1000	11.1174	50.8342
FE	−0.3315	0.4	1000	−23.9982	72.3928
AF	−0.5515	0.45	1000	−34.6945	62.9094
			Sum	−3.0300	233.1003
		Correction $\Delta_1 = -(-3.03)/(1.85 \times 233.1003) = 0.007$			
Bottom loop *BCDE*					
BC	0.3702	0.4	1200	35.3243	95.4194
CD	0.1402	0.35	1000	9.3525	66.7083
DE	−0.3598	0.4	1200	−33.5103	93.1360
EB	−0.2187	0.4	900	−10.0057	45.7508
			Sum	1.1607	301.0145
		Correction $\Delta_2 = -(-1.1607)/(1.85 \times 301.0145) = -0.0021$			

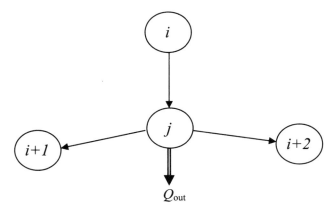

Figure 16-7 A WDN node with three pipe connections and an outflow.

The mass balance equation for node j can be written (with flow toward a node as positive) as

$$Q_{i,j} - Q_{j,i+1} - Q_{j,i+2} - Q_{out} = 0 \qquad (16.27)$$

or

$$\left(\frac{h_i - h_j}{K_{i,j}}\right)^{0.54} - \left(\frac{h_j - h_{i+1}}{K_{j,i+1}}\right)^{0.54} - \left(\frac{h_j - h_{i+2}}{K_{j,i+2}}\right)^{0.54} - Q_{out} = 0 \qquad (16.28)$$

Equation 16.28 can be written for each node. Thus, there will be a system of nonlinear equations with the same number of equations as the number of unknowns. These are then solved to obtain the unknown heads and thereby the flows in the pipes.

16.2.4 Node Formulation: Linear

This method is quite similar to the loop formulation of the Hardy–Cross method, but it has many advantages over the other traditionally used methods. Let Q_i be the discharge in pipe i. Rewriting the Hazen–Williams equation, we have

$$h_L = K\, Q^{1.85} \qquad (16.29)$$

If a link begins at node a and ends at node b, we have

$$h_b - h_a = K\, Q^{1.85} \qquad (16.30)$$

and $h_b > h_a$. Equation 16.30 can be linearized using the Taylor series expansion as

$$h_b - h_a = K\, Q_i^{1.85} + 1.85\, K\, Q_{io}^{0.85}\, q \qquad (16.31)$$

where Q_{io} is the estimated discharge in pipe i and q is the correction term. We write

$$Q_i = Q_{io} + q \qquad (16.32)$$

where Q_i is the updated discharge in pipe i.

Substituting for q from Eq. 16.32 and simplifying, one gets

$$Q_i = 0.46\, Q_{io} + 0.54\, [h_b - h_a\,]/(K\, Q_{io}^{0.85}) \qquad \textbf{(16.33)}$$

There will be n such equations for n nodes. Note that the equations are linear because the unknown head appears with a power of unity. This set of equations can be solved to get the unknowns in an iterative manner. The Hardy–Cross node method begins with a set of estimated heads. These form inputs for computing flows at the nodes and the residual discharge rates are determined by using the continuity equation. The heads are iteratively adjusted to get the solution. Many software packages are available to analyze a WDN. WADISCO (Water Distribution Simulation and Optimization) by Walski et al. (1990) is one such package. The software KYPIPE was developed by Wood (1980) and Rossman (2000) has described the software EPANET. The software FlowMaster (Meadows and Walski 1998) can be used for hydraulic analysis and design of pipes, ditches, and open channels.

Example 16.7 Consider the WDN shown in Fig. 16-8. The network has 30 nodes and 42 pipe links. The network receives water supply at node 1 whereas the other nodes are either demand nodes or connecting nodes. The demand nodes are shown as solid black dots and the connecting nodes are shown by double circles. Table 16-1 shows the demand (m^3) at various nodes along with node elevation. Pipe details, such as diameter (mm) and length (m), are given in Table 16-2. The Hazen–Williams coefficient (HWC) is 100 (metric units) for each pipe. Find the flow and head loss in each pipe. This example is excerpted from Kumar (1999).

Solution The WDN is hydraulically analyzed by assuming that all nodes are perfectly functional. The pressure at various nodes (in meters and tons per square meter, TPSM), and the energy grade line (in meters) are shown in Table E16-7a. The discharge in cubic meters per day (cmd), velocity, and head loss in different pipes are given in Table E16-7b.

16.3 Reliability of a WDN

The reliability of a WDN can be defined as "the ability of the WDN to meet the demands that are placed on it where such demands are specified in terms of (1) the flows to be supplied (total volume and flow rate); and (2) the range of pressure at which those flows must be provided" (Goulter 1995). Thus, the reliability is the ability of the system to provide adequate service with acceptable interruptions in spite of abnormal conditions.

Reliability analysis of WDNs is receiving much attention these days for many reasons. Many existing WDNs were designed and constructed several decades ago. Some of these are unable to meet current demands, which have increased considerably over the intervening time. Further, because of aging, the

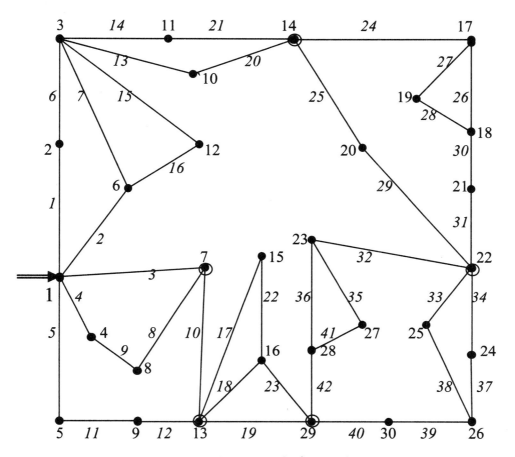

Figure 16-8 WDN network of Example 16.7.

capacity of components is declining and many of them are failing. Since water is a basic necessity, WDNs are required to have high reliability. Among other things, this requires that a certain amount of redundancy should be introduced into the network. A redundant network has an adequate residual capacity and alternate flow paths to provide uninterrupted water supply.

In the study of the reliability of a WDN, two categories are identified: mechanical reliability and hydraulic reliability. Mechanical reliability is concerned with the failure of the network components, such as pumps and pipes. It mainly depends on the design and manufacture of the components, age, and environment. Hydraulic reliability of a WDN refers to its ability to provide sufficient water to meet the demand at a required pressure. The hydraulic failure of a network could arise either from mechanical failure or when the demand outstrips its ability to supply the requisite quantity of water.

Table E16-7a *Pipe network analysis (node data and analysis) for Example 16.7.*

Node	Elevation (m)	Discharge (cmd)	Energy grade line (m)	Pressure head (m)	Pressure (TPSM)
1	65.0	−19700.0	100.0	35.0	35.0 (supply)
2	60.0	700.0	95.3	35.3	35.3
3	60.0	700.0	87.4	27.4	27.4
4	60.0	700.0	99.2	39.2	39.2
5	60.0	1000.0	95.8	35.8	35.8
6	60.0	1000.0	94.9	34.9	34.9
7	60.0		95.5	35.5	35.5
8	60.0	900.0	98.2	38.2	38.2
9	60.0	500.0	94.4	34.4	34.4
10	60.0	1000.0	83.7	23.7	23.7
11	60.0	1200.0	83.0	23.0	23.0
12	60.0	1000.0	86.0	26.0	26.0
13	60.0		92.9	32.9	32.9
14	60.0		76.3	16.3	16.3
15	60.0	1200.0	88.1	28.1	28.1
16	60.0	600.0	90.1	30.1	30.1
17	60.0	800.0	75.3	15.3	15.3
18	58.0	1000.0	72.7	14.7	14.7
19	56.0	500.0	72.0	16.0	16.0
20	57.0	700.0	73.9	16.9	16.9
21	57.0	500.0	70.4	13.4	13.4
22	55.0		70.8	15.8	15.8
23	56.0	1200.0	73.1	17.1	17.1
24	54.0	1000.0	70.1	16.1	16.1
25	53.0	1000.0	65.3	12.3	12.3
26	54.0	800.0	74.9	20.9	20.9
27	53.0	500.0	75.5	22.5	22.5
28	54.0	600.0	79.2	25.2	25.2
29	57.0		86.1	29.1	29.1
30	57.0	600.0	82.8	25.8	25.8

Table E16-7b *Pipe network analysis (pipe data and analysis) for Example 16.7.*

Pipe	Node		Diameter	Length	Flow	Velocity	Head loss
	From	To	(mm)	(m)	(cmd)	(m/s)	(m)
1	1	2	300.0	600.0	7018. 7	1.15	4.72
2	1	6	200.0	500.0	2791.3	1.03	5.14
3	1	7	150.0	900.0	887.6	0.58	4.50
4	1	4	300.0	400.0	3390.4	0.56	0.82
5	1	5	300.0	800.0	5611.9	0.92	4.16
6	2	3	250.0	500.0	6318.7	1.49	7.86
7	6	3	150.0	800.0	1242.2	0.81	7.45
8	8	7	200.0	600.0	1790.4	0.66	2.71
9	4	8	250.0	300.0	2690.4	0.63	0.97
10	7	13	250.0	800.0	2678.0	0.63	2.57
11	5	9	300.0	400.0	4611.9	0.76	1.45
12	9	13	300.0	500.0	4111.9	0.67	1.46
13	3	10	250.0	600.0	3836.2	0.90	3.75
14	3	11	200.0	500.0	2573.9	0.95	4.42
15	3	12	150.0	1000.0	450.9	0.30	1.43
16	6	12	100.0	600.0	549.1	0.81	8.87
17	13	15	150.0	900.0	926.1	0.61	4.86
18	13	16	200.0	600.0	1831.4	0.67	2.82
19	13	29	250.0	1000.0	4032.5	0.95	6.85
20	10	14	200.0	700.0	2836.2	1.04	7.40
21	11	14	150.0	600.0	1373.9	0.90	6.73
22	16	15	100.0	500.0	273.9	0.40	2.04
23	16	29	150.0	700.0	957.5	0.63	4.02
24	14	17	300.0	700.0	2678.9	0.44	0.92
25	14	20	200.0	700.0	1531.1	0.56	2.36
26	17	18	200.0	800.0	1519.4	0.56	2.66
27	17	19	100.0	500.0	359.5	0.53	3.37
28	18	19	100.0	600.0	140.5	0.21	0.71
29	20	22	150.0	700.0	831.1	0.54	3.10
30	18	21	100.0	300.0	378.9	0.56	2.23
31	22	21	100.0	400.0	121.1	0.18	0.36
32	23	22	100.0	800.0	224.0	0.33	2.25
33	22	25	100.0	600.0	425.6	0.63	5.54
34	22	24	150.0	400.0	508.4	0.33	0.71
35	27	23	150.0	1000.0	598.5	0.39	2.41
36	28	23	150.0	1400.0	825.5	0.54	6.11
37	26	24	100.0	400.0	491.6	0.72	4.82
38	26	25	100.0	600.0	574.4	0.85	9.64
39	30	26	150.0	400.0	1866.0	1.22	7.91
40	29	30	200.0	400.0	2466.0	0.91	3.27
41	28	27	150.0	500.0	1098.5	0.72	3.71
42	29	28	150.0	200.0	2524.0	1.65	6.92

16.3.1 Stochastic Hydraulic Reliability Analysis of a WDN

In the examples presented in the previous sections, it was assumed that the supply always equals or exceeds demand and that these two are deterministic. However, in many networks this assumption many not hold. Frequently, supply is less than demand because the demand either was underestimated at the time of planning or has overgrown beyond expectation while the network remains the same. Note that, beyond a certain limit, increasing supply does not help much. Because the sizes of pipes remain the same, forcing higher flows results in higher head losses. The result is insufficient pressure at the demand nodes and there may be insufficient or no flow at some of the nodes.

The procedure to compute reliability when supply and demand are random variables was explained in Chapter 14. When supply is S and demand is D, the reliability (R_s) can be computed by the probability (Pr) that $S - D$ exceeds zero:

$$R_s = P_r[(S - D) > 0]$$
$$= \int_0^\infty f_S(S)\left[\int_0^\infty f_D(D)dD\right]dS \tag{16.34}$$

where $P_r()$ stands for probability, and $f_S(S)$ and $f_D(D)$ are the probability density functions of supply and demand, respectively. Equation 16.34 is difficult to solve since the joint probability density function of supply and demand is difficult to derive. If supply and demand follow normal distribution, the reliability (R_s) can be computed as

$$R_S = \phi\left(\frac{\mu_S - \mu_D}{\sqrt{\sigma_S^2 + \sigma_D^2}}\right) \tag{16.35}$$

where ϕ represents the cumulative density function of a standard normally distributed variable, $N(0,1)$, μ and σ stand for mean and standard deviation, respectively, and subscripts S and D represent supply and demand, respectively.

Equation 16.35 gives the probability that the supply is greater than the demand. However, even when supply exceeds demand, it is possible that demand is not satisfied at some of the demand nodes because pressure is less than the required service head. This will reduce the nodal reliability and consequently the system reliability. To compute the system reliability, demand is bifurcated into two parts: $D < D_0$ and $D > D_0$, where D_o is the capacity of the system. This implies that nodal pressures at all the demand nodes will be greater than the service head when $D = D_o$. Thus, the expression for R_s becomes

$$R_s = P_r[(S - D) > 0, D < D_0] \times R_{h1} + P_r[(S - D) > 0, D > D_0] \times R_{ha} \tag{16.36}$$

where D_0 is the demand truncation level or design capacity of the system, R_{h1} is the network hydraulic reliability when $D < D_0$ (which equals one as the network is designed for D_0), R_{ha} is the network hydraulic reliability when $D > D_0$, and all the pipes in the network are operational.

Since the demand is less than the capacity of the system and all the pipelines are working, computations for the first component of Eq. 16.36 can be carried out analytically as discharge and pressure criteria will be satisfied at any node of the system. However, when the demand is greater than the network capacity, some of the nodes may not receive water with sufficient pressure head. To compute the second component of Eq. 16.36, demand above D_0 is divided into m discrete demand intervals and we get

$$P_r\{(S-D)>0, D>D_0\} \times R_{ha} = \sum_{i=1}^{m} P_r\{(S-D)>0, D_{i-1}<D>D_i\} \times (R_{ha})_{i-1,i} \quad \textbf{(16.37)}$$

where the number of discrete intervals (m) depends upon the accuracy of the results desired. For simplicity, each demand interval is assumed to be represented by the average demand of that interval. For example, demand D_k, which is the average of D_{i-1} and D_i, represents the kth demand interval as far as the hydraulic reliability is concerned. A hydraulic simulation of WDN having n demand nodes is carried out for this D_k and the corresponding pressure heads at all the demand nodes are computed. The water supplied to the jth node will depend on the head attainable at that node. For each node, two head limits must be given: a minimum head, H_{min}, and a service head, H_s.

For system performance to be termed satisfactory, all the imposed demands for each node should be met with heads above the service limit (H_s). If the available head (H) at a node is below H_s but above H_{min}, the system cannot supply the full demand. It can meet the reduced supply at that node. However, no supply is possible if the pressure head at the node is below H_{min}. Many relationships are available to estimate this reduced supply.

The reduced supply can be computed using the following equation:

$$Q_{j,supplied} = \begin{cases} Q_{j,required} \text{ (adequate flow) for } H \geq H_{min} \\ [(H-H_{min})/(H_S-H_{min})]^{0.5} \\ \times Q_{j,required} \text{ (partial flow) for } H_{min} \leq H < H_s \\ 0 \text{ (no flow) for } H < H_{min} \end{cases} \quad \text{(16.38)}$$

The rationale for Eq. 16.33 is that, according to hydraulic laws for pipe flow, flow in a pipe is proportional to the square root of the pressure head.

After computing the flow at various demand nodes, one can estimate the node hydraulic reliability (R_j), which is defined as the ratio of the discharge supplied to the required discharge at that node. In general, for m different demands with different probabilities of occurrence, the hydraulic reliability for node j can be computed as

$$R_j = \sum_{i=1}^{m} P_r\{(S-D)>0, D_{i-1}<D<D_i\} \times \frac{Q_{j,supplied}}{Q_{j,required}} \quad \text{(16.39)}$$

The network hydraulic reliability for the ith demand interval can be computed by taking the arithmetic mean or the weighted average of the nodal hydraulic reliability. If the arithmetic mean is taken, one gets

$$(R_{ha})_{i-1,i} = \frac{1}{n}\sum_{j=1}^{n}R_j \qquad (16.40)$$

where n is the number of demand nodes in the network.

Example 16.8 A WDN is shown in Fig. 16-9. The hydraulic equivalence of this WDN is shown in Fig. 16-10. The water demand for this network is 13,500 cubic meters per day (cmd). The HWC for all the pipes is 100. The pressure head at the supply node is 35 m. The service head (H_s) required to satisfy the full demand at a demand node is 16 m. No flow is possible if the residual pressure at a given demand node is less than 12 m, which is the minimum required head (H_{min}). If the pressure at a demand node is between 12 and 16 m, the demand will be satisfied only partially. The capacity of the network is 12,000 cmd. Therefore, all the demand nodes will receive water at a pressure of more than 16 m when the network demand is less than or equal to 12,000 cmd. Based upon the past data and the local climatic conditions, the following statistical data are available:

mean of supply series (μ_s) = 17,000 cmd, standard deviation (σ_s) = 2,000 cmd

mean of demand series (μ_D) = 15,000 cmd, standard deviation (σ_D) = 3,000 cmd

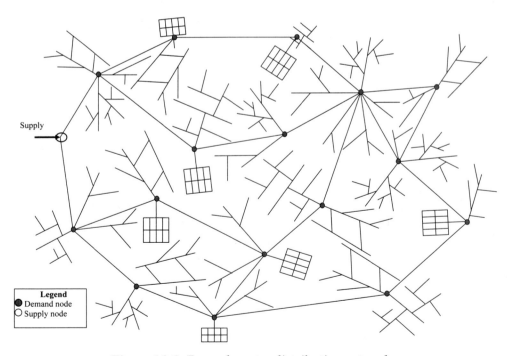

Figure 16-9 Example water distribution network.

Find the hydraulic reliability of this network when it is fully operational. Also, compute the hydraulic reliability when, owing to aging of the network, the HWC for all the pipes changes to 90 (metric units). This example is excerpted from Kansal (1996).

Solution To compute the static system reliability using Eq. 16.35, first the probability R_S of supply being greater than the demand is computed as

$$R_S = \phi\left(\frac{17000 - 15000}{\sqrt{2000^2 + 3000^2}}\right) = \phi(0.5547)$$

From the tables of the normal distribution, $R_s = 0.7105$. This will be the system reliability if there are no capacity constraints. The WDN is presented in Fig. 16-10 in terms of node and links.

Considering the demand data to follow a normal distribution, one can find the probabilities of various demand levels using the normal distribution tables. For example, in the present illustrative example, the probabilities of demands less than 12,000 cmd, between 12,000 and 15,000, between 15,000 and 18,000, and between 18,000 and 21,000 cmd are computed as follows:

$$P_r\{D < 12,000\} = 0.1587$$

$$P_r\{12,000 < D < 15,000\} = 0.3413$$

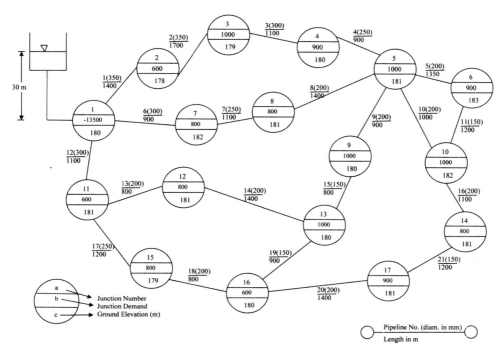

Figure 16-10 Line diagram of the water distribution network of Fig. 16-9.

$$P_r\{15,000 < D < 18,000\} = 0.3413$$

$$P_r\{18,000 < D < 21,000\} = 0.1359$$

$$P_r\{D > 21,000\} = 0.0228$$

Further, the probability of $(S - D) > 0$ is 0.7105. If supply and demand are independent and normally distributed, then the joint probabilities can be found as

$$P_r\{(S - D) > 0; D < 12,000\} = 0.1587 \times 0.7105 = 0.1128$$

$$P_r\{(S - D) > 0; 12,000 < D < 15,000\} = 0.2425$$

$$P_r\{(S - D) > 0; 15,000 < D < 18,000\} = 0.2425$$

$$P_r\{(S - D) > 0; 18,000 < D < 21,000\} = 0.0966$$

$$P_r\{(S - D) > 0; D > 21,000\} = 0.0161$$

The next step is the hydraulic analysis of the network to estimate pressures and flows in various pipelines for various demand intervals. The program WAD-ISO developed by Walski et al. (1990) was used for the purpose. It was assumed that a particular demand interval can be represented by its average demand. Thus, the demand intervals of 12,000–15,000, 15,000–18,000, and 18,000–21,000 are represented by the demands of 13,500, 16,500 and 19,500, respectively.

The results of hydraulic analysis for demand levels of 12,000, 13,500, 16,500, and 19,500 cmd are given in Table E16-8. As can be seen from Table E16-8, when the demand is 12,000 cmd, all the demand nodes will receive water more than the service head (16 m). However, when the demand is more than 12,000 cmd, some of the nodes have pressure heads below the minimum desired service head. As a result, some of the demand nodes may receive water in full, some may receive water at a reduced rate, and some may not receive any water. For example, when the demand of the network is 16,500 cmd, the demand node 5 will receive water at 13.2 m. This node will receive water in partial-flow mode and will get (from Eq. 16.33) 669 cmd against the requirement of 1,222 cmd. Similarly, node number 6 will receive water at a pressure of 7.8 m. Since this is below the minimum required head, no water will be withdrawn from this node.

The residual pressure heads at various nodes for different sets of demands can be utilized to compute the network hydraulic reliability using Eq. 16.40. After computing the R_{ha} values for all the demand intervals, the system reliability can be computed by using Eq. 16.36. Note that the probability of $(S - D) > 0$ and $D > 21,000$ cmd is very small (0.0161) and the corresponding hydraulic reliability will also be very small. Therefore, the hydraulic reliability computations for $D > 21,000$ cmd have been neglected. This gives the system hydraulic reliability

$$R_s = (0.1128 \times 1.0) + (0.2425 \times 0.9946) + (0.2425 \times 0.6604) + (0.0966 \times 0.4235) = 0.5550$$

If the value of the HWC changes (owing to aging of pipeline, etc.), the values of the residual pressures at various demand nodes will also change. The reliability can be computed by following the procedure just given. For the case when the HWC is 90 (metric units) for all the pipes, higher head loss at various demand nodes will reduce the system hydraulic reliability to 0.4716.

16.3.2 Reliability Analysis of a WDN with Unsatisfied Demands

In the previous section, hydraulic reliability was computed based on the node head analysis (NHA) of the WDN. In NHA, it is presumed that the demand of a node is always satisfied; that is, for node j, the available flow q_j^{avl} is always equal to the required demand q_j^{req} and the corresponding available nodal head is obtained. However, because the supply and demand at any node are related through Eq. 16.37, it is likely that outflow from some of the demand nodes may not be equal to the desired one. For instance, in the previous example, when the network demand is 16,500 cmd, demand at node 5 can be partially met and no demand can be met at node 6 (see Table E16-8). This deficiency in pressure and flow will change the flow conditions in the WDN, which, in turn, necessitates changes in the flow analysis. In Example 16.2, the hydraulic reliability was computed without modifying the flow conditions in the network owing to partial- or no-flow conditions at some of the demand nodes. However, it is mandatory to satisfy Eq. 16.37 at each and every demand node along with the usual flow analysis. In this study, a procedure for such modifications has been suggested. The steps of the modified methodology are as follows:

1. Carry out the NHA for a given set of demands.
2. Check the residual pressure of all the demand nodes. If all the nodes have residual pressures above the service head, the solution obtained is final. Otherwise, go to step 3.
3. For a deficient residual pressure node, modify the demand satisfied at that node by using Eq. 16.37.
4. Repeat the NHA to compute the new values of the residual pressures at various nodes.
5. Compute the demands satisfied by the new set of residual pressures at various nodes.
6. Compare the computed demand satisfied at the various demand nodes with that of the previously computed demand. If both the demands at all the demand nodes are the same, then the results obtained are final. Otherwise, repeat steps 4 to 6 until the two values are the same.

Example 16.9 Compute the hydraulic reliability of the WDN of Fig. 16-10 when the network demand is 16,500 cmd. This example is excerpted from Kansal (1996).

Solution In Table E16-8, some of the nodes are in partial-flow mode and some are in no-flow mode. For example, nodes 5, 9, 13, and 16 are in partial-flow mode and nodes 6, 10, 14, and 17 are in no-flow mode. Using Eq. 16.37, one can

Table E16-8 Pressure and flow data at various nodes for the WDN shown in Fig. 16-10 by node head analysis.

Node	Elevation (m)	Output (cmd)	Pressure head (m)	Output (cmd)	Pressure head (m)	Output (cmd)	Pressure head (m)	Output (cmd)	Pressure head (m)	R_j (Eq. 16.27)
1	180.0	−12000	35.0	−13500	35.0	−16500	35.0	−19500	35.0	—
2	178.0	533	31.6	600	30.3	733	27.3	867	23.7	1.0
3	179.0	889	28.4	1000	26.5	1222	22.3	1444	17.3	1.0
4	180.0	800	25.6	900	23.4	1100	18.1	1300	12.0[b]	0.8414
5	181.0	889	22.5	1000	19.6	*1222*	*13.2*[a]	*1444*	*5.6*[a]	0.6870
6	183.0	800	18.6	*900*	*15.3*[a]	*1100*	*7.8*[b]	*1300*	*−1.0*[a]	0.4688
7	182.0	711	26.4	800	24.9	978	21.2	1156	16.9	1.0
8	181.0	711	25.5	800	23.5	978	18.7	*1156*	*13.2*[a]	0.9158
9	180.0	889	23.1	1000	20.2	*1222*	*13.5*[a]	*1444*	*5.7*[b]	0.7091
10	182.0	889	19.4	1000	16.0	*1222*	*8.4*[b]	*1444*	*−0.5*[b]	0.5000
11	181.0	533	29.7	600	28.6	733	26.2	867	23.3	1.0
12	181.0	711	25.8	800	23.7	978	19.1	*1156*	*13.7*[a]	0.9300
13	183.0	889	20.9	1000	18.3	*1222*	*12.1*[a]	*1444*	*4.8*[b]	0.5540
14	181.0	711	20.0	800	16.6	*978*	*8.7*[b]	*1156*	*−0.4*[b]	0.5000
15	179.0	711	28.4	800	26.6	978	22.3	1156	17.4	1.0
16	180.0	533	24.1	600	21.4	*733*	*15.3*[a]	*867*	*8.1*[b]	0.8100
17	181.0	800	20.6	900	17.4	*1100*	*9.9*[b]	*1300*	*1.2*[b]	0.5000
Network hydraulic reliability (R_{ha})			1.0		0.9946		0.6604		0.4235	

a. Only partial flow can take place at these nodes.
b. No flow can take place at these nodes; flow shown in italics is not the actual flow.

Table E16-9 Modified pressure and flow data at various nodes for the WDN shown in Fig. 16-10.

Node	Elevation (m)	Output (cmd)	Pressure head (m)	Output (cmd)	Pressure head (m)	Output (cmd)	Pressure head (m)	Output (cmd)	Pressure head (m)	R_j (Eq. 16.7)
1	180.0	−12,000	35.0	−13450	35.0	−15200	35.0	−16383	35.0	—
2	178.0	533	31.6	600	30.3	733	28.7	867	27.4	1.0
3	179.0	889	28.4	1,000	26.6	1222	24.3	1444	22.7	1.0
4	180.0	800	25.6	900	23.5	1100	20.9	1300	19.1	1.0
5	181.0	889	22.5	1,000	19.8	1222	17.0	*1364*	15.5[a]	0.9686
6	183.0	800	18.6	850	15.6[a]	*658*	13.4[a]	*457*	12.5[a]	0.7325
7	182.0	711	26.4	800	24.9	978	22.9	1156	21.3	1.0
8	181.0	711	25.5	800	23.6	978	21.0	1156	19.2	1.0
9	180.0	889	23.1	1,000	20.3	1222	17.2	*1338*	15.5[a]	0.9686
10	182.0	889	19.4	1,000	16.2	*850*	14.0[a]	*686*	13.0[a]	0.8094
11	181.0	533	29.7	600	28.6	733	27.2	867	26.3	1.0
12	181.0	711	25.8	800	23.8	978	21.2	1156	19.6	1.0
13	183.0	889	20.9	1,000	18.3	*1120*	15.3[a]	*990*	14.0[a]	0.9062
14	181.0	711	20.0	800	16.7	*765*	14.4[a]	*710*	13.5[a]	0.8477
15	179.0	711	28.4	800	26.6	978	24.3	1156	22.9	1.0
16	180.0	533	24.1	600	21.5	733	18.6	866	17.1	1.0
17	181.0	800	20.6	900	17.5	*730*	14.8[a]	*870*	13.8[a]	0.8768
R_{ha}			1.0		0.9965		0.9274		0.8542	
R_v			1.0		0.9963		0.9212		0.8402	

a. Only partial flow is possible; figures in italics are the actual flows possible.

compute the flows at nodes 5, 9, 13, and 16 as 669, 748, 193, and 666 cmd, respectively. The flows at nodes 6, 10, 14, and 17 will be zero if the computed heads are stationary. In the next iteration, these new computed flows are considered; this will change the heads at these nodes, causing the possibility of increased flow at the pressure-deficient nodes. The procedure is repeated until the computed values of flow at all the demand nodes are the same in two consecutive iterations. The results have been tabulated in Table E16-9.

From Table E16-9, observe that of the network demand of 16,500 cmd, only 15,200 cmd can be met. Also, the demand nodes 6, 10, 14, and 17, which were in no-flow condition, will actually have flows of 658, 850, 765, and 930 cmd, respectively. The nodes 5, 9, 13, and 16, which were in partial-flow mode with flows of 669, 748, 193, and 666 cmd, will actually have flows of 1,222, 1,222, 1,120, and 930 cmd. Thus, nodes 5 and 9 will have full flow, whereas nodes 13 and 16 will still be under partial flow. Similarly, for a network demand of 13,500 and 19,500 cmd, the possible flows are 13,450 and 16,383 cmd, respectively. The actual nodal flows are shown in Table E16-9. Note that the nodal demands shown in Table E16-9 satisfy Eq. 16.37 at all the demand nodes.

Comparison of the last column of Table E16-9 with that of Table E16-8 shows that the nodal reliability has changed considerably.

Similarly, the network hydraulic reliability (R_{ha}) computed from Eq. 16.39 for most of the demand patterns has also gone up. Using the R_{ha} values with the modified outflows at various demand nodes (see Table 16-9), we can compute the system reliability from Eq. 16.35 as

$$R_s = (0.1128 \times 1.0) + (0.2425 \times 0.9965) + (0.2425 \times 0.9274) + (0.0966 \times 0.8542) = 0.6619$$

Thus, the hydraulic reliability after accounting for the actual flows is 0.6619 as compared to 0.555 from NHA. Of course, this value is less than the static hydraulic reliability of the system (0.7105), which was computed without the residual head criterion.

The nodal reliabilities can be plotted to get a reliability surface. Graphical representation of the results helps to visualize them and to identify the areas of low reliability. In turn, this helps in the operation and maintenance of a WDN.

16.3.3 Hydraulic Reliability Analysis Considering Correlation Between Demand and Supply

The hydraulic reliability analysis described in Section 16.3.1 did not consider the correlation between supply and demand. However, many times supply and demand have some correlation. To some extent, both of these depend upon variations in weather and other factors. Hence, it may be better to consider that supply and demand are correlated variables. The demand is likely to be inversely related with supply. If ρ is the correlation coefficient between supply and demand, Eq. 16.35 for the system reliability is modified to yield

$$R_S = \phi \left(\frac{\mu_S - \mu_D}{\sqrt{\sigma_S^2 + \sigma_D^2 - 2\rho\sigma_S\sigma_D}} \right) \tag{16.41}$$

where ϕ represents the cumulative density function of a standard normally distributed variable, $N(0,1)$, μ and σ stand for mean and standard deviation, subscripts S and D represent supply and demand, and ρ is the correlation coefficient between the variables. The correlation coefficient between the surplus $(S - D)$ series and demand series can be computed by

$$\rho_1 = \frac{\rho\mu_S - \mu_D}{\sqrt{\sigma_S^2 + \sigma_D^2 - 2\rho\sigma_S\sigma_D}} \tag{16.42}$$

If we denote the surplus series by X and the demand series by Y, the bivariate probability estimation based on the normal distribution can be computed as

$$P_r(X \le h, Y \le k, \rho_1) = \frac{1}{2\pi\sqrt{1-\rho_1^2}} \int_\infty^h \int_\infty^k \exp\left[\frac{x^2 - 2\rho_1 xy + y^2}{2(1-\rho_1^2)} \right] dx\,dy \tag{16.43}$$

The value of $P_r(X \le h, Y \le k, \rho_1)$ can be expressed using the T function (Kumar 1980) as

$$P_r(X \le h, Y \le k, \rho_1) = \frac{P_r(h)}{2} + \frac{P_r(k)}{2} - T(h, a_h) - T(k, a_k) - \begin{cases} 0 \\ 1/2 \end{cases} \tag{16.44}$$

where

$$a_h = \frac{k}{h\sqrt{1-\rho_1^2}} - \frac{\rho_1^2}{\sqrt{1-\rho_1^2}} \tag{16.45}$$

$$a_k = \frac{h}{k\sqrt{1-\rho_1^2}} - \frac{\rho_1^2}{\sqrt{1-\rho_1^2}} \tag{16.46}$$

In Eq. 16.44, the upper choice is made if $h \times k > 0$ and if $h \times k = 0$ but $h + k \ge 0$; the lower choice is made otherwise.

The T function is defined (Owen 1962) as

$$T(h, a) = \frac{1}{2\pi} \int_0^a \frac{\exp[-h^2(1+x^2)/2]}{1+x^2} dx \tag{16.47}$$

The values of the T function were tabulated by Owen (1962) for $0 \le a \le 1$ and ∞. To obtain the values for $1 < a < \infty$, the following formula may be used:

$$T(h, a) = 0.5P(h) + 0.5P(ah) - P(h)P(ah) - T(ah, 1/a) \tag{16.48}$$

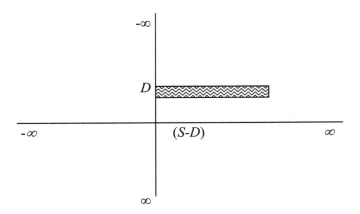

Figure 16-11 Graphical representation of Eq. 16.50.

To obtain the values for negative a and/or h, the formula is

$$T(h, -a) = -T(h, a) \text{ and } T(-h, a) = T(h, a)$$

Thus, $P_r(X \le h, Y \le k, \rho_1)$ can be computed using T functions. Further, for any two normally distributed random variables that fall in any region bounded by a polygon, the bivariate normal distribution can be computed using the T function. This gives

$$P_r(X \le h, Y \le k, \rho_1) = P_r(h, k, \rho) - P_r(h, b, \rho) - P_r(a, k, \rho) + P_r(a, b, k) \qquad \text{(16.49)}$$

Thus,

$$P_r[(S-D) > 0; D_0 < D \le D_1; \rho_1] = P_r(D \le D_1) - P_r(D \le D_0) - P_r[(S-D) \le 0, D < D_1; \rho_1]$$
$$+ P_r[(S-D) \le 0; D \le D_0; \rho_1] \qquad \text{(16.50)}$$

This is shown in Fig. 16-11.

Although the demand series D, in reality, is really a continuous function, to simplify the analysis it is discretized into m discrete demand intervals. The joint probability expression now can be expressed as

$$P_r(X \le h, Y \le k, \rho_1) = \sum_{i=1}^{m} P_r\{(S-D) > 0, D_{i-1} < D < D_i, \rho_1\} \qquad \text{(16.51)}$$

where D_0 is the design capacity of the WDN.

Example 16.10 For illustrating the methodology, consider again the WDN of Example 16.8, shown in Fig. 16-10. The hydraulic equivalent of this WDN was shown in Fig. 16-6. The network is then analyzed for pressures and flows in various pipelines for given demand intervals. This example has been excerpted from Kumar (1999).

Solution In the WDN of Example 16.8 with 17 nodes connected by 21 pipes, the head available in the reservoir is 30 m. The service head for all the demand nodes is 16 m and the minimum head for all the demand nodes is 12 m. The node will receive reduced supply if the pressure of water lies between 12 and 16 m and there will be no supply if the pressure goes below 12 m. In the present illustrative example, the following data have been used:

$$\text{supply series: mean } (\mu_S) = 17{,}000 \text{ cmd}$$

$$\text{standard deviation } (\sigma_S) = 2{,}000 \text{ cmd}$$

$$\text{demand series: mean } (\mu_D) = 15{,}000 \text{ cmd}$$

$$\text{standard deviation } (\sigma_S) = 3{,}000 \text{ cmd}$$

The coefficient of correlation between the supply and demand series $(\rho) = -0.4$. The coefficient of correlation ρ_1 between the $(S - D)$ series and the D series can be computed by using Eq. 16.30. For the given data, it turns out to be -0.9.

For computation of the system reliability as shown in Eq. 16.24, first the value of joint probabilities for various intervals of demand subject to the condition that supply is more than the demand are computed. These values have been computed using the T functions and are reported as follows:

$$P_r\{(S - D) > 0; D < 12{,}000\} = 0.158$$

$$P_r\{(S - D) > 0; 12{,}000 < D < 15{,}000\} = 0.325$$

$$P_r\{(S - D) > 0; 15{,}000 < D < 18{,}000\} = 0.192$$

$$P_r\{(S - D) > 0; D > 21{,}000\} \approx 0.00$$

The computations for the second demand interval $(12{,}000 < D < 15{,}000)$ are as follows: Using Eq. 16.50 one can express the probability $P_r\{(S - D) > 0; 12{,}000 < D < 15{,}000\}$ as

$$P_r\{(S - D) > 0; 12{,}000 < D < 15{,}000\} =$$

$$P_r\,(D \le 15{,}000) - P_r(D \le 12{,}000) - P_r\{(S - D) \le 0,$$

$$D < 15{,}000; -0.9\} + P_r\{(S - D) \le 0; D \le 12{,}000; -0.9\}$$

$$P_r\{D \le 15{,}000\} = \beta(\infty,0) = P_r\,(\infty)/2 + P_r\,(0)/2 - T(\infty,1.06) - T(0,\infty) - 0$$

$$= 0.5 + 0.25 - 0.0 - 0.25 = 0.5$$

Similarly,

$$P_r\{D \le 12{,}000\} = 0.1587$$

$$P_r\{(S - D) \le 0; D < 15{,}000; -0.9\} = 0.01126$$

$$P_r\{(S - D) \le 0; D < 12{,}000; -0.9) = 0.000562$$

Substituting these values in Eq. 16.39 gives

$$P_r\{(S - D) > 0; 12{,}000 < D < 15{,}000\} = 0.331$$

Similarly, for other intervals, joint probabilities can be evaluated.

As shown in Table E16-10, when the demand is less than 12,000 cmd, all the demand nodes receive water at more than the service head (16 m). However, when the demand is more than 12,000 cmd, some demand nodes may receive water in full, some may be in the reduced mode, and some may even not receive water at all, depending upon the availability of the pressure head. The results of pressures at various nodes for different sets of demand can then be utilized to compute the network hydraulic reliability from Eq. 16.29, as shown in the last row of Table E16-10. After computing R_h, one can compute the system reliability as

$$R_S = (0.158 \times 1.0) + (0.331 \times 0.9976) + (0.192 \times 0.6362) + (0.002 \times 0.439) = 0.611$$

If one does not consider the capacity constraint, the value of the system reliability (R_S), which is represented by $Pr\{(S - D) > 0\}$, would have been 0.683.

16.3.4 Reliability Analysis of a WDN Considering Pipeline Failures

The delivery of water in adequate quantity at the desired pressure to all the demand nodes is the primary goal of any WDN. As a system ages, its ability to transport water diminishes while the demand placed upon it typically increases. Thus, the rehabilitation, replacement, and/or expansion of an existing system to adequately meet the demand of flow at proper pressure head has always been of considerable interest to water utility engineers (Kim and Mays 1994). In the previous section, the hydraulic reliability of a WDN was estimated when the network was fully operational. However, this is not a realistic assumption: In real-life situations, a WDN is subject to pipeline failures. The computations of hydraulic reliability in such situations becomes much more difficult because it is very difficult to estimate the time and location of pipeline failure. Second, more than one pipeline may fail simultaneously and the number of such possible failure combinations may be very large even for a moderately sized WDN. The hydraulic simulation of all such failure events is complex. Thus, it has always been desired to suggest techniques that involve less computational effort and are robust and easy to comprehend (even at the cost of exactness) for computing the hydraulic reliability of a WDN when some of its pipelines are nonoperational.

The reliability that water will be available at various demand nodes of a distribution system may be expressed in terms of several measures. An earlier measure bases the reliability only on the connectivity of the demand point with the source of water. This measure is not representative of real systems since the adequacy of water supply requires not only connection to a source but also that a specified amount of flow must be delivered. To address this deficiency, a capacity-weighted reliability index was suggested by Wagner et al. (1988) and Quimpo et al. (1993). Two capacity-weighted measures are possible. In the reliability computation, a

Table E16-10 *Hydraulic reliability analysis of WDN.*

Node	Elevation (m)	Output (cmd)	Head (m)	Output (cmd)	Head (m)	Output (cmd)	Head (m)	Output (cmd)	Head (m)
1	185	–12000	30.0	–13500	30.0	–16500	30.0	–19500	30.0
2	179	720	30.5	810	29.1	990	26.0	1170	22.4
3	179	800	28.2	900	26.3	1100	21.9	1300	16.8
4	180	720	26.3	810	24.2	990	19.3	1170	13.6
5	181	960	22.7	1080	19.9	1320	13.6	1560	6.2
6	183	720	18.9	810	15.7	990	8.3	1170	–0.3
7	182	640	26.5	720	24.9	880	21.2	1040	16.9
8	181	640	25.7	720	23.7	880	19.0	1040	13.6
9	180	960	23.2	1080	20.3	1320	13.7	1560	5.9
10	182	960	19.4	1080	16.1	1320	8.5	1560	–0.4
11	181	480	29.8	540	28.8	660	26.4	780	23.7
12	181	640	26.0	720	24.0	880	19.5	1040	14.3
13	183	960	20.9	1080	18.2	1320	12.0	1560	4.8
14	178	800	23.0	900	19.5	1100	11.7	1300	2.5
15	179	640	28.6	720	26.8	880	22.7	1040	17.8
16	180	640	24.0	720	21.4	880	15.2	1040	8.1
17	181	720	20.8	810	17.6	990	10.2	1170	1.5
R_h			1.00		0.998		0.636		0.439

strict capacity measure excludes those paths that do not meet the desired demand fully. A more realistic measure is one that also takes into account the partial satisfaction of demand into consideration because two partially satisfying paths may combine to satisfy the required demand at a particular node. Another important dimension of the problem is that the intermediate nodes in any particular path may draw water and hence restrict the capacity of the path for the desired demand node.

16.4 Reliability Analysis of a WDN Using the Entropy Concept

Entropy theory, described in Chapter 9, can be employed for reliability analysis of a WDN. Recall that the redundancy is introduced in a WDN to increase its reliability. If there is only one path between source and sink in a network, it is perfectly ordered and therefore the entropy of this system will be zero. Redundancy at a node of a WDN can be considered as a measure of the disorder. One way to add redundancy to a WDN is by making looped networks rather than branched networks. In a redundant system, there will be many paths from a demand center to the source and the system will have nonzero entropy. As the redundancy increases, the uncertainty in the flow distribution in the system will increase and so will entropy. Hence, maximization of entropy is equivalent to maximization of redundancy.

In a water distribution network, redundancy is introduced so that demand points have alternative supply paths. This helps in delivery of water to demand centers even if a particular link becomes nonoperational. Redundancy is defined in terms of flow rather than any other variable, such as pressure, since flow is the most important variable of interest. Redundancy of a network is a measure of how well the network performs in terms of the total flow in the network when a link fails.

Redundancy of a WDN depends on its shape and is closely related to its reliability; a redundant network is more reliable. The redundancy of the entire network as a whole depends upon the redundancies of the individual nodes of the network. Note that summation of the redundancies of the individual nodes to get the redundancy of the whole network is not correct because redundancy is a measure of how well the network performs as a whole in moving the total flow when an individual link fails. The relative importance of a link to the local flow is not important; rather, the relative importance of a link to the total flow in the network should be used to assess the overall network performance.

Let S_j be the redundancy of the node j. To use the entropy concept, the definition of redundancy should have the following features:

1. S_j at node j should be a function of $X_{1j}, X_{2j}, ..., X_{n(j)j}$, where X_{ij} is the fraction of flow into node j derived from node i, and

$$X_{ij} = \frac{Q_{ij}}{Q_j}, \quad \text{and} \quad Q_j = \sum_{i=1}^{n(j)} q_{ij}$$

where $n(j)$ is the number of links incident on node j.

2. S_j should be zero if only one link is incident on node j or $n(j) = 1$.
3. For a given value of $n(j)$, the redundancy should have a maximum value when all X_j are equal.
4. The maximum value of S_j at a node should monotonically increase with $n(j)$.

Let the flow at node j be denoted by Q_j. If there are N nodes in the network, the total flow in the network Q_0 will be

$$Q_0 = \sum_{j=1}^{N} Q_j \tag{16.52}$$

Note that Q_0, which is the sum of flows in all links, is greater than the total demand. The redundancy for node j can be written in the same form as that for Shannon's entropy (see Chapter 9):

$$S_j = -\sum_{i=1}^{n(j)} \frac{q_{ij}}{Q_j} \ln \frac{q_{ij}}{Q_j} \tag{16.53}$$

where q_{ij} is the flow in the link from node i to node j, Q_j is the total flow into node j, and, in this equation, redundancy is measured by the extent to which the node receives water when a link that is incident to it fails. The redundancy S_j of a node will be a maximum when all q_{ij}/Q_j terms are equal (i.e., all links incident on the node carry the same flow). Awumah et al. (1991) argued that the relative importance of a link to the total flow is important in assessing the overall network performance and hence q_{ij}/Q_j in Eq. 16.53 should be replaced by q_{ij}/Q_0. With this replacement, Eq. 16.53 can be written as

$$S_j = -\sum_{i=1}^{n(j)} \frac{q_{ij}}{Q_j} \frac{Q_j}{Q_0} \ln \frac{q_{ij}}{Q_j} \frac{Q_j}{Q_0} = -\frac{Q_j}{Q_0} \left[\sum_{i=1}^{n(j)} \frac{q_{ij}}{Q_j} \ln \frac{q_{ij}}{Q_j} + \sum_{i=1}^{n(j)} \frac{q_{ij}}{Q_j} \ln \frac{Q_j}{Q_0} \right]$$

or

$$S_j = \frac{Q_j}{Q_0} S_j - \frac{Q_j}{Q_0} \ln \frac{Q_j}{Q_0} \tag{16.54}$$

The sum of redundancies at all nodes will give the entropic measure of the network redundancy:

$$S_N = \sum_{j=1}^{N} \left(S_j \right) \tag{16.55}$$

Substituting in Eq. 16.55 from Eq. 16.54 gives the entropic measure of network redundancy as

$$S_N = \sum_{j=1}^{N} \left[\frac{Q_j}{Q_0} S_j \right] - \sum_{j=1}^{N} \left[\frac{Q_j}{Q_0} \ln \frac{Q_j}{Q_0} \right] \qquad (16.56)$$

The first term on the right-hand side of this equation is the weighted sum of entropy at different nodes. The second term reflects the redundancy among nodes where uniformity among the nodes in terms of flow distribution (uniformity of Q_j/Q_0 value) indicates that the network has a better capacity to successfully overcome failure of any single link. Awumah et al. (1990) showed that maximization of the function given by Eq. 16.56 is equivalent to maximizing the ability of the network to supply water to each node. This is the same as maximizing the network redundancy.

A measure of network performance that reflects how well the network is able to supply flow under a range of failure conditions is the percentage of the total demand supplied at adequate pressure (PSPF). This parameter shows the performance of the network as a whole and therefore reflects the network-wide redundancy (Awumah et al. 1991).

The entropy approach can also be used to minimize the cost of the network in a formulation that includes, in addition to the hydraulic constraints, a set of constraints to ensure the minimum level of redundancy in the network. Constraining entropy of the network at each node individually has been suggested. This approach ensures that the network does not have a few unreliable nodes while maintaining good overall redundancy. The decision variables of this problem are the flows. The resulting problem is a nonlinear optimization problem since the entropy function is nonlinear.

Example 16.11 Figure 16-12 shows a WDN. The node numbers are shown in bold and the flows in the links (m^3/hour) are also shown. Compute the entropy of the WDN

Solution For the entire network, flow Q_0 obtained by summing the individual flows is 8,704 units. The entropy has been computed as shown in Table E16-11. For example, at node 1, the total flow is $840 + 760 = 1,600$ units. The last two columns give the first and the second terms of Eq. 16.56. Hence, $S_N = 2.9807$.

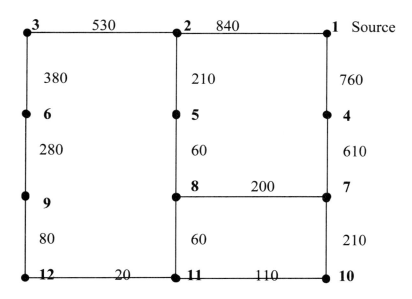

Figure 16-12 WDN of Example 16.11.

Table E16-11 *Computation of entropy for the WDN of Example 16.11.*

Node	Link 1	Link 2	Link 3	Node Q_j	S_j	$Q_j \times S_j/Q_0$	Entropy
1	840	760	0	1600	0.691897	0.127245	0.4385
2	840	530	210	1580	0.970512	0.176254	0.4859
3	530	380		910	0.6795	0.071074	0.3071
4	760	610		1371	0.68723	0.108298	0.3994
5	210	60		270	0.529706	0.016439	0.1242
6	380	280		660	0.681624	0.051709	0.2473
7	611	200	210	1022	0.952482	0.111889	0.3633
8	60	200	60	320	0.921493	0.033894	0.1553
9	280	80		360	0.529706	0.021919	0.1537
10	211	110		321	0.642797	0.023717	0.1454
11	60	110	20	190	0.917402	0.020035	0.1035
12	80	20		100	0.500402	0.005752	0.0571
			Sum	8704		0.768226	2.9807

16.5 Questions

16.1   The discharge passing through a horizontal pipe of diameter 80 mm is 0.01 m³/s. The pressures at the upstream and downstream sections are 20 and 15 kPa, respectively. What is the head loss in the pipe?

16.2   For the pipe of Question 16.1, let the elevation of the upstream end be 1.5 m higher than that of the downstream end. Further, at the downstream end, the pipe diameter is 70 mm (whereas it is 80 mm at the upstream end). Find the head loss between the two sections.

16.3   A 1000-m-long pipe with a diameter of 0.5 m carries water at a velocity of 3 m/s. Determine the head loss in the pipe if the relative roughness is k_s = 0.0001 m. Assume the kinematic viscosity $v = 1.0 \times 10^{-6}$ m²/s.

16.4   Tanks 1 and 2 are 1,000 meters apart and are connected through two pipes. The first pipe connected to pipe 1 is 600 m long and has a diameter of 300 mm. This is then connected to another pipe of 400-m length and a diameter of 200 mm, which is then connected to tank 2. The flow of water from the upper tank to the lower one is at 0.05m³/s and the Darcy–Weisbach friction factor is 0.025. Find the elevation of water in the lower tank if the elevation of water in the upper tank is 80 m.

16.5   A pipe of 0.15 m diameter is connected to a pipe of 0.20 m diameter. If the average velocity in the first pipe is 10 m/s, what is the average velocity in the second pipe? Also determine the discharge through the second pipe.

16.6   Four pipes of different diameters join at a junction as shown in the Fig. 16-13. The diameters of the pipes and the discharge passing through them are also shown in the figure. Find the value of Q_4 and the average velocity of flow in each pipe.

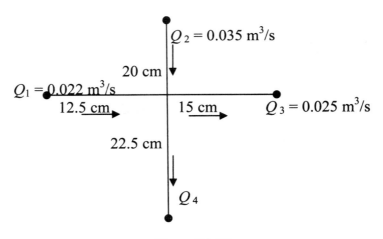

Figure 16-13

16.7   A 450-m-long cast-iron pipe of 15 cm diameter is to carry a flow of water ($v= 1.131 \times 10^6$ m^2/s) with $Q = 0.03$ m^3/s. Determine the friction factor of the pipe and head loss in the pipe.

16.8   A 15-cm PVC pipe transfers water between two reservoirs. The elevations of the water surfaces in the reservoirs differ by 50 m and the length of the pipe is 2,000 m. Find the flow rate through the pipe.

16.9   A 50-m-long cast-iron pipe of 30 cm diameter is connected with a 25-cm-diameter 75-m-long pipe in series. The discharge through the pipes is 0.2 m^3/s. Find the length of a 40-cm pipe that is equivalent to this system.

16.10 In Example 16.5, assume the diameter of pipe 1 to be 35 cm and the length of each pipe to be 90 m. Find the velocity of water in the pipes and discharge through pipe 2.

16.11 In Example 16.6, assume the roughness of all four pipes in the upper loop to be 125. Compute the flow in each pipe by the Hardy–Cross method. Compare the results with those obtained in Example 16.6 and explain the difference in flow in the various pipes.

References

Abramowitz, M., and Stegun, I. A. (1965). Handbook of mathematical functions, Dover Publications, New York.

Akaike, H. (1973). "Information theory and an extension of the maximum likelihood principle." Proc., 2nd International Symposium on Information Theory, B. N. Petrov and F. Csaki, eds., Publishing House of the Hungarian Academy of Sciences, Budapest, Hungary.

Ali, M. M., Mikhail, N. N. and Haq, M. S. (1978). "A class of bivariate distributions including the bivariate logistic." Journal of Multivariate Analysis, 8, 405–412.

Ang, A. H.-S., and Tang, W. H. (1984). Probability concepts in engineering planning and design, Vol. I: Basic principles, John Wiley & Sons, New York.

Aron, G. and White, E. L. (1982). "Fitting a gamma distribution over a synthetic unit hydrograph." Water Resources Bulletin, 18(1), 95–98.

ASCE. (1996). Hydrology handbook, American Society of Civil Engineers, New York.

Ashkar, F., Bobee, B., Lerous, D., and Morisette, D. (1988). "The generalized method of moments as applied to the generalized gamma distribution." Stochastic Hydrology and Hydraulics, 2, 161–174.

Awumah, K., Goulter, I. C., and Bhatt, S. K. (1990). "Assessment of reliability in water distribution networks using entropy-based measures." Stochastic Hydrology and Hydraulics, 4, 309–320.

Awumah, K., Goulter, I., and Bhatt, S. K. (1991). "Entropy-based redundancy measures in water distribution networks." Journal of Hydraulic Engineering, 117(5), 595–614.

Ayres, F., (1952). Theory and problems of differential equations, Schaum's Outline Series, McGraw-Hill Book Company, New York.

Bacchi, B., Becciu, G., and Kottegoda, N. T. (1994). "Bivariate exponential model applied to intensities and durations of extreme rainfall." Journal of Hydrology, 155, 225–236.

Barbe, D. E., Cruise, J. F., and Singh, V. P. (1991). "Solution of three-constraint entropy-based velocity distribution." Journal of Hydraulic Engineering, 117(10), 1389–1396.

Barbe, D. E., Cruise, J. F., and Singh, V. P. (1994). "Derivation of a distribution for the piezometric head in groundwater flow using stochastic and statistical methods." Hydrology and environmental engineering, 2, K. W. Hipel, ed., Kluwer Academic Publishers, Dordrecht, 151–160.

Barnes, H. H. (1962). "Roughness characteristics of natural channels." Water supply papers 1849, U.S. Geological Survey, Washington, D.C.

Bendat, J. S., and Piersol, A. G. (1980). Engineering applications of correlation and spectral analysis, Wiley-Interscience, New York.

Benjamin, J. R., and Cornell, C. A. (1970). Probability, statistics, and decision for civil engineers, McGraw-Hill Inc., New York.

Bobee, B., Pereault, L., and Ashkar, F. (1993). "Two kinds of moment ratio diagrams and their applications in hydrology." Stochastic Hydrology and Hydraulics, 7, 41–65.

Booy, C. (1990). Rational engineering decisions under conditions of uncertainty. Lecture notes for course 25-221, Department of Civil Engineering, University of Manitoba, Winnipeg, Canada.

Borgman, L. E. (1963). "Risk criteria." Journal of the Waterways and Harbor Division, Proc., ASCE, 89 (WW3), 1–35.

Bos, R. (1990). Aquifer parameter identification by the maximum entropy and Bayes' theorem. Unpublished report, Technion-Israel Institute of Technology, Haifa, Israel.

Bouchart, F. J.-C., and Goulter, I. C. (1998). "Is rational decision making appropriate for management of irrigation reservoirs?" Journal of Water Resources Planning and Management, 124 (6) 310–309.

Boudon, R. (2003). "Beyond rational choice theory." Annual Review of Sociology, 29, 1–29.

Box, G. E. P., and Muller, M. E. (1958). "A note on the generation of random normal deviates." Annals of Mathematical Statistics, 29, 610–611.

Burg, J. P. (1975). Maximum entropy spectral analysis. Ph.D. thesis, Stanford University, Palo Alto, Calif.

Burges, S. J., and Lettenmaier, D. P. (1982). "Reliability measures for water supply reservoirs and the significance of long-term persistence." Decision making for hydrosystems: Forecasting and operation, T. E. Unny and E. A. McBean, eds., Water Resources Publications, Littleton, Colo.

Carslaw, H. S. and Jaeger, J. C. (1959). Conduction of heat in solids, Clarendon Press, Oxford, England.

Cawlfield, J. D., and Wu, M. C. (1993). "Probabilistic sensitivity analysis for one dimensional reactive transport in porous media." Water Resources Research, 29(3), 661–672.

Chang, C. H., Tung, Y. K., and Yang, J. C. (1995). "Evaluation of probabilistic point estimate methods." Applied Mathematical Modelling, 19(2), 95–105.

Cheng, S. T. (1982). Overtopping risk evaluation for an existing dam. Ph.D. thesis, Department of Civil Engineering, University of Illinois at Urbana-Champaign, Urbana, Ill.

Chiu, C. L., and Murray, D. W. (1992). "Variation of velocity distribution along nonuniform open channel flow." Journal of Hydraulic Engineering, 118(1), 989–1001.

Chiu, C. L., Jin, W., and Chen, Y. C. (2000). "Mathematical models of distribution of sediment concentration." Journal of Hydraulic Engineering, 126(1), 16–23.

Choo, T. H. (2000). "An efficient method of the suspended sediment-discharge measurement using entropy concept." Water Engineering Research, 1(2), 95–105.

Chow, V. T. (1951). "A generalized formula for hydrologic frequency analysis." Transactions of the American Geophysical Union, 32(2), 231–237.

Chow, V. T. (1954). "The log-probability law and its engineering applications." Proc., ASCE, 80, 1–25.

Chow, V. T. (1979). "Risk and reliability analysis applied to water resources in practice." Reliability in water resources management, E. A. McBean, K. W. Hipel, and T. E. Unny, eds., Water Resources Publications, Fort Collins, Colo.

Cleary, R. W., and Adrian, D. D., 1973. "New analytical solutions for dye diffusion equations." Journal of the Environmental Engineering Division, ASCE, 99(EE3), 213–227.

Cochran, W. (1966). Sampling techniques, Second edition, John Wiley & Sons, New York.

Collins, M. A. (1983). "Discussion of 'Fitting a gamma distribution over a synthetic unit hydrograph, by G. Aron and E.L. White.'" Water Resources Bulletin, 19(2), 303–304.

Cornell, C. A. (1972). "First-order analysis of model and parameters uncertainty." Proc., International Symposium on Uncertainties in Hydrology and Water Resources Systems, University of Arizona, Tucson., 3, 1245–1272

Cox, D. R., and Miller, H. D. (1965). The theory of stochastic processes, Chapman and Hall, London.

Croley, T. E. (1980). "Gamma synthetic hydrographs." Journal of Hydrology, 47, 41–52.

Cross, H. (1936). "Analysis of flow in networks of conduits or conductors." Bulletin 286, University of Illinois Experiment Station, Urbana, Ill.

DeCoursey, D. (1966). A runoff hydrograph equation, U.S. Department of Agriculture, Agricultural Research Service, ARS-41-116, Washington, D.C.

De Michele, C., and Salvadori, G. (2003). "A generalized Pareto intensity-duration model of storm rainfall exploiting 2-copulas." Journal of Geophysical Research (Atmospheric), 108 (D2), ACL 15–11.

De Michele, C., Salvadori, G., Canossi, M., Petaccia, A., and Rosso, R. (2005). "Bivariate statistical approach to check adequacy of dam spillway." Journal of Hydrologic Engineering, 10(1), 50–57.

Deymie, P. (1939). "Propagation d'une intumescence allongee (Propagation of a long wave)." Revue Generale de l'Hydraulique, 3, 138–142.

Ditlevsen, O. (1981). Uncertainty modeling with applications to multidimensional civil engineering systems, McGraw-Hill Book Co., Inc., New York.

Dooge, J. C. I. (1959). "A general theory of the unit hydrograph." Journal of Geophysical Research, 64(2), 241–256.

Dooge, J. C. I. (1973). "Linear theory of hydrologic systems." Technical bulletin no. 1468, Agricultural Research Service, U.S. Department of Agriculture, Washington, D.C.

Dooge, J. C. I., and Harley, B. M. (1967). "Linear routing in uniform open channels." Proc., International Hydrology Symposium, Fort Collins, Colo., 1, 57–63.

Dooge, J. C. I., and Napiorkowski, J. J. (1987). "Applicability of diffusion analogy in flood routing." Acta Geophysica Polonica, 35(1), 66–75

Dooge, J. C. I., Napiorkowski, J. J., and Strupczeswki, W. G. (1987a). "Properties of the generalized channel response." Acta Geophysica Polonica, 35(4), 405–418.

Dooge, J. C. I., Napiorkowski, J. J., and Strupczeswki, W. G. (1987b). "The linear downstream response of a generalized uniform channel." Acta Geophysica Polonica, 35(3), 278.

Duffy, C. J., Gelhar, L. W., and Wierenga, P. J. (1984). "Stochastic models in agricultural watersheds." Journal of Hydrology, 69, 145–162.

Dwight, H. B. (1934). Tables of integrals and other mathematical data, Macmillan, New York.

Eagleson, P. S. (1978). "Climate, soil and vegetation. 2. The distribution of annual precipitation derived from observed storm sequences." Water Resources Research, 14(5): 713–721.

Edson, C. G. (1951). "Parameters for relating unit hydrographs to watershed characteristics." Transactions, American Geophysical Union, 32(4), 591–596.

Efron, B. (1977). Bootstrap methods: Another look at the jackknife, Technical Report No. 32, Division of Biostatistics, Stanford University, California.

Eilbert, R. F. and Christensen, R. A. (1983). "Performance of the entropy minimax hydrological forecasts for California, water years 1948–1977." Journal of Climate and Applied Meteorology, 22, 1654–1657.

Einstein H. A. (1942). "Formulas for the transportation of bed load." Transactions ASCE, Paper no. 2140, 561–597.

Favre, A.-C., El Adlouni, S., Perrault, L., Thiemonge, N., and Bobee, B. (2004). "Multivariate hydrological frequency analysis using copulas." Water Resources Research, 40, W01101-1–12.

Fiorentino, M, Claps, P., and Singh, V. P. (1993). "An entropy-based morphological analysis of river basin networks." Water Resources Research, 29(4), 1215–1224.

Folks, J. L., and Chhikara, R. S. (1978). "The inverse Gaussian distribution and its statistical application—A review." Journal of Royal Statistical Society B, 40(3), 263–289.

Fougere, P. F., Zawalick, E. J., and Radoski, H. R. (1976). "Spontaneous line splitting in maximum entropy power spectral analysis." Physics of the Earth and Planetary Interiors, 12, 201–207.

Frank, M. J. (1979). "On the simultaneous associativity of $F(x, y)$ and $x + y - F(x, y)$." Aequationes Mathematics, 19, 617–627.

Freeze, R. A. (1975). "A stochastic-conceptual analysis of one-dimensional groundwater flow in nonuniform homogenous media." Water Resources Research, 11(5) 725–741.

Fruedenthal, A. M. (1956). "Safety and probability of structural failure." Transactions of the ASCE, 121, 1337–1397.

Gelhar, L.W. (1974). "Stochastic analysis of phreatic aquifers." Water Resources Research, 10, 530–545.

Genest, C. and Ghoudi, K. (1994). "Une famille de lois bidimensionnelles insolite." Central Royal Academy of Science, Paris, Series I, 318, 351–354.

Genest, C., and Rivest, L.-P. (1993). "Statistical inference procedures for bivariate Archimedean copulas." Journal of the American Statistical Association, 88, 1034–1043.

Genest, C., Ghoudi, K., and Rivest, L.-P. (1995). "A semiprarametric estimation procedure of dependence parameters in multivariate families of distributions." Biometrika, 82(3), 543–552.

Gershenfeld, N. A., and Weigend, A. S. (1993). "The future of time series: Learning and understanding." Time series prediction: Forecasting the future and understanding the past, N. A. Gershenfeld and A. S. Weigend, eds., Proc., SFI Studies in the Sciences of Complexity, Col XV, Addison-Wesley, Reading, Mass.

Good, I. J. (1953). "The population frequencies of species and the estimation of population parameters." Biomerika, 40, 237–260.

Goulter, I. (1995). "Analytical and simulation models for reliability analysis in water distribution systems." Improving efficiency and reliability in water distribution systems, E. Cabrera and A. Vela, eds., Kluwer Academic Publishers, Dordrecht.

Gray, D. M. (1961). "Synthetic unit hydrographs fort small watersheds." Journal of Hydraulics Division, ASCE, 87(HY4), 33–54.

Greenwood, J. A., Landwehr, J. M., Matalas, N. C., and Wallis, J. R. (1979). "Probability-weighted moments: Definition and relation to parameters to several distributions expressible in inverse form." Water Resources Research, 15, 1049–1054.

Gumbel, E. J. (1958). Statistics of extremes, Columbia University Press, New York.

Gumbel, E. J. (1960). "Multivariate extreme distributions." Bulletin of the International Statistical Institute, 39(2), 471–475.

Gupta, V. L. and Moin, S. A. (1974). "Surface runoff hydrograph equation." Journal of Hydraulics Division, ASCE, 100(HY10), 1353–1368.

Gupta, V. L., Thongchareon, V. and Moin, S. A. (1974). "Analytical modeling of surface runoff hydrographs for major streams in northeast Thailand." Hydrological Sciences Bulletin, 19(4/12), 523–540.

Haan, C. T. (1977). Statistical methods in hydrology, The Iowa State University Press, Iowa City.

Haimes, Y. Y., and Stakhiv, E. Z., eds. (1985). Risk-based decision making in water resources, ASCE, New York.

Haldar, A., and Mahadevan, S. (2000). Probability, reliability, and statistical methods in engineering design, John Wiley & Sons, Inc., New York.

Haktanir, T. (1996). "Probability-weighted moments without plotting position formula." Journal of Hydrologic Engineering, 1(2), 89–91.

Hammersley, J. M., and Morton, K. W. (1956). "A new Monte Carlo technique: Antithetic variates." Proc., Cambridge Physics Society, 52, 449–474.

Harmancioglu, N. B., and Singh, V. P. (1998). "Entropy in environmental and water resources. Encyclopedia of hydrology and water resources, R. W. Hershey and R. W. Fairbridge, eds., Kluwer Academic Publishers, Boston, 225–241.

Harmancioglu, N. B., Fistikoglu, O., Ozkul, S. D., Singh, V. P., and Alpaslan, M. N. (1999). Water quality monitoring network design, Kluwer Academic Publishers, Dordrecht.

Harr, M. E. (1989). "Probability estimates for multivariate analyses." Applied Mathematical Modelling, 13, 313–318.

Hasofer, A. M., and Lind, N. C. (1974). "Exact and invariant second-moment code format." Journal of Engineering Mechanics Division, ASCE, 100, 111–121.

Hayami, S. (1951). On the propagation of flood waves, Bulletin 1, Disaster Prevention Research Institute, Kyoto University, Kyoto, Japan.

Hazen, A. (1914). "Storage to be provided in impounded reservoirs for municipal water supply." Transactions of the ASCE, 77, 1539–1640.

HEC. (1990). "Explaining flood risk." Training document no. 33, Hydrologic Engineering Center, U.S. Army Corps of Engineers, Davis, Calif.

Herr, H. D., and Krzystofowicz, R. (2005). "Generic probability distribution of rainfall in space: The bivariate model." Journal of Hydrology, 300, 234–263.

Hosking, J. R. M. (1986). "The theory of probability-weighted moments." Technical report RC 12210, Mathematics, IBM Thomas J. Watson Research Center, Yorktown Heights, N.Y.

Hosking, J. R. M. (1990). "L-moments: Analysis and estimation of distribution using linear combination of order statistics." Journal of Royal Statistical Society B, 52(1), 105–124.

Hosking, J. R. M., and Wallis, J. R. (1987). "Parameter and quantile estimation for the generalized Pareto distribution." Technometrics, 29(3), 339–349.

Houghton, J. C. (1978). "Birth of a parent: The Wakeby distribution for modeling flood flows." Water Resources Research, 14(6), 1105–1110.

Iman, R. L., and Helton, J. C. (1985). "An investigation of uncertainty and sensitivity analysis techniques for computer models." Risk Analysis, 8(1), 71–90.

ISWM Design Manual for Development and Redevelopment (2004). Prepared by Freese and Nichols, Inc., AMEC Earth and Environmental, Alan Plummer Associates, Inc., and Caffey Engineering, Inc., for the North Central Texas Council of Governments.

Jain, S. K., and Singh, V. P. (2003). Water resources systems planning and management, Elsevier, Amsterdam.

Jaynes, E. T. (1957). "Information theory and statistical mechanics, I." Physical Review, 106, 620–630.

Jaynes, E. T. (1982). "On the rationale of maximum entropy methods." Proc., IEEE, 70, 939–952.

Jenkinson, A. F. (1955). "The frequency distribution of the annual maximum (or minimum) values of meteorological elements." Quarterly Journal of the Royal Meteorological Society, 81, 251–261.

Jenkinson, A. F. (1969). "Estimation of maximum floods." World Meteorological Organization Technical Note, Geneva, Switzerland, 98(5), 183–227.

Jennings, M. E., and Benson M. A. (1969). "Frequency curve for annual flood series with some zero events or incomplete data." Water Resources Research, 5(1), 276–280.

Jeppson, R. W. (1977). Analysis of flow in pipe networks, Ann Arbor Science, Ann Arbor, Mich.

Joe, H. (1993). "Parametric families of multivariate distributions with given margins." Journal of Multivariate Analysis, 46(2), 262–282.

Johnson, N. L., and Kotz, S. (1985). "Moment ratios." Encyclopedia of Statistical Inferences, 5, 603–604.

Johnston, N. L., and Kotz, S. (1970). Distribution in statistics: Continuous univariate distributions 1, Houghton-Mifflin, Boston.

Jones, L. E. (1971). "Linearizing weight factors for least squares fitting." Journal of the Hydraulics Division, ASCE, 97(HY5), 665–675.

Jones, L. (1990). "Explicit Monte Carlo simulation head moment estimates for stochastic confined groundwater flow." Water Resources Research, 26(6), 1145–1153.

Kansal, M. L. (1996). Reliability analysis of water distribution system. Ph.D. thesis, Department of Civil Engineering, Delhi College of Engineering, Delhi, India.

Kaplan, S., and Garrick, B. J. (1981). "On the quantitative definition of risk." Risk Analysis, 1(1), 11–27.

Kelly, K. S., and Krzysztofowicz, R. (1997). "A bivariate meta-Gaussian density for use in hydrology." Stochastic Hydrology and Hydraulics, 11:17–31.

Kendall, M. G., and Stuart, A. (1969). The advanced theory of statistics, Vol. 1: Distribution theory, Charles Griffin & Company Limited, London.

Kendall, M. G., and Stuart, A. (1973). The advanced theory of statistics, Vol. 2: Functional and structural relationship, Charles Griffin & Company Limited, London.

Kim, H. J., and Mays, L. W. (1994). "Optimal rehabilitation model for water distribution systems." Journal of Hydraulic Engineering, ASCE, 120(5), 674–692.

Klemes, V. (1971). "Some problems in pure and applied stochastic hydrology." Proc., Symposium on Statistical Hydrology, University of Arizona, Tuscon, 2–15.

Klir, G. J. (1991). "Measures and principles of uncertainty and information. Information dynamics." Proc., NATO Advanced Study Institute on Information Dynamics, Plenum Press, New York.

Krasovskaia, R. (1997). "Entropy-based grouping of river flow regimes." Journal of Hydrology, 202, 173–1191.

Kroll, C. N., and Stedinger, J. R. (1996). "Estimation of moments and quantiles using censored data." Water Resources Research, 32(4), 1005–1012.

Kronmal, R. A., and Peterson, A. V. (1979). "On the alias method for generating random variables from a discrete distribution." American Statistician, 33, 214–218.

Krstanovic, P. F., and Singh, V. P. (1991a). "A univariate model for long-term streamflow forecasting: 1. Development." Stochastic Hydrology and Hydraulics, 5, 173–189.

Krstanovic, P. F., and Singh, V. P. (1991b). "A univariate model for long-term streamflow forecasting: 2. Application." Stochastic Hydrology and Hydraulics, 5, 189–205.

Krstanovic, P. F., and Singh, V. P. (1992a). "Evaluation of rainfall networks using entropy: 1. Theoretical development." Water Resources Management, 6, 279–293.

Krstanovic, P. F., and Singh, V. P. (1992b). "Evaluation of rainfall networks using entropy: II. Application." Water Resources Management, 6, 295–314.

Krstanovic, P. F., and Singh, V. P. (1993a). "A real-time flood forecasting model based on maximum entropy spectral analysis: 1. Development." Water Resources Management, 7, 109–129.

Krstanovic, P. F., and Singh, V. P. (1993b). "A real-time flood forecasting model based on maximum entropy spectral analysis: 2. Application." Water Resources Management, 7, 131–151.

Krzysztofowicz, R. (1999). "Probabilities for a period and its subperiods: Theoretical relations for forecasting." Monthly Weather Review, 127(2), 228–235.

Kubo, R. (1862). "Generalized cumulant expansion method." Journal of Physical Society of Japan, 17, 1100–1200.

Kuczera, G. (1982a). "Combining site-specific and regional information: An empirical Bayes approach." Water Resources Research, 18(2), 306–314.

Kuczera, G. (1982b). "Robust flood frequency models." Water Resources Research, 18(2), 315–324.

Kuczera, G. (1982c). "On the relationship between the reliability of parameter estimates and hydrologic time series data used in calibration." Water Resources Research, 18(1) 146–154.

Kuczera, G. (1982d). "Robust flood frequency models." Water Resources Research, 18(2), 315–324.

Kullback, S., and Leibler, R. A. (1951). "On information and sufficiency." Annals of Mathematical Statistics, 22, 79–86.

Kumar, A. (1980). Prediction and real-time hydrological forecasting. Ph.D. thesis, Indian Institute of Technology, Delhi.

Kumar, S. (1999). Some studies on hydrologic and hydraulic reliability analysis in water resource systems. Ph.D. thesis, Department of Civil Engineering, Delhi College of Engineering, Delhi.

Labadie, J., Fontane, D., Tabios, G., III, and Chou, N.-F. (1987). "Stochastic analysis of dependable hydropower capacity." Journal of Water Resources Planning and Management, ASCE, 113(3), 422–437.

Landwehr, J. M., Matalas, N. C., and Wallis, J. R. (1979b). "Estimation of parameters and quantiles of Wakeby distribution." Water Resources Research, 15, 1361–1369.

Law, A. M., and Kelton, W. D. (1991). Simulation modeling and analysis, McGraw Hill, Inc., New York.

Levine, R. D., and Tribus, M. (1979). The maximum entropy formalism, MIT Press, Cambridge, Mass.

Li, K. S. (1992). "Point estimate method for calculating statistical moments." Journal of Engineering Mechanics, 118(7), 1506–1511.

Lienhard, J. H., and Meyer, P. L. (1967). "A physical basis for the generalized gamma distribution." Quarterly of Applied Mathematics, 25(3), 330–334.

Lighthill, M. H., and Witham, G. B. (1955). "On kinematic waves. I. Flood movements in long rivers." Proc., Royal Society, London, Series A, 229, 281–316.

Lin, G. F., and Wang, Y. M. (1996). "General stochastic instantaneous unit hydrograph." Journal of Hydrology, 182, 227–238.

Loucks, D. P., and Dorfman, P. J. (1975). "An evaluation of some linear decision rules in chance-constrained models for reservoir planning and operation." Water Resources Research, 11(6), 777–782.

Loucks, D. P., Stedinger, J. R., and Haith, D. A. (1981). Water resources systems planning and analysis, Prentice Hall, N.J.

Low, B. K. (1996). "Practical probabilistic approach using spreadsheet." Proc., Uncertainty in Geologic Environment from Theory to Practice, ASCE Geotech. Spec. Publ. 58, 2, 1284–1302.

Low, B. K., and Tang, W. H. (1997). "Efficient reliability evaluation using spreadsheet." Journal of Engineering Mechanics, 123(7), 749–752.

Manache, G. (2001). Sensitivity of a continuous water-quality simulation model using Latin hypercube sampling. Ph.D. thesis, Vrje Universiteit Brussel, Brussels.

Maran, S. (2002). "A stochastic approach to the evaluation of residence times." Water Resources Research, 38(5), 4-1–4-9.

Markovic, R. D. (1965). "Probability functions of best-fit to distributions of annual precipitation and runoff." Hydrology Papers 8, Colorado State University, Fort Collins, Colo.

Marshall, A. W., and Ingram O. (1967). "A multivariate exponential distribution." Journal of American Statistical Association, 62(317), 30–44.

Matalas, N. C. (1967). "Mathematical assessment of synthetic hydrology." Water Resources Research, 3(4), 937–945.

Masse, R. (1937). "Des intumescences dans les torrents (Transitory waves in torrents)." Revue Generale de l'Hydraulique, Paris, 3(18), 305–306.

Mays, L. W., ed. (2000). Water distribution systems handbook, McGraw-Hill Book Company, New York.

McCuen, R. H., and Snyder, W. M. 1985. Hydrologic modelling, statistical methods and applications, Prentice Hall, Englewood Cliffs, N.J.

Mckay, M. D. (1988). "Sensitivity and uncertainty analysis using a statistical sample of input values." Chapter 4, Uncertainty analysis, Y. Ronen, ed., CRC Press, Boca Raton, Fla.

Meadows, M. E., and Walski, T. M. (1998). Computer applications in hydraulic engineering, Haestad Press, Waterbury, Conn.

Melching, C. S. (1995). "Reliability estimation." Chapter 3, Computer models of watershed hydrology, V. P. Singh, ed., Water Resources Publications, Littleton, Colo.

Melching, C. S. (2001). "Sensitivity measures for evaluating key sources of modeling uncertainty." Proc., International Symposium on Environmental Hydraulics, IAHR, Tempe, Ariz., 508.

Melching, C. S., and Anmangandla, S. (1992). "Improved first order uncertainty method for water quality modeling." Journal of Environmental Engineering, 118(5), 791–805.

Mitchell, W. D. (1948). Unit hydrographs in Illinois, Illinois Division of Waterways and U.S. Geological Survey, Ill.

Moharram, S. H., Gosain, A. K., and Kapoor, P. N. (1993). "A comparative estimation of the generalized Pareto distribution." Journal of Hydrology, 150, 169–185.

Moore, R. J. (1984). "A dynamic model of basin sediment yield." Water Resources Research, 20 (1), 89–103.

Moore, R. J., and Clarke, R. T. (1983). "A distributed function approach to modelling basin sediment yield." Journal of Hydrology, 65, 239–257.

Moramarco, T., and Singh, V. P. (2001). "Simple method for relating local stage and remote discharge." Journal of Hydrologic Engineering, 6(1), 78–81.

Morlat, G. (1956). "Note sur l'estimation des debits de crues." La Houille Blanche, No. Spec. B, 663–678.

Mukherjee, D., and Ratnaparkhi, M. V. (1986). "On the functional relationship between entropy and variance with related applications." Communications on Statistical Theory and Methods, 15(1), 291–311.

Nash, J. E. (1957). "The form of the instantaneous unit hydrograph." International Association of Scientific Hydrology Publication, 45(3), 114–121.

Nash, J. E. (1959). "Systematic determination of unit hydrograph parameters." Journal of Geophysical Research, 64(1), 111–115.

Nash, J. E. (1960). "A unit hydrograph study, with particular reference to British catchments." Proc., Institution of Civil Engineers, London, 17, 249–282.

Nash, J. E., and Sutcliffe, J. V. (1970). "River flow forecasting through conceptual models: 1. A discussion of principles." Journal of Hydrology, 10, 282–290.

Natale, L., and Todini, E. (1974). A constrained parameter estimation technique for linear models in hydrology, Publication no. 13, Institute of Hydraulics, University of Pavia, Pavia, Italy.

Natural Environment Research Council (NERC). (1975). Flood studies report, London, 1, 81–97

Nelson, R. B. (1999). An introduction to copulas, Springer-Verlag, New York.

Ochoa, I. D., Bryson, M. C., and Shen, H. W. (1980). "On the occurrence and importance of Paretian-tailed distribution in hydrology." Journal of Hydrology, 48, 53–62.

Ouarda, T. B. M. J, Ashkar, F., Ben Said, E. M., and Hournani, L. (1994). "Distribution statistiques utilisees en hydrologie: Transformation et proprietes asymptotiques." Rapport de recherché STAT-13, University of Moncton, New Brunswick, Canada.

Özisik, M. N. (1968). Boundary value problems of heat conduction, International Textbook Co., Scranton, Pa., 48–79.

Owen, D. B. (1962). Handbook of statistical tables, Addison-Wesley Publishing Company, Reading, Mass.

Ozkul, S., Harmancioglu, N. B., and Singh, V. P. (2000). "Entropy-based assessment of water quality monitoring networks." Journal of Hydrologic Engineering, 5(1), 90–100.

Padmanabhan, G., and Rao, A. R. (1986). "Maximum entropy spectra of some rainfall and river flow time series from southern and central India." Theoretical and Applied Climatology, 37, 63–73.

Padmanabhan, G., and Rao, A. R. (1988). "Maximum entropy spectral analysis of hydrologic data." Water Resources Research, 24(9), 1519–1533.

Pate-Cornell, M. E. (1996). "Uncertainties in risk analysis: Six levels of treatment." Reliability Engineering and System Safety, 54(2–3), 95–111.

Pebesma, E. J., and Heuvelink, G. B. M. (1999). "Latin hypercube sampling of Gaussian random fields." Technometrics, 41(4), 303–312.

Perreault, L., Bobee, B., and Rasmussen, P. F. (1999a). "Halpen distribution system. I: Mathematical and statistical properties." Journal of Hydrologic Engineering, 4(3), 189–199.

Perreault, L., Bobee, B., and Rasmussen, P. F. (1999b). "Halpen distribution system. II: Parameter and quantile estimation." Journal of Hydrologic Engineering, 4(3), 200–208.

Phien, H. N., and Jivajirajah, T. (1984). "Fitting the S_b curve by the method of maximum likelihood." Journal of Hydrology, 67, 67–75.

Plate, E. J. (2002). "Risk management for hydraulic systems under hydrological loads." Risk, reliability, uncertainty, and robustness of water resources systems, J. J. Bogardi and Z. W. Kundzewicz, eds., Cambridge University Press, Cambridge, U.K.

Quandt, R. E. (1966). "Old and new methods of estimation of the Pareto distribution." Biometrika, 10, 55–82.

Quimpo, R. G. (1993). "Reliability analysis of water distribution systems." Proc., Sixth Conference Sponsored by the Engineering Foundation on Risk-Based Decision Making in Water Resources, ASCE, 45–55.

Rackwitz, R. (1976). "Practical probabilistic approach to design." Comite European du Beton, Paris, Bulletin No. 112.

Rao, A. R., Padmanabhan, G., and Kashyap, R. L. (1984). "A comparative analysis of recently developed methods of spectral analysis." Frontiers in hydrology, V. Yevjevich, ed., Water Resources Publications, Fort Collins, Colo., 127–149.

Rao, A. R., and Hamed, K. H. (2000). Flood frequency analysis, CRC Press, Boca Raton, Fla.

Rescher, N. (1995). Satisfying reason: Studies in the theory of knowledge, Kluwer Acdemic Publishers, Dordrecht.

ReVelle, C., Joeres, E., and Kirby, W. (1969). "The linear decision rule in reservoir management and design, 1. Development of the stochastic model." Water Resources Research, 5(4), 767–777.

Rosenblueth, E. (1975). "Point estimates for probability moments." Proc., National Academy of Sciences U.S.A., 72(10), 3812–3814.

Rosenblueth, E. (1981). "Point estimates in probabilities." Applied Mathematical Modeling, 72(10), 3812–3814.

Rossman, L.A. (2000). "Computer models/EPANET." Water distribution systems handbook, L. W. Mays, ed., McGraw-Hill Book Company, New York.

Royal Society (1983). Risk assessment: A study group report, The Royal Society, London.

Rubinstein, R. Y. (1981). Simulation and the Monte Carlo method, John Wiley & Sons, New York.

Salas, J. D. (1993). "Analysis and modeling of hydrologic time series." Chapter 19, Handbook of hydrology, D. R. Maidment, ed., McGraw Hill, New York.

Salvadori, G., and De Michele, C. (2004). "Analytical calculation of storm volume statistics involving Pareto-like intensity-duration marginals." Geophysical Research Letters, 31, L04502.

Sarino and Serrano, S. E. (1990). "Development of the instantaneous unit hydrograph using stochastic differential equations." Stochastic Hydrology and Hydraulics, 4, 151–160.

Shrader, M. L., Rawls, W. J., Snyder, W. M., and McCuen, R. H. (1981). "Flood peak regionalization using mixed-mode estimation of the parameters of the log-normal distribution." Journal of Hydrology, 52, 229–237.

Scott, E. J. (1955). Transform calculus with an introduction to complex variables, Harper & Brothers, New York.

Shackle, G. L. S. (1961). Decision, order and time in human affairs, Cambridge University Press, Cambridge, U.K.

Shannon, C. E. (1948). "A mathematical theory of communications, I and II." Bell System Technical Journal, 27, 379–443.

Shih, S.-F., and Hamrick, R. L. (1975). "A modified Monte Carlo technique to compute Thiessen coefficients." Journal of Hydrology, 27, 339–356.

Shinozuka, M. (1983). "Basic analysis of structural safety." Journal of Structural Engineering, 109(3), 721–740.

Shore, J. E., (1979). Minimum cross-entropy spectral analysis, Naval Research Laboratory, NRL Memorandum Report 3921, Washington, D.C.

Shrader, M. L., Rawls, W. J., Snyder, W. M., and McCuen, R. H. (1981). "Flood peak regionalization using mixed-mode estimation of the parameters of the lognormal distribution." Journal of Hydrology, 52, 229–237.

Singh, K., and Singh, V. P. (1991). "Derivation of bivariate probability density functions with exponential marginals." Stochastic Hydrology and Hydraulics, 5, 55–68.

Singh, V. P. (1988). Hydrologic systems, Vol. I: Rainfall-runoff modeling, Prentice Hall, Englewood Cliffs, N.J.

Singh, V. P. (1989). Hydrologic systems, Vol. 2: Watershed modeling, Prentice Hall, Englewood Cliffs, N.J.

Singh, V. P. (1996). Kinematic wave modeling in water resources: Surface water hydrology, John Wiley & Sons, New York.

Singh, V. P. (1998a). "The use of entropy in hydrology and water resources." Hydrological Processes, 11, 587–626.

Singh, V. P. (1998b). Entropy-based parameter estimation in hydrology, Kluwer Academic Publishers, Boston.

Singh, V. P., Baniukiewicz, A., and Chen, V. J. (1982). "An instantaneous unit sediment graph study for small upland watersheds." Modeling components of hydrologic cycle, V. P. Singh, ed., Water Resources Publications, Littleton, Colo., 539–554.

Singh, V. P., Baniukiewicz, A., and Ram, R. S. (1982). "Some empirical methods for determining the unit hydrograph." Rainfall-runoff relationship, V. P. Singh, ed., Water Resources Publications, Littleton, Colo., 67–90.

Singh, V. P., and Fiorentino, M., eds. (1992). Entropy and energy dissipation in water resources, Kluwer Academic Publishers, Dordrecht.

Singh, V. P., and Rajagopal, A. K. (1986). "A new method of parameter estimation for hydrologic frequency analysis." Hydrological Science and Technology, 2(3), 33–40.

Singh, V. P., Rajagopal, A. K., and Singh, K. (1985). Application of the principle of maximum entropy (POME) to hydrologic frequency analysis, Completion report 06, Louisiana Water Resources Research Institute, Louisiana State University, Baton Rouge, La.

Singh, V. P., Singh, K., and Rajagopal, A. K. (1986). Derivation of some frequency distributions using the principle of maximum entropy. Advances in Water Resources, 9, 91–106.

Smith, O. E., Adelfang, S. I., and Tubbs, J. D. (1982). A bivariate gamma probability distribution with application to gust model, NASA technical memorandum 82483, National Aeronautics and Space Administration, Washington, D.C.

Snyder, W. M. (1972). "Fitting of distribution functions by nonlinear least squares." Water Resources Research, 8(6), 1423–1432.

Stedinger, J. R., and Tasker, G. D. (1985). "Regional hydrologic analysis: 1. Ordinary, weighted and generalized least squares compared." Water Resources Research, 21(9), 1421–1432.

Strupczewski, W. G., and Napiorkowski, J. J. (1989). "Properties of the distributed Muskingum model." ACTA Geologica Polonica, 37(3–4), 299–314.

Strupczewski, W. G., and Napiorkowski, J.J., (1990a). "Linear flood routing model for rapid flow." Hydrological Sciences Journal, 35 (1–2), 49–64.

Strupczewski, W. G., and Napiorkowski, J. J. (1990c). "Linear flood routing model for rapid flow." Hydrological Sciences Journal, 35(1–2), 49–64.

Strupczewski, W. G., Napiorkowski, J. J., and Dooge, J. C. I. (1989). "The distributed Muskingum model." Journal of Hydrology, 111, 235–257.

Strupczewski, W. G., Singh, V. P., and Weglarczyk, S. (2001). "Impulse response of linear diffusion analogy as a flood probability density function." Hydrology Science Journal, 46(5), 761–780.

Stuart, A., and Ord, J. K., (1987). Kendall's advanced theory of statistics, Vol. 1: Distribution theory, Fifth edition, Oxford University Press, New York.

Swamee, P. K., and Jain, A. K. (1976). "Explicit equations for pipe-flow problems." Journal of Hydraulics Division, ASCE, 102(HY5), 657–664.

Taha, H. A. (2003). Operations research: An introduction, Prentice Hall of India Private Limited, New Delhi.

Tribus, M. (1969). Rational description: Decision and designs, Pergamon Press, New York.

Tsai, C., and Franceschini, S. (2003). "An improved point estimate method for probabilistic risk assessment." Proc., World Water and Environmental Resources Congress, ASCE, Philadelphia.

Tung, Y. K. (1990). "Mellin transformation applied to uncertainty analysis in hydrology/hydraulics." Journal of Hydraulic Engineering, 116(5), 659–674.

Tweedie, M. C. K. (1957). "Statistical properties of the inverse Gaussian distributions, I." Annals of Mathematical Statistics, 28, 362–377.

Tyagi, A. (2000). A simple approach to reliability, risk, and uncertainty analysis of hydrologic, hydraulic, and environmental engineering systems. Ph.D. thesis, Oklahoma State University, Stillwater, Okla.

Ulrych, T. J., and Clayton, R. W. (1976). Time series modeling and maximum entropy. Physics of the Earth and Planetary Sciences, 12, 188–199.

USEPA (1989). Risk assessment guidance for superfund, Vol. 1: Human health evaluation manual (part A), Interim final, Rep. no. EPA/540/1-89/002, U.S. Office of Emergency and Remedial Response, Washington, D.C.

USEPA (1990). Guidance on remedial actions for superfund sites with PCB contamination, Rep. no. EPA/540/G-90/007, U.S. Office of Emergency and Remedial Response, Washington, D.C.

Venetis, C. (1970). "Finite aquifers: Characteristic response and applications." Journal of Hydrology, 12, 53–62.

Veneziano, D. (1974). Contributions to second moment reliability, Res. Rep. No. R74-33, Department of Civil Engineering, Massachusetts Institute of Technology, Cambridge, Mass.

Vrijling, J. K., and van Gelder, P. H. A. J. M. (2000). "Policy implications of uncertainty integration in design." Stochastic hydraulics 2000, Wang and Hu, eds., Balkema, Rotterdam., 633–646.

Vrijling, J. K., Van Hengel, W., and Houben, R. J. (1995). "A framework for risk evaluation." Journal of Hazardous Materials, 43, 245–261.

Wagner, J. M., Shamir, U., and Marks, D. H. (1988) "Water distribution reliability: Analytical methods." Journal of Water Resources Planning and Management, ASCE, 114(3), 253–275.

Walski, T. M., Gessler, J., and Sjostrom, J. W. (1990). Water distribution systems: Simulation and sizing, Lewis Publishers Inc., Chelsea, Mich.

Wang, Q. J. (1997). "LH moments for statistical analysis of extreme events." Water Resources Research, 33(12), 2841–2848.

Wang, S. X., and Adams, B. J. (1984). Parameter estimation in flood frequency analysis, Publication 84-02, Department of Civil Engineering, University of Toronto, Toronto, Canada.

Wang, S. X., and Singh, V. P. (1995). "Frequency estimation for hydrological samples with zero value." Journal of Water Resources Planning and Management, 121(1), 98–108.

Williams, B. J., and Yeh, W. W.-G. (1983). "Parameter estimation in rainfall-runoff models." Journal of Hydrology, 63, 373–393.

Willmott, C. J., Ackleson, S. G., Davis, R. E., Feddema, J. J., Klink, K. M., Legates, D. R., O'Donnell, J., and Rowe, C. M. (1985). "Statistics for the evaluation and comparison of models." Journal of Geophysics Research, 90, 8995–9005.

Woo, M. K., and Wu, K. (1989). "Fitting annual floods with zero flows." Canadian Water Resources Journal, 14(2), 10–16.

Wood, D. J. (1980). Computer analysis of flow in pipe networks including extended period simulations (KYPIPE)—Users manual, Department of Civil Engineering. University of Kentucky, Lexington, Ky.

Woodbury, A. D., and Ulrych, T. J. (1993). "Minimum relative entropy: Forward probabilistic modeling." Water Resources Research, 29(8), 2847–2860.

Wu, I. P. (1963). "Design hydrographs for small watersheds in Indiana." Journal of Hydraulics Division, ASCE, 89(HY6), 35–66.

Yang, C. T. (1994). "Variational theories in hydrodynamics and hydraulics." Journal of Hydraulic Engineering, 120(6), 737–756.

Yang, Y., and Burn, D. H. (1994). "An entropy approach to data collection network design." Journal of Hydrology, 157, 307–324.

Yen, B. C., Cheng, S.-T., and Melching, C. S. (1986). "First-order reliability analysis." Stochastic and risk analysis in hydraulic engineering, B.C. Yen, ed., Water Resources Publications, Littleton, Colo.

Yevjevich, V., and Obeysekera, J. T. B. (1984). "Estimation of skewness of hydrologic variables." Water Resources Research, 20(7), 935–943.

Yue, S. (2000). "The bivariate lognormal distribution to model a multivariate flood episode." Hydrological Processes, 14, 2575–2588.

Yue, S. (2001a). "The Gumbel logistic model for representing a multivarite storm event." Advances in Water Resources, 24, 179–185.

Yue, S. (2001b). "A review of bivariate gamma distributions for hydrological application." Journal of Hydrology, 246, 1–18.

Yue, S., Ouarda, T. B. M. J., and Bobee, B. (2001). "A review of bivariate gamma distribution for hydrological application." Journal of Hydrology, 246, 1–18.

Zhang, L., and Singh, V. P. (2006). "Bivariate flood frequency analysis using the copula method." Journal of Hydrologic Engineering, 11(2), 150–164.

Zhao, Y. G., and Ono, T. (1999). "New approximation for SORM: Part 2." Journal of Engineering Mechanics, 125(1), 86–93.

Index

About the Authors

Vijay P. Singh, Ph.D., D.Sc., P.E., P.H., D.WRE, holds the Caroline and William N. Lehrer Distinguished Chair in Water Engineering and is a professor of biological and agricultural engineering and a professor of civil and environmental engineering at Texas A&M University. He earned B.S., M.S., Ph.D., and D.Sc. degrees in engineering and has widely published in the areas of hydrology, hydraulics, irrigation engineering, environmental engineering, and water resources. He currently serves as editor-in-chief of ASCE's *Journal of Hydrologic Engineering*, editor-in-chief of the Water Science and Technology book series for Springer, and associate editor or member of 16 editorial boards. He has won more than 42 national and international awards for his contributions and professional service. Dr. Singh has been president and senior vice president of the American Institute of Hydrology and a member of numerous committees of ASCE, the Hydrology Section of the American Geophysical Union, and the American Water Resources Association.

Sharad K. Jain, Ph.D., is a senior scientist and head of the Water Resources Systems Division of the National Institute of Hydrology, Roorkee, India. He holds bachelor's, master's, and doctorate degrees in civil engineering. He has written textbooks on water resources systems planning and management and on hydrology and water resources of India; in addition, he published numerous journal articles and conference papers. Dr. Jain is a former editor of the *Journal of Indian Water Resources Society.* He has been involved in many research and consultancy projects that address real-life problems of water resources development and management.

Aditya Tyagi, Ph.D., P.E., is a senior water resources engineer and technologist in the water business group of CH2M HILL. Previously, he worked as a scientist in the National Institute of Hydrology in India. He obtained his B.S. in civil engineering and M.S. in environmental engineering from the Indian Institute of Technology and his Ph.D. in biosystems engineering from Oklahoma State University. His research interests encompass the application of various analytical, numerical, statistical, stochastic, and optimization techniques to solve hydrologic, hydraulic, and environmental engineering problems.